Chemistry and Technology of Silicones

Chemistry and Technology of Silicones

by WALTER NOLL
Leverkusen, Germany

With contributions to the chapters on technical applications by

OSKAR GLENZ, *Leverkusen*; WILFRIED KNIEGE, *Leverkusen*; WALTER KRAUSS, *Leverkusen*; HUBERT ROTHERT, *Berlin*; and BRUNO ZORN, *Leverkusen*

and to the medical part by

GERHARD HECHT, *Wuppertal-Elberfeld*

Translated from the second, revised and substantially expanded German edition by B. HAZZARD, *Tooting, London*, and M. LANDAU, *Cheadle Hulme, Cheshire, England* in collaboration with EXPRESS TRANSLATION SERVICE, *Wimbledon, London.*

1968

ACADEMIC PRESS New York San Francisco London
A Subsidiary of Harcourt Brace Jovanovich, Publishers

ACADEMIC PRESS INC.
111 Fifth Avenue, New York, New York 10003

United Kingdom Edition published by
ACADEMIC PRESS, INC. (LONDON) LTD.
24/28 Oval Road, London NW1

First published in the German language under the title
"Chemie und Technologie der Silicone," copyright 1960
by Verlag Chemie G.m.b.H., Weinheim/Bergstrasse, Germany

LIBRARY OF CONGRESS CATALOG CARD NUMBER: 67-22772

PRINTED IN THE UNITED STATES OF AMERICA

79 80 81 82 9 8 7 6 5 4 3 2

Preface to the Second German Edition

The rapid development of the field of silicones has made it desirable to bring out a new, amended edition of this book; I am grateful to Verlag Chemie for making it possible to achieve this aim.

The whole work has been thoroughly revised and updated, as a result of which its size has been increased by almost fifty percent. More than half of the text has been rewritten. Particularly heavily revised and expanded are those sections of Chapters 8–10 that deal with the technology of silicone rubber; Chapter 4, which deals with the chemistry of the organofunctional silicon compounds; and Chapter 7, which gives an account of copolymers and cocondensates. The necessity of rewriting Chapters 4 and 7, in particular, shows clearly the direction of the most intensive work in silicone chemistry.

In spite of its expanded scope, the book retains the nature of a monograph giving the reader in condensed form not only a wide-ranging but also a thorough review of this rapidly growing field. In contrast to some other monographs on organosilicon compounds that have appeared in the interim, the silicones occupy in this as in the first edition the central position, and the technological part of the work is entirely devoted to them. In the chemical part it was necessary to broaden the range in order to provide the reader with a basis for understanding the present-day technological situation and with information that could be useful for further work. In view of the great volume of published literature, such an aim made it necessary to omit many detailed aspects of the field, increasing the danger of disagreement concerning the best selection of material.

As in the first edition, some contributions to the chapters on technical applications and on medical questions have been written by the author's colleagues. The subject of silicone rubber has been rewritten by Wilfried Kniege.

In view of the wide range of material, the advice and information from my associates and colleagues were welcomed. For assistance of this type I thank Dr. Klaus Damm, Dr. Hans Dietrich Gölitz, Dr. Hans Niederprüm, Dr. Karl Schnurrbusch, Dr. Walter Simmler, Dr. Hans-Horst Steinbach, and Dr. Engelbert Walaschewski.

The enormous task of arranging and recording the literature was taken care of by Edith Geselle, to whom I am particularly grateful for her untiring, patient, and persevering cooperation.

WALTER NOLL

Leverkusen, Farbenfabriken Bayer AG,
Anorganische Abteilung
November, 1965.

Acknowledgment

The present work is based on the second German edition, the translation of which was done by B. J. Hazzard and M. Landau. I am very much obliged to both of them for their close and willing cooperation. I also wish to thank Express Translation Service for many useful discussions and for the preparation of the text in its final form.

WALTER NOLL

Preface to the First German Edition

In the present work an attempt has been made to describe concisely and yet thoroughly the entire field of silicone chemistry, from the monomers to the polymers and the final technical products, from the methods of their synthesis to their chemical and physical properties and the technology of their use and application. However, the book is not the result of a systematic accumulation of the available material, claiming encyclopedic completeness. The main object has been rather to select what is, in the opinion of the author and his colleagues, the essential information from the mass of individual data available, and to bring it together into a closely integrated whole. That subjective considerations could not be excluded from a work undertaken in this spirit is obvious, but we hope that the words once spoken at a presentation of the Gmelin-Beilstein medal also apply to this work: "The number of individual contributions will increase further. What is permanent and valuable can be passed on only by those into whose hands their daily work on the subject has placed the sole adequate yardstick of critical evaluation."

The literature dealing with organosilicon compounds has indeed grown enormously, probably amounting at the present time to between 6000 and 8000 publications. A selection of about 2000 of these is cited in this book as a guide to the reader who wishes to delve further into the subject. Part of this material stems from the personal experience of the author and his colleagues. Unfortunately there is probably much important information in the hands of various manufacturers.

The guidelines on which the book is based will be clear to the reader from the list of contents. Care has always been taken to show the basic and broad interrelationships: the role of the monomers as precursors of the polymers, the characteristics of the polymer molecules as a basis for the understanding of their physicochemical properties, and this again as a foundation for the many-sided technology of the silicones. The descriptions of the compounds are supplemented by a large number of newly compiled tables, which are by no means complete but which contain a selection of compounds well beyond the framework of the text and should thus help to convey some idea of the extent and versatility of each group of compounds.

The literature has been considered up to about October, 1958. More recent publications have been used only in a few exceptional cases. Patents have been cited where possible together with the name of the inventor and the applicant company. An exception had to be made in the case of countries in which the names of the inventors are not given on the patents. The date given in each case is the date of application. Where the actual patent cited is not from the country of invention, the priority date in the country of origin is cited.

The wide-ranging subject matter of this book will, I hope, show a most striking characteristic of silicone chemistry—its enormous versatility.

In the first part of the book the emphasis is placed on matter of interest to the chemist. The second part deals with the technological applications of silicones of interest to electrical, mechanical, and civil engineers, as well as to chemists and technologists engaged in a host of industries such as rubber, plastics, paints and varnishes, textiles, leather, paper, polishes and protective coatings, cosmetics, pharmaceuticals and dental materials. To make the relevant material more easily accessible to the reader interested purely in the end products and their application, Chapter 10 (on application techniques) has been divided into industries and fields of utilization. From there the reader may refer back to the more chemical part of the book; for this purpose the section references given throughout the text should prove useful.

Within the framework of the discussion of the properties conferred by molecular structure, I have gone to particular pains to bring out the relationship between the siloxanes and the silicates, since the knowledge gained in either field will undoubtedly prove fruitful in the other. These sections should show mineralogists and silicate chemists the boundary region that the silicones have opened up, from a point of view of particular interest to them.

My thanks go first of all to the authors who have enriched the technological part of this book by their contributions: Professor Hubert Rothert, AEG-Fabriken, Berlin, whose material is based on many years of experience in the laboratories and plants of AEG; Drs. Oskar Glenz, Walter Krauss, Julius Peter, and Bruno Zorn, Farbenfabriken Bayer, Leverkusen, whose reports are likewise based on their own profound work on the special applications of silicones; and Professor Gerhard Hecht, Farbenfabriken Bayer, Wuppertal-Elberfeld, who, ever since we have been dealing with them, has been responsible for the evaluation of the pharmacological and toxicological properties of silicones. Wilfried Kniege, Farbenfabriken Bayer, Leverkusen, placed his knowledge at my disposal for the chapter on rubber technology.

For stimulus, guidance, and criticism I am indebted to several of my colleagues at Farbenfabriken Bayer, in particular, Drs. Klaus Damm, Karl Schnurrbusch, and Walter Simmler to whom I am bound by many years of collaboration. In the collection of the literature and preparation of the tables I have been assisted by Dr. Dora Brück, and in the preparation of the manuscript by my faithful and indefatigable assistant Edith Geselle and my son Klaus Dieter Noll.

WALTER NOLL

Leverkusen, Farbenfabriken Bayer,
Spring, 1960

Contents

CHAPTER 11. **Esters of Silicic Acid (Organooxysilanes and Polyorganooxysiloxanes**

CHAPTER 12. **Analytical Methods**

CHAPTER 1

General Discussion

1.1 Chemistry and Molecular Structure of the Silicones

The term "silicone" is not by any means used consistently. It is sometimes employed generically to designate all monomeric and polymeric organosilicon compounds containing Si—C bonds and sometimes as a collective term for organosilicon polymers of all types; however, it is also used in the more restricted sense of organosilicon polymers containing Si—O—Si bonds, and this is the usage that we shall adopt in this book. Accordingly, we define silicones as organosilicon polymers in which the silicon atoms are bound to each other through oxygen atoms, the silicon valences not taken up by oxygen being saturated by at least one organic group. The prototype for the simple linear polymers is thus built up according to the following scheme:

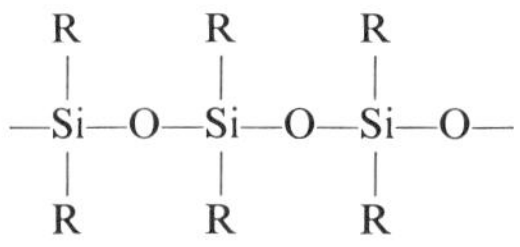

$(R_2SiO)_n$

where R is an organic radical.

The name "silicone" was coined at a time when it was thought that far-reaching analogies to the chemistry of carbon could be found in the chemistry of silicon. This idea is reflected in the occasional usage of terms like "silicochloroform" or "silicoformic acid". The term "silicone" was adopted by analogy with *ketone*, since the structural unit of the chain shown above, R_2SiO, appears to correspond to a ketone R_2CO. The analogy, however, is only formal. Although Stock believed he had isolated a monomeric pro-siloxane H_2SiO, it can now be accepted as certain that unlike the C=O double bond the Si=O double bond is unstable, at least at low temperatures. The tendency to form single bonds Si—O—Si causes silicon to combine with oxygen to give polymeric compounds, while carbon is capable of forming single molecules with oxygen. This also explains the completely different behavior of the two elements toward oxygen in nature. Silicon *builds up* with oxygen crystalline silicates with polymeric anions as stable end products, while carbon compounds are *degraded* by oxygen to the gaseous carbon dioxide.

"*Polyorganosiloxanes*" is a more precise term for these compounds than silicones. The name "siloxane" was based on the formulation of the Si—O—Si unit as *sil-oxane*, and this has found general acceptance in scientific nomenclature (compare Section 1.2). If the name "silicone" nevertheless continues to be used, it is for reasons of simplicity of speech and because of an adherence to the older usage which it would now be almost impossible to abandon. For this reason the rejection of "silicone" as a technical term would be unthinkable today. However, the convention adopted in this book, i.e., to use the term "silicone" only for *technical* products, could with advantage find general acceptance. Such a collective term would appear to be justified because as a rule the technical silicone products are not distinct chemical individuals. For scientific terminology, the nomenclature based on "siloxane" is more useful.

In our definition silicones are therefore characterized according to their general structural principles as follows:

1. Silicones are polymeric. They thus possess the typical structural features of organic macromolecules.

2. They contain silicon-oxygen bonds, and are thus fundamentally identical with the units forming the structure of silicic acids and silicates.

3. They contain hydrocarbon radicals combined directly with silicon and, therefore, structural units and bonds relate them to organic and, particularly, organometallic and organosilicon chemistry.

Silicones, therefore, occupy an intermediate position between organic and inorganic compounds, and in particular between silicates and organic polymers. It is precisely this dual nature which gives this class of compounds its particular fascination.

The silicon in polyorganosiloxanes can be combined with one, two, or three organic groups, the remaining valences being satisfied by oxygen. Accordingly, there are three groups containing silicon, oxygen, and organic radicals which can be considered *siloxane units* of the polymers. They belong to a series beginning with a group consisting only of silicon and oxygen and ending with one containing only silicon and hydrocarbon radicals:

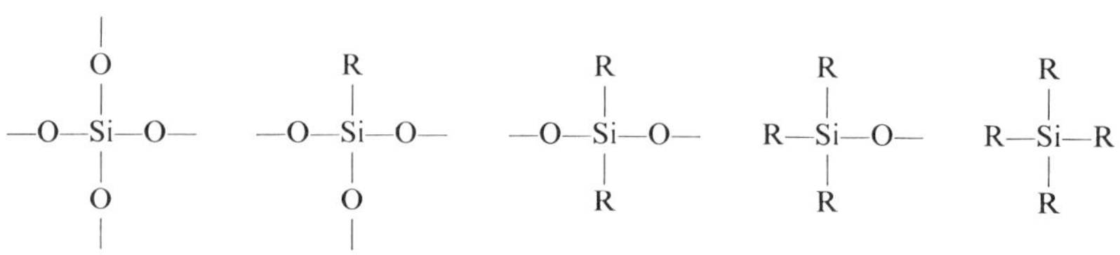

The complex ion $(SiO_4)^{4-}$ is the structural unit from which the three-dimensional lattices of the modifications of silica and the silicates are formed; R_4Si represents a nonfunctional organosilicon compound. The structural units of the silicon polymers lie between these extremes, and here again the dual nature of the polyorganosiloxanes as a link between silicate chemistry and organic chemistry is made evident.

The composition of a siloxane unit is determined by the fact that each oxygen atom functions as a link between two silicon atoms so that each silicon can be considered to have only a half-share in each oxygen atom. The general formula of members of the above series is therefore:

$$R_nSiO_{(4-n)/2}$$

The actual organosiloxane units are represented by this formula with n values of from 1 to 3.

The free valences on the oxygen determine the functionality of each siloxane unit. The organosiloxane units are therefore mono-, di-, or trifunctional. The nonfunctional molecule R_4Si cannot be used as a structural unit in polymers, but the tetrafunctional structural group $(SiO_{4/2})$ is occasionally used with the others in the synthesis of organosiloxane polymers, so that the basic units required for the production of silicones can be tabulated as shown below.

TABLE 1
STRUCTURAL UNITS OF THE POLYORGANOSILOXANES

Structural formula	Composition	Functionality	Symbol
$R_3Si—O—$	$R_3SiO_{1/2}$	monofunctional	M
R above, R below: $—O—Si—O—$	$R_2SiO_{2/2}$	difunctional	D
R above, O— below: $—O—Si—O—$	$RSiO_{3/2}$	trifunctional	T
—O above, O— below: $—O—Si—O—$	$SiO_{4/2}$	tetrafunctional	Q[a]

[a] The symbol "Q" is derived from "quadrifunctional" rather than "tetrafunctional" to distinguish it from T for "trifunctional".

* Page number of chapter references.

The symbols M, D, T, and Q in the last column of Table 1 have been introduced to represent the functionalities of the structural units and have proved very useful, since they make it possible to represent the structures of even complicated polymers rapidly and clearly.

The great diversity of the types of compounds found in polyorganosiloxane chemistry arises primarily from the fact that different *siloxane units* can be *combined* with one another in the same molecule. The routes already followed to this end are shown below by means of a number of typical examples.

1. The monofunctional siloxane unit can combine with itself only once, giving a hexaorganodisiloxane (I) of the type M_2:

$$R_3Si—O—SiR_3$$

I [M_2]

2. The difunctional units give closed rings on combination with one another. The smallest known ring (II) contains three siloxane units; rings with four (III) or five (IV) siloxane units are also easily obtainable.

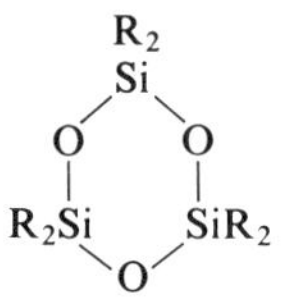

II [D_3]

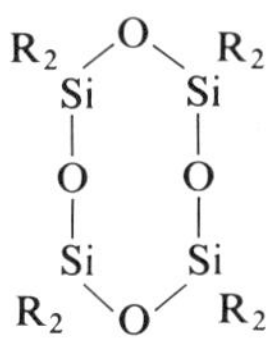

III [D_4]

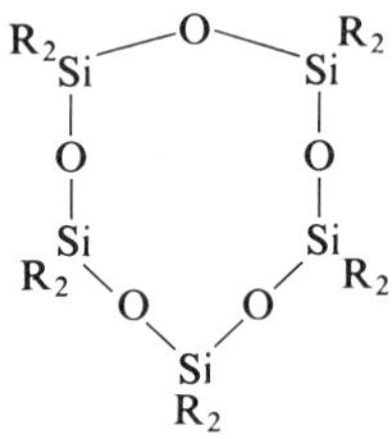

IV [D_5]

With certain polyorganosiloxanes, multimembered rings with a greater number of components, for example, eight or nine siloxane units, have also been iso

3. Trifunctional siloxane units combining with one another generally give molecules cross-linked randomly in three dimensions. Formula (V) gives a schematic cross section in plane projection.

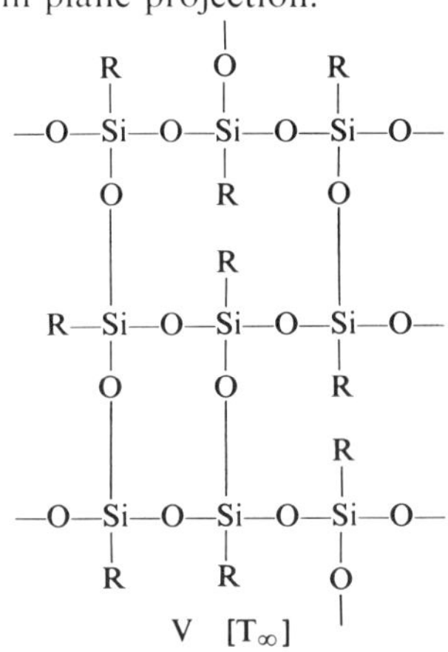

V [T_∞]

Under special circumstances, small cagelike structures, with four, six, eight, and twelve siloxane units, which could be interpreted as being polycyclic, have also been found.

4. The combination of mono- and difunctional siloxane units leads to linear siloxanes of various chain lengths, depending on the ratio of di- to monofunctional units

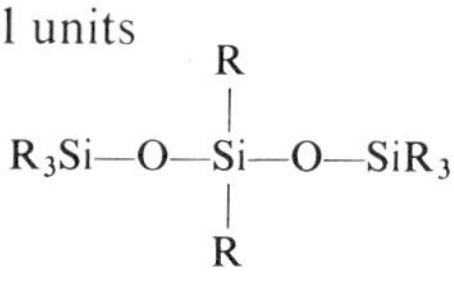

VI [M_2D]

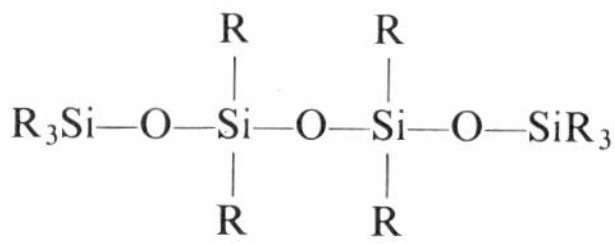

VII [M_2D_2]

and so on to linear high polymers.

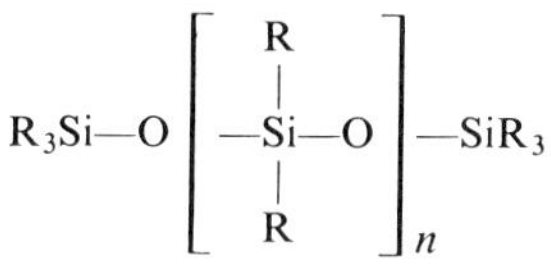

VIII [M_2D_n]

5. The combination of mono- and tri- or tetrafunctional siloxane units leads to low-molecular-weight structures such as

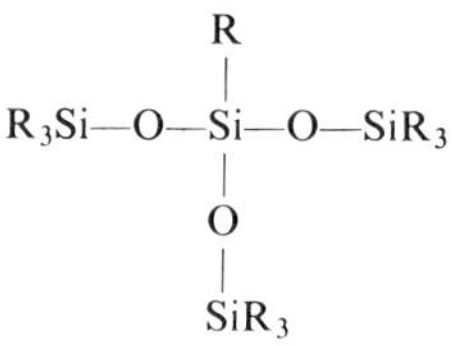

IX [M_3T]

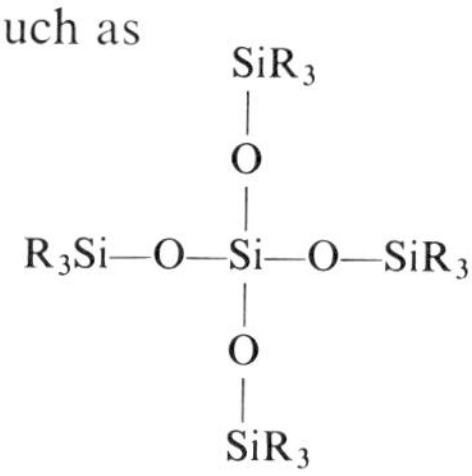

X [M_4Q]

6. Di- and trifunctional siloxane units in general yield macromolecules which have the nature of networks when the content of T units is high:

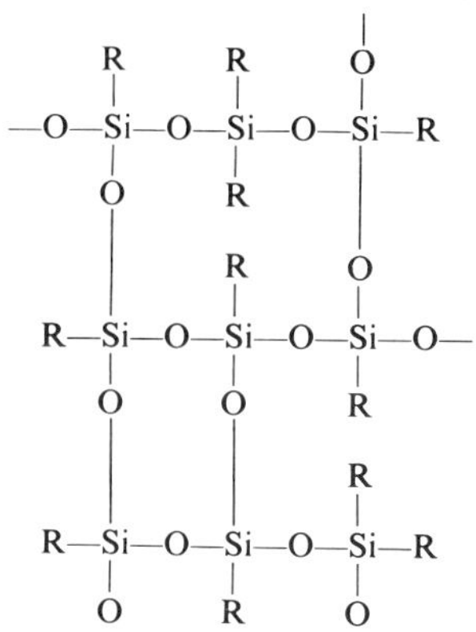

XI [D_nT_m $(m > n)$]

while an excess of D units gives chains which are cross-linked or branched at long intervals.

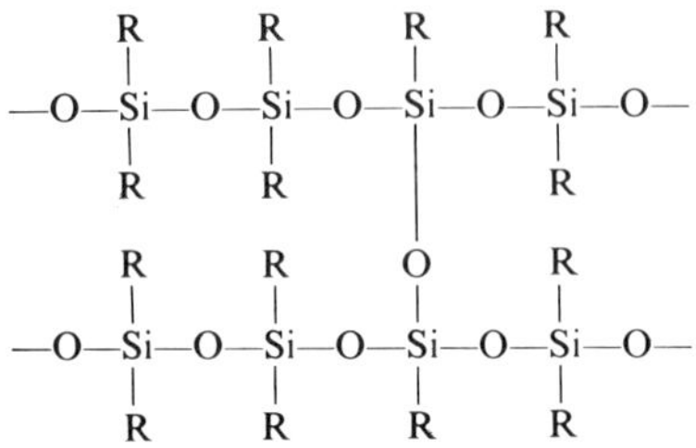

XII $[D_nT_m\ (n \gg m)]$

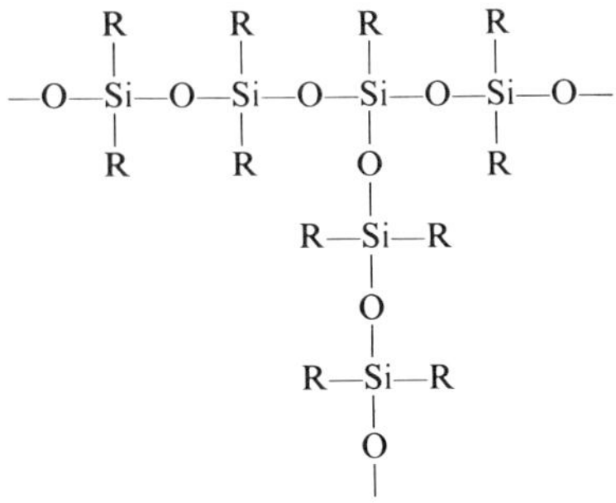

XIII $[D_nT_m\ (n \gg m)]$

On the other hand, limited numbers of D and T units can join together to form products of low molecular weight:

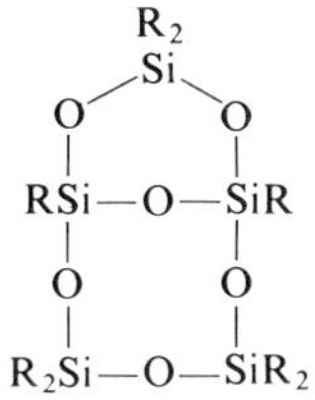

XIV $[D_3\ T_2]$

7. Finally, the union of di- and tetrafunctional units gives not only three-dimensional highly cross-linked molecular structures—like the combination of di- and trifunctional units—but also, under special conditions, spiro-compounds of the type (XV), whose structure is determined by the tetrahedral arrangement of the oxygen atoms around the silicon atom in the Q unit.

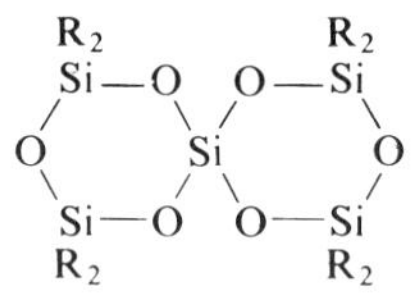

XV $[D_4\ Q]$

Given the same number of silicon atoms, it is therefore possible to obtain molecules of quite different structural types according to the nature of the participating siloxane units. This can be shown in principle by a series of polyorganosiloxanes each of which contains five silicon atoms (Table 2).

TABLE 2

POLYMETHYLSILOXANES WITH FIVE SILICON ATOMS IN THE MOLECULE[a]

Structural symbol	Structure classification	Empirical formula	B.p. (°C)	M.p. (°C)
$[D_5]$ (IV)[b]	cyclic	$(CH_3)_{10}Si_5O_5$	+210	−38
$[M_2D_3]$ (VIII)	linear	$(CH_3)_{12}Si_5O_4$	+229	ca. −80
$[M_4Q]$ (X)	branched	$(CH_3)_{12}Si_5O_4$	+91/9 mm Hg	−54
$[D_3T_2]$ (XIV)	cyclic and branched	$(CH_3)_8Si_5O_6$	+203	+118
$[D_4Q]$ (XV)	spiro	$(CH_3)_8Si_5O_6$	+204	+121

[a] After H. W. Kohlschütter (9).
[b] The numbers in parentheses refer to the corresponding structural formulas in the text.

Since, besides combinations of two different types of siloxane units, it is also possible to use combinations of three or four different types to construct a molecule, numerous other possibilities are available for synthesis. The siloxane units are incorporated according to their functionality and following the rules well-established in macromolecular chemistry: the D units for *chain* or *ring formation*, the M units as *chain stoppers* or *regulators*, and the T and sometimes the Q units for *cross-linking*.

A second method of influencing the properties of the molecules is offered by variation of the *organic radicals* on the silicon atoms. This approach seems to offer almost inexhaustible possibilities. The organic group may be a saturated or unsaturated hydrocarbon radical; it may also be substituted in various ways, and different organic groups may be attached simultaneously to one and the same silicon atom:

$$[R'R''R'''SiO_{1/2}] \quad [R'_2R''SiO_{1/2}]$$

The two methods of varying the molecular structure are not independent of each other, since whether a given siloxane structure can occur at all may also depend on the nature of the organic substituents on the silicon atom. This can be seen, for instance, with low-molecular-weight polyorganosiloxanes of the type T_n; the combination with $n=4$ occurs only when the organic groups are highly branched (for example, *t*-butyl or isopropyl groups), while configurations with $n=6$, 8, and 12 are found with siloxanes containing small organic groups, especially methyl radicals.

The possibilities of organic substitution are probably by no means so nearly exhausted today as those depending on the ability of the siloxane units to combine with each other. This is because, in the first place, the synthesis of the intermediates for the polyorganosiloxanes often encounters considerable difficulties, and, in the second place, attention has perforce been concentrated on the *methyl*-substituted compounds as a result of the discovery of the polymethylsiloxanes and their many unique and completely novel properties. Consequently, it is these compounds which, from the beginning of the technical development of silicone chemistry until today, have constituted the bulk of technical silicone products.

The special position occupied by the *methyl*-substituted polyorganosiloxanes will be constantly brought to the reader's attention and justified in detail in this book. At this point let us consider the following facts merely to gain a preliminary basic idea of this situation.

In the silicones, inorganic and organic structural units are linked together. It is obvious that the "inorganic" specifically silicatelike character of the silicones is more pronounced when the content of the "organic" constituents in the molecule is lower, i.e., it will be most marked in silicones containing the organic groups most deficient in carbon such as the methylsilicones. The latter are in fact the most unlike organic compounds, as is immediately shown by their low compatibility with them; on the other hand, their highly cross-linked forms resemble silica glass in their tendency to brittleness. In brief, they can claim to be the least "organophilic" and most strongly silicatelike members of this class.

It is also easily understandable that the organic character of the polyorganosiloxanes increases with increasing carbon content of the substituents on the silicon atoms. This pattern is again confirmed immediately by the compatibility of polyorganosiloxanes containing higher organic substituents with organic compounds.

Higher molecular weight liquid polydimethylsiloxanes, for example, are practically insoluble in the lower alcohols or in paraffin oils, while polymethylsiloxanes containing phenyl groups are miscible in all proportions with ethanol and special types even with methanol and paraffins.

Although *methyl*silicones completely dominate present-day silicone technology, it is nevertheless impossible to do without other organic groups for incorporation in the siloxane skeleton in certain products. *Phenyl* groups are the most important in this connection; their beneficial influence on the heat stability of silicones was recognized at an early stage. Methylsiloxane components containing *vinyl* groups, incorporated in straight-chain dimethylsiloxane polymers in small numbers to provide reactive centers in the otherwise almost unreactive chain, have also become important. In addition, interest has been devoted to organosiloxane units with higher aliphatic groups or, sometimes, with radicals substituted by fluorine or by amino, cyano, or other groups.

Corresponding to the variety of molecular structures, silicones are today known in various *states of aggregation*. They may be *crystalline*, *liquid*, *resinous*, or *elastomeric* under normal conditions. Furthermore, certain types have been produced which, like the famous bouncing putty, seems to mock all attempts at classifying their state. Depending on the mechanical stresses applied, they appear as viscous liquids, display rubberlike elasticity, or behave like brittle solids.

Technical interest has concentrated almost entirely on fluid, resinous, and rubberlike polymers. From the technological point of view, therefore, the silicone products are divided into *silicone oils*, *silicone resins*, and *silicone rubbers*. Silicone oils contain molecules of the type MD_nM (formula VIII); silicone resins consist, for example, of combinations of the type D_nT_m (formula XI). Silicone rubbers are produced from long-chain polymers MD_nM (formula VIII).

1.2 Nomenclature

1.2.1 General

The system of nomenclature used at present is based partly on nomenclature of the inorganic silicon compounds and partly on that of organic chemistry. No entirely consistent terminology has yet been evolved, and the structural formula is still *indispensable* in ambiguous cases.

Silicones belong to the class of organosilicon compounds. Their characteristic feature is that they have at least *one direct Si—C bond in the molecule.* Orthosilicic and polysilicic acid esters (organic silicates) are, therefore, not classified as organosilicon compounds, since the carbon in them is attached to silicon only through oxygen. What organosilicon compounds and silicic acid esters have in common is that they are both organic compounds of silicon. This latter term is therefore the most comprehensive one, covering also all those silicon compounds that contain organic groups but no Si—C bonds—not only the silicic acid esters but also isocyanosilanes, isocyanatosilanes, isothiocyanatosilanes, acyloxysilanes, etc. The scheme set out below is intended to illustrate these relationships and, in particular, the position of the silicones in the system of nomenclature of the organic compounds of silicon (see Table 3).

An early system was developed by Kipping, who called the silicon hydride SiH_4 "silicane" (*5*). With more complicated compounds this system led to unwieldy polysyllabic names, almost impossible to take in at one glance, and it therefore failed to gain acceptance.

The basis for present-day terminology has been provided by the proposals of Stock, who called SiH_4 "silane," by analogy with the alkanes (*22*). In addition, Stock coined terms such as "silanol" for H_3SiOH and "disiloxane" for $H_3SiOSiH_3$, which have also become part of current usage (*23*, *24*).

In its details, the terminology is based on the recommendations of Sauer (*19*) made in 1944; these were then systematized in 1946 by the American Chemical

TABLE 3

FAMILY TREE OF THE ORGANIC COMPOUNDS OF SILICON

Organosilicon compounds (containing Si—C bonds)		Organooxysilicon compounds (silicic acid and polysilicic acid esters or organosilicates) without Si—C bonds		Other organic compounds of silicon without Si—C bonds
Monomers	Polymers	Monomers	Polymers	(Predominantly monomeric)
Organosilanes Organohalo-silanes Organoalkoxy-silanes Organoamino-silanes, etc.	Polyorganosil-oxanes (Silicones) Polyorganosil-carbanes Polyorgano-silazanes Polyorgano-silthianes, etc.	Organooxy-silanes (Orthosilicic acid esters, Organic-orthosilicates)	Polyorganooxy-siloxanes (Polysilicic esters; organic poly-silicates)	Isocyanosilanes Isocyanato-silanes Isothiocyanato-silanes Acyloxysilanes, etc.

Society (*26*). In addition, the International Union of Pure and Applied Chemistry (IUPAC) published its rules for the nomenclature of organosilicon compounds in 1952 (*29*).

In the discussion of nomenclature which follows, it is intended to adhere to these rules by and large, although not in every respect, since it is proposed to supplement them with new suggestions.

1.2.2 MONOMERS

To name a monomer, we consider it as a substituted silane SiH_4. It follows, from the definition of organosilicon compounds that at least one hydrogen atom must be substituted by an organic group (R). The remaining hydrogen atoms can either still be present as such or be substituted by halogens, alkoxy, amino, isocyano groups, etc. (usually symbolized as X in general formulas).

The organic group can be a saturated, unsaturated, or substituted hydrocarbon radical. Organic chemical nomenclature is used for its characterization and particular differentiation is therefore made between the groups of the alkyl-, aryl-, and aralkylsilanes. The use of the terms "alkyl" instead of "organo" and "alphyl" for "alkyl," suggested by Krause and von Grosse (*10*) and by Rochow (*18*), has not been accepted.

If the organic group is a saturated unsubstituted hydrocarbon radical, there is a considerable difference between the reactivities of the R and X groups attached to silicon. The former are in general less easily attacked chemically than the latter. These differences are particularly striking in the case of their hydrolytic behavior. Organosilanes substituted exclusively with saturated hydrocarbon radicals (i.e., carrying no X groups) are called "nonfunctional," while organosilanes or silanes containing a *hydrolyzable* X group are called "functional" or, more specifically, "silicon functional," since the functional group is attached to a silicon atom. If the organic group is an *unsaturated* hydrocarbon radical or one *substituted* with reactive atoms or groups, it too becomes a functional group. In this case the silane is said to be "organofunctional." A silane can therefore be either silicon functional, organofunctional, or both silicon functional and organofunctional, e.g.,

Only silicon functional	$SiCl_4$, $CH_3Si(OC_2H_5)_3$
Only organofunctional	$CH_2{=}CH(CH_3)_3Si$, $ClCH_2(CH_3)_3Si$
Silicon functional and organofunctional	$CH_2{=}CH(CH_3)SiCl_2$, $ClC_2H_4SiCl_3$, $H_2N(CH_2)_3Si(OC_2H_5)_3$

Silanes in which X is a halogen atom, an alkoxy group, or an amino group are called "halosilanes," "alkoxysilanes," and "aminosilanes," respectively. The aminosilanes may also be called "silylamines." The Si—OH group is called the "silanol group." Metal salts of the silanols, for example, $(C_6H_5)_3SiONa$, are called "metal silanolates."

If the organic group is substituted, the normal nomenclature of organic chemistry for the characterization of its structure is used, e.g., γ-aminopropyltriethoxysilane, hydroxymethyltrimethoxysilane, tri-(4-chlorophenyl)-ethoxysilane.

The *Si—H bond* is regarded as one of the silicon-functional groups. The derivation of the nomenclature system from silane SiH_4 implies that only atoms and groups of atoms replacing hydrogen need be expressly named. The names "trimethylsilane" for $(CH_3)_3SiH$, "methyldichlorosilane" for CH_3SiHCl_2, and "trichlorosilane" for $SiHCl_3$ all characterize the substituted silanes unambiguously. This is not the case in group names. The term "organochlorosilanes" does not specify whether the silanes contain hydrogen atoms as well as organic groups and chlorine atoms. To avoid misunderstandings, and to make the terms unambiguous, the hydrogen must be specified in these and in similar cases.

The IUPAC rules provide the principle of indicated hydrogen for similar problems in organic chemistry (*30a*). It seems reasonable to apply this principle to organosilicon nomenclature. In the example given, therefore, organochlorosilanes that also contain Si—H bonds will be designated organo-*H*-chlorosilanes.*

* This is based on a suggestion by Dr. H. Grünewald, Verlag Chemie.

It seems highly desirable to use a definite system in the formation of *names of compounds* and in the representation of *formulas*. The following rules will be adopted in this book.

In names of compounds and formulas the organofunctional groups will be given first, then the nonfunctional groups, and finally the silicon-functional substituents, e.g.:

Bromomethyldimethylchlorosilane

$BrCH_2(CH_3)_2SiCl$

In the naming of compounds, the substituents within the three groups are given in alphabetical order. This is to facilitate the use of the subject index. In the formulas, on the other hand, the organic substituents are arranged in order of the number of carbon atoms, e.g.:

*M*ethyl*p*henyldichlorosilane $CH_3(C_6H_5)SiCl_2$

*E*thyl*m*ethyldichlorosilane $CH_3(C_2H_5)SiCl_2$

*B*romomethyl*c*hloromethyl*m*ethylchlorosilane

$BrCH_2(ClCH_2)(CH_3)SiCl$

The nomenclature is further supplemented by using radical names formed according to the following principles, which are borrowed from organic chemistry (all H atoms can be substituted).

silyl-	$—SiH_3$	siloxy-	$—OSiH_3$
silylene-	$>SiH_2$	silylthio-	$—SSiH_3$
silylidyne-	$—SiH<$ (three bonds)	silylamino-	$—NHSiH_3$

This increases the choice of names considerably, so that several names can be assigned to one and the same compound; for example, in the case of the compound $(H_3Si)_3N$ the possibilities are (*25*):

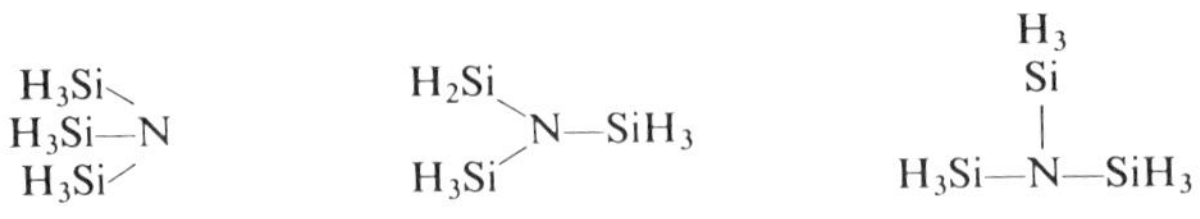

Trisilylamine — *N*,*N*-Disilylaminosilane — *N*-Silyldisilazane

The radical nomenclature has become firmly established with some monomers. Compounds in which the metal atoms are attached to Si are designated as silyl derivatives of the metal; e.g., triphenylsilyllithium, trimethylsilylpotassium. The same procedure is adopted with esters of sulfuric or phosphoric acid: trimethylsilyl sulfate, trimethylsilyl phosphate.

1.2.3 Polymers

In polymeric organosilicon compounds, the units forming the polymer skeleton are preceded by the prefix "poly." In this book the use of the prefix "oligo" to denote low-molecular-weight polymers is avoided, particularly since the boundary between oligomers and polymers is arbitrary.

If the silicon atoms in the polymers are linked directly with one another, the compounds are called "polysilanes."* Polymers with oxygen bridges between silicon atoms are called "polysiloxanes," those with NH bridges "polysilazanes," and those with S bridges "polysilthianes." The characteristic linkages are:

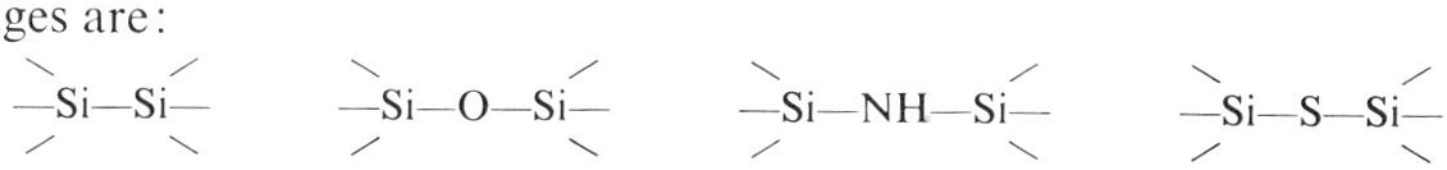

Polysilane | Polysiloxane | Polysilazane | Polysilthiane

Polymeric combinations of silicon and carbon atoms are generally classified under the generic names of "polysilcarbanes" or "polycarbosilanes." The usage varies considerably in detail. The following terms will be used in this book.

Polymers in which the silicon atoms are linked through alkylene bridges are "polysilalkylenes," and those with arylene bridges "polysilarylenes." If alkylene or arylene bridges and oxygen bridges alternate, the polymers are called "polysilalkylenosiloxanes" or "polysilarylenosiloxanes." Polymers in which alkylene or arylene bridges alternate with Si—Si bonds are called "polysilalkylenosilanes." This is illustrated below, the characteristic linkages being shown in each case.

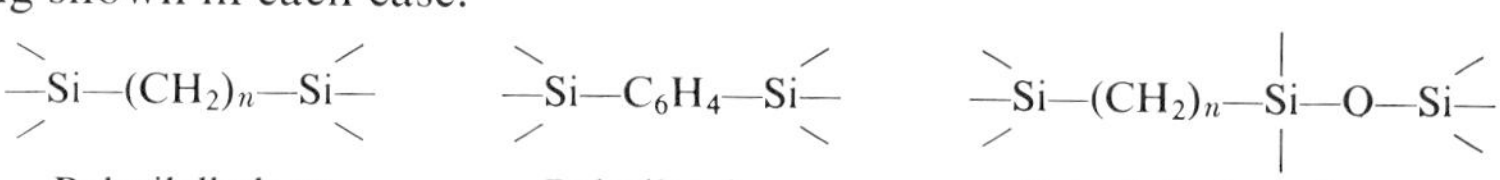

Polysilalkylene | Polysilarylene | Polysilalkylenosiloxane

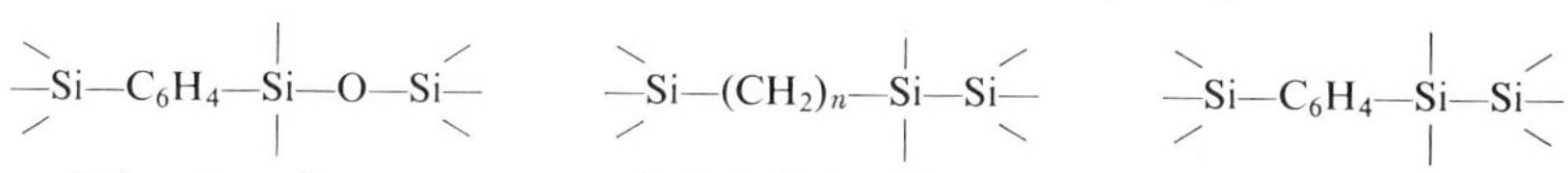

Polyarylenosiloxane | Polysilalkylenosilane | Polysilarylenosilane

Terminal groups in chain or cross-linked polymers are generally placed before the polymer-forming units. For example, α,ω-dichloropolydimethylsilane denotes

$$Cl\left[-\underset{CH_3}{\overset{CH_3}{\underset{|}{\overset{|}{Si}}}}\right]_n-Cl$$

* The IUPAC rules do not provide for the use of the prefix "poly." The difficulties that can arise because of this are very obvious in dealing with the polysilanes. If these were simply called silanes it would be impossible to distinguish them from the monomeric silanes. The use of the prefix "poly," moreover, corresponds to the usual practice in the chemistry of organic polymers.

The nomenclature is rather more complex in the case of polymers containing bridge atoms, as can be seen from the following examples:

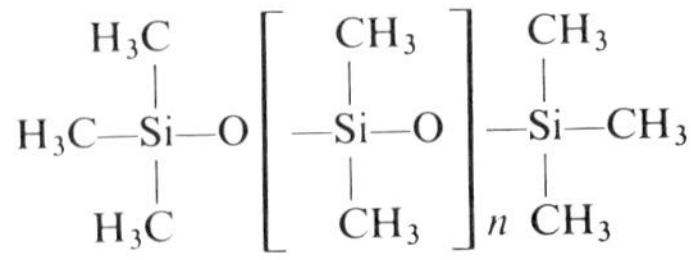

α,ω-Bis-trimethylsiloxy-polydimethylsiloxane

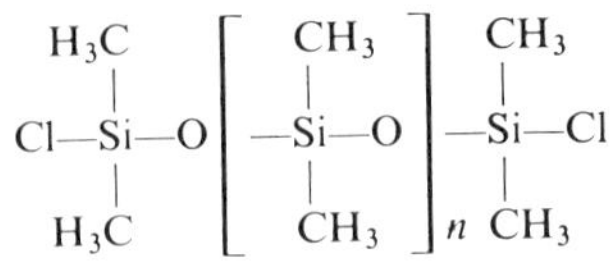

α,ω-Bis-dimethylchlorosiloxy-polydimethylsiloxane

In the case of siloxanes with terminal OH groups, the nomenclature may also be based on the silanol:

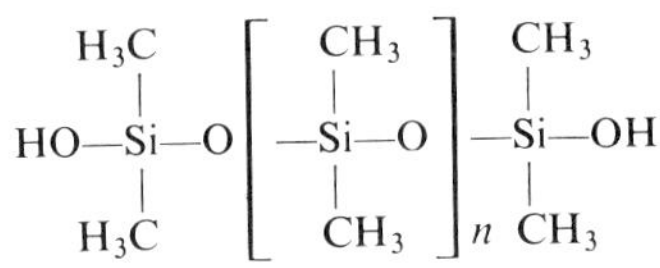

α,ω-Bis-dimethylhydroxysiloxy-polydimethylsiloxane or polydimethylsiloxane-α,ω-diol

In smaller polymer molecules with definite numbers of silicon atoms, these are counted and indicated in the usual manner by the prefixes "di," "tri," and "tetra." Cyclic polymers are denoted by the prefix "cyclo," e.g.:

$[(CH_3)_2SiO]_4$	$[(CH_3)_2SiNH]_4$
Octamethylcyclotetrasiloxane	Octamethylcyclotetrasilazane

The organic portion is in all cases designated by the prefix "organo" ("alkyl," "aryl"), while the presence of X groups is shown in the manner adopted with the monomers.

A special nomenclature must be used for *hydrogen* bonded to silicon, as was the case for the general group names of substituted silanes (Section 1.2.2). If the hydrogen in polymers were not named, serious ambiguity could arise. A polymer having the structural units

$$[CH_3SiHO]$$

would simply be polymethylsiloxane, and so could not be distinguished by its name from a siloxane having the structural groups

$$[CH_3SiO_{3/2}]$$

As in the special problems of monomer nomenclature, it is thought that the hydrogen in polymers should be denoted in accordance with the principle of indicated hydrogen. Accordingly, hydrogen bonded to silicon in the polymer will be indicated in this book by the insertion of an italicized -*H*- in the name of the compound. Thus the above polymer having the structural units $[CH_3SiHO]$ is polymethyl-*H*-siloxane.

To designate the *position of the substituents* in cyclic and linear polymers, the atoms are numbered starting from a key atom as is usual in organic chemistry. In chains, the counting begins from the terminal silicon atom. Each chain atom is included in the enumeration as recommended by the American Chemical Society. While this would appear to be an unnecessary burden in the case of the siloxane polymers, since the nonsubstituted oxygen atoms in the siloxane chain also have to be counted, in other cases, as with the silazanes or the silmethylenes, where the bridge atoms between the silicon atoms could also carry substituents, this method of counting is meaningful and it is therefore extended to the siloxanes in the interests of consistency. A typical example is 1,5-dihydroxy-1,1,5,5-tetramethyl-3,3-diphenyltrisiloxane:

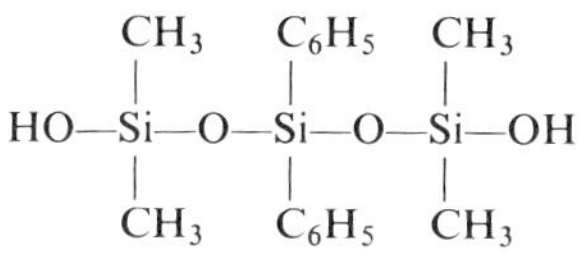

The use of *radical names* also broadens the possibilities with the polymers; in analogy with the radical nomenclature adopted for the monomers, the following, for example, are in current use:

$H_3Si{-}O{-}SiH_2{-}$	disiloxanyl-	$H_3Si{-}O{-}SiH_2O{-}$	disiloxanoxy-
$H_3Si{-}NH{-}SiH_2{-}$	disilazanyl-	$H_3Si{-}NH{-}SiH_2NH{-}$	disilazanylamino-

In a branched-chain compound, the chain carrying the largest number of atoms determines whether the parent compound is a silane, siloxane, silthiane, or silazane, etc., the branches being named according to the rules of radical nomenclature:

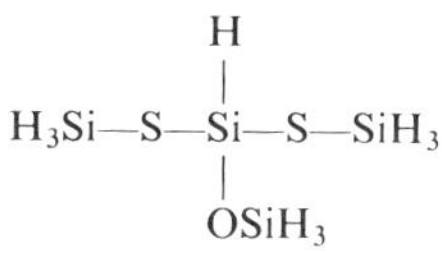

3-Siloxytrisilthiane

If the choice between two or more parent compounds cannot be made on the basis of numbers of atoms, the decision is made according to the following order of precedence: siloxane, silthiane, silazane, silane:

$$H_3Si{-}O{-}SiH_2{-}S{-}SiH_3$$

Silylthiodisiloxane

All these rules apply to mono- or polymeric compounds in which silicon is the predominant element or can reasonably be considered as such. The more pronounced the organic moiety of an organosilicon compound, the more useful it might be found to turn to organic nomenclature and use the so-called "oxa-aza convention" ("*a*" nomenclature) (*15*, *21*). This system can also be used with organosilicon compounds rich in silicon if it seems to offer the advantages of clarity and simplicity:

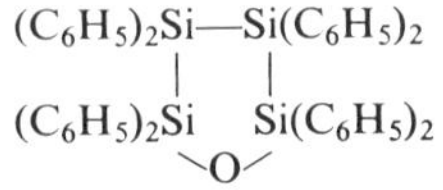

Octaphenylcyclooxapentasilane

While here the nomenclature is based on that of organic chemistry, it may sometimes—in accordance with the dual nature of the silicones—be useful to adhere to the *terminology of silicate chemistry*. This is done in naming metal salts of organosiloxanols; these can formally be related to the silicates, from which they differ by the partial replacement of the O atoms by hydrocarbon residues. For simplicity—and not quite in accordance with the sense of our definitions—the anions of such saltlike compounds will frequently or usually be called "siliconate" ions rather than "siloxanolate" ions. This gives rise to names like "sodium methylsiliconate"—as a counterpart to sodium silicate—and the group name "heterosiliconates" for saltlike compounds with anions containing both silicon and other metal atoms such as aluminum, gallium, and iron [e.g., (I) (*20*)] and following the structural principle of the alumosilicates [e.g., feldspar (II)]:

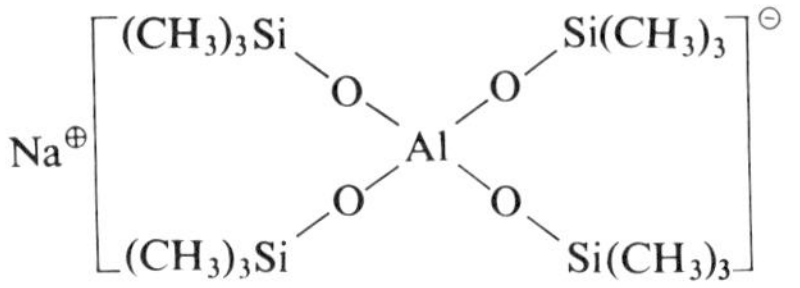

(I) Sodium tetrakis-trimethylsilylaluminate

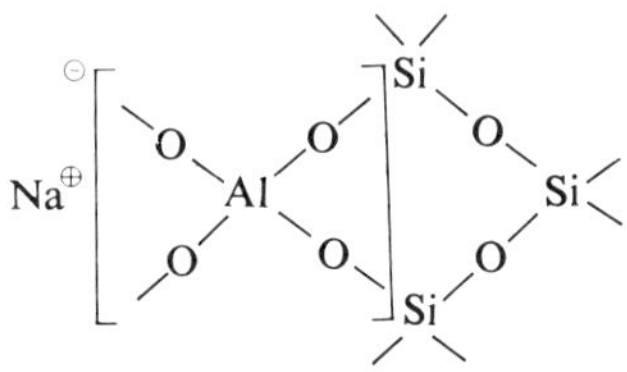

(II) Section of a feldspar structure

If metal heteroatoms are incorporated into a polymer skeleton as substituents of Si atoms, "heterosiloxanes" such as, for example, alumosiloxanes ferrosiloxanes, plumbosiloxanes, and stannosiloxanes are produced.

1.3 Historical Survey

Compounds of silicon, in the form of the naturally occurring silicates, have been of service to man for thousands of years and belong to the oldest of man's raw materials.

After the appropriate operating techniques had been developed, a chemical technology of considerable dimensions became established. It was based on the natural silicates as raw materials, and its products were still silicates, refined and adapted to the needs of man (hydraulic cements, glass in all its forms, ceramic products in their widest sense, etc.). A mass-production industry of specifically inorganic nature and based especially on silicates has thus developed up to very recent times. In addition, interest has arisen in elementary silicon, its alloys, and silicon carbide. Compared with the large variety of possible silicon compounds, however, the variety of products used in industry was relatively small. Other inorganic compounds of silicon, such as silicon tetrachloride, or organic compounds, such as silicic acid esters, have begun to arouse interest only in quite recent times.

On the other hand, there has never been any lack of effort in the field of pure scientific investigation to prove that silicon, in analogy with carbon, should be in a position to generate its own quasi-organic chemistry. Even if this particular hope was shattered by Stock's work, in which he established the instability of the higher silanes, there were always research workers eagerly in pursuit of a combined silicon-carbon chemistry. Their first aim was to find routes and methods for the preparation of compounds with silicon-carbon bonds, a type of bond having no known pattern in any natural substance.

The first tentative steps in this direction were made by Friedel and Crafts in the years 1863–1880, and later by Ladenburg. These authors found that organosilicon compounds could be prepared by heating dialkylzincs with silicon tetrachloride in sealed tubes at about 160°C (*3*):

$$2Zn(C_2H_5)_2 + SiCl_4 \rightarrow 2ZnCl_2 + (C_2H_5)_4Si$$

Pape (1884) carried out the reaction of trichlorosilane, as well as silicon tetrachloride, with dipropylzinc (*14*):

$$2SiHCl_3 + 4Zn(C_3H_7)_2 \rightarrow (C_3H_7)_3SiH + (C_3H_7)_4Si + 3ZnCl_2 + Zn + C_3H_8$$

Polis (1885) succeeded in synthesizing aromatic derivatives of silicon by means of the Wurtz reaction:

$$SiCl_4 + 4RCl + 8Na \rightarrow R_4Si + 8NaCl$$

and accordingly prepared tetraphenylsilane, tetratolylsilane, and tetrabenzylsilane (*16*).

Methods for the partial alkylation of silicon compounds were discovered at an early stage. Friedel and Ladenburg obtained alkylalkoxysilanes by the reaction of diethylzinc with triethoxychlorosilane and metallic sodium without the use of sealed tubes. The following reaction scheme can be postulated (*2*):

$$Zn(C_2H_5)_2 + 2Si(OC_2H_5)_3Cl + 2Na \rightarrow 2C_2H_5Si(OC_2H_5)_3 + 2NaCl + Zn$$

Ladenburg prepared the series of the various ethylchlorosilanes, ethylethoxychlorosilanes, ethylethoxysilanes, and ethoxychlorosilanes by this method and described their properties (Table 4) (*11*).

TABLE 4

SUBSTITUTED SILANES SYNTHESIZED BY A. LADENBURG, 1872

$SiCl_4$	$Si(OC_2H_5)Cl_3$	$Si(OC_2H_5)_2Cl_2$	$Si(OC_2H_5)_3Cl$	$Si(OC_2H_5)_4$
$C_2H_5SiCl_3$	$C_2H_5Si(OC_2H_5)Cl_2$	$C_2H_5Si(OC_2H_5)_2Cl$	$C_2H_5Si(OC_2H_5)_3$	
$(C_2H_5)_2SiCl_2$	$(C_2H_5)_2Si(OC_2H_5)Cl$	$(C_2H_5)_2Si(OC_2H_5)_2$		
$(C_2H_5)_3SiCl$	$(C_2H_5)_3Si(OC_2H_5)$			
$(C_2H_5)_4Si$				

By the reaction of diphenylmercury with silicon tetrachloride in sealed tubes at 300°C, Ladenburg (*12*) also succeeded in synthesizing phenylchlorosilanes

$$Hg(C_6H_5)_2 + SiCl_4 \rightarrow C_6H_5SiCl_3 + C_6H_5HgCl$$

Research in this sphere received a tremendous impulse from the work of Kipping, who in 1904 (*4*)—almost simultaneously with Dilthey (*1*)—found a route into the field of organosilicon compounds via the Grignard reaction:

$$2C_2H_5MgCl + SiCl_4 \rightarrow (C_2H_5)_2SiCl_2 + 2MgCl_2$$

Kipping, a professor at University College, Nottingham, England, worked for 45 years on the chemistry of the organosilicon compounds, publishing at least 57 papers, of which 53 appeared as a numbered series in the *Journal of the Chemical Society* (London) (*8*) and four elsewhere (*4*, *6*).

After partially alkyl-substituted silanes also containing hydrolyzable functional groups had become available, the route to the silanols was opened. Kipping discovered the principle of intermolecular condensation of silanediols and silanetriols to polysiloxanes, and recognized, as Stock had done in a different context, the essential differences between the chemistry of silicon and classical organic chemistry.

Oddly enough, Kipping overlooked the polymeric phenomena he encountered at every turn. What is now the focus of interest of silicon chemistry was, to him, a troublesome by-product. He and his co-workers were constantly at pains to isolate crystalline compounds from the perpetually occurring resinous, oily, or gelatinous products. It is thus certainly to Kipping's credit that he gained a knowledge of a large number of defined organosilicon compounds, and his work became the classical scientific foundation of the subject, but it was not the starting point for present-day silicone chemistry, because he failed to recognize the importance of polymerization in this field. This fact would appear to throw light on Kipping's remarks in his now famous Bakerian lecture in 1937, which he concluded with a certain scepticism (*7*):

"Most, if not all, of the known types of organic derivatives of silicon have now been considered and it may be seen how few they are in comparison with those which are entirely organic; as moreover the few which are known are very limited in their reactions, the prospect of any immediate and important advance in this section of organic chemistry does not seem to be very hopeful."

Shortly after that the picture changed radically. Toward the end of the thirties, two groups of workers in the United States, one in the laboratories of the Corning Glass Works under J.F. Hyde and the other at the General Electric Co. (W.J. Patnode and E.G. Rochow), who were tackling problems quite different from Kipping's, obtained a large number of interesting and technically promising results. The Russians (B.N. Dolgov and K.A. Andrianov) also turned their attention to organosilicon polymers at that time. The discovery of a process for the direct synthesis of the silicone intermediates, the organochlorosilanes, by Rochow (*17*), after Dow Corning's first step had still been based on the Grignard process, stimulated further large-scale technical development.

The need in industry, particularly in the electrical industry, for polymers more heat resistant than those which could be supplied from purely organic sources, provided the drive for this new departure. The aim of this work was to transfer to the synthesis of new polymers the structural principles of the silicon-oxygen skeleton known from the study of the natural silicates. The incorporation of organic groups could make it possible to arrive at linear polymers starting from the highly cross-linked glass structure. These working hypotheses made Kipping's organosiloxanes and the principle of their synthesis the starting point of new researches which were, of course, determined by the problems of polymer chemistry.

Silicone chemistry soon developed well beyond the original aim of finding heat-resistant polymers. Once the unique and versatile character of this type of compound was recognized, progress toward wider diversification became rapid. The American firms also had the problem of liberating their development from the original points of growth, and enabling it to continue along independent lines. Corning Glass Works and Dow Chemical founded the Dow Corning Corporation of Midland, Michigan, in 1943, and the General Electric Co. established its own Silicone Department in Waterford, New York, in 1947. The usefulness of silicone products in armaments, particularly in aircraft construction, resulted in an energetic drive for their development during World War II. In Germany, too, some silicone research was started at the beginning of the war (*13*). In contrast to the situation in the United States, however, the war and the control of research here exerted a retarding influence.

Interest was aroused in all the larger industrial countries after the war, and numerous firms engaged in research and production in this field. In the United States, Union Carbide entered the market alongside the other two producers in 1957 with its own Silicone Division at Long Beach, West Virginia. This development was based on preliminary work carried out by the Linde Company at Tonawanda, New York. In 1965, the Stauffer Chemical Co. also started up a silicone plant, at Adrian, Michigan. In the German Federal Republic, Farbenfabriken Bayer AG of Leverkusen, since 1952, and Wacker-Chemie of Munich, since 1951, have offered a complete range of silicone products. Special products are catered for by Th. Goldschmidt AG, Essen. In East Germany silicones are produced by Chemie-Werke, Nünchritz. Rhône-Poulenc and S.I.S.S. (Société Industrielle des Silicones et des Produits Chimiques du Silicium) are silicone producers in France, and Imperial Chemical Industries and Midland Silicones in Great Britain. Selected silicone products and fields of application are dealt with by the Union Chimique Belge in Belgium. Silicones are made and marketed in Japan by the Shin Etsu Chemical Industry Co., the Tokyo Shibaura Electrical Co. Ltd., Fuji Kobunshi Kogyo, and Toray Silicone. The U.S.S.R. has made silicones probably since 1947 and a smaller amount is produced in Czechoslovakia.

As a result of this intensive preoccupation with silicones, increased scientific efforts were simultaneously devoted to the entire field of organosilicon compounds. The interests of research workers also turned to silicon polymers with bridge atoms other than oxygen, with the result, however, that the unique position of the silicones was made even more abundantly clear. So far, no organosilicon polymers have been found in this related field the properties of which would have demanded anything like the same interest. On the other hand, stimulated by the technological successes scored by silicones, research on esters of polysilicic acids was taken up again. These were known in principle and show considerable analogies in their make-up to silicones themselves. A few classes of products have developed in this field which are an interesting complement and parallel to silicones.

The increasing scientific and technical interest lavished on silicones perforce stimulated research in the field of monomers. The wealth of knowledge gained as a result is reflected in the large proportion of this book devoted to monomers. Such work cannot be valued too highly for the future, since it provides the basis for the further development and modification of polymers. Many spectacular advances in the polymer field would be unthinkable without corresponding advances in monomers.

The most recent trend in monomer research is toward the organofunctional silanes. This appears particularly hopeful, since it would offer the possibility of a bridge between organosilicon chemistry and pure organic chemistry, in spite of the special position of the silicones proper.

1.4 Economics

The available statistical material on the production of silicones in individual countries and on a world scale is sparse and not without contradictions. The existing uncertainties may be due to the fact that no differentiation has been made between plant capacities and actual production figures, or between the production of finished products—which may contain water or solvent—and that of 100% silicone material.

In view of this, only estimates of the production figures will be possible.

American production, which in 1950 was only in the region of 1000 tons* per annum of 100% silicone material, is estimated to have risen to 14,000 tons per annum in 1965, corresponding to a turnover of about $90,000,000.

In the producing countries of western Europe—Germany, Great Britain, France, and Belgium—about 6000 tons are estimated to have been produced in 1965.

The production schedules of the producing firms in all these countries differ widely, especially in their emphasis on different products. It is thus particularly difficult to give a more precise account of what share different classes of products have of the total production. A forecast for 1958 (*28*) divided production in the United States among the major silicone products as in the accompanying tabulation (column 2) by tonnages. The author's own estimate for 1963 (column 3) shows an increase in the oils and emulsions produced.

Product	1958[a] (%)	1963[b] (%)
Oils, emulsions	33.3	~40
Rubbers	33.3	~30
Resins	20.0	20
Miscellaneous	13.4	10

[a] See ref. *28*.
[b] Author's estimate.

For the most part this would probably be due to the development of liquid polyether-siloxanes which have rapidly established themselves as stabilizers for synthetic foams. The wider application of silicone emulsions in textile finishing may also have made a larger contribution. The changes in the over-all picture are not very marked, and it has probably maintained its fundamental character throughout the last decade owing to the fact that every range of silicone products is made up of a host of individual products, each of which is

* All figures are in metric tons.

directed at a small or medium-sized market and used in an even greater variety of fields. The upward or downward movements in the demand for individual products generally compensate each other, unless the picture is radically altered by a special and particularly successful new development.

From these same considerations, it would appear that, given a certain stability in the over-all situation, the ratio of 100% silicone to the total amount of finished product, an interesting figure for market estimates, has probably not changed appreciably for years. If the market is divided between the four main classes of products as given above, this figure has probably remained around 0.5 in the West.

Japanese firms increased their production from a total of about 300 tons in 1956 to 1770 tons in 1962 (*30*). The capacities of the producing firms are said to have been increased to 2400 tons per annum at Shin Etsu and 900 tons per annum at Tokyo Shibaura in recent years (*27*).

Silicone production in the eastern block can only be estimated. The U.S.S.R. has a number of silicone-producing plants whose output has been quoted at 2200 tons per annum (*31*) for 1957. It is uncertain if this figure refers to the final products or their content of 100% silicone. Annual production is said to have risen considerably since then. The production in East Germany and Czechoslovakia on the other hand may be comparatively small. It appears that silicone resins form a very much larger proportion of the total output in the eastern block than in the West.

The production of silicones in the western world and in Japan can thus be estimated at approximately 22,000 to 24,000 tons for 1965.

The sales of silicone products have up to now risen steeply and steadily throughout the world. Not only are new applications constantly found for the "classical" products of silicone chemistry but also entirely new developments are continually being made available. This whole field thus has an extremely interesting future.

The progress achieved by technical improvements in production and by the extension of the market is also reflected by price changes. The present market price for silicones is only about one-third of the 1950 figure.

REFERENCES

(1) W. Dilthey, *Ber. dtsch. chem. Ges.* **37**, 319 (footnote 2) [1904]. (2) C. Friedel and A. Ladenburg, *Liebigs Ann. Chem.* **159**, 259 [1871]. (3) — and J. M. Crafts, ibid. **127**, 31 [1863]; idem "Siliciumtetramethyl," ibid. **136**, 302 [1865]. (4) F. S. Kipping, *Proc. Chem. Soc.* [London] **20**, 15 [1904]. (5) —, ibid. **28**, 243 [1912]; *J. Chem. Soc.* [London] **101**, 2106 [1912]. (6) — and co-workers, *Proc. Chem. Soc.* [London] **15**, 174 [1899]; **28**, 243 [1912]. (7) —, *Proc. Roy. Soc.* [London], Ser. A **159**, 139 [1937]. (8) — and co-workers, *J. Chem. Soc.* [London] **79**, 449 [1901]; **91**, 209, 717 [1907]; **93**, 198, 439, 457, 2004, 2090 [1908]; **95**, 69, 302, 489 [1909]; **97**, 142, 755 [1910]; **99**, 138 [1911]; **101**, 2106, 2108, 2125, 2142, 2156 [1912]; **105**, 40,

484, 679 [1914]; **107**, 459 [1915]; **119**, 647, 830, 848 [1921]; **123**, 2590, 2598, 2830 [1923]; **125**, 2291, 2616 [1924]; **1927**, 104, 2719, 2728, 2734; **1928**, 1427, 1431; **1929**, 357, 360, 1176, 1180, 2545; **1930**, 1020, 1029; **1931**, 1290, 2774, 2830; **1932**, 2200, 2205; **1933**, 1040; **1935**, 1085, 1088; **1944**, 81. (9) H. W. KOHLSCHÜTTER, *Fortschr. chem. Forsch.* **1**, 1 [1949].

(10) E. KRAUSE and A. v. GROSSE: *Chemie der metallorganischen Verbindungen*, published by Bornträger, Berlin, 1937, p. 254 ff. (11) A. LADENBURG, *Liebigs Ann. Chem.* **164**, 300 [1872]. (12) —, ibid. **173**, 151 [1874]. (13) R. MÜLLER, VEB Silikonchemie, German Pat. (East German) 5348 [6.6.1942]. (14) C. PAPE, *Liebigs Ann. Chem.* **222**, 370 [1884]. (15) A. M. PATTERSON, *J. Amer. Chem. Soc.* **50**, 3074 [1928]. (16) A. POLIS, *Ber. dtsch. chem. Ges.* **18**, 1540 [1885]. (17) E. G. ROCHOW, General Electric, U.S. Pat. 2380995 [26.9.1941]. (18) —: *An Introduction to the Chemistry of the Silicones*, 2nd Ed., New York, John Wiley and Sons Inc., 1951. (19) R. O. SAUER, *J. chem. Educat.* **1944**, 303.

(20) H. SCHMIDBAUR and M. SCHMIDT, *Angew. Chem.* **74**, 589 [1962]; *J. Amer. Chem. Soc.* **84**, 3600 [1962]. (21) R. STELZNER: *Literatur-Register der organischen Chemie*, Einführung zu Band V, Verlag Chemie, Berlin, Leipzig, 1926. (22) A. STOCK, *Ber. dtsch. chem. Ges.* **49**, 108 [1916]. (23) —, ibid. **50**, 169 [1917]. (24) —, ibid. **50**, 1769 [1917]. (25) — and K. SOMIESKI, ibid. **54**, 740 [1921]. (26) *Chem. Engng. News* **24**, 1233 [1946]. (27) *Chem. Ind.* **14**, June, 308 1962. (28) *Chemical Week*, **1958**, 85. (29) *J. Chem. Soc.* [London] **1952**, 5064.

(30) *Kunststoffe* **54**, 71 [1964]. (30a) *Nomenclature of organic chemistry*, Sect. A+B, 2nd Ed., published by IUPAC, London, Butterworths, 1966. (31) *Rubber and Plastics Age* **1959**, 136.

CHAPTER 2

Preparation of Silanes with Nonfunctional Organic Substituents

2.1 Monomers in Silicone Chemistry

The starting materials for the preparation of polyorganosiloxanes are usually monomeric organosilicon compounds simultaneously containing organic groups and silicon-functional groups. In the development of any polyorganosiloxane and, indeed, any polymeric organosilicon compound, the first problem is usually the synthesis of the monomeric intermediates. Frequently, as in the case of the polymethylsiloxanes, the production of the monomeric precursors is far more difficult and exacting than the subsequent conversion of the monomers into polymers. The technological development of the silicones themselves has, therefore, been decisively influenced by the discovery of suitable and efficient processes for the synthesis of the monomers.

The monomers are also of interest as starting materials for the preparation of silicones *in situ* when, for example, a silicone film is to be formed on the surface of a material in the position in which it will be finally used. The significance of the monomers for syntheses outside the immediate sphere of silicone chemistry, in particular the part played by organofunctional silanes, has already been emphasized.

The present chapter deals with methods of obtaining monomeric compounds in which unsubstituted saturated organic radicals are attached directly to silicon. Methods for the preparation of silanes with organic groups containing substituents or unsaturated bonds will be found in Chapter 4. In accordance with the nature of the book, particular emphasis has been placed on the treatment of those processes that are important in the *technical* production of silicones. Readers versed in pure chemistry are referred to the appropriate standard works (*102*, *177*) for processes of academic interest or suitable for laboratory preparations.

To appreciate the tasks and problems arising in the synthesis of the monomers it is of fundamental importance to understand that the process used must be capable of yielding organosilanes with different functionalities, since all functionalities are required for synthesis of silicones. The silanes of different functionalities that are required and the siloxane units to which they give rise are as follows:

Substituted silane		Organosiloxane unit	
R_3SiX	$\rightarrow$	$R_3SiO_{1/2}$	monofunctional
R_2SiX_2	$\rightarrow$	$R_2SiO_{2/2}$	difunctional
$RSiX_3$	$\rightarrow$	$RSiO_{3/2}$	trifunctional
SiX_4	$\rightarrow$	$SiO_{4/2}$	tetrafunctional

The steps between these are of quite different types. The processes most important technologically are based on the Si—Cl bond as the silicon-functional linkage. They thus comprise first the synthesis of organochlorosilanes and subsequently their conversion to organosiloxanes, via the organosilanols, after elimination of the chlorine by hydrolysis. Besides the Si—Cl group, the Si—OC_2H_5 group has found technical use as a functional unit. So far, all other functional groups on the Si atom have in general been of only academic interest.

However, the problem of synthesis is not simply to make the organosilanes of different functionalities, but also to produce the functionalities in the ratio in which they will finally be utilized for the synthesis of the actual polymers. It can be anticipated here—and this will be brought out in detail in later chapters on the polymers and their applications—that the need in technical silicone chemistry is primarily for *difunctional siloxane units*, to build up longer or shorter chains. *Trifunctional groups* come next in order of importance, while *mono-* and *tetrafunctional siloxane units* are third and fourth, respectively. An ideal synthesis should therefore be capable of direction toward one or the other of the silanes, as required. This ideal case is hardly realizable in its entirety. Almost all syntheses yield organosubstituted silanes of different functionalities side by side, in a ratio which in general depends on the ratio of organic groups to silicon atoms used. By varying this ratio, it is possible to influence the proportions of the silane functionalities to a certain extent. Catalysts also help to direct some syntheses, but by and large one is always dependent on a yield of products in a ratio *specific to a given process*. It has therefore become particularly important to find processes permitting subsequent transformation of the functionalities in order to adjust the disparity between the actual monomers recovered and the requirements of the polymer preparation. The most important of these is probably the *redistribution reaction*, a process which

* Page number of chapter references.

allows the synthesis of an *n*-functional silane from silanes with functionalities $n+1$ and $n-1$, as, for example,

$$R_3SiX + RSiX_3 \rightarrow 2R_2SiX_2$$

These reactions will be considered in Section 2.5 on substituent exchange.

The processes available for the production of the organosilanes R_nSiX_{4-n}, i.e., generally for the formation of Si—C bonds in monomers, can be divided into three major groups.

1. Direct syntheses—this is understood to mean the *direct* reaction of silicon with organic halides.
2. Organometallic syntheses—a group of processes in which organometallic compounds in stoichiometric quantities act as carriers in the transfer of the organic radical to the Si atom. The distinction from group 1 is not quite sharp, since there, too, metallic additives are employed to function as radical-carriers, though only in catalytic quantities.
3. Syntheses by reactions of hydrocarbons with silanes, in particular the reaction of olefins with the Si—H bond.

From the standpoint of their technical significance, none of these processes can generally be said to be preeminent. None is optimally suited for the synthesis of *all* organosilanes with all conceivable organic groups and at the same time superior economically and with respect to its ability to determine functionalities. Each process has its specific advantages and disadvantages, which can be understood in detail only from detailed discussion. Versatile industrial silicone production must therefore be based on a variety of processes, depending on which synthesis is found optimal for the required organosilane, or which synthesis is best able to fulfill the need for particular functionalities.

2.2 Direct Syntheses

2.2.1 Fundamentals

Following a suggestion by Rochow (*202*), in the terminology of organosilicon chemistry, reactions of organic halides with silicon are grouped together under the heading of *direct syntheses*. Rochow also regards the reactions of olefins with the Si—H bond as direct syntheses. These will be treated here as a distinct group (Section 2.4), and thus separated from the direct syntheses as defined above.

The reaction of organic halides with silicon and silicon alloys was first disclosed in a patent by Rochow (*196*), and the process became known in the literature as the "Rochow synthesis." Müller (*152*) also described the reaction of alkyl halides with silicon alloys, but his priority date was nine months later.

The simplest reaction of this type can be described by the equation:

$$2RX + Si \rightarrow R_2SiX_2 \tag{1}$$

This is, as it were, the idealized form of the direct synthesis which would permit the utilization of all the starting materials if the reaction went to completion. On the other hand, it is clear that in the ideal case the direct synthesis leads only to *one* kind of organochlorosilane, i.e., a diorganodichlorosilane.

The reaction never proceeds ideally in practice. A number of other processes take place in addition to reaction (1). In the main, these are

$$3RX + Si \rightarrow RSiX_3 + 2R\cdot \tag{2}$$

$$3RX + Si \rightarrow R_3SiX + X_2 \tag{3}$$

$$2X_2 + Si \rightarrow SiX_4 \tag{4}$$

Organic free radicals are generated during reaction (2), some of which react further as such, while some decompose. This phenomenon is responsible for the formation of organo-*H*-halosilanes such as $RSiHX_2$ and R_2SiHX as further by-products, as well as the deposition of finely divided carbon, which is frequently troublesome and which inhibits the reaction. Chlorine generated in (3) also participates further in the reaction and causes, *inter alia*, the formation of silicon tetrachloride.

The reaction as a whole is clearly very complex. Consequently, it yields a mixture containing the different organohalosilanes together with tetraalkylsilanes, organo-*H*-halosilanes, silicon tetrachloride, and small amounts of hydrocarbons (*222*), the latter being formed by recombination of the organic radicals. Table 5* gives the products obtained from the reaction of methyl chloride with silicon in the presence of a copper catalyst. The products of the reaction at approximately 500°C between chlorobenzene and silicon, with copper as a catalyst, included silicon tetrachloride and about 10% high-boiling silicon-free compounds, among them biphenyl, 1,3-diphenylbenzene, 1,4-diphenylbenzene, monochlorobiphenyl, and dichlorobiphenyl (*272*), in addition to the phenylchlorosilanes.

The actual direct synthesis is as difficult to carry out in practice as it appears easy in principle. Commercially available silicon commonly reacts only very sluggishly with organic halides. The use of catalysts to initiate the reaction and permit it to proceed with profitable yield was accepted as necessary at an early stage. The optimal conditions are, however, even then narrowly restricted. Of course, the reaction and the composition of the resulting crude product can always be modified by choosing suitable catalysts, controlling the temperature,

* A still more detailed list is given in ref. *292*.

TABLE 5

PRODUCTS OF THE REACTION OF METHYL CHLORIDE AND SILICON (DIRECT SYNTHESIS)[a]

Compound	Formula	Bp (°C/760 mm)
Dimethyldichlorosilane	$(CH_3)_2SiCl_2$	70.2 ± 0.1
Methyltrichlorosilane	CH_3SiCl_3	66.1 ± 0.1
Trimethylchlorosilane	$(CH_3)_3SiCl$	57.3 ± 0.1
Silicon tetrachloride	$SiCl_4$	57.6
Tetramethylsilane	$(CH_3)_4Si$	26.2
Methyldichlorosilane	CH_3SiHCl_2	40.4 ± 0.1
Dimethylchlorosilane	$(CH_3)_2SiHCl$	35.4
Trichlorosilane	$SiHCl_3$	31.8
Also:		
α,ω-Dichloropolydimethylsilanes	$Cl(CH_3)_2Si—[(CH_3)_2Si]_n—Si(CH_3)_2Cl$	
2- and 3-Methylhexanes (*222*)		
2- and 3-Methylpentanes (*222*)		
Hydrogen, methane, ethane, ethylene (*232*)		

[a] Numbers in parentheses indicate reference numbers.

using diluent gases, etc. In what follows, details of the method of operation will be treated from these aspects. It must, however, be mentioned at once that any single one of these measures alone is not likely to be strikingly successful, and that it is always a combination of several which leads to the best result.

The successful technical achievement of the synthesis is a matter of experience and rigid adherence to very special working conditions, as is frequently the case with reactions in heterogeneous gas/solid systems. These are more easily achieved on the industrial scale than in laboratory experiments, and this is why the direct synthesis has today been developed to its utmost refinement as a substantially automated process in a number of the world's great producing centers. The silanes thus produced have in fact established themselves as large-scale industrial raw materials for organosilicon chemistry.

2.2.2 Influence of the Type of Silicon

Among the types of commercial silicon there are many products which either do not react at all with organic halides, or can be made to do so only with difficulty. Much effort has therefore been devoted to attempts to increase the reactivity of the silicon by suitable pretreatment. In the final account, these processes, such as pretreatment with hydrogen fluoride (*138*) or with other highly corrosive acids (*147*), slurrying with heavy metal salts (*165*), with copper sulfate solution, and with hydrofluoric acid (*275*) amount to an activation of the metal surface.

It is likely that the formation of an oxide skin on the surface of the silicon inhibits at least the start of the reaction. This is borne out by the success of additions of metals with deoxidizing properties (Section 2.2.3).

Clearly, the development of maximum surface area of the metallic silicon also plays a very important part. It is preferred to use finely powdered material, and attempts have been made to establish optimum values for the particle size distribution (*91*).

2.2.3 Influence of Catalytically Active Metals

The direct synthesis cannot be achieved in practice without the addition of a catalyst.

In the past, no other aspect of the direct synthesis has received so much attention and effort as the search for suitable catalysts. It is difficult, if not impossible, for the unintiated to extract from the sometimes contradictory evidence in the literature a generally valid guide for the best method of working. The value of a large part of this work is limited by the fact that attention has been concentrated exclusively on the problem of catalysts, without at the same time considering the large number of other factors which also have a bearing on the result. At least, it is impermissible to compare the results of different authors directly for this reason.

Full-scale industrial plants have, in fact, operated the direct synthesis on a largely "classical" basis with considerable success, without the necessity for one of the many catalytic additives recommended in the more recent literature.

Copper has become the "classical" catalyst for the reaction of methyl chloride with silicon (*196*), and silver for that of chlorobenzene with silicon (*204*). Metallic copper can be replaced by copper(I) chloride, which increases the yield of CH_3SiHCl_2 (*200*). Under certain conditions, the presence of copper oxide increases the efficiency of the catalyst (*203*). Post-oxidized CuSi alloys are said to have proved to be particularly suitable for the synthesis of phenylchlorosilanes (*203*). The surface development of the catalyst is also of importance (*75*).

The optimum amounts of copper and silver are between about 5 and 10% of the silicon used. An increase of the amount of catalyst within this range or even beyond it does not improve the results of the synthesis.

A thoroughly homogenized mixture of silicon and catalyst is a prerequisite for the successful outcome of the reaction. In types of reactors in which the contact mass is in motion and is continuously mixed (Section 2.2.6) it is possible to work with silicon and copper powders (*194*). A more intimate contact, which is particularly necessary in a stationary bed, is achieved by the superficial alloying of the two metals, for example, by sintering the finely ground powders at about 1000 to 1100°C in a stream of hydrogen (*205*). It is desirable to use highly purified, oxygen-free hydrogen for such pretreatment in the synthesis of methylchlorosilanes (*174*) in order to prevent the formation of passivating layers of SiO_2. The addition of 10% of nitrogen, which is said to form nitrides on the silicon surface (*104*), is also recommended. The pretreatment of powdered silicon with copper(I) chloride at temperatures of 200 to 400°C has proved to be a particularly suitable means of superficially alloying silicon and copper and producing a substrate permitting reproducible results to be obtained even on the laboratory scale (*202*, *274*):

$$4CuCl + nSi \rightarrow (n-1)Si + 4Cu + SiCl_4$$

On the other hand, processes employing the wet precipitation of copper onto silicon metal are considerably more troublesome, since they require a technically undesirable drying process [e.g., treatment of ferrosilicon (*271*) with copper(II) salt solution].

The discovery of this specific catalytic action, particularly that of copper, immediately raised the question of the reaction mechanism concerned. Early experiments in this direction on the system $Si/Cu/CH_3Cl$ were conceived and carried out by Rochow (*112*, *130*). These suggested the following mechanism.

Copper reacts with methyl chloride to give methylcopper and copper(I) chloride:

$$2Cu + CH_3Cl \rightarrow CuCH_3 + CuCl$$

Methylcopper is unstable at the reaction temperature and breaks down at 250°C in approximately 3×10^{-3} second into metallic copper and methyl radicals:

$$CuCH_3 \rightarrow Cu + \cdot CH_3$$

These were detected by means of the Paneth test (*112*), i.e., by their ability to volatilize a lead mirror deposited on glass. The free methyl radicals then methylate the elementary silicon:

$$\cdot CH_3 + Si \rightarrow \overset{\diagdown}{\underset{\diagup}{-}}Si{-}CH_3$$

On the other hand, copper(I) chloride, the second product of the reaction of copper with methyl chloride, is reduced by elementary silicon at 250°C or above (*131*):

$$CuCl + Si \rightarrow Cu + \overset{\diagdown}{\underset{\diagup}{-}}Si{-}Cl$$

The original copper is re-formed by this reaction, as well as by the pyrolytic decomposition of methylcopper and can therefore re-enter the reaction cycle.

Copper thus plays the part of a genuine catalyst by participating in the reaction process but emerging from it again unchanged:

$$2CH_3Cl + Si \xrightarrow{Cu} (CH_3)_2SiCl_2$$

The part played by silver in the direct synthesis of phenylchlorosilanes from chlorobenzene and silicon can probably be interpreted analogously:

$$2C_6H_5Cl + Si \xrightarrow{Ag} (C_6H_5)_2SiCl_2$$

This view of the reaction mechanism has been opposed in recent years by a different concept developed in the work of a number of authors (*31*, *76*, *121*, *129*, *268*, *274*). This is based on the interesting experimental discovery that, of all the intermetallic phases existing in the copper/silicon system, the η'-phase corresponding approximately to the composition Cu_3Si displays particularly marked activity toward methyl chloride (*268*). Pure η'-phase yields a mixture of methylchlorosilanes with a dimethyldichlorosilane content of more than 80% (*274*) even under laboratory conditions, which always give a more unfavorable picture than the technical process. Mixtures of this phase with metallic silicon continue to yield products with a high dimethyldichlorosilane content until no more free silicon is available to fill the vacant lattice sites of the η'-phase. The ϵ-phase of the Cu/Si system, corresponding approximately to $Cu_{15}Si_4$, also yields dimethyldichlorosilane selectively (*274*).

This realization of the importance of an intermetallic phase as the catalytically active component and the assumption of a partially polar nature for the $Cu^{\delta+}—Si^{\delta-}$ bond led to the idea that the primary step in the reaction with methyl chloride was the dissociative adsorption of this molecule. The subsequent steps until a dimethyldichlorosilane molecule is detached from the surface are shown in the scheme below (*274*) (where M indicates the body of the metal).

Metallic copper is not merely inactive but is actually harmful, since it promotes the decomposition of methyl chloride, with the formation of highly

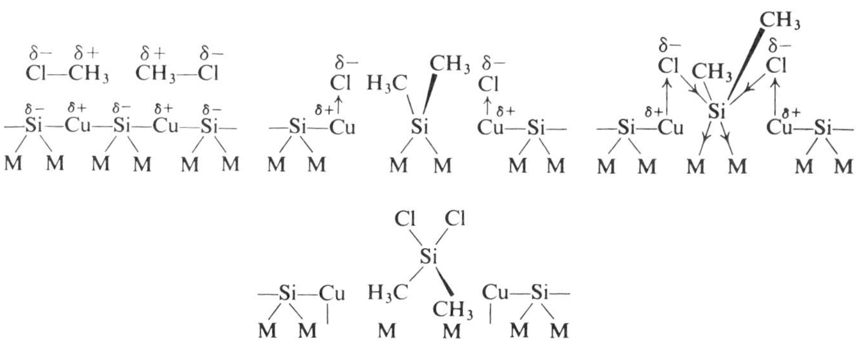

chlorinated hydrocarbons and the separation of elementary carbon. If metallic copper is continuously eliminated in the form of copper sulfide which is fine enough to be carried off as dust by the addition of hydrogen sulfide to the methyl chloride, at least 90% of the silicon can be converted into a mixture of chlorosilanes containing at least 80% of dimethyldichlorosilane (*274*).

Rochow's ideas also seem to have been shaken by other results. Methyl chloride does react with copper to give copper(I) chloride, but no volatile methylcopper compounds are formed. The transport of copper observed by Rochow can be explained by the sublimation of copper(I) chloride and reduction by hydrogen. Methyl radicals must thus be transferred directly to the silicon and not through methylcopper (*154*).

The influence of other metals has been the subject of numerous investigations. A few details follow below.

The effect of the addition of *zinc* to the copper/silicon mixture in the synthesis of methylchlorosilanes can be considered a reproducible and technically valuable one. Zinc assumes the role of a *promoter* of the reaction. It lowers the "striking" and reaction temperatures; consequently, it increases the yield of dimethyldichlorosilane and suppresses harmful side reactions (*90*, *259*). The use of *copper-zinc* alloys is said to have advantages under certain working conditions (*208*, *255*). *Aluminum* is reported to be active in a similar context (*109*). Very recent investigations even ascribe to it a most decisive influence on the rate of the reaction. It is said to accelerate both the reaction of methyl chloride with copper and that of copper(I) chloride with silicon (*155*). The addition of *cobalt* (*71*) or metals with a deoxidizing action, such as *calcium* or *titanium*, before the sintering of copper-silicon tablets (*176*) also causes an increase in the proportion of more highly methylated silanes in the reaction product.

Calcium-silicon alloys containing between 10 and 80% of calcium can react with methyl chloride at 300 to 375°C in the presence of hydrogen chloride. Under these conditions they give relatively high yields of trimethylchlorosilane and even tetramethylsilane (*35*).

Iron in the silicon (ferrosilicon alloys) acts in the opposite sense, in that it seems to favor the formation of more highly chlorinated derivatives (*223*).

Lead is reported to promote the selective formation of methyldichlorosilane. At temperatures above 400°C, yields of over 70% of this product are obtained in the reaction mixture (*273*). Its activity as a catalyst, however, is low. Both lead and *bismuth* act as catalyst poisons when added to the Si/Cu system (*269*). *Antimony* is a promoter of the reaction of methyl chloride and silicon in concentrations of 0.002 to 0.005%, inasmuch as it increases the yield of dimethyldichlorosilane. In higher concentrations it too becomes a catalyst poison (*269*). *Silver*, combined with a copper catalyst, results in an enrichment of CH_3SiHCl_2 (*239*) when used with gaseous mixtures of CH_3Cl and HCl.

The addition of zirconium halides to the Si—CuCl—CH_3Cl system (*168*) acts in a similar manner, but the gain with respect to methyldichlorosilane is obtained at the expense of a considerable increase in the amount of methyltrichlorosilane.

Other additives and pretreatments specifically favor the combination of certain hydrocarbon radicals with silicon. This should be compared with the discussion of the direct synthesis of vinyl- and allylchlorosilanes in Section 4.3.1.1.

However, a warning must be given against a too far-reaching generalization of these results, since the experiments were frequently carried out with different types of silicon and under not strictly comparable conditions. The structure, surface state, and purity of the metallic silicon are other parameters which can markedly affect the situation. A series of experiments on a silicon purified with hydrofluoric acid and undoubtedly surface-active showed that in this case the best yields of dimethyldichlorosilane were obtained when the pure metal was used (*120*).

2.2.4 Composition of the Gas Phase

Changes in the composition of the gas phase have been made for various reasons. Initially, it was merely to attempt to influence the heat balance of the reaction by adding nitrogen to the organic halide. Additions of 25 to 50% nitrogen have been recommended (*252*). This, however, makes the condensation of the reaction products more difficult, especially when their boiling points are low.

Numerous attempts have been made to influence the yield and composition of the reaction products with suitable additives. Mixing hydrogen with methyl chloride results in only a slight increase of the yield of CH_3SiHCl_2 (*251*). The addition of hydrogen chloride (*18*, *239*) is more effective; in this way the proportion of CH_3SiHCl_2 can be raised to almost 20% (*281*). The addition of HCl, moreover, frequently accelerates the reaction and increases the yield (*19*, *198*), particularly in the case of the reaction of chlorobenzene with silicon. The proportion of dimethyldichlorosilane in the reaction product is said to be increased by the addition of small amounts of tertiary alkyl halides (*36*) or dimethyl ether (*207*). However, with increasing amounts of dimethyl ether the yield of silanes, referred to methyl chloride, falls off (*53*). Benzene is a suitable additive for bringing about an increase in the yield for the chlorobenzene/HCl/Si system (*51*). The addition of chlorine to the alkyl halide is also said to favor the formation of organohalosilanes (*110*). Nitric oxide, on the other hand, inhibits the reaction of methyl chloride with silicon (*233*).

Instead of passing an organic halide over silicon or specially prepared contact masses, it is possible to start with mixtures of olefins and hydrogen chloride (*65*). The reaction mechanism in this case may be different from that when organic halides are used, since the first step must now be the formation of chlorosilanes, e.g.,

$$3HCl + Si \rightarrow SiHCl_3 + H_2$$

which then add to the olefin

$$C_2H_4 + SiHCl_3 \rightarrow C_2H_5SiCl_3$$

The yield of diorganodihalosilanes in this process can be considerably increased by working with 40 to 100 times the amount of hydrogen, under pressures of approximately 50 atm (*283*). The deposition of carbon on the solid phase of the system, which frequently inhibits or terminates the reaction, is here minimized or virtually suppressed. It is thus possible to achieve almost quantitative conversions of the silicon used in the reaction (*285*). The practical performance of this process, however, involves appreciable difficulties with respect to apparatus. A significant decrease in the carbon deposition is said to result from the addition of hydrogen with the simultaneous use of hydrogenation catalysts such as nickel or cobalt (*126*).

For the sake of completeness, it should be mentioned that the reaction of the lower aliphatic hydrocarbons, particularly methane, and hydrogen chloride with powdered silicon has also been studied. Small yields (15 to 20%) of methyltrichlorosilane in addition to silicon tetrachloride and trichlorosilane (*58*) have been obtained at 600 to 900°C, in some cases with the application of pressure.

Besides alkyl halides, dialkyl ethers can be used as the donors of alkyl groups for this reaction. If dimethyl ether is mixed with dry hydrogen chloride in a molecular ratio of approximately 1 : 0.1 to 3 and passed over a silicon-copper alloy at temperatures between 200 and 500°C, methyl-substituted chlorosilanes are obtained (*201*). On the other hand, if lower alcohols are used as the carriers of alkyl groups, the reaction does not proceed according to the following equation, as expected:

$$2ROH + Si \xrightarrow{Cu} R_2Si(OH)_2 \xrightarrow{-H_2O} [R_2SiO]$$

but yields tetraalkoxysilanes (*199*):

$$4CH_3OH + Si \xrightarrow[250^\circ]{Cu} Si(OCH_3)_4 + 2H_2$$

The hydrogen formed in this reaction, only part of which is liberated in the molecular state, gives rise to side reactions in the course of which silanes, hydrocarbons, and water are formed. The latter again causes hydrolytic cleavage, at least in part, of the tetraalkoxysilane originally formed. Ethanol reacts more sluggishly and gives poorer yields than methanol. Phenol reacts at 300 to 350°C with silicon and copper in analogous manner, with evolution of hydrogen and formation of tetraphenoxysilane (*135*).

2.2.5 Influence of Temperature and Pressure

The reaction temperature is determined primarily by the nature of the organic halide used. The optimum in the CH_3Cl/Si system lies between 250 and 300°C, but in the C_6H_5Cl/Si system it is appreciably higher (400 to 600°C). The catalyst employed has, moreover, a considerable influence on the optimum reaction temperature; the data given above relate to copper and silver as the catalysts. A somewhat higher temperature is usually required to start up the reaction than to keep it running.

The composition of the crude mixture of silanes produced is extremely sensitive to the reaction temperature. The proportion of halogen-rich silanes rises continuously with increasing temperature. In principle, the reaction of methyl chloride and silicon can, for example, be carried out at 400°C. Under those conditions, however, it yields mainly CH_3SiCl_3. The deposition of carbon, which inhibits the reaction, increases considerably at the same time, because of the increased cracking of the organic residues and, in turn, this leads to a catalytic acceleration of the pyrolytic decomposition of the organic halides. If the highest possible yields of dimethyldichlorosilane, which is so important for silicone chemistry, are to be obtained, it is essential to work at the lowest possible temperature.

Since the reaction is strongly exothermic, local hot spots can have a very unfavorable effect on the result. Problems of heat technology must therefore play a considerable part in any attempt to carry out this process on the industrial scale. Experience has shown that the lowest possible temperature at which the reaction can be maintained has a tendency to rise, which is probably due in part to the passivating action of deposited carbon on the surfaces. Suitable control of the temperature during the process is essential for the successful synthesis.

The "striking" temperature is reduced by an increase in pressure. If the reaction starts at 300°C under normal pressure, it can be expected to commence at about 240°C at 4 to 8 atm (*128*). At the same time, an increase in pressure has a favorable influence on the yield of dimethyldichlorosilane (*39*, *128*), particularly in the region 0 to 4 atm. It is additionally significant from the purely technical viewpoint that with a correspondingly higher pressure the condensation of excess methyl chloride leaving the reactor can be achieved with water cooling.

2.2.6 Performance on the Industrial Scale

The problems arising in the industrial-scale operation of this process derive from the conditions under which reaction proceeds optimally. These problems are typical for an exothermic reaction in a gas-solid system in which the heat has to be removed rapidly from a poorly conducting solid.

In principle, this process can be carried out in static (*254*) or weakly (*175*) or strongly agitated beds. Operation with strongly agitated beds has the advantage that the heat of reaction can be abstracted more easily and that the movement of the powdered solid causes fresh surface to be continually exposed. Under static conditions, this can be all too easily covered with carbon and thus passivated. The significance of the continuous renewal of the surface area during the reaction is also made plain by investigations which prove that silicon is capable of reacting at normal temperature with chlorine or carbon tetrachloride when ground continuously in contact with these reagents (*105*).

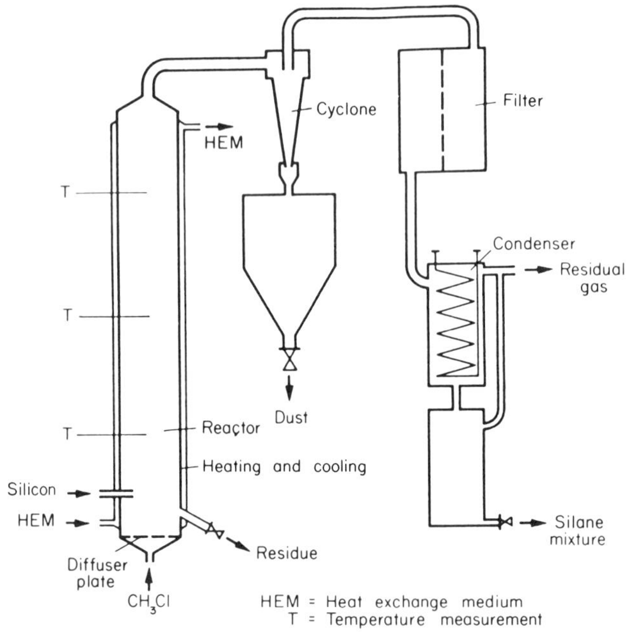

FIG. 1.

A modern variant of the process, which takes the above factors into consideration, is carried out in a fluidized bed (*36*, *194*). The mixture of silicon and copper powder is treated with the organic halide which at the same time acts as the fluidizing medium in comparatively tall reactors. Special devices are employed for the removal of dust. The heat of reaction is removed through the walls of the vessel by means of a heat-transfer liquid. The same medium and device are also used to start the reaction by heating. The principle of this process, which has in recent years been developed to a considerable level of refinement, is shown in Fig. 1.

The fluidized-bed process can be further improved by continuously or intermittently removing a portion of the solid from the reaction space and grinding it in a jet mill before returning it to the reactor. This ensures that the silicon surface, which has been gradually passivated by carbon deposition and other factors, is constantly renewed. As a result, the conversion of the silicon can be raised to 80 or 90% and the yield of organotrihalosilane can be minimized in favor of diorganodihalosilane (*64*).

According to an older proposal a mechanically stirred fluidized bed can be used (*230*). The metal powder is agitated in an up-and-down motion in a vertical reactor by means of a spiral band rotated by a central shaft, while a stream of the organic halide is passed upward through it (Fig. 2).

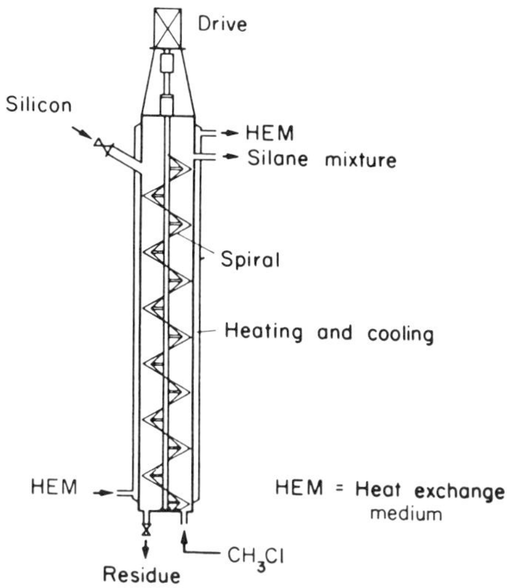

FIG. 2.

Numerous other proposals show the difficulty of the problem. Thus, a device has been described which consists of a system of horizontal tubular reactors through which the powdered solid reactants are continuously shoveled by means of a special device. Part-streams of the gas phase are fed in at different locations (*123*) and piecemeal catalyst addition has been considered advisable (*124*).

To remove the heat of reaction more satisfactorily, it has been proposed to suspend the metal powder in high-boiling oils or eutectic melts of metal chlorides and to pass the gas phase into this system in finely divided form, e.g., through perforated plates, with continuous stirring (*225*, *226*); however, this method of working has the disadvantage that silanes with a low degree of alkylation are formed as reaction products.

To achieve strong mechanical agitation of the silicon-catalyst mixture it has been proposed to carry out the reaction in a combined reactor and mill (*38*), or with vibration (*256*) and continuous recirculation of the solid reactants (*73*).

2.2.7 Silanes Produced

In the course of years, the direct synthesis has been used for the preparation of the most diverse types of silanes. In principle, all the work is based on the classical investigations of Rochow (*197*), which concentrated primarily on the methylchlorosilanes. Ethylchlorosilanes were prepared analogously by the reaction of ethyl chloride (*166*), and also that of ethylene and HCl (*65*, *283*), with silicon. The direct synthesis has also been used to obtain chlorosilanes with higher alkyl groups, such as butyl- and amylchlorosilanes (*68*), and also alkylbromosilanes, namely, methyl- and ethylbromosilanes (*92*, *261*, *265*) and propyl- and butylbromosilanes (*156*).

In the field of the aromatically substituted chlorosilanes the synthesis of the phenylchlorosilanes has been thoroughly investigated (*195*, *204*). The phenylbromosilanes $C_6H_5SiBr_3$, $(C_6H_5)_2SiBr_2$, and $(C_6H_5)_3SiBr$ have also been prepared by the direct synthesis (copper catalyst, 410 to 420°C) (*205*, *262*, *265*). With α-chloronaphthalene and hydrogen chloride silicon-copper alloy with 20% of added silver oxide gives a 24% yield of naphthyltrichlorosilane at 500 to 540°C (*217*).

Polychloroalkanes generally yield complicated mixtures of different silicon compounds in the direct synthesis, but usually in poor yields.

Dichloromethane, when passed over copper-silicon contact masses at 300 to 400°C, yields compounds with Si—CH_2—Si bonds, such as $Cl_3SiCH_2SiCl_3$, $Cl_3SiCH_2SiHCl_2$, and cyclic silanes of the type of $[CH_2SiCl_2]_n$ with $n \geqslant 3$ (*82*, *179*, *264*). 1,1-Dichloroethane yields a complex mixture of reaction products in which it has been possible to detect 4.5% of $SiCl_4$, 6.4% of $CH_2{=}CHSiHCl_2$, 16% of $CH_2{=}CHSiCl_3$, 6.5% of $CH_3CH(SiHCl_2)_2$, 18.5% of $CH_3CH(SiCl_3)$-$(SiHCl_2)$, and 18.5% of $CH_3CH(SiCl_3)_2$ (*182*). The products of the reaction of 2,2-dichloropropane with silicon-copper alloys were found to contain 2% of $SiCl_4$, 7.6% of $CH_2{=}C(CH_3)SiHCl_2$, 10.1% of $CH_2{=}C(CH_3)SiCl_3$, 14% of $(CH_3)_2C(SiHCl_2)_2$, 11.5% of $(CH_3)_2C(SiHCl_2)(SiCl_3)$, and 8.5% of $(CH_3)_2C(SiCl_3)_2$ (*182*). With silicon and silicon alloys, halogen-substituted olefins yield alkenylhalosilanes, generally in moderate yields (see Section 4.3.1.1). Perfluorinated organic halides are unsuitable for the direct synthesis (see Section 4.4.1.1).

In the reaction of carbon tetrachloride with copper-silicon mixtures, $SiCl_4$, C_2Cl_4, Si_2Cl_6, $C(SiCl_3)_4$, $Cl_3Si{-}C{\equiv}C{-}SiCl_3$, and $Cl_3Si{-}CCl{=}CCl{-}SiCl_3$ were found as reaction products. The formation of the last two compounds mentioned can be explained by dechlorination, due to copper, of trichloromethyltrichlorosilane $Cl_3C{-}SiCl_3$, which is not found in the products (*153*).

2.2.8 Working Up of the Reaction Products

Working up begins with a fractional distillation in which, in a first stage, the distillable portion is separated from the heavy residue, and, in a second stage, the components of the former are separated from one another. This part of the process can be very laborious, since the crude reaction product almost always consists of a number of components in a very narrow boiling range, and since the highest purity is frequently demanded of the individual organosilanes. This applies particularly to the diorganodihalosilanes which are required for the synthesis of the difunctional siloxane unit, and thus for the production of linear or cyclic molecular structures. The slightest contamination with mono- or trifunctional siloxane units could be of considerable disadvantage, since they act as chain-terminating and cross-linking agents. It is for the methylchlorosilanes—the compounds of greatest industrial interest—that the problems of separation are particularly difficult. The boiling points of dimethyldichlorosilane and methyltrichlorosilane differ by only about 4°C. With the help of present-day process and distillation technology it is nevertheless possible to recover dimethyldichlorosilane containing less than 0.1% of methyltrichlorosilane from mixtures of chlorosilanes (cf. Section 3.2.1). The fact that mixtures of different methylchlorosilanes, such as those of dimethyldichloro- and methyltrichlorosilanes, are required for the production of certain types of silicone products does facilitate production methods to a certain extent. In such a case quantitative separations of the constituents are of course not required.

Although the undistillable residues rarely constitute more than 10% of the total crude synthesis product, it is still economically desirable to work these up into usable materials. In the case of the methylchlorosilane synthesis these consist of chlorine-containing polymethylsilanes and also polymers with Si—O—Si or Si—CH_2—Si bonds. It is possible to obtain distillable methylchlorosilanes [especially CH_3SiHCl_2 but also CH_3SiCl_3 and $(CH_3)_2SiCl_2$ (*26*)] from them by treatment with hydrogen chloride at elevated temperatures, in certain cases with the addition of organic amines as catalysts (*34*). Cleavage of the Si—Si bonds takes place in accordance with the following equation:

$$\equiv Si-Si\equiv + HCl \rightarrow \equiv SiCl + HSi\equiv$$

Conversion of the polysilane components into methylchlorosilanes can also be achieved by heating the residue with catalytic amounts of quaternary ammonium or phosphonium halides, e.g., tetraethylphosphonium iodide (0.5 to 10%). The reaction concerned is probably a disproportionation of the polysilanes (*87*).

Efforts have also been made to utilize the solid residues remaining after the reaction. It has already been mentioned that the silicon used in direct synthesis can be consumed completely only if special measures, such as operation under a high pressure of hydrogen, are adopted. Otherwise, the silicon reacts only partially, and its reaction is limited by passivation of its surface as a result of carbon deposition. The solid residues contain therefore silicon and carbon, and they are also enriched in the catalyst (copper or silver). In spite of its passivation by carbon, the silicon is capable of further reaction with dry hydrogen chloride or chlorine, so that up to 90% of the silicon used can be converted into trichlorosilane or silicon tetrachloride (*125*). For the recovery of copper, it has been proposed to disperse the residues in water or dilute hydrochloric acid and to treat them with gaseous chlorine at 75 to 80°C. The silicon, which is not attacked, except for the smallest particles, is separated, washed, and dried. The $CuCl_2$ in solution is reduced to CuCl and separated after precipitation (*170*).

2.2.9 Advantages and Disadvantages of the Direct Synthesis

The attractive feature of the direct synthesis is its great simplicity: as the following flowsheet shows, it is possible to obtain the Si—C bond in a *single* step from a hydrocarbon derivative and elementary silicon.

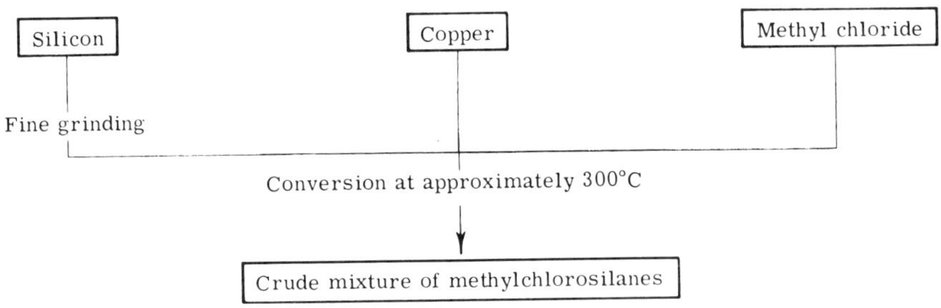

The catalyst is the only additional substance required for this process. Cyclic processes for the recovery and return of by-products are generally unnecessary unless the hydrochloric acid formed during the hydrolysis of the organohalosilanes is to be reused for the production of organic halides, particularly methyl chloride, by well-known processes. Consequently, the discovery and technical development of this process undoubtedly represented a decisive advance for silicone chemistry. The methyl- and phenylchlorosilanes required by technology, which constitute the greater part of silicone products, are today manufactured preponderantly by this process.

The above description of the conditions and results of the direct synthesis also indicates its limitations. They may in fact be considered narrow in view of the fact that it has not been possible with organic halides other than methyl and phenyl halides to find effective catalysts with sufficiently specific action or other modifications which would increase the yields or direct the reaction as required, i.e., without too great a yield of undesirable by-products.

Alkylhalosilanes with higher alkyl groups and vinylsilanes are now synthesized by other routes, since the direct synthesis has been found unsatisfactory in yield and reaction velocity.

Besides this lack of universal application, a further disadvantage of the direct synthesis is the difficulty of influencing the ratio of the end products. The ideal case, which would give a material balance according to which all the organic halide used was consumed with the formation of diorgano-substituted silanes alone, has so far not been realized. This result, however, is not of particularly practical interest, since silanes with one and with three organic substituents are also required for the synthesis of polyorganosiloxanes, and these would otherwise have to be manufactured by different routes. Their inevitable appearance in the products of the synthesis must be accepted from the standpoint of economics. Seen from the viewpoint of silicone chemistry as a whole, this could be considered fortunate, since the synthesis is thus capable of supplying *all* the necessary chlorosilanes.

Attempts to influence the course of the reaction toward a different ratio of the substituted silanes have met with considerable difficulties. Certain metal additives may favor the production of more highly alkylated or arylated silanes, but it is in no way possible to direct the reaction so that it proceeds almost exclusively in accordance with the equations

$$6CH_3Cl + 2Si \rightarrow 2(CH_3)_3SiCl + 2Cl_2 \quad \text{and} \quad 2Cl_2 + Si \rightarrow SiCl_4$$

It is equally impossible, by adding chlorine, to cause the reaction to take place solely in the following way:

$$CH_3Cl + Cl_2 + Si \rightarrow CH_3SiCl_3$$

Such measures achieve only a relative increase in the proportion of the desired derivative, while all the other silanes also occur—in each case in a somewhat altered quantitative ratio. In this respect the direct synthesis differs essentially and unfavorably from the other two groups of syntheses which will be discussed in the following sections.

2.3 Organometallic Syntheses

Let us now turn to those processes for the production of organosilanes in which organometallic compounds in stoichiometric amounts function as agents transferring the organic group to the silicon compounds. The Si—C bond is formed by substitution of a silicon-functional group. Reactions with organofunctional groups, therefore, do not belong to the subject matter of this section and are treated in Chapter 4.

The reactions proceed according to the scheme

$$RMe + X{-}Si\Big\langle \rightarrow R{-}Si\Big\langle + MeX$$

It makes no difference in this reaction whether the pre-formed organometallic compound intended for radical transfer is added to the system or is synthesized in the course of the over-all reaction.

2.3.1 GRIGNARD SYNTHESES

The Grignard reaction is of fundamental importance for the whole field of organic chemistry, but particularly for that of organosilicon chemistry where, as is generally known, it utilizes organomagnesium compounds to transfer organic groups to silicon:

$$RMgX + XSi\Big\langle \rightarrow RSi\Big\langle + MgX_2$$

Ever since Kipping (*127*) and Dilthey (*61*) originally used it for the production of organic derivatives from silicon tetrachloride, it has remained the preferred method for the production of Si—C bonds for more than six decades. Even at the present time it has at least maintained its value as a widely applicable laboratory method.

As normally carried out, the Grignard synthesis is a two-stage reaction. The first stage yields the corresponding alkyl- or arylmagnesium halide or its ethereal solution as a result of the reaction of an alkyl or aryl halide with magnesium turnings suspended in a lower aliphatic ether, preferably diethyl ether. In the solution, the Grignard compound is in equilibrium with the dialkylmagnesium and the magnesium halide.

$$2RMgX \rightleftharpoons R_2Mg + MgX_2$$

The ethereal solution of the mixture is added with stirring to the silicon compound—usually a silicon halide or a silicic acid ester. The reactions are exothermic and cooling is essential. Initiation of the reaction can cause difficulties, and must be promoted by special techniques (*209*). A magnesium halide or magnesium alkoxide precipitates, and the substituted silicon compounds are recovered from the solution by distillation. The yields lie in general in the region of 30 to 70% calculated on the initial silicon compound used.

The synthesis can also be carried out in a single step by adding the magnesium and the silicon compound to the solvent and running in the organic halide in the required amount. The mixture of the halides can also be added to magnesium suspended in the solvent. This then allows continuous operation (*212, 213*). Finally, it has been recommended to carry out the process continuously by arranging layers of magnesium turnings in a tower and irrigating them from above with the solution of a silicon tetrahalide or silicic acid ester in a sufficiently high-boiling solvent while passing the organic halide countercurrently from below. The reactions can be distributed over a number of towers (*141*).

In principle, all halo- or alkoxysilanes can react with organometallic compounds. The ability of silicon halides to decompose ethers is a disadvantage; this is particularly evident in the presence of traces of aluminum chloride and iron(III) chloride, and can lead to a considerable decrease in the yield (*229*). Consequently, the purest silicon halides should be used and iron vessels should be avoided.

The first choice for silicon components would fall on silicon tetrachloride, which is readily available on the industrial scale. It has been used for the majority of the classical organic substitutions but SiF_4 can also be employed with Grignard reagents (*47*, *69*, *86*, *150*, *183*, *289*). Even salts of fluorosilicic acid, such as sodium hexafluorosilicate, react with Grignard reagents at elevated temperatures (160 to 170°C) (*245*):

$$4RMgX + Na_2SiF_6 \rightarrow R_4Si + 2NaF + 4MgXF$$

The reaction can be carried out continuously, by running the Grignard reagent into ether or some other suitable solvent and adding silicon tetrafluoride simultaneously at a rate which ensures that the reaction mixture is always saturated with this reagent (*236*).

Silicon tetrabromide reacts less readily. No reaction takes place with methylmagnesium bromide below 160°C (*288*). With phenylmagnesium bromide it reacts at 180 to 200°C (*227*).

Silanes with 1 to 3 organic groups per silicon atom can be further substituted by means of Grignard compounds just like silicon halides, if such reaction is not opposed by steric hindrance, which will be discussed later. The possibility of this reaction makes organosilanes with differing types of substituents accessible.

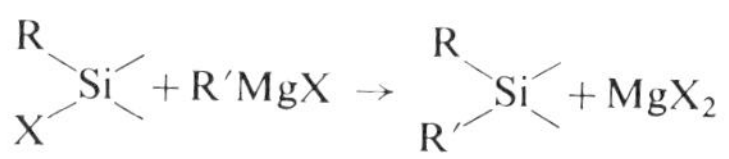

Silicon-hydrogen bonds do not react with Grignard compounds, so that trichlorosilane can be converted into the corresponding triorganosilane (*192*):

$$SiHCl_3 + 3RMgX \rightarrow R_3SiH + 3MgXCl$$

This also makes possible the synthesis of organo-*H*-halosilanes by partial halogen substitution in trichlorosilane (*13*, *66*, *72*, *139*, *214*):

$$CH_3MgCl + SiHCl_3 \rightarrow CH_3SiHCl_2 + MgCl_2 \quad (72)$$

$$C_6H_5MgBr + SiHCl_3 \rightarrow C_6H_5SiHCl_2 + MgBrCl \quad (13, 214)$$

However, the reactivity of the Si—H bond with a Grignard reagent depends also on the nature of the solvent (*100*). While triphenylsilane does not react with phenylmagnesium bromide in ether, xylene, or a mixture of ether and dioxane (*98*), it can be converted to tetraphenylsilane in 14% yield with tetrahydrofuran as solvent (*100*). On being boiled under reflux with an excess of phenylmagnesium bromide in tetrahydrofuran, diphenylsilane gives triphenylsilane in 79% yield (*100*).

It must be realized that if an organic halide compound or a Grignard compound is used in which the halogen is different from that of the silicon halide, a certain degree of exchange can take place between the halogen atoms during Grignard reaction (*139*).

Of the alkoxysilanes, the derivatives of the lower aliphatic alcohols, particularly $Si(OCH_3)_4$ and $Si(OC_2H_5)_4$, are most suitable for the Grignard reaction. Steric hindrance can considerably slow down the reaction with silicic acid esters of the higher alcohols. It is particularly interesting that many silicic acid esters, such as tetraethoxysilane, can act as their own solvent so that the use of ether, inconvenient in technical processes, becomes unnecessary (*6*, *146*, *263*). This allows single-stage operation; the Grignard reagent is formed from magnesium and alkyl chloride in the presence of the tetraethoxysilane and immediately alkylates it in the nascent state. This method of working has gained considerable interest since ethoxysilanes, particularly methylethoxysilanes, are more easily separated by fractional distillation than chlorosilanes because their boiling points are further apart (compare Section 3.3.2).

Even small additions of tetraethoxysilane have a catalytic effect in the reaction of silicon halides with Grignard reagents, so that it is possible to work without solvents or in solvents which are technically more easy to handle than diethyl ether (*5*). Like tetraethoxysilane, chlorobenzene can also play a dual role as reagent and solvent (*93*, *284*).

Experience has shown that the organic substitution of silicon halides or ethoxysilanes becomes more difficult with increasing size of the organic groups (*94*, *163*). The reason for this is to be found in steric hindrance, which already becomes manifest in the sequence methyl, ethyl, propyl (*85*, *241*). The effect of steric hindrance increases with increasing number of organic groups already present on the silicon atom, and becomes most evident when it is desired to synthesize tetraorganosilanes. Whether the substituent available on the silicon is chlorine or an alkoxy group can also make a difference in such cases: alkoxy groups are more difficult to substitute than the halogens (*85*). The Grignard compound must be used in considerable excess if more highly substituted silanes are to be obtained. It is frequently necessary to increase the reaction temperatures, using solvents with higher boiling points (*184*, *206*), or to distill off the solvent and heat the reactants for a relatively long time at temperatures up to 200°C (*52*, *60*, *96*, *227*). One to two percent of cobalt chloride acts as a catalyst in the reaction of arylmagnesium chlorides with silicon tetrachloride (*193*).

The value of the Grignard synthesis lies primarily in its broad applicability. Almost all organic groups can be bound to the silicon atom by means of Grignard compounds. If it is further considered that these substitutions can be carried out one, two, three, and four times, and that the nature of the organic groups can be different, it is understandable that the synthesis of a multitude of organosilanes is possible.

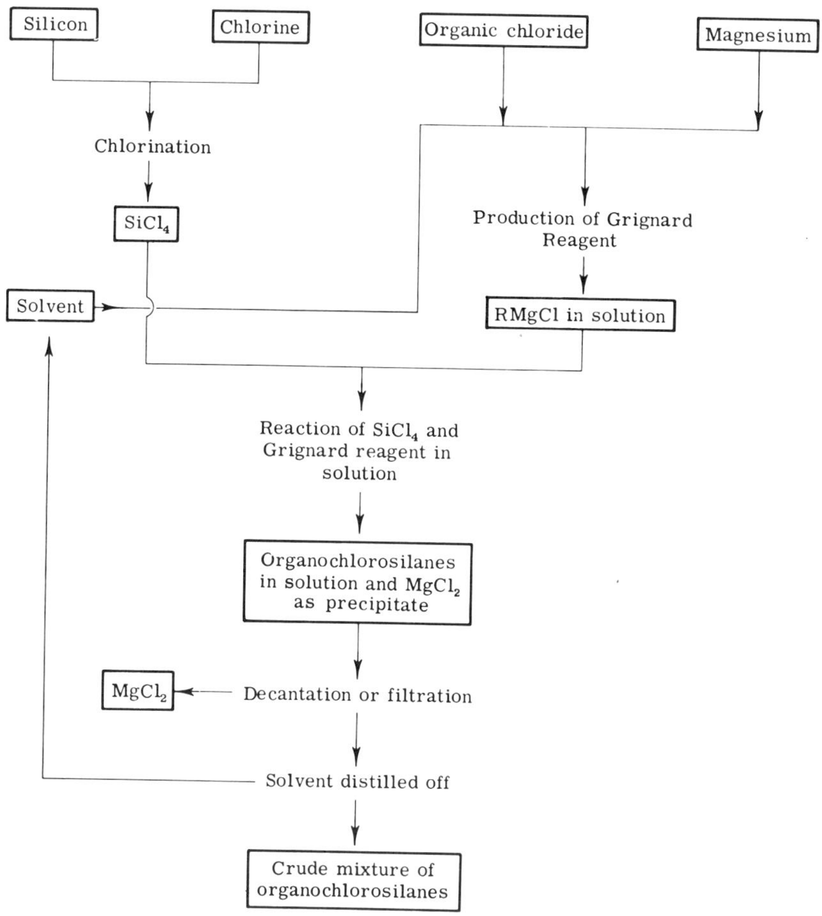

The Grignard process is, however, not only unusually *versatile* but also more easily directed than the direct synthesis. In general, even adjustment of the excess of Grignard reagent makes it possible to enforce the preferential formation of mono-, di-, tri-, or even tetrasubstituted silanes. Of course, *singly* substituted products are never formed alone, but are always accompanied by organosilanes with higher and lower degrees of substitution. Attempts to analyze the distribution functions mathematically have shown that the number of reactive atoms on the silicon is less important than the symmetry of the molecules in influencing their reactivity with respect to Grignard reagents (*84*).

If the Grignard process has been largely displaced for technical applications by the much more specialized direct synthesis, it is because of its greater complication and the larger ballast of side and intermediate products, as will be evident from a comparison of the above flowsheet with that of the direct synthesis (Section 2.2.9). In spite of this, it is likely to maintain a certain importance, even beyond the purely academic sphere, for the production of special organosilanes together with certain addition processes in silicone manufacture (compare Section 2.4.1).

2.3.2 Syntheses with Organic Compounds of the Alkali Metals

Organic compounds of the alkali metals can also fulfill the role of carriers of organic groups to the silicon atom, e.g.,

$$\mathrm{RLi + Cl{-}Si{\equiv} \rightarrow R{-}Si{\equiv} + LiCl}$$

Reactions of this kind are formally analogous to the reaction of a Grignard compound with a functional silicon compound. The organic group can, however, also be transferred by treating an organic halide with a halosilane in the presence of an alkali metal. In other words, the reaction is carried out in the manner of a Wurtz synthesis. In this form, it is one of the oldest reactions of organosilicon chemistry (see Section 1.3).

However, this group of reactions has not become even remotely as important for actual silicone chemistry as the Grignard synthesis or even the direct synthesis. The reason for this lies principally in the fact that it is here more difficult to direct the course of the reaction toward specific stages of substitution. Generally, the reaction proceeds as far as the organically tetrasubstituted silanes, and partially substituted derivatives occur only in small quantities. This strongly alkylating and arylating action is probably due primarily to the energy gain in the formation of the alkali-metal halide. It also manifests itself in the fact that steric hindrance plays a considerably smaller part in the reaction of organic compounds of the alkali metals with silicon compounds already containing organic substituents than is the case with Grignard compounds. From organolithium compounds and tetrahalosilanes, tetraorganosilanes can be obtained that, because of steric hindrance, are not accessible or are accessible only with difficulty in any other way—e.g., tetraisobutyl-, tetraisopropyl-, and tetranaphthylsilanes. While silicon tetrachloride does not react with Grignard reagents containing tertiary organic groups, *t*-butylhalosilanes such as *t*-butyltrichlorosilane and di-*t*-butyldichlorosilane can be synthesized with organolithium compounds (*270*). A further substitution with *t*-butyllithium, even under severe conditions, is however impossible. With a large excess of cyclohexyllithium, silicon tetrachloride gives only tricyclohexylchlorosilane (*163*).

The reactions can be directed better than is the case with the Wurtz synthesis if two-stage operation is employed. This can be shown in the case of phenyl transfer (*228*). The organometallic compound is prepared in the first stage from the organic halide and sodium in a suitable solvent, and this is treated with the silicon compound in the second stage.

Alkoxysilanes (*260*) and silanes with Si—H bonds can function as reactants in addition to halosilanes. The reaction of trichlorosilane with an organolithium compound in diethyl ether therefore proceeds as far as the tetraorganosilane (*97*, *142*):

$$SiHCl_3 + 4RLi \rightarrow R_4Si + LiH + 3LiCl$$

With an excess of phenyllithium or ethyllithium in diethyl ether, monophenylsilane can be converted into tetraphenylsilane or triethylphenylsilane (*159*); triphenylsilane reacts with phenyllithium to give tetraphenylsilane (*98*).

The choice of solvent is limited by the ability of the organic compounds of the alkali metals to react with many organic compounds. Aromatic hydrocarbons, such as toluene, which react with phenylsodium to give benzylsodium (*151*), are therefore excluded.

The reactions of difunctional organolithium compounds with chlorosilanes are interesting. The reaction of pentamethylenedilithium with silicon tetrachloride gives a spiro-compound in approximately 60% yield (*286*).

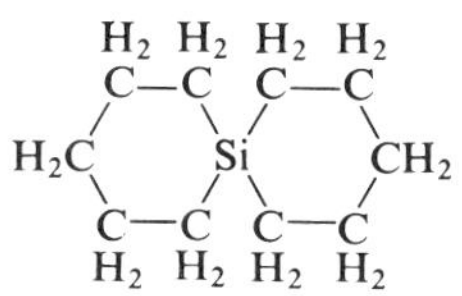

Similarly, a silicon compound corresponding to 9,9-diphenylfluorene can be obtained (*95*):

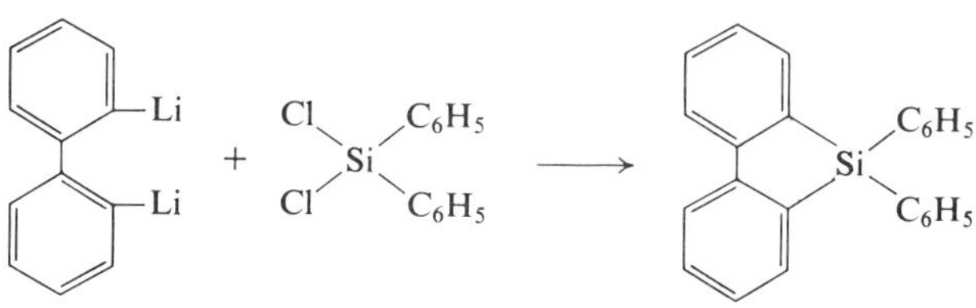

2.3.3 Syntheses with Other Organometallic Compounds

This group of syntheses includes the oldest known processes for the formation of Si—C bonds. Friedel and Crafts already worked with organozinc compounds (*101*), and Ladenburg used organomercury compounds. Halo- and alkoxysilanes were used as starting materials. Such reactions of the organic compounds of the heavy metals are today of only historical interest, since they have been superseded by the processes already described.

In very recent times, however, interest has also been directed to organoaluminum compounds, particularly after the partially organically substituted compounds $RAlCl_2$ and R_2AlCl, as well as R_3Al (*291*), had become industrially available; $RAlCl_2$ and R_2AlCl react with Si—Cl bonds with the formation of organosilicon compounds (*172*, *173*).

$$RAlCl_2+RSiCl_3 \rightarrow R_2SiCl_2+AlCl_3; \quad R_2AlCl+RSiCl_3 \rightarrow R_3SiCl+AlCl_3$$

This reaction proceeds with particular ease and smoothness if one starts with silanes that already have at least one organic substituent. Methyltrichlorosilane will already react rapidly on boiling under reflux with methylaluminum chlorides to give more highly methylated silanes. The mechanisms of the reactions of the methylaluminum chlorides both with silicon tetrachloride (*136*) and with organochlorosilanes (*32*) have recently been studied in detail.

At elevated temperatures, trialkylaluminums or their ether adducts likewise yield alkylfluorosilanes with SiF_4 or alkali-metal hexafluorosilicates (*114*, *115*). With silicon tetrachloride, they react in sealed vessels at 175 to 220°C to give alkylchlorosilanes (*253*). Orthosilicic acid esters can also be converted into organoalkoxysilanes with trialkylaluminums or alkylaluminum halides (*118*) or the ether adducts of trialkylaluminums (*116*). With halosilanes or alkoxysilanes, alkylaluminum hydrides give organosilanes containing Si—H bonds (*119*).

While pre-formed organometallic compounds are used in all these syntheses, it is also possible to transfer organic groups to silicon by passing silicon halides together with gaseous alkyl halides through a heated bed of granulated metallic zinc (*107*, *108*). Dimethyldichlorosilane and methyl chloride react in this way at 375°C to give trimethylchlorosilane, and methyltrichlorosilane gives dimethyldichlorosilane. Aluminum can be used as the carrier metal in the same way. Hurd presumed that organozinc or organoaluminum compounds are formed as intermediates in this reaction, which then transfer their organic groups to the silicon halide:

$$2Zn+2CH_3Cl \rightarrow Zn(CH_3)_2+ZnCl_2$$

$$Zn(CH_3)_2+2CH_3SiCl_3 \rightarrow 2(CH_3)_2SiCl_2+ZnCl_2$$

$$2Al+3CH_3Cl \rightarrow CH_3AlCl_2+(CH_3)_2AlCl$$

$$CH_3AlCl_2+CH_3SiCl_3 \rightarrow (CH_3)_2SiCl_2+AlCl_3, \text{ etc.}$$

Noncrystalline forms of silicic acid, as well as certain silicates, also react with alkylaluminum halides either when introduced into the reaction as such (*1*, *2*), or in the form of their complexes with alkali-metal halides (*240*), or when they are synthesized in the reaction mixture from aluminum and alkyl halide (*3*). Silica gel, diatomaceous earth, and bentonite in the dehydrated state are preferred as silicon-containing starting material because of their reactivity. The temperatures correspond approximately to those of the direct synthesis (Section 2.2), i.e., they are in the range of 250 to 350°C; pressure increases the reaction rate. In contrast to direct synthesis this process tends to yield highly alkylated silanes. One thus obtains from bentonite and methylaluminum sesquichloride at 300°C in sealed vessels a mixture containing 22% of tetramethylsilane, 70% of trimethylchlorosilane, and 8% of dimethyldichlorosilane (*2*).

2.4 Syntheses by the Reaction of Hydrocarbons with Silanes

A group of processes for the production of Si—C bonds will now be considered which in part display the character of addition reactions and in part that of substitution reactions. In addition reactions, the Si—H bond is attached to an unsaturated hydrocarbon. Addition to alkenes gives silanes with saturated organic groups:

$$\equiv SiH + R_2C{=}CR_2 \rightarrow R_2CH{-}CR_2{-}Si{\equiv} \qquad (1)$$

Addition to alkynes yields silanes with unsaturated organic radicals (see Section 4.3.1.3):

$$\equiv SiH + RC{\equiv}CR \rightarrow RCH{=}CR{-}Si{\equiv} \qquad (2)$$

The addition of the Si—H bond to chloroolefins gives chloroorganosilanes (see Section 4.5.1.2):

$$\equiv Si{-}H + ClCH_2{-}CH{=}CH_2 \rightarrow ClCH_2{-}CH_2{-}CH_2{-}Si{\equiv} \qquad (3)$$

The *substitution reactions* comprise the interaction of saturated or halogenated hydrocarbons with the Si—H or more rarely the Si—Cl bond according to the following reaction schemes:

$$\equiv SiH + C_6H_6 \rightarrow H_2 + C_6H_5Si{\equiv} \qquad (4)$$

$$\equiv SiH + RCl \rightarrow HCl + RSi{\equiv} \qquad (5)$$

$$\equiv SiCl + C_6H_6 \rightarrow HCl + C_6H_5Si{\equiv} \qquad (6)$$

Because of its lower bond strength, the Si—H bond is in general found to be more reactive than the Si—Cl bond, which has to be activated by the supply of energy.

The major part of the work carried out in this field has, for this reason, been concerned with the Si—H bond. The addition reactions of the Si—H bond have been investigated with particular thoroughness. If, in spite of this, the addition of the Si—Cl bonds is continually being investigated, it is doubtless because these are industrially more accessible either in the form of silicon tetrachloride or of already organically substituted chlorosilanes. In contrast to this, there are only two compounds for reactions with the Si—H bond available today from technical sources: trichlorosilane $SiHCl_3$, as a product of the reaction of hydrogen chloride with elementary silicon, and methyldichlorosilane CH_3SiHCl_2 as a by-product of the direct synthesis. The amounts of dimethylmonochlorosilane $(CH_3)_2SiHCl$ arising in the latter process play only a subordinate role. The production of dichlorosilane SiH_2Cl_2 which would broaden this basis of raw material in a very welcome fashion has not so far been achieved on a large scale in satisfactory yields.

2.4.1 Addition of the Si—H Bond to Unsaturated Hydrocarbons

The reactions with olefins according to Eq. (1) are catalyzed by uv light and peroxides. This led to the conclusion even at an early stage that they proceed by a radical mechanism. The reaction of trichlorosilane with pent-1-ene, for instance, proceeds as follows (*247*) (A· is the initiating radical):

$$A\cdot + SiHCl_3 \rightarrow AH + \cdot SiCl_3$$

$$CH_3{-}CH_2{-}CH_2{-}CH{=}CH_2 + \cdot SiCl_3 \rightarrow CH_3{-}CH_2{-}CH_2{-}\dot{C}H{-}CH_2{-}SiCl_3$$

$$CH_3{-}CH_2{-}CH_2{-}\dot{C}H{-}CH_2{-}SiCl_3 + SiHCl_3 \rightarrow CH_3(CH_2)_4SiCl_3 + \cdot SiCl_3$$

A self-polymerization of the olefins is often a more or less pronounced side-reaction which is in kinetic competition with the addition. Practically all olefins appear to be capable of adding the Si—H bond. This reactivity appears to be limited only in isolated cases. Steric factors then appear to play a part, as can be deduced from the fact that trichlorosilane can be added to the double bond in diisobutylene (formula I), but not to that in triisobutylene (formula II) (*55*):

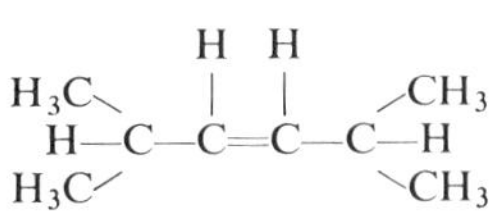
I

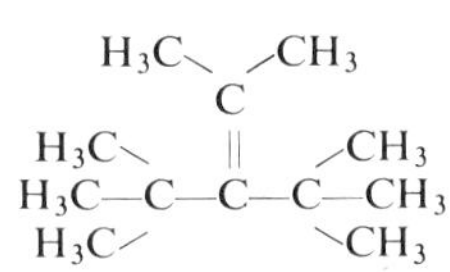
II

In general, the reactivity of the Si—H bonds increases with increasing substitution of electronegative atoms or groups on the Si atom. Thus, trichlorosilane can be added to olefins more easily than methyldichlorosilane. Triethylsilane did not add to vinyltrichlorosilane in the presence of benzoyl peroxide even after several weeks, while methyldichlorosilane did undergo this reaction, although very slowly (*234*). Diethylsilane and dibutylsilane add to oct-1-ene in a molar ratio of 1:2 by only one Si—H bond:

$$R_2SiH_2 + 2CH_2{=}CH{-}(CH_2)_5{-}CH_3 \rightarrow R_2[CH_3(CH_2)_7]SiH + CH_2{=}CH{-}(CH_2)_5{-}CH_3$$

On the other hand, with dec-1-ene, diphenylsilane gives decyldiphenylsilane in 61% yield and didecyldiphenylsilane in 10% yield (*267*).

A very interesting observation is that the addition to olefins with a double bond not in the terminal position can yield *n*-alkylsilanes if carried out in the presence of catalysts. The addition of methyldichlorosilane or trichlorosilane to pent-2-ene in the presence of a chloroplatinic acid catalyst yields *n*-pentylsilanes (*248*), and the addition of trichlorosilane to hept-3-ene leads to *n*-heptyltrichlorosilane (*216*).

The great advantage of the addition reactions with the Si—H bond is that they are capable of leading very precisely to the theoretically expected product and that the number and amount of by-products are relatively small. This is in very marked contrast to the direct synthesis as well as to the organometallic processes. Thus, for instance, trichlorosilane and an olefin give the organotrichlorosilane, and methyldichlorosilane and an olefin give the methylorganodichlorosilane:

$$SiHCl_3 + C_2H_4 \rightarrow C_2H_5SiCl_3$$

$$CH_3SiHCl_2 + C_2H_4 \rightarrow CH_3(C_2H_5)SiCl_2$$

Only the addition reactions with monosilane SiH_4 are more complicated. Not only the higher polymeric products with Si—Si and Si—C bonds are obtained in this case, but also various alkylation stages (*79–81*, *111*, *287*).

In practice, these reactions can be carried out under various conditions (*20*, *21*, *23*, *24*). One can work in autoclaves *without catalysts* in the temperature region between 200 and 600°C at different pressures depending on the nature of the olefin. Trichlorosilane reacts with octadecene at 300°C to give octadecyltrichlorosilane in 94% yield after 18 hours (*24*). Cyclopentenyltrichlorosilane can be obtained from the reaction of dicyclopentadiene and trichlorosilane in sealed vessels at 240 to 290°C in approximately 30% yield (*231*). A survey by Barry, DePree, Gilkey, and Hook shows the versatility of this method and the products synthesized by these authors (see Table 6).

In addition to the monomers, telomerization products are easily formed when the reactions are carried out at high temperatures and pressures. With methyldichlorosilane at 260 to 275°C and 560 atm gage, ethylene yields telomeric compounds $(CH_3)H(CH_2CH_2)_nSiCl_2$ with $n=2$ to 6, and at 270 to 285°C and 200 atm trichlorosilane gives the compounds $H(CH_2CH_2)_nSiCl_3$

TABLE 6

PRODUCTS OF THE NONCATALYZED REACTION OF OLEFINS WITH ORGANO-*H*-CHLOROSILANES (*24*)

Starting materials		Final product	Bp (°C) at	p (mm Hg)
$SiHCl_3$	But-1-ene	*n*-Butyltrichlorosilane	64	40
$SiHCl_3$	But-2-ene	*s*-Butyltrichlorosilane	145–146	736
$SiHCl_3$	*n*-Hexene	*n*-Hexyltrichlorosilane	103–140	50
$SiHCl_3$	*n*-Hexadecene	*n*-Hexadecyltrichlorosilane	194–196	7.5
$SiHCl_3$	*n*-Octadecene	*n*-Octadecyltrichlorosilane	185–199	2–3
$SiHCl_3$	Hexa-1,5-diene	Bis-trichlorosilyl-hexane	148–153	10
$SiHBr_3$	Propylene	*n*-Propyltribromosilane	183	756
CH_3SiHCl_2	Ethylene	Ethylmethyldichlorosilane	100	744
CH_3SiHCl_2	Propylene	Methylpropyldichlorosilane	123–124	747
CH_3SiHCl_2	*n*-Butene	*n*-Butylmethyldichlorosilane	147.5–148	744
CH_3SiHCl_2	*n*-Hexene	*n*-Hexylmethyldichlorosilane	192	743
CH_3SiHCl_2	*n*-Octene	Methyl-*n*-octyldichlorosilane	100–116	20
CH_3SiHCl_2	Octadecene	Methyloctadecyldichlorosilane	200–210	6
CH_3SiHCl_2	Cyclohexene	Cyclohexylmethyldichlorosilane	204	745
$C_6H_5SiHCl_2$	Propylene	Phenylpropyldichlorosilane	140–144	44–47
$(C_2H_5)_2SiHCl$	Propylene	Diethylpropylchlorosilane	164–166	742
$(CH_3)C_6H_5SiHCl$	Propylene	Methylphenylpropylchlorosilane	124–126	30
$(C_6H_5)_2SiHCl$	Propylene	Diphenylpropylchlorosilane	174–176	10

with $n=2$ to 5 (*160*, *161*). The compounds $H(CH_2CH_2)_n(C_6H_5)SiCl_2$ with $n = 1, 2,$ and 3 have been obtained by heating ethylene with phenyldichlorosilane at 280°C and 90 atm gage (*162*).

By *activation with uv light or with peroxides*, it is possible to carry out the reaction under considerably milder conditions, e.g., by boiling under reflux. A selection of the yields obtainable with various olefins under given conditions is presented in Table 7.

Other peroxides suitable as catalysts are dibenzoyl peroxide, *t*-butyl perbenzoate, *t*-butyl peracetate, di-*t*-butyl peroxide, and methyl amyl ketone peroxide. An exhaustive investigation of their influence on the reaction of trichlorosilane and pentene showed (*247*) that dibenzoyl peroxide was by far the most effective in this system. Ultraviolet irradiation appears to be generally less successful than catalysis by peroxides. This is very strikingly demonstrated by the example of the sterically strongly hindered triphenylsilane $(C_6H_5)_3SiH$. This compound reacts with oct-1-ene to give octyltriphenylsilane $(C_6H_5)_3$-$(C_8H_{17})Si$ in the presence of dibenzoyl peroxide but not under irradiation with uv light (*83*). The effect of ozone is similar to that of an organic peroxide.

TABLE 7

CONDITIONS AND PRODUCTS OF THE REACTION OF TRICHLOROSILANE AND OLEFINS WITH ACTIVATION (*188*)

Olefin	Reaction conditions: Activated with[a]	T (°C)	Duration (hr)	Yield (%)	End product
Oct-1-ene	AP	50–63	9	99	$CH_3(CH_2)_7SiCl_3$
Oct-1-ene	uv	44–74	40	47	$CH_3(CH_2)_7SiCl_3$
Oct-2-ene	AP	50–62	18	99	$C_8H_{17}SiCl_3$
2-Methylhept-1-ene	AP	49–58	25	70	$CH_3(CH_2)_4CH(CH_3)CH_2SiCl_3$
2-Methylbut-2-ene	AP	41–45	20	24.6	$C_5H_{11}SiCl_3$
2-Methylbut-2-ene	uv	42–52	45	64	$C_5H_{11}SiCl_3$
2-Methylprop-1-ene	uv	25–30	135	5	$CH_3CH(CH_3)CH_2SiCl_3$
2,3-Dimethylbut-2-ene	AP	45–53	23	59.3	$CH_3CH(CH_3)C(CH_3)_2SiCl_3$
2,4,4-Trimethylpent-1-ene	AP	49–50	15	9.1	$C_8H_{17}SiCl_3$
2,4,4-Trimethylpent-2-ene	AP	49–52	42	9.3	$C_8H_{17}SiCl_3$
1,1-Dineopentylethylene	AP	50–60	27	1.6	$C_{12}H_{25}SiCl_3$
Allyl chloride	AP	43–47	27	20.4	$ClC_3H_6SiCl_3$
Allyl chloride	uv	41–43	45	8.4	$ClC_3H_6SiCl_3$

[a] Here AP indicates diacetyl peroxide and uv indicates ultraviolet light.

It has been applied with considerable success in the addition of perfluoro-olefins to the Si—H bond (*290*). The reactions of a large number of unsaturated aliphatic hydrocarbons with Si—H bonds have been brought about largely by the use of peroxidic catalysts (*44*, *133*, *134*, *188*, *243*, *250*, *266*). In addition, cyclohexene (*157*, *250*), methylcyclopentene (*157*), dicyclopentadiene (*106*), terpenes (*103*) and many other olefins have been used as reactants. The radical donor *N*,*N*′-azo-bis-isobutyronitrile can be used for initiating the addition reaction (*140*, *247*).

When olefins having a strong tendency toward self-polymerization, like acrylonitrile or styrene, are used, polymer formation cannot be prevented, even with peroxidic catalysts. For example, the reaction of styrene with trichlorosilane in the presence of benzoyl peroxide yields compounds of the type of $H(C_6H_5CHCH_2)_nSiCl_3$ with $n_{mean} \sim 59$ (*122*).

A third group of suitable catalysts is represented by certain *metals and metal halides*. Foremost among those are, according to recent investigations, platinum and hexachloroplatinic(IV) acid (*10*, *158*, *180*, *191*, *246*, *248*, *276*, *279*, *280*), besides compounds of iridium and rhodium. For this purpose the platinum is preferably used in an active form, i.e., as platinum black, or precipitated on charcoal, silica gel or γ-alumina. Hexachloroplatinic(IV) acid, in particular, is a highly effective catalyst which appears outstandingly suitable for alkyldichlorosilanes; trichlorosilane frequently gives better yields with peroxidic catalysts (*181*). Methyldichlorosilane can be added almost quantitatively to ethylene and propylene in the presence of hexachloroplatinic(IV) acid (*191*). In comparison with platinum on silica gel, hexachloroplatinic(IV) acid increases the yield of addition products from the reaction of CH_3SiHCl_2 with $CH_2{=}CF_2$ by a factor of 3 or 4 (*191*). Even cyclohexene, which reacts only with difficulty, adds the Si—H bond better with hexachloroplatinic(IV) acid than with peroxides (*248*).

Finally, the action of the platinum catalysts is distinguished from that of the peroxidic catalysts and uv irradiation in that, with them, the rate of the addition reaction exceeds the tendency of the olefinic compounds toward self-polymerization. Acrylonitrile and acrylic acid esters, which polymerize with peroxidic catalysts, will add methyldichlorosilane with platinum-on-charcoal catalysts (*242*). The reaction of styrene with trichlorosilane in the presence of a peroxidic catalyst (diacetyl peroxide) gave high-boiling products and polystyrene exclusively (*44*); with a platinum/γ-alumina catalyst in tetrahydrofuran as the solvent (*189*), or with hexachloroplatinic(IV) acid (*185*), on the other hand, it was possible to synthesize β-phenylethyltrichlorosilane from styrene and trichlorosilane.

$$C_6H_5{-}CH{=}CH_2 + SiHCl_3 \rightarrow C_6H_5{-}CH_2{-}CH_2{-}SiCl_3$$

In the case of the platinum-catalyzed addition of methyldichlorosilane to styrene, thorough investigation showed that two isomers (I and II) were formed side by side, in a ratio of 33:53 (*215*):

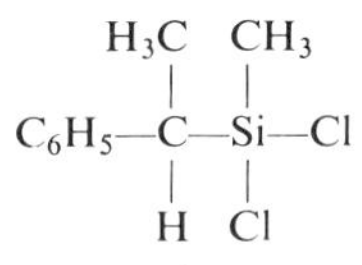
I

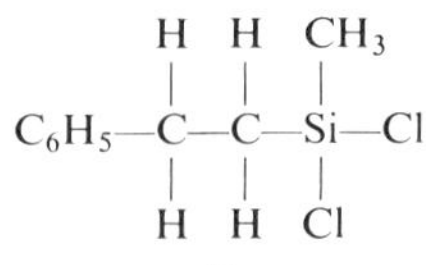
II

while the addition of the same silane to α-methylstyrene gave only one isomer, methyl-(2-phenylpropyl)-dichlorosilane (*215*):

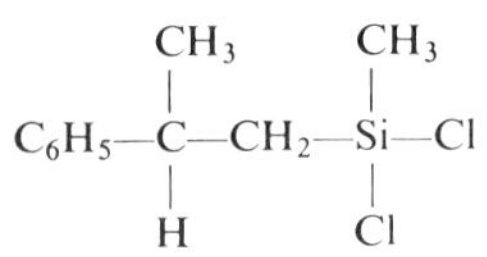

Catalytic activity toward the addition reaction is also displayed by copper (*224*), nickel (*167*, *180*, *242*), titanium tetrachloride (*78*), tin and tin tetrachloride (*248*), and iron pentacarbonyl (*77*). Elementary tin and its di- or tetravalent oxides and chlorides promote the addition reaction catalyzed by organic peroxides (*282*).

In addition, trichlorosilane can be added to many alkenes, sometimes in high yield, under the *influence of γ-rays*. Methyldichlorosilane reacts under these conditions with somewhat greater difficulty (*70*). Certain addition reactions have been carried out with *tertiary organic bases*, preferably pyridine, as catalysts (*178*, *190*).

Silane addition reactions can also be carried out with *alkynes and polyolefins*; they yield alkenylsilanes and are treated together with these compounds in Section 4.3.1.3. The addition reaction has also become important for the production of other organofunctional silanes, and is of particular interest for the synthesis of fluororganosilanes (Section 4.4), epoxyorganosilanes (Section 4.9), and cyanoalkylsilanes (Section 4.11).

2.4.2 Substitution Reactions of the Si—H Bond

The best known reaction of this kind is the reaction of benzene with an Si—H bond. It is achieved in autoclaves at temperatures of around 400°C and pressures between 88 and 130 atm (*15, 20, 22, 27, 29, 46, 278*). The reaction conditions can be made milder and the yield can be increased by the use of Friedel-Crafts catalysts, especially BCl_3 (*14*). Russian authors have also used Raney nickel (*46*) as well as boric acid (*143*) as the catalysts. Zinc halides (*219*) and compounds with Si—O—B bonds such as $[(CH_3)_2C_2H_5OSiO]_3B$ (*17*) are also mentioned in the literature.

The yields that can be obtained are always only moderate. They can under certain optimal conditions be increased to about 45 to 55% of the silane used. For the BCl_3-catalyzed reaction of trichlorosilane and benzene

$$SiHCl_3 + C_6H_6 \rightarrow C_6H_5SiCl_3 + H_2 \qquad (1)$$

this optimum lies at 280°C and approximately 60 atm (*27*). Methyldichlorosilane reacts more easily and gives the best yields of methylphenyldichlorosilane at temperatures as low as about 200°C:

$$CH_3SiHCl_2 + C_6H_6 \rightarrow CH_3(C_6H_5)SiCl_2 + H_2 \qquad (2)$$

However, the reactions yield numerous by-products, particularly compounds of lower and higher molecular weight with $\equiv Si—C_6H_4—Si \equiv$ linkages. Other silanes are formed in addition by disproportionation under the influence of the Lewis acids such as, in the case of (1), $(C_6H_5)_2SiCl_2$ and $SiCl_4$ and, in the case of (2), $SiHCl_3$, $C_6H_5SiCl_3$, CH_3SiCl_3, and $(CH_3)_2SiCl_2$ (*27*).

Toluene, mesitylene (*15*, *42*), biphenyl (*27*), naphthalene, dihydronaphthalene, methylnaphthalene, anthracene, phenanthrene, and fluorene (*30*) can be converted into the corresponding organochlorosilanes with trichlorosilane and methyldichlorosilane in a similar manner. The reaction of methane with trichlorosilane in the presence of boron trichloride as a catalyst has also been described (*235*).

Organic substitution of the hydrogen bound to silicon can also, in principle, be brought about by reaction with halogenated hydrocarbons:

$$SiHCl_3 + C_6H_5Cl \rightarrow C_6H_5SiCl_3 + HCl \quad (3)$$

$$CH_3SiHCl_2 + C_6H_5Cl \rightarrow CH_3(C_6H_5)SiCl_2 + HCl \quad (4)$$

but the reactions occurring are more complex and obscure than in the case of those mentioned above. This is because the hydrogen chloride formed in the reaction can cause cleavage of the R—Si bond, reactions (1) and (2) are in competition with reactions (3) and (4), respectively, and disproportionation of the silanes can take place at the high temperatures required.

Investigation of the purely thermal reaction of trichlorosilane with chlorobenzene in closed systems indicated that optimal yields of phenyltrichlorosilane were obtained at 387°C, but these amounted to only 38% of the trichlorosilane used (*27*); $(C_6H_5)_2SiCl_2$, $C_6H_5C_6H_4SiCl_3$, $Cl_3SiC_6H_4SiCl_3$, and some $ClC_6H_4SiCl_3$ were found as by-products. Reaction at normal pressure can be promoted by a palladium catalyst in the range from 450 to 575°C (*149*). It has also been suggested that the mixture of the starting materials in the vapor phase should be led over a copper catalyst at 400 to 600°C and that silicon should be added in the form of the element or as a copper-silicon alloy to remove the hydrogen chloride formed during the reaction. The $SiHCl_3$ so formed is utilized in the reaction, and the $SiCl_4$ which also arises does not interfere. Even under these conditions, however, the yields were only about 18 to 40% (*169*). The temperatures can be reduced to 275–350°C by the use of an alumina catalyst containing copper, iron, or zinc (*164*).

If the reaction is carried out in the presence of BCl_3 or $AlCl_3$, reactions (1) or (2) preponderate and chlorophenylchlorosilanes are formed preferentially (*16*).

Dichlorobenzene, chlorobiphenyl (*16*), α- and β-chloronaphthalene (*28*, *186*), and propyl chloride (*59*), have also been investigated as possible reactants. Vinyl chloride and olefins react under special conditions according to the same pattern (see Section 4.3.1.4).

2.4.3 Substitution Reactions of the Si—Cl Bond

Substitution of the Si—Cl bond would undoubtedly be of particular technical interest, since the easily available silicon tetrachloride could then be used as the silicon component. But, just as in the processes discussed in the above section, the yields are poor and the reaction conditions difficult and demanding.

It has been reported that phenylchlorosilanes can be produced from silicon tetrachloride and benzene by a reaction in the vapor phase between 840 and 1000°C (*54*, *148*). Methylchlorosilanes are said to be obtained with methane (*54*). The yields can be increased to 50% in the presence of finely divided silicon, which binds the hydrogen chloride formed during the reaction (*57*).

$$CH_4 + SiCl_4 \rightarrow CH_3SiCl_3 + HCl$$

$$4HCl + Si \rightarrow SiCl_4 + 2H_2$$

The reaction conditions recommended are: temperatures between 275 and 600°C, pressures between 5 and 180 atm, and boron trichloride or aluminum chloride as the catalyst. Other metals are also suitable for the trapping of hydrogen chloride (*56*).

The efforts to pursue this theoretically interesting route are reflected in the discovery that a reaction between silicon tetrachloride and hydrocarbons can be brought about by a silent electric discharge (*4*), as well as in attempts to apply electron-beam irradiation to overcome the activation threshold of these reactions (*43*). On irradiation with the 800 kV electron beam of a resonance transformer, solutions of various hydrocarbons in cold silicon tetrachloride gave, among other compounds, phenyltrichlorosilane (42%) and cyclohexyltrichlorosilane (33%).

Silicon tetrachloride can also be converted into alkylchlorosilanes by reaction with olefins and hydrogen in the presence of silicon, or an alloy of silicon, and $SnCl_2$, CuCl, or $CuCl_2$ and under pressures of 10 to 300 atm at 325 to 400°C. Under these conditions again, however, the yields are moderate and organotrihalosilanes are obtained as the main reaction products (*67*).

2.4.4 Syntheses with Diazomethane

As a reactant with the Si—halogen bond, diazomethane is used primarily for the preparation of halomethylsilanes (Section 4.5.1.3). It has been shown recently that diazomethane can also react with the Si—H bond to give a methyl—Si bond in the presence of copper catalysts or light (*132*).

$$C_6H_5SiH_3 + CH_2N_2 \rightarrow (CH_3)C_6H_5SiH_2 + N_2$$

$$(C_6H_5)_2SiH_2 + CH_2N_2 \rightarrow CH_3(C_6H_5)_2SiH + N_2$$

2.5 Substituent Exchange

Under certain circumstances various substituents on the silicon atom can be exchanged for one another. The reactions lead to equilibria which are established by *redistribution* of the substituents of two or more differently substituted silanes or by the *disproportionation* of a single molecular species. Both organic and inorganic substituents can be mutually exchanged in this way.

The mobility of hydrocarbon radicals bound to silicon brings a flexibility to silane synthesis which is frequently very desirable. Synthetic by-products with a degree of alkylation making them of little use in practice can be converted into useful higher alkylated compounds, and silanes with various hydrocarbon radicals on the same silicon atom can be made. Since new Si—C bonds are formed in this case, such reactions are also interesting from the point of view of those methods which aim at the production of these bonds.

Substituent exchange can frequently be achieved simply under the influence of increased temperature. In practice, catalytic additives, which lower the reaction temperatures and shorten the reaction times, sometimes considerably, are generally used.

The examples quoted below show that aluminum chloride is most frequently used as the catalyst. The effectiveness of Friedel-Crafts catalysts with reference to the regrouping of hydrocarbon radicals as studied in the case of the disproportionation of ethyltrimethylsilane decreases as follows (*210*):

$$AlBr_3 > AlCl_3 > AlI_3 > GaBr_3 > GaCl_3, BCl_3, FeCl_3$$

In general, an equilibrium distribution of all possible substitution products is obtained. Some catalysts are, however, able to act selectively and cause only certain substituents to migrate. The equilibria established are, for the sake of simplicity, formulated in such a way that only the most frequently occurring substitution products are taken into consideration, as has been done below.

2.5.1 Exchange of R with R′

Hydrocarbon radicals in mixtures of tetraorganosilanes R_4Si, R'_4Si can be exchanged in such a way that tetraorganosilanes $R_nR'_{4-n}Si$ (n being 1, 2, or 3) with mixed substituents are formed. Thus after equimolecular amounts of tetraethylsilane and tetra-*n*-propylsilane had been heated for 5 hours at 175 to 180°C with the addition of 2.5 mole-% of aluminum chloride, the following mixture of ethyl-*n*-propylsilanes was obtained, besides residues of the original silanes (*45*):

$$(C_2H_5)_4Si + (n\text{-}C_3H_7)_4Si \rightleftharpoons \begin{cases} (C_2H_5)_4Si\ (5\%) \\ (C_2H_5)_3(n\text{-}C_3H_7)Si\ (21\%) \\ (C_2H_5)_2(n\text{-}C_3H_7)_2Si\ (40\%) \\ C_2H_5(n\text{-}C_3H_7)_3Si\ (24\%) \\ (n\text{-}C_3H_7)_4Si\ (10\%) \end{cases}$$

In organohalosilane systems in which different organic groups are substituted on the same silicon atom, the organic groups can undergo exchange with halogen atoms, and organic groups can undergo exchange with one another. Such reactions are interesting *inter alia* for the production of the industrially otherwise not easily accessible methylphenylchlorosilanes (*7*, *48*, *49*, *220*):

$$C_6H_5SiCl_3 + (CH_3)_3SiCl \xrightleftharpoons{AlCl_3} (CH_3)C_6H_5SiCl_2 + (CH_3)_2SiCl_2$$

$$CH_3(C_6H_5)_2SiCl + (C_6H_5)_2SiCl_2 \xrightleftharpoons{AlCl_3} (CH_3)C_6H_5SiCl_2 + (C_6H_5)_3SiCl$$

2.5.2 Exchange of R with X

As an example of the possibility of exchanging hydrocarbon radicals and halogen atoms, the mobility of the methyl groups and chlorine atoms in the methylchlorosilanes is of primary interest in view of the technical problems posed, and it has been studied by various workers (*25*, *33*, *137*, *144*, *145*, *218*, *221*). Aluminum chloride or one of its complexes with alkali metal chlorides is used as the catalyst, and has the effect, in particular, that the pressures required for the reaction can be considerably reduced (*33*, *37*, *145*). Organoaluminum compounds are also recommended as catalysts (*113*). The amounts of aluminum chloride added are usually in the region of 5 to 10%. Equilibrium is established after heating under pressure to approximately 200 to 400°C. The reactions require many hours under these conditions. The reaction rate is considerably increased by the addition of organo-*H*-halosilanes, particularly methyldichlorosilane (*144*). By means of these compounds, the temperatures can be reduced to as low as 50°C and quite acceptable reaction rates still be obtained.

It is of practical interest that trimethylchlorosilane and methyltrichlorosilane, which perforce occur as by-products in the direct synthesis, can in this manner be converted into dimethyldichlorosilane. The reaction mixture obtained after heating equimolecular amounts of trimethylchlorosilane and methyltrichlorosilane for 23 hours with the addition of 6% of methyldichlorosilane and 11% of aluminum chloride at 150°C in sealed vessels contains 79% of dimethyldichlorosilane (*144*):

$$(CH_3)_3SiCl + CH_3SiCl_3 \rightleftharpoons 2(CH_3)_2SiCl_2$$

Methyl groups and fluorine substituents on the Si atom are also exchangeable. By heating equimolecular amounts of SiF_4 and $(C_2H_5)_4Si$ for 3 hours in the autoclave at 300 to 350°C, a mixture of ethylfluorosilanes consisting of 22% of $C_2H_5SiF_3$, 46.6% of $(C_2H_5)_2SiF_2$, 23.9% of $(C_2H_5)_3SiF$, and 7.5% of $(C_2H_5)_4Si$ was obtained (*117*).

As an example of the exchange of phenyl groups and chlorine atoms, it may be mentioned that tetraphenylsilane gives phenylchlorosilanes on being heated at 300 to 400°C in sealed with silicon tetrachloride (*171*, *244*):

$$SiCl_4 + (C_6H_5)_4Si \rightleftharpoons (C_6H_5)_nSiCl_{4-n} \qquad (n = 1, 2, 3)$$

The reaction is catalyzed by Cu(I) chloride (*244*), so that 34% of $(C_6H_5)_3SiCl$ could be identified in the reaction product within 12 hours at 300 to 360°C. Aluminum chloride cleaves the phenyl-silicon bond and thus leads to losses of the reaction product on distillation. These can be reduced to a certain extent by complexing with alkali-metal chlorides (see Section 3.2.1). Zirconium tetrachloride, which specifically promotes the exchange of phenyl groups and chlorine atoms, does not exhibit these disadvantages (*7*).

2.5.3 EXCHANGE OF H WITH R OR X

Migrations of the substituents on *organosilanes having Si—H bonds* take place very easily. Trimethylsilane disproportionates with $AlBr_3$ at 40 to 60°C in 162 hours to give a mixture of silanes containing 2.15 mole-% of SiH_4, 6.15 mole-% of CH_3SiH_3, 17.3 mole-% of $(CH_3)_2SiH_2$, 39.2 mole-% of $(CH_3)_3SiH$, and 35.0 mole-% of $(CH_3)_4Si$ (*211*). Diphenylsilane disproportionates at temperatures around 100°C in the presence of platinum or hexachloroplatinic(IV) acid, and at higher temperatures in the absence of a catalyst (*99*), e.g., at 300°C, as follows:

$$(C_6H_5)_2SiH_2 \rightleftharpoons SiH_4(?) + C_6H_5SiH_3\ (11.2\%) + (C_6H_5)_2SiH_2\ (45\%) + (C_6H_5)_3SiH\ (11.8\%) + (C_6H_5)_4Si\ (0.1\%)$$

Monophenylsilane disproportionates in benzene solution in the presence of aluminum chloride as follows (*249*):

$$4C_6H_5SiH_3 \rightleftharpoons (C_6H_5)_4Si + 3SiH_4$$

In methylphenylsilanes, phenyl and methyl groups are exchanged with one another under similar conditions:

$$4CH_3(C_6H_5)SiH_2 \rightleftharpoons (C_6H_5)_4Si + 2CH_3SiH_3 + (CH_3)_2SiH_2$$

The general rule for the $AlBr_3$-catalyzed disproportionation of organosilanes of the type $(CH_3)_3SiR$ appears to be that the mobility of R decreases in the order

$$H > C_6H_5 > C_2H_5, CH_3$$

This order corresponds to the gradation of the reactivities of the Si—R bonds toward electrophilic reagents (*210*).

The mobility of the hydrogen bound to silicon is also manifest in organo-*H*-halosilanes. Much evidence for this is available (*40, 41, 50, 62, 63, 88, 249, 277*); monophenyldichlorosilane, for example, disproportionates under the influence of aluminum chloride basically as follows (*277*):

$$2C_6H_5SiHCl_2 \rightleftharpoons (C_6H_5)_2SiCl_2 + SiH_2Cl_2$$

The redistribution reaction of methyltrichlorosilane with methyl-*H*-silanes catalyzed with aluminum trichloride gives yields of over 80% of methyldichlorosilane even under mild conditions. In this reaction, the ease of exchange between various pairs of substituents on the Si atom decreases in the sequence (*89*):

$$\text{H-Cl} > \text{H-alkyl} \gg \text{alkyl-alkyl, alkyl-Cl}$$

Methyl-*H*-chlorosilanes can also be obtained in good yields from methylchlorosilanes and silanes containing Si—H bonds by heating with activated alumina as the catalyst (*237, 238*).

Secondary or tertiary amines, or their salts or complexes, as well as nitriles

(*11*, *187*) and aliphatic cyanamides (*12*), act *specifically* on the exchange of *hydrogen and halogen atoms* (*187*):

$$2C_2H_5SiHCl_2 \rightleftharpoons C_2H_5SiH_2Cl + C_2H_5SiCl_3$$

The exchange likewise remains limited to hydrogen and halogen atoms if methylchlorosilanes or silicon tetrachloride is treated with silanes or methylsilanes in salt melts consisting of eutectic mixtures of alkali-metal chlorides and aluminum chloride. The specific hydrogen-chlorine exchange can be explained in this case by the fact that reactions in fused salts will take place preferentially on bonds with at least partial polar character (*257*, *258*).

Finally, under the catalytic influence of sodium ethoxide, organosilanes containing ethoxy groups also undergo a redistribution reaction in which hydrocarbon and alkoxy groups migrate (*9*, *10*):

$$C_2H_5(C_6H_5)Si(OC_2H_5)_2 \rightleftharpoons \begin{cases} C_2H_5Si(OC_2H_5)_3 \\ C_2H_5(C_6H_5)_2Si(OC_2H_5) \\ (C_6H_5)_3Si(OC_2H_5) \end{cases}$$

Under these conditions, in organo-*H*-alkoxysilanes hydrogen atoms and alkoxy groups are exchanged predominantly (*8*):

$$3C_2H_5SiH(OC_2H_5)_2 \rightleftharpoons C_2H_5SiH_3 + 2C_2H_5Si(OC_2H_5)_3$$

REFERENCES

(1) R. C. ANDERSON and G. J. SLEDDON, ICI, Brit. Pat. 805674 [28.9.1956]. (2) —, —, —, U.S. Pat. 2949481 [Brit. Prior. 16.8.1957]. (3) —, —, *Chem. and Ind.* **1960**, No. 43, 1335. (4) D. N. ANDREEV, *Doklady Akad. Nauk SSSR* **100**, 263 [1955]. (5) K. A. ANDRIANOV, *Zhurnal Obshcheĭ Khimii* **16**, 487 [1946]. (6) — and M. A. KAMENSKAYA, ibid. **8**, 969 [1938]. (7) B. A. ASHBY, General Electric, U.S. Pat. 2888478 [9.9.1957]. (8) D. L. BAILEY, Union Carbide, U.S. Pat. 2723983 [16.10.1952]. (9) —, —, U.S. Pat. 2723984 and 2723985 [17.10.1952].

(10) — and A. N. PINES, *Ind. Engng. Chem.* **46**, 2363 [1954]. (11) —, P. W. SHAFER and G. H. WAGNER, Union Carbide, German Pat. (West German) 956405 [U.S. Prior. 25.3.1953]. (12) — and G. H. WAGNER, —, U.S. Pat 2732280 [25.3.1953]. (13) A. J. BARRY, Dow Chemical, Brit. Pat. 618403 [U.S. Prior. 4.10.1945]. (14) —, Dow Corning, U.S. Pat. 2572302 [16.10.1948]. (15) —, —, U.S. Pat. 2626266 [16.10.1948]. (16) —, —, Brit. Pat. 671710 [U.S. Prior. 23.2.1949]. (17) —, —, U.S.Pat. 2611775 [10.4.1950]. (18) — and L. DEPREE, Dow Chemical, U.S. Pat. 2488487 [7.2.1946]. (19) —, —, Dow Corning, German Pat. (West German) 870257 [U.S. Prior. 15.1.1945].

(20) —, — and D. E. HOOK, —, U.S. Pat. 2510853 [6.6.1946]. (21) —, —, —, —, U.S. Pat. 2626271 [6.6.1946]. (22) —, —, —, —, U.S. Pat. 2546330 [11.12.1948]. (23) —, —, —, —, U.S. Pat. 2626268 [11.12.1948]. (24) —, —, J. W. GILKEY and D. E. HOOK, *J. Amer. Chem. Soc.* **69**, 2916 [1947]. (25) — and J. W. GILKEY, Dow Chemical, U.S. Pat. 2647912 [19.7.1947]. (26) —, —, Dow Corning, U.S. Pat. 2681355 [20.2.1953]. (27) —, — and D. E. HOOK, *Advances in Chemistry Series* **23**, 246 [1959]. (28) —, D. E. HOOK and L. DEPREE, Dow Corning, U.S. Pat. 2511820 [13.9.1947]. (29) —, —, —, —, U.S. Pat. 2591668 [4.11.1947].

(30) —, —, —, —, U.S. Pat. 2556462 [5.1.1948]. (31) V. BAŽANT, *Rozprav. Českoslov. akad. věd* **71**, 11 [1961] Prague. (32) L. BERANEK and V. BAŽANT, *Chem. Listy* **50**, 1250 [1956]. (33) B. A. BLUESTEIN, General Electric, U.S. Pat. 2717257 [9.9.1952]. (34) —, —, U.S. Pat. 2709176 [12.3.1954]. (35) —, —, U.S. Pat. 2887501 [26.12.1956]. (36) —, —, U.S. Pat. 2887502 [13.6.1957]. (37) — and H. R. MCENTEE, *Advances in Chemistry Series* **23**, 233 [1959]. (38) E. BONITZ, Bad. Anilin- u. Sodafabrik, German Appl. (West German) 1100006 [4.12.1958]. (39) E. F. BOON, F. C. J. BRANDT, H. KRAMERS, H. I. WATERMAN and N. P. M. DE WIT, *De Ingenieur* ('s-Gravenhage) **70**, Ch. 89 [1958].

(40) S. N. BORISOV, M. G. VORONKOV and B. N. DOLGOV, *Bull. Acad. Sci. USSR, Div. Chem. Sci.* (*Eng. transl.*) **1957** (II), 1416. (41) —, —, —, *Proc. Acad. Sci. USSR, Sect. Chem.* (*Engl. transl.*) **114**, 433 [1957]. (42) S. D. BREWER, General Electric, U.S. Pat. 2600198 [21.9.1949]. (43) J. F. BROWN JR., 130th Meeting, Amer. Chem. Soc., Atlantic City, N.J., 1956, Abstr. 55-0, No. 93; German Appl. (West German) 1048920 [U.S. Prior. 22.6.1956]. (44) C. A. BURKHARD and R. H. KRIEBLE, *J. Amer. Chem. Soc.* **69**, 2687 [1947]. (45) G. CALINGAERT, H. SCROOS and V. HNIZDA, ibid. **62**, 1107 [1940]. (46) E. A. CHERNYSHEV and A. D. PETROV, *Bull. Acad. Sci. USSR, Sect. Chem. Sci.* (*Engl. transl.*) **1956**, 637. (47) V. S. CHUGUNOV, ibid. **1953**, 761. (48) H. A. CLARK, Dow Corning, German Pat. (West German) 836649 [U.S. Prior. 23.2.1949]. (49) —, —, German Pat. (West German) 842057 [U.S. Prior. 23.2.1949].

(50) —, —, German Pat. (West German) 845198 [U.S. Prior. 23.2.1949]. (51) J. T. COE and W. A. SCHWENKER, General Electric, U.S. Pat. 2595767 [10.8.1950]. (52) N. W. CUSA and F. S. KIPPING, *J. Chem. Soc.* [London] **1933**, 1040. (53) A. CZERNOTZKY, *Int. Symp. Organosilicon Chem. Prague*, 1965, p. 381. (54) G. A. DALIN, Plax Corp., U.S. Pat. 2405019 [5.10.1943]. (55) K. DAMM, Farbenfabriken Bayer, 1960, unpublished. (56) W. H. DAUDT, Dow Corning, U.S. Pat. 2576448 [3.10.1949]. (57) —, —, U.S. Pat. 2710875 [24.10.1951]. (58) —, —, U.S. Pat. 2672475 [1.12.1952]. (59) L. DEPREE, A. J. BARRY and D. E. HOOK, Dow Corning, U.S. Pat. 2469355 [13.11.1946].

(60) P. A. DIGIORGIO, W. A. STRONG, L. H. SOMMER and F. C. WHITMORE, *J. Amer. Chem. Soc.* **68**, 1380 [1946]. (61) W. DILTHEY, *Ber. dtsch. chem. Ges.* **37**, 319 [1904] (footnote 2). (62) B. N. DOLGOV, S. N. BORISOV and M. G. VORONKOV, *J. Gen. Chem. USSR* (*Engl. transl.*) **27**, 2117 and 2733 [1957]. (63) —, —, —, ibid. **27**, 781 [1957]. (64) J. M. DOTSON, General Electric, French Pat. 1311472 [U.S. Prior. 28.11.1960]. (65) Dow Chemical, Brit. Pat. 609507 [U.S. Prior. 15.1.1945]. (66) Dow Corning, Brit. Pat. 647940 [U.S. Prior. 2.10.1944]. (67) J. J. DRYSDALE, Du Pont, U.S. Pat. 2917529 [10.7.1956]. (68) P. K. DUTT, *Sci. and Cult.* **20**, 559 [1955]. (69) C. EABORN, *J. Chem. Soc.* [London] **1952**, 2840.

(70) A. M. EL-ABBADY and L. C. ANDERSON, *J. Amer. Chem. Soc.* **80**, 1737 [1958]. (71) H. G. EMBLEM and R. C. DE FLEMING, Unilever, Brit. Pat. 907161 [24.6.1960]. (72) H. J. EMELÉUS and S. R. ROBINSON, *J. Chem. Soc.* [London] **1957**, 1592. (73) E. ENK, J. NICKL, O. HARBICH and W. SCHINDELIN, Wacker-Chemie, German Appl. (West German) 1102119 [22.4.1959]. (74) Farbwerke Hoechst, Belg. Pat. 612921 [German Prior. (West German) 20.1.1961]. (75) CH. S. FERGUSON and J. E. SELLERS, General Electric, U.S. Pat. 2443902 [27.6.1947]. (76) V. S. FIKHTENGOLTS and A. L. KLEBANSKY, *J. Gen. Chem. USSR* (*Engl. transl.*) **27**, 2535 [1957]. (77) R. KH. FREIDLINA, E. TS. CHUKOVSKAYA and TS'AO-I, *Proc. Acad. Sci. USSR, Sect. Chem.* (*Engl. transl.*) **127**, 551 [1959]. (78) —, —, — and A. N. NESMEYANOV, ibid. **137**, 361 [1961]. (79) G. FRITZ, *Z. Naturforsch.* **7b**, 207 [1952].

(80) —, ibid. **7b**, 379 [1952]. (81) —, *Z. anorg. allg. Chem.* **273**, 275 [1953]. (82) — and H. Thielking, ibid. **306**, 39 [1960]. (83) R. Fuchs and H. Gilman, *J. Org. Chem.* **22**, 1009 [1957]. (84) R. M. Fuoss, *J. Amer. Chem. Soc.* **65**, 2406 [1943]. (85) P. D. George, L. H. Sommer and F. C. Whitmore, ibid. **77**, 6647 [1955]. (86) J. A. Gierut, F. J. Sowa and J. A. Nieuwland, ibid. **58**, 897 [1936]. (87) A. R. Gilbert and G. D. Cooper, General Electric, U.S. Pat. 2842580 [22.11.1955]. (88) —, —, —, U.S. Pat. 2902506 [18.4.1956]. (89) —, — and R. W. Shade, *Ind. Engng. Chem.* **51**, 665 [1959].

(90) W. F. Gilliam, General Electric, U.S. Pat. 2464033 [27.2.1947]. (91) — and R. N. Meals, —, U.S. Pat. 2466412 [21.2.1946]. (92) —, — and R. O. Sauer, *J. Amer. Chem. Soc.* **68**, 1161 [1946]. (93) H. Gilman and R. E. Brown, ibid. **52**, 3330 [1930]. (94) — and R. N. Clark, ibid. **69**, 1499 [1947]. (95) — and R. D. Gorsich, ibid. **77**, 6380 [1955]. (96) — and R. K. Ingham, ibid. **77**, 1680 [1955]. (97) — and S. P. Massie jr., ibid. **68**, 1128 [1946]. (98) — and H. W. Melvin jr., ibid. **71**, 4050 [1949]. (99) — and D. H. Miles, *J.Org.Chem.* **23**, 326 [1958].

(100) — and E. A. Zuech, *J. Amer. Chem. Soc.* **79**, 4560 [1957]. (101) B. Gluck, C. Shaw and W. E. Smith, Union Chimique, Brit. Pat. 601938 [7.12.1945]. (102) *Gmelins Handbuch der anorganischen Chemie*, 8. Aufl., System-Nr. 15, Teil C, Organische Silicium-Verbindungen, Verlag Chemie, Weinheim/Bergstr. 1958. (103) L. A. Goldblatt and D. M. Oldroyd, U.S.A. Secretary of Agriculture, U.S. Pat. 2533240 [16.1.1948]. (104) A. Grimm, Chem. Werke Buna, German Pat. (East German) 9956 [21.3.1952]. (105) H. Grohn and R. Paudert, *Chem. Techn.* **10**, 307 [1958]. (106) D. B. Hatcher, Allied Chemical and Dye, U.S. Pat. 2665287 [9.8.1948]. (107) D. T. Hurd, *J. Amer. Chem. Soc.* **67**, 1545 [1945]. (108) —, General Electric, U.S. Pat. 2403370 [15.3.1945]. (109) —, —, U.S. Pat. 2427605 [15.3.1945].

(110) —, —, U.S. Pat. 2466429 [18.3.1946]. (111) —, —, U.S. Pat. 2537763 [8.2.1950]. (112) — and E. G. Rochow, *J. Amer. Chem. Soc.* **67**, 1057 [1945]. (113) ICI, French Pat. 1204830 [Brit. Prior. 4.10.1957]. (114) H. Jenkner, Kali-Chemie, German Pat. (West German) 972855 [11.6.1953]. (115) —, —, German Pat. (West German) 948975 [2.11.1953]. (116) —, —, German Pat. (West German) 974764 [6.5.1954]. (117) —, —, German Pat. (West German) 965328 [21.10.1954]. (118) — and H. W. Schmidt, —, German Pat. (West German) 974338 [11.6.1953]. (119) —, —, —, German Appl. (West German) 1038553 [12.11.1954].

(120) J. Joklík and V. Bažant, *Coll. Czechoslov. Chem. Commun.* **26**, 417 [1961]. (121) —, M. Kraus and V. Bažant, ibid. **27**, 974 [1962]. (122) M. Kadonaga and K. Jino, Jap. Pat. 5645 [1954]; *C.A.* **49**, 14377 [1955]. (123) M. Kahlert, VEB Silikonchemie, German Pat. (East German) 9253 [19.4.1951]. (124) —, —, German Pat. (East German) 9754 [19.4.1951]. (125) —, —, German Pat. (East German) 6872 [24.4.1951]. (126) —. —. German Pat. (East German) 7651 [18.2.1953]. (127) F. S. Kipping, *Proc. Chem. Soc.* [London] **20**, 15 [1904]. (128) A. L. Klebansky and V. S. Fikhtengolts, *J. Gen. Chem. USSR (Engl. transl.)* **27**, 2693 [1957]. (129) —, —, ibid. **26**, 2795 [1956].

(130) H. W. Kohlschütter, *Fortschr. chem. Forsch.* **1**, 1 [1949]. (131) —, H. Schifferdecker and O. Klump, *Z. anorg. allg. Chem.* **283**, 257 [1956]; **286**, 193 [1956]. (132) K. Kramer and A. Wright, *Angew. Chem.* **74**, 468 [1962]. (133) R. H. Krieble, General Electric, U.S. Pat. 2479374 [8.4.1947]. (134) —, —, U.S. Pat. 2524529 [8.4.1947]. (135) I. Kuis and H. I. Waterman, *Research Corresp., Suppl. to Research* **8**, No. 1, S 2 [1955]. (136) B. Lengyel and T. Szekely, *Z. anorg. allg. Chem.* **287**, 273 [1956]. (137) —, —, S. Jenei and G. Garzó, ibid. **323**, 65 [1963]. (138) A. A. Levine, DuPont, U.S. Pat. 2478493 [21.2.1946]. (139) R. N. Lewis, *J. Amer. Chem. Soc.* **69**, 717 [1947].

(140) R. D. W. LIPSCOMB and M. L. ERNSBERGER, DuPont, U.S. Pat. 2570462 and 2570463 [30.1.1948]. (141) A. E. MEADOWCROFT, C. SHAW and W. E. SMITH, Brit. Pat. 609841 [14.4.1946]. (142) R. N. MEALS, *J. Amer. Chem. Soc.* **68**, 1880 [1946]. (143) E. P. MECHEEV, A. L. KLEBANSKY, G. N. MEL'NOVA and K. K. POPKOV, *Plast. Massy* **1961**, No. 1, p. 19. (144) H. R. McENTEE, General Electric, U.S. Pat. 2786861 [26.4.1955]. (145) —, —, German Pat. (West German) 1018062 [U.S. Prior. 13.7.1955]. (146) R. R. McGREGOR and E. L. WARRICK, Corning Glass Works, U.S. Pat. 2380057 [23.10.1941] (147) W. MERTENS, Siemens & Halske, German Appl. (West German) p 3092 [31.5.1949]. (148) H. C. MILLER and R. S. SCHREIBER, DuPont, U.S. Pat. 2379821 [30.12.1942]. (149) D. MOHLER and J. E. SELLERS, General Electric, U.S. Pat. 2598436 [21.9.1949].

(150) G. L. MOORE, Air Reduction, U.S. Pat. 2802851 [14.3.1955]. (151) A. A. MORTON and J. T. MASSENGALE, *J. Amer. Chem. Soc.* **62**, 120 [1940]. (152) R. MÜLLER, VEB Silikon-chemie, German Pat. (East German) 5348 [6.6.1942]. (153) — and H. BEYER, *Chem. Ber.* **92**, 1018 [1959]. (154) — and H. GÜMBEL, *Z. anorg. allg. Chem.* **327**, 302 [1964]. (155) —, —, ibid. **327**, 286 and 293 [1964]. (156) N. S. NAMETKIN, A. V. TOPCHIEV and L. I. KARTASHEVA, *Doklady Acad. Nauk SSSR* **101**, 885 [1955]. (157) —, — and T. I. CHERNYSHEVA, *Proc. Acad. Sci. USSR, Sect. Chem. (Engl. transl.)* **111**, 767 [1956]. (158) —, —, — and L. I. KARTASHEVA, Chem. and Practical Application of Silicon Organic Compounds, Proceedings of Conference; Leningrad **1958**, No. 1, 241 (*Chem. Zbl.* **1961**, 5771). (159) W. H. NEBERGALL, *J. Amer. Chem. Soc.* **72**, 4702 [1950].

(160) A. N. NESMEYANOV, R. KH. FREIDLINA and E. TS. CHUKOVSKAYA, *Tetrahedron* **1**, 248 [1957]. (161) —, —, —, *Proc. Acad. Sci. USSR, Sect. Chem. (Engl. transl.)* **112**, 41 [1957]. (162) —, —, —, ibid. **115**, 785 [1957]. (163) W. H. NEBERGALL and O. H. JOHNSON, *J. Amer. Chem. Soc.* **71**, 4022 [1949]. (164) K. NISHIKAWA, I. SHIIHARA and R. SHIMAMOTO, U.S. Pat. 2759960 [Jap. Prior. 15.4.1950]. (165) S. NITZSCHE, Wacker-Chemie, German Pat. (West German) 899352 [26.1.1949]. (166) —, —, German Pat. (West German) 906455 [23.2.1949]. (167) —, —, German Pat. (West German) 916529 [14.2.1949]. (168) — and P. BUCHHEIT, —, German Appl. (West German) 1134671 [22.6.1961]. (169) — and R. RIEDLE, —, German Appl. (West German) 1046619 [7.12.1954].

(170) — and E. SCHMIDT, —, German Pat. (West German) 901889 [5.3.1952]. (171) W. NOLL, Farbenfabriken Bayer, German Pat. (West German) 825087 [15.8.1949]. (172) — and H. JONAS, —, German Pat. (West German) 888852 [10.6.1950]. (173) —, —, —, German Pat. (West German) 908019 [10.6.1950]. (174) — and P. SIMONS, —, German Appl. (West German) p 40075 [16.4.1949]. (175) —, —, —, German Pat. (West German) 848953 [16.4.1949]. (176) —, —, —, German Pat. (West German) 970283 [19.5.1950]. (177) —, W. SIMMLER and A. ZAPPEL, in Houben-Weyl: *Methoden der organischen Chemie*, Georg-Thieme-Verlag Stuttgart, 4. Aufl., Bd. 12, in the press. (178) S. NOZAKURA, *Bull. chem. Soc. Japan* **29**, 784 [1956]. (179) W. I. PATNODE and R. W. SCHIESSLER, General Electric, U.S. Pat. 2381000 [4.3.1942].

(180) A. D. PETROV, KH. M. MINACHEV, V. A. PONOMARENKO, B. A. SOKOLOV and G. V. ODABASHYAN, *Proc. Acad. Sci. USSR, Sect. Chem. (Engl. transl.)* **112**, 45 [1957]. (181) —, V. F. MIRONOV, V. A. PONOMARENKO, S. I. SADYKH-ZADE and E. A. CHERNYSHEV, *Bull. Acad. Sci. USSR, Sect. Chem. Sci. (Engl. transl.)* **1958**, No. 8, 926. (182) —, S. I. SADYKH-ZADE and I. L. TSETLIN, *Proc. Acad. Sci. USSR, Sect. Chem. (Engl. transl.)* **107**, 147 [1956]. (183) — and T. I. CHEMYSHOVA, *Izvest. Akad. Nauk. SSSR, Ser. Khim.* **1951**, 820. (184) — and V. S. CHUGUNOV, *Doklady Akad. Nauk. SSSR*, **73**, 323 [1950]; **77**, 815 [1951]. (185) —, E. A. CHERNYSHEV, M. E. DOLGAYA, YU. P. EGOROV and L. A. LEITES *J. Gen. Chem. USSR (Engl. transl.)* **30**, 400 [1960]. (186) —, — and LI KUANG-LIANG, *Proc. Acad. Sci. USSR, Sect. Chem.*

(*Engl. transl.*) **132**, 679 [1960]. (187) — and V. M. VDOVIN, *Bull. Acad. Sci. USSR, Div. Chem. Sci.* (*Engl. transl.*) **1960** (3), 486. (188) E. W. PIETRUSZA, L. H. SOMMER and F. C. WHITMORE, *J. Amer. Chem. Soc.* **70**, 484 [1948]. (189) R. A. PIKE and R. C. BORCHERT, Union Carbide, U.S. Pat. 2954390 [5.12.1957].

(190) —, *J. Org. Chem.* **27**, 2186 [1962]. (191) V. A. PONOMARENKO, V. G. CHERKAEV, A. D. PETROV and N. A. ZADOROZHNYI, *Bull. Acad. Sci. USSR, Sect. Chem. Sci.* (*Engl. transl.*) **1958**, 238. (192) F. P. PRICE, *J. Amer. Chem. Soc.* **69**, 2600 [1947]. (193) H. E. RAMSDEN, Metal and Thermit, U.S. Pat. 2813887 [20.11.1953]. (194) C. E. REED and J. T. COE, General Electric, U.S. Pat. 2389931 [27.9.1943]. (195) L. RICCOBONI and M. ZOTTA, *J. Soc. Chem. Ind.* **67**, 235 [1948]. (196) E. G. ROCHOW, General Electric, U.S. Pat. 2380995 [26.9.1941]. (197) —, *J. Amer. Chem. Soc.* **67**, 963 [1945]. (198) —, General Electric, U.S. Pat. 2483373 [16.4.1946]. (199) —, *J. Amer. Chem. Soc.* **70**, 2170 [1948]; U.S. Pat. 2473260 [26.6.1946].

(200) —, General Electric, U.S. Pat. 2447873 [28.6.1946]. (201) —, —, U.S. Pat. 2459539 [22.5.1947]. (202) —: *An Introduction to the Chemistry of the Silicones*, 2nd Ed., New York, John Wiley and Sons Inc., 1951. (203) — and W. F. GILLIAM, General Electric, U.S. Pat. 2383818 [26.7.1943]. (204) —, —, *J. Amer. Chem. Soc.* **67**, 1772 [1945]. (205) — and W. I. PATNODE, General Electric, U.S. Pat. 2380996 [26.9.1941]. (206) S. D. ROSENBERG, J. J. WALBURN and H. E. RAMSDEN, *J. Org. Chem.* **22**, 1606 [1957]. (207) G. ROSSMY, TH. GOLDSCHMIDT, German Appl. (West German) 1062700 [13.5.1957]. (208) —, —, German Appl. (West German) 1088051 [2.1.1959]. (209) F. RUNGE: *Organometallverbindungen*, Wiss. Verlagsgesellschaft, Stuttgart 1944, 2. Teil, p. 202.

(210) G. A. RUSSEL, *J. Amer. Chem. Soc.* **81**, 4815 [1959]. (211) —, ibid. **81**, 4825 [1959]. (212) J. B. RUST and C. A. MACKENZIE, Montclair Research, U.S. Pat. 2426121 [3.10.1944]. (213) —, —, —, U.S. Pat. 2426122 [17.10.1944]. (214) —, —, —, U.S. Pat. 2698334 [27.3.1945]. (215) J. W. RYAN and J. L. SPEIER, *J. Org. Chem.* **24**, 2052 [1959]. (216) J. C. SAAM and J. L. SPEIER, *J. Amer. Chem. Soc.* **80**, 4104 [1958]. (217) P. S. SANIN and A. D. PETROV, *J. Gen. Chem. USSR* (*Engl. transl.*) **27**, 1007 [1957]. (218) R. O. SAUER, General Electric, U.S. Pat. 2647136 [22.1.1948]. (219) —, —, German Pat. (West German) 886308 [U.S. Prior. 21.9.1949].

(220) —, —, U.S. Pat. 2730540 [26.6.1953]. (221) — and E. M. HADSELL, *J. Amer. Chem. Soc.* **70**, 3590 [1948]. (222) —, W. J. SCHEIBER and E. M. HADSELL, ibid. **70**, 4254, 4259 [1948]. (223) W. SCHLENK, Bad. Anilin u. Sodafabrik, German Appl. (West German) p 1169 [1.10.1948]. (224) —, —, German Pat. (West German) 899500 [1.10.1948]. (225) W. SCHMIDT and K. JOST, —, German Pat. (West German) 887343 [18.12.1951]. (226) —, —, —, German Pat. (West German) 920187 [26.7.1952]. (227) W. C. SCHUMB and C. M. SAFFER JR., *J. Amer. Chem. Soc.* **61**, 363 [1939]. (228) —, —, ibid. **63**, 93 [1941]. (229) R. SCHWARZ and W. KUCHEN, *Ber. dtsch. chem. Ges.* **89**, 169 [1956].

(230) J. E. SELLERS and J. L. DAVIS, General Electric, U.S. Pat. 2449821 [8.12.1945]. (231) I. W. SELLERS and M. P. WAGNER, Columbia Southern, German Appl. (West German) 1060398 [U.S. Prior. 24.12.1956]. (232) K. SETINEK, V. BAŽANT and F. ŠORM, *Chem. Listy* **50**, 1954 [1956]. (233) —, —, —, ibid. **51**, 639 [1957]. (234) D. SEYFERTH and E. G. ROCHOW, *J. Org. Chem.* **20**, 250 [1955]. (235) P. W. SHAFER and G. H. WAGNER, Union Carbide, U.S. Pat. 2660597 [17.12.1948]. (236) C. SHAW, W. E. LANGRISH-SMITH and G. A. PAGE, Brit. Pat. 637739 [21.5.1948]. (237) Shinetsu Kagaku Kogyo K.K., Jap. Pat. 23172 [8.4.1959]. (238) —, Jap. Pat. 9272 [24.5.1960]. (239) Shinetsu Chemical Industry Co. Ltd., French Pat. 1278517 [30.1.1961].

(240) G. J. SLEDDON, ICI, U.S. Pat. 2961453 [Brit. Prior. 13.9.1957]. (241) L. H. SOMMER, G. T. KERR and F. C. WHITMORE, *J. Amer. Chem. Soc.* **70**, 434 [1948]. (242) —, F. P. MACKAY, O. W. STEWARD and P. G. CAMPBELL, ibid. **79**, 2764 [1957]. (243) —, E. W. PIETRUSZA and F. C. WHITMORE, ibid. **69**, 188 [1947]. (244) E. M. SOSHESTVENSKAYA, *J. Gen. Chem. USSR* (*Engl. transl.*) **26**, 247 [1956]. (245) —, *Zhurnal Obshcheĭ Khimii SSSR* **8**, 295 [1938]; **10**, 1689 [1940]; *J. Gen. Chem. USSR* (*Engl. transl.*) **22**, 1163 [1952]. (246) J. L. SPEIER and D. E. HOOK, Dow Corning, U.S. Pat. 2823218 [5.12.1955]. (247) — and J. A. WEBSTER, *J. Org. Chem.* **21**, 1044 [1956]. (248) —, — and G. H. BARNES, *J. Amer. Chem. Soc.* **79**, 974 [1957]. (249) — and R. E. ZIMMERMAN, ibid. **77**, 6395 [1955].

(250) —, — and J. WEBSTER, ibid. **78**, 2278 [1956]. (251) M. M. SPRUNG and W. F. GILLIAM, General Electric, U.S. Pat. 2380998 [4.3.1942]. (252) —, —, —, U.S. Pat. 2380999 [4.3.1942]. (253) Stamicarbon, Belg. Pat. 554994 [Dutch Prior. 22.2.1956]. (254) H. STAMM and H. ARENS, Th. Goldschmidt, German Pat. (West German) 967044 [29.8.1950]. (255) —, —, —, German Pat. (West German) 970341 [29.11.1950]. (256) P. O. STELLING and R. B. M. EKLUND, Swed. Pat. 138857 [1953]; *C.A.* **47**, 6714 [1953]. (257) W. SUNDERMEYER, European Research S.A., German Appl. (West German) 1058476 [3.8.1957]. (258) — and O. GLEMSER, *Angew. Chem.* **70**, 625 [1958]. (259) Y. TAKAMI and T. ARAKI, U.S. Pat. 2903473 [Jap. Prior. 19.3.1954].

(260) L. TIGANIK, Uddeholms Aktiebolag, U.S. Pat. 2521267 [Swed. Prior, 8.9.1947]. (261) A. V. TOPCHIEV and N. S. NAMETKIN, *Doklady Akad. Nauk. SSSR* **78**, 295 [1951]. (262) —, — and N. M. SHMYKOVA, ibid. **78**, 497 [1951]. (263) —, —, ibid. **80**, 897 [1951]. (264) —, — and V. I. SETKIN, *Chem. Techn.* **4**, 508 [1952]. (265) —, —, *J. Gen. Chem. USSR* (*Engl. transl.*) **23**, 73 [1953]. (266) —, — and T. I. CHERNYSHEVA, *Proc. Acad. Sci. USSR, Sect. Chem.* (*Engl. transl.*) **115**, 741 [1957]. (267) —, —, —, ibid. **118**, 87 [1958]. (268) P. TRAMBOUZE, *Bull. Soc. chim. France, Mém.* **1956**, 1756. (269) I. V. TROFIMOVA, N. P. LOBUSEVICH, S. A. GOLUBTSOV and K. A. ANDRIANOV, *J. Gen. Chem. USSR* (*Engl. transl.*) **32**, 835 [1962].

(270) L. J. TYLER, L. H. SOMMER and F. C. WHITMORE, *J. Amer. Chem. Soc.* **70**, 2876 [1948]. (271) Unilever Ltd., Brit. Pat. 929024 [2.6.1960]. (272) M. VAVRUŠKA, *Chem. Listy* **51**, 319 [1957]. (273) R. J. H. VOORHOEVE and J. C. VLUGTER, *Rec. trav. chim. Pays-Bas* **82**, 605 [1963]. (274) —, Diss. Techn. Hochschule Delft, Holland, 1964; —, et al., *J. Catalysis* **3**, 414 [1964]; **4**, 43 [1965]; **4**, 123 [1965]; **4**, 220 [1965]. (275) G. H. WAGNER, Union Carbide, U.S. Pat. 2657114 [21.6.1949]. (276) —, —, U.S. Pat. 2637738 [17.9.1949]. (277) — and M. M. BURNHAM, —, U.S. Pat. 2746981 [26.9.1952]. (278) — and P. W. SHAFER, —, U.S. Pat. 2775606 [22.3.1954]. (279) — and C. O. STROTHER, —, U.S. Pat. 2632013 [9.10.1946].

(280) — and W. G. WHITEHEAD, —, U.S. Pat. 2851473 [23.12.1955]. (281) E. WALASCHEWSKI, Farbenfabriken Bayer, 1962, unpublished. (282) J. A. WEBSTER, Dow Corning, U.S. Pat. 2894968 [26.3.1956]. (283) J. WEISE and H. WELZ, Farbenfabriken Bayer, German Pat. (West German) 859164 [27.2.1951]. (284) A. WEISSENBORN, I.G. Farbenindustrie, German Pat. 660075 [24.2.1932]; German Pat. 697420 [3.5.1932]. (285) H. WELZ, Farbenfabriken Bayer, German Appl. (West German) H 1419 [28.1.1950]. (286) R. WEST and E. G. ROCHOW, *Naturwissenschaften* **40**, 142 [1953]. (287) D. G. WHITE and E. G. ROCHOW, *J. Amer. Chem. Soc.* **76**, 3897 [1954]. (288) C. J. WILKINS, N. P. H. BROWN and A. D. STEVENS, *J. Chem. Soc.* [London] **1950**, 163. (289) A. YA. YAKUBOVICH and V. A. GINSBURG, *Uspekhi Khimiĭ* **18**, 46 [1949].

(290) A. ZAPPEL, Farbenfabriken Bayer, German Pat. (West German) 949659 [5.2.1954]. (291) K. ZIEGLER, *Angew. Chem.* **68**, 721 [1956]. (292) J. J. ZUCKERMAN, in Emeléus, *Advances of Inorganic Chemistry* **6**, 383 [1964].

CHAPTER 3

Monomeric Organosilicon Compounds R_nSiX_{4-n}

This chapter will be devoted to a discussion of selected compounds of the type R_nSiX_{4-n}, where R is a saturated unsubstituted hydrocarbon radical but where the silicon-functional substituents may vary. Organosilanes with unsaturated or substituted hydrocarbon radicals will be examined in Chapter 4.

The selection made takes into account both the technological and preparative importance of the compounds. The monomers considered are therefore primarily those of importance in the production of the silicones, but others which may play a special part in technological applications or could be of interest as reagents have also been included.

The Si—C bond is formed as a result of processes described fully in Chapter 2. The Si—X bonds formed at the same time are predominantly Si—Cl, Si—OR (R is mainly C_2H_5) or Si—H. Silanes with such bonds are available today as highly purified tonnage raw materials. Other Si—X bonds are mostly

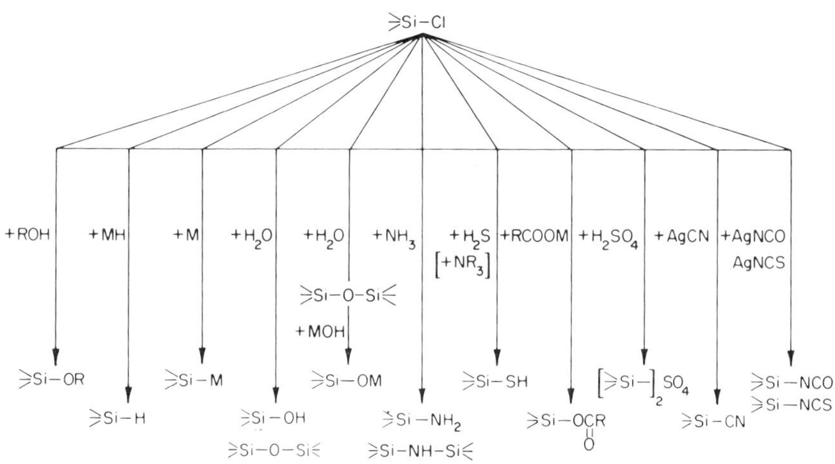

prepared from these basic products of organosilicon chemistry by changing the silicon function. Thus, starting with organochlorosilanes the "family tree" shown in the accompanying insert is obtained.

3.1 Tetraorganosilanes

Tetraorganosilanes have acquired relatively little importance in the field of silicone chemistry, unless they happen to contain unsaturated or substituted organic groups and are thus capable of participating in organic reactions.

Tetramethyl- and tetraethylsilanes are the oldest known organosilanes. They were prepared in the last century by Friedel and Crafts and by Ladenburg (see Section 1.3). Tetramethylsilane is today obtained in small quantities as a by-product from the preparation of methylchlorosilanes by the direct synthesis. All other tetraorganosilanes must be obtained by special processes. Organometallic—particularly organolithium—syntheses are the most suitable (Section 2.3.3).

Tetraorganosilanes containing only saturated hydrocarbon groups are without exception poorly reactive and occasionally display remarkable stability. Depending on the nature of the substituents, they are either liquid or solid at normal temperature: tetramethylsilane boils at 26.2°C and tetraethylsilane as high as 153.7°C; tetraphenylsilane is crystalline at normal temperatures, melts at 234 to 235°C, and boils at 428°C without decomposition (Table 8). The boiling points of some other aromatically substituted silanes are well above 500°C. Compounds with the following groups possess increasing thermal stability (*121*):

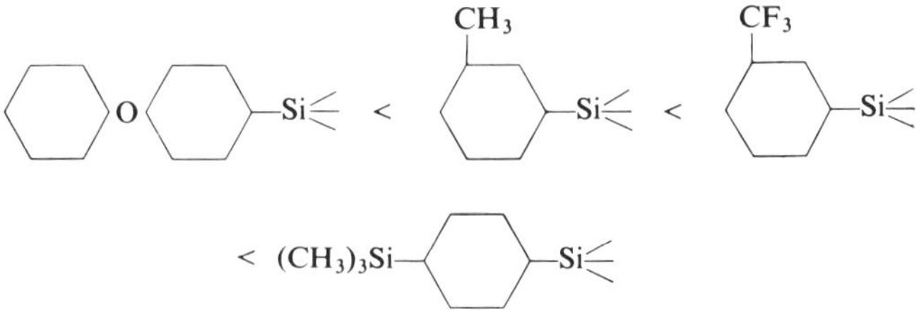

The Si—C bonds in tetraorganosilanes containing lower alkyl groups are not easily attacked by nucleophilic or electrophilic reagents (but cf. Section 5.3). The organic groups can be chlorinated, brominated, or nitrated without cleavage of the Si—C bond. On the other hand, tetraphenylsilane and tetraethylsilane are subject to stepwise degradation by aluminum chloride; the ultimate product is silicon tetrachloride:

$$(C_6H_5)_4Si + AlCl_3 \rightarrow (C_6H_5)_3SiCl + C_6H_5AlCl_2$$
$$(C_6H_5)_3SiCl + AlCl_3 \rightarrow (C_6H_5)_2SiCl_2 + C_6H_5AlCl_2, \text{ etc.}$$

In tetraorganosilanes containing unsaturated groups, such as vinyltrimethylsilane, however, the alkenyl—Si bond is ruptured completely by concentrated sulfuric acid (*187*).

Tetraorganosilanes are thermally decomposed at very high temperatures. Such cracking processes have been investigated very thoroughly in the case of tetramethyl- and tetraethylsilanes. Tetramethylsilane decomposes between 580 and 700°C, in a very complicated manner, to give H_2, CH_4, SiH_4, C_2H_4, C_2H_6, simple methylsilanes like $(CH_3)_3SiH$, and higher boiling products among which there are cyclic compounds with Si—C bonds (*108*) (cf. Section 7.2.4).

Since, in principle, any of the enormous variety of organic groups can be selected for the composition of a tetraorganosilane and the four groups concerned may be all the same, some the same, or all different (R′R″R‴R⁗Si), a very large number of such silanes is theoretically possible, although the number that can actually be synthesized is limited by steric effects. In the reaction of silicon tetrachloride with cyclohexyllithium, for instance, only three chlorine atoms can be substituted; the tricyclohexylchlorosilane no longer takes up any methyl, ethyl, or phenyl groups if it is treated with the corresponding organolithium compound (*248*). Tricyclohexylchlorogermane is capable of accepting a phenyl or benzyl group as the fourth substituent, but not a cyclohexyl group. Tetracyclohexylsilane is accessible by quite a different route, the hydrogenation of tetraphenylsilane (*188*).

Table 8 (p. 70) gives a selection of tetraorganosilanes and their physical constants. X-ray crystallographic investigations have been carried out on tetraphenylsilane (*169*, *249*).

3.2 Organohalosilanes and Tetrahalosilanes

Organohalosilanes of the general type R_nSiX_{4-n} (with $n=1$, 2, and 3) are the basic products of silicone chemistry, both for the preparation of the important siloxanes (M, D, T units, cf. Section 1.1) and for the synthesis of the silanes with other silicon-functional groups (see the "family tree" at the beginning of Chapter 3).

Tetrahalosilanes are required in limited quantities for the preparation of Q units.

3.2.1 Preparation

Organohalosilanes are accessible by the methods described in Chapter 2. Almost all syntheses give mixtures of R_nSiX_{4-n} with $n=1$, 2, and 3.

Fractional distillation is necessary to obtain the different silanes in the pure state. Isolation of the technologically very important methylchlorosilanes is

* Page number of chapter references.

TABLE 8

TETRAORGANOSILANES R_4Si[a]

Compound	Mp (°C)	Bp (°C)	at p (mm Hg)	n_D^t	at t (°C)	d_4^t (gm/cc)	at t (°C)
$(CH_3)_4Si$	α: −101.7 (*357*) β: −99.5 (*357*)	26.2 (*357*)	760	1.3582 (*357*)	20	0.6495 (*298*)	10
$(C_2H_5)_4Si$	−82.54 (*357*)	153.70 (*357*)	760	1.4267 (*357*)	20	0.7662 (*357*)	20
$(n\text{-}C_3H_7)_4Si$	−46 (*269*)	213–214 (*246*)	742	1.4378 (*246*)	20	0.7870 (*246*)	20
$(n\text{-}C_4H_9)_4Si$	—	156–157 (*118*)	22	1.4465 (*118*)	20	0.8008 (*118*)	20
$(C_6H_5)_4Si$	234–235 (*133, 267*)	428 (*213*)	—	—	—	1.0780 (*272*)	20
$(CH_3)_3C_2H_5Si$	—	62 (*382*)	760	1.3820 (*382*)	20	0.6849 (*382*)	20
$(CH_3)_3n\text{-}C_3H_7Si$	—	90 (*382*)	760	1.3929 (*382*)	20	0.7020 (*382*)	20
$(CH_3)_3n\text{-}C_4H_9Si$	—	115 (*382*)	760	1.4030 (*382*)	20	0.7181 (*382*)	20
$(CH_3)_3C_6H_5Si$	—	169–171 (*228*)	n.d.	1.4886 (*228*)	25	—	—
$(CH_3)_2(n\text{-}C_3H_7)_2Si$	—	141.69 (*357*)	760	1.4146 (*357*)	20	0.7399 (*357*)	20
$CH_3(n\text{-}C_3H_7)_3Si$	—	181–181.5 (*380*)	760	1.4291 (*380*)	20	0.7648 (*380*)	20
$CH_3(n\text{-}C_3H_7)(C_2H_5)(C_6H_5)Si$	—	228–230 (*194*)	n.d.	—	—	—	—
$(CH_3)_2(CH_2)_4Si$[b]	—	107 (*377*)	760	1.4335 (*377*)	25	0.780 (*377*)	25
$(CH_3)_2(CH_2)_5Si$[b]	—	133 (*377*)	760	1.4380 (*377*)	25	0.798 (*377*)	25
$(CH_3)_2(CH_2)_6Si$[b]	—	161 (*377*)	760	1.4335 (*377*)	25	0.780 (*377*)	25
$(CH_2)_4(CH_2)_4Si$[c]	< −70 (*377*)	178.5 (*377*)	760	1.4860 (*377*)	25	0.899 (*377*)	25
$(CH_2)_4(CH_2)_5Si$[c]	−35 (*377*)	203 (*377*)	760	1.4860 (*377*)	25	0.899 (*377*)	25
$(CH_2)_5(CH_2)_5Si$[c]	−18 (*377*)	227 (*377*)	760	1.4869 (*377*)	25	0.900 (*377*)	25

[a] Numbers in parentheses indicate reference numbers.
[b] Compound with one heterocyclic ring.
[c] Spiro-compound with two heterocyclic rings.

particularly difficult, since their boiling points are close to one another. Columns with a large number of plates are required and these must be operated at a high reflux ratio (*116*). In the case of the high-boiling phenylchlorosilanes, the differences between the boiling points are large but they cannot be fully utilized in practice since the distillation must be carried out in vacuum. The recovery of pure phenylchlorosilanes is rendered even more difficult by the interference of by-products, particularly chlorinated biphenyls, with boiling points near that of diphenyldichlorosilane. Furthermore, small amounts of aluminum chloride or ferric chloride tend to catalyze the decomposition of diphenyldichlorosilane at the temperatures occurring in the still bottoms. This silane reacts with $AlCl_3$ and $FeCl_3$ with cleavage of the aromatic Si—C bond, and formation of organoaluminum or organoiron compounds (*387*), e.g.,

$$(C_6H_5)_2SiCl_2 + AlCl_3 \rightarrow C_6H_5SiCl_3 + C_6H_5AlCl_2$$

$$(C_6H_5)_2SiCl_2 + 2AlCl_3 \rightarrow SiCl_4 + 2C_6H_5AlCl_2$$

$SbCl_3$ does not attack diphenyldichlorosilane, but $SbCl_5$ reacts as follows:

$$(C_6H_5)_2SiCl_2 + SbCl_5 \rightarrow C_6H_5SiCl_3 + C_6H_5SbCl_4$$

The organoantimony compound is unstable and decomposes rapidly (*386*):

$$C_6H_5SbCl_4 \rightarrow C_6H_5Cl + SbCl_3$$

The halides mentioned can be arranged in order of their dearylating activities as follows (*386*):

$$AlCl_3 > FeCl_3 \gg SbCl_5$$

The ability of $AlCl_3$ to cause scission is decreased by the addition of NaCl, which forms the complex $NaAlCl_4$ with $AlCl_3$ (*271*).

Further difficulties in the separation of the organosilanes by distillation are caused by formation of azeotropes. The azeotrope consisting of trimethylchlorosilane and silicon tetrachloride (Table 9) is particularly unwelcome. New methods for its separation are continually being investigated.

TABLE 9

BINARY AZEOTROPIC SYSTEMS (*196, 301*)

Component A	Component B	B in azeotrope (%)	Bp_{760}(°C)
$SiCl_4$	$(CH_3)_3SiCl$	35.2	54.5
$SiCl_4$	Acetonitrile	9.6	49.1
$(CH_3)_3SiCl$	Acetonitrile	7.3	56.5
$SiCl_4$	Acrylonitrile	11.0	51.2
$(CH_3)_3SiCl$	Acrylonitrile	7.0	57.0

The two chlorosilanes give azeotropes with acetonitrile and acrylonitrile whose boiling points lie sufficiently far apart to allow satisfactory separation (*301*). The boiling points of the $SiCl_4$/acetonitrile and $SiCl_4$/acrylonitrile azeotropes are also sufficiently far from that of trimethylchlorsilane. To recover this in the pure state, i.e., without a solvent phase, only sufficient acetonitrile is added for the $SiCl_4$/acetonitrile azeotrope to pass over, the $(CH_3)_3SiCl$ remaining behind as the higher boiling fraction.

If the azeotrope of $SiCl_4$ and $(CH_3)_3SiCl$ is treated with an equivalent amount of an amine or pyridine, the corresponding adduct, e.g., $SiCl_4 \cdot 2C_2H_5N$ (*372*) is formed. Pure $(CH_3)_3SiCl$ is obtained when the adduct is distilled (*370*). The addition compound can be decomposed with anhydrous HCl and worked up to the tetrahalosilane.

Other possibilities for the separation of the azeotropic mixture are offered by changing the silicon-functional groups of either one or both silanes. Hydrolysis of the azeotrope with a mixture of water and dioxane containing less water than theoretically required for the hydrolysis of the silicon tetrachloride results in selective attack on the tetrachloride, which is converted into high-molecular-weight, nondistillable chlorosiloxanes. The $(CH_3)_3SiCl$ can be recovered by distillation and fractional separation from the dioxane (*53*).

Hydrolysis of the azeotrope in a solvent with an excess of water at +1 to +5°C gives hexamethyldisiloxane in a yield of 65 to 70% (*183*).

By esterifying the azeotrope with absolute ethanol, both chlorosilanes are converted into ethoxy derivatives, which can be separated by distillation (*375*). If, in the liquid phase and in the presence of an acid catalyst such as HCl, the azeotrope is treated with trimethylethoxysilane, the silicon tetrachloride component reacts to give ethoxychlorosilane and trimethylchlorosilane

$$SiCl_4 + (CH_3)_3Si(OC_2H_5) \rightarrow Si(OC_2H_5)Cl_3 + (CH_3)_3SiCl$$

The latter is separated from the ethoxytrichlorosilane by fractional distillation, together with the trimethylchlorosilane already present in the azeotrope (*157*).

Treatment of the azeotropic mixture with ethylene oxide yields the β-chloroethoxy derivative, which can be separated by distillation (*295*). The action of phenol on the azeotrope at an elevated temperature gives the silanes $(CH_3)_3SiOC_6H_5$ and $Si(OC_6H_5)_4$, which can again be separated by fractional distillation (*69*). With catechol, the tetrahalo component gives a derivative insoluble in organic solvents, while trimethylchlorosilane forms the soluble 1,2-bis-(trimethylsiloxy)-benzene, which can be converted by hydrolysis into hexamethyldisiloxane (*279*).

Other proposals aim at facilitating the separation of the azeotrope by converting the Si—Cl bond into an Si—F bond. With zinc fluoride, trimethylchlorosilane gives the corresponding fluorosilane faster than $SiCl_4$. The $(CH_3)_3SiF$ formed boils at 17°C and can easily be separated from the $SiCl_4$

by distillation (*252*). If the azeotrope is added to an aqueous solution of hydrofluoric acid, $SiCl_4$ is converted into H_2SiF_6, which remains in solution, while $(CH_3)_3SiF$ is evolved as a gas (*237*, *238*).

The separation of chlorosilanes that do not form azeotropic mixtures can also be facilitated by suitable measures. With at least two moles of formamide or dimethylformamide per mole of silane, the more highly chlorinated silanes (i.e., methyltrichlorosilane and $SiCl_4$) give crystalline nonvolatile complexes, while dimethyldichlorosilane and trimethylchlorosilane do not. The solid amide complexes can be separated from the solution of the non-complex-forming silanes by filtration or decantation. Among other things, the process has also been applied to the separation of phenyltrichlorosilane and methylphenyldichlorosilane, which is also particularly difficult because of the closeness of their boiling points (cf. Table 10). It must naturally be borne in mind that, when the solutions are heated to distill off the solvent, the dimethyldichlorosilane can react with dimethylformamide (*164*) (cf. Section 3.2.3).

Certain dinitriles, particularly adiponitrile, have a greater solvent power for dichlorosilanes than for other chlorosilanes, and can therefore be used for selective extraction in the liquid or vapor phase. This process is also of interest for the separation of phenyltrichlorosilane and methylphenyldichlorosilane (*217*).

Finally, organotrihalosilanes can be separated from diorganodihalosilanes, particularly CH_3SiCl_3 from $(CH_3)_2SiCl_2$, by treating the mixture with equivalent quantities of a diorganodiacyloxysilane (acetate or propionate), so that at least one halogen atom of the organotrihalosilane is replaced by an acyloxy group. The reaction is catalyzed by triethanolamine triacetate. The diorganodihalosilane can then be recovered by fractional distillation (*218*).

3.2.2 Physical Constants (Tables)

Tables 10, 11, and 12 (p. 74) give some physical constants for a number of organohalosilanes with saturated organic groups.

In addition to the data given above, the reader's attention is drawn to the following investigations: dipole moments (*61*, *280*), bond refractivities (*323*), nuclear magnetic resonance spectra (*307*), microwave spectra (*145*, *316*), Raman spectra (*67*, *142*, *242*, *318*), surface tension (*281*), and thermal conductivity (*172*).

3.2.3 Chemical Properties

The special importance of the organohalosilanes for silicone chemistry is ultimately determined by the often widely differing reactivities of the substituents attached to the silicon; i.e., the organic groups on the one hand and the halogen atoms on the other. If the organic group is not substituted, various

TABLE 10

ORGANOHALOSILANES AND TETRAHALOSILANES[a]

Compound	Mp (°C)	Bp (°C)	at p (mm Hg)	n_D^t	at t (°C)	d_4^t (gm/cc)	at t (°C)
$(CH_3)_3SiCl$	−57.7 ± 0.2 (*44*)	57.3 ± 0.1 (*44*)	760	1.3884 (*276*)	20	0.8581 ((*276*)	20
$(CH_3)_2SiCl_2$	−76.1 ± 0.2 (*44*)	70.2 ± 0.1 (*116*)	760	—	—	1.067 (*116*)	27[b]
CH_3SiCl_3	−77.5 (*321*)	66.1 ± 0.1 (*116*)	760	—	—	1.270 (*116*)	27[b]
$SiCl_4$	−70 (*288*)	57.6 (*288*)	760	—	—	1.48 (*288*)	20
$(C_2H_5)_3SiCl$	—	145.8–146 (*61*)	~745	1.4299 (*61*)	25	0.8977 (*21*)	20
$(C_2H_5)_2SiCl_2$	—	129.0 (*61*)	744	1.4291 (*61*)	25	1.0472 (*61*)	25
$C_2H_5SiCl_3$	—	100–100.1 (*61*)	746	1.4257 (*52*)	19.8	1.2342 (*61*)	24
$(n\text{-}C_3H_7)_3SiCl$	—	202 (*79*)	760	1.442 (*79*)	16.5	0.886 (*366*)	20
$(i\text{-}C_3H_7)_3SiCl$	—	202 (*83*)	760	1.4515 (*119*)	20	0.9017 (*119*)	20
$(n\text{-}C_3H_7)_2SiCl_2$	—	171.5 (*270*)	753.5	1.4411 (*270*)	20	1.0174 (*270*)	20
$(i\text{-}C_3H_7)_2SiCl_2$	—	67–69 (*159*)	11	—	—	1.06 (*159*)	20
$(n\text{-}C_3H_7)SiCl_3$	—	122.2 (*270*)	740	1.4290 (*270*)	20	1.185 (*270*)	20
$(i\text{-}C_3H_7)SiCl_3$	—	120.3 (*270*)	747.8	1.4319 (*270*)	20	1.1934 (*270*)	20
$(n\text{-}C_4H_9)_3SiCl$	—	93–94 (*269*)	4.5	1.4478 (*269*)	20	0.8795 (*269*)	20
$(n\text{-}C_4H_9)_2SiCl_2$	—	62–63.5 (*269*)	4.5	1.4448 (*269*)	20	0.9915 (*269*)	20
$(i\text{-}C_4H_9)_2SiCl_2$	—	93 (*159*)	16	—	—	1.00 (*159*)	20
$(t\text{-}C_4H_9)_2SiCl_2$	−15 (*365*)	190 (*365*)	729	1.4561 (*365*)	20	1.009 (*365*)	20
$CH_3(n\text{-}C_4H_9)SiCl_2$	—	147.5–148 (*27*)	744	—	—	—	—
$(n\text{-}C_4H_9)SiCl_3$	—	146.5 (*270*)	741	1.4379 (*270*)	20	1.1603 (*270*)	20
$(i\text{-}C_4H_9)SiCl_3$	—	139.5 (*270*)	744.7	1.4358 (*270*)	20	1.1467 (*270*)	20
$(sec\text{-}C_4H_9)SiCl_3$	—	145–146 (*27*)	736	—	—	1.188 (*26*)	20
$(t\text{-}C_4H_9)SiCl_3$	98–99 (*365*)	132 (*365*)	730	—	—	—	—

$C_{18}H_{37}SiCl_3$	—	159–162 (*160*)	13	—	—	0.95 (*160*)	22
$(CH_3)C_{18}H_{37}SiCl_2$	—	200–210 (*27*)	6	—	—	—	—
$(C_6H_5)_3SiCl$	96 ± 1(*373*)	378 (*289*)	760	—	—	—	—
$(C_6H_5)_2SiCl_2$	—	305.2 (*289*)	760	1.5814 (*255*)	20	1.220 (*255*)	20
$C_6H_5SiCl_3$	—	201.5 (*289*)	760	—	—	1.3256 (*52*)	18.8
$(CH_3)C_6H_5SiCl_2$	—	82.5 (*165*)	13	—	—	1.1876 (*165*)	25[c]
$(CH_3)_2C_6H_5SiCl$	—	191 (*65*)	740–60	1.5082 (*65*)	20	1.032 (*65*)	20
$CH_3(C_6H_5)_2SiCl$	—	295 (*65*)	740–60	1.5740 (*65*)	20	1.110 (*65*)	20
$(C_6H_{11})_3SiCl$	100–101 (*248*)	—	—	—	—	—	—
$(C_6H_{11})_2SiCl_2$	—	150 (*262*)	4	—	—	—	—
$C_6H_{11}SiCl_3$	—	208–211 (*262*)	760	—	—	—	—
$(CH_3)_3SiF$	−74.3 ± 0.2 (*44*)	16.4 ± 0.1 (*44*)	760	—	—	—	—
$(CH_3)_3SiBr$	−43.5 to −43.2 (*229*)	77.3 (*229*)	735	1.4211 (*229*)	25	1.1727 (*229*)	25
$(CH_3)_3SiI$	—	106.5 (*276*)	734	—	—	1.47 (*276*)	20
$(CH_3)_2SiF_2$	−87.5 ± 0.2 (*44*)	2.7 ± 0.1 (*44*)	760	—	—	—	—
$(CH_3)_2SiBr_2$	−58.5 to −58.0 (*229*)	110.2–110.3 (*229*)	748	1.4696 (*229*)	25	1.6952 (*229*)	25
$(CH_3)_2SiI_2$	—	170 ± 1 (*5*)	760	—	—	2.203 (*5*)	20
CH_3SiF_3	−72.8 ± 0.2 (*41*)	−30.2 ± 0.1 (*41*)	760	—	—	—	—
CH_3SiBr_3	−28.4 to −28.1 (*229*)	131.2–131.3 (*229*)	744	1.5152 (*229*)	25	2.2130 (*229*)	25
CH_3SiI_3	—	229 ± 1 (*5*)	760	—	—	2.946 (*5*)	20
SiF_4	−95.7 (*288*)	−65 (*288*)	1810	—	—	—	—
$SiBr_4$	5 (*288*)	153 (*288*)	760	—	—	2.81 (*288*)	20
SiI_4	121 (*288*)	290 (*288*)	760	—	—	—	—

[a] Numbers in parentheses indicate reference numbers.
[b] d_{27}.
[c] d_{25}

TABLE 11

VAPOR PRESSURES OF A NUMBER OF ORGANOHALOSILANES[a]

	Bp (°C) at p (mm Hg)							
	10	20	40	60	100	200	400	760
$SiCl_4$ (*346*)	−35.1	−24.4	−12.6	−5.0	5.3	20.9	38.6	57.3
CH_3SiCl_3 (*346*)	−27.3	−16.5	−4.6	3.1	13.5	29.3	47.3	66.4
$(CH_3)_2SiCl_2$ (*346*)	−24.8	−13.7	−1.4	6.5	17.2	33.2	51.4	70.3
$(CH_3)_3SiCl$ (*171*)	−34.9	−24.2	−12.3	−4.7	5.6	21.2	38.9	57.6
$C_2H_5SiCl_3$ (*346*)	−1.6	9.7	22.3	30.6	41.5	58.4	77.9	98.8
$(C_2H_5)_2SiCl_2$ (*346*)	21.0	33.7	47.9	56.9	69.1	87.5	108.4	130.4
$(C_2H_5)_3SiCl$ (*346*)	32.0	45.5	60.2	69.5	82.3	101.6	123.6	146.3
$C_6H_5SiCl_3$ (*346*)	75.3	89.8	105.9	116 3	130.3	151.5	175.5	201.0
$(C_6H_5)_2SiCl_2$ (*346*)	158.4	175.2	194.0	205.9	222.2	246.8	274.8	304.4

[a] Numbers in parentheses indicate reference numbers.

TABLE 12

IGNITION POINTS, FLASH POINTS, AND EXPLOSIVE LIMITS OF A NUMBER OF IMPORTANT SILANES (*392*) (CF. ALSO REFS. *23*, *24*)

Silane	Ignition point[a] (°C)	Flash point[a] (°C)	Explosive limit[a] in air 760 mm Hg, 20°C	
			Lower (vol-%)	Upper (vol-%)
Trimethylchlorosilane $(CH_3)_3SiCl$	400	Lower than −20	3.6	6.4
Dimethyldichlorosilane $(CH_3)_2SiCl_2$	425	−1	5.5	10.4
Methyltrichlorosilane CH_3SiCl_3	455	−1	7.2	11.9
Methyldichlorosilane CH_3SiHCl_2	230	Lower than −20	4.5	>70[b]
Trichlorosilane $SiHCl_3$	215	Lower than −20	6.9	>70[b]

[a] Methods: Ignition point, DIN 51794; flash point, DIN 51755; explosive limit, steady supply of substance, varying addition of air; lower and upper inflammable mixture determined after ignition with pilot flame.

[b] The upper explosive limit was outside the range of the instrument and could not be determined accurately.

TABLE 13

C—X AND Si—X INTERNUCLEAR DISTANCES IN Å[a,b]

C—F		C—Cl		Si—Cl						Si—F		
CH_3F	1.42	CH_3Cl	1.77	SiH_3Cl	2.05	(*345*)	$(CH_3)_3SiCl$	2.09		$(CH_3)_3SiF$	1.55	(*145*)
CH_2F_2	1.36	CH_2Cl_2	1.77	SiH_2Cl_2	2.02		$(CH_3)_2SiCl_2$	1.99				
		$CHCl_3$	1.77	$SiHCl_3$	2.01		CH_3SiCl_3[c]	2.01	(*181*)	CH_3SiF_3	1.555	(*316*)
CF_4	1.36	CCl_4	1.76	$SiCl_4$	2.02		$SiCl_4$	2.02		SiF_4	1.54	
Theor.	1.41		1.76		2.16			2.16			1.81	

[a] Numbers in parentheses indicate reference numbers.
[b] Data from Haul (*150*), with some additions.
[c] In $C_2H_5SiCl_3$, $CH_3ClCHSiCl_3$, and $ClCH_2CH_2SiCl_3$ the Si—Cl distance was also found to be 2.00 ± 0.02 Å (*181*).

ways are available for bringing about reaction of the halogen without appreciable involvement of the Si—C bond. Substitution of the organic group may increase the reactivity of the Si—C bond—in certain circumstances quite appreciably. The high reactivity of the silicon-halogen bond strikingly demonstrates the difference between the chemistry of silicon and carbon; carbon tetrachloride is stable in water, while silicon tetrachloride reacts very vigorously.

The reactivity of the silicon-halogen bond is undoubtedly due primarily to its strongly polar character. Unlike the carbon-halogen bond, the measured interatomic distances in Si—X bonds differ considerably from the sum of the Pauling's atomic radii (Table 13). To interpret these reductions in the distances it is assumed that the silicon-halogen bond involves resonance between an ionic bond (I), a covalent bond (II), and a double-bond structure (III) (*266*).

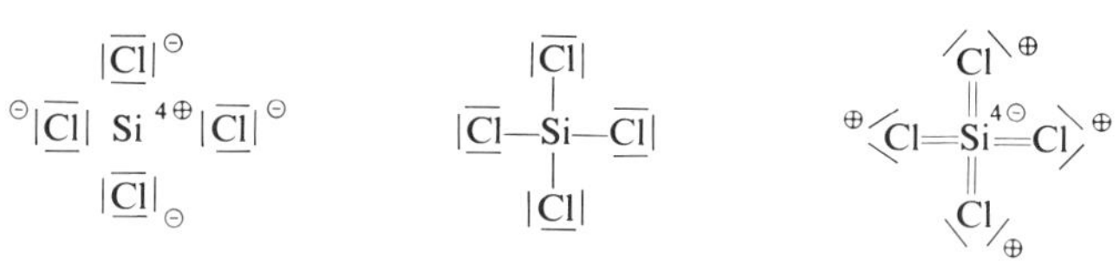

I II III

Evidence from Raman spectra can also be taken to confirm that the Si—Cl bond has a partial double-bond character (*137*).

Further information contributing to an understanding of the different behavior of the Si—C and the C—C bond can be obtained from the electron configurations and the stereochemical relationships of the atoms concerned. Unlike carbon, silicon with its unfilled $3d$ levels has the ability to extend its electron shell and attach nucleophilic groups, with the formation of an activated *pentacovalent transition state*.

Moreover, the radius of a silicon atom for single bonds, 1.17 Å, is appreciably larger than that of carbon, 0.77 Å. An atom, group, or molecule—particularly the OH group—attacking nucleophilically would have much easier access in $SiCl_4$ than in CCl_4. All these viewpoints are of the utmost importance in understanding the difference between the chemistry of carbon and silicon. They will be discussed in more detail later (Section 6.4). The Si—C bond, unlike the Si—halogen bond, is virtually covalent (see Section 6.4).

The reactivity of the halogens attached to silicon increases in the sequence:

$$\equiv\mathrm{Si{-}F} < \equiv\mathrm{Si{-}Cl} < \equiv\mathrm{Si{-}Br} < \equiv\mathrm{Si{-}I}$$

With substitution by the same halogen, it decreases as the number of organic groups is increased (cf. Chapter 5):

$$\mathrm{SiCl_4} > \mathrm{RSiCl_3} > \mathrm{R_2SiCl_2} > \mathrm{R_3SiCl}$$

Trialkylfluorosilanes react slowly with water, and are decomposed rapidly when alkalis are added.

The Si—halogen bond is capable of undergoing many types of reactions. Because of its accessibility through synthetic processes (Chapter 2), it occupies a key position in silane and silicone chemistry. Details are discussed in different contexts in the following sections: Reactions with alcohols, Section 3.3.1; with alkoxysilanes, Section 5.2.1; with alcohols and carboxylic acids, Section 5.2.2; with carboxylic acids, Section 3.10; with organoacyloxysilanes, Section 5.2.3; with aldehydes, Section 7.4.2; with diazomethane, Section 4.5.1.3; with water, Sections 3.6 and 5.1.1; with organosilanols, Section 5.2.4; with metals, Section 3.5; with metal oxides, Section 5.2.6; with organometal silanolates, Section 5.2.5; with metal hydrides, Section 3.4; with ammonia and amines, Section 3.8.1; with hydrogen sulfide, Section 3.9; with sulfuric, phosphoric, and perchloric acids, Section 3.11; with cyanides, cyanates, and thiocyanates, Section 3.12; with Grignard and other organometallic compounds, Sections 2.3, 4.3.1.2, 4.4.1.2, and 4.5.1.1.

The following facts are also pertinent.

Organohalosilanes react with ethyl orthoformate (*317*):

$$\equiv\mathrm{SiX} + \mathrm{HC(OC_2H_5)_3} \rightarrow \equiv\mathrm{SiOC_2H_5} + \mathrm{C_2H_5X} + \mathrm{HCOOC_2H_5}$$

This method has the advantage over the conventional esterification of halosilanes with alcohols in that no hydrogen halide capable of giving rise to interfering side reactions is liberated.

Si—O—C bonds are also produced by the reactions of organochlorosilanes with aldehydes (*88*) and epoxides (*220*, *221*, *264*, *300*, *333*):

$$\geq SiCl + H_2C{=}O \rightarrow \geq Si{-}O{-}CH_2Cl$$

$$\geq SiCl + H_2C{-}CH_2 (\text{-O-}) \rightarrow \geq Si{-}O{-}CH_2CH_2Cl$$

If the first reaction is carried out with equivalent amounts of water and in the presence of pyridine, the products are silylformals (cf. Section 7.4.2):

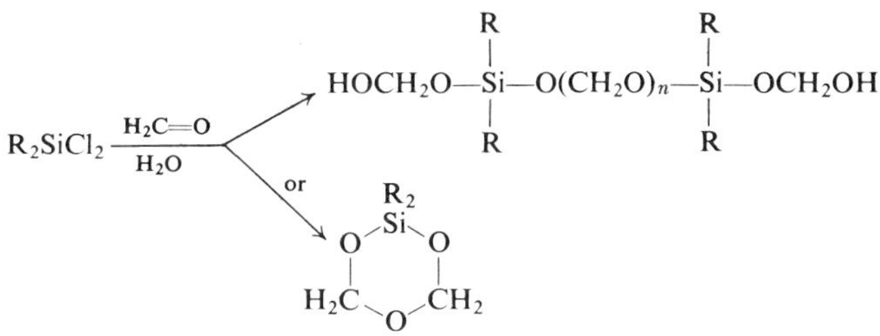

However, no compounds with Si—O—C bonds have been found in the products of the reaction of silicon tetrachloride with benzaldehyde. They probably occur only as unstable intermediates. The final products are benzylidene chloride and polydichlorosiloxanes (*391*):

$$SiCl_4 + C_6H_5CHO \rightarrow \frac{1}{n}[SiCl_2O]_n + C_6H_5CHCl_2$$

A reaction that has been known for an even longer time is the cleavage of ethers by silicon halides (*312*); $SiCl_4$ reacts slowly with diethyl ether to form $Si(OC_2H_5)Cl_3$ and ethyl chloride—in principle, therefore, according to the equation:

$$\geq Si{-}X + ROR \rightarrow \geq Si{-}OR + RX$$

However, the cleavage of ethers is so slight under the reaction conditions normally used that it presents no obstacle to the utilization of the good solvent properties of diethyl ether in many organometallic reactions with chlorosilanes.

At higher temperatures (~125 to 350°C) and with the use of Friedel-Crafts catalysts, the reaction of organochlorosilanes and dialkyl ethers gives organosiloxanes and alkyl chlorides since, under such conditions, the $\geq SiOR$ groups formed by the cleavage of the ethers react further with $\rangle$SiX bonds in accordance with the equation (*296*) (cf. Section 5.2.1):

$$\rangle Si{-}OR + X{-}Si\langle \rightarrow \rangle Si{-}O{-}Si\langle + RX$$

Solutions of organochlorosilanes in acetone become deep red on standing at room temperature. Under these conditions, dimethyldichlorosilane is converted into cyclic and high-molecular-weight dimethylsiloxanes, while $SiCl_4$ and CH_3SiCl_3 give gels. Methyl ethyl ketone behaves similarly, while acetophenone does not react with the chlorosilanes (*355*).

Organochlorosilanes react with amides to form *N*-silyl-substituted compounds. With acetamide, trimethylchlorosilane gives the mono- or disubstituted derivative (*35*)

$$(CH_3)_3SiCl + CH_3CONH_2 \xrightarrow[\text{pyridine}]{-HCl} (CH_3)_3SiNHCOCH_3 \xrightarrow[\text{pyridine}]{(CH_3)_3SiCl} [(CH_3)_3Si]_2NCOCH_3$$

the first of which forms a good starting material for the synthesis of silanols (Section 3.6.1). The behavior of dimethylformamide is interesting. On boiling, solutions of organochlorosilanes in dimethylformamide react to form organosiloxanes; the halogen is bound as dimethylammonium chloride, and carbon monoxide is formed as a by-product. The optimum ratio of the amounts of the reactants appears to be about 1.9 moles of dimethylformamide to one Si—Cl bond (*287*).

Interhalogenation of the organohalosilanes can easily be carried out in the sequence

$$\rangle Si{-}I \rightarrow \rangle Si{-}Br \rightarrow \rangle Si{-}Cl \rightarrow \rangle Si{-}F$$

i.e., a halosilane on the right can be obtained from one of those to the left, but not the other way around. Even the pseudohalogens can be included in this sequence (Section 3.12) (*80*). In particular, $\rangle$Si—F bonds can be obtained from $\rangle$Si—Cl, —Br, or —I bonds by reaction with aqueous alcoholic (*82*) or aqueous hydrofluoric acid (*223*) or with anhydrous hydrogen fluoride (*268*), ammonium fluoride (*384*), zinc fluoride (*86, 251*), antimony trifluoride (*38–43*) even in aqueous solution (*239*), or sulfur tetrafluoride (*240*), as well as by passing the vapors of the silanes through eutectic melts of alkali-metal fluorides (*350*).

The organic group can be halogenated in all organohalosilanes. The haloorganohalosilanes are again starting materials for substitutions which can be carried out by the methods of classical organic chemistry; amino, hydroxyl, carboxyl, and other groups can thus be introduced into the organic radical (see Chapter 4).

The pyrolysis of methylchlorosilanes around 700°C gives a complicated mixture of linear and cyclic silmethylenes with methyl and chloro substituents on the Si atoms (*103*, *106*).

3.3 Organo(organooxy)silanes

Organo(organooxy)silanes have the general formula $R_nSi(OR)_{4-n}$ ($n=1$, 2, or 3). After the halosilanes, they are the most interesting silanes in silicone chemistry. Like the corresponding halosilanes, they lead to M, D, and T siloxane units (cf. Section 1.1). In principle, Q units are produced by all the possible tetra(organooxy)silanes $Si(OR)_4$. In practice only those are of interest which can be hydrolyzed relatively fast, i.e., tetramethoxy- and tetraethoxysilanes. These two silanes may be mentioned here as limiting members of the series. Detailed treatment of the tetra(organooxy)silanes and polyorganooxysiloxanes will be found in Chapter 11.

3.3.1 Preparation

The processes available for the preparation of the organically substituted organooxysilanes are very similar to those for the tetra(organooxy)silanes (orthosilicic acid esters).

These compounds are usually obtained by the reaction of organohalosilanes with alcohols or alkoxides:

$$\equiv Si{-}X + ROH \rightarrow \equiv Si{-}OR + HX \tag{1}$$

$$\equiv Si{-}X + ROM \rightarrow \equiv Si{-}OR + MX \tag{2}$$

where pyridine (*92*, *243*) or tertiary amines (*320*) can be added as acceptors for the acid formed in (1). Alternatively, the reaction of tetra(organooxy)silanes with organometallic compounds, preferably Grignard compounds and organic derivatives of the alkali metals can be employed (Section 2.3.1).

Of all the Si—X bonds the Si—F bond is the most inert. The reaction with alcohol takes place only if anhydrous ammonia is passed into the mixture of organofluorosilane and alcohol (*176*):

$$R_{4-x}SiF_x + xNH_3 + xROH \rightarrow R_{4-x}Si(OR)_x + xNH_4F$$

In addition, organofluorosilanes can be converted into organooxysilanes with aluminum, magnesium, or calcium alkoxides at 200 to 260°C, e.g.:

$$3R_3SiF + Al(OR)_3 \rightarrow 3R_3SiOR + AlF_3$$

in yields of up to 90% (*175*).

Silanes with Si—Cl bonds are generally used for alcoholysis. Their reactivity decreases with increasing loading of the silicon atom with methyl or ethoxy groups, as can be seen from the following sequence (*388*):

$$CH_3SiCl_3 > CH_3Si(OC_2H_5)Cl_2 > (CH_3)_2SiCl_2 > CH_3Si(OC_2H_5)_2Cl$$
$$> (CH_3)_2Si(OC_2H_5)Cl > (CH_3)_3SiCl$$

This phenomenon is due at least in part to steric factors. The yields of distillable alkoxysilanes are diminished by side reactions; in the case of alcoholysis with methanol and ethanol, they are between 50 and 60% when HCl acceptors are used. With the higher alcohols, such as *n*-butanol, yields of 70 to 80% are obtained (*109*). On the other hand, the yields in the reaction of $SiCl_4$ with the same alcohols are up to about 20% higher (*319*). The side reactions include, in particular, the condensation of alkoxysilanes already formed with the chlorosilanes still present:

$$\equiv Si{-}OR + Cl{-}Si\equiv \rightarrow \equiv Si{-}O{-}Si\equiv + RCl$$

which may take place not only intermolecularly but also intramolecularly:

$$>Si{<}^{Cl}_{OR} \rightarrow [{-}\overset{|}{\underset{|}{Si}}{-}O]{-} + RCl$$

Very high yields of alkoxysilanes are obtained by the reaction of chlorosilanes with alkyl orthoformates (*109*) in the presence of $AlCl_3$:

$$(CH_3)_2SiCl_2 + 2HC(OC_2H_5)_3 \rightarrow (CH_3)_2Si(OC_2H_5)_2 + 2C_2H_5Cl + 2HCOOC_2H_5$$

The Si—H bond can also be converted into an Si—OR bond by catalyzed reactions with alcohols or alkoxides, and with aldehydes, ketones, and ethers (Section 3.4.3). It remains intact when Si—OR bonds are obtained from Si—Cl bonds by means of ethyl orthoformate (Section 3.4.1).

Organooxysilanes in which the OR group is derived from a higher monohydric alcohol, a polyhydric alcohol, a polyglycol ether, or a phenol are easily obtained by transesterification from methoxy- and ethoxysilanes. Not only do these contain very reactive alkoxy groups, but in the transesterification process they yield alcohols which can easily be removed continuously from the reaction system because of their low boiling points. In this sense, therefore, methoxy- and ethoxysilanes occupy a key position in the preparation of other organooxysilanes.

The transesterification process is accelerated by catalysts. Suitable catalysts are sodium alkoxides (*74*), titanic acid esters (*149*), sulfuric acid (*128*), trifluoroacetic acid (*20*), amines and quaternary ammonium hydroxides such as tetramethylammonium hydroxide (*381*).

3.3.2 Physical Constants (Table)

Physical constants for a number of organomethoxy- and organoethoxysilanes of interest as starting materials for the production of organosiloxanes have been collected in Table 14.

3.3.3. Chemical Properties

The Si—OR bond is subject to hydrolytic cleavage. In conformity with the more pronounced covalent character of the bond as compared with that of the Si—X bond, its sensitivity to hydrolysis is considerably lower. In addition, it depends strongly on the nature of the radical R in the organic group. The sensitivity of alkoxy groups to hydrolysis decreases with an increasing number of carbon atoms in the group; aryloxy groups are less easily split off by hydrolysis than lower alkoxy groups. Organomethoxy- and organoethoxysilanes (i.e., silanes with the most reactive alkoxy groups) have gained interest mainly as intermediates for the production of silicones. They are particularly reactive not only toward water but also in alcoholysis and in the wide variety of reactions which the alkoxy group can undergo.

The organoalkoxysilanes are among the most important intermediates and auxiliary products of silicone chemistry, although the Si—OR bond is less reactive than the Si—X bond. This is due primarily to the fact that organoalkoxysilanes can be converted to siloxanes without the production of corrosive by-products; this may be of interest, for example, for an *in situ* production of silicone films on surfaces where the liberation of acid might be harmful. Of no less importance is the fact that in reactions with higher alcohols it is only an alcohol, and that the lower alcohol formed during transesterification, which is liberated. This enables reactions with hydroxy compounds sensitive to acids to be carried out. Organoalkoxysilanes and particularly organoethoxysilanes have therefore become the preliminary stages in the production of *silicone-modified organic resins* (cf. Section 8.1.2.3).

The reactivity of the alkoxy groups is, however, also affected by the nature and extent of the organic substitution of the silicon atoms. If several relatively large organic groups are bound to the silicon, the reactivity of a Si—OR bond can be considerably reduced by steric hindrance. Triphenylethoxysilane is unaffected by boiling with water, 5% hydrochloric acid, or 30% aqueous alkali (*244*). This resistance to hydrolysis can be further increased by attaching bulky groups to the silicon. Trioctylphenoxysilane, diethyldioctyloxysilane, and phenyltridodecyloxysilane are very resistant toward hydrolysis. Since these compounds are remarkable for their low freezing point and the relatively small temperature-dependence of their viscosity, they have been recommended as *hydraulic fluids* (*259*). Phenylaryloxysilanes combine relatively good resistance to hydrolysis with good heat stability and high boiling points. They can therefore be used as *heat-transfer media* (*62, 63, 168, 236*).

TABLE 14

ORGANO(ORGANOOXY)SILANES [a]

Compound	Mp (°C)	Bp (°C) at	p (mm Hg)	n_D^t at	t (°C)	d_4^t (gm/cc) at	t (°C)
$(CH_3)_3SiOCH_3$	—	56.5–56.7 (*297*)	747	1.3678 (*297*)	20	—	—
$(CH_3)_2Si(OCH_3)_2$	—	80–80.5 (*189*)	760	—	—	—	—
	—	82 (*258*)	760	1.3708 (*258*)	20	0.8646 (*258*)	20
$CH_3Si(OCH_3)_3$	—	103–103.5 (*189*)	760	1.3687 (*356*)	25	—	—
$Si(OCH_3)_4$	2 (*286*)	121.4 (*199*)	760	1.3681 (*199*)	20	1.0339 (*323*)	20
$(CH_3)C_2H_5Si(OCH_3)_2$	—	105.5 (*59*)	750	1.3854 (*59*)	20	0.8731 (*59*)	20
$(CH_3)n\text{-}C_3H_7Si(OCH_3)_2$	—	126 (*59*)	750	1.3931 (*59*)	20	0.8689 (*59*)	20
$(CH_3)n\text{-}C_4H_9Si(OCH_3)_2$	—	149 (*59*)	750	1.4007 (*59*)	20	0.8682 (*59*)	20
$(CH_3)C_6H_5Si(OCH_3)_2$	—	199 (*59*)	750	1.4694 (*59*)	20	0.9934 (*59*)	20
$(C_2H_5)_2Si(OCH_3)_2$	—	130 (*87*)	760	—	—	—	—
$C_2H_5Si(OCH_3)_3$	—	123 (*87*)	760	—	—	0.9747 (*203*)	0
$(n\text{-}C_3H_7)_3SiOCH_3$	—	63 (*209*)	4	—	—	—	—
$(n\text{-}C_3H_7)_2Si(OCH_3)_2$	—	168 (*59*)	760	1.4082 (*59*)	20	—	—
$(n\text{-}C_4H_9)_2Si(OCH_3)_2$	—	205.4 (*59*)	760	1.4184 (*59*)	20	—	—
$(C_6H_5)_3SiOCH_3$	54.5–55 (*31*)	164–165 (*31*)	0.5	—	—	—	—
$(C_6H_5)_2Si(OCH_3)_2$	—	286 (*59*)	760	1.5385 (*59*)	20	—	—
		145–152 (*156*)	8				
$C_6H_5Si(OCH_3)_3$	—	130 (*189*)	45	1.47331 (*323*)	20	1.0641 (*323*)	20
$(CH_3)_3SiOC_2H_5$	—	75.5–76 (*352*)	760	1.3737 (*352*)	25	0.755 (*352*)	25
$(CH_3)_2Si(OC_2H_5)_2$	—	112–113.5 (*352*)	760	1.3840 (*352*)	25	0.827 (*352*)	25
$CH_3Si(OC_2H_5)_3$	—	143–143.5 (*352*)	760	1.3844 (*352*)	25	0.899 (*352*)	25
$Si(OC_2H_5)_4$	−77 (*60*)	168.5 (*346*)	760	1.3831 (*368*)	20	0.9346 (*368*)	20
		153 (*203*)	760	—	—	0.8403 ((*203*)	0

[illegible]		[illegible] (199)	760	1.4980 (*199*)	20	0.8750 (*199*)	20
$C_2H_5Si(OC_2H_5)_3$	—	158–160 (*274*)	760	1.3853 (*274*)	22	0.928 (*274*)	22
$(n\text{-}C_3H_7)_3SiOC_2H_5$	—	109 (*223*)	40	1.4245 (*223*)	20	0.8155 (*223*)	20
$(i\text{-}C_3H_7)_3SiOC_2H_5$	—	200 (*119*)	738	1.4560 (*119*)	20	0.8657 (*119*)	20
$(n\text{-}C_3H_7)_2Si(OC_2H_5)_2$	—	185 (*205*)	760	1.4100 (*205*)	20	0.8544 (*205*)	20
$(i\text{-}C_3H_7)_2Si(OC_2H_5)_2$	—	186–187 (*79*)	760	1.4130 (*79*)	25.8	—	—
$n\text{-}C_3H_7Si(OC_2H_5)_3$	—	179–180 (*275*)	760	1.4076 (*275*)	26	—	—
$i\text{-}C_3H_7Si(OC_2H_5)_3$	—	177–179 (*14*)	760	—	—	0.9194 (*14*)	—
$(n\text{-}C_4H_9)_3SiOC_2H_5$	—	145 (*223*)	34	1.4341 (*223*)	20	0.8246 (*223*)	20
$(n\text{-}C_4H_9)_2Si(OC_2H_5)_2$	—	125 (*223*)	38	1.4190 (*223*)	20	0.8500 (*223*)	20
$n\text{-}C_4H_9Si(OC_2H_5)_3$	—	190–193 (*275*)	740	1.3976 (*275*)	29	0.895 (*275*)	29
$(C_6H_5)_3SiOC_2H_5$	65 (*244*)	207 (*244*)	3	—	—	—	—
$(C_6H_5)_2Si(OC_2H_5)_2$	—	109–110 (*338*)	0.7	1.5250 (*338*)	25	—	—
$C_6H_5Si(OC_2H_5)_3$	—	133 (*317*)	25	1.4580 (*317*)	25	0.988 (*317*)	25
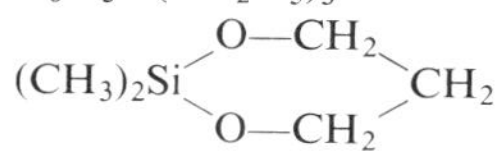	—	121.7 (*200*)	760	1.4120 (*200*)	20	0.9703 (*200*)	20
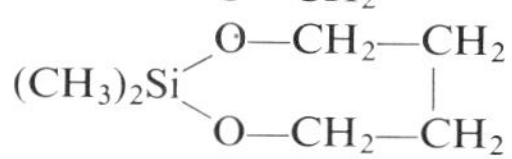	—	142 (*200*)	760	1.4218 (*200*)	20	1.0059 (*195*)	17
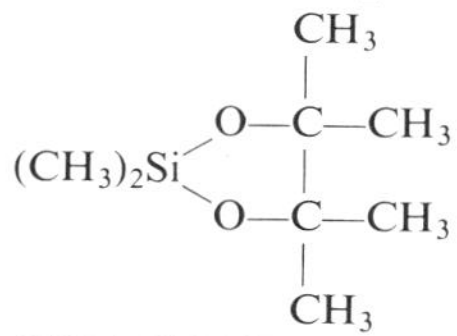	—	152–153 (*148*)	760	—	—	—	—
$(CH_3)_3SiO(CH_2)_4OSi(CH_3)_3$	—	203 (*195*)	759	1.413 (*195*)	15	0.8579 (*195*)	15
$p\text{-}(CH_3)_3SiOC_6H_4OSi(CH_3)_3$.	—	132 (*339*)	25	—	—	—	—

[a] Numbers in parenthese indicate reference numbers.

The fact that certain organoalkoxysilanes prepared by the reaction of organoalkoxysilanes with polyhydric alcohols [e.g., ethylene glycol (*112*), glycerol (*341*)] are soluble in water is of considerable interest. The reaction product of dimethyldiethoxysilane with glycerol, for instance, probably has the structure (*341*):

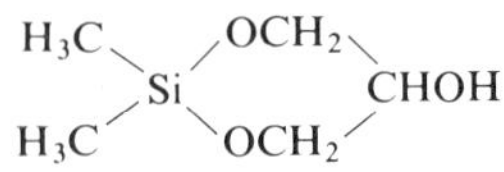

The stability of the aqueous solutions of these compounds is of course limited, since the Si—OC bond will hydrolyze sooner or later. Water-soluble alkoxysilanes of this type containing the vinyl radical, are suitable for the surface-finishing of glass fibers (Section 10.7.2) and for the protection of masonry (Section 10.12.2.1).

The Si—OR bond is capable of numerous other reactions.

Trialkylalkoxysilanes readily react with anhydrous hydrogen chloride to give the corresponding chlorosilanes (*91*). Depending on the type of silane used, larger or smaller amounts of siloxanes are formed by side processes during this reaction [reaction of the Si—Cl bonds formed with Si—OR bonds (see Section 5.2.1)]. A smooth conversion into the chlorosilanes with yields of over 90% is achieved by the reaction of alkoxysilanes with acyl chlorides (*65*):

$$CH_3(C_6H_5)_2SiOC_2H_5 + CH_3COCl \rightarrow CH_3(C_6H_5)_2SiCl + CH_3COOC_2H_5$$

or with thionyl chloride (*285*), and also with phosphorus pentachloride (*211*, *230*), phosphorus trichloride (*91*), boron trichloride (*97*, *383*), and aluminum trichloride (*170*), and by substituent exchange with silicon tetrachloride (*114*, *211*).

Conversion into fluorosilanes has been accomplished with aqueous or aqueous-alcoholic solutions of hydrogen fluoride (*82*), with fluorosulfonic acid (*336*), and with thionyl fluoride (*241*).

The reaction with alkali-metal hydroxides gives alkali-metal silanolates. The reaction with triorganoalkoxysilanes clearly proceeds according to the equation (*162*, *167*):

$$R_3SiOR + KOH \rightarrow R_3SiOK + ROH$$

Other reactions of the Si—OR bond are described in detail elsewhere: self-condensation of alkoxysilanes to give siloxanes (Section 5.2.9), and the reactions of alkoxysilanes with carboxylic acids (Section 5.2.10), with acyloxysilanes (Section 5.2.11), and with organosilanols (Section 5.2.12).

Diethyldiethoxysilane gives the corresponding acetal and cyclic as well as linear siloxanes with anhydrous aldehydes in the presence of sulfuric acid:

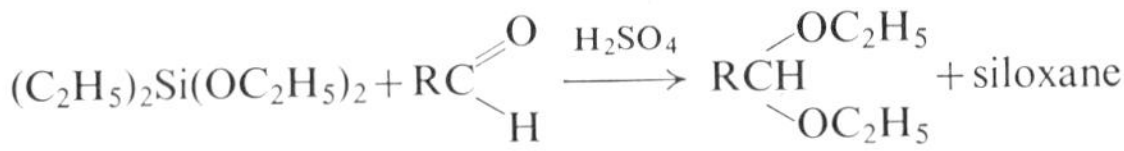

The reaction with acetone also gives siloxanes, but the mechanism is obscure (*214*).

3.4 Organo-*H*-silanes

Besides the organohalosilanes and organo(organooxy)silanes, the organosilanes still containing Si—H bonds form a third group that has acquired importance for the chemistry of the silicones, even though it is not so much the capacity of the Si—H bond for undergoing hydrolytic cleavage as its other reactions that are of interest.

Among the compounds belonging to this group are organosilanes having only organic groups and hydrogen atoms as substituents, and corresponding to the general formula

$$R_nSiH_{4-n} \qquad (n = 1, 2, \text{and } 3)$$

and also organo-*H*-halosilanes and organo-*H*-alkoxysilanes of the general formula

$$R_nSiH_mX_{4-(m+n)} \quad (m+n = 2, 3; m = 1, 2; n = 1, 2)$$

3.4.1 PREPARATION

Methyldichlorosilane CH_3SiHCl_2, an organo-*H*-halosilane, is a technical product which is obtained as a by-product of the direct synthesis (Chapter 2). Trichlorosilane is also a product of technical processes, and is used as a starting material for the preparation of ultrapure (metallic) silicon. The reaction of silicon, ferrosilicon, or calcium silicide with hydrogen chloride, carried out most frequently in the presence of a copper catalyst at 300–450°C, proceeds theoretically as follows:

$$Si + 3HCl \rightarrow SiHCl_3 + H_2$$

but silicon tetrachloride and higher silicon chlorides are always obtained as by-products.

Organic groups can be introduced into trichlorosilane by the Grignard process; only the Si—Cl (but not the Si—H bond) is involved in this reaction. Organo-*H*-chlorosilanes can be produced by suitable methods (Section 2.3.1):

$$SiHCl_3 + RMgCl \rightarrow RSiHCl_2 + MgCl_2$$

Conversely, the chlorine in chlorosilanes already containing organic substituents can be replaced by hydrogen. Reactions with metal hydrides, such as lithium tetrahydroaluminate (*31*, *94*, *219*, *267*), lithium hydride (*273*), sodium hydride (*111*), and the reactive aluminum tetrahydroborate $Al(BH_4)_3$ (*136*) are most suitable for this reaction, which proceeds in principle as follows:

$$\geq Si\text{—}Cl + MH \rightarrow \geq Si\text{—}H + MCl$$

The action of aluminum tetrahydroborate on trimethylchlorosilane can be represented schematically as follows:

$$2Al(BH_4)_3 + 6(CH_3)_3SiCl \rightarrow Al_2Cl_6 + 3B_2H_6 + 6(CH_3)_3SiH$$

The reaction with sodium hydride dispersed in mineral oil or another inert high-boiling hydrocarbon requires temperatures of 175 to 350°C (*111*). The addition of catalysts can lower the reaction temperature or increase the yield. Alkylborons, boric acid esters, and triethylaluminum are suitable catalysts (*178*, *389*). An interesting embodiment of the substitution reaction consists in passing the halogen compound through a eutectic melt of an alkali-metal halide containing lithium hydride. In this way, at 360 to 400°C dimethyldichlorosilane yields dimethylsilane $(CH_3)_2SiH_2$ with 95% conversion and 89% yield (*347*).

Lithium tetrahydroaluminate also breaks the Si—O—Si bond with the formation of silanes (see Section 5.5.7).

The reaction of a chlorosilane with calcium hydride at 200 to 320°C gives only moderate yields of silanes containing hydrogen attached to silicon. With alkylaluminum halides as catalysts trimethylchlorosilane has been converted into trimethylsilane in 46% yield (*322*).

Attempts to convert the Si—Cl into the Si—H bond by passing organochlorosilane vapors with hydrogen or hydrogen halide over aluminum, magnesium, or zinc at 200 to 500°C also gave only moderate yields. A liquid mixture which contained 5% of methyldichlorosilane was produced from methyltrichlorosilane (*158*).

At elevated temperatures the Si—F bond has been converted into the Si—H bond with sodium hydride (*177*) and likewise with the aluminum compounds R_2AlH (*179*).

Triarylsilanes are formed rapidly and quantitatively by the reaction of triarylsilylmetal compounds with triarylmethanes (*50*):

$$R_3SiM + R_3CH \rightarrow R_3SiH + R_3CM$$

The replacement of halogen by hydrogen invariably runs through to the perhydrogenated silanes; organosilanes simultaneously containing both Si—halogen and Si—H bonds can be obtained only by redistribution between the *H*-silane and the halosilane in a second operation (cf. Section 2.5) (*115*).

$$CH_3SiCl_3 \xrightarrow{NaH} CH_3SiH_3$$

$$CH_3SiCl_3 + CH_3SiH_3 \xrightarrow{AlCl_3} CH_3SiHCl_2 + CH_3SiH_2Cl$$

Organosilanes containing both Si—OR and Si—H bonds can be obtained correspondingly (Section 2.5). The latter are also produced in very good yield by the reaction of the *H*-chlorosilanes with ethyl orthoformate. In this way, methyldichlorosilane is 90% converted into methyldiethoxysilane (*109*):

$$CH_3SiHCl_2 + 2HC(OC_2H_5)_3 \rightarrow CH_3SiH(OC_2H_5)_2 + 2C_2H_5Cl + 2HCOOC_2H_5$$

3.4.2 Physical Constants (Table)

The physical constants of some selected compounds of the group of silanes discussed here have been collected in Table 15.

TABLE 15

SUBSTITUTED SILANES WITH Si—H BONDS[a]

Compound	Mp (°C)	Bp (°C) at	p (mm Hg)	n_D^{20}	d_4^t (gm/cc) at	t (°C)
SiH_4	−185 (*288*)	−112 (*288*)	760	—	0.68 (*288*)	−185
SiH_3Cl	−118 (*288*)	−30 (*288*)	760	—	1.15 (*288*)	−113
SiH_2Cl_2	−122 (*288*)	8.3 (*288*)	760	—	1.42 (*288*)	−122
$SiHCl_3$	−127 (*288*)	31.8 (*232*)	760	1.4020 (*232*)	1.3417 (*232*)	20
$(CH_3)_3SiH$	−135.89 (*357*)	6.7 (*357*)	760	—	0.6375 (*357*)	6.7
$(CH_3)_2SiH_2$	−150.22 (*357*)	−19.6 (*357*)	760	—	0.6377 (*357*)	−19.6
CH_3SiH_3	−156.81 (*357*)	−57.5 (*357*)	760	—	0.6277 (*357*)	−57.5
$(CH_3)_2SiHCl$	—	35.4 (*348*)	760	—	—	—
CH_3SiH_2Cl	−134.8 to −134.1 (*344*)	7–8 (*344*)	760	—	0.935 (*344*)	−80
CH_3SiHCl_2	− 90.6 ± 0.5 (*40*)	40.4 ± 0.1 (*40*)	760	—	1.135 (*40*)	0
$(C_2H_5)_3SiH$	−156.90 (*357*)	108.77 (*357*)	760	1.4119 (*357*)	0.7318 (*357*)	20
$(C_2H_5)_2SiH_2$	−134.39 (*357*)	55.99 (*357*)	760	1.3920 (*357*)	0.6832 (*357*)	20
$C_2H_5SiH_3$	−179.7 (*357*)	−13.7 (*357*)	760	—	—	—
$(C_2H_5)_2SiHCl$	−143 (*232*)	99.7 (*232*)	760	1.4133 (*232*)[b]	0.8895 (*232*)	20
$C_2H_5SiHCl_2$	−107 (*232*)	74.9 (*232*)	760	1.4148 (*232*)	1.0926 (*232*)	20
$(C_6H_5)_2SiH_2$	—	75–76 (*31*)	0.5	1.5795 (*31*)	0.9969 (*31*)	20
$C_6H_5SiH_3$	—	120 (*31*)	760	1.5125 (*94*)	0.8681 (*94*)	20
$(C_6H_5)_2SiHCl$	—	143 (*25*)	10	—	1.118 (*25*)	20
$C_6H_5SiHCl_2$	—	184 (*212*)	760	—	—	—

[a] Numbers in parentheses indicate reference numbers.
[b] n_D^{25}.

3.4.3 Chemical Properties

The Si—H bond is silicon functional and, in particular, subject to hydrolytic cleavage. The ease of hydrolysis is however very much dependent on the pH, being considerably less in the acid than in the alkaline region. Halosilanes with Si—H bonds can thus be converted into the corresponding siloxanes by hydrolysis without liberation of appreciable quantities of hydrogen, particularly if the lowest possible temperatures are employed. Decomposition in the alkaline region however is so fast that it forms the basis for a method of quantitative estimation of Si—H bonds (see Chapter 12). The alkaline hydrolysis of alkyl-*H*-silanes has been the subject of detailed investigations (*22*, *120*, *378*). The rate-determining step is the formation of a pentacovalent intermediate according to Eq. (1), which rapidly decomposes into the end products Eq. (2) (*325*, *329*).

(1)

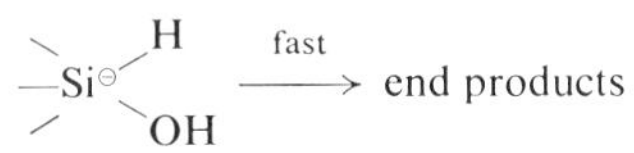

(2)

The ability of the Si—H bond to add to unsaturated hydrocarbons (Section 2.4.1) is of considerable practical importance, as it gives access to organically substituted silanes, particularly those with a high number of carbon atoms, which are frequently difficult to obtain by other routes.

For such reactions, proceeding via a radical mechanism, the bond character and especially the polar component of the Si—H bond appears to be decisive. The reactivity of different carbon and silicon compounds toward olefins can be qualitatively arranged in a sequence which is the inverse of the sequence of the proportion of polar component in the bond concerned, as shown in the accompanying tabulation. There appears to be no correlation with the individual bond energies (*390*).

Bond	C—Br	Si—H	C—H	C—Cl	Si—Br	Si—Cl
		Decreasing reactivity of olefins ⟶				
Polar component (%)	3	3	4	6	22	30

The practical application of the silanes with Si—H bonds, or the siloxanes produced from them, is frequently based on their reactivity with oxygen. A considerable energy drop from the Si—H (70.4 kcal/mole) to the Si—O bond (88.2 kcal/mole) can be calculated from Pauling's single-bond energies, and this drop must be responsible for the great sensitivity of the Si—H bond to oxidation.

The simple silanes SiH_4, Si_2H_6, Si_3H_8, etc., ignite spontaneously in air. An explosion takes place, probably as a result of an intermediate liberation of hydrogen (*288*)

$$n\mathrm{SiH_4} + \tfrac{n}{2}\mathrm{O_2} \rightarrow [\mathrm{H_2SiO}]_n + n\mathrm{H_2}$$

which gives an explosive mixture with air. The introduction of a chlorine atom or a methyl group into the silane molecule is sufficient to suppress this tendency to spontaneous ignition. The hydrogen atom in a compound like CH_3SiHCl_2 is not attacked by atmospheric oxygen at room temperature.

All Si—H bonds can however be easily oxidized at elevated temperatures and can thus be used as reducing agents. Diphenylsilane gives thioxanthene in 63% yield on being heated with 10-thioxanthone, and anthracene in 45% yield on being heated with anthraquinone (*68*). A very active reduction system can be made up from silanes, dibutyltin laurate, and alcohol, which acts simultaneously as hydrogen donor and solvent (*254*); carbonyl groups are converted into hydroxyl groups and nitro to amino groups. The reactions proceed only slowly at room temperature, but rapidly at 80 to 180°C.

Treatment of a silane containing Si—H bonds with oxygen in the presence of copper salts and amines at moderately elevated temperatures has been recommended for the production of Si—OH and Si—O—Si bonds (*152*).

Among the other reactions of the Si—H bond the following may be mentioned.

The reaction of organo-*H*-silanes with bromine, iodine, or acid chlorides gives the corresponding organohalosilanes (*66*, *173*, *229*, *362*, *367*):

$$2\mathrm{C_6H_5SiH_3} + 3\mathrm{Br_2} \rightarrow \mathrm{C_6H_5SiH_2Br} + \mathrm{C_6H_5SiHBr_2} + 3\mathrm{HBr}\ (229)$$

$$\mathrm{R_3SiH} + \mathrm{I_2} \rightarrow \mathrm{R_3SiI} + \mathrm{HI}\ (66)$$

$$(\mathrm{C_6H_5CH_2})_3\mathrm{SiH} + \mathrm{C_6H_5COCl} \rightarrow (\mathrm{C_6H_5CH_2})_3\mathrm{SiCl} + \mathrm{C_6H_5CHO}\ (173)$$

$$2(\mathrm{C_2H_5})_3\mathrm{SiH} + 2\mathrm{CH_3COCl} \xrightarrow{\mathrm{AlCl_3}} 2(\mathrm{C_2H_5})_3\mathrm{SiCl} + \mathrm{CH_3COOC_2H_5}\ (313)$$

Explosions have been reported in the bromination of ethylsilane by bromine (*229*). Reports that hydrogen iodide also reacts with the Si—H bond to give Si—I bonds (*367*) can perhaps be explained by the fact that some free iodine is always present in equilibrium with hydrogen iodide.

Hydrogen halides split the Si—C bond in phenylsilanes with the formation of Si—halogen bonds, e.g.:

$$\mathrm{C_6H_5SiH_3} + \mathrm{HX} \rightarrow \mathrm{C_6H_6} + \mathrm{SiH_3X}$$

In the presence of copper salts at 0°C, HF will fluorinate the Si—H bond without splitting the Si—C bond (*107*).

Aryl-*H*-silanes react with alkali metals to give arylsilyl compounds of these metals (*29–31*); for example, triphenylsilane reacts as follows:

$$(\mathrm{C_6H_5})_3\mathrm{SiH} + 2\mathrm{K} \rightarrow (\mathrm{C_6H_5})_3\mathrm{SiK} + \mathrm{KH}$$

while diphenylsilane and phenylsilane ultimately give tetraphenylsilane on reaction with Na—K alloys; the following intermediate steps are assumed:

$$(C_6H_5)_2SiH_2 \xrightarrow{K} C_6H_5SiH_2K + C_6H_5K$$

$$(C_6H_5)_2SiH_2 \xrightarrow{C_6H_5K} (C_6H_5)_3SiH \xrightarrow{C_6H_5K} (C_6H_5)_4Si$$

Triethylsilane does not give this reaction.

The reaction of the Si—H bond with alcohols has long been known (*75*, *98*, *101*, *151*, *253*, *267*):

$$\geq Si{-}H + HOC{\leq} \rightarrow \geq Si{-}O{-}C{\leq} + H_2$$

As early as 1867, Friedel and Ladenburg obtained not only triethoxysilane but also tetraethoxysilane from the reaction of trichlorosilane with ethanol (*98*). The possibility of esterification of siloxane-bonded Si—H groups with methanol in the presence of ammonia was confirmed in 1926 (*101*). For practical purposes, alkali-metal oxides (*253*, *267*), metal halides (*72*, *75*), or hydrogen halides (*151*) are used as catalysts. The reaction with alkali-metal alkoxides is even more vigorous (*74*):

$$\geq Si{-}H + M{-}O{-}C{\leq} \rightarrow \geq Si{-}O{-}C{\leq} + MH$$

Sodium phenoxide, on the other hand, hardly reacts with the Si—H bond in trichlorosilane (*224*).

The Si—H bond undergoes addition reactions with aldehydes (*58*) and ketones (*54*, *55*, *153*) with the formation of Si—O—C bonds:

$$\geq Si{-}H + \underset{O}{\overset{R}{}}{\gt}C{-}R' \rightarrow \geq Si{-}O{-}\underset{H}{\overset{R}{C}}{-}R'$$

Anhydrous zinc chloride acts as a catalyst (*55*). As in the case of the Si—X bond, ethers are decomposed with the formation of Si—O—R bonds (*104*):

$$\geq Si{-}H + ROR \rightarrow \geq Si{-}OR + RH$$

Furthermore, like halosilanes, triorgano-*H*-silanes react with carboxylic acids or their mercury salts:

$$R_3SiH + RCOOH \rightarrow R_3Si{-}O{-}\underset{O}{\overset{}{C}}{-}R + H_2$$

Yields of 80 to 90% of trialkylacetoxysilanes have been achieved with the use of strong mineral acids, aluminum, iodine, or colloidal nickel as catalysts (*73*, *260*).

On reaction with an excess of diazomethane monophenylsilane gave methylphenylsilanes of which monomethylphenylsilane (I) was the major product, together with dimethylphenylsilane (II); trimethylphenylsilane was not formed (*197*):

$$C_6H_5SiH_3 \xrightarrow{CH_2{-}N_2} \underset{\text{I}}{(CH_3)C_6H_5SiH_2} + \underset{\text{II}}{(CH_3)_2C_6H_5SiH}$$

This reaction, too, is analogous to the behavior of the Si—Cl bond (Section 4.5.1.3).

The analogy with the Si—halogen bond has finally been made complete by investigations on the reactivity of the Si—H bond toward ethylene oxide (*100*) (cf. Section 3.2.3). With ethylene oxide at 450°C, SiH_4 gives compounds which include $Si(OC_2H_5)H_3$ and probably also $Si(OC_3H_7)H_3$. The latter silane also appears to be formed in the reaction of SiH_4 with acetone at 450°C.

The behavior of organo-*H*-silanes on heating has been examined in detail with the three ethylsilanes, $C_2H_5SiH_3$, $(C_2H_5)_2SiH_2$, and $(C_2H_5)_3SiH$. At 440 to 460°C, gaseous products (H_2, hydrocarbons, SiH_4, CH_3SiH_3, and similar compounds), liquid organosilicon polymers with Si—Si bonds, and a solid of composition $(SiCH_3)_x$ were found (*102*, *105*).

It was impossible to effect an exchange of hydrogen in the Si—H bond with deuterium from C_2H_5OD, D_2O, and $DN(C_2H_5)_2$ (*47*, *193*).

3.5 Organosilanes with Silicon-Metal Bonds

Organosilylmetal compounds take part as intermediates in silicon chemistry, both in organoalkali-metal and in Grignard syntheses. Organosilyl compounds of the alkali metals can be isolated and their reactions studied.

Triphenylsilyl derivatives can be obtained by the reaction of the Si—halogen, the Si—alkoxy, or the Si—H bond with alkali metals:

$$(C_6H_5)_3SiCl + 2K \rightarrow (C_6H_5)_3SiK + KCl \ (130)$$

$$(C_6H_5)_3SiOCH_3 + 2K \rightarrow (C_6H_5)_3SiK + KOCH_3 \ (31)$$

$$(C_6H_5)_3SiH + 2K \rightarrow (C_6H_5)_3SiK + KH \ (29)$$

Triphenyl(phenylisopropyl)silane reacts with an Na-K alloy with the elimination of a phenylisopropyl group (*33*).

$$(C_6H_5)_3Si{-}\overset{\displaystyle CH_3}{\underset{\displaystyle CH_3}{C}}{-}C_6H_5 + 2K \rightarrow (C_6H_5)_3SiK + C_6H_5{-}\overset{\displaystyle CH_3}{\underset{\displaystyle CH_3}{C}}{-}K$$

A useful route is through the cleavage of hexaphenyldisilane with a sodium-potassium alloy in ether. It yields particularly pure triphenylsilyl-potassium (*130*, *131*).

$$(C_6H_5)_3Si{-}Si(C_6H_5)_3 + 2K \rightarrow 2(C_6H_5)_3SiK$$

Compounds of lithium (*48*, *124*, *225*), rubidium, and cesium (*131*) can also be prepared in this way. Diglyme and tetrahydrofuran have been found particularly useful as solvents (*48*, *124*, *225*).

Unlike hexaphenyldisilane, hexaethyldisilane cannot be split by potassium to form potassium compounds (*123*).

On the other hand, disilanes with both aromatic and aliphatic substituents can be converted into the corresponding silylpotassium compounds (*123*):

$$(C_6H_5)_3Si{-}Si(CH_3)_3 \xrightarrow{Na/K} (C_6H_5)_3SiK + (CH_3)_3SiK$$

Triphenylsilyl compounds of the alkali metals can undergo a number of interesting reactions:

$$(C_6H_5)_3SiK + C_6H_5Br \rightarrow (C_6H_5)_4Si + KBr \ (33, 130)$$

$$(C_6H_5)_3SiK + HCl \rightarrow (C_6H_5)_3SiH + KCl \ (33)$$

$$(C_6H_5)_3SiK + (C_6H_5)_3SiH \rightarrow (C_6H_5)_4Si + \text{other products} \ (49)$$

$$(C_6H_5)_3SiK + (C_6H_5)_3CH \rightarrow (C_6H_5)_3SiH + (C_6H_5)_3CK \ (50)$$

$$(C_6H_5)_3SiK + HCHO + H_2O \rightarrow (C_6H_5)_3SiCH_2OH + KOH \ (134)$$

$$(C_6H_5)_3SiK + (C_6H_5)_3SiOH \rightarrow (C_6H_5)_4Si + ? \ (133)$$

$$(C_6H_5)_3SiK + (CH_3)_3SiCl \rightarrow (C_6H_5)_3Si{-}Si(CH_3)_3 + KCl \ (33, 48, 124)$$

$$2(C_6H_5)_3SiK + (C_6H_5)_2SiCl_2 \rightarrow (C_6H_5)_3Si{-}\underset{\underset{C_6H_5}{|}}{\overset{\overset{C_6H_5}{|}}{Si}}{-}Si(C_6H_5)_3 + 2KCl \ (135)$$

$$(C_6H_5)_3SiLi + R_{ar}OCH_3 \xrightarrow{H^{\oplus}} (C_6H_5)_3SiCH_3 + R_{ar}OH + Li^{\oplus} \ (129)$$

$$(C_6H_5)_3SiLi + C_6H_5C\begin{matrix}{\nearrow}O\\ {\searrow}H\end{matrix} \rightarrow (C_6H_5)_3Si{-}O{-}\underset{\underset{Li}{|}}{CH}{-}C_6H_5 \xrightarrow{H_2O} (C_6H_5)SiOCH_2C_6H_5 \ (385)$$

Using *trans*-stilbene as an example, it has been shown that the Si—K bond can attach itself to double bonds (*132*):

$$(C_6H_5)_3SiK + C_6H_5CH{=}CHC_6H_5 \rightarrow (C_6H_5)_3SiCH(C_6H_5)CH(C_6H_5)K$$

3.6 Organosilanols

Monomeric *organosilanols* play an important part as intermediate products in silicone chemistry. They are formed in every hydrolysis of silanes with silicon-functional groups and are then transformed more or less rapidly into *siloxanes* by spontaneous or constrained condensation:

$$\geq Si{-}OH + HO{-}Si\leq \rightarrow \geq Si{-}O{-}Si\leq + H_2O$$

The tendency of the silanols to undergo spontaneous condensation depends on their molecular structure as well as on the conditions of the hydrolysis in which they are formed. These factors will be set forth in detail during the discussion of the condensation of silanes with silicon-functional groups (Section 5.4.1); it need only be pointed out here that the tendency of the silanols to condense decreases with decreasing number of OH groups bound to the Si atom and with increasing steric shielding of these OH groups by large and bulky organic groups. The tendency to condensation also depends on the electronegativity of the R groups, which may be substituted, the pH of the medium, and the nature of the catalyst (cf. Section 5.4.1).

Monomeric silanols can therefore be obtained only if their structure satisfies certain requirements and very special conditions are observed during their production.

3.6.1 Preparation

The simplest process, the hydrolysis of organohalosilanes—especially organochlorosilanes—in water, gives the desired result with silanols relatively insensitive to condensation. The hydrochloric acid formed during hydrolysis generally promotes condensation so vigorously that the monomeric silanols can either not be obtained at all or only in very moderate yield. A number of routes are available for obtaining useful yields of relatively sensitive silanols:

1. Hydrolysis of silanes giving weak acids on hydrolysis. Acyloxysilanes, especially acetoxysilanes, are generally better suited for silanol synthesis than halosilanes (*99*, *334*). Triethylsilanol has been prepared from triethylacetoxysilane in 91% yield (*334*).

2. Hydrolysis of silanes giving no acids or bases on hydrolysis. Dimethylsilanediol, which is sensitive to condensation, can be obtained from dimethyldimethoxy- and dimethyldiethoxysilanes (*163*, *189*, *190*), and phenylsilanetriol from phenyltrimethoxysilane (*364*). It is also appropriate to include here the route followed by Sauer (*297*) for the preparation of trimethylsilanol through the cleavage of the dimethylsiloxane bond with methylmagnesium iodide and subsequent hydrolysis of the Grignard complex:

$$[(CH_3)_2SiO]_x + xCH_3MgI \rightarrow x(CH_3)_3SiOMgI$$

$$(CH_3)_3SiOMgI + H_2O \rightarrow (CH_3)_3SiOH + MgIOH$$

A good synthesis for trimethylsilanol starts with *N*-trimethylsilylacetamide, which can be prepared from trimethylchlorosilane and acetamide and which can be hydrolyzed without change of pH in the reaction mixture (*35*):

$$(CH_3)_3Si—NH—COCH_3 \xrightarrow{H_2O} (CH_3)_3SiOH + CH_3CONH_2$$

3. Hydrolysis under conditions ensuring immediate neutralization of acids or bases formed. This principle is very widely used. The hydrolysis of trimethyl fluorosilane by slow addition of a 1*N* solution of caustic soda to an ethereal solution of the silane is an excellent method for preparing trimethylsilanol (*328*). Triethylsilanol can be made from triethylchlorosilane or triethylsilyl sulfate (*328*) and diethylsilanediol from diethyldichlorosilane (*70*) in the same way. Using aniline as an HCl acceptor, dimethylsilanediol was obtained in 71% and diethylsilanediol in 76% yield (*351*). Trimethylsilanol can be obtained in 90% yield by the acid hydrolysis of hexamethyldisilazane (*297*, *334*). Even the yields from synthesis of relatively stable silanols like diphenylsilanediol are increased (up to 95%) if the acid is neutralized by, e.g., the addition of sodium carbonate, calcium carbonate, calcium hydroxide, pyridine, or aniline (*351*, *353*, *354*).

Organic solvents sparingly miscible or immiscible with water are often used with these hydrolytic processes. The success of this measure can be explained by various factors: Dilution of the silanols soluble in the organic solvent would inhibit self-condensation; by taking up the silanol in the organic phase it is removed from the aqueous environment and the danger of condensation, particularly with an acid or alkaline pH, is decreased; finally, solid silanols remain in solution and can easily be recovered by phase separation. Suitable solvents include, among others, diethyl ether, dibutyl ether, toluene, and xylene.

A further variant of the hydrolysis consists in using an alcohol in addition to the organic solvents. Aliphatic alcohols with 3 to 5 carbon atoms in combination with toluene or butyl acetate are the preferred solvents. Diphenylsilanediol has been obtained from diphenyldichlorosilane in 93% yield, by hydrolysis in the presence of toluene and *t*-amyl alcohol (*51*).

Sodium hydrogen carbonate or sodium hydrogen sulfite can be used to supply water for the hydrolysis of halosilanes (*77*, *353*). Anhydrous solvents are used, and atmospheric moisture is excluded (*353*):

$$(C_6H_5)_2SiCl_2 + 2NaHCO_3 \xrightarrow{\text{acetone}} (C_6H_5)_2Si(OH)_2 + 2NaCl + 2CO_2$$

3.6.2 Physical Constants (Table)

The physical constants for a number of silanols have been collected in Table 16.

TABLE 16

ORGANOSILANOLS[a]

Compound	Mp (°C)	Bp (°C) at	p (mm Hg)	n_D^t at	t (°C)	d_4^t (gm/cc) at	t (°C)
$(CH_3)_3SiOH$	−4.5 (*28*)	98.9 (*144*)	760	1.3889 (*144*)	20	0.8141 (*328*)	20
$(CH_3)_2Si(OH)_2$	96–98 (*163*)	—	—	$n_1 = 1.454 \pm 0.002$ (*163*) $n_2 = 1.446 \pm 0.002$ (*163*)	25	1.097 ± 0.002 (*163*)	20
$CH_3(C_6H_5)Si(OH)_2$	74–75 (*331*)	—	—	—	—	—	—
$(C_2H_5)_3SiOH$	—	63 (*144*)	12	1.4329 (*144*)	20	0.8638 (*328*)	20
$(C_2H_5)_2Si(OH)_2$	96 (*113*)	140 (decomp.) (*113*)	760	α: 1.413 β: 1.493 (*113*) γ: 1.517	~20	1.134 (*184*)	20
$(n\text{-}C_3H_7)_3SiOH$	—	206–208 (*263*)	760	—	—	—	—
$(i\text{-}C_3H_7)_3SiOH$	—	196 (*81*)	750	1.4532 (*81*)	21	—	—
$(n\text{-}C_3H_7)_2Si(OH)_2$	99–100 (*113*)	—	—	—	—	—	—
$(i\text{-}C_3H_7)_2Si(OH)_2$	114 (*81*)	—	—	—	—	—	—
$(n\text{-}C_4H_9)_3SiOH$	—	99–99.5 (*126*)	1	1.4470 (*126*)	20	0.8505 (*126*)	20
$(n\text{-}C_4H_9)_2Si(OH)_2$	95–97 (*113*)	—	—	—	—	—	—
$(i\text{-}C_4H_9)_2Si(OH)_2$	100–101 (*81*)	—	—	α: 1.476 β: 1.487 (*84*) γ: 1.489	~20	—	—
$(t\text{-}C_4H_9)_2Si(OH)_2$	151–153 (*324*)	—	—	—	—	—	—
$(C_6H_5)_3SiOH$	155 (*71*)	—	—	—	—	1.1777 (*180*)	—
$(C_6H_5)_2Si(OH)_2$	137–141 (*247*) (decomp.)	—	—	α: 1.564 ± 0.004 β: 1.648 ± 0.004 (*247*) γ: 1.656 ± 0.004	—	—	—

[a] Numbers in parentheses indicate reference numbers.

3.6.3 Chemical Properties

Organosilanols can undergo a large variety of reactions. Those leading to the formation of Si—O—Si bonds are treated in detail in Section 5.2. The following reactions are involved:

$$\equiv Si{-}OH + HO{-}Si\equiv \rightarrow \equiv Si{-}O{-}Si\equiv + H_2O$$

$$\equiv Si{-}OH + Cl{-}Si\equiv \rightarrow \equiv Si{-}O{-}Si\equiv + HCl$$

$$\equiv Si{-}OH + RO{-}Si\equiv \rightarrow \equiv Si{-}O{-}Si\equiv + ROH$$

$$\equiv Si{-}OH + H_2N{-}Si\equiv \rightarrow \equiv Si{-}O{-}Si\equiv + NH_3$$

$$\equiv Si{-}OH + H{-}Si\equiv \rightarrow \equiv Si{-}O{-}Si\equiv + H_2$$

$$\equiv Si{-}OH + AcO{-}Si\equiv \rightarrow \equiv Si{-}O{-}Si\equiv + HOAc$$

The reactions of the Si—OH bond have been studied mainly with the triorganosilanols, since these are sufficiently stable to permit the preparative working up and isolation of the reaction products.

Although triethylchlorosilane readily reacts with water, hydrogen chloride can convert triethylsilanol into triethylchlorosilane, i.e., the equilibrium

$$(C_2H_5)_3SiOH + HCl \rightleftharpoons (C_2H_5)_3SiCl + H_2O$$

can be strongly displaced to the right. Triethylchlorosilane is obtained in 77% yield by treating the silanol with 38% hydrochloric acid, cooling in ice, drying, and distilling the separated silane phase (*328*). Triethylsilanol gives the acetoxysilane with acetic anhydride and the silyl sulfate with fuming sulfuric acid (*328*):

$$(C_2H_5)_3SiOH + (CH_3CO)_2O \rightarrow (C_2H_5)_3SiOOCCH_3 + CH_3COOH$$

$$(C_2H_5)_3SiOH + H_2SO_4 \rightarrow (C_2H_5)_3SiOSO_3H + H_2O$$

Diethylsilanediol has been converted into the chlorosilane in 50% yield with thionyl chloride and quinoline (*70*, *113*).

The course of the reaction of silanols with acetyl chloride depends on the nature of the organic substituents on the Si atom, the tendency of the silanol groups to condense with one another, and the working conditions. Silanols with aromatic or aralkyl substituents give the chlorosilane or the acetoxysilane (*226*, *245*), while dimethylsiloxanols readily condense under the action of acetyl chloride to give polysiloxanes (*308*).

Organosilanols can be converted into organofluorosilanes with dilute aqueous hydrofluoric acid (*82*) or thionyl fluoride (*241*).

Unlike the analogous carbinols, trimethyl- and triethylsilanols react with alkali metals, liberating hydrogen and forming the alkali-metal silanolates (*328*, *361*):

$$(CH_3)_3SiOH + M \rightarrow (CH_3)_3SiOM + \tfrac{1}{2}H_2$$

Trimethylsilanol gives the silanolate with 12*N* NaOH (*328*):

$$(CH_3)_3SiOH + NaOH \rightarrow (CH_3)_3SiONa + H_2O$$

Organosilanols can be condensed with alcohols if they do not exhibit an excessive tendency toward self-condensation (*342*):

$$R_3SiOH + ROH \rightarrow R_3SiOR + H_2O$$

Triethylsilanol reacts with lower aliphatic alcohols in this way to give triethylalkoxysilanes with yields between 10 and 60% (*12*). With methanol in the presence of $ZnCl_2$, $(p\text{-}C_6H_5\text{—}C_6H_4)_3SiOH$ gives $(p\text{-}C_6H_5\text{—}C_6H_4)_3SiOCH_3$ (*127*), and with *t*-butanol in the presence of iodine, pyridine, and SO_2 (Karl Fischer reagent) $(C_6H_5)_3SiOH$ gives $(C_6H_5)_3SiOC(CH_3)_3$ (*117*).

An acetal is obtained by the action of vinyl ethyl ether on trimethylsilanol (*334*):

$$(CH_3)_3SiOH + CH_2{=}CH\text{—}OC_2H_5 \rightarrow CH_3\text{—}\underset{OSi(CH_3)_3}{\overset{OC_2H_5}{\underset{|}{\overset{|}{C}}H}}$$

and trialkylvinyloxysilane by reaction with an excess of acetylene (*335*):

$$R_3SiOH + CH{\equiv}CH \rightarrow R_3SiOCH{=}CH_2$$

The reactions of the Si—OH group with isocyanates are particularly interesting. Studies of the reactions of numerous aromatically and aliphatically substituted *diols* with mono- and diisocyanates have shown that siloxanes and not silylurethanes are formed. In other words, the tendency toward self-condensation outweighs that for the formation of urethanes. The isocyanate removes water of condensation from the system and can thus accelerate the reaction:

$$2\ {\gt}SiOH \rightleftharpoons {\gt}Si\text{—}O\text{—}Si{\lt} + H_2O$$

$$H_2O + 2C_6H_5NCO \rightarrow C_6H_5NHCONHC_6H_5 + CO_2 \quad (64, 332)$$

It is possible that a silylurethane is formed as an intermediate:

$${\gt}SiOH + C_6H_5NCO \rightarrow {\gt}SiOCONHC_6H_5$$

$$2\ {\gt}SiOCONHC_6H_5 \rightarrow {\gt}SiOSi{\lt} + C_6H_5NHCONHC_6H_5 + CO_2 \quad (64)$$

but this has not been isolated.

On the other hand, certain triorgano*monosilanols* appear to be able to react with isocyanates to form the corresponding silylurethanes:

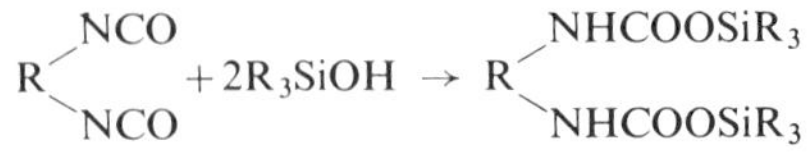

This has been found with ethyl, propyl, and butyl substituents (*18*, *19*), but triphenylsilanol again gave siloxane and urethane (*64*). The silylurethanes are decomposed by water (*19*) and alcohols (*12*).

Unlike alcohols, the silanols react with the Karl Fischer reagent (*127*) (apart from compounds with pronounced steric hindrance) so that their OH content can be quantitatively determined in this way (cf. Chapter 12).

Many of these reactions, particularly those with isocyanates, illustrate the distinctive dual nature of the Si—OH bond in silanols, which in its properties is analogous partly to silicic acid and partly to the alcohols.

The similarity to the alcohols decreases, and the acidity of the OH protons increases, on passing from the trialkylsilanols to silicic acid itself. However, trialkylsilanols are still more acidic than the corresponding tertiary alcohols (*302*). Electronegative substituents such as phenyl groups increase the acidity over that of the corresponding alkylsilanols. Triphenylsilanol is a considerably stronger acid than trialkylsilanols, or even trialkylmethanols, but on the other hand it is less acidic than phenol (*379*).

Corresponding to the greater or lesser polar character of the Si—OH bond, silanols are more or less strongly associated in the liquid or solid phase and in solution (*144*). The association is due to the formation of hydrogen bonds, possibly in the following way:

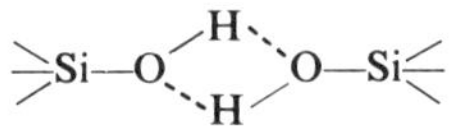

This can be observed in the Raman and infrared spectra (*64*, *186*, *284*, *376*).

The absorption band at 2.7 μ, which is characteristic of the free stretching vibrations of the OH groups in the infrared region, can be detected unambiguously only at high dilutions (*64*).

In the crystals of diethylsilanediol (*184*, *185*), diallylsilanediol (*191*), and diisobutylsilanediol (*84*), the tetrahedral molecules are arranged as chains in the direction of the *b*-axes, the linkage of the tetrahedra being through hydrogen bonds:

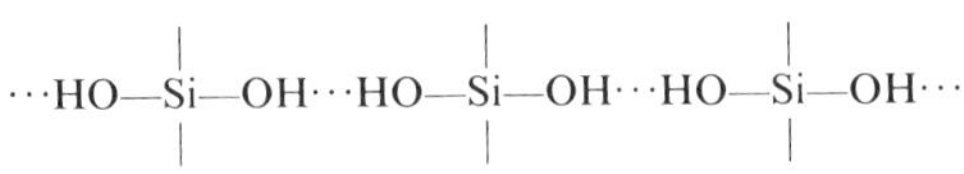

In diethylsilanediol and diallylsilanediol these chains are arranged in layers parallel to (001), and therefore lead to an excellent cleavage in this plane.

Needlelike habit with the *b*-axis as the needle axis and good cleavage parallel to the needle are observed in diisobutylsilanediol.

The boiling points of the silanols are relatively high, because of their polarity and ability to form hydrogen bonds. Trimethylsilanol, with a molecular weight of 90, boils at the same temperature (100°C) as hexamethyldisiloxane with a molecular weight of 162. On the other hand, their polarity and capacity for forming hydrogen bonds with water molecules is responsible for the more or less pronounced water solubility of the silanols. Diethylsilanediol is soluble in water.

The solubility in water decreases and the solubility in organic solvents increases with increasing length of the alkyl groups. Di-*n*-butylsilanediol is already almost insoluble in cold water. Triorganosilanols are insoluble in water; diphenylsilanediol is insoluble and phenylsilanetriol only very slightly soluble in water.

By reaction with NaOD in analogy with the preparation of $(CH_3)_3SiOH$ (*215*, *216*), $(CH_3)_3SiOD$ can be prepared from $(CH_3)_3SiF$.

3.7 Organosilanolates

The alkali-metal salts are the most interesting of all the metal derivatives of the organosilanols. Among these are water-soluble compounds stable in aqueous solution, from which silanols or siloxanes can be liberated by reaction with acids. They can be used as impregnating agents.

The sodium salts of the triorganosilanols can be obtained from the corresponding hexaorganodisiloxanes and sodium hydroxide in alcoholic solution (*162*, *167*, *360*):

$$R_3Si{-}O{-}SiR_3 + 2NaOH \rightarrow 2R_3Si{-}ONa + H_2O$$

However, the formation of methane, which indicates cleavage of the Si—C bond, cannot be avoided in the reaction of hexamethyldisiloxane with sodium hydroxide in benzene or ether solution (*360*). The Si—C bond has proved less sensitive in the following reactions, in which yields of up to about 80% were achieved (*167*).

$$[(C_6H_5)_3Si]_2O \xrightarrow{NaOH} (C_6H_5)_3SiONa$$

$$[CH_3(C_6H_5)_3Si]_2O \xrightarrow{NaOH} CH_3(C_6H_5)_2SiONa$$

$$[(CH_3)_2C_6H_5Si]_2O \xrightarrow{NaOH} (CH_3)_2C_6H_5SiONa$$

Trimethylsilanolates are obtained in good yield (about 80%) by the reaction of hexamethyldisiloxane with sodamide or potassamide in liquid ammonia or absolute ether (*166*, *167*):

$$2(CH_3)_3Si{-}O{-}Si(CH_3)_3 + 2NaNH_2 \rightarrow 2(CH_3)_3SiONa + [(CH_3)_3Si]_2NH + NH_3$$

High yields are also obtained by the cleavage of hexamethyldisiloxane with methyllithium or phenyllithium (*292*, *315*):

$$(CH_3)_3Si—O—Si(CH_3)_3 + LiCH_3 \rightarrow (CH_3)_3SiOLi + (CH_3)_4Si$$

$$(CH_3)_3Si—O—Si(CH_3)_3 + LiC_6H_5 \rightarrow (CH_3)_3SiOLi + (CH_3)_3C_6H_5Si$$

The corresponding silanolates can also be obtained by the reaction of triorganoalkoxysilanes with alkali-metal hydroxides (Section 3.3.3) and of triorganosilanols with alkali metals or alkali-metal hydroxides (Section 3.6.3).

Alkali-metal salts of tri- or diorganosilanols are rapidly hydrolyzed in water. Alkali-metal salts of organosilanetriols, on the other hand, are stable in aqueous solution at certain concentrations of alkali. If methyltrichlorosilane is hydrolyzed and the precipitate is washed free of acid and dissolved in caustic soda, it is possible to obtain water-soluble products with a ratio of one mole of NaOH to one gram-atom of Si. Water solubility decreases rapidly below this ratio. The structure

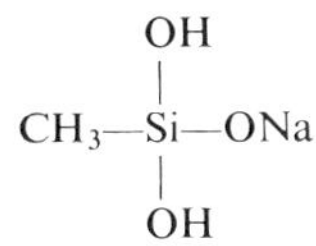

is usually ascribed to the compound in solution (*192*), but this is probably a mixture of partially condensed molecules rather than the monomer (hence the name *sodium methylsiliconate*, cf. Sections 1.2 and 10.12.2.1). A group separation of siloxanes of different functionalities can be carried out on the basis of the different behavior of the alkali-metal silanolates with water (Section 5.5.4) (*161*).

The water-soluble metal compounds of the monoorganosilanols have found an important application as *water-repellency agents*, particularly for the protection of masonry (*110*). Silicone films are formed from solutions by reaction with atmospheric carbon dioxide (Section 10.12).

The alkali-metal derivatives of organosilanols are suitable as catalysts for the production of macromolecular linear siloxanes from low-molecular cyclic siloxanes (*166*, *182*).

On the other hand, interest in triorganosilanolates as intermediates for the synthesis of siloxanes has remained limited. The anhydrous salts, which are not easy to prepare, can react with chlorosilanes (*167*):

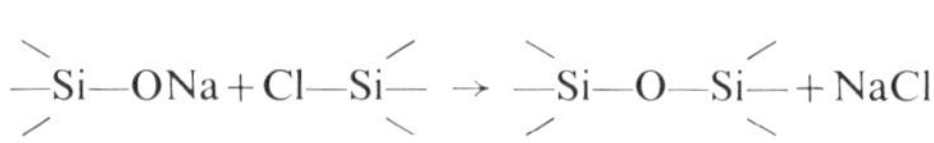

They can also be made to react with heavy-metal halides in solvents like dioxane or carbon tetrachloride to form heavy-metal derivatives of the triorganosilanols (*361*):

$$4(CH_3)_3SiOK + SnCl_4 \rightarrow [(CH_3)_3SiO]_4Sn + 4KCl$$
$$2(CH_3)_3SiONa + HgCl_2 \rightarrow [(CH_3)_3SiO]_2Hg + 2NaCl$$
$$2(CH_3)_3SiOK + (CH_3)_2SnCl_2 \rightarrow (CH_3)_2[(CH_3)_3SiO]_2Sn + 2KCl$$

In an analogous manner to the production of silica sols from solutions of alkali-metal silicates, it is possible to prepare dispersions of organosilanols or organosiloxanes by treating aqueous solutions of alkali-metal silanolates with ion-exchangers (*293*). Special precautions must be taken if such dispersions are to be obtained by simple acidification of the alkaline solutions. The critical region between pH 9 and 11 must be traversed quickly and a pH between 4 and 7 established (*234*, *294*).

3.8 Organoaminosilanes (Organosilylamines)

3.8.1 Preparation

Compounds of this group belong to series of the general composition:

$$R_nSi(NH_2)_{4-n}, \quad R_nSi(NHR)_{4-n}, \quad R_nSi(NR_2)_{4-n}$$

They are generally accessible by the reaction of halosilanes with ammonia or amines in the presence of an inert solvent—usually diethyl ether, benzene, or carbon tetrachloride—or, in the case of those with an unsubstituted amino group, by the addition of the halosilanes to liquid ammonia. Ammonolysis with Divers' liquid—a solution of ammonium salts in anhydrous ammonia or amines—is particularly useful. It can be carried out at room temperature and permits simple isolation of the reaction products by phase separation (*290*).

The extent and nature of the reaction process depends, in each case, on the organic substituents on the Si atom and the character of the amine. Steric hindrance can make reactions of this type more difficult; on the other hand, the condensibility of the silylamines, which is analogous to that of the silanols:

$$\equiv Si{-}NH_2 + H_2N{-}Si\equiv \rightarrow \equiv Si{-}NH{-}Si\equiv + NH_3$$

makes the isolation of certain silylamines impossible. Thus, the reaction of *trimethylchlorosilane* with *ammonia* under various conditions always yields hexamethyldisilazane but no trimethylaminosilane (*297*, *299*).

$$2(CH_3)_3SiCl + 3NH_3 \rightarrow (CH_3)_3Si{-}NH{-}Si(CH_3)_3 + 2NH_4Cl$$

Similarly, dimethylethylchlorosilane and diethylmethylchlorosilane still yield the corresponding disilazanes (*330*) and triethylchlorosilane (*21*), tri-*n*-propylchlorosilane (*209*), and tri-*n*-butylchlorosilane (*206*) are the first compounds that form monomeric aminosilanes with ammonia (cf. also refs. *57*, *146*, *147*, *210*, *340*).

Di- and *monoorganohalosilanes* generally react with ammonia to form polysilazanes (Section 7.2.2). Di- and monoorganohalosilanes with particularly voluminous organic groups are an exception, since they, like the triorganohalosilanes, give monomeric aminosilanes. This can be shown with di-α-naphthyldiaminosilane, which can be prepared from di-α-naphthyldifluorosilane and ammonia (*363*):

$$(C_{10}H_7)_2SiF_2 \xrightarrow{NH_3} (C_{10}H_7)_2Si(NH_2)_2$$

The considerable influence on the reaction of organic substitution on the nitrogen is shown if *amines* are used in place of the ammonia. The formation of a monomeric aminosilane from trimethylchlorosilane is made possible by replacement of even *one* H atom on the nitrogen by CH_3 (*299*).

$$(CH_3)_3SiCl + 2CH_3NH_2 \rightarrow (CH_3)_3SiNHCH_3 + [CH_3NH_3]Cl$$

While *n*-propyltrichlorosilane reacts rapidly and vigorously in ether or benzene with an amine having a small R group, the addition of an ethereal or benzene solution of the silane to an excess of *t*-butylamine or diethylamine in the same solvent results in the replacement of only two chlorine atoms even on boiling or heating to 160°C under pressure. In the case of diisopropylamine or dicyclohexylamine the reaction is limited to *one* chlorine atom in the silane molecule. The residual chlorine atoms can react only with ammonia or isopropylamine (*358*). Ethyltri-(*t*-butylamino)-silane and propyltri-(diethylamino)-silane can, however, be prepared by heating the corresponding alkyltribromosilanes and -triiodosilanes in benzene solution with the appropriate amine under reflux (*359*).

Useful yields of silazanes with relatively pronounced steric hindrance, like triphenyl-(di-*n*-butylamino)-silane, can be obtained when alkylsilyllithium compounds are used as starting materials instead of halosilanes (*125*):

$$R_3SiLi + R'_2NH \rightarrow R_3SiNR'_2 + LiH$$

or when triphenylsilane is treated with lithium dialkylamides (*122*):

$$(C_6H_5)_3SiH + LiNR_2 \rightarrow (C_6H_5)_3SiNR_2 + LiH$$

or when the Si—halogen or the Si—H bond is caused to react with alkali-metal amides in ether (X = Cl or H) (*122*):

$$(C_6H_5)_3SiX + LiN(n\text{-}C_4H_9)_2 \rightarrow (C_6H_5)_3SiN(n\text{-}C_4H_9)_2 + LiX$$

The observation that aniline reacts completely, but not as rapidly as cyclohexylamine, with all three chlorine atoms in *n*-propyltrichlorosilane cannot be explained by steric effects, and neither can the fact that diphenylamine does not react at all while dicyclohexylamine reacts rapidly with one chlorine atom. Differences in the reactivities of diethylamine and piperidine toward *n*-propyltrichlorosilane are paralleled by their behavior toward $SiCl_4$. The former replaces only two chlorine atoms in *n*-propyltrichlorosilane and three in $SiCl_4$ while the latter always reacts completely (*358*).

Aminosilanes derived from high-boiling amines can be obtained from those derived from amines with lower boiling points by transamination (*204*):

$$\geq SiNHR' + R''NH_2 \rightarrow \geq SiNHR'' + R'NH_2$$

Triethylaminosilane exchanges its amino group with the hydrazine radical on being heated with *N*,*N*-dimethylhydrazine in the presence of ammonium sulfate:

$$(C_2H_5)_3Si{-}NH_2 + H_2N{-}N(CH_3)_2 \rightarrow (C_2H_5)_2SiNH{-}N(CH_3)_2 + NH_3$$

With an excess of hydrazine the H atom in amino-*H*-silanes is also substituted (*314*):

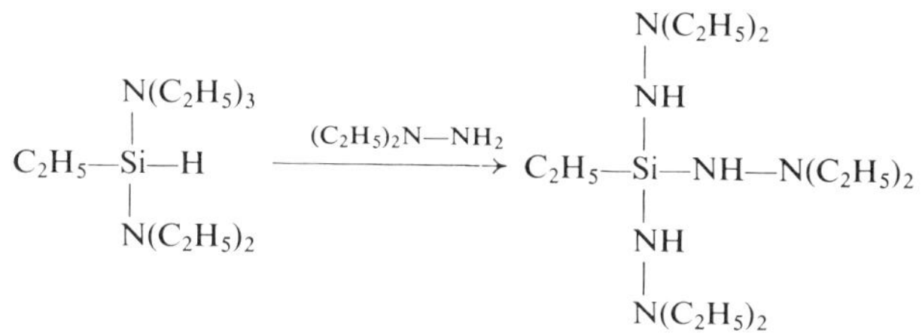

Polymers have been obtained from the transamination of diaminosilanes with diamines (*15*).

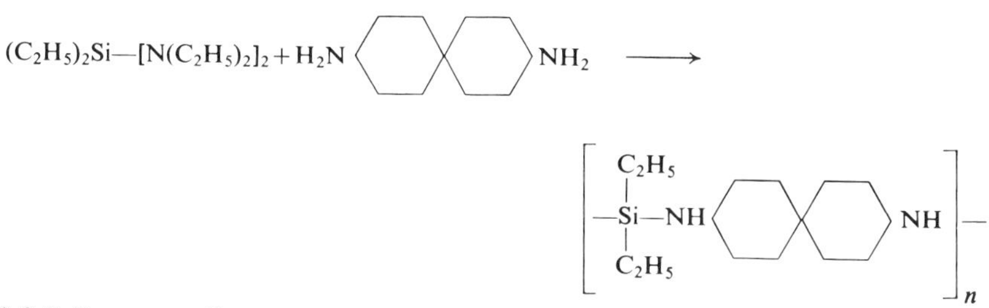

3.8.2 Physical Constants (Table)

Physical constants of a number of organoaminosilanes are given in Table 17 (p. 106).

3.8.3 Chemical Properties

The Si—N bond can be cleaved by water, but its sensitivity to hydrolysis is far less pronounced than that of the Si—Cl or the Si—OCH_3 bond. Hexamethyldisilazane is decomposed only slowly even by boiling water. However,

TABLE 17

ORGANOAMINOSILANES[a]

Compound	Mp (°C)	Bp (°C) at	p (mm Hg)	n_D^t at	t (°C)	d_4^t (gm/cc) at	t (°C)
$(CH_3)_3SiNHCH_3$	—	71 (*299*)	755	1.3905 (*299*)	20	0.7395 (*299*)	20
$(CH_3)_2Si(NHCH_3)_2$	—	105 (*210*)	776	—	—	—	—
$(CH_3)_3SiN(CH_3)_2$	—	85–86 (*235*)	760	—	—	—	—
$(CH_3)_2Si[N(CH_3)_2]_2$	−98 (*6*)	128.4 (*6*)	760	1.4169 (*6*)	22	0.809 (*6*)	22
$CH_3Si[N(CH_3)_2]_3$	−11 (*6*)	161 (*6*)	760	1.4324 (*6*)	22	0.850 (*6*)	22
$Si[N(CH_3)_2]_4$	−2 (*6*)	180 (*6*)	760	—	—	0.973 (*6*)	22
$(CH_3)_3SiNHC_2H_5$	—	90.1–90.8 (*299*)	760	1.3912 (*299*)	20	—	—
$(CH_3)_2Si(NHC_2H_5)_2$	—	139 (*210*)	775	1.4151 (*210*)	20	0.8067 (*210*)	20
$(CH_3)_3SiN(C_2H_5)_2$	—	126.1–126.4 (*299*)	750	1.4112 (*299*)	20	—	—
$(C_6H_5)_3SiNH_2$	55–56 (*198*)	—	—	—	—	—	—
$(C_6H_5)_3SiN(CH_3)_2$	80–81 (*122*)	—	—	—	—	—	—
$(C_6H_5)_2Si[NH(CH_3)]_2$	—	155–156 (*207*)	3	—	—	—	—
$(C_6H_5)_3SiNHC_2H_5$	47–48 (*32*)	163–165 (*32*)	0.5	—	—	—	—
$(C_6H_5)_3SiN(C_2H_5)_2$	84–85 (*122*)	—	—	—	—	—	—

[a] Numbers in parentheses indicate reference numbers.

the hydrolysis can be accelerated by the addition of small amounts of acid. As with the Si—X or Si—OR bonds, the behavior on hydrolysis in individual cases is strongly dependent on steric factors, particularly the space requirements of the organic substituents not only on the Si but also on the N atom; the rate of hydrolysis decreases with increasing steric hindrance.

The proximity of electron-attracting groups, such as the carbonyl group, increases the sensitivity of the Si—N bond toward hydrolytic attack. *N*-trimethylsilylacetamide, which can be obtained from trimethylchlorosilane and acetamide in the presence of triethylamine:

$$(CH_3)_3SiCl + H_2NCOCH_3 \xrightarrow{(C_2H_5)_3N} (CH_3)_3SiNHCOCH_3 + (C_2H_5)_3N \cdot HCl$$

hydrolyzes smoothly to trimethylsilanol (*35*) (Section 3.6.1). Polyfunctional compounds in this group, i.e. those having a number of Si—N bonds on the Si atom, are suitable for cross-linking in room-temperature-vulcanizing silicone rubber single-component systems (Section 8.1.2.3.2).

The reaction of the Si—N bond with alcohols gives alkoxysilanes. Tri-(ethylamino)-silane readily reacts with alcohols from ethanol to *n*-octanol (*13*). The ability of $\diagdown\!\!\!\diagup$—Si—NH_2 groups to condense with each other to form polysilazanes has already been discussed in Section 3.8.1. Aminosilanes and silanols can condense to form polysiloxanes (Section 5.2.13).

Aminosilanes react with phenyl isocyanate as follows:

$$\begin{matrix} C_6H_5 \diagdown \\ R_3Si \diagup \end{matrix} NH + C_6H_5NCO \rightarrow \begin{matrix} C_6H_5 \diagdown \\ R_3Si \diagup \end{matrix} N\text{—}CO\text{—}NH\text{—}C_6H_5 \ (201)$$

$$(CH_3)_3SiN(C_2H_5)_2 + C_6H_5NCO \rightarrow (CH_3)_3Si\text{—}N(C_6H_5)\text{—}CO\text{—}N(C_2H_5)_2 \ (257)$$

while silazanes are decomposed at the Si—N bond (*93*) (cf. Section 7.2.2).

The following reactions are also interesting:

$$(CH_3)_3SiN(C_2H_5)_2 + CO_2 \rightarrow (CH_3)_3Si\text{—}O\text{—}CO\text{—}N(C_2H_5)_2 \ (46)$$

$$(CH_3)_2Si[N(C_2H_5)_2]_2 + 2CO_2 \rightarrow (C_2H_5)_2N\text{—}CO\text{—}O\text{—}Si(CH_3)_2\text{—}O\text{—}CO\text{—}N(C_2H_5)_2 \ (256)$$

$$(CH_3)_3SiN(C_2H_5)_2 + CS_2 \rightarrow (CH_3)_3Si\text{—}S\text{—}CS\text{—}N(C_2H_5)_2 \ (45)$$

$$(CH_3)_3SiN(C_2H_5)_2 + SO_3 \rightarrow (CH_3)_3Si\text{—}O\text{—}SO_2\text{—}N(C_2H_5)_2 \ (303)$$

Hydrogen halides can reconvert the organoaminosilanes into halosilanes (*21*). Boron trifluoride can add to aminosilanes, giving addition compounds stable for short periods at low temperatures, but decomposing into fluorosilanes at higher temperatures (*85*, *143*):

$$(CH_3)_2Si[N(CH_3)_2]_2 \xrightarrow{BF_3} (CH_3)_2SiF_2 + 2(CH_3)_2NBF_2$$

3.9 Organosilanethiols

These compounds form a group of little practical importance. They are mentioned only in connection with the (polymeric) organosilthianes (Section 7.2.3). Literature references are scant. Organosilanethiols can be prepared by the reaction of organoaminosilanes with hydrogen sulfide (*208*):

$$\geq Si—NH_2 \xrightarrow{H_2S} \geq Si—SH$$

or from halosilanes and hydrogen sulfide in the presence of tertiary bases which combine with the hydrogen chloride formed during the reaction (*90*):

$$\geq Si—X + H_2S + R_3N \rightarrow \geq Si—SH + R_3NHX$$

The route through the reaction of a halosilane and lithium hydrogen sulfide in toluene solution has also been used (*56*):

$$(CH_3)_3SiCl + LiHS \rightarrow (CH_3)_3Si—SH + LiCl$$

Silanethiols undergo self-condensation as readily as silanols and aminosilanes, so that polymers with Si—S—Si bonds are always formed side by side with the monomers; the reaction of trialkylsilanethiols thus gives hexaalkyldisilthianes.

The physical constants of some compounds in this group—which, like all silicon-functional silanes, are sensitive to hydrolysis—are given in Table 18.

TABLE 18

ORGANOSILANETHIOLS[a]

Compound	Mp (°C)	Bp (°C) at	p (mm Hg)
$(CH_3)_3SiSH$	—	77–78 (*56*)	760
$(C_2H_5)_3SiSH$	—	158 (*208*)	760
$(C_3H_7)_3SiSH$	—	83–84 (*208*)	7
$(C_6H_5)_3SiSH$	103 (*90*)	—	—

[a] Numbers in parentheses indicate reference numbers.

3.10 Organoacyloxysilanes

Of this group of compounds, the general composition of which is $R_nSi(OOCR')_{4-n}$, the organoacetoxysilanes, in particular, became technically interesting at an early stage. They hydrolyze easily and give off a less corrosive acid than, e.g. the halosilanes. Acyloxysilanes are used as cross-linking agents in the room-temperature vulcanization of silicone rubber in single-component systems (Section 8.1.2.3.2). See Table 19.

TABLE 19

ORGANOACYLOXYSILANES[a]

Compound	Mp (°C)	Bp (°C)	at p (mm Hg)	n_D^t	at t (°C)	d_4^t (gm/cc)	at t (°C)
$(CH_3)_3SiOOCCH_3$	−32 (*89*)	30–31 (*311*)	35	1.388 (*34*)	[b]	0.8961 (*311*)	25
$(CH_3)_2Si(OOCCH_3)_2$	—	44–45 (*311*)	3	1.403 (*34*)	[b]	1.0485 (*311*)	25
$CH_3SiH(OOCCH_3)_2$	—	83–84 (*311*)	45	—	—	1.0761 (*311*)	25
$CH_3Si(OOCCH_3)_3$	—	94–95 (*311*)	9	1.407 (*34*)	[b]	1.1677 (*311*)	25
$Si(OOCCH_3)_4$	110 (*311*)	—	—	—	—	—	—
$(C_2H_5)_3SiOOCCH_3$	—	168 (*202*)	760	—	—	0.9039 (*202*)	0
$(C_2H_5)_2Si(OOCCH_3)_2$	—	70–72 (*311*)	4	—	—	1.0190 (*311*)	25
$C_2H_5Si(OOCCH_3)_3$	—	97–99 (*311*)	4	—	—	1.1426 (*311*)	25
$(C_6H_5)_3SiOOCCH_3$	97 (*245*)	—	—	—	—	—	—
$(C_6H_5)_2Si(OOCCH_3)_2$	—	176–178 (*7*)	3	—	—	—	—

[a] Numbers in parentheses indicate reference numbers.
[b] Temperature not given

Organoacetoxysilanes can be obtained by the reaction of the corresponding chlorosilanes with anhydrous sodium acetate, with acetic anhydride, or with acetic acid (*233*, *283*).

Tetraacetoxysilane is made from $SiCl_4$ and acetic acid in low-boiling organic solvents such as ether (*141*, *231*). The presence of tertiary bases can increase the yields from reactions of chlorosilanes with carboxylic acids (up to 75% in the case of trimethylacetoxysilane) (*89*). Numerous other carboxylic acids and their sodium or silver salts have been used in reactions with halosilanes (*7*, *9–11*, *154*). Fluoroacetoxysilanes have been prepared from chlorosilanes and sodium fluoroacetate in benzene suspension (*310*).

It is also possible to use the Si—H bond as the starting point for the synthesis of acyloxysilanes, by its reaction with carboxylic acids or their mercury salts (Section 3.4.3).

Exchange of the carboxyl groups on the silicon atom frequently permits the synthesis of new acyloxysilanes like those of chloroacetic acid (*8*, *309*). Acyloxysilanes can also be synthesized by the reaction of organosilanols (Section 3.6.3) and alkoxysilanes (Section 5.2.10), and of hexaalkyldisiloxanes (Section 5.5.3) and hexaalkyldisilazanes, with acid anhydrides.

The acetoxysilanes are relatively sensitive compounds. Depending on their substituents, they decompose more or less rapidly at elevated temperatures. Tetraacetoxysilane decomposes easily on heating, with liberation of acetic anhydride and formation of siloxane polymers. With silicon tetrachloride it forms acetyl chloride and polymeric Si—O—Si compounds (*141*). Thermal decomposition, which commences at 50°C, can be completely avoided by the presence of ether.

Organoacyloxysilanes can be used as starting materials for the preparation of siloxanes. They can react with organoalkoxysilanes (*155*):

$$\geq Si{-}OAc + RO{-}Si{\leq} \rightarrow \geq Si{-}O{-}Si{\leq} + ROAc$$

With alcohols, they react either similarly, i.e., with the elimination of a carboxylic acid ester (*76*):

$$\geq Si{-}OAc + ROH \rightarrow \geq Si{-}OH + ROAc$$

or with the formation of a carboxylic acid (*17*, *76*):

$$\geq Si{-}OAc + ROH \rightarrow \geq Si{-}OR + AcOH$$

Similarly, depending on the reaction conditions, aminolysis gives either a silanol and an acid amide:

$$\geq Si{-}OAc + R_2NH \rightarrow \geq Si{-}OH + R_2NAc$$

or an aminosilane and a free acid *(291)*:

$$\geq Si{-}OAc + R_2NH \rightarrow \geq Si{-}NR_2 + AcOH$$

3.11 Organosilyl Sulfates, Phosphates, and Perchlorates

Organosilyl sulfates, phosphates, and perchlorates (see Table 20) can be considered esters of silanols and inorganic acids.

$$R_3Si{-}O{-}ClO_3 \qquad (R_3Si{-}O)_2SO_2 \qquad (R_3Si{-}O)_3P{=}O$$

TABLE 20

Organosilyl Sulfates, Phosphates, and Perchlorates[a]

Compound	Mp (°C)	Bp (°C) at	p (mm Hg)	n_D^t at	t (°C)
$[(CH_3)_3Si]_2SO_4$	56–58 *(78)*	118 *(78)*	18	—	—
$[(C_2H_5)_3Si]_2SO_4$	—	170 *(327)*	12	1.4442 *(327)*	20
$[(CH_3)_3Si]_3PO_4$	—	85–87 *(297)*	4	1.4090 *(297)*	20
	—	110–112 *(261)*	10	1.4080 *(261)*	20
$[(C_6H_5)_3Si]_3PO_4$	244 *(261)*	—	—	—	—
$[(CH_3)_3Si]ClO_4$	—	35–38 *(371)*	14	—	—
$[(C_2H_5)_3Si]ClO_4$	—	45–46 *(371)*	1	1.4256 *(371)*	25

[a] Numbers in parentheses indicate reference numbers.

Trimethylchlorosilane and sulfuric acid give bis-(trimethylsilyl)sulfate *(265)*:

$$2(CH_3)_3SiCl + H_2SO_4 \rightarrow [(CH_3)_3Si]_2SO_4 + 2HCl$$

and dimethyldichlorosilane and sulfuric acid form dimethylsilyl sulfate *(305)*:

$$(CH_3)_2SiCl_2 + H_2SO_4 \rightarrow (CH_3)_2SiSO_4 + 2HCl$$

Silyl sulfates are formed by cleavage of the Si—O—Si bond with sulfuric acid or SO_3:

$$(CH_3)_3Si{-}O{-}Si(CH_3)_3 \xrightarrow{H_2SO_4} [(CH_3)_3Si]_2SO_4 \ (326\text{–}328)$$

$$[(CH_3)_2SiO]_n \xrightarrow{nSO_3} n(CH_3)_2SiSO_4 \ (305)$$

Silyl hydrogen sulfates have not yet been isolated, but chlorosulfuric and methylsulfuric esters have been produced from methylsilanols *(304)*.

Silyl sulfates and silyl hydrogen sulfates are formed as intermediates during the polymerization and equilibration of polyorganosiloxanes with sulfuric acid (see Section 5.4.2.2). They can be readily hydrolyzed. In concentrated sulfuric acid, the methylsilyl sulfates appear to be dissociated into HSO_4^- and siliconium cations (*277*).

Tris-(trimethylsilyl) phosphate was originally prepared by cleavage of hexamethyldisiloxane with phosphorus pentoxide (*297*):

$$3[(CH_3)_3Si]_2O + P_2O_5 \rightarrow 2[(CH_3)_3Si]_3PO_4$$

Later, however, numerous other routes were found to be feasible:

$$3R_3SiOSiR_3 + 2H_3PO_4 \xrightarrow{H^\oplus} 2[R_3Si]_3PO_4 + 3H_2O \; (261)$$

$$6R_3SiOH + P_2O_5 \rightarrow 2[R_3Si]_3PO_4 + 3H_2O \; (261)$$

$$3R_3SiOH + H_3PO_4 \rightarrow [R_3Si]_3PO_4 + 3H_2O \; (261)$$

$$3R_3SiCl + H_3PO_4 \rightarrow [R_3Si]_3PO_4 + 3HCl \; (369)$$

$$3R_3SiCl + Ag_3PO_4 \rightarrow [R_3Si]_3PO_4 + 3AgCl \; (306)$$

Trimethylsilyl dihydrogen phosphate $(CH_3)_3SiH_2PO_4$ has been prepared in 80–90% yield from trimethylchlorosilane and potassium dihydrogen phosphate in ethereal solution (*16*):

$$(CH_3)_3SiCl + KH_2PO_4 \rightarrow [(CH_3)_3Si]H_2PO_4 + KCl$$

Trialkylsilyl phosphorodichloridates have also been described (*306*).

Trialkylchlorosilanes yield trialkylsilyl perchlorates with silver perchlorate (*371*):

$$R_3SiCl + AgClO_4 \rightarrow R_3SiClO_4 + AgCl.$$

These decompose explosively at elevated temperatures, but not as unpredictably or dangerously as the alkyl perchlorates.

The silyl phosphates and perchlorates, too, are easily cleaved by hydrolysis. The triorganosilyl derivatives always form the corresponding hexaorganodisiloxane by reaction with water (*16*, *371*).

3.12 Organocyanosilanes, Organoisocyanatosilanes, and Organoisothiocyanatosilanes

3.12.1 Preparation

In trimethylcyanosilane the cyano and isocyano isomers are in equilibrium. The equilibrium is predominantly on the side of the normal structure at ordinary temperatures, but is displaced toward the iso-form at higher temperatures (*36*). The iso-form seems to predominate at normal temperatures in a number of phenyl-substituted silanes (*227*).

Trimethylcyanosilane $(CH_3)_3SiCN$ and dimethyldicyanosilane $(CH_3)_2Si(CN)_2$ were originally prepared by the reaction of the methylhalosilanes with silver cyanide (*228*). Compounds with more than two CN groups on the Si atom were not formed. The synthesis of phenyl-substituted isocyanosilanes from phenylbromosilanes (*227*) can be carried out similarly. It was later shown that trialkylcyanosilanes can also be obtained by other routes; e.g., from chlorosilanes and $K_2Hg(CN)_4$ in dimethylformamide, from hexaalkyldisilazanes and HCN, and from trialkylmethoxysilanes and trimethylacetyl cyanide in the presence of $AlCl_3$ (*36*). The possibility of obtaining cyano compounds of silicon by passing halosilane vapors through eutectic LiCl-KCl melts containing potassium cyanide is particularly interesting (*349*).

The iso-structure of the four isocyanatosilanes $Si(NCO)_4$, $CH_3Si(NCO)_3$, $(CH_3)_2Si(NCO)_2$, and $(CH_3)_3SiNCO$ has been established by infrared and Raman spectra (*138*). The same is true of the corresponding isothiocyanates, as shown by their infrared spectra (*337*).

The reaction of the appropriate organosilanes with lithium, sodium, potassium, silver, or lead salts has been used to prepare the organoisocyanatosilanes and organoisothiocyanatosilanes (*1–4*, *37*, *95*, *96*, *140*, *174*, *222*). The isothiocyanates are also obtained rapidly and in good yield from the halosilanes and ammonium thiocyanate in acetonitrile or liquid SO_2 (*337*).

Of other methods of preparation, the following may be mentioned: trimethylchlorosilane reacts with urea at 240°C in the autoclave, giving a 65 to 75% yield of trimethylisocyanatosilane:

$$(CH_3)_3SiCl + CO(NH_2)_2 \rightarrow (CH_3)_3SiNCO + NH_4Cl$$

However, this process fails with dimethyldichlorosilane (*140*).

Isocyanic acid (*343*, *374*) gives isocyanatosilanes with chlorosilanes in the presence of amines as HCl acceptors.

Isocyanatosilanes are prepared by the reaction of hexamethyldisilazane with phosgene (*278*).

$$4(CH_3)_3Si{-}NH{-}Si(CH_3)_3 + 3COCl_2 \rightarrow NH_4Cl + 3(CH_3)_3SiNCO + 5(CH_3)_3SiCl$$

A particularly elegant process is the reaction of halosilanes with alkali-metal cyanates and thiocyanates dissolved in eutectic mixtures of fused salts. The question here is whether the alkali pseudo-halides which definitely possess the normal structure also give the normal form of the silicon compounds (*349*).

3.12.2 Physical Constants (Table)

Physical constants for a number of organocyanosilanes, organoisocyanatosilanes, and organoisothiocyanatosilanes are given in Table 21.

TABLE 21

ORGANOCYANOSILANES, ORGANOISOCYANATOSILANES, AND ORGANOISOTHIOCYANOTOSILANES[a]

Compound	Mp (°C)	Bp (°C)	at p (mm Hg)	n_D^t	at t (°C)	d_4^t (gm/cc)	at t (°C)
$(CH_3)_3SiCN$	10.5–11.5 (*228*)	117.8 (*228*)	760	1.3891 (*228*)	23	0.7834 (*228*)	30
$(CH_3)_2Si(CN)_2$	84–85 (*228*)	168–169 (*228*)	762	—	—	—	—
$(CH_3)_3SiNCO$	−49.0 ± 3 (*96*)	91.0 ± 0.3 (*96*)	760	1.3960 (*96*)	20	0.867 (*96*)	20
$(CH_3)_2Si(NCO)_2$	−31.2 ± 3 (*96*)	139.2 ± 0.3 (*96*)	760	1.4221 (*96*)	20	1.076 (*96*)	20
$CH_3Si(NCO)_3$	2.7 ± 0.5 (*96*)	170.8 ± 0.3 (*96*)	760	1.4430 (*96*)	20	1.267 (*96*)	20
$Si(NCO)_4$	26 (*96*)	185.6 ± 0.5 (*96*)	760	1.4610 (*96*)	20	1.434 (*96*)	25
$(CH_3)_3SiNCS$	−32.8 (*1*)	143.1 ± 0.3 (*1*)	760	1.4820 (*1*)	20	0.931 (*1*)	20
$(CH_3)_2Si(NCS)_2$	18.0 (*1*)	217.3 ± 1 (*1*)	760	1.5677 (*1*)	20	1.142 (*1*)	20
$CH_3Si(NCS)_3$	72.4 (*1*)	266.8 ± 1 (*1*)	760	—	—	1.304 (*1*)	20
$Si(NCS)_4$	143.8 (*282*)	313.0 (*1*)	760	—	—	1.409 (*1*)	20
$(C_6H_5)_3SiNC$	136–138 (*227*)	210 (*227*)	5	—	—	—	—
$(C_6H_5)_2Si(NC)_2$	46–48 (*227*)	142 (*227*)	2	1.5599 (*227*)	32	1.0904 (*227*)	32
$(C_6H_5)_3SiNCO$	95.0 ± 1 (*95*)	372 ± 1 (*95*)	760	—	—	—	—
$(C_6H_5)_2Si(NCO)_2$	22.9 ± 0.3 (*95*)	319.6 ± 1 (*95*)	760	1.5675 (*95*)	20	1.188 (*95*)	20
$C_6H_5Si(NCO)_3$	—	251.9 ± 1 (*95*)	760	1.5210 (*95*)	20	1.273 (*95*)	20
$(C_6H_5)_3SiNCS$	76 ± 1 (2)	396 ± 1.5 (2)	760	—	—	—	—
$(C_6H_5)_2Si(NCS)_2$	46 ± 1 (2)	376 ± 1.5 (2)	760	—	—	1.188 (2)	30
$C_6H_5Si(NCS)_3$	52 ± 1 (2)	339.6 ± 1 (2)	760	—	—	1.291 (2)	31

[a] Numbers in parentheses indicate reference numbers.

3.12.3 Chemical Properties

The Si—X bonds can be arranged in a sequence according to their reactivities with silver salts (and lead or cadmium salts), which shows the possibilities for (irreversible) conversions in the series. For ethyl-substituted silanes (*80*):

$$(C_2H_5)_3SiI \rightarrow [(C_2H_5)_3Si]_2S \rightarrow (C_2H_5)_3SiBr \rightarrow (C_2H_5)_3SiCN$$
$$\rightarrow (C_2H_5)_3SiCl \rightarrow (C_2H_5)_3SiNCS$$

analogously, for the methylsilanes:

$$(CH_3)_3SiI \rightarrow [(CH_3)_3Si]_2S \rightarrow (CH_3)_3SiCl \rightarrow (CH_3)_3SiNCS$$

The organosilicon pseudo-halides generally have high thermal stability; phenyl-substituted isothiocyanatosilanes have boiling points of around 400°C (Table 21) at normal pressure. However, they are easily cleaved by water to give silanols or siloxanes. The sensitivity to hydrolysis in dioxane solution of silanes of the type R_3SiX increases in the sequence (*80*):

$$(C_2H_5)_3SiH, (C_2H_5)_3SiF < (C_2H_5)_3SiNCO < (C_2H_5)_3SiNCS$$
$$< (C_2H_5)_3SiCN, (C_2H_5)_3SiCl$$

The influence of the organic substituents is shown by the sequence (*96*):

$$(CH_3)_3SiNCO < (CH_3)_2Si(NCO)_2 < CH_3Si(NCO)_3$$

Of the remaining reactions, the addition of the isocyanatosilanes to amines is of particular interest. All methylisocyanatosilanes, as well as $Si(NCO)_4$, have been smoothly converted into urea derivatives with primary or secondary amines (*139*, *250*):

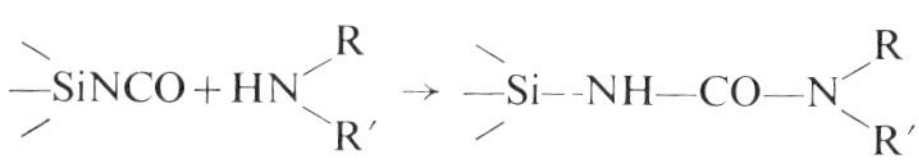

Only aromatic amines with steric hindrance (e.g., due to *t*-butyl groups) or electron-attracting substituents (*p*-nitro group) failed to react (*250*). The corresponding reaction of the isocyanatosilanes with silanols did not take place (*139*).

The silylureas obtained are sensitive toward water, hydrolyzing to give ureas and siloxanes.

No silylurethane is formed from the reaction of trimethylisocyanatosilane with methanol; it goes almost quantitatively to trimethylmethoxysilane and methyl carbamate (*140*):

$$R_3SiNCO + 2CH_3OH \rightarrow R_3SiOCH_3 + H_2NCOOCH_3$$

REFERENCES

(1) H. H. ANDERSON, *J. Amer. Chem. Soc.* **69**, 3049 [1947]. (2) —, ibid. **70**, 1220 [1948]. (3) —, ibid. **72**, 193 [1950]. (4) —, ibid. **72**, 196 [1950]. (5) —, ibid. **73**, 2351 [1951]. (6) —, ibid. **74**, 1421 [1952]. (7) —, ibid. **74**, 2371 [1952]. (8) — and H. FISCHER, *J. Org. Chemistry* **19**, 1296 [1954]. (9) D. N. ANDREEV, B. N. DOLGOV and E. V. KUKHARSKAYA, *Bull. Acad. Sci. USSR, Div. Chem. Sci.* (*Engl. transl.*) **1955**, 465.

(10) —, L. L. SHCHUKOVSKAYA, ibid. **1953**, 121. (11) — and G. I. ZAVOROTNOVA, ibid. **1954**, 605. (12) K. A. ANDRIANOV and V. V. ASTAKHIN, *J. Gen. Chem. USSR* (*Engl. transl.*) **29**, 2665 [1959]. (13) —, S. A. GOLOBTSOV and E. A. SEMENOVA, *Bull. Acad. Sci. USSR, Div. Chem. Sci.* (*Engl. transl.*) **1958**, 41. (14) — and O. GRIBANOVA, *Zhurnal obshchei Khimii* **8**, 558 [1938] (*Chem. Zbl.* **1938**, II, 3674). (15) —, G. A. KURAKOV, L. M. KHANANASHVILI and T. A. LOMONOSOVA, *J. Gen. Chem. USSR* (*Engl. transl.*) **33**, 1265 [1963]. (16) —, B. N. RUTOVSKY and A. A. KAZAKOVA, ibid. **26**, 285 [1956]. (17) —, A. A. ZHDANOV and A. A. BOGDANOVA, ibid. **27**, 2128 [1957]. (18) V. V. ASTAKHIN, I. P. LOSEV and K. A. ANDRIANOV, *Proc. Acad. Sci. USSR, Sect. Chem.* (*Engl. transl.*) **113**, 247 [1957]. (19) —, —, —, *J. Gen. Chem. USSR* (*Engl. transl.*) **29**, 887 [1959].

(20) D. L. BAILEY and F. M. O'CONNOR, Union Carbide, German Appl. (West German) 1040251 [U.S. Prior. 22.3.1954]. (21) —, L. H. SOMMER and F. C. WHITMORE, *J. Amer. Chem. Soc.* **70**, 435 [1948]. (22) J. E. BAINES and C. EABORN, *J. Chem. Soc.* [London] **1955**, 4023. (23) E. W. BALIS and H. A. LIEBHAFSKY, *Ind. Engng. Chem.* **38**, 583 [1946]. (24) —, — and D. H. GETZ, ibid. **41**, 1459 [1949]. (25) A. J. BARRY, Dow Chemical, Brit. Pat. 622970 [U.S. Prior. 4.2.1946]. (26) —, D. J. HOOK and L. DEPREE, —, U.S. Pat. 2626271 [6.6.1946]. (27) —, L. DEPREE, J. W. GILKEY and D. E. HOOK, *J. Amer. Chem. Soc.* **69**, 2916 [1947]. (28) M. J. BATUEV, M. F. SHOSTAKOVSKII, V. I. BELYAEV, A. D. MATVEEVA and E. V. DUBROVA, *Doklady Akad. Nauk. SSSR* **95**, 531 [1954]. (29) R. A. BENKESER and D. I. FOSTER, *J. Amer. Chem. Soc.* **74**, 4200 [1952].

(30) —, —, ibid. **74**, 5314 [1952]. (31) —, H. LANDESMAN and D. J. FOSTER, ibid. **74**, 648 [1952]. (32) —, R. E. ROBINSON and H. LANDESMAN, ibid. **74**, 5699 [1952]. (33) — and R. G. SEVERSON, ibid. **73**, 1424 [1951]. (34) A. F. BIDAUD and P. DUMONT, Rhône-Poulenc, U.S. Pat. 2573302 [French Prior. 23.7.1947]. (35) L BIRKOFER, A. RITTER and H. DICKOPP, *Chem. Ber.* **96**, 1473 [1963]. (36) T. A. BITHER, W. H. KNOTH, R. V. LINDSEY JR. and W. H. SHARKEY, *J. Amer. Chem. Soc.* **80**, 4151 [1958]. (37) B. A. BLUESTEIN, General Electric, U.S. Pat. 2559340 [21.6.1947]. (38) G. S. BOOTH and P. H. CARNELL, *J. Amer. Chem. Soc.* **68**, 2650 [1946]. (39) — and H. S. HALBEDEL, ibid. **68**, 2652 [1946].

(40) — and R. L. JARRY, ibid. **71**, 971 [1949]. (41) — and W. F. MARTIN, ibid. **68**, 2655 [1946]. (42) — and A. A. SCHWARTZ, ibid. **68**, 2662 [1946]. (43) — and D. R. SPESSARD, ibid. **68**, 2660 [1946]. (44) — and J. F. SUTTLE, ibid. **68**, 2658 [1946]. (45) H. BREEDERVELD, *Rec. Trav. Chim. Pays-Bas* **79**, 1126 [1960]. (46) —, ibid. **81**, 276 [1962]. (47) A. I. BRODSKII and I. G. KHASKIN, *Doklady Akad. Nauk. SSSR* **74**, 299 [1950]. (48) A. G. BROOK and H. GILMAN, J. *Amer. Chem. Soc.* **76**, 278 [1954]. (49) —, —, ibid. **76**, 2333 [1954].

(50) —, —, ibid. **76**, 2338 [1954]. (51) C. A. BURKHARD, ibid. **67**, 2173 [1945]; also *Inorg. Syntheses* **3**, 62 [1950]. (52) A. BYGDÉN, *Dissert.*, Uppsala 1916. (53) R. P. CAHOY, General Electric, German Appl. (West German) 1087601 [U.S. Prior. 8.11.1958]. (54) R. CALAS and N. DUFFAUT, *C.R. hebd. Séances Acad. Sci.* **245**, 906 [1957]. (55) —, E. FRAINET and J. BONASTRE, ibid. **251**, 2987 [1960]. (56) G. CHAMPETIER, Y. ÉTIENNE and R. KULLMANN, ibid.

234, 1985 [1952]. (57) N. D. CHERONIS, U.S. Pat. 2564674 [7.8.1947]. (58) P. CHEVALIER, Rhône-Poulenc, German Appl. (West German) 1031520 [French Prior. 13.7.1956]. (59) S. CHRZCZONOWICZ and Z. LASOCKI, *Roczniki Chem.* **34**, 1667 [1960].

(60) H. D. COGAN and C. A. SETTERSTROM, *Chem. Engng. News* **24**, 2499 [1946]. (61) C. CURRAN, R. M. WITUCKI and P. A. MCCUSKER, *J. Amer. Chem. Soc.* **72**, 4471 [1950]. (62) E. DAFANO and H. I. STERNHEIM, J. B. Pierce Foundation, French Pat. 1082863 [U.S. Prior. 22.10.1952]. (63) —, —, J. I. CASE, U.S. Pat. 2813389 [15.10.1954]. (64) K. DAMM and W. NOLL, *Kolloid-Z.* **158**, 97 [1958]. (65) W. H. DAUDT and J. F. HYDE, *J. Amer. Chem. Soc.* **74**, 386 [1952]. (66) D. R. DEANS and C. EABORN, *J. Chem. Soc.* [London] **1954**, 3169. (67) M. L. DELWAULLE, M. B. BUISSET and M. DELHAYE, *J. Amer. Chem. Soc.* **74**, 5768 [1952]. (68) J. W. DIEHL and H. GILMAN, *Chem. and Ind.* **1959** No. 35, 1095. (69) P. A. DIGIORGIO and A. E. NEWKIRK, General Electric, U.S. Pat. 2485366 [15.12.1948].

(70) —, L. H. SOMMER and F. C. WHITMORE, *J. Amer. Chem. Soc.* **68**, 344 [1946]. (71) W. DILTHEY and F. EDUARDOFF, *Ber. dtsch. chem. Ges.* **37**, 1139 [1904]. (72) B. N. DOLGOV, N. P. KHARITONOV, N. E. GLUSHKOVA and YU. I. KHUDOBIN, *J. Gen. Chem. USSR (Engl. transl.)* **28**, 2737 [1958]. (73) —, — and M. G. VORONKOV, ibid. **24**, 859 [1954]. (74) —, —, —, ibid. **24**, 1169 [1954]. (75) —, YU. I. KHUDOBIN and N. P. KHARITONOV, *Bull. Acad. Sci. USSR, Div. Chem. Sci. (Engl. transl.)* **1958**, 112. (76) —, V. P. DAVYDOVA and M. G. VORONKOV, *J. Gen. Chem. USSR (Engl. transl.)* **27**, 1666 [1957]. (77) J. J. DUANE, Union Carbide, U.S. Pat. 2646441 [17.12.1949]. (78) N. DUFFAUT, R. CALAS and J. DUNOGUÈS, *Bull. Soc. Chim. France* **1963**, 512. (79) C. EABORN, *J. Chem. Soc.* [London] **1949**, 2755.

(80) —, ibid. **1950**, 3077. (81) —, ibid. **1952**, 2840. (82) —, ibid. **1952**, 2846. (83) —, ibid. **1953**, 494. (84) — and N. H. HARTSHORNE, ibid. **1955**, 549. (85) E. A. V. EBSWORTH and H. J. EMELÉUS, ibid. **1958**, 2150. (86) H. J. EMELÉUS and C. J. WILKINS, ibid. **1944**, 454. (87) — and S. R. ROBINSON, ibid. **1947**, 1592. (88) H. H. ENDER, *J. appl. Polymer Sci.* **7**, 1133 [1963]. (89) Y. ÉTIENNE, *C.R. hebd. Séances Acad. Sci.* **235**, 966 [1952].

(90) —, *Bull. Soc. chim. France* **1953**, 791. (91) J. FERTIG, W. GERRARD and H. HERBST, *J. Chem. Soc.* [London] **1957**, 1488. (92) R. FILLER, *J. Org. Chem.* **19**, 544 [1954]. (93) W. FINK, *Chem. Ber.* **97**, 1433 [1964]. (94) A. E. FINHOLT, A. C. BOND JR, K. E. WILZBACH and H. I. SCHLESINGER, *J. Amer. Chem. Soc.* **69**, 2692 [1947]. (95) G. S. FORBES and H. H. ANDERSON, ibid. **70**, 1043 [1948]. (96) —, —, ibid. **70**, 1222 [1948]. (97) M. J. FRAZER, W. GERRARD and J. A. STRICKSON, *J. Chem. Soc.* [London] **1960**, 4701. (98) C. FRIEDEL and A. LADENBURG, *Liebigs Ann. Chem.* **143**, 118 [1867]. (99) K. C. FRISCH, P. A. GOODWIN and R. E. SCOTT, *J. Amer. Chem. Soc.* **74**, 4584 [1952].

(100) G. FRITZ, *Z. Naturforsch.* **6b**, 116 [1951]. (101) —, Discussion comments on S. Nitzsche, *Angew. Chem.* **63**, 490 [1951]. (102) —, *Z. anorg. allg. Chem.* **273**, 275 [1953]. (103) —, D. HABEL, D. KUMMER and G. TEICHMANN, ibid. **302**, 60 [1959]. (104) — and H. KAUTSKY, German Pat. (West German) 964859 [16.3.1951]. (105) —, —, German Appl. (West German) 1004180 [17.1.1953]. (106) — and D. KSINSIK, *Z. anorg. allg. Chem.* **322**, 46 [1963]. (107) — and D. KUMMER, ibid. **308**, 105 [1961]; **310**, 327 [1961]; *Chem. Ber.* **94**, 1143 [1961]. (108) — and B. RAABE, *Z. anorg. allg. Chem.* **286**, 149 [1956]. (109) A. GANCBERG and J. VANDEVELDE, Industrie Chimique Belge No. 6, 591 [1964].

(110) General Electric, Brit. Pat. 736561 [U.S. Prior. 27.5.1952]. (111) —, Belg. Pat. 553496 [U.S. Prior. 19.12.1955]. (112) —, Belg. Pat. 557059 [U.S. Prior. 30.4.1956]. (113) P. D. GEORGE, L. H. SOMMER and F. C. WHITMORE, *J. Amer. Chem. Soc.* **75**, 1585 [1953].

(114) W. GERRARD and K. D. KILBURN, *J. Chem. Soc.* [London] **1956**, 1536. (115) A. R. GILBERT, G. D. COOPER and R. W. SHADE, *Ind. Engng. Chem.* **51**, 665 [1959]. (116) W. F. GILLIAM, E. M. HADSELL, H. A. LIEBHAFSKY and M. M. SPRUNG, *J. Amer. Chem. Soc.* **73**, 4252 [1951]. (117) H. GILMAN, A. G. BROOK and L. S. MILLER, ibid. **75**, 3757 [1953]. (118) — and R. N. CLARK, ibid. **69**, 967 [1947]. (119) —, —, ibid. **69**, 1499 [1947].

(120) — and G. E. DUNN, ibid. **73**, 3404 [1951]. (121) — and J. J. GOODMAN, J. *Org. Chem.* **22**, 45 [1957]. (122) —, B. HOFFERTH, H. W. MELVIN and G. E. DUNN, *J. Amer. Chem. Soc.* **72**, 5767 [1950]. (123) —, R. K. INGHAM and A. G. SMITH, *J. Org. Chem.* **18**, 1743 [1953]. (124) — and G. D. LICHTENWALTER, *J. Amer. Chem. Soc.* **80**, 608 [1958]. (125) —, —, ibid. **82**, 3319 [1960]. (126) — and F. J. MARSHALL, ibid. **71**, 2066 [1949]. (127) — and L. S. MILLER, ibid. **73**, 2367 [1951]. (128) — and G. N. R. SMART, *J. Org. Chem.* **19**, 441 [1954]. (129) — and W. J. TREPKA, *J. Organometall. Chem.* **1**, 222 [1964].

(130) — and T. C. WU, *J. Amer. Chem. Soc.* **73**, 4031 [1951]. (131) —, —, *J. Org. Chem.* **18**, 753 [1953]. (132) —, —, *J. Amer. Chem. Soc.* **75**, 234 [1953]. (133) —, —, ibid. **75**, 2509 [1953]. (134) —, —, ibid. **76**, 2502 [1954]. (135) —, —, H. A. HARTZFELD, G. A. GUTER, A. S. SMITH, J. J. GOODMAN and S. H. EIDT, ibid. **74**, 561 [1952]. (136) N. C. GOODSPEED and R. T. SANDERSON, *J. Inorg. Nucl. Chem.* **2**, 266 [1956]. (137) J. GOUBEAU, IUPAC-Colloquium, Münster 1954, Verlag Chemie, Weinheim/Bergstr., p. 69. (138) —, E. HEUBACH, D. PAULIN and I. WIDMAIER, *Z. anorg. allg. Chem.* **300**, 194 [1959]. (139) —, —, *Chem. Ber.* **93**, 1117 [1960].

(140) — and D. PAULIN, ibid. **93**, 1111 [1960]. (141) — and R. MUNDIEL, *Z. anorg. allgem. Chem.* **272**, 313 [1953]. (142) —, H. SIEBERT and M. WINTERWERB, ibid. **259**, 240 [1949]. (143) H. GROSSE RUYKEN and R. KLEEZAAT, *Z. Chemie* **1**, 27 [1960). (144) W. T. GRUBB and R. C. OSTHOFF, *J. Amer. Chem. Soc.* **75**, 2230 [1953]. (145) R. C. GUNTON, J. F. OLLOM and H. N. REXROAD, *J. Chem. Physics* **22**, 1942 [1954]. (146) E. L. GUSTUS and N. D. CHERONIS, U.S. Pat. 2579416 [25.1.1946]; U.S. Pat. 2579417 [7.8.1947]. (147) —, —, U.S. Pat. 2579418 [24.1.1949]. (148) R. A. HALL, ICI, German Appl. (West German) 1044080 [Brit. Prior. 17.8.1955 and 22.5.1956]. (149) J. H. HASLAM, Du Pont, U.S. Pat. 2820806 [26.10.1956].

(150) R. HAUL, *Angew. Chem.* **60**, 227 [1948]. (151) M. E. HAVILL, H. JOFFE and H. W. POST, *J. Org. Chem.* **13**, 280 [1948]. (152) A. S. HAY, General Electric, French Pat. 1309761 [U.S. Prior. 11.1.1961]. (153) R. N. HAZELWOOD, McGraw Electric, U.S. Pat. 2727880 [22.1.1953]. (154) F. A. HENGLEIN, G. ABELSNES, H. HENEKA, K. LIENHARD, P. NAKHRE and K. SCHEINOST, *Makromolekulare Chem.* **24**, 1 [1957]. (155) —, P. SCHMULDER and H. NIEBERGALL, German Pat. (West German) 947739 [27.11.1953]. (156) K. HIRATSUKA, Jap. Pat. 5330 [1951], (C.A. **1953**, 9345). (157) M. J. HUNTER and P. C. SERVAIS, Dow Corning, French Pat. 970668 [U.S. Prior. 13.10.1947]. (158) D. T. HURD, General Electric, U.S. Pat. 2406605 [15.3.1945]. (159) C. D. HURD and W. A. YARNALL, *J. Amer. Chem. Soc.* **71**, 755 [1949].

(160) J. F. HYDE, Corning Glass Works, U.S. Pat. 2413050 [21.4.1945]. (161) —, —, U.S. Pat. 2574265 [15.11.1945]. (162) —, —, U.S. Pat. 2472799 [25.11.1946]. (163) —, *J. Amer. Chem. Soc.* **75**, 2166 [1953]. (164) —, Dow Corning, French Pat. 1141735 [U.S. Prior. 23.2.1955]. (165) — and R. C. DE LONG, *J. Amer. Chem. Soc.* **63**, 1194 [1941]. (166) — and O. K. JOHANNSON, Corning Glass Works, U.S. Pat. 2453092 [5.8.1947]. (167) —, —, W. H. DAUDT, R. F. FLEMING, H. B. LAUDENSLAGER and M. P. ROCHE, *J.Amer. Chem. Soc.* **75**, 5615 [1953]. (168) R. M. ISMAIL, *Z. Naturforschung* **19b**, No. 10, 873 and 964 [1964]. (169) I. G. ISMAILZADE and G. S. ZHDANOV, *Zhurnal fizicheskoi Khimii* **26**, 1619 [1952].

(170) E. Itsugi, Nippon Silicone Resin, Jap. Pat. 3861 [1957] (C.A. **52**, 8182 [1958]). (171) A. C. Jenkins and G. F. Chambers, *Ind. Engng. Chem.* **46**, 2367 [1954]. (172) — and A. J. Reid, ibid. **46**, 2566 [1954]. (173) I. W. Jenkins and H. W. Post, *J. Org. Chem.* **15**, 552, 556 [1950]. (174) L. H. Jenkins and D. S. Sears, Virginia-Carolina Chemical, U.S. Pat. 2873171 [3.7.1956]. (175) H. Jenkner, Kali-Chemie, German Pat. (West German) 950636 [23.7.1953]. (176) —, —, German Pat. (West German) 959910 [17.12.1953]. (177) —, —, German Appl. (West German) 1028575 [2.11.1955]. (178) —, —, German Appl. (West German) 1055511 [15.12.1956]; Brit. Pat. 823483 [German Prior. (West German) 3.11., 12.12. and 15.12.1956]. (179) — and H. W. Schmidt, —, French Pat. 1130998 [German Prior. (West German) 18.9.1954 and 12.11.1954].

(180) G. Jerusalem, *J. Chem. Soc.* [London] **97**, 2190 [1910]. (181) C. Jida, M. Yokoi and K. Yamasaki, *J. Chem. Soc. Japan* **73**, 882 [1952]. (182) O. K. Johannson and J. F. Hyde, Dow Corning, German Pat. (West German) 886075 [U.S. Prior. 10.1.1947]. (183) M. Kahlert and H. Clauss, German Pat. (East German) 4955 [8.8.1951]. (184) M. Kakudo and T. Watase, *J. Chem. Physics* **21**, 167 [1953]. (185) —, —, *Technol. Rep. Osaka Univ.* **2**, 247 [1952]. (186) —, P. N. Kasai and T. Watase, *J. Chem. Physics* **21**, 1894 [1953]. (187) M. Kanazashi, *Bull. Chem. Soc. Japan* **28**, 44 [1955]. (188) — and M. Takakusa, ibid. **27**, 441 [1954]. (189) S. W. Kantor, *J. Amer. Chem. Soc.* **75**, 2712 [1953].

(190) —, General Electric, U.S. Pat. 2717258 [25.2.1953]. (191) N. Kasai and M. Kakudo, *Bull. Chem. Soc. Japan* **27**, 605 [1954]. (192) W. S. Kather and A. Torkelson, *Ind. Engng. Chem.* **46**, 381 [1954]. (193) I. G. Khaskin, *J. Gen. Chem. USSR* (*Engl. transl.*) **23**, 29 [1953]. (194) F. S. Kipping, *Proc. Chem. Soc.* [London] **23**, 9 [1907]; *J. Chem. Soc.* [London] **91**, 209 [1907]. (195) H. W. Kohlschütter and G. Jaekel, *Z. anorg. allg. Chem.* **271**, 185 [1953]. (196) T. B. Korablina, M. V. Barinova, A. I. Kurakina and G. I. Ryabova, *Plast. massy* 1963, No. 12, 18. (197) K. Kramer and A. Wright, *Angew. Chem.* **74**, 468 [1962]. (198) C. A. Kraus and R. Rosen, *J. Amer. Chem. Soc.* **47**, 2739 [1925]. (199) A. P. Kreshkov, *Zhurnal fizicheskoi Khimii* **28**, 537 [1954] (*Chem. Zbl.* 1955, 2630).

(200) R. H. Krieble and C. A. Burkhard, *J. Amer. Chem. Soc.* **69**, 2689 [1947]. (201) C. Krüger, Diss. T.H. Aachen, 1961. (202) A. Ladenburg, *Ber. dtsch. chem. Ges.* **5**, 319 [1872]; *Liebigs Ann. Chem.* **164**, 300 [1872]. (203) —, *Ber. dtsch. chem. Ges.* **5**, 1081 [1872]; *Liebigs Ann. Chem.* **173**, 143 [1874]. (204) E. Larsson, *Svensk. kem. Tidskr.* **61**, 59 [1949]. (205) —, *Kungl. Fysiografiska Sällskap. i. Lund, Hand.* **63**, No. 12, 8 [1952]; (C.A. 1954, 10545). (206) —, ibid. *Förh.* **28**, 1 [1958]; C. A. **53**, 21622i [1959]. (207) — and L. Bjellerup, *J. Amer. Chem. Soc.* **75**, 995 [1953]. (208) — and R. Mårin, *Acta chem. scand.* **5**, 964 [1951]. (209) —, —, ibid. **5**, 1173 [1951].

(210) — and B. Smith, ibid. **3**, 487 [1949]. (211) —, —, *Svensk. kem. Tidskr.* **62**, 87 [1950]. (212) R. N. Lewis, *J. Amer. Chem. Soc.* **69**, 717 [1947]. (213) — and A. E. Newkirk, ibid. **69**, 701 [1947]. (214) N. S. Leznov, L. A. Sabun and K. A. Andrianov, *J. Gen. Chem. USSR* (*Engl. transl.*) **29**, 1488 [1959]. (215) K. Licht and H. Kriegsmann, *Z. anorg. allg. Chem.* **325**, 1 [1963]. (216) —, —, ibid. **323**, 190 [1963]. (217) R. C. Linville and H. R. McEntee, General Electric, U.S. Pat. 3007956 [1.7.1960]. (218) G. R. Lucas, General Electric, U.S. Pat. 2594946 [4.1.1951]. (219) M. Maienthal, M. Hellmann, C. P. Haber, L. A. Hymo, S. Carpenter and A. J. Carr, *J. Amer. Chem. Soc.* **76**, 6392 [1954].

(220) M. S. Malinovsky and M. K. Romantsevich, *J. Gen. Chem. USSR* (*Engl. transl.*) **27**, 1749 [1957]. (221) —, —, ibid. **27**, 1935 [1957]. (222) H. Manami and S. Nishizaki, *J.*

Chem. Soc. Japan, Ind. Chem. Sect. **61**, 214 [1958]. (223) N. S. MARANS, L. H. SOMMER and F. C. WHITMORE, *J. Amer. Chem. Soc.* **73**, 5127 [1951]. (224) M. MARCHAND, German Pat. (East German) 12115 [19.4.1951]. (225) G. MARR and D. E. WEBSTER, *J. Organometall. Chem.* **2**, 93 [1964]. (226) G. MARTIN and F. S. KIPPING, *J. Chem. Soc.* [London] **95**, 302 [1909]. (227) J. J. MCBRIDE JR., *J. Org. Chem.* **24**, 2029 [1959]. (228) — and H. C. BEACHELL, *J. Amer. Chem. Soc.* **74**, 5247 [1952]. (229) P. A. MCCUSKER and E. L. REILLY, ibid. **75**, 1583 [1953].

(230) R. R. MCGREGOR and E. L. WARRICK, Corning Glass Works, U.S. Pat. 2386488 [1.9.1943]. (231) C. A. MCKENZIE, Montclair Research, U.S. Pat. 2566347 [17.4.1946]. (232) —, A. P. MILLS and J. M. SCOTT, *J. Amer. Chem. Soc.* **72**, 2032 [1950]. (233) — and M. SCHOFFMAN, Montclair Research, U.S. Pat. 2537073 [16.7.1946]. (234) C. W. MCMULLEN and A. MARZOCCHI, Cowles Chemical, U.S. Pat. 2723211 [13.4.1953]. (235) O. MJÖRNE, *Svensk. kem. Tidskr.* **72**, 120 [1950]. (236) C. R. MORGAN and W. F. OLDS, California Research, U.S. Patent 2674579 [3.1.1950]. (237) R. MÜLLER and C. DATHE, German Appl. (West German) 1142364 [5.8.1960]. (238) —, —, *J. prakt. Chemie* **22**, 232 [1963]. (239) —, —, *Z. anorg. allg. Chem.* **330**, 195 [1964].

(240) — and D. MROSS, ibid. **324**, 78 [1963]. (241) —, —, ibid. **324**, 86 [1963]. (242) H. MURATA, *J. Chem. Soc. Japan, Pure Chem. Sect.* **73**, 465 [1952]. (243) R. NAGEL, C. TAMBORSKI and H. W. POST, *J. Org. Chem.* **16**, 1768 [1951]. (244) N. S. NAMETKIN, A. V. TOPCHIEV and F. F. MACHUS, *Doklady Akad. Nauk. SSSR* **83**, 705 [1952]. (245) —, —, —, ibid. (2) **87**, 233 [1952]. (246) —, —, and L. I. KARTASHEVA, ibid (2) **93**, 667 [1953]. (247) D. J. NEAL and R. D. BLAUMANIS, *Analytic. Chem.* **32**, 139 [1960]. (248) W. H. NEBERGALL and O. H. JOHNSON, *J. Amer. Chem. Soc.* **71**, 4022 [1949]. (249) W. D. NEFEDOV and N. A. VARSCHAV, *Zhurnal fizicheskoi Khimii* **28**, 981 [1954].

(250) R. G. NEVILLE and J. J. MCGEE, *Canad. J. Chem.* **41**, 2123 [1963]. (251) A. E. NEWKIRK, *J. Amer. Chem. Soc.* **68**, 2736 [1946]. (252) —, General Electric, U.S. Pat. 2449815 [23.5.1946]. (253) S. NITZSCHE and E. PIRSON, Wacker-Chemie, German Pat. (West German) 896414 [23.5.1950]. (254) — and M. WICK, *Angew. Chem.* **69**, 96 [1957]. (255) W. NOLL, K. MENNICKEN and E. WALASCHEWSKI, Farbenfabriken Bayer, unpublished. (256) G. OERTEL, H. HOLTSCHMIDT and H. MALZ, —, German Appl. (West German) 1157226 [28.10.1961]. (257) —, —, — and E. DEGENER, —, German Appl. (West German) 1154475 [10.11.1961]. (258) R. OKAWARA, *J. Chem. Soc. Japan, Ind. Chem. Sect.* **61**, 690 [1958]. (259) B. A. ORKIN, Socony-Vacuum Oil, U.S. Pat. 2701803 [20.10.1948].

(260) N. F. ORLOV, R. A. BOGATKIN, Z. I. SERGEEVA and M. G. VORONKOV, *J. Gen. Chem. USSR (Engl. transl.)* **32**, 2526 [1962]. (261) — and M. G. VORONKOV, ibid. **30**, 2206 [1960]. (262) K. W. PALMER and F. S. KIPPING, *J. Chem. Soc.* [London] **1930**, 1020. (263) C. PAPE, *Ber. dtsch. chem. Ges.* **14**, 1872 [1881]; *Liebigs Ann. Chem.* **222**, 354 [1884]. (264) W. J. PATNODE and R. O. SAUER, General Electric, U.S. Pat. 2381138 [14.5.1942]. (265) — and F. C. SCHMIDT, *J. Amer. Chem. Soc.* **67**, 2272 [1945]. (266) L. PAULING: *The Nature of the Chemical Bond*, Cornell Univ. Press, Ithaca, N.Y. (USA) 1960, third Ed. (267) J. S. PEAKE, W. H. NEBERGALL and Y. T. CHEN, *J. Amer. Chem. Soc.* **74**, 1526 [1952]. (268) W. H. PEARLSON, T. J. BRICE and J. H. SIMONS, ibid. **67**, 1769 [1945]. (269) A. D. PETROV and E. A. CHERNYSHEV, *Doklady Akad. Nauk. SSSR* (2) **86**, 737 [1952].

(270) —, N. P. SMETANKINA and G. I. NIKISHIN, *J. Gen. Chem. USSR (Engl. transl.)* **25**, 2305 [1955]; (*Chem. Zbl.* **1956**, 10216). (271) E. P. PLUEDDEMANN, Allied Chemical, U.S. Pat. 2739165 [15.2.1952]. (272) A. POLIS, *Ber. dtsch. chem. Ges.* **19**, 1012 [1886]. (273) V. A.

PONOMARENKO and V. F. MIRONOV, *Bull. Acad. Sci. USSR, Div. Chem. Sci. (Engl. transl.)* **1954**, 423. (274) H. W. POST and C. H. HOFRICHTER JR., *J. Org. Chem.* **4**, 363 [1939]; **5**, 443 [1940]. (275) —, —, ibid. **5**, 572 [1940]. (276) B. O. PRAY, L. H. SOMMER, G. M. GOLDBERG, G. T. KERR, P. A. DIGIORGIO and F. C. WHITMORE, *J. Amer. Chem. Soc.* **70**, 433 [1948]. (277) F. P. PRICE, ibid. **70**, 871 [1948]. (278) J. PUMP and U. WANNAGAT, *Monatshefte Chem.* **93**, 352 [1962]. (279) S. D. RADOSAVLJEVIĆ, M. D. DRAGOJEVIĆ and M. S. JAĆOVIĆ, *Trans. Chem. Soc. Belgrade* **21**, 51 [1956].

(280) E. L. REILLY, C. CURRAN and P. A. MCCUSKER, *J. Amer. Chem. Soc.* **76**, 3311 [1954]. (281) H. REUTHER, *Chem. Technik* **6**, 302 [1954]; see also correction ibid. **6**, 426 [1954]. (282) J. E. REYNOLDS, *J. Chem. Soc.* [London] **89**, 397 [1906]. (283) Rhône-Poulenc, Brit. Pat. 622985 [French Prior. 29.11.1946]. (284) R. E. RICHARDS and H. W. THOMPSON, *J. Chem. Soc.* [London] **1949**, 124. (285) J. D. ROBERTS and R. H. MAZUR, *J. Amer. Chem. Soc.* **73**, 2509 [1951]. (286) E. G. ROCHOW, ibid. **70**, 2170 [1948]. (287) — and K. GINGOLD, ibid. **76**, 4852 [1954]. (288) —: *An Introduction to the Chemistry of the Silicones.* German Edition, Verlag Chemie, Weinheim/Bergstr. 1952, 2nd Ed., p. 28. (289) — and W. F. GILLIAM. *J. Amer. Chem. Soc.* **67**, 1772 [1945].

(290) K. RÜHLMANN, VEB-Berlin-Chemie, German Appl. (West German) 1163826 [23.3.1962]. (291) —, *J. prakt. Chem.* [4], **16**, 172 [1962]. (292) I. RUIDISCH and M. SCHMIDT, *Angew. Chem.* **75**, 575 [1963]. (293) T. R. SANTELLI, Libbey-Owens, French Pat. 1075762 [U.S. Prior. 29.3.1952]. (294) —, —, French Pat. 1079899 [U.S. Prior. 31.3.1952]. (295) R. O. SAUER, General Electric, U.S. Pat. 2381139 [14.5.1942]. (296) —, —, U.S. Pat. 2534149 [22.5.1947]. (297) —, *J. Amer. Chem. Soc.* **66**, 1707 [1944]. (298) —, ibid. **68**, 954 [1946]. (299) — and R. H. HASEK, ibid. **68**, 241 [1946].

(300) — and W. PATNODE, ibid. **67**, 1548 [1945]. (301) — and C. E. REED, General Electric, U.S. Pat. 2388575 [23.6.1943]. (302) H. SCHMIDBAUR, *Chem. Ber.* **97**, 830 [1964]. (303) M. SCHMIDT and H. SCHMIDBAUR, *Angew. Chem.* **70**, 657 [1958]. (304) —, —, ibid. **70**, 469 [1958]; *Chem. Ber.* **95**, 47 [1962]. (305) —, —, ibid. **93**, 878 [1960]. (306) —, — and A. BINGER, ibid. **93**, 872 [1960] (307) E SCHNELL and E. G. ROCHOW, *J. Amer. Chem. Soc.* **78**, 4178 [1956]. (308) K. SCHNURRBUSCH and W. KNIEGE, Farbenfabriken Bayer, German Pat. (West German) 1106080 [20.3.1959]. (309) G. SCHOTT and H. KELLING, *Z. Chem.* **3**, 107 [1963].

(310) — and R. SCHILD, ibid. **3**, 108 [1963]. (311) H. A. SCHUYTEN, J. W. WEAVER and J. D. REID, *J. Amer. Chem. Soc.* **69**, 2110 [1947]. (312) R. SCHWARZ and W. KUCHEN, *Chem. Ber.* **89**, 169 [1956]. (313) Z. I. SERGEEVA, D. D. TSITOVICH and M. G. VORONKOV, *Proc. Acad. Sci. USSR, Sect. Chem. (Engl. transl.)* **134**, 1195 [1960]. (314) — and SE TSZYAN-LAN', *J. Gen. Chem. USSR (Engl. transl.)* **32**, 1968 [1962]. (315) D. SEYFERTH and D. L. ALLESTON, *Inorg. Chem.* **2**, 418 [1963]. (316) J. SHERIDAN and W. GORDY, *Physic Rev.* **77**, 719 [1950]. (317) L. M. SHORR, *J. Amer. Chem. Soc.* **76**, 1390 [1954]. (318) E. R. SHULL, R. A. THURSACK and C. M. BIRDSALL, *J. Chem. Physics* **24**, 147 [1956]. (319) W. SIMMLER, Houben-Weyl, *Methoden der Organischen Chemie*, 4 Aufl. Band IV/2, Sauerstoff-Verbindungen I, 1963, p. 71.

(320) W. SIMPSON, Thomson-Houston, Brit. Pat. 635726 [21.7.1947]. (321) H. H. SISLER and R. MATTAIR, *J. Amer. Chem. Soc.* **70**, 3826 [1948]. (322) G. J. SLEDDON, ICI, Brit. Pat. 870102 [10.4.1959]. (323) B. E. F. SMITH, *Svensk. kem. Tidskr.* **61**, 213 [1949]. (324) L. H. SOMMER, Dow Corning, U.S. Pat. 2626270 [8.7.1949]. (325) —, O. F. BENNET and P. G. CAMPBELL, *Amer. Chem. Soc.*, 135th Meeting, Boston 1959, 22 M, No. 62 (cf. *Angew. Chem.* **71**, 532 [1959]). (326) —, G. T. KERR and F. C. WHITMORE, *J. Amer. Chem. Soc.* **70**, 445 [1948]. (327) —, E. W. PIETRUSZA, G. T. KERR and F. C. WHITMORE, ibid. **68**, 156 [1946].

(328) —, — and F. C. Whitmore, ibid. **68**, 2282 [1946]. (329) —, D. R. Weyenberg and P. B. Campbell, *Amer. Chem. Soc.*, 135th Meeting, Boston, April 1959, 23 M, No. 63 (cf. *Angew. Chem.* **71**, 532 [1959]).

(330) M. F. Shostakovsky, D. A. Kochkin, I. A. Shikhiev and W. M. Vlasov, *J. Gen. Chem. USSR* (*Engl. transl.*)**25**, 593 [1955]. (331) —, —, Kh. I. Kondratyev and V. M. Rogov, ibid. **26**, 3721 [1956]. (332) —, V. N. Kotrelev, D. A. Kochkin, S. P. Kalinina and V. V. Borisenko, *J. applied Chem. USSR* (*Engl. transl.*) **31**, No. 6, 944 [1958]. (333) —, M. S. Malinovsky, M. R. Romantsevich and D. A. Kochkin, *Bull. Acad. Sci. USSR, Div. Chem. Sci.* (*Engl. transl.*) **1956**, 639. (334) —, I. A. Shikhiev, D. A. Kochkin and V. I. Belyaev, *J. Gen. Chem. USSR* (*Engl. transl.*) **24**, 2175 [1954]. (335) — and V. M. Vlasov, ibid. **31**, 626 [1961]. (336) F. J. Sowa, U.S. Pat. 2477704 [27.1.1947]. (337) D. B. Sowerley, *J. Inorg. Nucl. Chem.* **22**, 205 [1961]. (338) S. B. Speck, *J. Org. Chem* **18**, 1689 [1953] (339) J. L. Speier, Dow Corning, U.S. Pat. 2611781 [Brit. Prior. 29.11.1949].

(340) —, —, German Pat. (West German) 938845 [U.S. Prior. 3.9.1952]. (341) — and L. M. Shorr, —, U.S. Pat. 2811542 [5.8.1953]. (342) M. M. Sprung and F. O. Guenther, *J. Org. Chem.* **26**, 552 [1961]. (343) P. R. Steyermark, ibid. **28**, 586 [1963]. (344) A. Stock and C. Somieski, *Ber. dtsch. chem. Ges.* **52**, 695 [1919]. (345) H. A. Stuart: *Die Physik der Hochpolymeren*, Bd. 1: Die Struktur des freien Moleküls, Springer-Verlag, Berlin-Göttingen-Heidelberg 1952. (346) D. R. Stull, *Ind. Engng. Chem.* **39**, 517 [1947]. (347) W. Sundermeyer European Research Associates S.A., German Appl. (West German) 1080077 [8.8.1957]. (348) —, Farbenfabriken Bayer, 1961, unpublished. (349) —, *Z. anorg. allg. Chem.* **313**, 290 [1962].

(350) —, ibid. **314**, 100 [1962]. (351) T. Takiguchi, *J. Amer. Chem. Soc.* **81**, 2359 [1959]. (352) —, *J. Chem. Soc. Japan, Ind. Chem. Sect.* **62**, 1262 [1959]. (353) —, *Bull. Chem. Soc. Japan* **32**, 556 [1959]. (354) —, *J. Chem. Soc. Japan, Ind. Chem. Sect.* **62**, 1261 [1959]. (355) —, *J. Org. Chem.* **23**, 1216 [1958]. (356) C. Tamborski and H. W. Post, ibid. **17**, 1400 [1952]. (357) S. Tannenbaum, S. Kaye and G. F. Lewenz, *J. Amer. Chem. Soc.* **75**, 3753 [1953]. (358) L. Tansjö, *Acta chem. scand.* **11**, 1613 [1957]. (359) —, ibid. **13**, 35 [1959].

(360) W. S. Tatlock and E. G. Rochow, *J. Amer. Chem. Soc.* **72**, 528 [1950]. (361) —, —, *J. Org. Chem.* **17**, 1555 [1952]. (362) A. G. Taylor and B. V. de G. Walden, *J. Amer. Chem. Soc.* **66**, 842 [1944]. (363) V. S. Chugunov, J. *Gen. Chem. USSR* (*Engl. transl.*) **23**, 811 [1953]. (364) L. J. Tyler, *J. Amer. Chem. Soc.* **77**, 770 [1955]. (365) —, L. H. Sommer and F. C. Whitmore, ibid. **70**, 2876 [1948]. (366) M. G. Voronkov, B. N. Dolgov and B. N. Dimitrieva, *Doklady Akad. Nauk. SSSR* (2) **84**, 959 [1952]. (367) — and Yu. I. Khudobin, *Bull. Acad. Sci. USSR, Div. Chem. Sci.* (*Engl. transl.*) **1956**, 823. (368) — and B. N. Dolgov, *J. applied Chem. USSR* (*Engl. transl.*) **24**, 103 [1951]. (369) — and V. N. Zgonnik, *J. Gen. Chem. USSR* (*Engl. transl.*) **27**, 1557 [1957].

(370) G. H. Wagner and W. G. Whitehead jr., Union Carbide, Brit. Pat. 746866 [U.S. Prior. 29.10.1952]. (371) U. Wannagat and W. Liehr, *Angew. Chem.* **69**, 783 [1957]. (372) — and F. Vielberg, *Z. anorg. allg. Chem.* **291**, 310 [1957]. (373) —, H. Bürger and E. Ringel, *Monatshefte Chem.* **93**, 1363 [1962]. (374) G. K. Weisse and R. M. Thomas, Olin Mathieson Chemical, French Pat. 1311033 [U.S. Prior. 26.1.1961]. (375) M. Weist, German Pat. (East German) 10599 [28.9.1952]. (376) —, *Chem. Techn.* **6**, 568 [1954]. (377) R. West, *J. Amer. Chem. Soc.* **76**, 6012 [1954]. (378) —, ibid. **76**, 6015 [1954]. (379) — and R. H. Baney, *J. Inorg. Nucl. Chem.* **7**, 297 [1958].

(380) H. WESTERMARK, *Svensk. kem. Tidskr.* **64**, 283 [1952]. (381) M. L. WHEELER JR., Union Carbide, French Pat. 1323121 [U.S. Prior. 7.4.1961]. (382) F. C. WHITMORE, L. H. SOMMER, P. A. DIGIORGIO, W. A. STRONG, R. E. VAN STRIEN, D. L. BAILEY, H. K. HALL, E. W. PIETRUSZA and G. T. KERR, *J. Amer. Chem. Soc.* **68**, 475 [1946]. (383) E. WIBERG and U. KRÜERKE, *Z. Naturforsch.* **8b**, 608 [1953]. (384) C. J. WILKINS, *J. Chem. Soc.* [London] **1951**, 2726. (385) D. WITTENBERG, T. C. WU and H. GILMAN, *J. Org. Chem.* **24**, 1349 [1959]. (386) A. YA. YAKUBOVICH and G. W. MOTSAREV, *J. Gen. Chem. USSR (Engl. transl.)* **23**, 1481 [1953]. (387) —, —, *Doklady Akad. Nauk. SSSR* **88**, 87 [1953]. (388) CHIANG YIN-YEN, LIN I and WANG PAO-JEN, *Acta chim. sinica* **25**, 197 [1959]. (389) L. I. ZAKHARKIN, *Bull. Acad. Sci. USSR, Div. Chem. Sci. (Engl. transl.)* **1960**, 2079.

(390) A. ZAPPEL, Tagungsbericht der Chem. Ges. i. d. DDR, Jahrestagung Leipzig 1954, Akademie Verlag, Berlin, p. 83. (391) A. ZAPPEL, *J. Amer. Chem. Soc.* **77**, 4228 [1955]. (392) M. ZIMMERMANN, Farbenfabriken Bayer, 1962, unpublished.

CHAPTER 4

Organosilanes with Organofunctional Groups

4.1 Organofunctional Silanes in Silicone Chemistry

For many years scientific and technical research in the field of silicones was concerned almost exclusively with silanes or siloxanes in which only alkyl, aryl, or aralkyl groups were attached to the silicon atom. More recently, interest has turned increasingly to organofunctional compounds, i.e., compounds with *unsaturated* or *functionally substituted* hydrocarbon radicals.

Organofunctional groups not only confer their specific reactivities on silane or siloxane molecules, but can also impart to them interesting physical properties, differing from those of the classical silicones.

The modification of *chemical* properties by organofunctional groups has been of great benefit to silicone chemistry. The fact that vinyl groups facilitate the cross-linking of silicone rubber molecules has been recognized almost from the beginning of silicone rubber technology (Section 8.1.1). Likewise, silicon-functional and organofunctional silanes and siloxanes have long been known for their adhesive properties in glass-resin bonding (Section 10.7.2). Vinylsilanes or vinylsiloxanes can be cross-linked with siloxanes containing Si—H bonds in the presence of platinum catalysts (Section 8.1.2.3.1). A particularly interesting possibility is the combination of silanes or siloxanes with organic components by means of organofunctional groups to form relatively stable structures linked through Si—C bonds. This approach has led to some surprising discoveries, such as the fact that as little as 0.5% of a hydroxy-functional siloxane can increase considerably the aging resistance of polyurethane elastomers (Section 10.6). The latent potential or organofunctional silicon compounds in plastics, dyestuffs, pharmaceuticals, and insecticides should not be overlooked.

The following examples illustrate the considerable influence that organic substituents can exert on *physical* properties: The lubricating action of polymethylphenylsiloxane oils is increased by chlorosubstitution of the phenyl radicals (Section 9.2.10), and the swelling resistance of silicone rubber is improved (Section 9.6.6) by the introduction of γ-trifluoropropyl groups. The dielectric properties of oils are considerably modified by nitrile groups. Condensates of siloxanes with polyethers in which the two partners are linked via Si—C bonds give stable aqueous solutions. These substances display interesting surface-active properties and can act both as foam stabilizers (Section 10.6) and foam suppressors, as emulsifiers (Section 8.3.2), and as emulsion-breakers (Section 10.4.3).

Processes leading to organofunctional silicon compounds are frequently rather complicated, and economic considerations could therefore militate against their use. On the other hand, there are undoubtedly applications where the performance of the products is more important than their price. It must be borne in mind that organofunctional silanes and siloxanes are active in relatively small concentrations when they are used as surfactants or to modify surfaces. The example quoted shows how their effect can be observed even when small concentrations are distributed throughout the matrix of, for example, a polymer composition.

To keep this account within the framework of silicone chemistry proper, the following discussion will survey only a selection of organofunctional silicon compounds in which consideration is devoted mainly to the organofunctionalities of current or possibly future technological interest.

The following should be borne in mind from the standpoint of silicone chemistry: Many investigations of possible syntheses, reactivities, and transformations have been carried out on silanes whose structure corresponds formally to that of neopentyl derivatives, particularly on methylsilanes of the type:

$$(CH_3)_3Si{-}CH_2{-}X$$

Since these contain only *one* functional group, they are clearly ideal model compounds for organofunctional silane chemistry. For practical applications, however, silanes containing either a number of organofunctional groups or both silicon-functional and organofunctional groups are of far greater interest. Syntheses and transformations of these types of compounds are frequently made more difficult, if not impossible, by the reactivity of the silicon-functional groups present. It is therefore often necessary to carry out syntheses and conversions of the organic functionalities on low-molecular-weight siloxanes. Their incorporation into a polymer skeleton or their conversion into individual polymers can be achieved with suitable polymerization

* Page number of chapter references.

and equilibration catalysts (Section 5.4), if the Si—O—Si bond has not been made too stable by the organofunctional substituents.

Syntheses of organofunctional siloxanes and their reactions are therefore included in the following sections.

4.2 Industrial Availability

Various methods have been used in laboratory-scale investigations on organofunctional silanes to obtain the desired compounds and study their properties (*66*, *239*). As individual substances or classes of compounds acquired technological interest, the problem of their technically rational and economically acceptable availability became more acute. It is natural that in this situation a search was begun for methods permitting as many different derivatives as possible to be obtained, in the manner of a family tree, from a *single* parent compound or by means of a *single* process, this compound and this process therefore being analogous to the methylchlorosilanes as parent compounds and the direct synthesis as a standard process, respectively.

In the present state of development it seems clear that the direct synthesis is far less important for the technical preparation of organofunctional silanes than for the synthesis of methyl- and phenylchlorosilanes. On the other hand, processes depending on the *addition of the Si—H bond* to multiple bonds, or based on *replacement of halogen atoms in chloro- and bromo-substituted organic groups* are gaining greater technical importance.

These processes represent two basically different possibilities for the synthesis of organofunctional silanes: either the organofunctional group is attached to the Si atom *with the formation of an Si—C bond* or the desired functionality is imparted to the organic group in a preformed organosilane *without increasing the number of Si—C bonds*. The Si—H addition processes represent the first route and the halogen substitution processes the second.

The following examples will illustrate this further:

1. *Addition processes*

(a) The addition of methyldichlorosilane, a by-product of the direct synthesis of methylchlorosilanes (Section 2.2), to acrylonitrile in the presence of certain catalysts gives β-cyanoethylmethyldichlorosilane preferentially (cf. Section 4.11).

$$CH_3SiHCl_2 + CH_2{=}CH{-}CN \rightarrow NCCH_2CH_2(CH_3)SiCl_2$$

which can be the parent compound for amino- and carboxy-functional silanes.

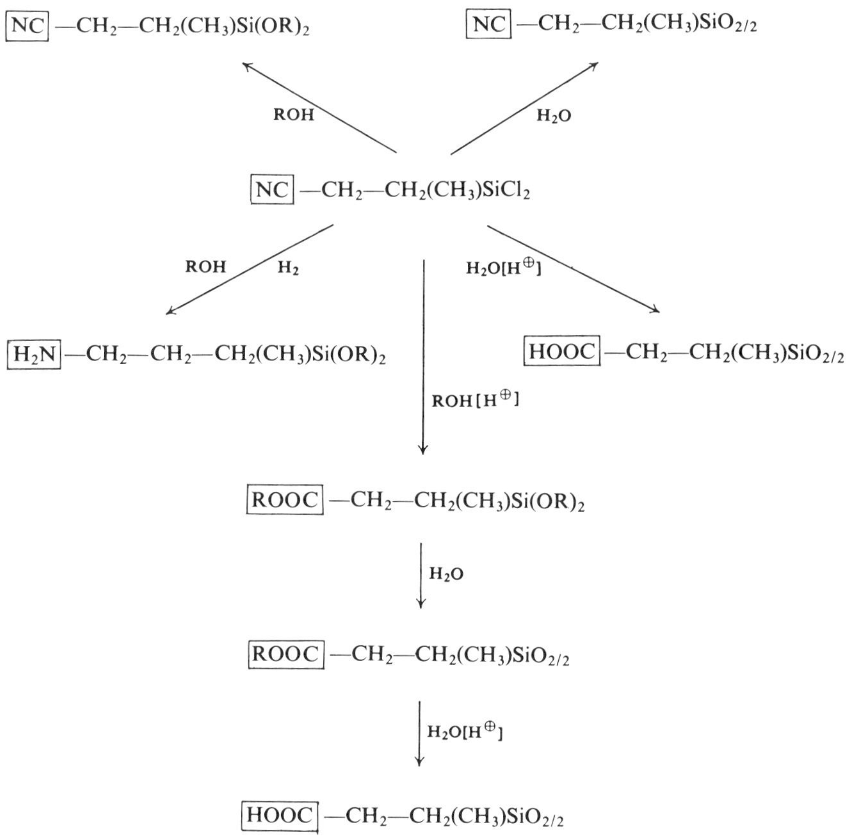

(b) The addition of a methyldialkoxysilane by its Si—H bond to allyl glycide ether gives the γ-glycidoxypropylmethyldialkoxysilane

$$CH_3{-}SiH(OR)_2 + CH_2{=}CH{-}CH_2{-}O{-}CH_2{-}\underbrace{CH{-}CH_2}_{O} \rightarrow$$

$$\underbrace{CH_2{-}CH}_{O}{-}CH_2{-}O{-}(CH_2)_3(CH_3)Si(OR)_2$$

which displays the well-known reactivities of the epoxy group. For example, it can react with alcohols, amines, and acids and thus give access to further functional compounds.

(c) The addition of methyldichlorosilane to allyl acetate gives γ-acetoxypropylmethyldichlorosilane:

$$CH_3SiHCl_2 + CH_2{=}CH{-}CH_2{-}O{-}CO{-}CH_3 \xrightarrow{Pt} CH_3COO(CH_2)_3(CH_3)SiCl_2,$$

which again occupies a key position in numerous further reactions:

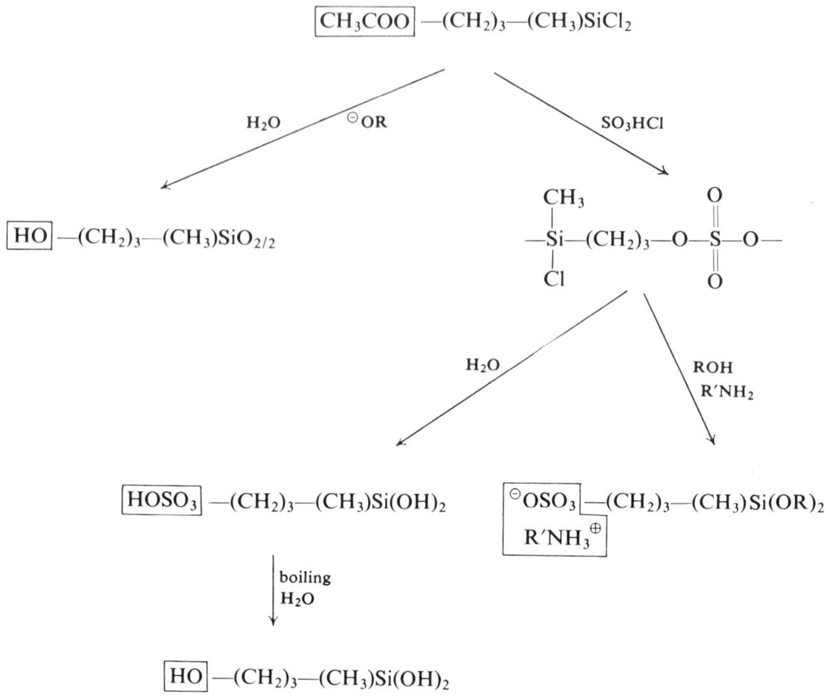

2. *Substitution processes*

The chlorobromination of trimethylchlorosilane gives bromomethyldimethylchlorosilane (Section 4.6):

$$(CH_3)_3SiCl + \tfrac{1}{2}Cl_2/Br_2 \rightarrow BrCH_2(CH_3)_2SiCl + HCl$$

The organically bound bromine can easily be replaced in various ways, as shown in the reaction scheme on p. 129:

It should finally be mentioned that the Grignard reaction is another widely applicable method for the attachment of organofunctional hydrocarbon radicals to silicon. However, in general, it is used industrially only if products not obtainable by the above processes are required.

Other methods, largely specific to individual products, will be mentioned in the following sections.

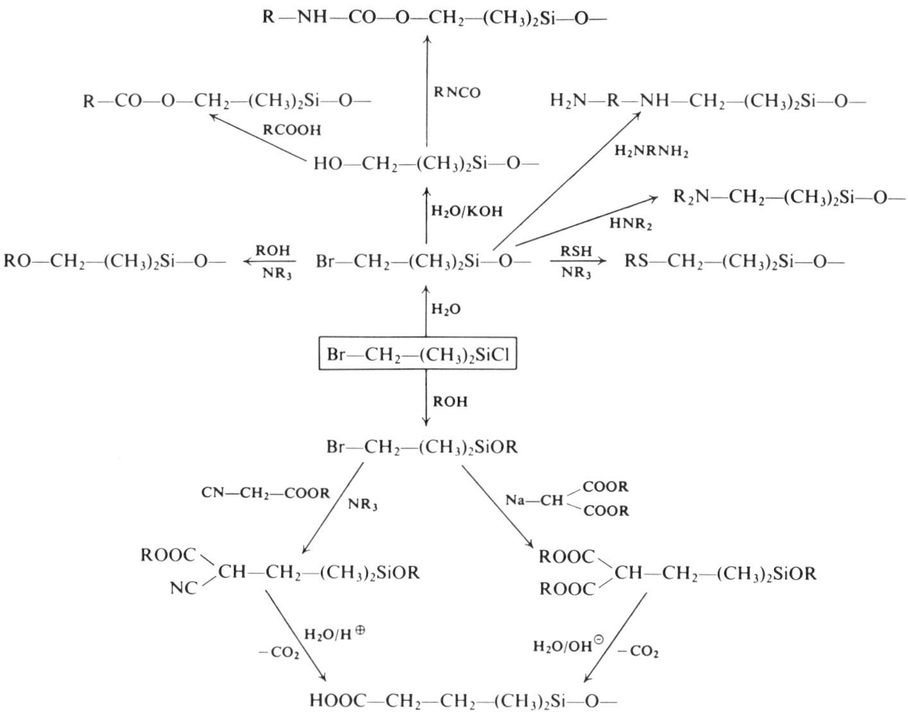

4.3 Silanes with Unsaturated Organic Groups

Vinyl- and allyl-substituted silanes are the main compounds of interest in this group. They can be produced either by reactions in which the vinyl-silicon or allyl-silicon bond is newly formed or by the transformation of organic groups in pre-formed organosilanes, in which case the number of Si—C bonds per silicon atom is not increased.

Syntheses of the first kind include direct syntheses and organometallic and other methods of substitution, as well as special addition reactions of the Si—H bond. Among the latter are dehydrohalogenation reactions.

4.3.1 Preparation by the Formation of New Si—C Bonds

4.3.1.1 *Direct Syntheses*

Vinyl chloride reacts only slightly with silicon-copper alloys at 300 to 350°C in the direct method of synthesis described in Section 2.2. Vinyltrichlorosilane and divinyldichlorosilane are formed in a ratio of about 3:1, and the yield is poor (*102*, *103*).

Alloys of silicon and nickel (*237*) or silicon and tin (*290*) are said to be more useful than the alloys of silicon and copper. Silicon-tin alloys are obtained by heating metallic silicon and tin(II) chloride (*290*). Other contact masses are prepared by the precipitation of copper oxide onto silicon (*9*). Although many variations in the experimental conditions have been tried, it has not been possible to raise the yields of vinylchlorosilanes above 30 to 35%. The direct synthesis is therefore not attractive for the production of vinylsilanes.

Allyl chloride reacts more easily than vinyl chloride with silicon-copper alloys in the 230 to 300°C region, and gives allyldichlorosilane, allyltrichlorosilane, and diallyldichlorosilane. These compounds, particularly diallyldichlorosilane, undergo pyrolysis or polymerization at the required reaction temperature (*102*, *103*). The allylchlorosilanes therefore make up only about 60% of the crude products derived from the direct synthesis. However, with silicon activated by precipitating copper sulfide onto its surface, total yields of over 80% are said to have been achieved (*179*).

With dichloroolefins as organic halide components, direct synthesis leads to mixtures of different chlorosilanes. The reaction of 1,3-dichlorobut-2-ene and silicon at 250°C gives a product from which it has been possible to isolate chlorosilanes with chloro-substituted unsaturated organic groups such as:

$$CH_3—\underset{\displaystyle Cl}{\underset{|}{C}}=CH—CH_2—SiHCl_2 \quad \text{and} \quad CH_3—\underset{\displaystyle Cl}{\underset{|}{C}}=CH—CH_2—SiCl_3$$

The chlorine atom remote from the double bond thus reacts preferentially under the conditions of the reaction. The proportion of silanes such as:

$$Cl_3Si—CH_2—CH=\underset{\displaystyle CH_3}{\underset{|}{C}}—SiHCl_2 \quad \text{and} \quad Cl_3Si—CH_2—CH=\underset{\displaystyle CH_3}{\underset{|}{C}}—SiCl_3$$

formed by the reaction of both chlorine atoms increases with increasing reaction temperature (*226*).

The following compounds:

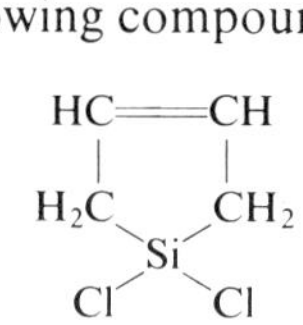

and $Cl_3Si—CH_2—CH=CH—CH_2—SiCl_3$

have been shown to be products of the reaction between 3,4-dichlorobut-1-ene and silicon-copper alloys (*191*).

1,3-Dichloroprop-1-ene does not react with a silicon-copper alloy at 270 to 280°C. Increasing the temperature to 380°C resulted in an apparently complicated reaction process, and the products contained small proportions of allyltrichlorosilane $CH_2=CH—CH_2—SiCl_3$ and propenyltrichlorosilane

$CH_3—CH═CH—SiCl_3$. 2,3-Dichloroprop-1-ene gives products containing allyltrichlorosilane and isopropenyltrichlorosilane even at low temperatures (275 to 290°C), in addition to compounds with Si—C(—C)—C—Si bonds (*197*).

4.3.1.2 *Organometallic Syntheses*

The second group of processes comprises the organometallic syntheses. The versatility of the Grignard process is again shown in this case: various substitutions can be carried out on the Si—X bond in good yield.

The synthesis of vinyl-substituted silanes is best achieved by the reaction of vinylmagnesium chloride or bromide with the Si—Cl bond, using tetrahydrofuran as the solvent (*140*, *176*, *221*).

Allylmagnesium halides have frequently been used to prepare allylsilanes from chlorosilanes and ethoxysilanes (*18*, *34*, *36*, *106*, *166*, *229*, *268*, *318*, *319*). A variant of this process consists in treating the allyl halide with magnesium and the silicon compound without first forming the Grignard compound as such (*6*, *126*, *184*, *185*, *186*). Methallyl halides can be used similarly (*34*, *190*, *292*).

p-Styryltrimethylsilane can be obtained in 76% yield by a Grignard reaction between *p*-styrylmagnesium chloride or bromide and trimethylchlorosilane in tetrahydrofuran (*182*, *202*, *230*).

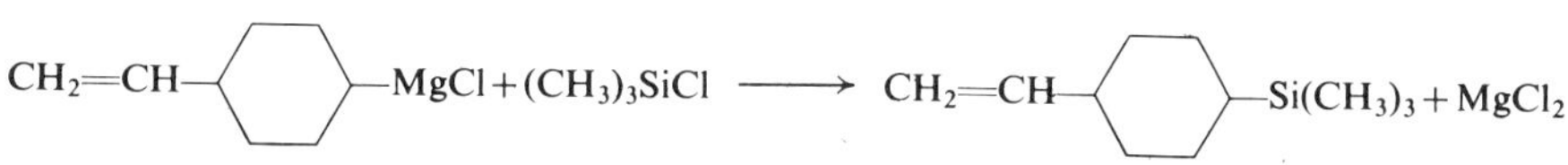

β-Styryltrimethylsilane (*256*) and cyclopentadienyltrimethylsilane (*60*):

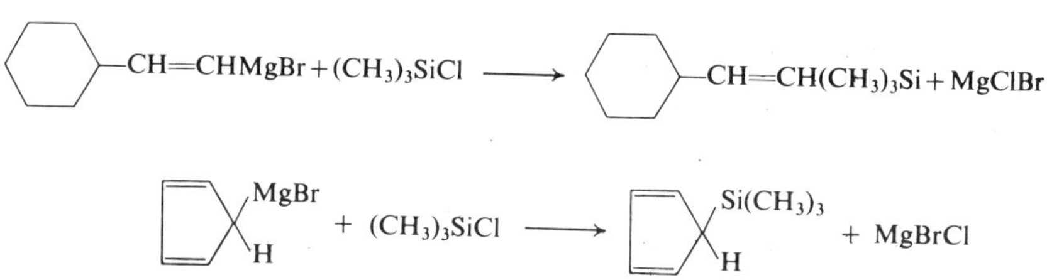

have also been obtained by Grignard reactions.

In addition to Grignard reagents, organo-alkali-metal compounds have been found suitable for attaching unsaturated hydrocarbon radicals to silicon. β-Styryltriphenylsilane has been produced from β-styryllithium and triphenylchlorosilane (*77*), and indenylsilane from indenylsodium and indenyllithium (*72*, *264*).

4.3.1.3 *Addition Reactions of the Si—H Bond*

Silanes with unsaturated organic groups can be obtained in an elegant manner by the addition of the Si—H bond to *alkynes* or *polyolefins*. These are typical cases of radical reactions which can be catalyzed by platinum, chloroplatinic acid, palladium, peroxides, uv light and azodinitriles.

Trichlorosilane can add to aliphatic alk-1-ynes:

$$SiHCl_3 + RC{\equiv}CH \rightarrow RCH{=}CHSiCl_3$$

Yields of 50 to 60 mole-% referred to the trichlorosilane used were obtained with *acetylene* at 175°C under pressure and using platinized asbestos (*311*) or platinized charcoal (*308*) as a catalyst. Homologs (*1*, *21*) and derivatives (*37*, *123*) of acetylene react analogously, e.g.,

$$CH_3{-}CH_2{-}CH_2{-}C{\equiv}CH + SiHCl_3 \rightarrow CH_3{-}CH_2{-}CH_2{-}CH{=}CH{-}SiCl_3$$

$$ClC{\equiv}CCl + SiHCl_3 \rightarrow ClCH{=}CCl{-}SiCl_3$$

Platinum catalysts are the most suitable promoters for the addition reaction with homologs of acetylene and give yields of 70 to 90%, while peroxides give only moderate yields (40 to 50%). Moreover, the addition reactions follow different stereochemical routes, platinum catalysts giving *cis*-addition with the formation of *trans*-products (*22*):

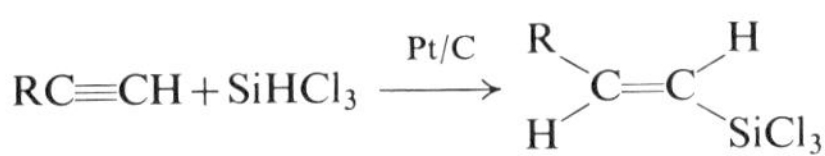

while peroxides give *trans*-addition with the formation of *cis*-products (*22*):

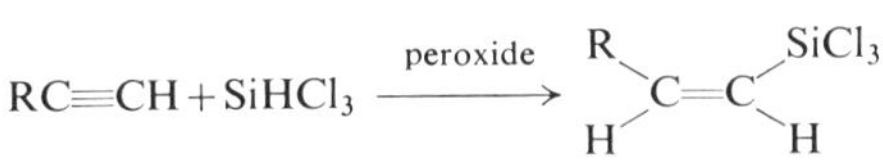

1,2-Bis(trichlorosilyl)-ethane, which is formed from vinyltrichlorosilane and trichlorosilane:

$$CH_2{=}CH{-}SiCl_3 + SiHCl_3 \rightarrow Cl_3Si{-}CH_2{-}CH_2{-}SiCl_3$$

was obtained in only negligible yield (a few per cent) by the peroxide-catalyzed reaction of acetylene with trichlorosilane (*39*).

Platinum (*210*) and palladium catalysts (*234*, *235*) have also proved useful for the addition of *methyldichlorosilane* to acetylene according to the following equation:

$$CH_3SiHCl_2 + HC{\equiv}CH \rightarrow (CH_2{=}CH)CH_3SiCl_2$$

Using chloroplatinic acid as the catalyst, methyldichlorosilane has been added to *vinylacetylene* in 50% yield and trichlorosilane in a yield of 49% on the silane (*195*):

$$CH_3SiHCl_2 + HC{\equiv}C{-}CH{=}CH_2 \rightarrow CH_2{=}CH{-}CH{=}CH{-}(CH_3)SiCl_2$$

$$SiHCl_3 + HC{\equiv}C{-}CH{=}CH_2 \rightarrow CH_2{=}CH{-}CH{=}CH{-}SiCl_3$$

Under these conditions, partial homopolymerization of the reaction products cannot be avoided (*195*).

The reaction of trichlorosilane with *butadiene* in the presence of a platinized charcoal catalyst resulted in 1,4-addition to give crotyldichlorosilane in 44% yield [referred to the trichlorosilane (*15*, *308*)]:

$$SiHCl_3 + CH_2{=}CH{-}CH{=}CH_2 \rightarrow CH_3{-}CH{=}CH{-}CH_2{-}SiCl_3$$

However, the addition of another molecule of $SiHCl_3$ gives 1,3-bis(trichlorosilyl)butane, $CH_3{-}\underset{\displaystyle SiCl_3}{\underset{|}{CH}}{-}CH_2{-}CH_2{-}SiCl_3$, as well. Trichlorosilane also adds to allyl acrylate in the presence of platinum catalysts (*208*):

$$CH_2{=}CH{-}COO{-}CH_2{-}CH{=}CH_2 + SiHCl_3 \rightarrow CH_2{=}CH{-}COO{-}(CH_2)_3{-}SiCl_3$$

4.3.1.4 *Condensation Reactions of the Si—H and Si—Cl Bonds*

Besides substitutions by means of organometallic compounds, a number of other possible substitutions have recently been examined. These lead to compounds with unsaturated organic groups by the formation of Si—C bonds and involve the Si—H or Si—Cl bond.

The *Si—H bond*, which usually undergoes addition reactions with *olefins* or *chloroolefins* (Sections 2.4.1, 4.2, and 4.3.1.3), can condense with these reactants under certain conditions. These reactions proceed at considerably higher temperatures than the addition reactions.

In this way the reaction of trichlorosilane with a slight excess of *vinyl chloride* for a short time (30 seconds) at 560 to 580°C has given a 60 to 80% yield (referred to trichlorosilane) of vinyltrichlorosilane (*41*, *83*, *146*, *313*).

$$CH_2{=}CHCl + SiHCl_3 \rightarrow CH_2{=}CH{-}SiCl_3 + HCl$$

Optimum conditions have also been found for the reaction of methyldichlorosilane with vinyl chloride; with a residence time of 30 seconds at 530 to 540°C, methylvinyldichlorosilane is obtained in 75 to 80% yield, referred to the original silane (*83*):

$$CH_2{=}CHCl + CH_3SiHCl_2 \rightarrow CH_2{=}CH{-}(CH_3)SiCl_2 + HCl$$

The reactivity decreases with the increasing number of methyl groups on the Si atom. Dimethylchlorosilane and vinyl chloride gave dimethylvinylchlorosilane in only 45% yield at 550°C (*212*). Trialkylsilanes do not undergo this reaction (*41*, *146*); there is only an exchange of H and Cl. Triethylsilane gave ethylene and triethylchlorosilane (*146*):

$$(C_2H_5)_3SiH + CH_2{=}CHCl \rightarrow (C_2H_5)_3SiCl + CH_2{=}CH_2$$

The yields for the corresponding reaction of allyl chloride with silanes containing Si—H bonds are lower (between 5 and 40%) (*41*, *55*, *212*).

The Si—H bond also condenses with olefins at elevated temperatures under conditions similar to those for the chloroolefins (*41*, *146*, *148*), for example:

$$CH_2{=}CH_2 + SiHCl_3 \rightarrow CH_2{=}CH{-}SiCl_3 + H_2$$

Widely differing yields—between 10 and 80%—are quoted for this reaction (*41*, *146*, *148*). Styrene and trichlorosilane give only a 15% yield of β-styryltrichlorosilane at 600°C (*149*).

Another condensation reaction taking place with halogens bound to silicon consists in the reaction of alkylchlorosilanes with sodium acrylate or methacrylate, for example (*4*):

$$(C_2H_5)_3SiCl + CH_2{=}CH{-}COONa \rightarrow (C_2H_5)_3SiOOC{-}CH{=}CH_2 + NaCl$$

A similar reaction can be carried out on a halogen atom which is a substituent of an alkyl group on the silane (see Section 4.3.2.2).

4.3.2 Preparation by the Transformation of Organic Groups

4.3.2.1 *Dehydrohalogenation of Haloalkylsilanes*

The dehydrohalogenation of haloalkyl groups in haloalkylsilanes is particularly important for the preparation of vinyl- and allylsilanes. The reaction is carried out in the presence of a base which will act as a hydrogen-halide acceptor, or a catalyst promoting the elimination of hydrogen halide.

Quinoline and isoquinoline are suitable bases for binding the hydrogen halide. Hydrogen halide is eliminated from β-haloethylsilanes by quinoline (*258*):

$$ClCH_2{-}CH_2{-}SiCl_3 \xrightarrow{C_9H_7N} CH_2{=}CH{-}SiCl_3$$

A number of γ-chloroalkylsilanes have been dehydrohalogenated with isoquinoline (*206*):

$$ClCH_2{-}CH_2{-}CH_2{-}SiCl_3 \xrightarrow{\text{iso-}C_9H_7N} CH_2{=}CH{-}CH_2{-}SiCl_3$$

On the other hand, chloromethylalkylsilanes are converted into alkylchlorosilanes, with cleavage at the Si—C bond, on boiling with quinoline (*53*):

$$ClCH_2(CH_3)_3Si \rightarrow (CH_3)_3SiCl$$

In general, the dehydrohalogenation reaction competes with the so-called "β-decomposition." The former will preponderate if the silicon carries strongly electronegative substituents as in the case of chloroethyltrichlorosilane (cf. Sections 4.5.4 and 6.4).

While acid acceptors must of course be used in stoichiometric quantities, catalysts—the most important is aluminum chloride—are effective even in small amounts. Their optimum range of concentrations is given as 0.1 to 10% (*15*, *310*). Not only vinyltrichlorosilane (*310*) but also allyl- or β-methylvinyltrichlorosilane have been obtained from chloropropyltrichlorosilanes (*15*) in the presence of aluminum chloride as catalyst.

Side reactions that may take place with an $AlCl_3$ catalyst are the addition of HCl to the unsaturated organic group (*310*), i.e., a back-displacement of the equilibrium, on the one hand, and a cleavage of the Si—C bond (β-decomposition, Section 4.5.4) on the other. The latter was established by using stoichiometric quantities of aluminum chloride in the following reaction as an example (*258*):

$$ClCH_2{-}CH_2(C_2H_5)_2SiCl \xrightarrow{AlCl_3} CH_2{=}CH_2 + (C_2H_5)_2SiCl_2$$

The back-addition of HCl can be prevented by the continuous removal by distillation of the alkenylsilanes formed.

In principle, dehydrohalogenation can be carried out on silanes containing organic groups with a *number of chlorine atoms*, by methods similar to those employed with monochlorinated organic groups. Bases (*252*, *309*) and catalysts (*187*, *188*) can again be used. Because of the low reactivity of the α-Cl, quinoline converts α,β-dichloroethyltrichlorosilane into α-chlorovinyltrichlorosilane (*2*, *252*):

$$ClCH_2{-}CHCl{-}SiCl_3 \xrightarrow{\text{quinoline}} CH_2{=}CCl{-}SiCl_3$$

The conversion of β,β-dichloroethyltrichlorosilane into β-chlorovinyltrichlorosilane has been achieved with aluminum chloride catalysts (*187*, *188*):

$$Cl_2CH{-}CH_2{-}SiCl_3 \xrightarrow{AlCl_3} ClCH{=}CH{-}SiCl_3$$

Aliphatic dinitriles have a catalytic effect on the elimination of hydrogen chloride from polychloroalkylhalosilanes containing a β-chlorine atom and at least one other attached to a neighboring carbon atom (*309*).

Powdered silicon, ferrosilicon, copper, and iron have also been used as hydrogen-halide acceptors (*164*, *165*). The reaction takes place even under mild conditions near or below the boiling point of the chloroalkylsilanes (*164*):

$$\begin{aligned} CH_3{-}CHCl{-}CH_2{-}SiCl_3 \xrightarrow{Fe/Si} & \; CH_2{=}CH{-}CH_2{-}SiCl_3 \; (89\%) \\ & + CH_3{-}CH{=}CH{-}SiCl_3 \; (7.6\%) \end{aligned}$$

The process is particularly suitable for the preparation of allyl- and crotyltrichlorosilanes. It is again evident here how the ease of elimination of the hydrogen chloride is affected by the other substituents on the silicon atom. Dehydrochlorination proceeds most easily in the case of the chloroalkyltrichlorosilanes, and is made more difficult by increasing replacement of the chlorine atoms on the silicon by methyl groups. Methylvinyldichlorosilane could not be prepared by this method (*165*).

A variant of the synthesis of alkenylsilanes by dehydrohalogenation is the *pyrolytic* cleavage of monochloroorganosilanes at elevated temperatures. The vaporized starting material is heated to about 500°C and passed over unglazed porcelain (*97*). The principle of pyrolysis can be combined with that of displacing the position of the equilibrium by means of hydrogen-halide acceptors, by passing haloalkylsilanes at about 500°C over silicon, a silicide, or a silicon alloy which binds the hydrogen halide arising from the pyrolysis by the formation of halosilanes (*162*, *163*).

4.3.2.2 *Substitutions in the Organic Groups*

Another way of producing silanes with unsaturated groups is to make substitutions in functional organic groups, In practice, chloroalkyl- or bromoalkylsilanes are preferred. A number of examples are given below.

Silanes or siloxanes containing chloromethyl groups react with acrylic acid in the presence of triethylamine as acid acceptor and hydroquinone as polymerization inhibitor. Acryloxymethylmethyldiethoxysilane (*138*, *139*) can be obtained in this way:

$$ClCH_2(CH_3)Si(OC_2H_5)_2 + CH_2{=}CH{-}COOH \xrightarrow{(C_2H_5)_3N} CH_2{=}CH{-}COO{-}CH_2(CH_3)Si(OC_2H_5)_2$$

The same result can be achieved by reaction with the alkali-metal salts of the carboxylic acids (*7*, *139*):

$$ClCH_2Si(OC_2H_5)_3 + CH_2{=}C(CH_3)COOK \rightarrow CH_2{=}C(CH_3)COOCH_2Si(OC_2H_5)_3 + KCl$$

$$[ClCH_2(CH_3)_2Si]_2O + 2CH_2{=}C(CH_3)COOK \rightarrow [CH_2{=}C(CH_3)COOCH_2(CH_3)_2Si]_2O + 2KCl$$

The Grignard synthesis is naturally also available for substitutions on the Si—R—X bond (*190*):

$$ClMgCH_2(CH_3)_3Si + CH_2{=}CH{-}CH_2Br \rightarrow CH_2{=}CH{-}CH_2{-}CH_2(CH_3)_3Si + MgClBr$$

but its application is limited. β-Halogenated compounds cannot be used because of the sensitivity of the Si—C bond to cleavage (Section 4.3.4), nor can siloxanes, because of the similar tendency of the Si—O bond (Section 5.5.5) nor many silanes also having a silicon-functional group because of the reactivity of the silicon-functional atoms toward the Grignard reagent.

Starting with hydroxymethyl groups on the Si atom, substitutions can also be achieved by reactions with maleic anhydride (*170*) or methacrylyl chloride (*5*) in the presence of tertiary amines.

4.3.3 Physical Constants (Table)

Physical constants of a number of selected silanes with unsaturated organic groups are given in Table 22.

TABLE 22

SILANES WITH UNSATURATED ORGANIC GROUPS[a, b]

Compound	Bp (°C) at	p (mm Hg)	n_D^t at	t (°C)	d_4^t (gm/cc) at	t (°C)
$CH_2{=}CH(CH_3)_3Si$	54.4 (*166*)	745	1.3880 (*166*)	25	0.6865 (*166*)	25
$(CH_2{=}CH)_2(CH_3)_2Si$	80.0–81.6 (*111*)	760	1.4176 (*111*)	20	0.7337 (*111*)	20
$(CH_2{=}CH)_2SiCl_2$	118–119 (*102*)	760	—	—	1.088[c] (*102*)	27
$CH_2{=}CHSiCl_3$	92 (*102*)	760	—	—	1.264[c] (*102*)	27
$CH_2{=}CHSiHCl_2$	66.5 (*290*)	760	—	—	—	—
$CH_2{=}CH(CH_3)SiCl_2$	92–93 (*102*)	760	—	—	1.085[c] (*102*)	25
$CH_2{=}CHSi(OC_2H_5)_3$	62.5–63 (*167*)	20	1.3960 (*167*)	25	0.9027 (*167*)	25
$CH_2{=}CH(CH_3)Si(OC_2H_5)_2$	133–134 (*45*)	760	1.4000 (*45*)	20	0.858 (*45*)	20
$CH_2{=}CH(CH_3)_2SiOC_2H_5$	99 (*45*)	760	1.3983 (*45*)	20	0.790 (*45*)	20
$CH_2{=}CH{-}CH_2(CH_3)_3Si$	85.4 (*20*)	752	1.4075 (*20*)	20	0.7195 (*20*)	20
$(CH_2{=}CH{-}CH_2)_2(CH_3)_2Si$	135.5 (*20*)	760	1.4420 (*20*)	20	0.7679 (*20*)	20
$(CH_2{=}CH{-}CH_2)_3CH_3Si$	180.3 (*20*)	745	1.4662 (*20*)	20	0.8055 (*20*)	20
$(CH_2{=}CH{-}CH_2)_4Si$	103.4 (*20*)	15	1.4864 (*20*)	20	0.8353 (*20*)	20
$(CH_2{=}CH{-}CH_2)_3SiCl$	97–98 (*229*)	30	1.4779 (*229*)	25	0.941 (*229*)	25
$CH_2{=}CH{-}CH_2(CH_3)_2SiCl$	111–113 (*102*)	760	—	—	0.922[c] (*102*)	27
$(CH_2{=}CH{-}CH_2)_2SiCl_2$	65–66 (*229*)	26	—	—	—	—
$(CH_2{=}CH{-}CH_2)CH_3SiCl_2$	119–120 (*102*)	760	—	—	1.057[c] (*102*)	27
$CH_2{=}CH{-}CH_2SiCl_3$	117.5 (*102*)	760	—	—	1.211[c] (*102*)	27
$CH_2{=}CH{-}CH_2SiHCl_2$	97 (*102*)	760	—	—	1.086[c] (*102*)	27
$C_6H_5CH{=}CH(CH_3)_3Si$	98 (*256*)	10.5	1.5270 (*256*)	20	0.888 (*256*)	20
p-$CH_2{=}CHC_6H_4(CH_3)_3Si$	91 (*202*)	10	1.5250 (*182*)	20	0.8913 (*182*)	20
$(CH_2{=}CH{-}CH_2)_2CH_3SiH$	122.5 (*185*)	745	1.4430 (*185*)	20	0.7630 (*185*)	20
$(CH_2{=}CH{-}CH_2)_3SiOC_2H_5$	127–129 (*293*)	84	1.4569 (*293*)	25	0.8543 (*293*)	25
$(CH_2{=}CH{-}CH_2)_2Si(OC_2H_5)_2$	189.5 (*293*)	741	1.4316 (*293*)	25	0.8787 (*293*)	25
$CH_2{=}CH{-}CH_2Si(OC_2H_5)_3$	175.8 (*293*)	740	1.4063 (*293*)	25	0.9032 (*293*)	25

(*continued on p.* 138)

TABLE 22 *(Contd.)*

SILANES WITH UNSATURATED ORGANIC GROUPS[a, b]

Compound	Bp (°C) at	p (mm Hg)	n_D^t at	t (°C)	d_4^t (gm/cc) at	t (°C)
$CH_2{=}CH{-}CH_2(CH_3)Si(OC_2H_5)_2$	155 (*129*)	760	1.4104 (*129*)	20	0.8629 (*129*)	20
$(CH_2{=}CH{-}CH_2)_2CH_3SiOC_2H_5$	165 (*129*)	760	1.4368 (*129*)	20	0.8350 (*129*)	20
$CH_2{=}CH{-}CH_2(CH_3)_2SiOC_2H_5$	121 (*129*)	760	1.4094 (*129*)	20	0.8065 (*129*)	20
$(CH_2{=}CH{-}CH_2)_2CH_3SiOH$	34.5–35.5 (*232*)	1	1.4598 (*232*)	20	0.8803 (*232*)	22
$(CH_2{=}CH{-}COOCH_2)CH_3Si(OC_2H_5)_2$	111 (*138*)	2.5	1.4214 (*138*)	25	0.986 (*138*)	25

[a] Numbers in parentheses indicate reference numbers.
[b] For siloxanes see Tables 36 and 45.
[c] d_{27}^t.

4.3.4 Chemical Properties

Silanes with unsaturated organic groups have acquired technical importance in a number of different ways. By incorporating siloxane units with unsaturated organic groups into otherwise poorly reactive organosiloxane polymers, "labeled" reactive sites can be created upon which, for example, cross-linking can readily be initiated. The *reactivity* of alkenyl groups on the Si atom has long been utilized in silicone rubber technology, by incorporating small amounts of methylvinylsiloxane units in a random distribution into polydimethylsiloxane chains by copolymerization (Section 5.4.3.1). On the other hand, silanes with unsaturated groups can also undergo *homopolymerization*. Silane- or siloxane-modified hydrocarbon polymers can thus be formed on the following principle:

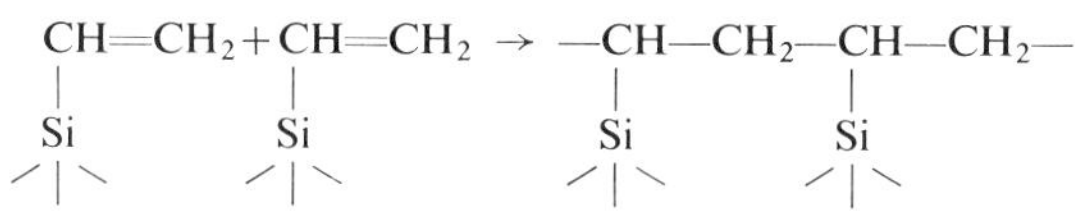

The possibility of *copolymerization* with unsaturated organic compounds has also aroused interest at various times.

The *ability of the Si—H bond to add to the vinyl—or allyl—Si bond*, which is promoted by peroxidic (*39*) or platinum catalysts (*308*) has acquired technological importance (see also Section 7.2.4):

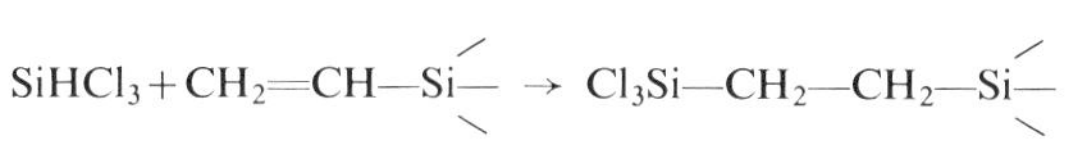

If the Si—H and Si—vinyl bonds are components of siloxane units belonging to siloxane molecules, cross-linking reactions which take place at relatively low temperatures are made possible and these have led to the development of certain low-temperature-vulcanizing (LTV) types of rubber (Section 8.1.2.3.1).

Vinyl- and allylsilanes undergo further reactions well known from organic chemistry. They *add halogens* and *hydrogen halides* (*310*). The addition of hydrogen halide to allylsilanes and higher alkenylsilanes follows Markovnikov's rule, i.e., the halogen atom adds to the carbon atom with the smaller number of hydrogen atoms. In contrast, with vinylsilanes the halogen atom adds to the β-carbon having the greater number of hydrogen atoms.

With *ozone* vinyl- and allylsilanes form ozonides (*15, 256, 310*), which break down on hydrolysis into silanols and formaldehyde.

The vinyl group in trialkylvinylsilanes can be converted into an *epoxy group* with peracetic or perphthalic acid (*52*):

Vinylchlorosilanes undergo *Diels-Alder reactions* with, for example, butadiene, cyclopentadiene (*310*), and hexachlorocyclopentadiene (Section 4.5.2.2).

$$CH_2{=}CH{-}CH{=}CH_2 + CH_2{=}CH{-}SiCl_3 \longrightarrow$$

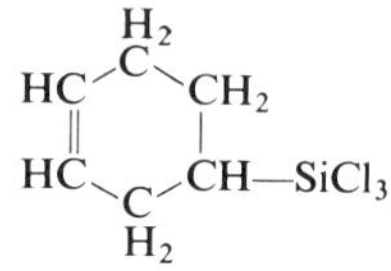

A *Friedel-Crafts reaction* takes place with benzene in the presence of aluminum chloride (*10*, *310*):

$$CH_2{=}CH(CH_3)SiCl_2 + C_6H_6 \xrightarrow{AlCl_3} C_6H_5CH_2CH_2(CH_3)SiCl_2$$

The reactivity of the silanes here decreases in the sequence:

$$CH_2{=}CHSiCl_3 > CH_2{=}CH(CH_3)SiCl_2 > CH_2{=}CH(CH_3)_2SiCl$$

Trimethylvinylsilane gave no adduct (*10*). Details of other reactions of vinylsilanes will be found in ref. (*236*).

The capacity of the alkenylsilanes for undergoing *homopolymerization* depends on the nature of the alkenyl groups, their number, the nature of the other substituents on the Si atom, and the particular type of catalyst. The relationships are not always absolute clear. In general, the capacity of homopolymerization becomes greater with increasing distance of the double bonds from the silicon atom (*165*). Allylsilicon compounds generally undergo homopolymerization more readily than the corresponding vinyl compounds. The tendency to homopolymerization rises with increasing number of alkenyl groups on the Si atom. The homopolymerization reaction is promoted by radical and ionic catalysts.

Vinyltriethylsilane polymerized to only a slight extent in the presence of *t*-butyl peroxide at normal pressure (*120*, *306*), but gave an approximately 50% yield at 6000 atm with the same peroxide (*120*). Polymer yields of around 75% were achieved with *vinyltrimethylsilane* under the same conditions (*120*). Ziegler catalysts have hardly any influence on vinyltrimethylsilane, but polymer yields of 80 to 90% have been achieved at 0 to 50°C in the presence of 2 to 10% of ethyllithium dissolved in *n*-heptene (*298*).

Vinyltrichlorosilane gives only polymers of very low molecular weight both in the absence and in the presence of catalysts on heating to 250 to 350°C while vinyltriethoxysilane can be converted easily and almost completely in the presence of peroxides into polymers having molecular weights of 5000 to 8000 (*310*). It is interesting to note that *vinylsilanes* containing *Si—H bonds* polymerize in the presence of alkali-metal alkoxides in the complete absence of water according to the equation:

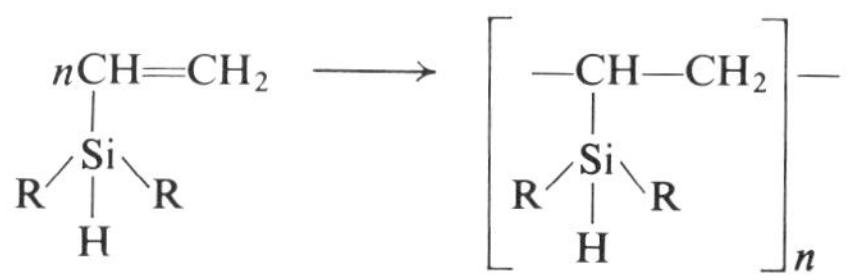

without rupture of the Si—H bond or its addition to the double bond (*48*).

Allyltrimethylsilane (*144*) and allyltri-*n*-butylsilane (*119*) hardly polymerize with peroxides under normal conditions. The latter compound is also unaffected by ionic catalysts of the type $AlCl_3$ and BF_3 (*119*), while allyltrimethylsilane gives only low molecular polymers in their presence (*268*). It has, however, been polymerized with Ziegler catalysts in inert solvents between 30 and 150°C to give high-molecular-weight products which were linear and isotactic and had a crystallinity of 40% (*168*).

The polymerization of *diallyldimethylsilane* has been investigated more thoroughly (*27*, *118*, *133*). Ziegler catalysts (triethylaluminum or triisobutylaluminum and titanium tetrachloride in hexane or heptane at 30 to 80°C) gave polymers consisting of a series of linearly linked rings.

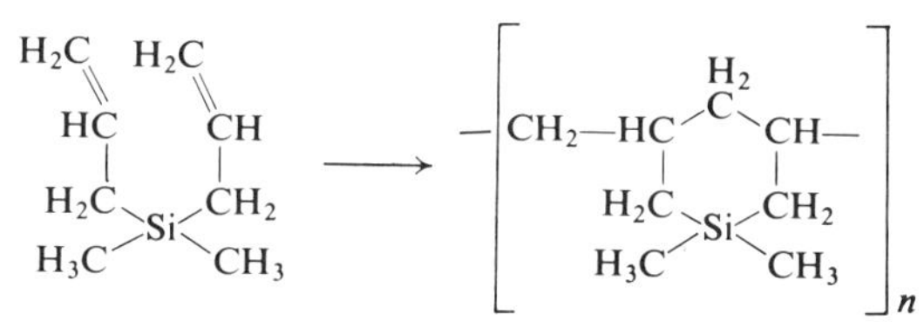

Diallyldimethylsilane thus behaves analogously to diallyldiethylammonium bromide

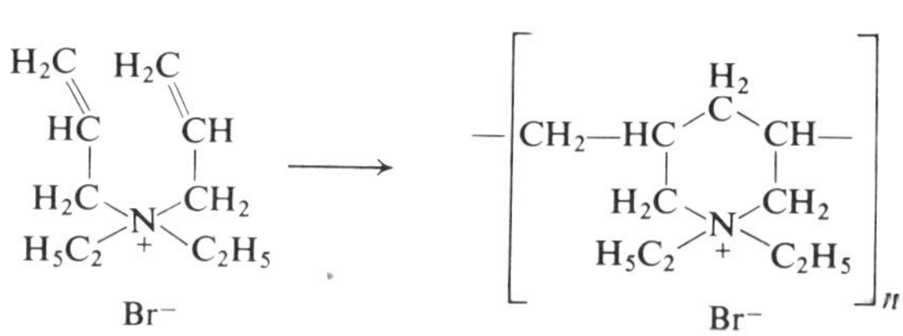

or hexa-1,5-diene and hepta-1,6-diene (*133*).

Allyltriethoxysilane, *diallyl-* and *allylmethyldiethoxysilanes*, and *polyallylsiloxanes* (*104*, *293*, *294*, *302*, *318*) have been polymerized with peroxide catalysts. Polydiallylsiloxanes behave like drying oils (*104*). The three allylethoxysilanes $(CH_2{=}CH{-}CH_2)_nSi(OC_2H_5)_{4-n}$ ($n = 1, 2,$ or 3) even polymerize without catalysts on prolonged storage at room temperature (*128*). Polymeric residues are always obtained on distillation or redistillation, since polymerization is accelerated by elevated temperatures (*128*).

Silanes or siloxanes containing acryloxy or methacryloxy groups undergo homopolymerization on heating (*4*). *p-Styryl derivatives* have been homopolymerized using azodiisobutyronitrile (*89*, *121*), *t*-butyl peroxide (*121*), or diacetyl peroxide (*202*) as catalyst. Even when the *p*-styrylsilanes simultaneously contain vinyl or allyl groups, polymerization will proceed only through the styryl group, so that polymers with side-chain vinyl or allyl groups are obtained (*89*):

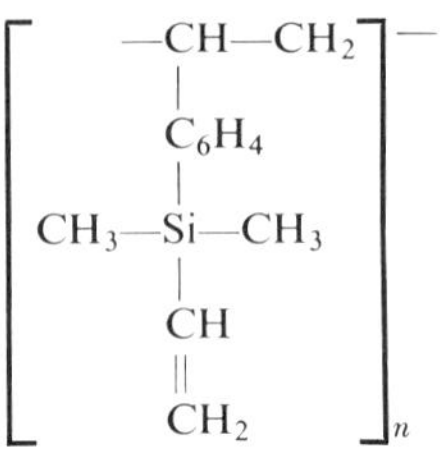

In addition to homopolymerization, the question of *copolymerization* with unsaturated organic monomers or prepolymers has gained particular interest. This will be discussed further in Section 7.4.6.

The *silicon-functional* groups in alkenyl-substituted silanes behave basically like those in alkylsilanes. Vinyltrichlorosilane has been converted into vinyltrifluorosilane with HF (*107*), ZnF_2 (*228*), and antimony trifluoride (*147*). The corresponding acetoxy- and isocyanatosilanes have been obtained from vinyltrichlorosilane with silver salts of carboxylic acids or silver isocyanate (*3*).

Alkenylchlorosilanes and alkenylethoxysilanes can, like the corresponding methyl- or ethylsilanes be converted into siloxanes by hydrolysis. The allyl—Si bond is however sensitive toward halogen halides and sulfuric acid (*268*). In the hydrolysis of allyltrichlorosilane, therefore, a weak base is used as an acid acceptor (*15*).

4.4 Fluoroorganosilanes

The desire to combine in one and the same compound the chemical stability of the C—F bond, well known from the polyfluoroethylenes, with the thermal stability of the Si—O bond has repeatedly led to the preparation of silanes and siloxanes with fluoroorganic substituents.

4.4.1 Preparation by the Formation of New Si—C Bonds

4.4.1.1 *Direct Syntheses*

The transfer of the process of direct synthesis from methylsilanes to trifluoromethylsilanes, has not been successful; trifluoromethyl chloride does not react with silicon or a silicon-copper alloy below 400°C; an exothermic reaction between 500 and 1000°C leads only to chlorofluorosilanes (*105*). The passage of 1,1,2-trichlorotrifluoroethane over a 75:25 mixture of silicon and copper at 300°C gives trifluorovinyltrichlorosilane $CF_2{=}CF{-}SiCl_3$ in about 25% yield. However, more than 50% of chlorotrifluoroethylene, besides other halohydrocarbons and $SiCl_4$, are formed simultaneously through dehalogenation (*160*).

4.4.1.2 *Organometallic Syntheses*

The Grignard synthesis can be used very widely for the preparation of fluoroorganosilanes. Trifluoromethylhalosilanes have been prepared from trifluoromethylmagnesium iodide and silicon tetrachloride (*93*, *94*). Other examples of compounds obtained in this way are $(CF_3{-}CH_2{-}CH_2)_nSiCl_{4-n}$ (*312*), $(CF_3{-}CF_2{-}CF_2{-}CH_2{-}CH_2)_3SiCl$, and $(CF_3{-}CF_2{-}CF_2{-}CH_2{-}CH_2)_2Si(OC_2H_5)_2$ (*200*), trifluoromethylphenylsilanes (*62*, *115*, *116*, *219*), and 4-chloro-3-trifluoromethylphenylsilanes (*216*). Trifluoromethylphenylsilanes have also been prepared by an organolithium synthesis (*117*).

4.4.1.3 *Addition Reactions of the Si—H Bond*

Addition reactions of the Si—H bond to fluoro-substituted olefins are very suitable for the synthesis of fluorosilanes, and have therefore received much attention. Because of the nature of this reaction—one hydrogen atom always remains in the alkyl group—it is impossible to produce silanes with perfluoroalkyl groups and these are made by the Grignard reaction.

Catalysts such as platinized charcoal (*193*, *211*, *297*), chloroplatinic acid (*210*) or its salts, tertiary amines (*213*), ozone (*326*), or irradiation with light (*67–70*, *95*) are used, as in the addition reactions of the corresponding fluorine-free olefins. Heating from 400 to 600°C without catalysts is also said to give the desired result (*90*, *91*), but yields for this method have not been disclosed.

The following are examples of the addition reactions discussed above:

$$CF_2{=}CF_2 + CH_3SiHCl_2 \rightarrow HCF_2{-}CF_2(CH_3)SiCl_2 \quad (69) \tag{1}$$

$$CFCl{=}CF_2 + SiHCl_3 \rightarrow HCFCl{-}CF_2SiCl_3 \quad (96) \tag{2}$$

$$CF_2{=}CH_2 + CH_3SiHCl_2 \rightarrow HCF_2{-}CH_2(CH_3)SiCl_2 \quad (210) \tag{3}$$

$$CF_3{-}CH{=}CH_2 + SiHCl_3 \rightarrow CF_3{-}CH_2{-}CH_2SiCl_3 \quad (70, 297) \tag{4}$$

$$CF_3{-}CH{=}CH_2 + CH_3SiHCl_2 \rightarrow CF_3{-}CH_2{-}CH_2(CH_3)SiCl_2 \quad (296) \tag{5}$$

$$(CF_3)_2C{=}CH_2 + SiHCl_3 \rightarrow (CF_3)_2CH{-}CH_2SiCl_3 \quad (142) \tag{6}$$

Telomerization is observed with these reactions to a greater or smaller extent, e.g. (*69*):

$$HC_2F_4(CH_3)SiCl_2 \xrightarrow{nC_2F_4} H(C_2F_4)_{n+1}(CH_3)SiCl_2$$

This can be interpreted as a radical mechanism:

$$CH_3SiHCl_2 \xrightarrow{h\nu} H\cdot + CH_3\underset{\bullet}{Si}Cl_2 \xrightarrow{+C_2F_4} \begin{array}{l} CH_3\diagdown \\ \qquad\qquad SiCl_2 \\ \cdot CF_2{-}CF_2\diagup \end{array}$$

This course of such reactions can be influenced by the ratio of the starting materials: an excess of chlorosilanes suppresses telomerization and produces mainly monomeric adducts (*69*):

$$\begin{array}{l} CH_3\diagdown \\ \qquad\qquad SiCl_2 \\ \cdot CF_2{-}CF_2\diagup \end{array} \xrightarrow{CH_3SiHCl_2} \begin{array}{l} \quad CH_3\diagdown \\ \qquad\qquad\quad SiCl_2 + CH_3\underset{\bullet}{Si}Cl_2 \\ HCF_2{-}CF_2\diagup \end{array}$$

4.4.1.4 *Condensation Reactions of the Si—H Bond*

The possibilities of preparing fluoroorganosilanes by a condensation reaction of the Si—H bond have been mentioned, but they are hardly of interest for preparative purposes. Thus, methyldichlorosilane reacts with fluorobenzene at 255 to 265°C in the presence of 0.3 % of boric acid as a catalyst to yield fluorophenylmethyldichlorosilane:

$$CH_3SiHCl_2 + C_6H_5F \rightarrow FC_6H_4(CH_3)SiCl_2 + H_2$$

However, the yield is only 18% (*o*-compound 25%, *m*-compound 60%, *p*-compound 15%) (*132*). Trifluoromethylphenylsilanes are obtained analogously, with liberation of hydrogen, from trifluoromethyl-substituted benzenes and Si—H-containing silanes, again in only moderate yield (*92*). The reaction

$$m\text{-}CF_3{-}C_6H_4{-}Cl + CH_3SiHCl_2 \rightarrow m\text{-}CF_3{-}C_6H_4{-}(CH_3)SiCl_2 + HCl$$

for instance, gives a conversion of only 25% if the reactants are passed with short residence times (15 seconds) through a reaction zone at 560 to 580°C (*92*).

4.4.2 Preparation by the Transformation of Organic Groups

4.4.2.1 *Addition Reactions of Alkenylsilanes*

Fluoroorganosilanes with partially fluorinated organic groups are obtained by the addition of an alkenylsilane to a fluoroolefin. The following reactions, for example, have been carried out in this way.

$$(CH_2{=}CH{-}CH_2)_2Si(OC_2H_5)_2 + 2CF_2{=}CF_2 \rightarrow \begin{array}{l} (CH_2{-}CH{-}CH_2)_2Si(OC_2H_5)_2\ (61) \\ \;\;\;| \qquad | \\ \;\;CF_2{-}CF_2 \end{array} \qquad (7)$$

$$CH_2{=}CH{-}SiCl_3 + CF_2{=}CFCl \rightarrow \begin{array}{l} CH_2{-}CH{-}SiCl_3\ (85, 86) \\ \;| \qquad | \\ CF_2{-}CFCl \end{array} \qquad (8)$$

$$CH_2{=}CH(CH_3)SiCl_2 + CF_2{=}CFCl \rightarrow \begin{array}{l} \phantom{CH_2{-}CH{-}}CH_3 \\ \phantom{CH_2{-}CH{-}}| \\ CH_2{-}CH{-}SiCl_2 \; (99) \\ |\phantom{CH_2{-}}| \\ CF_2{-}CFCl \end{array} \qquad (9)$$

Another interesting synthesis by addition consists in the reaction of vinylchlorosilanes or vinylethoxysilanes with a perfluoroalkanesulfonyl chloride. The reaction proceeds in the presence of a radical-forming catalyst, e.g., a peroxide. The halogen and the perfluoroalkyl radical add to the alkenyl group:

$$CH_2{=}CH(CH_3)Si(OC_2H_5)_2 + CF_3{-}SO_2Cl \rightarrow CF_3{-}CH_2{-}CHCl(CH_3)Si(OC_2H_5)_2 + SO_2 \qquad (10)$$

$$CH_2{=}CH(CH_3)SiCl_2 + C_8F_{17}SO_2Cl \rightarrow C_8F_{17}{-}CH_2{-}CHCl(CH_3)SiCl_2 + SO_2 \; (145) \qquad (11)$$

Greater or smaller yields of telomerization products are also obtained, of course, in addition to the adducts described above.

4.4.2.2 *Substitutions in Functional Organic Groups*

Fluoroorganosilanes have been prepared by halogen exchange between chloroalkyl- or chloroarylsilanes and hydrogen fluoride in the presence of antimony pentachloride or mixtures of antimony(III) fluoride and antimony pentachloride. The Si—Cl bonds are simultaneously converted to Si—F bonds (*141*):

$$CCl_3{-}C_6H_4{-}SiCl_3 \xrightarrow[SbCl_5]{HF} CF_3{-}C_6H_4{-}SiF_3$$

Silanes with partially fluoro-substituted organic groups can also be obtained by the reaction of carboxyorganosilicon compounds with fluoroalcohols (*227*)

$$\equiv Si(CH_2)_nCOOH + HOCH_2(CF_2)_mCF_3 \rightarrow \equiv Si(CH_2)_nCOOCH_2(CF_2)_mCF_3 + H_2O$$

or by reactions between hydroxyorganosilanes and perfluorocarboxylic acids (*244*):

$$\equiv Si(CH_2)_nOH + HOOC(CF_2)_mCF_3 \rightarrow \equiv Si(CH_2)_nOOC(CF_2)_mCF_3 + H_2O$$

4.4.3 PHYSICAL CONSTANTS (TABLE)

Table 23 lists the physical constants of a number of selected fluoroorganosilanes.

TABLE 23

FLUOROORGANOSILANES[a, b]

Compound	Mp (°C)	Bp (°C) at	p (mmHg)	n_D^t at	t (°C)	d_4^t (gm/cc) at	t (°C)
$F_2CH—CH_2—SiCl_3$	—	104–105.5 (*308*)	760	1.4050 (*308*)	20	1.43 (*308*)	25
$F_2CH—CF_2(CH_3)SiCl_2$	—	95.5–96 (*69*)	770	—	—	—	—
$F_3CCH_2CH_2(CH_3)_3Si$	—	94.3–95.1 (*135*)	750	1.3576 (*135*)	20	0.896 (*135*)	20
$(F_3CCH_2CH_2)_2SiCl_2$	—	180 (*312*) 162–163 (*70*)	760 760	1.3715 (*312*)	35	1.364 (*312*)	35
$F_3CCH_2CH_2(CH_3)SiCl_2$	—	121–122 (*135*)	737	—	—	—	—
$F_3CCH_2CH_2SiCl_3$	—	113 (*312*)	760	1.3780 (*312*)	35	1.381 (*312*)	35
$F_3CCH_2CH_2(CH_3)Si(OH)_2$	82–85 (*98*)	—	—	—	—	—	—
$(F_3C)_2CH—CH_2SiCl_3$	—	122 (*142*)	742	1.3661 (*142*)	25	—	—
$(F_3C)_2CH—CH_2SiH(OC_2H_5)_2$	—	67.5 (*142*)	38	1.3433 (*142*)	25	1.178 (*142*)	25
$(F_3CCF_2CF_2CH_2CH_2)_3SiCl$	34–35 (*200*)	133–134 (*200*)	15	—	—	—	—
$(F_3CCF_2CF_2CH_2CH_2)_2Si(OC_2H_5)_2$	—	117.5–118 (*200*)	25	—	—	—	—
$F_3CCF_2CF_2CH_2CH_2Si(OC_2H_5)_3$	—	92.5–93 (*200*)	25	—	—	—	—
$CH_2—CH—CH_2(CH_3)Si(OC_2H_5)_2$ | | $F_2C——CF_2$	< −70 (*61*)	201 (*61*)	760	—	—	1.096 (*61*)	25
$CH_2—CH—CH_2(CH_3)_2SiOC_2H_5$ | | $F_2C——CF_2$	< −75 (*61*)	176 (*61*)	760	—	—	1.092 (*61*)	25
$(CH_2—CH—CH_2)_2Si(OC_2H_5)_2$ | | $F_2C——CF_2$	< −75 (*61*)	117 (*61*)	3	—	—	1.259 (*61*)	25
p-F—$C_6H_4SiCl_3$	—	194.5–196.5 (*43*)	738	—	—	—	—
$(p$-F—$C_6H_4)_2SiCl_2$	—	293–294 (*43*)	738	—	—	—	—
p-F—$C_6H_4(CH_3)Si(OC_2H_5)_2$	—	87.5–88 (*64*)	4–5	1.4558 (*64*)	25	1.0149 (*64*)	22
p-F—$C_6H_4(CH_3)_2SiOC_2H_5$	—	69 (*64*)	6	1.3639 (*64*)	25	0.9846 (*64*)	28

[a] Numbers in parentheses indicate reference numbers.

[b] For siloxanes see Tables 36 and 45.

4.4.4 Chemical Properties

The high electronegativity of fluorine makes the C—F bond relatively stable. The reactivity of fluorine bound to carbon is lower than that of the other halogens, the difference between I and Br being only small (*243*):

$$\rangle C{-}I \sim \rangle C{-}Br > \rangle C{-}Cl > \rangle C{-}F$$

Substitution reactions of fluorine in fluoroorganosilanes are thus of no interest. The practical importance of the fluoroorganosilanes lies in their ability to produce siloxanes with unusual technical properties.

Fluorinated substituents on the α- and, especially, on the β-C atom weaken the Si—C bond and make it particularly sensitive to nucleophilic agents (see Section 6.4). Polymeric $[CHF_2{-}CF_2{-}SiO_{3/2}]_n$ is rapidly decomposed by dilute alkali with the liberation of tetrafluoroethane, $CHF_2{-}CHF_2$ (*95*). The heat resistance of α- and β-fluoroalkyl silicon compounds is also low, which is in part due to the high affinity of fluorine for silicon; $CF_2Cl{-}CF_2{-}SiCl_3$ decomposes readily in the gas phase into $CF_2{=}CFCl$ and $SiFCl_3$ (*95*).

Practical interest has therefore largely been concentrated on fluoroorganosilanes in which the fluoro substituents are sufficiently far from the Si—C bond—at least on the γ-C atom*. The compounds obtained in accordance with Eqs. (5), (6), (10), and (11) as given in Sections 4.4.1.3 and 4.4.2.1 can therefore be hydrolyzed and worked up to silanols and the corresponding siloxanes (*98*). The silanes and siloxanes of the type obtained by reaction (9) are hydrolytically and thermally unstable. Siloxane polymers with fluorocyclobutyl radicals can however be obtained if the additions are carried out to cyclosiloxanes and not to chlorosilanes and if the products are then subjected to acid polymerization (*99*).

One of the most important applications of fluoroorganosilanes is the incorporation of methyl-γ-trifluoropropylsiloxane units into siloxane rubber, which considerably improves the latter's swelling resistance (Sections 8.1.1 and 9.6.6). Oils based on siloxane molecules with these groups have enhanced lubricant action (Section 9.2.10). Fluoroorganosiloxanes with fluoroorganic substituents based on reaction (6) or (11), have been recommended for the simultaneous water- and oil-repellency impregnation of textiles (*142*, *145*).

4.5 Chloroorganosilanes

The compounds of this group, like those described in Section 4.6 are of interest for other purposes besides the modification of siloxane polymers. They are the gateway to the entire field of organofunctional silanes, since the

* However a CF_3 group does still exert a detectable inductive influence even when it is on the γ-C atom of a normal aliphatic hydrocarbon residue, as can be demonstrated by the susceptibility of an Si—H bond present in the molecule to hydrolysis (*288*).

halogen atom in the organic radical can be replaced by hydroxyl, amino, carboxyl, and other groups, by means of classical organic reactions.

4.5.1 Preparation by the Formation of New Si—C Bonds

The methods of synthesis for these compounds, like those for other organofunctional silanes, can be divided into two groups: in the first, *Si—C bonds* are actually *formed*, while in the second *changes in the organic group* are involved. The first group comprises organometallic syntheses, addition reactions of the Si—H bond with haloolefins, and the reaction of diazomethane with chlorosilanes; the second group includes the halogenation of the organic group by substitution or addition, which has acquired much importance for the further extension of organosilicon chemistry.

4.5.1.1 *Organometallic Syntheses*

Organometallic syntheses have again proved to be very versatile; *p*-ClC_6H_4MgBr reacts well with trimethylchlorosilane (*35*), trichlorosilane (*16*), vinyltrichlorosilane, and allyltrichlorosilane (*229*). More highly chlorinated silicon compounds can be obtained by similar means (*38*); *p*-ClC_6H_4Li reacts smoothly with chlorosilanes and ethoxysilanes (*76*).

4.5.1.2 *Addition and Substitution Reactions of the Si—H Bond*

The Si—H bond can be added to chloroolefins to give chloroorganosilanes. Chloroplatinic acid (*28*, *223*, *251*) and platinum on charcoal (*308*) have been used as catalysts; a pyridine/ethylene/$PtCl_2$ complex has also been recommended as particularly suitable (*114*).

The following are examples of such additions:

$$ClCH_2—CH{=}CH_2 + SiHCl_3 \rightarrow ClCH_2—CH_2—CH_2SiCl_3 \quad (308)$$

$$ClCH_2—CH{=}CH_2 + CH_3SiHCl_2 \rightarrow ClCH_2—CH_2—CH_2(CH_3)SiCl_2 \quad (114)$$

$$ClCH_2—\underset{\displaystyle CH_3}{\underset{|}{C}}{=}CH_2 + CH_3SiHCl_2 \rightarrow ClCH_2—\underset{\displaystyle CH_3}{\underset{|}{CH}}—CH_2(CH_3)SiCl_2 \quad (223)$$

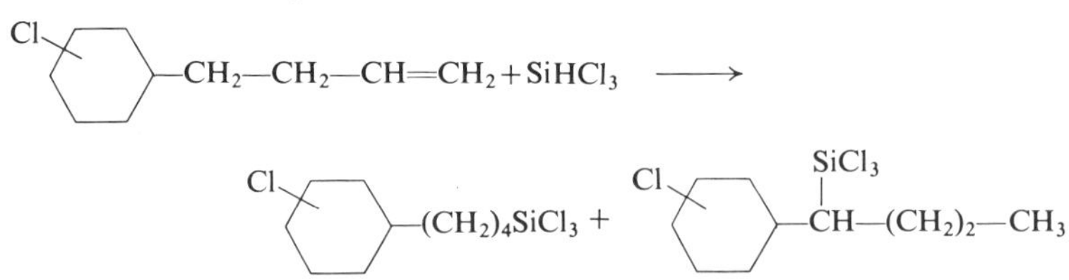

(28)

$$o\text{-}Cl—C_6H_4—CH{=}CH_2 + SiHCl_3 \rightarrow o\text{-}Cl—C_6H_4—CH_2—CH_2—SiCl_3 \quad (28)$$

As has been mentioned earlier, at temperatures above 250°C in the presence of $AlCl_3$, the Si—H bond is capable of undergoing a *condensation reaction* with chloroaromatic hydrocarbons, which leads to arylsilanes (Section 2.4.2); a side reaction yields chloroarylsilanes:

$$SiHCl_3 + C_6H_5Cl \xrightarrow{AlCl_3} C_6H_5SiCl_2 + HCl \qquad (1)$$

$$SiHCl_3 + C_6H_5Cl \xrightarrow{AlCl_3} ClC_6H_4SiCl_3 + H_2 \qquad (2)$$

Starting with CH_3SiHCl_2, it is possible to favor reaction (2) somewhat by lowering the temperature to about 150 to 250°C (boron trichloride catalyst, 200°C for 16 hours, 29% yield) (*17*):

$$CH_3SiHCl_2 + C_6H_5Cl \xrightarrow{BCl_3} ClC_6H_4(CH_3)SiCl_2 + H_2$$

On the other hand, *p*-dichlorobenzene mixed with benzene reacts with trichlorosilane above 600°C according to (1); *p*-Cl—$C_6H_4SiCl_3$ (yield 30%) is formed, with other products (*183*):

$$SiHCl_3 + p\text{-Cl—}C_6H_4Cl \rightarrow p\text{-Cl—}C_6H_4SiCl_3 + HCl$$

Numerous side reaction, such as allyl rearrangements, double-bond isomerizations, and olefin eliminations take place during the addition reaction with allyl chloride (*223*, *251*). 3-Chlorobut-l-ene and CH_3SiHCl_2 give a mixture consisting of butenes, methyltrichlorosilane, butylmethyldichlorosilane, and 3-chlorobutylmethyldichlorosilane (*251*).

4.5.1.3 *Syntheses with Diazomethane*

These syntheses are, in principle, reactions of the Si—Cl bond with diazomethane.

$$ClSi\lessdot + CH_2N_2 \rightarrow ClCH_2Si\lessdot + N_2$$

Diazoethane reacts analogously (*231*, *233*, *320*, *321*).

The reaction yields derivatives of methyl- or ethylsilane *mono*chloro-substituted in the organic group.

Diazomethane and silicon tetrachloride gave $(ClCH_2)_2SiCl_2$ and $(ClCH_2)_3SiCl$, and $ClCH_2(CH_3)SiCl_2$ was obtained with methyltrichlorosilane. The reaction must be carried out in ether at 50°C and then gives yields of 40 to 80%. At room temperature it proceeds very vigorously and gives only polymethylenes. Copper bronze and anhydrous copper sulfate markedly accelerate the formation of chloromethylsilanes.

Phenyltrichlorosilane and diazomethane give the chloromethylphenylsilanes $ClCH_2(C_6H_5)SiCl_2$ and $(ClCH_2)_2C_6H_5SiCl$. However, diphenyldichlorosilane and triphenylchlorosilane do not react (*122*).

4.5.2 Preparation by the Transformation of Organic Groups

4.5.2.1 *Chloro Substitution in the Organic Group*

The chlorination of *methylsilanes* is carried out with chlorine or sulfuryl chloride under uv irradiation (*50*, *280*, *307*, *314*, *325*) in the presence of azodiisobutyronitrile (*42*, *154*) or small amounts of fluorine (*161*), or with sulfuryl chloride in the presence of benzoyl peroxide (*136*). The photochemical chlorination can be accelerated by the addition of small amounts of PCl_5 (*314*). The various stages of chlorination are usually obtained side by side (*124*). Catalysis with azodiisobutyronitrile, however, leads predominantly to monochlorinated methylsilanes. Methyl groups already substituted by chlorine undergo further reaction with chlorine more readily than methyl groups without previous chlorine substitution (*124*). This makes the synthesis of monochlorosilanes difficult. If, however, a side stream is continuously withdrawn from the reaction mixture and passed into a fractionation column the mean temperature of which is above that of the nonchlorinated silane and below that of the singly chlorinated silane, and the former is recycled to the reactor while the latter is removed, it is possible to raise the yield of monochlorosilane to 90% (*327*).

Increasing halogen substitution on the Si atoms again makes the chlorination more difficult. It proceeds most readily with trimethylchlorosilane and is most difficult with methyltrichlorosilane (*154*). Azodiisobutyronitrile is effective with silanes rich in Si—Cl bonds even when benzoyl peroxide fails (*325*).*

*Ethyl*silanes and silanes with *higher alkyl groups* are also chlorinated photochemically (*143*), or, more effectively, by boiling with sulfuryl chloride in the presence of dibenzoyl peroxide (*270*). Tetraethylsilane can be chlorinated with carbon tetrachloride under the action of uv irradiation (*217*):

$$(C_2H_5)_4Si + CCl_4 \xrightarrow{h\nu} CH_3CHCl(C_2H_5)_3Si + CHCl_3$$

In the chlorination of tetraethylsilane with chlorine under the action of uv light, or with SO_2Cl_2 in the presence of peroxides, the chlorine enters mainly in the α-position. The proportion of chlorine substituents on the β-C atom however increases with increasing chlorine substitution on the Si atom, so that in the case of diethyldichlorosilane the chlorine substitutes the α- and β-C atoms equally, and in the case of ethyltrichlorosilane mainly the β-C atom (*19*).

With *aryl-substituted* silanes, chlorination of the nucleus occurs only in the presence of catalysts. Phenyltrichlorosilane and diphenyldichlorosilane cannot be halogenated without catalysts even with a large excess of chlorine at 170 to 180°C (*322*, *323*). Suitable catalysts are $AlCl_3$ (*100*), $FeCl_3$ (*58*), $SbCl_3$ (*322*),

* Explosions have occurred with chlorination of methylsiloxanes; moreover, the more highly chlorinated methylchlorosilanes tend to decompose (*328*).

$SbCl_5$ (*323*), PCl_5 (*323*), iron powder (*181*, *322*, *323*), iodine (*322*, *323*), and uv light (*194*).

The aromatic Si—C bond is sensitive toward the hydrogen chloride formed in the reaction, so that side reactions can easily take place which lead to cleavage of the Si—C bond. The extent of these reactions will depend on the type of catalyst used. The greatest degree of dephenylation in the chlorination of *phenyltrichlorosilane* is caused by $AlCl_3$; $FeCl_3$ is less harmful; no side reactions have been observed with $SbCl_3$ and iodine (*322*).

For high yields of chlorophenylchlorosilanes from the chlorination of phenyltrichlorosilane with $AlCl_3$ as catalyst, the reaction must be carried out in solution in carbon tetrachloride. A sixty percent yield of tetrachlorophenyltrichlorosilane can then be obtained (*100*, *127*).

Without this solvent the products are essentially the trichloro-substitution derivatives, obtained in 55% yield, while 43% of the original silane is split at the Si—C bond (*322*). The action of the carbon tetrachloride is probably due to its poor solvent action for hydrogen chloride, which is responsible for the Si—C cleavage, so that its concentration in the reaction mixture is kept low (*127*).

This type of chlorination substitutes the chlorine atoms into the ring in the sequence 2, 5, 3, 6, so that the *p*-position remains free (*127*).

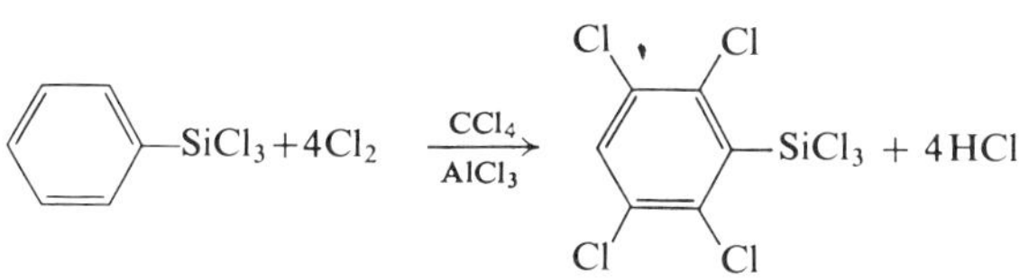

The chlorination of phenyltrichlorosilane at 50 to 60°C in the absence of a solvent and in the presence of iron(III) chloride gives mainly (up to 75%) *m*-chlorophenyltrichlorosilane with only 21% of the *o*-isomer and 4% of the *p*-isomer (*324*). Chlorination reactions with iron powder as catalyst proceed similarly (*181*). The $SiCl_3$ group also proved to be *m*-directing in the photochemical chlorination of phenyltrichlorosilane (*194*). It is interesting that below 20°C the substitution reaction is increasingly replaced by the addition of chlorine to the ring, so that in diffuse daylight below 0°C only hexachlorocyclohexyltrichlorosilane is obtained, in 67% yield (*156*); $SbCl_3$ gives mainly tri- and iodine mainly mono-substitution products. Considerable concentrations of catalysts (approximately 3%) must however be used in the latter case if yields of over 80% are to be achieved (*322*).

The use of a mixed $FeCl_3$—$SbCl_5$ catalyst enables yields of up to 86% of tetrachlorophenyltrichlorosilane to be obtained with only slight cleavage of the Si—C bond (*101*).

Diphenyldichlorosilane is even more sensitive toward side reactions involving cleavage of the Si—C bond than phenyltrichlorosilane. During the chlorobromination of diphenyldichlorosilane in the presence of iron powder, well over 60% of the Si—C bonds are broken. Only mono- to tribromophenyltrichlorosilanes (*238*) are obtained as reaction products: $SbCl_3$ is said to have the least cleaving action (*323*).

Silanes containing both *methyl* and *phenyl* groups are chlorinated either in the ring or in the methyl group, depending on the catalyst used and on the operating conditions (*155*). As in the case of phenyltrichlorosilane, chlorination in diffuse daylight between 0 and +5°C results in addition of chlorine, to yield a hexachlorocyclohexyl derivative.

The chlorination of *p-tolyltrichlorosilane* and *benzyltrichlorosilane* with chlorine in the presence of azodiisobutyronitrile as catalyst at 85 to 140°C produces all the possible *p*-tolylchlorosilanes chlorinated in the side chain and α-chloro- and α,α-dichlorobenzyltrichlorosilanes, respectively, in yields of 80 to 90% (*159*). The Si—C bond is not attacked in this case, and neither is the aromatic nucleus. On the other hand halogenation of *p*-tolyltrichlorosilane *in the ring* takes place in the presence of iron or $SbCl_3$ and products containing two or three chlorine atoms in the ring are obtained, depending on the ratio of the reactants. Still more intensive chlorination breaks the Si—C bond, forming tetra- and pentachlorotoluenes together with $SiCl_4$ (*158*). The monochloro-*p*-tolyltrichlorosilane has the structure:

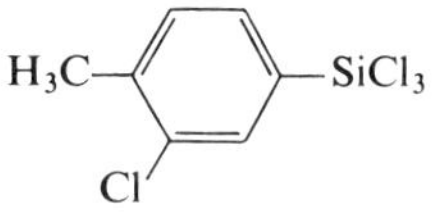

while the more highly chlorinated products are mixtures of isomers.

The ring-substituted mono- (4-), di- (2,4-), tri- (2,4,6-), tetra-, and pentachlorobenzyltrichlorosilanes can be obtained smoothly from benzyltrichlorosilane and chlorine. In this case, $SbCl_3$ and iron powder are useful catalysts. The $—CH_2SiCl_3$ group is thus first *para*- and then *ortho*-directing in halogen substitution on the nucleus (*157*).

4.5.2.2 *Addition of Chlorine to the Organic Radical*

As well as by substitution reactions on the organic group, haloorganosilanes are also accessible through a number of addition reactions of olefinic organic groups attached to silicon. Among these is the addition of halogen or hydrogen halide to vinyl- or allysilanes, as, for example, in (*2*, *295*):

$$CH_2{=}CH{-}SiCl_3 + Cl_2 \rightarrow ClCH_2{-}ClCH{-}SiCl_3$$

Low temperatures must be employed to preserve the Si—C bond. Vinylsilanes react with hydrogen halides only in the presence of catalysts. Hydrogen chloride was added in the presence of aluminum chloride (*310*):

$$CH_2{=}CH{-}SiCl_3 + HCl \rightarrow ClCH_2{-}CH_2{-}SiCl_3$$

Hydrogen bromide was added using dibenzoyl peroxide as a catalyst (*256*):

$$CH_2{=}CH(CH_3)_3Si + HBr \rightarrow BrCH_2{-}CH_2(CH_3)_3Si$$

Markovnikov's rule, according to which the hydrogen adds to the C atom richer in hydrogen and the halogen to the C atom poorer in hydrogen, is not followed in the addition of hydrogen halides to vinylsilanes. It holds in other cases, such as the addition of hydrogen iodide to allyltrimethylsilanes (*256*):

$$CH_2{=}CH{-}CH_2(CH_3)_3Si + HI \rightarrow CH_3{-}CHI{-}CH_2(CH_3)_3Si$$

The Diels-Alder reaction between vinylsilanes and hexachlorocyclopentadiene can also be considered an addition reaction to unsaturated organic groups attached to the silicon atom (*113*, *199*, *240*). Thus, vinyltrichlorosilane yields 1,4,5,6,7,7-hexachloro[2.2.1]bicyclohepten-5-yl-2-trichlorosilane:

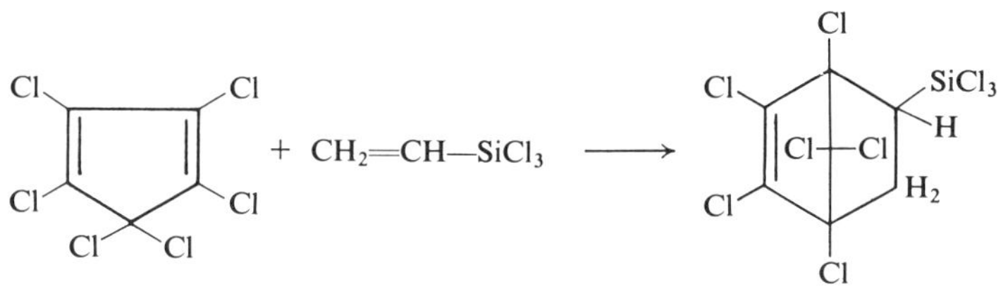

4.5.3 Physical Constants (Table)

Physical constants for a number of selected chloroorganosilanes are given in Table 24 (pp. 154–155).

4.5.4 Chemical Properties

Silanes containing chloroalkyl groups can undergo numerous reactions. The reactivity of the chlorine substituted in the alkyl group is highly dependent on its position in relation to the Si atom and decreases in the sequence:

$$\beta \gg \gamma \sim \alpha > \delta$$

(cf. Section 6.4).

Thus, the chlorine in β-chloroethyltrichlorosilane, β-chloroethyldiethylchlorosilane, and β-chloro-*n*-propyltrichlorosilane can be titrated with caustic soda (*257*, *259*, *270*):

$$ClCH_2{-}CH_2{-}SiCl_3 + 4NaOH \rightarrow CH_2{=}CH_2 + 4NaCl + Si(OH)_4$$

β-Chloroethyltriethylsilane reacts quantitatively on standing for one hour with caustic soda even at room temperature, while α-chloroethyltriethylsilane

TABLE 24

CHLOROORGANOSILANES[a, b]

Compound	Mp (°C)	Bp (°C)	at p (mm Hg)	n_D^t	at t (°C)	d_4^t (gm/cc)	at t (°C)
$ClCH_2(CH_3)_2Si$	—	97.1 (*314*)	734	1.4180 (*314*)	20	0.8765 (*280*)	25
$(ClCH_2)_2(CH_3)_2Si$	—	160–161 (*280*)	749	1.4573 (*280*)	25	1.08 (*280*)	25
$(ClCH_2)_3CH_3Si$	—	205 (*280*)	749	1.4857 (*280*)	25	1.24 (*280*)	25
$(ClCH_2)_3SiCl$	—	70 (*320*)	4	—	—	1.4377[c] (*320*)	20
$ClCH_2(CH_3)_2SiCl$	—	114 (*280*)	737	1.4360 (*66*)	20	1.0865 (*66*)	20
$(ClCH_2)_2CH_3SiCl$	—	172 (*280*)	760	—	—	—	—
$ClCH_2(CH_3)SiCl_2$	—	121–122 (*320*)	760	1.4503 (*299*)	20	1.2778[c] (*320*)	20
$(ClCH_2)_2SiCl_2$	—	58.5 (*320*)	16	—	—	1.4624[c] (*320*)	20
$ClCH_2SiCl_3$	—	116.5 (*320*)	750	—	—	1.4776[c] (*320*)	20
$Cl_2CH(CH_3)_3Si$	—	133–134 (*280*)	749	1.4430 (*280*)	25	1.04 (*280*)	25
$Cl_2CH(CH_3)_2SiCl$	—	148 (*280*)	734	1.4588 (*280*)	25	—	—
$Cl_2CH(CH_3)SiCl_2$	—	107 (*124*)	225	—	—	—	—
$Cl_3C(CH_3)_3Si$	60–66 (*280*)	146–156 (*280*)	749	—	—	—	—
$Cl_3C(CH_3)_2SiCl$	95–96 (*280*)	161 (*280*)	739	—	—	—	—
$Cl_3C(CH_3)SiCl_2$	99 (*124*)	109 (*124*)	150	—	—	—	—
Cl_3CSiCl_3	115–116 (*50*)	155–156 (*50*)	740	—	—	—	—

	—	132 (*270*)	734	—	—	—	—
$CH_3CHClSiCl_3$	—	66 (*320*)	75	—	—	1.3919[c] (*320*)	20
$ClCH_2CHClSiCl_3$	—	63–65 (*2*)	13	1.4876 (*2*)	20	1.53 (*2*)	27
$ClCH_2CCl_2SiCl_3$	56 (*2*)	91 (*2*)	17	—	—	—	—
$Cl_2C{=}CCl(CH_3)SiCl_2$	—	105.8–106.6 (*134*)	42	—	—	—	—
$Cl_2C{=}CClSiCl_3$	—	193 (*134*)	748	—	—	—	—
$Cl_2CH(CH_3)_2SiOH$	—	89 (*125*)	40	—	—	—	—
o-$ClC_6H_4(CH_3)_3Si$	< 75 (*44*)	207–208 (*44*)	740	1.512 (*44*)	25	1.0266 (*44*)	25
o-$ClC_6H_4SiCl_3$	—	240.5 (*194*)	742.5	1.5510 (*194*)	20	1.4629 (*194*)	20
m-$ClC_6H_4SiCl_3$	—	230.5 (*194*)	738	1.5421 (*194*)	20	1.4384 (*194*)	20
p-$ClC_6H_4SiCl_3$	—	232.5 (*194*)	745.5	1.5418 (*194*)	20	1.4316 (*194*)	20
(*p*-$ClC_6H_4)_2SiCl_2$	71.2 (*43*)	234 (*43*)	36	—	—	—	—
p-$ClC_6H_4(CH_3)_3Si$	0.5 (*44*)	206–207 (*44*)	740	1.5067 (*44*)	25	0.9959 (*44*)	25
$Cl_2C_6H_3SiCl_3$[d]	—	155 (*151*)	27	—	—	1.537 (*151*)	26
$Cl_3C_6H_2SiCl_3$[d]	69–71 (*57*)	113 (*57*)	1	—	—	—	—
$Cl_4C_6HSiCl_3$[d]	—	125–126 (*322*)	4	—	—	1.6340 (*322*)	20
$Cl_5C_6SiCl_3$	59.5 (*322*)	146–147 (*322*)	8	—	—	—	—
p-$CCl_3C_6H_4SiCl_3$	66–67 (*159*)	162–163 (*159*)	14	—	—	—	—

[a] Numbers in parentheses indicate reference numbers.
[b] For siloxanes see Tables 36, 40, and 45.
[c] d_{20}^{20}.
[d] Position of Cl in ring not known or not sufficiently certain.

reacts only on heating to 145°C (*306*). Of the monochloro derivatives of *n*-propyltrichlorosilane, only the β-chloro compound can be titrated with alkali; the γ- and α-derivatives do not react (*262*). Results of work on γ-chloro-*n*-propyltrichlorosilane, on the one hand, and α-chloro-*n*-propyltrichlorosilane and α-chloroethyltrichlorosilane, on the other, appear to indicate a higher reactivity of chlorine in the γ-position than in the α-position (*259*). Chlorine in the β-position always has the highest reactivity, while the reactivity of chlorine on a δ-C atom is similar to that of normal aliphatic halides.

Chloro substitution of the organic group weakens the Si—C bond increasingly with increasing degree of substitution; Cl_3CSiCl_3 is readily decomposed by cold water, with the liberation of chloroform (*50*). On the other hand, α-mono- or α,α-dichlorosilanes with $ClCH_2$, Cl_2CH, or CH_3CHCl groups can be hydrolyzed to siloxanes without cleavage of the Si—C bond (*50*, *270*).

The weakening of the Si—C bond also depends on the position of the substituent. Chloro substituents on the β-carbon increase the sensitivity of the Si—C bond toward nucleophilic and electrophilic attack; chlorine on the α-carbon increases sensitivity toward nucleophilic agents but decreases ease of cleavage by electrophilic agents (see Section 6.4).

The particularly strong influence of the β-halogen is called the "β-effect." It is responsible for the so-called "β-decomposition" (or "β-elimination"), which is accompanied by liberation of an olefin.

$$ClCH_2{-}CH_2(C_2H_5)_3Si + H_2O \xrightarrow{OH^\ominus} (C_2H_5)_3SiOH + CH_2{=}CH_2 + HCl$$

Other reactions also demonstrate the consequences of the "β-effect." It can be observed merely on heating; β-chloroethyltriethylsilane splits off ethylene on heating under reflux (*258*):

$$ClCH_2{-}CH_2(C_2H_5)_3Si \xrightarrow{\text{heat}} CH_2{=}CH_2 + (C_2H_5)_3SiCl$$

(For the conditions of a dehydrohalogenation reaction, see Section 4.3.2.1.)

An analogous β-elimination is observed in reactions with Grignard compounds (*262*):

$$ClCH_2{-}CH_2{-}SiCl_3 + 4RMgBr \rightarrow R_4Si + CH_2{=}CH_2 + 4MgBrCl$$

The reactivity of the β-halogen toward Grignard reagents is again in the strongest contrast to the lower reactivity of the halogen on the α-carbon. α-Chloroethyltrichlorosilane and methylmagnesium bromide undergo only the substitution reaction, replacing the Cl atoms attached to silicon by methyl radicals and yielding α-chloroethyltrimethylsilane (*262*):

$$CH_3CHClSiCl_3 + 3CH_3MgBr \rightarrow CH_3CHCl(CH_3)_3Si + 3MgBrCl$$

An "α-decomposition" corresponding to the thermal β-decomposition can be brought about only under considerably more severe conditions. Heating α-chloroethyltrichlorosilane to 600°C gave considerable quantities of ethylene and silicon tetrachloride in addition to vinyltrichlorosilane (*2*).

The reactivity of the organofunctional halogen atom and the ease of cleavage of the Si—C bond depend on the nature of the substituents on the Si atom, as well as on the type of attack—purely thermal decomposition or nucleophilic or electrophilic attack—and on the position of the halogen substituent (cf. Section 6.4). The siloxane oxygen also acts as an electronegative substituent in this case. This will be demonstrated by the following examples.

As just mentioned, β-chloroethyltriethylsilane readily undergoes β-decomposition on heating. β-Chloroethyltrichlorosilane does so only on heating to 600°C and then undergoes both possible reactions—β-decomposition (1) and the dehydrohalogenation reaction (2) (cf. Section 4.3.2.1) (*2*):

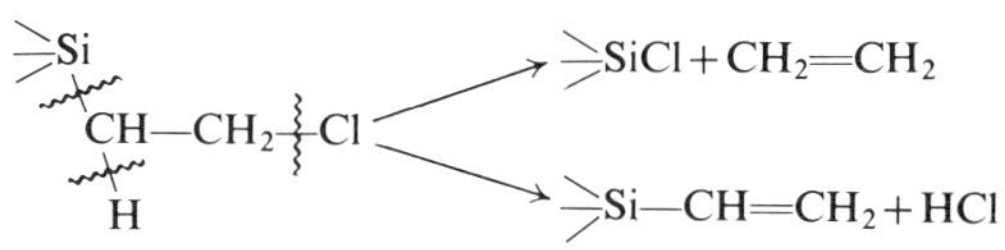
(1)
(2)

In the presence of quinoline, the dehydrochlorination (2) is favored to such an extent that the β-decomposition (1) is suppressed almost completely (*192*). The tendency toward β-decomposition decreases in the order (*192*):

$$ClCH_2—CH_2—R_3Si > ClCH_2—CH_2—R_2SiCl > ClCH_2—CH_2—RSiCl_2 > ClCH_2—CH_2—SiCl_3$$

Analogous rules hold for the electrophilic attack of $AlCl_3$ on β-haloorganosilanes: β-Chloroethyltriethylsilane preferentially undergoes β-decomposition with scission of the Si—C and Si—X bonds, while β-chloroethyltrichlorosilane is dehydrohalogenated (*310*) (Section 4.3.2.1).

Here, therefore, the electronegative silicon-functional substituent has the effect that cleavage of the Si—C bond gives way to that of the C—X bond. Conversely, with nucleophilic attack on halomethylsilanes substituted on the α-C, Si—C cleavage is facilitated. The following reactions may compete:

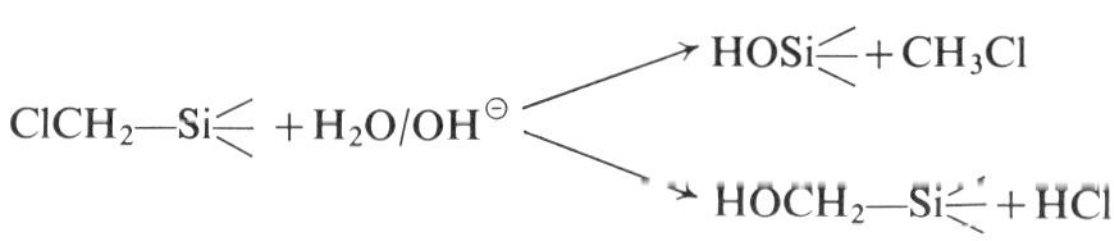

The Si—C bond in $ClC_2(p\text{-}ClC_6H_4)(CH_3)_2Si$ is decomposed faster by the nucleophilic attack of sodium ethoxide in ethanol, than the Si—C bond in $ClCH_2(C_6H_5)(CH_3)_2Si$ (*51*). The influence of the siloxane oxygen is confirmed by the fact that the Si—C bond in $Cl—CH_2\,(CH_3)_2Si—O—Si(CH_3)_3$ can be split more easily by alkalis than that in $ClCH_2(CH_3)_3Si$ (*125*, *220*).

The doubly chloro-substituted bis-dichloromethyl-tetramethyldisiloxane can no longer be converted into the corresponding hydroxymethyl derivatives with alkalis, since cleavage of the Si—C bond is by far the predominant reaction (*125*, *242*, *250*).

The reaction of chloromethyltrimethylsilane with alkoxides has given the interesting result that the number of carbons in the alkoxide also affects the extent to which the two reactions

$$ClCH_2{-}Si\equiv \xrightarrow[ROH]{NaOR} ROSi\equiv \qquad (1)$$

$$ClCH_2{-}Si\equiv \xrightarrow[ROH]{NaOR} ROCH_2{-}Si\equiv \qquad (2)$$

proceed (*279*). While sodium methoxide reacts almost exclusively according to (2), even sodium ethoxide gives 10% and the butoxide around 30% of reaction (1). Dichloromethyltrimethylsilane is decomposed quantitatively by sodium butoxide (*283*).

Mercaptides react analogously to the alkoxides, but more readily (*175*):

$$ClCH_2(CH_3)_3Si + NaS(n\text{-}C_4H_9) \rightarrow (n\text{-}C_4H_9)SCH_2(CH_3)_3Si$$

The greater reactivity of the mercaptans as compared with the alcohols also enables them to undergo reactions with chloromethylsilanes in the presence of triethylamine. Chloromethyldimethylchlorosilane, dithioglycol, and triethylamine in ethereal solution gave 2,2-dimethyl-2-sila-1,4-dithiacyclohexane (*316*):

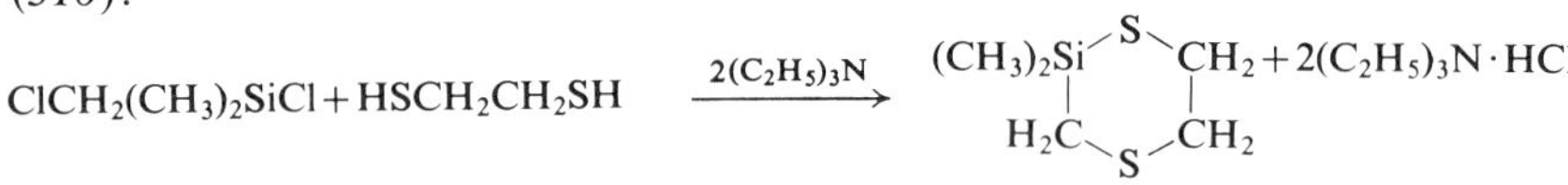

Chloroalkylsilanes react with ammonia and amines, preferably when the halogen is present as a halomethyl group (*174*):

$$ClCH_2(CH_3)_2(C_6H_5)Si + NH_3 \rightarrow H_2NCH_2(CH_3)_2(C_6H_5)Si\ (68\%) + HN[CH_2(CH_3)_2(C_6H_5)Si]_2\ (28\%)$$

$$ClCH_2Si(OC_2H_5)_3 + NH_3 \rightarrow H_2NCH_2Si(OC_2H_5)_3\ (51\%) + HN[CH_2Si(OC_2H_5)_3]_2\ (25\%)$$

$$ClCH_2(CH_3)_3Si + C_6H_{11}NH_2 \rightarrow C_6H_{11}NHCH_2(CH_3)_3Si$$

Cleavage of the Si—C and Si—O groups is again unavoidable in the corresponding reaction of the $ClCH_2$ group in disiloxanes.

The condensation of haloalkylsilicon compounds with compounds having active CH groups (such as cyanoacetic ester or malonic ester) has acquired some interest for the chemistry of the organofunctional silicones. It proceeds basically as follows:

$$\equiv Si(CH_2)_nHal + {}^{\ominus}CH(COOR)R' \rightarrow \equiv Si(CH_2)_nCH(COOR)R' + Hal^{\ominus}$$

Chloromethylsilicon compounds can only partly undergo this reaction, under special conditions; the reactions with cyanoacetic ester can be carried out with their sodium derivatives (*266*) or by using a sodium alkoxide as the condensation agent (*289*). The same holds for the reaction with malonic ester (*266*). The bromo compounds are more suitable for some of these syntheses (Section 4.6).

Finally, chloroalkylsilanes can be converted to Grignard compounds, which then give access to other organic functions such as hydroxy (Section 4.7) or carboxy derivatives (Section 4.8).

Thus, chloroalkylsilanes are versatile starting materials for organosilicon chemistry, but the situation with chloroarylsilanes is different. Up to now, practical interest has centered on only a few chlorophenylsilanes, because their incorporation into siloxanes is capable of imparting useful properties to the latter. Thus, the lubricant action of polydimethylsiloxanes is improved by incorporation of tetrachlorophenylsiloxane units (see Section 9.2.10).

4.6 Bromoorganosilanes

Bromoalkylsilanes have become of interest because of the high reactivity of the bromine in these compounds. They are accessible through many processes which are analogous to those used in the preparation of chloro derivatives. Only a few methods will be considered here.

4.6.1 Preparation

The action of bromine alone leads to brominated compounds only with higher tetraalkylsilanes (*130*, *180*). Trimethylchlorosilane, dimethyldichlorosilane, and hexamethyldisiloxane do not react with bromine alone, but will do so when exposed to the action of a mixture of chlorine and bromine with uv irradiation (*281*). Bromine chloride, which is in equilibrium with the two halogens, may be the actual brominating agent:

$$(CH_3)_3SiCl + BrCl \rightarrow BrCH_2(CH_3)_2SiCl + HCl$$

Chlorobromination gives preponderately monobromo-substitution products; the proportion of chloroalkylsilanes is small.

Bromoalkylsilanes are also obtained by the addition of bromine (*256*, *295*) or hydrogen bromide (*256*) to vinylsilanes:

$$CH_2{=}CHSiCl_3 + Br_2 \xrightarrow{Cl_2} BrCH_2{-}CHBrSiCl_3 \ (295)$$

$$CH_2{=}CH(CH_3)_3Si + HBr \xrightarrow{\text{benzoyl peroxide}} BrCH_2{-}CH_2(CH_3)_3Si \ (256)$$

Hydrogen bromide can also be added to the allyl group (*268*), Markovnikov's rule being followed:

$$CH_2{=}CH{-}CH_2(CH_3)_3Si + HBr \rightarrow CH_3{-}CHBr{-}CH_2(CH_3)_3Si$$

However, the reaction of even allyltrimethylsilane with bromine leads to cleavage of the Si—C bond (*268*):

$$CH_2{=}CH{-}CH_2(CH_3)_3Si + Br_2 \rightarrow CH_2{=}CH{-}BrCH_2 + (CH_3)_3SiBr$$

The possibility of the addition of the Si—H bond to bromoolefins demonstrates a further analogy between preparations of the bromo- and chloroorganosilanes. Triorganosilanes can add to allyl bromide under uv irradiation, but the bromine will go to the silicon atom in the presence of $AlCl_3$ (*54*):

$$R_3SiH + CH_2{=}CH{-}CH_2{-}Br \xrightarrow{AlCl_3} R_3SiBr + CH_2{=}CH{-}CH_3$$

$$R_3SiH + CH_2{=}CH{-}CH_2{-}Br \xrightarrow{uv} BrCH_2CH_2CH_2R_3Si$$

4.6.2 Physical Constants (Table)

The constants of a number of selected bromosilanes are given in Table 25.

4.6.3 Chemical Properties

The bromo compounds occupy a key position at least as interesting as that of the chloroorganosilanes with respect to subsequent conversion into other organofunctional silanes. On the one hand, the lower electronegativity of bromine generally permits it to be more easily exchanged in the organic group than chlorine and, on the other hand, it also weakens the Si—C bond less than a chlorine substituent.

Reactions of the bromo compounds usually proceed more readily and under less severe conditions than those of the chloro compounds. The danger of cleavage reactions is smaller, so that siloxanes can also be used in this case. Electronegative substituents on the Si atom—of which siloxane oxygen is one—again increase the reactivity of the C_α—Br bonds, as with the chloro derivatives. Bis-bromomethyl-tetramethyldisiloxane reacts more rapidly with carboxylic-acids and potassium thiocyanate than bromomethyltrimethylsilane (*171*).

The reaction-promoting influence of oxygen and the reaction-inhibiting influence of amine nitrogen is shown very strikingly by the following sequence of decreasing reactivities of the Br—CH_2 bond with respect to diethylamine:

$$BrCH_2(CH_3)_2SiOC_2H_5 > BrCH_2(CH_3)_2C_2H_5Si > BrCH_2(CH_3)_2SiN(C_2H_5)_2$$

The differences are so marked that the first compound mentioned reacts

TABLE 25

BROMOORGANOSILANES[a, b]

Compound	Bp (°C)	at p (mm Hg)	n_D^t	at t (°C)	d_4^t (gm/cc)	at t (°C)
$BrCH_2(CH_3)_3Si$	115.5 (*281*)	742	1.4423 (*281*)	25	1.170 (*281*)	25
$BrCH_2(CH_3)_2SiCl$	130 (*281*)	740	1.4630 (*281*)	25	1.375 (*281*)	25
$BrCH_2(CH_3)SiCl_2$	140–141 (*281*)	740	1.4750 (*281*)	25	1.57 (*281*)	25
$Br_2CH(CH_3)SiCl_2$	86–91 (*281*)	25	1.5185 (*281*)	25	—	—
$(BrCH_2)_3SiBr$	123 (*321*)	5	—	—	2.3440 (*321*)	20
$(BrCH_2)_2SiBr_2$	107 (*321*)	9	—	—	2.4614 (*321*)	20
$BrCH_2SiBr_3$	70 (*321*)	6	—	—	2.5730 (*321*)	20
$BrCH_2(CH_3)_2SiOC_2H_5$	55 (*250*)	30	1.4409 (*250*)	20	—	—
$BrCH_2(CH_3)Si(OC_2H_5)_2$	79 (*249*)	22	1.4350 (*249*)	20	—	—

[a] Numbers in parentheses indicate reference numbers.

[b] For siloxanes see Tables 36 and 40.

exothermically even at room temperature, while the last-mentioned reacts very slowly even at the boil (*171*).

The following reactions also illustrate this point; these have been chosen to parallel reactions discussed in connection with the chloroalkylsilanes.

Bromine in the bromomethyl group of siloxanes can be replaced directly by the hydroxyl group, using KOH as the reagent. This reaction does not proceed satisfactorily with $ClCH_2{-}Si{\equiv}$ (Section 4.5.4).

$$BrCH_2Si{\equiv} + OH^{\ominus} \rightarrow HOCH_2Si{\equiv} + Br^{\ominus}$$

Unlike chloro compounds, the bromo compounds can yield ethers with alcohols if tertiary amines are present to act as hydrogen-bromide acceptors (*242*):

$$BrCH_2Si{\equiv} + HOR \xrightarrow{NR_3} ROCH_2Si{\equiv} + [HNR_3]Br$$

$$BrCH_2(CH_3)_2Si{-}O{-}Si(CH_3)_2CH_2Br + 2HOR \xrightarrow{2NR_3} ROCH_2(CH_3)_2SiOSi(CH_3)_2CH_2OR + 2[HNR_3]Br$$

The reaction with 1,2-diols also proceeds very smoothly. Monosiladioxanes are formed as a result of ring closure if an Si—OR group is present (*242*):

$$BrCH_2(CH_3)_2SiOC_2H_5 + HOCH_2{-}CH_2OH \xrightarrow{NR_3}$$

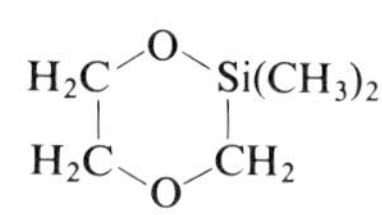

Bromomethylsilicon compounds react rapidly and in excellent yield with mercaptans even at low temperatures in the presence of tertiary amines (*242*, *245*):

$$BrCH_2Si{\equiv} + HSR \xrightarrow{NR_3} RSCH_2Si{\equiv} + [HNR_3]Br$$

This reaction also takes place with siloxanes without side reactions.

Correspondingly, the reactions of bromomethylsilanes with ammonia or amines also proceed more readily than those of the chloromethylsilicon compounds, and these reactions are particularly applicable to the siloxanes (*241*).

Many reactions with compounds containing active CH groups also proceed more readily. Chloromethylsilicon compounds hardly condense with cyanoacetic ester in the presence of triethylamine, but bromomethylsilicon compounds react readily (see Section 4.5.4) (*248*). In the malonic ester synthesis, however, as with chloro derivatives, sodium ethoxide must be used as the condensing agent, or sodiomalonic ester must be employed as the starting material (*266*).

4.7 Hydroxyorganosilanes and Their Esters

Compounds belonging to this group are characterized by the partial formulas:

$$\geq Si-R-OH \quad \text{and} \quad \geq Si-R-O-\underset{\underset{O}{\parallel}}{C}-R'$$

This includes cases where the carbon chain of the aliphatic R radicals is interrupted by the insertion of other atoms such as O or S, for example:

$$\geq Si-R-O-R'-OH, \qquad \geq Si-R-S-R'-OH$$

4.7.1 Preparation

The formally simplest route to the hydroxyorganosilanes, the addition of the Si—H bond to unsaturated alcohols, is not feasible, because of the reactivity of the Si—H bond even toward the C—OH groups. The simple conversion of Si—CH_2—Cl into Si—CH_2—OH groups by the action of bases is impracticable as has been shown in the preceding sections. The only possible substitution reaction of this kind is the replacement of bromine in siloxanes containing bromomethyl groups by the action of KOH (*242*):

$$\geq Si-CH_2-Br+KOH \rightarrow Si-CH_2-OH+KBr$$

A far more important route for the preparation of hydroxyorganosilanes is the roundabout way through their carboxylic acids or silicic acid esters. These yield the desired compounds after saponification:

$$\geq Si-R-OOC-R' \xrightarrow{H_2O} \geq Si-R-OH+HOOCR'$$

$$\geq Si-R-O-Si\leq \xrightarrow{H_2O} \geq Si-R-OH+HOSi\leq$$

The earliest known process used this method (*59*, *173*):

$$ClC_2H_4(C_2H_5)_3Si \xrightarrow{CH_3COOK} CH_3COOC_2H_4(C_2H_5)_3Si \xrightarrow[KOH]{\text{alcohol}} HOC_2H_4(C_2H_5)_3Si$$

The individual steps can be modified in various ways, as will appear from the following examples:

$$ClCH_2(CH_3)_3Si \xrightarrow{CH_3COOK} CH_3COOCH_2(CH_3)_3Si \xrightarrow[H_2SO_4]{CH_3OH} HOCH_2(CH_3)_3Si \quad (247, 284, 285)$$

$$[ClCH_2(CH_3)_2Si]_2O \xrightarrow[(C_2H_5)_3N]{CH_3COOH} [CH_3COOCH_2(CH_3)_2Si]_2O \xrightarrow[HCl]{CH_3OH} [HOCH_2(CH_3)_2Si]_2O \quad (138)$$

$$(CH_3)_3Si{-}Cl \xrightarrow[Na]{(CH_3)_3Si{-}O{-}R{-}Cl} (CH_3)_3Si{-}O{-}R{-}(CH_3)_3Si \xrightarrow{H_2O} (CH_3)_3Si{-}R{-}OH \ (275, 282)$$

$$CH_3SiHCl_2 \xrightarrow[Pt]{CH_2{=}CH{-}CH_2OOCCH_3} CH_3COO(CH_2)_3(CH_3)SiCl_2\ (I) \xrightarrow{H_2O} HO(CH_2)_3(CH_3)SiO_{2/2}\ (222)$$

or

$$(I) \xrightarrow{SO_3HCl} {-}\underset{Cl}{\overset{CH_3}{Si}}{-}(CH_2)_3O{-}\underset{O}{\overset{O}{S}}{-}O{-} \xrightarrow{H_2O} HOSO_3(CH_2)_3(CH_3)Si(OH)_2 \xrightarrow[{[H^+]}]{H_2O} HO(CH_2)_3(CH_3)Si(OH)_2\ (222)$$

Other syntheses are possible through the reaction of carboxyalkylsilanes or carboxyalkylsiloxanes or their esters with tetrahydroborates. In this process, Si—C bonds are simultaneously converted into Si—H bonds and these then form siloxanes during the hydrolytic working-up process (*169*).

$$ROOC(CH_2)_3(CH_3)_2SiCl \xrightarrow[H_2O]{NaBH_4} \tfrac{1}{2}[HO(CH_2)_4(CH_3)_2Si]_2O$$

The reactions of organosilyl Grignard compounds with epoxides (*269*) or aldehydes (*315*) have been employed for the preparation of hydroxyorganosilanes, e.g.:

$$ClMgCH_2(CH_3)_3Si \xrightarrow[H_2O]{\overset{O}{CH_2{-}CH_2}} HO(CH_2)_3(CH_3)_3Si$$

It is immaterial for their further application whether the hydroxy-functional compounds have other bridge atoms, such as oxygen or sulfur, in the alkyl residues, so that simple condensation processes are possible.

A bromomethyl group at the end of a siloxane chain can be converted into a hydroxyl function very smoothly and in good yields by reaction with monothioglycol in the presence of tertiary amines (*245*) or bases (*246*).

$$\sim{-}\underset{CH_3}{\overset{CH_3}{Si}}{-}CH_2{-}Br \xrightarrow[NR_3,\ KOH]{HS(CH_2)_2OH} \sim{-}\underset{CH_3}{\overset{CH_3}{Si}}{-}CH_2{-}S(CH_2)_2OH$$

Bis-(β-hydroxyethoxymethyl)-tetramethyldisiloxane has been obtained in relatively low yield (35 %) from chloromethyldimethylchlorosilane and sodium glycollate with subsequent hydrolysis (*8*):

$$2ClCH_2(CH_3)_2SiCl \xrightarrow[H_2O]{NaOCH_2CH_2OH} [HOCH_2CH_2OCH_2(CH_3)_2Si]_2O$$

4.7.2 Physical Constants (Table)

The physical constants of a number of hydroxyorganosilanes are given in Table 26.

TABLE 26

HYDROXYORGANOSILANES[a, b]

Compound	Mp (°C)	Bp (°C)	at p (mm Hg)	n_D^t	at t (°C)	d_4^t (gm/cc)	at t (°C)
$HOCH_2(CH_3)_3Si$	—	121.7–121.9 (*284*)	751	1.4169 (*284*)	25	0.8261 (*284*)	25
$(HOCH_2)_2(CH_3)_2Si$	—	130 (*283*)	27	1.4611 (*283*)	25	0.993 (*283*)	11
$HOCH_2(CH_2)_2(CH_3)_3Si$	—	82 (*282*)	24	1.4250 (*282*)	24	0.822 (*282*)	25
$CH_3CH(OH)CHCH_2(CH_3)_3Si$	—	48 (*315*)	10	1.4281 (*315*)	20	—	—
$(HOCH_2CH_2CH_2)_2(CH_3)_2Si$	—	176 (*282*)	24	1.4649 (*282*)	25	0.940 (*282*)	25
$HOCH_2(CH_2)_3(CH_3)_3Si$	—	96 (*282*)	25	1.4315 (*282*)	25	0.830 (*282*)	25
$HOCH_2(CH_2)_4(CH_3)_3Si$	—	110 (*282*)	25	1.4358 (*282*)	25	0.841 (*282*)	25
p-$HOC_6H_4(CH_3)_3S$	74–74.2 (*282*)	—	—	—	—	—	—
m-$HOC_6H_4(CH_3)_3Si$	—	70 (*31*)	1	1.5190 (*31*)	20	0.977 (*31*)	20
o-$HOC_6H_4(CH_3)_3S$	8.6–9.2 (*282*)	113 (*282*)	26	1.515 (*282*)	25	0.966 (*282*)	25
$(p$-$HOC_6H_4)_2(CH_3)_2Si$	170–171 (*282*)	—	—	—	—	—	—

[a] Numbers in parentheses indicate reference numbers.
[b] For siloxanes see Tables 36 and 45.

4.7.3 Chemical Properties

What has been said with respect to the chloroalkylsilanes obviously also applies to the reactivities of the hydroxyalkylsilanes. If the OH group is attached to the β-C atom, β-decomposition takes place readily under the influence of both bases and acids. On being heated with sulfuric acid, β-hydroxypropyltrimethylsilane is decomposed at the Si—C bond with the liberation of an olefin (*315*).

An OH group on the α-carbon is more reactive than the corresponding alcohol because of the proximity of the electropositive Si atom. Hydroxymethyltrimethylsilane reacts three times faster with phenyl isocyanate than methanol, and six times faster than neopentanol (*284*). The similarity to the alcohols is increased when the hydroxyl group moves away from the Si atom beyond the β-carbon. Hydroxyalkylsilanes with OH in the δ-position generally behave as alcohols. Hydroxyalkylsilicon compounds form polyurethanes with diisocyanates (*272*). The acetic acid esters of bis-hydroxyorgano-organodisiloxanes give polyesters with dibasic acids (*273*).

Hydroxyarylsilanes (*282*, *291*) have the properties of phenols; for example, they can condense with formaldehyde (*276*). All these reactions are interesting for the incorporation of siloxanes into organic synthetic resins and elastomers (cf. also Section 7.4).

4.8 Carboxyorganosilanes and Their Esters

The compounds of this group are characterized by the partial formulas:

$$\equiv\mathrm{Si{-}R{-}\overset{\displaystyle O}{\overset{\|}{C}}{-}OH} \quad \text{and} \quad \equiv\mathrm{Si{-}R{-}\overset{\displaystyle O}{\overset{\|}{C}}OR'}$$

The aliphatic hydrocarbon chain of the R groups may also be interrupted by other atoms, such as oxygen, or by the carbonyl group.*

* It is customary to name the position of the carboxyl in carboxyalkylsilicon compounds as:

$$[\alpha]\ \equiv\mathrm{Si{-}CH_2{-}COOH} \qquad [\beta]\ \equiv\mathrm{Si{-}CH_2{-}CH_2{-}COOH}.$$

If the carbon of the carboxyl group is also counted, one could consider the carbonyl oxygen as a substituent. The following would then hold:

$$[\alpha]\ \equiv\mathrm{Si{-}\overset{\displaystyle O}{\overset{\|}{C}}OH} \qquad [\beta]\ \equiv\mathrm{Si{-}CH_2{-}\overset{\displaystyle O}{\overset{\|}{C}}OH} \qquad [\gamma]\ \equiv\mathrm{Si{-}CH_2{-}CH_2{-}\overset{\displaystyle O}{\overset{\|}{C}}OH}$$

The latter nomenclature has the advantage of maintaining consistency with the lettering of, for example, halogen derivatives. β-Carboxy compounds could correspond to β-halogen compounds to which they in fact approximate in their chemical behavior. The accepted method, however, means that α-carboxyl compounds must be compared chemically with β-halogen compounds, and β-carboxyl with γ-halogen compounds. The same holds for the cyanoalkylsilanes (Section 4.11).

4.8.1 PREPARATION

Carboxyalkylsilicon compounds can be prepared by the addition of Si—H bonds to unsaturated carboxylic acids (*63*). A better method is the reaction with esters, since the Si—H bond can react with the carboxyl group (Section 3.4.3); peroxides or platinized charcoal are usually employed as catalysts:

$$(C_6H_5)_3SiH + CH_2{=}CH(CH_2)_8COOH \xrightarrow{(C_6H_5COO)_2} HOOC(CH_2)_{10}(C_6H_5)_3Si\ (63)$$

$$C_6H_5SiH_3 + CH_2{=}CH(CH_2)_8COOCH_3 \xrightarrow[\text{perbenzoate}]{t\text{-butyl}} CH_3OOC(CH_2)_{10}(C_6H_5)SiH_2\ (277)$$

$$SiHCl_3 + CH_2{=}CHCOOCH_3 \xrightarrow{Pt/C} CH_3OOCCH_2CH_2SiCl_3\ (196)$$

$$CH_3SiHCl_2 + CH_2{=}CHCOOCH_3 \xrightarrow{Pt/C} CH_3OOCCH_2CH_2(CH_3)SiCl_2\ (84)$$

A synthesis starting with partially substituted silicon hydrides and unsaturated nitriles proceeds via an addition reaction (Section 4.11.1)

$$CH_3SiHCl_2 + CH_2{=}CHCN \rightarrow NC{-}CH_2{-}CH_2(CH_3)SiCl_2$$

and produces the carboxyl group by saponification of the nitrile (*24*):

$$NC{-}CH_2{-}CH_2(CH_3)SiCl_2 \xrightarrow{H_2O\ [H^+]} HOOC{-}CH_2{-}CH_2(CH_3)SiO_{2/2}$$

The corresponding esters are obtained if the solvolysis is carried out with acidified alcohols. Halogen atoms on the silicon are of course simultaneously converted into alkoxy groups (*23*):

$$NC{-}CH_2{-}CH_2{-}SiCl_3 \xrightarrow[HCl]{C_2H_5OH} C_2H_5OOC{-}CH_2{-}CH_2{-}Si(OC_2H_5)_3$$

Other preparations are based on substitution reactions and start with haloalkylsilicon compounds. The most interesting is the reaction of halomethyl groups with compounds containing active CH groups:

$$\gt Si(CH_2)_nHal + {}^{\ominus}CH(COOR)R' \rightarrow \gt SiCH_{2n}CH(COOR)R' + Hal^{\ominus}$$

Bromo derivatives react more readily than chloro compounds, but β-derivatives cannot be used because of the sensitivity of the Si—C bond in an alkaline medium to cleavage (Section 4.5.4). Details of some syntheses are given below.

(*a*) *Malonic ester syntheses*

Chloro-, bromo-, and iodomethylmethylsilanes react with sodiomalonic ester in ethanol to give the silalkyl-substituted malonic esters (*253*, *261*, *266*):

$$ClCH_2(CH_3)_3Si + Na{-}CH(COOR)_2 \rightarrow (ROOC)_2CH{-}CH_2(CH_3)_3Si$$

The carboxyorganosilane is then obtained by hydrolysis and decarboxylation (*261*):

$$\begin{array}{l}ROOC\diagdown\\ \qquad CH{-}CH_2(CH_3)_3Si\\ ROOC\diagup\end{array} \xrightarrow[-CO_2]{H_2O} HOOC(CH_2)_2(CH_3)_3Si$$

Hydrolytic cleavage of an Si—C bond with sulfuric acid and water gives the disiloxane with the liberation of methane (cf. Section 5.3) (*254*):

$$2HOOC(CH_2)_2(CH_3)_3Si \xrightarrow[H_2O]{H_2SO_4} [HOOC(CH_2)_2(CH_3)_2Si]_2O + 2CH_4$$

Likewise, haloalkylalkoxysilanes can react with sodiomalonic esters in about 60% yield (*266*):

$$\begin{array}{l}\quad\diagup COOC_2H_5\\ Na{-}CH\\ \quad\diagdown COOC_2H_5\end{array} + \begin{array}{l}ClCH_2(CH_3)Si(OC_2H_5)_2 \rightarrow (C_2H_5OOC)_2CH{-}CH_2(CH_3)Si(OC_2H_5)_2\\ ClCH_2(CH_3)_2SiOC_2H_5 \rightarrow (C_2H_5OOC)_2CH{-}CH_2(CH_3)_2SiOC_2H_5\end{array}$$

Heating the latter reaction product with a mixture of concentrated hydrochloric acid and glacial acetic acid then gives a 95% yield of the bis-(carboxyalkyl)-disiloxane:

$$2(C_2H_5OOC)_2CH{-}CH_2(CH_3)_2SiOC_2H_5 \xrightarrow[HOAc]{HCl} [HOOCCH_2{-}CH_2(CH_3)_2Si]_2O$$

Chloromethylpentamethyldisiloxane is analogously converted into the carboxy derivative with a yield of 70 to 74% (*266*).

(*b*) *Cyanoacetic ester syntheses*

Chloro-, bromo-, and iodoalkylsilicon compounds react with sodiocyanoacetic ester (*266*) or cyanoacetic ester itself in the presence of a sodium alkoxide (*289*):

$$ClCH_2(CH_3)Si(OC_2H_5)_2 + NaCHCN(COOC_2H_5) \rightarrow (C_2H_5OOC)CN{-}CH{-}CH_2(CH_3)Si(OC_2H_5)_2$$

Bromo derivatives can also be condensed with cyanoacetic ester in the presence of triethylamine (*248*):

$$BrCH_2(CH_3)_2SiOC_2H_5 + CNCH_2COOR \xrightarrow{(C_2H_5)_3N} \begin{array}{l}ROOC\diagdown\\ \qquad CH{-}CH_2(CH_3)_2SiOC_2H_5\\ NC\diagup\end{array}$$

Hydrolysis and subsequent decarboxylation and saponification then yield the carboxyalkylsiloxane (*248*):

$$\begin{array}{l}ROOC\diagdown\\ \qquad CH{-}CH_2(CH_3)_2SiOC_2H_5\\ NC\diagup\end{array} \xrightarrow[-CO_2]{H_2O\,[H^+]} HOOC{-}CH_2{-}CH_2(CH_3)_2SiO_{1/2}$$

(*c*) *Acetoacetic ester syntheses*

The basis for the acetoacetic ester syntheses is less broad—only iodomethyl derivatives have been found useful as starting materials (*263*):

$$ICH_2(CH_3)_3Si + CH_3COCH_2COOC_2H_5 \xrightarrow[\text{excess}]{NaOC_2H_5}$$

$$C_2H_5OOCCH_2CH_2(CH_3)_3Si \xrightarrow{H_2O\ [OH^-]} HOOCCH_2CH_2(CH_3)_3Si$$

Further possible syntheses are as follows.

The reaction of chloromethylsilanes with magnesium gives the Grignard compounds, which can be converted to the organosilylcarboxylic acid compounds with CO_2 (*260*):

$$ClMgCH_2(CH_3)_3Si \xrightarrow[H_2O]{CO_2} HOOCCH_2(CH_3)_3Si$$

Carboxyarylsilicon compounds are also accessible in this way, but the expected dicarboxylic acids are obtained in very small yields when α,ω-bis-(chloromethyl)-polydimethylsiloxanes are used as the starting materials, since Grignard compounds can cleave siloxane bonds. The disiloxane still gives a yield of about 20% (*88*).

Chloromethylsiloxanes have been caused to react with sodium salts of dicarboxylic acids:

$$[ClCH_2(CH_3)_2Si]_2O + 2HOOC(CH_2)_4COONa \rightarrow$$

$$[HOOC(CH_2)_4COOCH_2(CH_3)_2Si]_2O + 2NaCl$$

In the products of this reaction, just as in the corresponding reaction of alkali-metal glycollates with the $ClCH_2Si{\lessgtr}$ group (Section 4.7.1), the aliphatic chain is modified by O atoms, and here by carbonyl groups as well (*137*).

Silanes containing carboxyaryl groups can also be obtained by the oxidation of suitable substituents in the phenyl group. Triphenyl-*p*-tolylsilane gives 4-carboxyphenyltriphenylsilane (*71*) in 81% yield on oxidation with CrO_3:

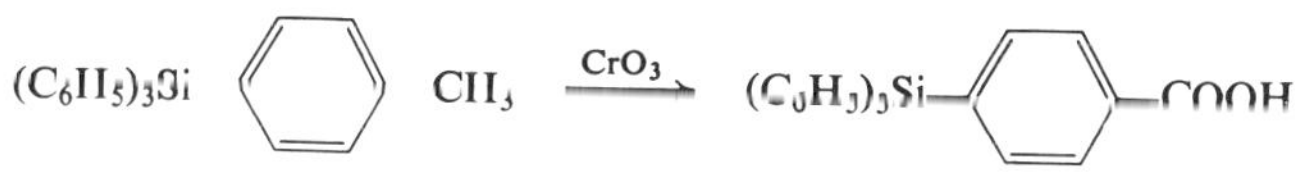

The corresponding dicarboxy compound is obtained similarly (*300*).

4.8.2 PHYSICAL CONSTANTS (TABLE)

Physical constants for some compounds of the types discussed above are given in Table 27 (p. 170).

TABLE 27

CARBOXYORGANOSILANES AND THEIR ESTERS[a, b]

Compound	Mp (°C)	Bp (°C)	at p (mm Hg)	n_D^t	at t (°C)	d_4^t (gm/cc)	at t (°C)
$HOOCCH_2(CH_3)_3Si$	40 (*260*)	—	—	—	—	—	—
$HOOCCH_2(CH_3)_2C_6H_5Si$	90 (*260*)	—	—	—	—	—	—
$HOOCCH_2CH_2(CH_3)_3Si$	22 (*260*)	147 (*260*)	65	1.4279 (*260*)	20	0.9196 (*263*)	20
$HOOCCH_2CH_2CH_2(CH_3)_3Si$	4 (*253*)	91 (*253*)	5	1.4325 (*253*)	20	0.9098 (*253*)	20
$HOOC(CH_2)_{10}(C_6H_5)_3Si$	58.5–59 (*63*)	250–270 (*63*)	7×10^{-5}	—	—	—	—
o-$HOOC—C_6H_4—(CH_3)_3Si$	97.5–98.5 (*32*)	—	—	—	—	—	—
m-$HOOC—C_6H_4—(CH_3)_3Si$	113.7–114.2 (*218*)	—	—	—	—	—	—
p-$HOOC—C_6H_4—(CH_3)_3Si$	117–118 (*32, 218*)	—	—	—	—	—	—
(*p*-$HOOC—C_6H_4)_2(CH_3)_2Si$	288 (decomp.) (*300*)	—	—	—	—	—	—
$CH_3OOC(CH_2)_2(CH_3)SiCl_2$	—	80.0 (*84*)	9.6	—	—	—	—
$C_2H_5OOCCH_2(CH_3)_3Si$	—	157 (*81*)	730	1.4149 (*81*)	20	0.8762 (*81*)	20
$C_2H_5OOC(CH_2)_2(CH_3)_3Si$	—	93 (*263*)	40	1.4198 (*263*)	20	0.8763 (*263*)	20
$C_2H_5OOC(CH_2)_2(CH_3)_2C_6H_5Si$	—	106 (*263*)	2	1.4972 (*263*)	20	0.9856 (*263*)	20
$C_2H_5OOC(CH_2)_2(CH_3)Si(OC_2H_5)_2$	—	95–95.5 (*23*)	5	1.4121 (*23*)	25	—	—
$C_2H_5OOC(CH_2)_2Si(OC_2H_5)_3$	—	95 (*23*)	2.5	1.4125 (*23*)	25	0.982 (*23*)	25
$(C_2H_5OOC)_2CHCH_2(CH_3)_3Si$	—	119 (*261*)	13	1.4312 (*261*)	20	0.9680 (*261*)	20
$C_2H_5OOC(CH_2)_3(CH_3)Si(OC_2H_5)_2$	—	127–129 (*23*)	8	1.4208 (*23*)	25	—	—
$C_2H_5OOC(CH_2)_3Si(OC_2H_5)_3$	—	86–89 (*23*)	0.5	1.4152 (*23*)	25	—	—
$(C_2H_5OOC)_2CHCH_2(CH_3)_2C_6H_5Si$	—	132 (*261*)	1	1.4900 (*261*)	20	1.042 (*261*)	20

[a] Numbers in parentheses indicate reference numbers.

[b] For siloxanes see Table 36.

4.8.3 Chemical Properties

The properties of carboxyalkylsilicon compounds depend on the position of the carboxyl group relative to the silicon atom. The acid strength of the carboxyl group is depressed by the proximity of a trimethylsilyl group (for further information see Section 6.4).

A carboxyl group directly attached to Si is labile: on heating to about 170°C $(C_6H_5)_3SiCOOH$ decomposes into triphenylsilanol with the liberation of CO (*30*); $(CH_3)_3SiCH_2COOH$ is sensitive to hydrolysis (*315*). α-Carboxyorganosilanes or their esters having the carbonyl oxygen in the β-position behave like β-halogen or β-hydroxy compounds (see footnote, Section 4.8). They readily undergo β-decomposition (*81*) (cf. Section 4.5.4); not only acids and bases, but also ethanol and bromine cause cleavage (*81*) in these compounds, e.g.:

$$C_2H_5OOCCH_2(CH_3)_3Si + HCl \rightarrow (CH_3)_3SiCl + CH_3COOC_2H_5$$

$$C_2H_5OOCCH_2(CH_3)_3Si + C_2H_5OH \rightarrow (CH_3)_3SiOC_2H_5 + CH_3COOC_2H_5$$

$$C_2H_5OOCCH_2(CH_3)_3Si + Br_2 \rightarrow (CH_3)_3SiBr + BrCH_2COOC_2H_5$$

If the carboxyl group moves to the β-atom (oxygen in the γ-position), it no longer appreciably weakens the Si—C bond. Silanes of the type $HOOC(CH_2)_n(CH_3)_3Si$ with $n > 2$ can thus be converted into disiloxanes (*254*) by means of sulfuric acid and water, with the liberation of methane (compare Sections 4.8.1 and 5.3).

The polarity and reactivity of the carboxyl group make carboxyorganosilanes or carboxyorganosiloxanes interesting for technical purposes. The addition of small amounts of carboxyundecyltriphenylsilane can considerably enhance the lubricant action of silicone oils. It has also been shown that dicarboxyorganosilanes can be worked up to polyesters with ethylene glycol (*271*), glycerol (*301*), and bis-hydroxymethyl-dimethylsilane (*301*). The general behavior pattern of these products is comparable with that of their hydrocarbon analogs (cf. Section 7.4).

4.9 Epoxyorganosilanes

Among the ether-functional monomeric and other low-molecular-weight silicon compounds, those containing epoxy groups have become particularly interesting for silicone chemistry.

These compounds can be obtained readily and in good yield by the addition to unsaturated epoxides of silanes or siloxanes containing hydrogen attached to the silicon. This reaction has also proved of value for the synthesis of other ether-functional compounds. Addition reactions of allyl glycidyl ether to silanes (*207*, *209*) and siloxanes (*11*, *207*, *209*, *286*) have been described repeatedly in the literature. Like other reactions of this type, they are catalyzed by ultraviolet light, peroxides, and platinum compounds:

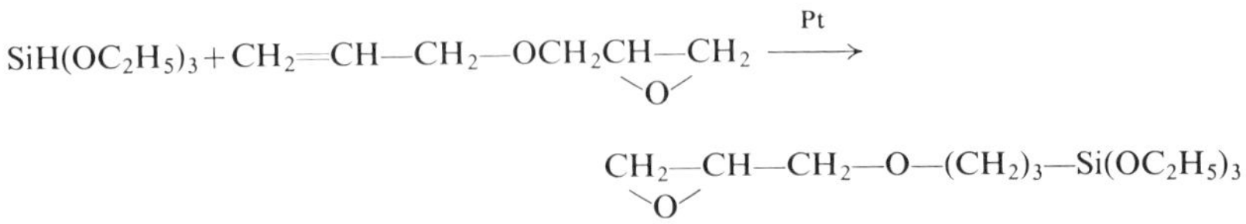

Such compounds can also sometimes be made by the elimination of hydrogen halide from organofunctional silanes containing adjacent halogen and hydroxy groups, which can be obtained by the Grignard reaction with epichlorohydrin (yield ca. 30%) or chloroacetone (yield ca. 70%):

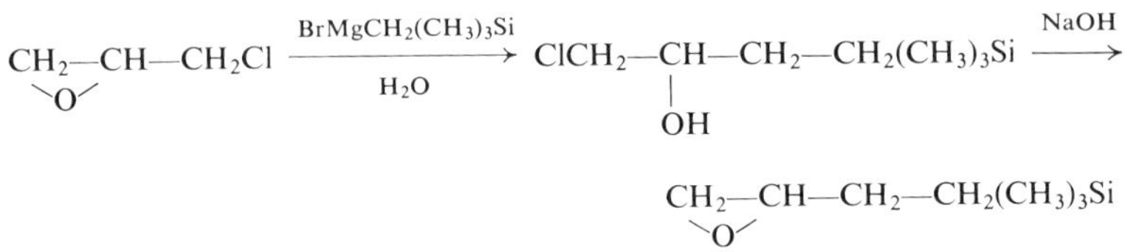

(*33*)

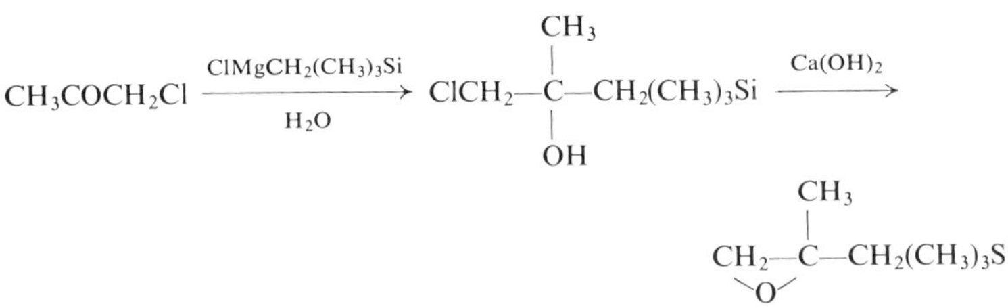

(*225*)

Silanes and siloxanes with epoxy groups as substituents in the organic groups are interesting starting materials for the preparation of modified siloxanes. The reaction of the epoxy groups of epoxyalkylsiloxanes with the OH groups of polyethers yields siloxanyl polyethers. These are interesting because of their solubility in water or alcohol (*49*) or their surface-active properties (*287*). Epoxyalkylsilanes or epoxyalkylsiloxanes can be incorporated directly into epoxy resins as such and have a favorable influence on certain properties (Section 10.6). Resinous materials are also obtained by the reaction of the diglycidyl ether of bisphenol A and and γ-aminopropyltriethoxysilane, followed by thermal condensation (*112*). Finally, epoxyalkylsilanes containing silicon-functional groups, particularly —OC_2H_5 groups, are adhesives for the bonding of epoxy and phenol resins to glass (see Table 28 and Section 10.7.2).

TABLE 28

EPOXYORGANOSILANES[a, b]

Compounds	Mp (°C)	Bp (°C)	at p (mm Hg)	n_D^t	at t (°C)	d_4^t (gm/cc)	at t (°C)
$\underbrace{CH_2\text{—}C}_{O}(CH_3)\text{—}CH_2(CH_3)_3Si$	—	67.0 (*225*)	10	1.4243 (*225*)	20	0.8338 (*225*)	20
$\underbrace{CH_2\text{—}C}_{O}(CH_3)\text{—}CH_2(CH_3)_2(C_2H_5)Si$	—	154.0 (*225*)	750	1.4393 (*225*)	20	0.8539 (*225*)	20
$\underbrace{CH_2\text{—}CH}_{O}\text{—}(CH_2)_2(CH_3)_3Si$	53–53.5 (*33*)	58–60 (*33*)	20	—	—	—	—
$\underbrace{CH_2\text{—}C}_{O}(CH_3)\text{—}(CH_2)_3(CH_3)(C_2H_5)_2Si$	—	85–86 (*225*)	3	1.4452 (*225*)	20	0.8581 (*225*)	20
$\underbrace{CH_2\text{—}C}_{O}(CH_3)\text{—}(CH_2)_3(CH_3)_2(C_2H_5)Si$	—	207.5 (*225*)	755	1.4382 (*225*)	20	0.8484 (*225*)	20
$\underbrace{CH_2\text{—}CH}_{O}\text{—}CH_2O(CH_2)_3(CH_3)_2SiOC_2H_5$	—	100 (*209*)	3	1.4337 (*209*)	25	0.947 (*209*)	25
$\underbrace{CH_2\text{—}CH}_{O}\text{—}CH_2O(CH_2)_3(CH_3)Si(OC_2H_5)_2$	—	122–126 (*209*)	5	1.4305 (*209*)	25	0.978 (*209*)	25
$\underbrace{CH_2\text{—}CH}_{O}\text{—}CH_2O(CH_2)_3Si(OC_2H_5)_3$	—	122–125 (*209*)	3	1.4256 (*209*)	25	1.003 (*209*)	25
$\underbrace{CH_2\text{—}CH}_{O}\text{—}CH_2O(CH_2)_3(C_6H_5)Si(OC_2H_5)_2$	—	137–141 (*209*)	0.1	1.4769 (*209*)	25	1.044 (*209*)	25

[a] Numbers in parentheses indicate reference numbers.
[b] For siloxanes see Table 36.

4.10 Aminoorganosilanes

4.10.1 Preparation

These compounds can be prepared relatively simply from haloorganosilanes and ammonia or amines (*65*, *174*, *274*). This method is mainly applicable to haloalkylsilanes, particularly halomethylsilanes. If the substitution is to be carried out on siloxanes, it is best to use bromo derivatives as the starting materials, since cleavage of the Si—C or Si—O—Si bonds could occur in the chloro compounds.

Primary, secondary, and tertiary amines react with halomethyltrimethylsilane as follows (*174*):

$$ClCH_2{-}R_3Si \xrightarrow{R'NH_2} R'NH{-}CH_2{-}R_3Si$$

$$ClCH_2{-}R_3Si \xrightarrow{R'_2NH} R'_2N{-}CH_2{-}R_3Si$$

$$BrCH_2{-}R_3Si \xrightarrow{NR'_3} [R'_3N{-}CH_2{-}R_3Si]^+Br^-$$

A 20-fold excess of ammonia must be used to obtain the primary amine as the main product [yield between 50 and 70% (*174*)]:

$$ClCH_2{-}R_3Si \xrightarrow{NH_3} H_2N{-}CH_2{-}R_3Si$$

From 20 to 30% of the secondary amine $R_3Si{-}CH_2NHCH_2{-}SiR_3$ is also formed:

$$H_2N{-}CH_2{-}R_3Si \xrightarrow{ClCH_2{-}R_3Si} R_3Si{-}CH_2{-}NH{-}CH_2{-}SiR_3$$

If the silane contains another silicon-functional group in addition, the reaction with primary amine yields either disiloxanes (*171*)

$$BrCH_2(CH_3)_2SiOC_2H_5 \xrightarrow{n\text{-}C_4H_9NH_2} [n\text{-}C_4H_9NHCH_2(CH_3)_2Si]^{\oplus}Br^{\ominus} \xrightarrow{KOH} [n\text{-}C_4H_9NHCH_2(CH_3)_2Si]_2O$$

or disilazanes:

$$2BrCH_2(CH_3)_2SiCl \xrightarrow{CH_3NH_2} BrCH_2(CH_3)_2SiN(CH_3)Si(CH_3)_2CH_2Br$$

which are then cyclized to diazadisilacyclohexanes with excess of amine (*172*):

$$BrCH_2(CH_3)_2Si{-}\underset{}{\overset{CH_3}{\overset{|}{N}}}{-}Si(CH_3)_2CH_2Br \xrightarrow{CH_3NH_2}$$

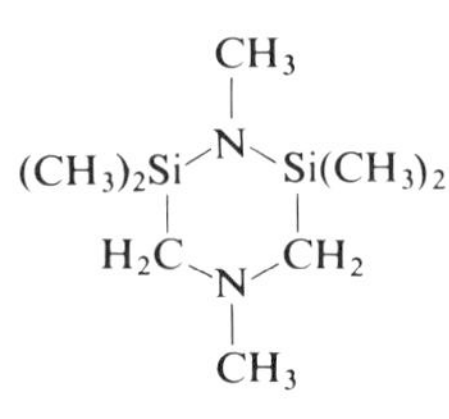

+small amounts of the 2,5-isomer

An oxazadisilacyclohexane, which is analogous to a diazadisilacyclohexane, is formed from bis-(bromomethyl)-tetramethyldisiloxane and an alkylamine (*172*, *241*):

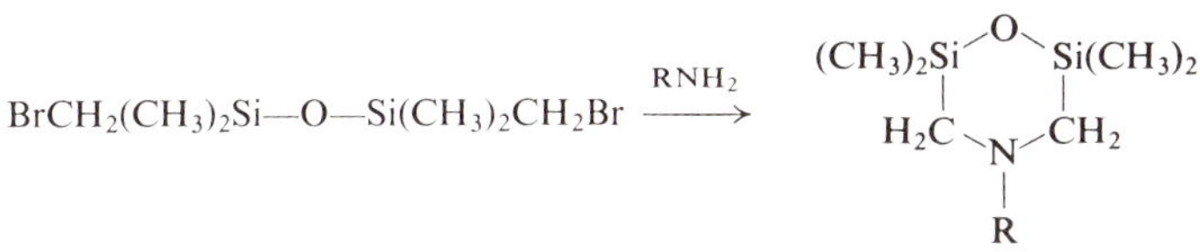

while bromomethyldimethylethoxysilane gives an oxazasilacyclohexane with *N*-methylethanolamine (*242*):

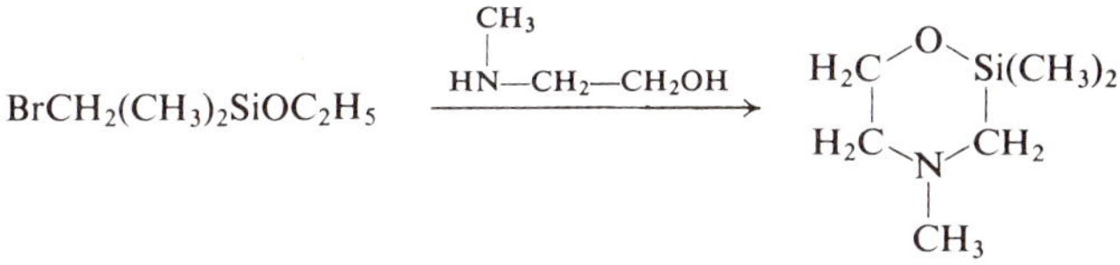

Of the metal amides, sodamide in liquid ammonia will split the Si—C bond in chloromethyltrimethylsilane (*174*), so that a silicon-functional amino compound is formed in place of an organofunctional one:

$$ClCH_2(CH_3)_3Si + NaNH_2 \rightarrow (CH_3)_3Si—NHCH_3 + NaCl$$

p-Haloaryltriorganosilanes can be substituted with amino groups using lithium dimethylamide, the substituent changing its position in the nucleus (*74*, *75*):

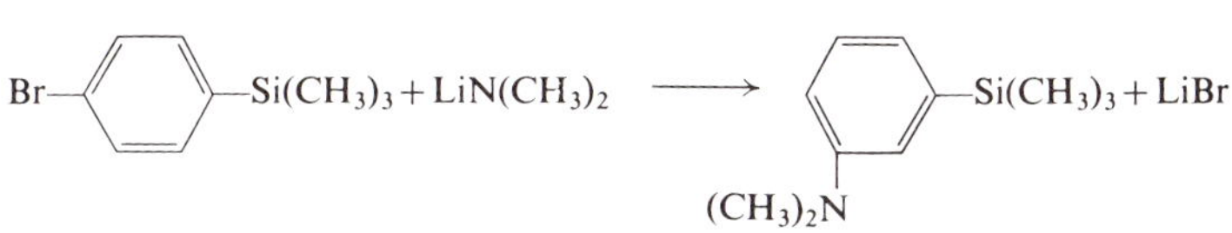

Chloroalkylsilanes can also be converted into aminomethylsilanes by means of phthalimide (*267*):

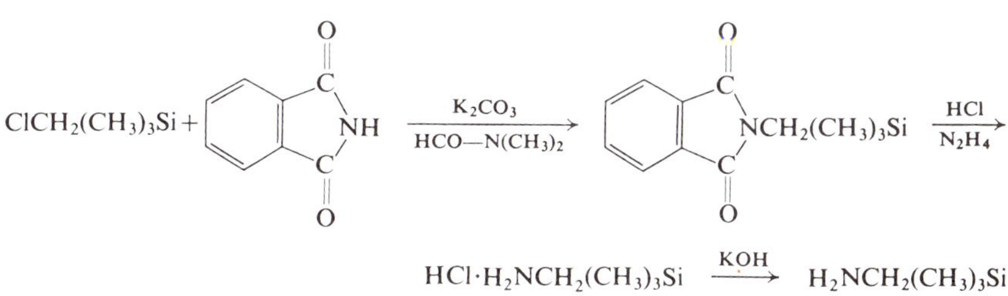

Bis-aminomethyl-dimethylsilane has been obtained in this way from bis-(chloromethyl)-dimethylsilane. The product was easily converted into bis-(aminomethyl)-tetramethyldisiloxane (*87*).

Aminoalkylfunctional silanes with Si—C—N—C bridges are obtained by the condensation of diamines with halopropylsilanes, e.g., as follows (*278*):

$$ClCH_2{-}CH_2{-}CH_2{-}Si(OCH_3)_3 \xrightarrow{H_2NCH_2CH_2NH_2} H_2N{-}CH_2{-}CH_2{-}NH(CH_2)_3\,Si(OCH_3)_3$$

Chlorine or bromine atoms on the α-C atom react more readily than in the γ-position. With diamines, however, they readily lead to cyclic or cross-linked products (*242*).

The reduction of the cyanoalkylsilanes, which can be synthesized by addition reactions, is of considerable interest in the preparation of aminoalkylsilanes (Section 4.11.1). Reduction is carried out either under pressure in the presence of hydrogenation catalysts such as nickel or platinum (*108*, *131*) or, very smoothly, with hydroborates (*170*), which, unlike lithium tetrahydroaluminate, can also be used with siloxanes since they do not cleave the siloxane bond:

$$NCCH_2CH_2(CH_3)SiCl_2 \xrightarrow[CH_3OH]{NaBH_4} H_2N(CH_2)_3(CH_3)Si(OCH_3)_2$$

Aminoarylsilanes have been obtained by the reduction of nitro compounds (*12*).

4.10.2 Physical Constants (Table)

The physical constants of a number of aminoorganosilanes are given in Table 29.

4.10.3 Chemical Properties

The basicity of the amino group in aminoalkylsilanes is determined in part by the substituents attached to the silicon atom. The trimethylsilyl group increases the basicity, an effect which is still noticeable with an amino group in the γ-position. Phenyl substituents or siloxane oxygen, on the other hand, reduce the basicity. However, the part played by the trimethylsilyl group is different in aromatic *p*-amino-substituted silanes (Section 6.4).

All aminoorganosilicon compounds form stable salts with acids (*29*, *174*, *267*). Sulfuric acid splits off a methyl group as methane from aminomethyl- and β-aminoethyltrimethylsilanes. Treating the solution with alkali gives the bis-(aminoalkyl)-disiloxane in almost quantitative yield (*174*, *265*) (cf. Section 5.3).

TABLE 29

AMINOORGANOSILANES[a, b]

Compound	Mp (°C)	Bp (°C) at	p (mm Hg)	n_D^t at	t (°C)	d_4^t (gm/cc) at	t (°C)
$H_2NCH_2(CH_3)_3Si$	—	94 (*267*)	729	1.4168 (*267*)	20	0.7697 (*267*)	20
$(H_2NCH_2)_2(CH_3)_2Si$	−44 (*87*)	70–72 (*87*)	15	—	—	—	—
$H_2NCH_2(CH_3)_2SiOC_2H_5$	—	131.8 (*174*)	740	1.411 (*174*)	25	0.849 (*174*)	25
$H_2NCH_2(CH_3)Si(OC_2H_5)_2$	—	67.5 (*174*)	24	1.4120–1.4126 (*174*)	25	0.914–0.916 (*174*)	25
$H_2NCH_2Si(OC_2H_5)_3$	—	93 (*174*)	26	1.4080 (*174*)	25	0.955 (*174*)	25
$(CH_3)_2NCH_2(CH_3)_3Si$	—	110.1 (*174*)	746	1.4102 (*174*)	25	0.746 (*174*)	25
$H_2NCH_2CH_2(CH_3)_3Si$	—	121 (*267*)	735	1.4241 (*267*)	20	0.7807 (*267*)	20
$H_2NCH_2CH_2CH_2(CH_3)_3Si$	—	145.5 (*267*)	732	1.4295 (*267*)	20	0.7866 (*267*)	20
$H_2NCH_2CH_2CH_2Si(OC_2H_5)_3$	—	217 (*304*)	760	1.4198 (*304*)	25	0.942 (*304*)	25
$H_2N(CH_2)_4(CH_3)Si(OC_2H_5)_2$	—	233 (*305*)	760	1.4293 (*305*)	25	0.905 (*305*)	25
$H_2NCH_2CH_2CH_2CH_2Si(OC_2H_5)_3$	—	123–124 (*108*)	15	1.4222 (*108*)	25	0.934 (*108*)	25
$H_2NCH_2CH_2NH(CH_2)_3Si(OCH_3)_3$	—	140.5 (*278*)	15	1.4416 (*278*)	25	1.01 (*278*)	25
$H_2NCH_2CH_2NH(CH_2)_3Si(OC_2H_5)_3$	—	115–124 (*131*)	0.1–1.1	1.4365 (*131*)	25	—	—
o-$H_2NC_6H_4(CH_3)_3Si$	—	101 (*29*)	10	1.5388 (*29*)	20	0.952 (*29*)	20
m-$H_2NC_6H_4(CH_3)_3Si$	—	111 (*29*)	10	1.5362 (*29*)	20	0.947 (*29*)	20
p-$H_2NC_6H_4(CH_3)_3Si$	—	113 (*29*)	10	1.5393 (*29*)	20	0.947 (*29*)	20
p-$H_2NC_6H_4Si(OC_2H_5)_3$	—	145–150 (*56*)	14	—	—	—	—
m-$(CH_3)_2NC_6H_4(CH_3)_3Si$	—	59–60 (*75*)	0.4	1.5312 (*75*)	25	0.9484 (*75*)	20
p-$(CH_3)_2NC_6H_4(CH_3)_3Si$	—	252–253 (*73*)	760	1.5362 (*75*)	20	0.9254 (*75*)	20
[*p*-$(CH_3)_2NC_6H_4]_3SiOC_2H_5$	125–126 (*78*)	—	—	—	—	—	—
[*p*-$(CH_3)_2NC_6H_4]_2Si(OC_2H_5)_2$	—	229–231 (*79*)	3	1.5765 (*79*)	25	1.102 (*79*)	25
p-$(CH_3)_2NC_6H_4Si(OC_2H_5)_3$	—	144–146 (*79*)	2	1.5012 (*79*)	25	1.012 (*79*)	25

(*continued on p.* 178)

TABLE 29 *(Contd.)*

AMINOORGANOSILANES[a, b]

Compound	Mp (°C)	Bp (°C)	at p (mm Hg)	n_D^t	at t (°C)	d_4^t (gm/cc)	at t (°C)
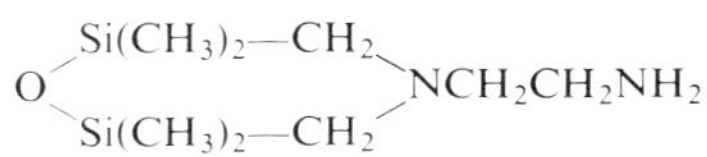	—	65 (*241*)	0.3	1.4519 (*241*)	20	—	—
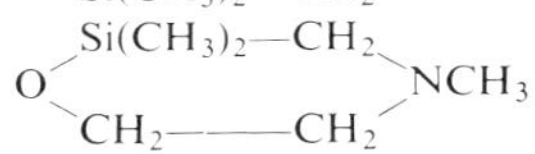	—	28 (*241*)	1.0	1.4339 (*241*)	20	—	—
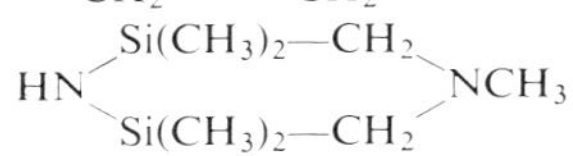	—	66 (*172*)	15	1.4503 (*172*)	20	—	—

[a] Numbers in parentheses indicate reference numbers.
[b] For siloxanes see Tables 36 and 45.

Aminoorganosilanes add normally to acrylonitrile and methyl acrylate:

$$C_2H_5NHCH_2(CH_3)_3Si + CH_2{=}CHCN \rightarrow NC{-}CH_2{-}CH_2{-}\underset{\displaystyle C_2H_5}{\underset{|}{N}}{-}CH_2(CH_3)_3Si \quad (150)$$

$$NH_2(CH_2)_3Si(OC_2H_5)_3 + CH_2{=}CH{-}COOCH_3 \rightarrow CH_3OOCCH_2CH_2NH(CH_2)_3Si(OC_2H_5)_3 \quad (204)$$

They also add to isocyanates (*201*):

$$NH_2(CH_2)_3Si(OC_2H_5)_3 + C_6H_5NCO \rightarrow C_6H_5NHCONH(CH_2)_3Si(OC_2H_5)_3$$

Use is made of the reactivity or simply the polarity of the aminoalkylsilanes when they are employed, in the simultaneous presence of silicon-functional groups, as adhesives for bonding certain resins to glass (see Section 10.7.2).

Siloxanes containing triazine groups, which can be obtained by the reaction of cyanuric chloride with aminoalkylethoxysilanes, have been recommended for the modification of melamine resins (*13*, *80*).

Aminoarylsilanes can be diazotized and coupled, and numerous azo compounds have been prepared in this way (*66*). The coupling can be carried out on fibers. The azo components can be adsorbed on the fibers and then coupled with a diazonium salt of an aminoarylalkoxysilane. This colors the fibers and makes them hydrophobic (*14*). *N*-Aryl-substituted aminoalkylsilanes containing —OC_2H_5 groups as silicon-functional substituents are bonding agents for azo dyes on glass. They improve the fastness to light, washing, and abrasion of dyed glass fibers. The silicon-functional groups are anchored to the silicate surface and the arylamino group provides the bond to the dye as shown (*40*):

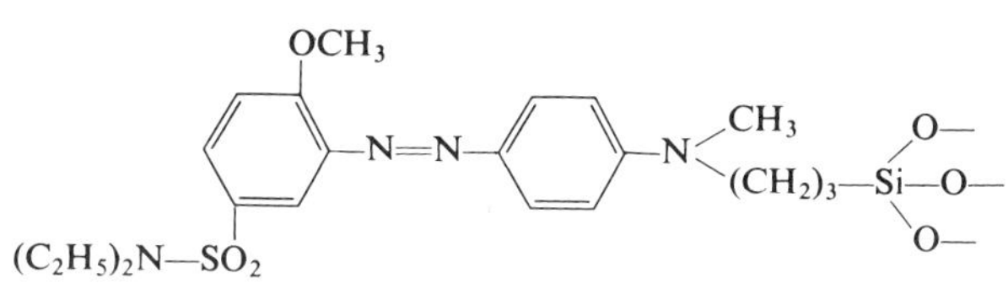

An interesting observation concerning the aminoorganosilanes is that they —particularly γ-aminopropyltriethoxysilane—are capable of protecting metals lying above magnesium in the electrochemical series against corrosion even when used in very small concentrations. Such surface treatment can, for example, prevent the tarnishing of silver (*153*). The corresponding amino-functional silicon compounds are also effective corrosion inhibitors if they are added to aqueous liquors in contact with metals of the type mentioned above (*152*).

4.11 Cyanoorganosilanes

4.11.1 Preparation

Organofunctional silanes with CN groups on the organic radical are usually made by addition of the Si—H bond to unsaturated nitriles, generally acrylonitrile. The reaction is promoted by radical as well as ionic catalysts. Many such catalysts have been proposed.

The addition of acrylonitrile can produce α- or β-cyanoethylsilanes:

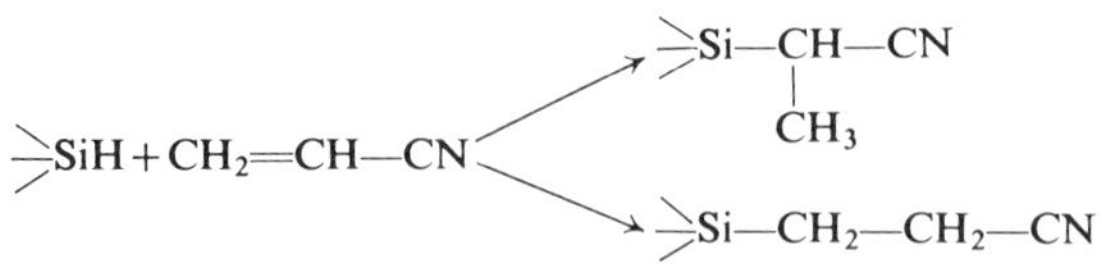

α-form*

β-form*

Since the strongly electron-attracting nitrile group considerably weakens the Si—C bond in the α-C position, interest has largely centered on methods for the preparation of the β-cyano form. This direction of the reaction is favored by catalysts based on amines. A combination consisting of tri-*n*-butylamine, tetramethylethylenediamine and copper(I) chloride or oxide (*25*) has been found useful for the addition of methyldichlorosilane to acrylonitrile:

$$CH_3SiHCl_2 + CH_2{=}CH{-}CN \rightarrow CNCH_2CH_2(CH_3)SiCl_2 \ (79\%)$$

Amines and phosphines, particularly triethylamine and triphenylphosphine (*110*, *203*), trimethylaminosilanes (*205*) such as trimethyldiethylaminosilane (*205*), dimethylformamide (*82*), and other amides (*224*) are suitable catalysts for the addition of trichlorosilane to acrylonitrile (70 to 75% yields can be achieved):

$$SiHCl_3 + CH_2{=}CH{-}CN \rightarrow NC{-}CH_2{-}CH_2{-}SiCl_3$$

α-Cyanoethylmethyldichlorosilane has been obtained in 60 to 75% yield by using a mixed catalyst consisting of palladium(II) chloride, tributylamine, and tetramethylethylenediamine (*26*).

γ-Cyanopropylsilanes can be prepared in 80 to 90% yield from the silanes containing hydrogen attached to silicon and allyl or methallyl cyanide using a platinum-on-alumina catalyst (*109*).

Compounds of this type can also be obtained in good yield by the reaction of dicyanogen with organosilicon Grignard compounds (*214*). Chloroorganosilanes and sodium cyanide react sluggishly in the absence of a solvent (*214*), but more readily in organic solvents (*303*). The reaction of trimethylchlorosilane with acetonitrile and sodium yields a complex mixture (*215*). Finally, the methods for cyanoethylating alcohols and amines have been applied to hydroxy- and aminoorganosilanes (*150*, *189*, *255*).

* For the lettering of the α- and β-positions in cyanoorganosilanes compare the footnote in Section 4.8.

4.11.2 Physical Constants (Table)

Physical constants of a number of the compounds described above are given in Table 30 (p. 182).

4.11.3 Chemical Properties

In the α-position, the strongly electron-attracting cyano group considerably weakens the Si—C bond (this effect formally corresponds to the action of α-carbonyl compounds or β-halogen substitution; see footnote, Section 4.8). Consequently, the cyanomethyl group in $NCCH_2(CH_3)_3Si$ is easily split off by boiling water (*214*). With the corresponding β- and γ-cyano compounds, on the other hand, hydrolysis causes only saponification of the CN group and not cleavage of the Si—C bond (*177*):

$$NC(CH_2)_2(CH_3)_3Si \xrightarrow[C_2H_5OH]{NaOH} HOOC(CH_2)_2(CH_3)_3Si$$

The position of the cyano group on the alkyl radical has a marked influence on the thermal behavior of the compound: the β-functional "α-cyanoethyl" compounds are strongly subject to β-decomposition. The accompanying tabulation gives percentages of the original carbon content found with siloxanes substituted in different ways on heating for 100 hours to 250°C in air (*317*). β-Cyanoethyl- and γ-cyanopropylsilanes are therefore suitable

Substituent	C(%)
α-Cyanoethyl	< 1
β-Cyanoethyl	88.3
β-Cyanopropyl	75.6
γ-Cyanopropyl	94.4

starting materials for the preparation of carboxy-functional silanes (see Section 4.8.1). They are also important as a preliminary stage in the production of amino-functional silanes; these can be obtained from them by reduction (see Section 4.10.1):

$$NCCH_2CH_2(CH_3)SiCl_2 \xrightarrow[CH_3OH]{NaBH_4} H_2N(CH_2)_3(CH_3)Si(OCH_3)_2$$

Besides their interesting key position in the preparation of other organofunctional silanes, cyanoorganosilanes have been used to modify the properties of classical silicone products. β-Cyanoethyl- and γ-cyanopropylmethylsiloxane units inserted into dimethylsiloxane chains improve the swelling resistance of silicone rubber (Section 8.1.1). The former have also been used to increase the dielectric constant of silicone oils (Section 9.2.9).

TABLE 30

CYANOORGANOSILANES[a, b]

Compound	Mp (°C)	Bp (°C)	at p (mm Hg)	n_D^t	at t (°C)	d_4^t (gm/cc)	at t (°C)
$CNCH_2(CH_3)_3Si$	—	84–85 (*214*)	54	1.4203 (*214*)	20	0.827 (*214*)	20
$CH_3CH(CN)(CH_3)_3Si$	—	91.5–92.5 (*178*)	58	1.4232 (*178*)	25	0.8254 (*178*)	25
$CH_3CH(CN)SiCl_3$	—	96–98 (*178*)	41.5	—	—	—	—
$CH_3CH(CN)Si(OC_2H_5)_3$	—	119–120 (*178*)	26.5	1.4070 (*178*)	25	0.9630 (*178*)	25
$CNCH_2CH_2(CH_3)_3Si$	—	97.5–98 (*177*)	64–65	1.4202 (*177*)	25	0.8226 (*177*)	25
$CNCH_2CH_2(CH_3)_2SiCl$	—	119–120 (*47*)	42	1.4442 (*47*)	20	—	—
$CNCH_2CH_2(CH_3)SiCl_2$	—	111–113 (*198*)	40	1.4518 (*198*)	20	1.1722 (*198*)	20
$CNCH_2CH_2(C_2H_5)SiCl_2$	—	122–124 (*198*)	40	1.4557 (*198*)	20	1.1561 (*198*)	20
$CNCH_2CH_2SiCl_3$	34.6–35.1 (*177*)	109 (*177*)	30	—	—	—	—
$CNCH_2CH_2(CH_3)_2SiOC_2H_5$	—	106–108 (*47*)	30	—	—	—	—
$CNCH_2CH_2(CH_3)Si(OC_2H_5)_2$	—	123–125 (*47*)	30	—	—	—	—
$CNCH_2CH_2Si(OC_2H_5)_3$	—	111.5–112.5 (*177*)	10	1.4103 (*177*)	25	0.9699 (*177*)	25
$CNCH_2CH_2CH_2(CH_3)SiCl_2$	—	79–82 (*109*)	1	1.4568 (*109*)	25	1.14 (*109*)	25
$CNCH_2CH_2CH_2(C_6H_5)SiCl_2$	—	145–148 (*109*)	2	1.5312 (*109*)	25	1.19 (*109*)	25
$CNCH_2CH_2CH_2SiHCl_2$	—	43–45 (*109*)	0.3	1.4602 (*109*)	25	1.21 (*109*)	25
$CNCH_2CH_2CH_2SiCl_3$	—	72–75 (*109*)	1.1	1.4638 (*109*)	25	1.28 (*109*)	25
$CNCH_2CH_2CH_2(CH_3)Si(OC_2H_5)_2$	—	83.5–85 (*109*)	3.7	1.4206 (*109*)	25	0.929 (*109*)	25
$CH_3CH(CN)CH_2SiCl_3$	—	88 (*213*)	8	1.4595 (*213*)	20	—	—
$CN{-}C(CH_3)_2(CH_3)SiCl_2$	—	98–107 (*46*)	49	—	—	—	—

[a] Numbers in parentheses indicate reference numbers.

[b] For siloxanes see Tables 36 and 45.

REFERENCES

(1) C. L. AGRE, *J. Amer. Chem. Soc.* **71**, 300 [1949]. (2) — and W. HILLING, ibid. **74**, 3895 [1952]. (3) H. H. ANDERSON, *J. Org. Chemistry* **26**, 276 [1961]. (4) D. N. ANDREEV and E. V. KUKHARSKAYA, *J. Gen. Chem. USSR (Engl. transl.)* **30**, 2763 [1960]. (5) K. A. ANDRIANOV and A. K. DABAGOVA, ibid. **30**, 1947 [1960]. (6) — and M. KAMENSKAYA, *Zhurnal obshchei Khimii* **8**, 969 [1938]. (7) —, N. S. LEZNOV and A. K. DABAGOVA, *Bull. Acad. Sci. USSR, Div. Chem. Sci. (Engl. transl.)* **1957**, 467. (8) — and L. I. MAKAROVA, *Proc. Acad. Sci. USSR (Engl. transl.)* **127**, 653 [1957]. (9) —, I. V. TROFIMOVA, G. S. POPELEVA and S. A. GOLUBTSOV, *Plast. Massy* **1959**, No. 3, 25.

(10) —, A. A. ZHDANOV and V. A. ODINETS, *J. Gen. Chem. USSR (Engl. transl.)* **31**, 3764 [1961] and **32**, 1102 [1962]. (11) D. L. BAILEY, Union Carbide, U.S. Pat. 2970150 [17.12.1957]. (12) —, E. J. PEPE and B. KANNER, —, U.S. Pat. 3007957 [18.9.1959]. (13) — and R. M. PIKE, —, German Appl. (West German) 1120144 [U.S. Prior. 12.10.1956]. (14) —, —, —, French Pat. 1205606 [U.S. Prior. 29.4.1957]. (15) — and A. N. PINES, *Ind. Engng. Chem.* **46**, 2363 [1954]. (16) A. J. BARRY, Dow Chemical, Brit. Pat. 618403 [U.S. Prior. 4.10.1945]. (17) —, Dow Corning, U.S. Pat. 2499561 [23.2.1949]. (18) E. A. BASSETT, H. G. EMBLEM, M. FRANKEL and D. RIDGE, *J. Soc. Chem. Ind.* **67**, 177 [1948]. (19) M. I. BATUEV, A. D. PETROV, V. A. PONOMARENKO and A. D. MATVEEVA, *Bull. Acad. Sci. USSR, Div. Chem. Sci. (Engl. transl.)* **1956**, 1087.

(20) P. A. BAZHULIN, YU. P. EGOROV and V. F. MIRONOV, *Doklady Akad. Nauk. SSSR* (2) **92**, 515 [1953]. (21) R. A. BENKESER and R. A. HICKNER, *J. Amer. Chem. Soc.* **80**, 5298 [1958]. (22) —, M. L. BURROUS, L. E. NELSON and J. V. SWISHER, ibid. **83**, 4385 [1961]. (23) W. T. BLACK, D. L. BAILEY and V. B. JEX, Union Carbide, French Pat. 1189990 [U.S. Prior. 12.10.1956]. (24) B. A. BLUESTEIN, General Electric, French Pat. 1205232 [U.S. Prior. 20.3.1957]. (25) —, *J. Amer. Chem. Soc.* **83**, 1000 [1961]; General Electric, U.S. Pat. 2971970 [29.12.1958]. (26) —, —, German Appl. (West German) 1092917 [U.S. Prior. 29.12.1958]. (27) V. YA. BOGOMOL'NYI, *Polymer Sci. (USSR)* **2**, 6 [1959]. (28) R. W. BOTT, C. EABORN and K. LEYSHON, *J. Chem. Soc.* [London] **1964**, 1548. (29) R. A. BENKESER and P. E. BRUMFIELD, *J. Amer. Chem. Soc.* **74**, 253 [1952].

(30) — and R. G. SEVERSON, ibid. **73**, 1424 [1951]. (31) — and H. R. KRYSIAK, ibid. **75**, 2421 [1953]. (32) —, —, ibid. **76**, 599 [1954]. (33) S. BRYNOLF, *Acta Chem. Scand.* **10**, 883 [1956]. (34) R. H. BUNNELL, Libbey-Owens, U.S. Pat. 2632755 [2.10.1947]. (35) C. A. BURKHARD, *J. Amer. Chem. Soc.* **68**, 2103 [1946]. (36) —, ibid. **72**, 1078 [1950]. (37) —, ibid. **72**, 1402 [1950]. (38) —, ibid. **74**, 6275 [1952]. (39) — and R. H. KRIEBLE, ibid. **69**, 2687 [1947].

(40) P. C. CANOVAI JR., Union Carbide, U.S. Pat. 2934459 [2.8.1957]. (41) E. A. CHERNYSHEV, V. F. MIRONOV and A. D. PETROV, *Bull. Acad. Sci. USSR, Div. Chem. Sci. (Engl. transl.)* **1960**, 1989. (42) H. CHIH-TANG and W. PAO-JEN, *Acta Chim. Sinica* **22**, 127 [1956]. (43) V. CHVALOVSKY and V. BAŽANT, *Chem. Listy* **46**, 158 [1952]. (44) H. A. CLARK, A. F. GORDON, C. W. YOUNG and M. J. HUNTER, *J. Amer. Chem. Soc.* **73**, 3798 [1951]. (45) M. COHEN and J. R. LADD, ibid. **75**, 988 [1953]. (46) Q. P. COLE, General Electric, U.S. Pat. 2776306 [31.12.1953]. (47) G. D. COOPER and M. PROBER, *J. Org. Chem.* **25**, 240 [1960]. (48) J. W. CURRY, Dow Corning, U.S. Pat. 2811541 [20.6.1955]. (49) K. DAMM, K. OTT and H. STEINBACH, Farbenfabriken Bayer, French Pat. 1345921 [German Prior. (West German) 19.1.1962].

(50) P. A. DiGiorgio, L. H. Sommer and F. C. Whitmore, *J. Amer. Chem. Soc.* **70**, 3512 [1948]. (51) C. Eaborn and J. C. Jeffrey, *J. Chem. Soc.* [London] **1957**, 137. (52) J. J. Eisch and J. T. Trainov, *J. Org. Chem.* **28**, 487 [1963]. (53) R. J. Fessenden and J. S. Fessenden, ibid. **28**, 3490 [1963]. (54) E. Fischer, G. Schott and A. D. Petrow, *J. prakt. Chem.* [4] **21**, 149 [1963]. (55) — and A. D. Petrov, *Proc. Acad. Sci. USSR* (*Engl. transl.*) **138**, 462 [1961]. (56) R. F. Fleming, Corning Glass Works, U.S. Pat. 2386452 [25.9.1942]. (57) H. J. Fletcher and H. D. Dingman, Dow Corning, Brit. Pat. 673322 [U.S. Prior. 14.9.1949]. (58) —, —, —, German Pat. (West German) 861561 [U.S. Prior. 14.9.1949]. (59) C. Friedel and I. M. Crafts, *Liebigs Ann. Chem.* **138**, 19 [1866]; *Ann. Chimie* 4. Série, **19**, 334 [1870].

(60) K. C. Frisch, *J. Amer. Chem. Soc.* **75**, 6050 [1953]. (61) L. W. Frost, Westinghouse, U.S. Pat. 1596967 [19.11.1948]. (62) —, *J. Amer. Chem. Soc.* **78**, 3855 [1956]. (63) G. N. Gadsby, *Research* **3**, 338 [1950]. (64) G. C. Gainer and D. W. Lewis, Westinghouse, French Pat. 1039652 [U.S. Prior. 24.3.1950]. (65) P. D. George and J. R. Elliott, *J. Amer. Chem. Soc.* **77**, 3493 [1955]. (66) —, M. Prober and J. R. Elliott, *Chem. Reviews* **56**, 1065 [1956]. (67) A. M. Geyer and R. N. Haszeldine, *Nature* [London] **178**, 808 [1956]. (68) —, —, *J. Chem. Soc.* [London] **1957**, 1038. (69) —, —, ibid. **1957**, 3925.

(70) —, —, K. Leedham and R. I. Marklow, ibid. **1957**, 4472. (71) H. Gilman, C. G. Brannen and R. K. Ingham, *J. Amer. Chem. Soc.* **78**, 1689 [1956]. (72) —, A. G. Brook and L. S. Miller, ibid. **75**, 3757 [1953]. (73) — and F. J. Marshall, ibid. **71**, 2066 [1949]. (74) — and H. W. Melvin jr., ibid. **72**, 995 [1950]. (75) —, — and J. J. Goodman, ibid. **76**, 3219 [1954]. (76) — and L. S. Miller, ibid. **73**, 968 [1951]. (77) — and J. F. Nobis, ibid. **72**, 2629 [1950]. (78) —, M. A. Plunkett and G. E. Dunn, ibid. **73**, 1686 [1951]. (79) — and G. N. R. Smart, *J. Org. Chem.* **15**, 720 [1950].

(80) I. C. Goan, S. H. Smith jr. and R. R. Miller, ibid. **27**, 2657 [1962]. (81) J. R. Gold, L. H. Sommer and F. C. Whitmore, *J. Amer. Chem. Soc.* **70**, 2874 [1948]. (82) S. A. Golubtsov, Z. V. Belyakova and T. M. Yakusheva, *Soviet Plastics* [London] No. 12, 19 [1961]. (83) —, G. S. Popeleva, K. A. Andrianov and N. I. Zaslavskaya, ibid. No. 10, 21 [1962]. (84) L. Goodman, R. M. Silverstein and A. Benitez, *J. Amer. Chem. Soc.* **79**, 3073 [1957]. (85) A. F. Gordon, Dow Corning, French Pat. 1116316 [U.S. Prior. 21.12.1953]. (86) —, —, French Pat. 1143366 [U.S. Prior. 14.3.1955]. (87) J. Goubeau and H. D. Fromm, *Z. anorg. allg. Chem.* **317**, 41 [1962]. (88) G. Greber and L. Metzinger, *Makromol. Chem.* **39**, 226 [1960]. (89) — and E. Reese, ibid. **77**, 13 [1964].

(90) L. A. Haluska, Dow Corning, German Appl. (West German) 1022225 [U.S. Prior. 17.11.1952 and 14.8.1953]. (91) —, —, U.S. Pat. 2800494 [22.10.1954]. (92) —, —, German Appl. (West German) 1047782 [U.S. Prior. 6.4.1955]. (93) R. N. Haszeldine, *Nature* [London] **168**, 1028 [1951]. (94) —, *Angew. Chem.* **66**, 693 [1954]. (95) — and R. J. Marklow, *J. Chem. Soc.* [London] **1956**, 962. (96) — and J. C. Young, ibid. **1960**, 4503. (97) D. B. Hatcher, Libbey-Owens, U.S. Pat. 2574390 [27.3.1948]. (98) G. W. Holbrook and P. L. Brown, Dow Corning, U.S. Pat. 2915544 [11.12.1957]. (99) —, A. F. Gordon and O. R. Pierce, *J. Amer. Chem. Soc.* **82**, 825 [1960].

(100) N. G. Holdstock, General Electric, U.S. Pat. 2803638 [18.8.1954]. (101) —, —, U.S. Pat. 2887503 [19.8.1957]. (102) D. T. Hurd, *J. Amer. Chem. Soc.* **67**, 1813 [1945]. (103) —, General Electric, U.S. Pat. 2420912 [11.6.1945]. (104) — and G. F. Roedel, *Ind. Engng. Chem.* **40**, 2078 [1948]. (105) E. F. Izard and S. L. Kwolek, *J. Amer. Chem. Soc.* **73**, 1156 [1951]. (106) J. W. Jenkins, N. L. Lavery, P. R. Guenther and H. W. Post, *J. Org. Chem.* **13**, 862 [1948]. (107) H. Jenkner, Kali-Chemie, German Appl. (West German) 1091566

[13.8.1958]. (108) V. B. Jex and D. L. Bailey, Union Carbide, U.S. Pat. 2930809 [12.10.1956]. (109) — and J. E. McMahon, —, French Pat. 1169052 [U.S. Prior. 23.12.1955].

(110) —, —, —, U.S. Pat. 2907784 [28.6.1957, cip 23.12.1955]. (111) M. Kanazashi, *Bull. Chem. Soc. Japan* **26**, 493 [1953]. (112) B. Kanner, Union Carbide, French Pat. 1213000 [U.S. Prior. 21.10.1957]. (113) M. Kleiman, Arvey, U.S. Pat. 2697089 [16.3.1953]. (114) G. Koerner, Th. Goldschmidt, German Pat. (West German) 1165028 [14.11.1962]. (115) C. F. Kohl jr., Dow Corning, Brit. Pat. 655105 [U.S. Prior. 17.11.1947]. (116) —, —, U.S. Pat. 2530202 [2.12.1949]. (117) —, —, French Pat. 1024248 [U.S. Prior. 27.10. and 2.12.1949]. (118) G. S. Kolesnikov, S. L. Davydova and T. I. Ermolaeva, *Polymer Sci. USSR* **2**, 20 [1959]. (119) V. V. Korshak, A. D. Petrov, N. G. Matveeva, V. F. Mironov, G. I. Nikitin and S. I. Sadykh-Zade, *J. Gen. Chem. USSR* (*Engl. transl.*) **26**, 1367 [1956].

(120) —, A. M. Polyakova, A. A. Sakharova, A. D. Petrov, V. F. Mironov, V. G. Glukhovtsev and G. I. Nikishin, ibid. **27**, 2507 [1957]. (121) —, —, —, — and E. A. Chernyshev, *Proc. Acad. Sci. USSR* (*Engl. transl.*) **119**, 207 [1958]. (122) K. Kramer and N. Wright, *Chem. Ber.* **96**, 1877 [1963]. (123) R. H. Krieble, General Electric, U.S. Pat. 2510642 [8.4.1947]. (124) — and J. R. Elliott, *J. Amer. Chem. Soc.* **67**, 1810 [1945]. (125) —, —, ibid. **68**, 2291 [1946]. (126) E. L. Kropa, American Cyanamid, U.S. Pat. 2465731 [1.8.1944]. (127) H. F. Lamoreaux and N. G. Holdstock, General Electric, Amer. chem. Soc. 125th Meeting, Kansas City 1954, Abstr. 17 N. (128) E. Larsson, *Kungl. fysiografiska Sällskapets i Lund Förhandlingar* **25**, No. 7, 1 [1955]. (129) —, ibid. **26**, No. 4, 1 [1956].

(130) — and L. O. Knopp, *Acta Chem. Scand.* **1**, 268 [1947]. (131) R. J. Lisanke and D. L. Bailey, Union Carbide, U.S. Pat. 3046295 [21.6.1960]. (132) G. N. Mal'nova, E. P. Mikheev, A. L. Klebanskii and N. P. Filimonova, *Proc. Acad. Sci. USSR* (*Engl. transl.*) **123**, 909 [1958]. (133) C. S. Marvel and R. G. Woolford, *J. Org. Chem.* **25**, 1641 [1960]. (134) E. T. McBee, C. W. Roberts and G. W. R. Puerckhauer, *J. Amer. Chem. Soc.* **79**, 2326 [1957]. (135) —, —, —, ibid. **79**, 2329 [1957]. (136) J. J. McBride jr. and H. C. Beachell, ibid. **70**, 2532 [1948]. (137) R. L. Merker, Dow Corning, U.S. Pat. 2770632 [20.5.1954]. (138) —, —, U.S. Pat. 2833802 [27.2.1956]. (139) — and J. E. Noll, *J. Org. Chem.* **21**, 1537 [1956].

(140) Metal and Thermit, Brit. Pat. 795772 [U.S. Prior. 1.2.1955]. (141) Midland Silicones, Brit. Pat. 769500 [U.S. Prior. 11.1. and 24.5.1954]. (142) —, Brit. Pat. 864848 [U.S. Prior. 19.1.1959]. (143) E. P. Mikheev, *Proc. Acad. Sci. USSR* (*Engl. transl.*) **108**, 285 [1956]. (144) D. Mikulášova and A. Hrivik, *Chemické zvesti* **11**, 641 [1957]. (145) Minnesota Mining, Manufacture, Brit. Pat. 850228 [U.S. Prior. 6.9.1955]. (146) V. F. Mironov, *Coll. Czechoslov. chem. Commun.* **25**, 2167 [1960]. (147) —, *Bull. Acad. Sci. USSR, Div. Chem. Sci.* (*Engl. transl.*) **1962**, 1797. (148) — and N. G. Maksimova, ibid. **1962**, 1223. (149) — and V. V. Nepomnina, ibid. **1961**, 1759.

(150) —, A. D. Petrov and N. A. Pogonkina, ibid. **1955**, 685. (151) D. Mohler and J. E. Sellers, General Electric, U.S. Pat. 2598436 [21.9.1949]. (152) F. L. Morehouse, Union Carbide, German Appl. (West German) 1171235 [U.S. Prior. 2.12.1958]. (153) — and A. N. Pines, —, French Pat. 1207724 [U.S. Prior. 19.7.1957]. (154) G. V. Motsarev, A. L. Englin, A. Ya. Yakubovich, I. N. Uspenskaya and N. G. Ivanova, *J. Gen. Chem. USSR* (*Engl. transl.*) **28**, 1395 [1958]. (155) — and V. R. Rozenberg, ibid. **30**, 2983 [1960]; **31**, 1874 [1961]; **32**, 3654 [1962]; **33**, 247 [1963]; **33**, 2874 [1963]. (156) —, A. Ya. Yakubovich and V. R. Rozenberg, *Proc. Acad. Sci. USSR* (*Engl. transl.*) **148**, 33 [1963]. (157) —, —, *J. Gen. Chem. USSR* (*Engl. transl.*) **26**, 2927 [1956]. (158) —, —, ibid. **27**, 1404 [1957]. (159) —, —, ibid. **27**, 2822 [1957].

(160) R. Müller and M. Dressler, *J. prakt. Chem.* [4] **22**, 29 [1963]. (161) — and R. Köhne, ibid. [4] **21**, 163 [1963]. (162) — and K. Schnurrbusch, *Chem. Ber.* **91**, 1805 [1958]. (163) —, M. Weist and K. Schnurrbusch, Institut für Silikon- and Fluorkarbonchemie, German Appl. (West German) 1029377 [24.2.1956]. (164) S. Munkelt and R. Müller, —, German Appl. (West German) 1056609 [24.8.1957]. (165) —, —, *Chem. Ber.* **92**, No. 5, 1012 [1959]. (166) R. Nagel and H. W. Post, *J. Org. Chem.* **17**, 1379 [1952]. (167) —, C. Tamborski and H. W. Post, ibid. **16**, 1768 [1951]. (168) G. Natta, G. Mazzanti, P. Longi and F. Bernardini, Montecatini, German Appl. (West German) 1091760 [Ital. Prior. 23.12.1957 and 3.4.1958]; *La Chimica e l'Industria* **40**, No. 10, 813 [1958]. (169) H. Niederprüm, E. M. Horn and W. Simmler, Farbenfabriken Bayer, German Appl. (West German) 1216873 [14.5.1964].

(170) —, —, —, —, German Appl. (West German) 1195753 [14.3.1964]. (171) — and W. Simmler, —, unpublished. (172) —, —, *Chem. Ber.* **96**, 965 [1963]. (173) E. L. Niedzielski, *J. Amer. Chem. Soc.* **62**, 3519 [1940]. (174) J. E. Noll, J. L. Speier and B. F. Daubert, ibid. **73**, 3867 [1951]. (175) D. C. Noller and H. W. Post, *J. Org. Chem.* **17**, 1393 [1952]. (176) H. Normant, *C. R. hebd. Séances Acad. Sci.* **239**, 1510 [1954]. (177) S. Nozakura and S. Konotsune, *Bull. Chem. Soc. Japan* **29**, 322 [1956]. (178) —, —, ibid. **29**, 326 [1956]. (179) Office National d'Études et de Recherches Aéronautiques, French Pat. 1132611 [13.7.1955].

(180) C. Pape, *Liebigs Ann.* **222**, 354 [1884]. (181) A. D. Petrov, M. I. Batuev, V. A. Ponomarenko, A. D. Snegova, A. D. Matveeva and B. A. Sokolov, *J. Gen. Chem. USSR* (*Engl. transl.*) **27**, 2112 [1957]. (182) —, E. A. Chernyshev and T. L. Krasnova, *Proc. Acad. Sci. USSR* (*Engl. transl.*) **140**, 975 [1961]. (183) —, — and Li Kuang-Liang, ibid. **132**, 679 [1960]. (184) — and V. F. Mironov, *Doklady Akad. Nauk. SSSR* **75**, 707 [1950]. (185) —, —, ibid. (2) **80**, 761 [1951]. (186) —, —, *Bull. Acad. Sci. USSR, Div. Chem. Sci.* (*Engl. transl.*) **1952**, 591. (187) —, — and D. Komanich, ibid. **1957**, 1411. (188) —, — and D. Mashantsker, ibid. **1956**, 547. (189) —, — and N. A. Pogonkina, *Doklady Akad. Nauk. SSSR* **100**, 81 [1955].

(190) — and G. I. Nikishin, *Bull. Acad. Sci. USSR, Div. Chem. Sci.* (*Engl. transl.*) **1952**, 989. (191) —, —, N. P. Smetankina and Y. P. Egorov, ibid. **1955**, 861. (192) —, V. A. Ponomarenko, B. A. Sokolov and Y. P. Egorov, ibid. **1957**, 323. (193) —, —, — and G. V. Odabashyan, ibid. **1957**, 1230. (194) —, —, — and V. U. Roshal, *J. Gen. Chem. USSR* (*Engl. transl.*) **26**, 1391 [1956]. (195) — and S. I. Sadykh-Zade, *Bull. Acad. Sci. USSR, Div. Chem. Sci.* (*Engl. transl.*) **1958**, 496. (196) A. D. Pétrov and S. I. Sadikh-Sadé, *Bull. de la Société Chim. de France* (5. Série) **1959**, 1932. (197) —, S. I. Sadykh-Zade, N. P. Smetankina and Y. P. Egorov, *J. Gen. Chem. USSR* (*Engl. transl.*) **26**, 1419 [1956]. (198) — and V. M. Vdovin, *Bull. Acad. Sci. USSR, Div. Chem. Sci.* (*Engl. transl.*) **1957**, 1513. (199) O. R. Pierce, Dow Corning, French Pat. 1132606 [U.S. Prior. 28.6.1954].

(200) —, E. T. McBee and R. E. Cline, *J. Amer. Chem. Soc.* **75**, 5618 [1953]. (201) R. M. Pike, Union Carbide, French Pat. 1184200 [U.S. Prior. 12.10.1956]. (202) —, *J. Polymer Sci.* **40**, 577 [1959]. (203) —, J. E. McMahon, V. B. Jex, W. T. Black and D. L. Bailey, *J. Org. Chem.* **24**, 1939 [1959]. (204) — and E. L. Morehouse, Union Carbide, French Pat. 1184198 [U.S. Prior. 12.10.1956]. (205) R. A. Pike and R. L. Schank, *J. Org. Chem.* **27**, 2190 [1962]. (206) A. N. Pines and E. R. York, Union Carbide, U.S. Pat. 2735860 and U.S. Pat. 2736736 [12.8.1954]. (207) E. P. Plueddemann, Dow Corning, U.S. Pat. 2946701 [12.11.1957]. (208) — and H. A. Clark, —, Belg. Pat. 613466 [U.S. Prior. 6.2. and 19.5.1961]. (209) — and G. Fanger, *J. Amer. Chem. Soc.* **81**, 2632 [1959].

(210) V. A. PONOMARENKO, N. G. CHERKAEV, A. D. PETROV and N. A. ZADOROZHNYI, *Bull. Acad. Sci, USSR, Div. Chem. Sci.* (*Engl. transl.*) **1958**, 238. (211) —, B. A. SOKOLOV and A. D. PETROV, ibid. **1956**, 633. (212) G. S. POPOLEVA, K. A. ANDRIANOV, A. A. LARINOVA and S. A. GOLUBTSOV, *J. Gen. Chem. USSR* (*Engl. transl.*) **34**, 1101 [1964]. (213) M. PROBER, General Electric, French Pat. 1118500 [U.S. Prior. 31.12.1953]. (214) —, *J. Amer. Chem. Soc.* **77**, 3224 [1955]. (215) —, ibid. **78**, 2274 [1956]. (216) L. W. PROBST, G. C. GAINER and D. W. LEWIS, Westinghouse, French Pat. 1103476 [U.S. Prior 21.4.1953]. (217) G. A. RAZUVAEV and O. S. DYACHKOVSKAYA, *J. Gen. Chem. USSR* (*Engl. transl.*) **26**, 1255 [1956]. (218) J. D. ROBERTS, E. A. MCELHILL and R. ARMSTRONG, *J. Amer. Chem. Soc.* **71**, 2923 [1949]. (219) E. G. ROCHOW, General Electric, Brit. Pat. 659439 [U.S. Prior 4.3.1948].

(220) G. F. ROEDEL, *J. Amer. Chem. Soc.* **71**, 269 [1949]. (221) S. D. ROSENBERG, I. I. WALBURN, T. E. STANKOVICH, A. E. BALINT and H. E. RAMSDEN, *Amer. Chem. Soc.* 130th Meeting, Atlantic City, N.J., 1956, Abstr. of Papers, 81-O. (222) G. ROSSMY, *Chimia* **16**, 226 [1962]. (223) J. W. RYAN, G. K. MENZIE and J. L. SPEIER, *J. Amer. Chem. Soc.* **82**, 3601 [1960]. (224) I. C. SAAM, Dow Corning, U.S. Pat. 2860153 [25.4.1957]. (225) S. I. SADYKH-ZADE, L. V. NOZDRINA and A. D. PETROV, *Proc. Acad. Sci. USSR* (*Engl. transl.*) **118**, 107 [1958]. (226) — and A. D. PETROV, *Doklady Akad. Nauk. SSSR* **96**, 765 [1954]. (227) G. A. SCHMIDT, General Electric, French Pat. 1269856 [U.S. Prior. 19.8.1959]. (228) E. SCHNELL, *Monatshefte Chem.* **88**, 1004 [1957]. (229) R. E. SCOTT and K. C. FRISCH, *J. Amer. Chem. Soc.* **73**, 2599 [1951].

(230) A. E. SENEAR, J. WIRTH and R. G. NEVILLE, *J. Org. Chem.* **25**, 807 [1960]. (231) D. SEYFERTH and E. G. ROCHOW, *J. Amer. Chem. Soc.* **77**, 907 [1955]. (232) K. SEYFRIED, Farbenfabriken Bayer, 1964, unpublished. (233) R. A. SHAW, *J. Chem. Soc.* [London] **1957**, 2831. (234) M. F. SHOSTAKOVSKY and D. A. KOCHKIN, *Proc. Acad. Sci. USSR* (*Engl. transl.*) **109**, 363 [1956]. (235) —, —, *Bull. Acad. Sci. USSR, Div. Chem. Sci.* (*Engl. transl.*) **1956**, 1175. (236) —, —, *J. Applied Chem. USSR* (*Engl. transl.*) **31**, 746 [1958]. (237) —, E. M. SAVITSKY, D. A. KOCHKIN and L. V. MUSATOVA, *Bull. Acad. Sci. USSR, Div. Chem. Sci.* (*Engl. transl.*) **1957**, 1519. (238) W. SIMMLER, Farbenfabriken Bayer, 1958, unpublished. (239) —, in: W. SIMMLER, W. NOLL and A. ZAPPEL: *Herstellung und Umwandlung von Organosilicium-verbindungen*, Houben-Weyl: *Methoden der organischen Chemie*, Verlag Thieme, Stuttgart, 13.3 in preparation.

(240) —, Farbenfabriken Bayer, German Appl. (West German) 1093506 [3.4.1959]. (241) —, *Chem. Ber.* **94**, 1585 [1961]. (242) —, ibid. **96**, 349 [1963]. (243) —, Farbenfabriken Bayer, 1963, unpublished. (244) —, —, 1964, unpublished. (245) — and H. NIEDERPRÜM, —, Belg. Pat. 634126 [German Prior. (West German) 29.6.1962]. (246) —, — and H. JONAS, —, German Appl. (West German) 1193046 [31.12.1962]. (247) —, — and H. SATTLEGGER, *Chem. Ber.* **99**, 1368 [1966]. (248) —, — and H. WALZ, ibid. **97**, 1047 [1964]. (249) —, H. SATTLEGGER and H. NIEDERPRÜM, 1965, unpublished.

(250) —, H. WALZ and H. NIEDERPRÜM, *Chem. Ber.* **96**, 1495 [1963]. (251) A. G. SMITH, J. W. RYAN and J. L. SPEIER, *J. Org. Chem.* **27**, 2183 [1962]. (252) L. H. SOMMER, Dow Corning, U.S. Pat. 2512390 [3.6.1949]. (253) —, —, U.S. Pat. 2589445 [3.6.1949]. (254) —, —, U.S. Pat. 2589446 [Brit. Prior. 30.6.1949]. (255) —, —, U.S. Pat. 2687424 [31.3.1953]. (256) —, D. L. BAILEY, G. M. GOLDBERG, C. E. BUCK, T. S. BYE, F. J. EVANS and F. C. WHITMORE, *J. Amer. Chem. Soc.* **76**, 1613 [1954]. (257) —, —, W. A. STRONG and F. C. WHITMORE, ibid. **68**, 1881 [1946]. (258) —, — and F. C. WHITMORE, ibid. **70**, 2869 [1948]. (259) —, E. DORFMAN, G. M. GOLDBERG and F. C. WHITMORE, ibid. **68**, 488 [1946].

(260) —, J. R. GOLD, G. M. GOLDBERG and N. S. MARANS, ibid. **71**, 1509 [1949]. (261) —, G. M. GOLDBERG, G. H. BARNES and L. S. STONE JR., ibid. **76**, 1609 [1954]. (262) —, —, E. DORFMAN and F. C. WHITMORE, ibid. **68**, 1083 [1946]. (263) — and N. S. MARANS, ibid. **72**, 1935 [1950]. (264) —, —, ibid. **73**, 5135 [1951]. (265) —, —, G. M. GOLDBERG, J. ROCKETT and R. P. PIOCH, ibid. **73**, 882 [1951]. (266) —, J. M. MASTERSON, O. W. STEWARD and R. H. LEITHEISER, ibid. **78**, 2010 [1956]. (267) — and J. ROCKETT, ibid. **73**, 5130 [1951]. (268) —, L. J. TYLER and F. C. WHITMORE, ibid. **70**, 2872 [1948]. (269) —, R. E. VAN STRIEN and F. C. WHITMORE, ibid. **71**, 3056 [1949].

(270) — and F. C. WHITMORE, ibid. **68**, 485 [1946]. (271) S. B. SPECK, *J. Org. Chem.* **18**, 1689 [1953]. (272) J. L. SPEIER JR., Dow Corning, U.S. Pat. 2527590 and 2527591 [Brit. Prior. 26.8.1947]. (273) —, —, U.S. Pat. 2576486 [Canad. Prior. 13.5.1949]. (274) —, —, German Pat. (West German) 881654 [U.S. Prior. 30.6.1950]. (275) —, —, U.S. Pat. 2629727 [26.7.1950]. (276) —, —, U.S. Pat. 2611777 and 2645630 [18.9.1950]. (277) —, —, U.S. Pat. 2723987 [18.10.1954]. (278) —, —, French Pat. 1217009 [U.S. Prior. 23.12.1957 and 26.3.1958]. (279) —, *J. Amer. Chem. Soc.* **70**, 4142 [1948].

(280) —, ibid. **73**, 824 [1951]. (281) —, ibid. **73**, 826 [1951]. (282) —, ibid. **74**, 1003 [1952]. (283) — and B. F. DAUBERT, ibid. **70**, 1400 [1948]. (284) —, — and R. R. MCGREGOR, ibid. **70**, 1117 [1948]. (285) —, —, —, ibid. **71**, 1474 [1949]. (286) H. STEINBACH and W. NOLL, Farbenfabriken Bayer, German Appl. (West German) F35802 [18.1.1962]. (287) —, K. OTT and K. DAMM, —, French Pat. 1345922 [German Prior. (West German) 19.1.1962]. (288) O. W. STEWARD and O. R. PIERCE, *J. Amer. Chem. Soc.* **81**, 1983 [1959]. (289) — and L. H. SOMMER, *J. Org. Chem.* **26**, 4132 [1961].

(290) C. O. STROTHER and G. H. WAGNER, Linde Air Products, U.S. Pat. 2532430 [3.1.1947]. (291) S. V. SUNTHANKAR and H. GILMAN, *J. Amer. Chem. Soc.* **72**, 4884 [1950]. (292) J. SWISS and C. E. ARNTZEN, Westinghouse, U.S. Pat. 2595727 [9.3.1945]. (293) —, —, —, U.S. Pat. 2595728 [9.3.1945]. (294) —, —, —, U.S. Pat. 2595730 [9.3.1945]. (295) C. TAMBORSKI and H. W. POST, *J. Org. Chem.* **17**, 1397 [1952]. (296) P. TARRANT, Dow Corning, French Pat. 1182153 [U.S. Prior. 10.9.1956]. (297) —, G. W. DYCKES, R. DUNMIRE and G. B. BUTLER, *J. Amer. Chem. Soc.* **79**, 6536 [1957]. (298) A. V. TOPCHIEV, N. S. NAMETKIN, TSYU SYAO-PEI, S. G. DURGAR'YAN and N. A. KUZ'MIN, *Bull. Acad. Sci. USSR, Div. Chem. Sci. (Engl. transl.)* **1962**, 1417. (299) J. TSUCHIYA and M. A. KANAZASHI, *Bull. Electrotechn. Lab. Tokyo* **16**, 897 [1952].

(300) L. J. TYLER, Dow Corning, U.S. Pat. 2517146 [5.7.1949]. (301) —, —, U.S. Pat. 2601646 [5.7.1949]. (302) L. W. TYRAN, DuPont, U.S. Pat. 2532583 [24.7.1945]. (303) Union Carbide, Belg. Pat. 553606 [U.S. Prior. 23.12.1955]. (304) —, Technical Information Leaflet for Product X 1901. (305) —, Technical Information Leaflet for Product X 1902. (306) S. N. USHAKOV and A. M. ITENBERG, *Zhurnal obshchei Khimii SSSR* **7**, 2495 [1937] (*C. A.* **32**, 2083 [1938]). (307) M. G. VORONKOV and V. P. DAVYDOVA, *Proc. Acad. Sci. USSR (Engl. transl.)* **125**, 230 [1959]. (308) G. H. WAGNER, Union Carbide, U.S. Pat. 2637738 [17.9.1949]. (309) —, —, U.S. Pat. 2735859 [29.6.1954].

(310) —, D. L. BAILEY, A. N. PINES, M. L. DUNHAM and D. B. MCINTIRE, *Ind. Engng. Chem.* **45**, 367 [1953]. (311) — and C. O. STROTHER, Linde Air Products, Brit. Pat. 670617 [U.S. Prior. 9.10.1947]. (312) E. L. WARRICK, Dow Corning, Belg. Pat. 560351 [U.S. Prior. 28.8.1956]. (313) D. R. WEYENBERG, —, German Pat. (West German) 936445 [U.S. Prior. 17.7.1953]. (314) F. C. WHITMORE and L. H. SOMMER, *J. Amer. Chem. Soc.* **68**, 481 [1946].

(315) —, —, J. GOLD and R. E. VAN STRIEN, ibid. **69**, 1551 [1947]. (316) M. WIEBER and M. SCHMIDT, *Chem. Ber.* **96**, 1019 [1963]. (317) T. C. WILLIAMS, R. A. PIKE and F. FEKETE, *Ind. Engng. Chem.* **51**, 939 [1959]. (318) B. I. YAKOVLEV, *J. Gen. Chem. USSR* (*Engl. transl.*) **19**, a-425 [1949]. (319) — and N. V. VINOGRADOVA, ibid. **22**, 1507 [1952].

(320) A. YA. YAKUBOVICH and V. A. GINSBURG, ibid. **22**, 1821 [1952]. (321) —, S. P. MAKAROV, V. A. GINSBURG, G. I. GAVRILOV and E. N. MERKULOVA, *Doklady Akad. Nauk. SSSR* **72**, 69 [1950]; *C. A.* **45**, 2856 [1951]. (322) — and G. V. MOTSAREV, *J. Gen. Chem. USSR* (*Engl. transl.*) **26**, 611 [1956]. (323) —, —, ibid. **26**, 1593 [1956]. (324) —, —, ibid. **29**, 2360 [1959]. (325) C. YAO-TSO, C. SAN MEI and W. PAO JEN, *Acta chim. sinica* **23**, 124 [1957] (326) A. ZAPPEL, Farbenfabriken Bayer, German Pat. (West German) 949659 [6.2.1954]. (327) —, —, in W. NOLL and A. ZAPPEL, *Herstellung und Umwandlung organischer Silicium-verbindungen:* Houben-Weyl, *Methoden der organischen Chemie*, in the press (328) W. ZIMMERMANN, *Chem. Ber.* **87**, 887 [1954].

CHAPTER 5

Preparation of Polyorganosiloxanes

A very large number of methods is available for the production of polyorganosiloxanes from the monomeric silicon-functional organosilanes described in the preceding chapters. The process used most frequently is the transformation of the functional groups on the silicon atom by *hydrolysis*; this step is employed in virtually all modern large-scale manufacturing processes because of its simplicity and economy.

The hydrolysis, as such, of the various functional silanes normally leads to silanols. It has however already been emphasized during the discussion of the monomeric silanols, that the instability of the silanol bond requires special measures if the monomeric silanols are to be isolated. All silanols tend to *condense spontaneously* to *siloxanes* to a greater or smaller extent, so that as a rule hydrolysis of silicon-functional silanes leads rapidly and directly through the intermediate silanol stage to polymeric siloxanes. The M, D, T, and Q siloxane units are obtained from the functional silanes as given in the accompanying tabulation.

Silane		Generally unstable intermediate		Siloxane unit
R_3SiX	→	$R_3Si(OH)$	→	$R_3SiO_{1/2}$
R_2SiX_2	→	$R_2Si(OH)_2$	→	$R_2SiO_{2/2}$
$RSiX_3$	→	$RSi(OH)_3$	→	$RSiO_{3/2}$
SiX_4	→	$Si(OH)_4$	→	$SiO_{4/2}$

The method of working adopted for the hydrolysis of a silane has been shown by experience to have a decisive influence on the structure and size of the siloxane molecule. This is particularly true if mixtures of silanes with different functionalities are used, i.e., if M and D or M and T or D and T units, etc., are to be linked together to give polysiloxanes.

Viewed in this light, hydrolysis becomes a particularly important step on the path leading from the silanes to the siloxanes. The polymer acquires its ultimate structure only as a result of a *final condensation*, i.e., removal of the last condensable components, or by further construction and reconstruction of the polymer skeleton by *polymerization methods.*

In addition to these methods (hydrolysis, condensation, polymerization) for the production of siloxane bonds, there are a large number of others which depend only partially (or not at all) on hydrolysis. In these, the oxygen required for the formation of the siloxane bond is taken from alcohols, inorganic oxides, alkoxysilanes, acyloxysilanes, silanols, etc., or the reaction of the X group on the silane can be considered a hydrolysis in which the necessary water is supplied by a parallel reaction. These processes also lead directly to polysiloxanes, but are, on the whole, used mainly for special purposes.

5.1 Hydrolytic Processes

The choice of any particular method of hydrolysis will be influenced by the rate of hydrolysis of the individual silane used or of the components present in a silane mixture.

The rate of hydrolysis increases with the polarity of the Si—X bond and with the number of X atoms attached to each silicon in the molecule. It is also affected by the type and number of the organic groups, since the larger and bulkier these are the more do they oppose by steric hindrance the nucleophilic attack by water molecules and the possibility of scission of the Si—X bond. The greater their electronegativity, the more do they strengthen the Si—X bond and lower its reactivity with respect to water. Because of the combined effects of the size and electronegativity of phenyl groups, phenyl-substituted chlorosilanes, for example, are more difficult to hydrolyze than methylchlorosilanes of the same structural type.

Systematic investigations of such factors have been carried out in particular on alkyl- and arylsilanes R_nSiH_{4-n} (*25*, *100*, *249*, *272*, *313*). Since the Si—H bond can be cleaved by water with the evolution of hydrogen, the course of the hydrolysis cannot be affected by the by-products arising during the reaction, and can be satisfactorily followed by measuring the volume of H_2 evolved. For the kinetics of the hydrolysis of chlorosilanes and of triphenylfluorosilane the reader is referred to refs. *4* and *290*.

* Page number of chapter references.

While these factors must be taken into consideration in all attempts to convert a silicon-functional silane into a silanol, they are particularly important for the choice of the method when a mixture of silanes is simultaneously to be hydrolyzed and condensed to a copolymer. If the different components of a siloxane mixture have very different tendencies toward hydrolysis and self-condensation, it may be found impossible to achieve cocondensation simultaneously with hydrolysis, and subsequent molecular rearrangement by a special condensation step or equilibration is then unavoidable.

The characteristics of the Si—C bond must also be considered for the performance of the hydrolysis. Because of its weakly polar character, it is more sensitive toward acids arising from the hydrolysis of halosilanes than a normal C—C bond. In particular, the behavior of the Si—C bond during acid hydrolysis will be determined by the nature of the organic group. Bonds of saturated aliphatic hydrocarbons with silicon are generally not broken; the aromatic Si—C bond is more sensitive, and it is particularly important that electronegative substituents in the organic group considerably increase its sensitivity to hydrolysis (cf. Section 6.4).

5.1.1 Hydrolysis of Organohalosilanes

In its simplest form, the hydrolysis consists of allowing the organohalosilanes to run into an excess of water. The excess is usually adjusted to allow aqueous hydrochloric acid no stronger than 20% to be formed. The reaction is exothermic owing especially to the heat evolved by dissolution of the hydrogen chloride in water, so that either the reaction mixture is cooled or, more simply, the halosilanes are run onto ice.

The self-condensation of the silanols formed is accelerated by aqueous hydrochloric acid to an extent which depends on the latter's concentration. Siloxanes are therefore the main product of this method of hydrolysis.

Trimethylchlorosilane gives hexamethyldisiloxane:

$$2(CH_3)_3SiCl + H_2O \rightarrow (CH_3)_3Si—O—Si(CH_3)_3 + 2HCl$$

Dimethyldichlorosilane gives with water a mixture of polymers containing either between 20 and 50% of polydimethylcyclosiloxanes or between 80 and 50% of linear polydimethylsiloxane-α,ω-diols, depending on the method of working (*214*).

$$(CH_3)_2SiCl_2 \xrightarrow{H_2O} [(CH_3)_2SiO]_n \qquad n = 3, 4, 5, \text{ or higher}$$

$$(CH_3)_2SiCl_2 \xrightarrow{H_2O} HO(CH_3)_2SiO[(CH_3)_2SiO]_nSi(CH_3)_2OH$$

Hydrolysis with 6*N* aqueous hydrochloric acid instead of water, can increase the proportion of oligomeric cyclosiloxanes to about 70%. Conversely, the hydrolysis of dimethyldichlorosilane with 50 to 85% sulfuric acid yields high-molecular-weight siloxanes with only small amounts of cyclosiloxanes, since the sulfuric acid promotes polymerization in this case (*227*).

The aqueous hydrolysis of *methyltrichlorosilanes* gives highly cross-linked gel-like or pulverulent polymers. A highly disperse powder of the composition $(CH_3SiO_{3/2})_n$ with a particle size below 1 μ can be obtained by atomizing methyltrichlorosilane and water in spray towers (*57*).

The conditions of hydrolysis and condensation change with changing concentration of the aqueous hydrochloric acid during batchwise hydrolysis of halosilanes. The pH can be kept constant and the reaction can therefore be made more uniform when the process is carried out continuously, the halosilanes and water being mixed in metered amounts (*28*, *57*, *79*, *101*, *165*, *229*).

In one industrial form of this process, the organochlorosilane is continuously added with water or acid in metered amounts to a recirculating system, and the corresponding amount of the resulting mixture of polyorganosiloxane and aqueous acid is continuously withdrawn (*251*).

If a large excess of water (at least five times the amount of chlorosilane) is used, it is possible in the presence of suitable solvents (see below) to obtain polyorganosiloxanes rich in silanol groups even from the hydrolysis of easily condensable diorganodichlorosilanes and organotrichlorosilanes (*312*).

The foremost technique designed to influence the outcome of the hydrolysis is the use of *organic solvents*. These have a different effect according to whether they are fully or partially miscible or totally immiscible with water, and whether or not they can react with halosilanes.

While the hydrolysis of *dimethyldichlorosilane* in water gives the mixture of cyclosiloxanes and linear siloxanediols mentioned above, the addition of dimethyldichlorosilane to a mixture of water and inert *water-miscible solvents* such as tetrahydrofuran and dioxane gives a high yield of low-molecular-weight polycyclosiloxanes (*261*).

Organic solvents either *immiscible* or only *slightly miscible* with water and inert toward the halosilanes are frequently used. These are, in particular, toluene, xylene, diethyl ether, dibutyl ether, and trichloroethylene (*312*). They are excellent solvents for both the organohalosilanes and the polyorganosiloxanes formed from them. It is thus possible, on the one hand, for the organohalosilanes to react with water in greater dilution and therefore with better distribution and, on the other hand, for the hydrolysis products to be taken up by the solvent and so be protected from the action of the aqueous acid. Because of the dilution of the siloxane phase, the tendency to *intramolecular condensation* predominates over intermolecular condensation and the difunctional siloxane units yield ring compounds preferentially. The lower polycyclosiloxanes are formed preferentially when dimethyldichlorosilane is hydrolyzed in ethereal solution (*214*).

Dilution of the silane and siloxane phase with inert solvents also promotes *intramolecular* condensation in the case of the hydrolysis of *organotrichlorosilanes*. Organotrichlorosilanes with a bulky organic group, such as *t*-butyl or isopropyl, in a molar ratio of $RSiCl_3:H_2O = 1:1.5$, hydrolyze in the presence of a large excess of ether to give low-molecular-weight polymers $(RSiO_{3/2})_4$ with an interesting structure resembling that of hexamethylenetetramine (*316*).

Still other conditions are created if a solvent *partially miscible* with water is used. In the case of water-insoluble organic solvents, hydrolysis takes place at the phase boundary between the silanol solution and the water, and with partially miscible solvents the reaction is also to some extent transferred *into* the organic phase. Butyl acetate is particularly suitable for this method (*200*).

In none of the methods described so far does the solvent participate in the reaction. Special effects can be achieved if solvents capable of becoming involved in the reaction process are used. *Alcohols*, alone or in combination with nonreactive solvents of the kind discussed previously, are most suitable for this purpose. Organosilanols which tend to condense spontaneously on hydrolysis with water alone can be prepared in good yield by the addition of alcohol to the hydrolysis medium (*41*). Diethyldichlorosilane gives relatively good yields of low-molecular-weight cyclic polymers of the type $[D_3]$ or $[D_4]$ (*66*) on hydrolysis in mixtures of alcohol and water. It is, however, particularly important that still *soluble*, highly cross-linked siloxanes can be obtained in this way without gel formation, from mixtures of di-,tri-, or even tetrahalosilanes which would give gels on hydrolysis with water alone (*80*, *201*, *239*). Such results are probably due to the fact that, in the mixture of water and alcohol, hydrolysis and alcoholysis of the halosilanes take place side by side and that the growth of the polymer molecules is hindered by the blocking of even a relatively small proportion of the reaction centers by the less readily hydrolyzable Si—OR groups. The differences in the hydrolysis rates of different silanes appear to be evened out by the addition of alcohol, and the conditions necessary for uniform cohydrolysis and cocondensation are thus established.

The employment of organic solvents in hydrolysis has been found particularly useful for the preparation of mixtures of highly cross-linked siloxanes such as are present in silicone resins. Here, the different methods of working have been combined or made to follow one another in a number of stages. For example, the organohalosilanes are diluted with a polyether such as dioxane, bis-(β-ethoxyethyl) ether or ethylene glycol diethyl ether, and this mixture is added to a boiling mixture of solvent naphtha and a polyether containing slightly more water than that necessary for hydrolysis (*156*). In another method, the organochlorosilanes are first treated with alcohols and the reaction products are partially saponified with water or aqueous hydrochloric acid; the product of this reaction is completed with this solution (*193*).

The marked influence exerted by the solvent on the outcome of the hydrolysis

5.1.2 Hydrolysis of Organoalkoxysilanes

Since no cleavage products that could accelerate the condensation of the silanols are liberated during the hydrolysis of organoalkoxysilanes, these products can be employed very widely and are especially useful for the production of silanols sensitive to condensation. If all traces of alkali and acid are excluded from the system, dimethylsilanediol can be prepared from dimethyldiethoxysilane and dimethyldimethoxysilane (*131*, *141*, *142*), and the hydrolysis of phenyltrimethoxysilane gives good yields of phenylsilanetriol (*295*). It must, however, be borne in mind that the hydrolysis of the methoxy- and ethoxysilanes proceeds more slowly than that of the chlorosilanes. The latter are direct products of most organosilane syntheses, so that their conversion first into organoalkoxysilanes represents a detour associated with an over-all loss in yield. The hydrolysis of alkoxysilanes has therefore been of more specialized interest in silicone chemistry than that of the chlorosilanes.

The method of working is similar to that with organochlorosilanes; the organoalkoxysilane or a mixture of different silanes of this type is run into water. The hydrolysis can be greatly accelerated by small amounts of acids or alkalis. Acid catalysts which can be easily removed by washing, such as hydrochloric, oxalic, acetic, and trichloroacetic acids (*178*), are preferred. Addition of water to the alkoxysilane, the reverse of the normal method, is preferable if silanes with different functional groups are to be cohydrolyzed (*124*).

Inert solvents can also be used here, as in the case of the hydrolysis of the halosilanes, if the formation of cyclic organosiloxanes is to be favored. Small amounts of low-molecular-weight sublimable siloxanes having the compositions $[(CH_3)SiO_{3/2}]_6$ and $[(CH_3)SiO_{3/2}]_8$ have been isolated from the products of the complete hydrolysis of methyltriethoxysilane in benzene (*286*). Analogous products can be recovered from the hydrolysis of methyltri-*n*-butoxysilane (*288*) and ethyltriethoxysilane (*287*).

If the amount of water is reduced, unchanged methoxy and ethoxy groups will of course remain in the siloxane. If dimethyldiethoxysilane is hydrolyzed with an amount of water insufficient to replace all the ethoxy groups, linear polydimethylsiloxanes with terminal ethoxy groups are formed (α,ω-bis-dimethylethoxysiloxy-polydimethylsiloxanes) (*83*, *112*).

$$C_2H_5O-\underset{CH_3}{\overset{CH_3}{\underset{|}{\overset{|}{Si}}}}-O-\left[\underset{CH_3}{\overset{CH_3}{\underset{|}{\overset{|}{Si}}}}-O-\right]_n\underset{CH_3}{\overset{CH_3}{\underset{|}{\overset{|}{Si}}}}-OC_2H_5$$

Polymers with 2 to 11 Si atoms in the chain have been isolated from this series. Partial hydrolysis of methyltriethoxysilane can yield α,ω-bis-methyldiethoxysiloxy-polymethylethoxysiloxanes in the same way,

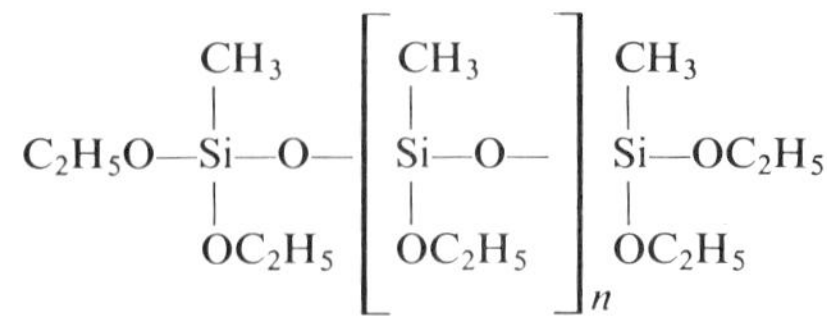

of which the polymer-homologous members with 2 to 5 Si atoms in the chain have been prepared (*83*).

As can be seen from the above examples, the outcome of the hydrolysis of any single silane can vary considerably according to the method of working.

Table 31 gives a review of the relationship between the conditions and the results of the hydrolysis for the silanes that have been investigated most thoroughly, namely, dimethyldichlorosilane, dimethyldimethoxysilane, and dimethyldiethoxysilane. This will illustrate and establish some of the possible variants of the process.

TABLE 31

SURVEY OF THE PRODUCTS OF THE HYDROLYSIS OF DIMETHYLDICHLOROSILANE (A), DIMETHYLDIMETHOXYSILANE (B), AND DIMETHYLDIETHOXYSILANE (C) UNDER DIFFERENT CONDITIONS

Silane	Conditions of hydrolysis	Result	Reference
A	Water	Cyclic polymers and polydimethylsiloxane-α,ω-diols	*214*
A	Salt hydrates	Mainly higher polymers	*320*
A	Strong HCl	Mainly lower polymers	*214*
A	50–85% sulfuric acid	Mainly higher polymers	*227*
A	Alkaline medium (excess of alkali)	Mainly higher polymers	*214*
A	Water and ammonia pH 6.5–8.5	Tetramethyldisiloxane-1,3-diol (60% of theoretical)	*169*
A	Water and water-miscible solvents	Mainly cyclic lower polymers	*261*
A	Water and diethyl ether	Mainly cyclic lower polymers	*214*
A	Deficiency of water, dioxane, and ether	Linear α,ω-bis-dimethylchlorosiloxypolydimethylsiloxanes	*211*
A	Alcohol and carboxylic acid	Mainly high-molecular-weight siloxanes	*263*
B, C	Water, free of acid and alkali	Dimethylsilanediol	*131*, *141*
C	Deficiency of water	α,ω-bis-Dimethylethoxysiloxypolydimethylsiloxanes	*84*

5.1.3 Hydrolysis of Other Silanes

Relatively little is known about the hydrolysis of *silicon-functional silanes*, other than organosilanes with Si—H groups (see Section 5.1). Acetoxysilanes are readily hydrolyzed; if the reaction is carried out in the homogeneous phase with mixtures of acetone and water, immediately after the preparation of the reaction mixture a specific electrolytic conductivity is observed which corresponds to that of an acetic acid solution of the same concentration in the same solvent mixture (*157*). Since the acetic acid formed during hydrolysis is not as powerful a promotor of the condensation as a hydrohalic acid, acetoxysilanes have been found to be better than halosilanes for the preparation of silanols, particularly those containing unsaturated R groups sensitive toward hydrogen halide attached to silicon (*86*, *87*). Acetoxysilanes are also better for the production of silicone films on substrates, such as textiles or paper, that would be destroyed by the attack of a hydrohalic acid. It is also possible to obtain polyorganosiloxanes from organoacetoxysilanes, just as with silanes containing other silicon-functional groups (*228*).

Organoaminosilanes appear to be less reactive. Only 10% of hexamethyldisilazane is hydrolyzed on boiling with water for 4 hours, but the reaction can be accelerated by the addition of acid (*240*). Trimethylsilanol can be prepared from hexamethyldisilazane (*240*).

Trimethylcyanosilane reacts slowly with water to give hexamethyldisiloxane; dimethyldicyanosilane hydrolyzes rapidly at first and then more slowly. The final product is an oily polymer (*176*). Acids and alkalis accelerate the reaction. Triethylisothiocyanatosilane hydrolyzes more slowly than triethylcyanosilane, and triethylisocyanatosilane more slowly than either (*5*, *72*); moreover, in the case of the organoisocyanatosilanes, in analogy with other cases, the rate of hydrolysis increases with increasing number of NCO groups, and decreasing number of organic groups substituted on the silicon atom (*85*).

The Si—H bond in silanes is frequently not regarded as a functional group for the purposes of hydrolysis, since the problem is often to keep the bond intact. Consequently, the times of contact with the hydrolyzing medium are usually kept as short as possible, and the temperatures low. Such measures enable, e.g., only the chlorine to be split off hydrolytically from methyldichlorosilane, while the Si—H bond remains unaffected.

The processes for the hydrolysis of silanes containing *organofunctional* as well as silicon-functional groups are basically analogous to those described above. Of course if the organofunctional group is intended to survive the hydrolysis unchanged, a method must be chosen which takes into account its reactivity and, in certain circumstances, also the altered strengths of the Si—C bonds (compare also Section 6.4). Trichloromethyltrichlorosilane is hydrolyzed by water even at room temperature (*65*):

$$Cl_3C{-}SiCl_3 + 4H_2O \rightarrow CHCl_3 + 3HCl + Si(OH)_4$$
$$\downarrow \text{condensation}$$

Dichloromethyl groups remain substantially unaffected during hydrolysis in ice water, and monochloromethyl groups during normal hydrolysis (*153*); both are removed from silicon in an alkaline medium on heating (*153, 327*):

$$Cl_2CH(CH_3)SiCl_2 + 3H_2O \xrightarrow{OH^{\ominus}} CH_3Si(OH)_3 + CH_2Cl_2 + 2HCl$$

$$ClCH_2(CH_3)_2SiCl + 2H_2O \xrightarrow{OH^{\ominus}} (CH_3)_2Si(OH)_2 + CH_3Cl + HCl$$

The CHF_2 and CF_3 groups are detached from silicon even with cold water (*231*), while γ-trifluoropropyl groups survive acid hydrolysis (*292*).

The influence of other substituents on the cleavage of the $Cl_3C-Si\equiv$ bond is shown by the following compounds, in which the sensitivity of the bond decreases in the given sequence (*89*):

$$(CCl_3)_2SiCl_2 > CCl_3SiCl_3 > (CCl_3)(C_6H_5)SiCl_2 > (CCl_3)(CH_3)SiCl_2$$

Similar rules hold for alkenylchlorosilanes.

β-Cyanoethylmethyldichlorosilane can be converted into the corresponding siloxane by acid hydrolysis, while the α-substituted silane decomposes at the Si—C bond under similar treatment (*53*). The hydrolysis of γ-cyanopropylmethyldichlorosilane proceeds readily and without such cleavage.

Reactions of the chlorosilanes at the silicon-functional group can be combined with those at the organofunctional group. Thus, the acid alcoholysis of γ-cyanopropylmethyldichlorosilane yields γ-ethoxycarbonylpropylmethyldiethoxysilane (*23*).

If a reactive group in an organic radical is to remain unchanged during hydrolysis, the system should remain neutral during the process and only those silicon-functional groups that can be hydrolyzed in a neutral medium should be used. The ethoxysilanes appear to be the most useful in this connection. Allyltrichlorosilane, which decomposes during hydrolysis according to the following equation:

$$CH_2{=}CH{-}CH_2SiCl_3 \xrightarrow{2H_2O} CH_2{=}CH{-}CH_3 + SiO_2 + 3HCl$$

can be converted into the siloxane if the system is kept neutral during the process by the continuous addition of a weak base (*24*). γ-Ethoxycarbonylpropylmethyldiethoxysilane hydrolyzes smoothly to give the corresponding cyclosiloxanes (*23*). The hydrolysis of the aminoalkylmethyldiethoxysilanes yields the expected siloxanes (*137*).

5.2 Nonhydrolytic Processes

The term chosen for the group of processes to be treated below is very arbitrary; it is however primarily intended to indicate their difference from the purely hydrolytic processes. All those processes for the production of Si—O—Si bonds from monomeric silanes in which water as such is not introduced into the reaction will be classified as *nonhydrolytic*. Processes in which side reactions yield water that can then take part in the reaction are also regarded as nonhydrolytic. Among these are the reactions of *halosilanes* with *alcohols* and *carboxylic acids* and of *alkoxysilanes* with *carboxylic acids*. However, the greater part of the processes discussed here work in the complete absence of water. A survey of the reactions concerned is given in Table 32.

TABLE 32

NONHYDROLYTIC REACTIONS LEADING TO THE FORMATION OF SILOXANES

Reactants		End products
≡Si—X + RO—Si≡	→	≡Si—O—Si≡ + RX
2≡Si—X + ROH + R′COOH	→	≡Si—O—Si≡ + 2HX + R′COOR
≡Si—X + RCOO—Si≡	→	≡Si—O—Si≡ + RCOX
≡Si—X + HO—Si≡	→	≡Si—O—Si≡ + HX
≡Si—X + MO—Si≡	→	≡Si—O—Si≡ + MX
2≡Si—X + MO	→	≡Si—O—Si≡ + MX_2
≡Si—H + HO—Si≡	→	≡Si—O—Si≡ + H_2
2≡Si—OR + 2HX	→	≡Si—O—Si≡ + 2RX + H_2O
≡Si—OR + RO—Si≡	→	≡Si—O—Si≡ + ROR
2≡Si—OR + 2R′COOH	→	≡Si—O—Si≡ + 2R′COOR + H_2O
≡Si—OR + R′COO—Si≡	→	≡Si—O—Si≡ + R′COOR
≡Si—OR + HO—Si≡	→	≡Si—O—Si≡ + ROH

(*continued on p.* 202)

TABLE 32 *(Contd.)*

NONHYDROLYTIC REACTIONS LEADING TO THE FORMATION OF SILOXANES

Reactants		End products
$\geq Si—NH_2 + HO—Si \leq$	→	$\geq Si—O—Si \leq + NH_3$
$\geq Si—OAc + HO—Si \leq$	→	$\geq Si—O—Si \leq + HOAc$
$\geq Si—OAc + KO—Si \leq$	→	$\geq Si—O—Si \leq + KOAc$

5.2.1 REACTIONS OF ORGANOHALOSILANES WITH ORGANOALKOXYSILANES

If an organohalosilane is heated with an organoalkoxysilane in the absence of a catalyst, an *exchange* of the halogen atoms and the alkoxy groups takes place. However, in the presence of Friedel-Crafts catalysts reactions take place with the elimination of alkyl halide, which could be interpreted as *intermolecular*,

$$RO—\overset{|}{\underset{|}{Si}}—X + RO—\overset{|}{\underset{|}{Si}}—X \rightarrow RO—\overset{|}{\underset{|}{Si}}—O—\overset{|}{\underset{|}{Si}}—X + RX$$

or *intramolecular* reactions (*90*, *106*, *188*, *253*, *257*, *302*).

$$>Si\begin{smallmatrix} OR \\ X \end{smallmatrix} \rightarrow \geq Si—O— + RX$$

The unwelcome effect of these reactions is shown in reduced yields during the syntheses of pure alkoxysilanes from halosilanes and alcohols.

The reaction has frequently aroused industrial interest. It has, *inter alia*, been used for the production of methylphenylsiloxanes (*253*) and for the production of mixed polymers from di- and tetrafunctional (*298*) as well as di- and trifunctional silanes (*9*). It is also suitable for the stepwise synthesis of linear polymers of *definite chain length* from di- and monofunctional silanes (*90*, *297*). In a first stage, for example, dimethyldichlorosilane and dimethyldiethoxysilane in a molar ratio of 1:1 react in the presence of a catalyst as follows:

$$(CH_3)_2SiCl_2 + (CH_3)_2Si(OC_2H_5)_2 \rightarrow Cl—\overset{CH_3}{\underset{CH_3}{Si}}—O—\overset{CH_3}{\underset{CH_3}{Si}}—OC_2H_5 + C_2H_5Cl$$

The resulting disiloxane then undergoes further reaction with similar molecules on heating, and the extent of the reaction can be followed by means of the content of hydrolyzable chlorine. The chains are finally terminated by addition of monofunctional silane. Trimethylchlorosilane, which reacts with the C_2H_5O—Si bond, trimethylethoxysilane, which reacts with the Cl—Si bond (*297*), or trimethylsilanol which can react with both bonds (*90*) are used for this purpose in the example given.

As a variant of this reaction, it is possible to transfer simultaneously the alkyl radical from the alkyl halide to the silicon by means of suitable metals (e.g., Mg or Zn) so that the *condensation reaction* is combined with an *alkylation* (*315*):

$$\geq\mathrm{Si{-}X} + \mathrm{RO{-}Si}\leq \;\rightarrow\; \geq\mathrm{Si{-}O{-}Si}\leq + \mathrm{RX}$$

$$\geq\mathrm{Si{-}OR} + \mathrm{RX} + \mathrm{M} \;\rightarrow\; \geq\mathrm{Si{-}R} + \mathrm{M(OR)X}$$

Condensation with the elimination of alkyl halide is also realized in a process characterized by the reaction of organohalosilanes with *t*-butanol (*35*, *68*) or methanol (*95*, *224*)—in the latter case, in the presence of a $ZnCl_2$ catalyst. The reaction products in both instances are not organoalkoxysilanes but polyorganosiloxanes, alkyl halides, and some hydrogen chloride. The formation of the latter can be easily interpreted by assuming that the chlorosilane is first partially esterified by the alcohol. The $\mathrm{RO{-}Si}\leq$ group then reacts with the $\mathrm{Cl{-}Si}\leq$ group, with elimination of alkyl halide and production of a siloxane bond.

5.2.2 Reactions of Organohalosilanes with Alcohols and Carboxylic Acids

The formation of siloxanes from organohalosilanes by heating with alcohols and carboxylic acids (*263*) proceeds according to the following over-all equation:

$$2\geq\mathrm{Si{-}X} + \mathrm{ROH} + \mathrm{R'COOH} \;\rightarrow\; \geq\mathrm{Si{-}O{-}Si}\leq + 2\mathrm{HX} + \mathrm{R'COOR}$$

The formation of the alkoxysilane from the halosilane, its reaction with the carboxylic acid, and the reaction of the silanol formed with the halosilane could proceed side by side as part of the over-all reaction

$$\geq\mathrm{Si{-}X} + \mathrm{ROH} \;\rightarrow\; \geq\mathrm{Si{-}OR} + \mathrm{HX} \qquad (1)$$

$$\equiv Si{-}OR + R'COOH \rightarrow \equiv Si{-}OH + R'COOR \qquad (2)$$

$$\equiv Si{-}OH + X{-}Si\equiv \rightarrow \equiv Si{-}O{-}Si\equiv + HX \qquad (3)$$

but the reactions given below are also conceivable, particularly after all the Si—X bonds have been consumed:

$$\equiv Si{-}OH + HO{-}Si\equiv \rightarrow \equiv Si{-}O{-}Si\equiv + H_2O \qquad (4)$$

$$\equiv Si{-}OH + RO{-}Si\equiv \rightarrow \equiv Si{-}O{-}Si\equiv + ROH \qquad (5)$$

$$ROH + R'COOH \rightarrow R'COOR + H_2O \qquad (6)$$

$$\equiv Si{-}OR + H_2O \rightarrow \equiv Si{-}OH + ROH \qquad (7)$$

On reaction in this way with glacial acetic acid and methanol, dimethyldichlorosilane yields mainly ($>90\%$) polydimethylsiloxanes.

5.2.3 Reactions of Organohalosilanes with Organoacyloxysilanes

The reactions are of the type:

$$\equiv Si{-}X + R'COO{-}Si\equiv \rightarrow \equiv Si{-}O{-}Si\equiv + R'COX$$

Dimethyldiacetoxysilane and methylphenyldiacetoxysilane have been made to react with phenyltrichlorosilane in the presence of iron(III) chloride as the catalyst (*17*). Diphenyldiacetoxysilane reacts in boiling methyl acetate with diphenyldichlorosilane to give hexaphenylcyclotrisiloxane in 44% yield (*191*).

5.2.4 Reactions of Organohalosilanes with Organosilanols

Organohalosilanes react with organosilanols on heating as follows (*82*, *159*, *175*, *208*, *265*, *296*, *321*):

$$\equiv Si{-}X + HO{-}Si\equiv \rightarrow \equiv Si{-}O{-}Si\equiv + HX$$

It is best to work in the presence of pyridine.

This can also take place as a part-reaction during the hydrolysis of organohalosilanes if their rate of hydrolysis is low or if insufficient water is used for complete hydrolysis.

5.2.5 Reactions of Organohalosilanes with Metal Silanolates

Reactions of this kind proceed according to the following scheme (M = Na, K) (*276*):

$$\equiv Si{-}X + MO{-}Si\equiv \rightarrow \equiv Si{-}O{-}Si\equiv + MX$$

This type of reaction is employed to prepare mixed polymers built up of siloxane units with different substituents, as well as to synthesize polysiloxanes which can be obtained only with difficulty (or not at all) by the cohydrolysis of different halosilanes (*20*, *52*, *126*, *276*).

5.2.6 Reactions of Organohalosilanes with Metal Oxides

Organohalosilanes, especially those such as $(CF_3C_6H_4)_2SiCl_2$ in which the reactivity of the halogen is very low, can be converted into the corresponding polyorganosiloxanes with metal oxides (*128*, *255*, *291*):

$$2\ {\equiv}\mathrm{Si{-}X} + \mathrm{MO} \rightarrow {\equiv}\mathrm{Si{-}O{-}Si}{\equiv} + \mathrm{MX_2}$$

Calcium, magnesium, zinc, lead, and copper oxides can be used in up to 10% excess. The mixture of organohalosilane and metal oxide is heated in a saturated aliphatic nitrile or a nitro compound, especially acetonitrile, as a solvent. Small amounts of $CuCl_2$ or $HgCl_2$ are effective catalysts (approximately 0.1 gm/mole of silane). Polyorganohalosiloxanes are obtained if a deficiency of oxide is used (*129*).

5.2.7 Reactions of Organo-*H*-silanes with Organosilanols

Organo-*H*-silanes can react with organosilanols in the presence of small quantities of alkalis or amines (*197*) or of colloidal nickel (*207*) as catalysts:

$${\equiv}\mathrm{Si{-}H} + \mathrm{HO{-}Si}{\equiv} \rightarrow {\equiv}\mathrm{Si{-}O{-}Si}{\equiv} + \mathrm{H_2}$$

The reaction can be initiated even at room temperature by small amounts of dibutyltin dilaurate (*197*) and thus gives the possibility of room-temperature vulcanization of silicone rubber systems of suitable composition (see Section 8.1.2.3).

5.2.8 Reactions of Organoalkoxysilanes with Hydrogen Chloride

The process can be considered an acidolysis of the Si—OR bond:

$$2\ {\equiv}\mathrm{Si{-}OR} + 2\mathrm{HCl} \rightarrow {\equiv}\mathrm{Si{-}O{-}Si}{\equiv} + 2\mathrm{RCl} + \mathrm{H_2O}$$

This part-reaction takes place to a greater or smaller extent with every esterification of a chlorosilane with alcohols. However, the method is of little interest for the preparation of siloxanes since the yields remain small even when catalysts are used. Boiling dimethyldiethoxysilane with $FeCl_3$ and HCl gave only a 15.7% yield of polysiloxane oil (*49*).

5.2.9 Reactions of Organoalkoxysilanes with One Another

Organoalkoxysilanes can be self-condensed to polysiloxanes with the splitting out of an ether:

$$\equiv Si-OR + RO-Si\equiv \rightarrow \equiv Si-O-Si\equiv + ROR$$

This type of condensation is frequently encountered as a side-reaction during the synthesis of alkoxysilanes; it proceeds the more readily in the above sense, the lower the number of carbons in the alkoxy radical. Ethoxy- and methoxysilanes undergo this reaction most readily. When appreciable amounts of higher siloxanes form in the still during the distillation of methylmethoxy- or, especially, phenylmethoxysilanes, this can be ascribed mainly to self-condensation with the elimination of an ether.

Boron and aluminum halides catalyze this reaction. Thus, heating diethyldiethoxysilane in the autoclave for 24 hours at 300°C with about 2% of $AlCl_3$ yields siloxanes which include 1,3-diethoxytetraethyldisiloxane, hexaethylcyclotrisiloxane, 1,5-diethoxyhexaethyltrisiloxane, and others (*302*). The reaction temperatures are generally so high and the yields so unsatisfactory, that other types of reaction are usually preferred for the production of polyorganosiloxanes from organoalkoxysilanes.

5.2.10 Reactions of Organoalkoxysilanes with Carboxylic Acids

This type of reaction is characterized by the equation

$$2\ \equiv Si-OR + 2R'COOH \rightarrow \equiv Si-O-Si\equiv + 2R'COOR + H_2O$$

and corresponds to a stage in the reaction of organohalosilanes with alcohols and carboxylic acids (Section 5.2.2). Catalytic amounts of H ions are used, which can be introduced, for example, by the addition of HCl, sulfuric acid, or *p*-toluenesulfonic acid (*163*, *282*).

Organoalkoxysilanes can similarly be made to react with carboxylic acid anhydrides (*110*), particularly when the reaction is catalyzed by small amounts of sulfuric acid.

5.2.11 Reactions of Organoalkoxysilanes with Organoacyloxysilanes

Organoacyloxysilanes react with organoalkoxysilanes like the carboxylic acids or their anhydrides (*7*, *17*, *108*, *109*):

$$\equiv Si-OR + R'COO-Si\equiv \rightarrow \equiv Si-O-Si\equiv + R'COOR$$

Acid or base catalysis is again employed, as, for example, with sulfuric acid; *p*-toluenesulfonic acid; iron, zinc, or aluminum chlorides; aluminum isopropoxide; or sodium ethoxide.

5.2.12 Reactions of Organoalkoxysilanes with Organosilanols

The course of this reaction is shown schematically below (*264*, *265*):

$$\equiv Si{-}OR + HO{-}Si\equiv \rightarrow \equiv Si{-}O{-}Si\equiv + ROH$$

The reaction of triethylsilanol with alkoxysilanes can be catalyzed by sodium ethylsilanolate (*264*).

This method, like many of the nonhydrolytic syntheses mentioned in this chapter, is also suitable for the preparation of mixed polymers with siloxane units bearing different substituents, as well as those which are either not accessible at all through cohydrolysis, or only in poor yield. Thus ethylmethyl- (*265*), methylpropyl- (*264*, *265*), and methylphenylsiloxane copolymers (*232*, *293*, *294*) have been prepared in this way.

The ability of hydroxyl, methoxy, or ethoxy groups bound to silicon to react even at room temperature with the aid of Sn compounds of higher carboxylic acids or other catalysts is utilized in the room-temperature vulcanization of silicone rubber (see Sections 5.4.1 and 8.1.2.3).

5.2.13 Reactions of Organoaminosilanes with Organosilanols

The reaction proceeds schematically according to the following equation (*220*, *264*, *283*):

$$\equiv Si{-}NR_2 + HO{-}Si\equiv \rightarrow \equiv Si{-}O{-}Si\equiv + NR_2H$$

where R = H or alkyl. The rate of the reaction of different aminosilanes with one and the same silanol decreases in the sequence (*220*):

$$(CH_3)_3SiN(C_2H_5)_2 > (CH_3)_3SiNH\text{-}n\text{-}C_4H_9 > (CH_3)_2Si[N(C_2H_5)_2]_2$$

An investigation of the condensation of hexamethyltrisiloxanediol and *N*,*N*-diethylaminotrimethylsilane has shown that there was no self-condensation of the silanediol to siloxanes. Consequently, it appears that this reaction is suitable, when certain working conditions are maintained, for building up or linking polymers in a definite manner and without side reactions (*220*).

5.2.14 Reactions of Organoacyloxysilanes with Organosilanols and Silanolates

Diphenyldiacetoxysilane reacts with dipotassium diphenylsilanediolate in methyl acetate according to the general scheme:

$$\equiv Si{-}O{-}Ac + KO{-}Si\equiv \rightarrow \equiv Si{-}O{-}Si\equiv + KOAc$$

The reaction products are hexaphenylcyclotrisiloxane and octaphenylcyclotetrasiloxane (*191*).

The terminal silanol group in linear polyorganosiloxanes is also easily accessible to reaction with acyloxysilanes (*42*):

$$\equiv Si-O-Ac + HO-Si\equiv \rightarrow \equiv Si-O-Si\equiv + HOAc$$

If both compounds are very thoroughly dried, the mix can be stored at room temperature. Under such conditions the reaction is initiated only by moisture. This makes it possible to produce systems which are stable during storage at room temperature and which can be vulcanized to rubber-elastic products on the admission of atmospheric moisture without any other catalytic additive or increase in temperature (*43*) (cf. Section 8.1.2.3.2).

α,ω-Bis-methyldiacetoxysiloxy-polydimethylsiloxanes, which can be obtained from the partial reaction of methyltriacetoxysilane with polydimethylsiloxane-α,ω-diols (*38*)

$$(AcO)_2(CH_3)Si-O-[Si(CH_3)_2-O-]_nSi(CH_3)(OAc)_2,$$

have the same property.

5.2.15 Other Methods

The heat sensitivity of the acetoxy—silicon bond has been used to obtain polyorganosiloxanes from organoacetoxysilanes

$$\equiv Si-OOCCH_3 + CH_3COO-Si\equiv \rightarrow \equiv Si-O-Si\equiv + (CH_3CO)_2O$$

The reaction is very rapid above 200°C (*13*).

The oxidation with mercury(II) oxide or silver oxide of organosilanes containing Si—H or Si—Si bonds has also been proposed (*51*). The reaction is carried out in an inert organic solvent such as toluene or carbon tetrachloride.

The course of the oxidation of alkoxysilanes in air at high temperatures is very obscure (*233*): alkoxy groups are said to be split off as aldehydes, acids, etc., and, in particular, even an attack on the Si—C bond would appear unavoidable under these conditions.

5.3 Syntheses by Cleavage of the Si—C Bond

All the reactions treated so far, whether hydrolytic or nonhydrolytic, take place on the silicon-functional group of the organosilane. The Si—C bond is involved only slightly or not at all.

However, there is a special method, leading from silanes to polysiloxanes by the hydrolytic cleavage of a saturated unsubstituted organic group from the Si atom, which can be applied to Si—methyl bonds. It depends on the fact that a single methyl radical can be removed from the $(CH_3)_3Si$— group by the combined action of sulfuric acid and water (*277*, *280*):

$$(CH_3)_3Si— \xrightarrow[H_2O]{H_2SO_4} \left[\begin{array}{c} CH_3 \\ | \\ —Si—O— \\ | \\ CH_3 \end{array} \right] + CH_4$$

The sulfuric acid reacts with the methylsilane, in the first stage, to form a monosilyl sulfate which is hydrolyzed with water and then condensed in a second stage. The methyl group is liberated as methane.

This method is particularly important for preparative purposes, since it can also be used when the silane contains, in addition to methyl groups, substituted organic radicals such as those containing keto (*271*), carboxy (*269*, *273–275*, *277*) or amino groups (*203*, *270*). The following are two of the possible reactions:

$$2HOOC(CH_2)_2(CH_3)_3Si \xrightarrow[H_2O]{H_2SO_4} \left[\begin{array}{c} \qquad\qquad\qquad\quad CH_3 \\ \qquad\qquad\qquad\quad | \\ HOOC—CH_2—CH_2—Si— \\ \qquad\qquad\qquad\quad | \\ \qquad\qquad\qquad\quad CH_3 \end{array} \right]_2 O + 2CH_4$$

$$2H_2NCH_2(CH_3)_3Si \xrightarrow[KOH]{H_2SO_4/H_2O} [H_2NCH_2(CH_3)_2Si]_2O + 2CH_4$$

The organofunctional disiloxanes obtained can then be equilibrated with polymethylsiloxanes (Section 5.4.2.3).

The question of how far other alkyl groups linked to the Si atom can be cleaved has been examined for a series of carboxyalkylmethylsilanes of the general formula

$$HOOC—CH_2—CH_2(CH_3)R_2Si$$

A decreasing stability of the Si—C bond within the series (*6*)

$$C_2H_5,\ n\text{-}C_3H_7 > CH_3 > n\text{-}C_5H_{11} > n\text{-}C_4H_9,\ i\text{-}C_5H_{11}$$

is obvious.

5.4 Polymerization and Polycondensation

It was pointed out during the discussion of the general principles of hydrolytic processes that condensation reactions can take place even during the hydrolysis of silicon-functional silanes because of their tendency to undergo *inter-* or *intramolecular condensation*. More or less condensed products with siloxane and silanol bonds or even siloxanes free from silanol groups, such as, in particular, cyclosiloxanes, are thus formed.

These products are of course seldom the final products in a technical sense. They are mostly *polymerized* or *condensed* finally in a third stage—if the production of the organosilanes is considered the first and their hydrolysis the second stage.

While the process occurring during condensation can be defined with considerable precision, the term "polymerization" will be used very loosely here to characterize a number of reactions whose mechanism is not as yet understood in detail. Judging by the nature of the catalysts known to be effective for the reaction it would appear that they are predominantly of *ionic* character, and proceed in part through silanol condensation as an intermediate stage. It is thus difficult, if not impossible, to differentiate between polymerization and polycondensation on a kinetic basis. A phenomenological differentiation which makes reference only to the starting and final products is more promising. *Polymerization* will be taken in this sense to mean the conversion of low-molecular-weight siloxanes free from silanol groups, such as the cyclic siloxanes D_n, into high-molecular-weight siloxanes. The starting materials and final products can thus be represented by the scheme

$$x[R_2SiO]_n \rightarrow [R_2SiO]_{nx}$$

The term "*condensation*" will be reserved for processes in which silanols or siloxanols participate in the reaction. For example, phosphoronitrile halides do not cause the splitting and polymerization of pure octamethylcyclotetrasiloxane, and therefore should not be called polymerization catalysts. Their action is exerted only with siloxanes containing silanol groups, and must therefore be ascribed to a condensation effect (*246*). Unfortunately the available literature does not always permit a clear classification into one type of reaction or the other, since the starting products are often insufficiently defined.

Special cases of polymerization are processes in which mixtures of siloxanes of different molecular sizes are *equilibrated*, i.e., brought into molecular equilibrium. The reaction has the effect of converting a mixture of silanes whose molecular weight distribution curve has a number of maxima into one having a *single* maximum in its molecular weight distribution curve. The curve is Gaussian (Fig. 124). The rate at which molecular distribution equilibrium is reached depends on the conditions (catalyst). *Equilibration* therefore differs from simple polymerization inasmuch as a decrease in the size of larger molecules must take place simultaneously with an increase in the size of smaller ones.

Another special case is the process of *copolymerization*. This comprises all processes in which two organosiloxanes with different substituents are polymerized together to form a single polymer. The following scheme would represent a copolymerization process for two siloxanes:

$$x[R_2SiO]_n + x[R'_2SiO]_n \rightarrow \left[\begin{array}{c} \quad R \qquad\; R' \\ \quad | \qquad\;\; | \\ -Si-O-Si-O \\ \quad | \qquad\;\; | \\ \quad R \qquad\; R' \end{array} \right]_{nx}\!\!-$$

Changes in the size of the molecules with all these methods are achieved by the breaking and re-forming of siloxane bonds. It is thus possible to break siloxane rings and re-form the fragments into linear macromolecules, a fact which has acquired considerable technical importance. In such processes it is thus possible, in principle, to blend the various functionalities and bring them into the molecular distribution equilibrium by suitable methods.

Although condensation and polymerization can be initiated by heat alone, *catalysts* are nearly always used to accelerate the reaction and to moderate the reaction conditions. The mode of action differs widely; additives designed to accelerate condensation operate partly by removing the water formed in condensation from the system, and partly by entering the polymer with hetero-atoms, thus increasing its molecular weight. Polymerization catalysts cleave the siloxane bond to give intermediate complexes with a higher energy state, and these tend to change spontaneously into polymers having a lower energy. However, these processes overlap in the sphere of condensation and polymerization, which makes a rigid classification of the catalysts difficult or well-nigh impossible.

The number of ways in which polymerization in the most general sense can take place is still further increased by the fact that, in addition to the growth of molecules by condensation reactions on silicon-functional groups or by the breaking and re-forming of siloxane bonds, the organic groups may also be involved in the reaction. Thus, reactions involving the methyl groups in linear polydimethylsiloxanes can be brought about by means of peroxides or high-energy radiations, resulting in the formation of methylene, ethylene, or Si—Si bridges and thus increasing the functionality of the siloxane unit concerned. This method is used in silicone rubber technology to achieve further *cross-linking* or *curing* of linear macromolecular siloxanes.

The last but by no means unimportant group of processes which enables the polymer skeleton to be modified comprises reactions by means of which *heteroatoms* are incorporated into the siloxane chain. Since the entire siloxane skeleton can be fundamentally altered by this method, its discussion will be reserved for special later sections (Section 7.1).

5.4.1 Condensations of Silanols and Siloxanols

The same rules apply to the spontaneous condensation of silanols taking place even during hydrolysis and to those condensation processes that require heat or catalysts (cf. Section 3.6). The tendency to condensation increases with the number of hydroxyl groups bonded to the silicon atom and decreases with the number, size, and bulkiness of the organic groups. The tendency to undergo condensation has also been found to depend on the electronegativity of the substituents, on the pH value of the system and on the type of catalyst used.

The importance of steric hindrance to the stability of organosilanols can be very clearly seen from the fact that silanediols containing *t*-butyl groups are extremely resistant to thermal condensation as compared with di-*n*-alkylsilanediols (*281*). The same inference can be drawn from the fact that silanols containing branched aliphatic groups are more stable than those with unbranched groups. Methyl and ethylsilanols are thermally less stable than the corresponding substituted phenyl and naphthyl compounds (*256*). However, these relationships can change in the presence of catalysts, as will be seen below.

Many catalysts capable of promoting the reaction are known. They probably act in very different ways. Some are merely dehydrating agents and displace the equilibrium

$$\geqslant Si-OH + HO-Si \leqslant \;\rightleftharpoons\; \geqslant Si-O-Si \leqslant + H_2O$$

toward the formation of the siloxane bond. This action is characteristic of the following substances: sulfuric acid (*230*), phosphoric acid (*230*), phosphorus halides (*230*), phosphorus pentoxide (*230*), phosphonitrile halides (*195*), phenyl phosphorodichloridate and phenylphosphonic dichloride (*284*, *285*), organically substituted nitrogen derivatives of phosphorous and phosphoric acid (*195*), dialkyl phosphites (*2*), boron trifluoride (*179*), organic acid halides, especially acetyl chloride (*247*), trifluoroacetic acid (*218*), isocyanates (*55*, *76*), zeolites (*219*), and others.

The probability of a rearrangement of the siloxane skeleton taking place simultaneously cannot be excluded with many of these catalysts. Selective catalysts for the condensation reaction which do not cause cleavage of the Si—O—Si bonds and equilibration of the siloxane molecules are tetramethylguanidine 2-ethylcaproate and *n*-hexylamine 2-ethylcaproate (*77*, *183*, *184*). Using these, it is possible to obtain block copolymers containing phenylenesiloxane and polydimethylsiloxane blocks by the cocondensation of prepolymers containing the requisite silanol groups (*183*, *184*).

Other catalysts may operate by a reaction of their functional groups with the silanol group leading to their incorporation into the siloxane chains as cross-linking or end-to-end linking units. The action of boric acid (*230*), titanic acid (*107*, *146*, *266*) and silicic acid esters (*266*) can be interpreted in this way:

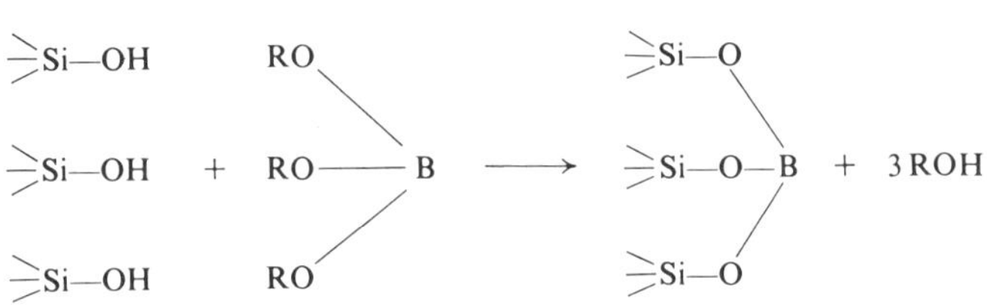

The action of acids such as HCl (*122*), HBr (*36*), or bases, especially amines (*230*), could be interpreted as a catalysis of the condensation of silanol groups.

Considerable differences are observed in the behavior of differently substituted silanols toward acidic or basic catalysts. The rate of condensation in the presence of hydrochloric acid decreases in the order

$$(CH_3)_2Si(OH)_2 > CH_2{=}CH(CH_3)Si(OH)_2 > ClCH_2(CH_3)Si(OH)_2 > Cl_2CH(CH_3)Si(OH)_2$$

and
$$(CH_3)_2Si(OH)_2 > CH_3(C_6H_5)Si(OH)_2 > (C_6H_5)_2Si(OH)_2$$

This may be regarded as resulting from the increase in the number of electronegative groups attached to the Si atom; these groups decrease the electron density on the siloxane oxygen and so make an attack by $H^{\oplus}$ more difficult (cf. Section 6.3.3.2).

In catalysis with bases, such as triethylamine, the situation is naturally reversed. The sequences in this case are

$$(CH_3)_2Si(OH)_2 < CH_2{=}CH(CH_3)Si(OH)_2 < ClCH_2(CH_3)Si(OH)_2 < Cl_2CH(CH_3)Si(OH)_2$$

and
$$(CH_3)_2Si(OH)_2 < CH_3(C_6H_5)Si(OH)_2 < (C_6H_5)_2Si(OH)_2$$

The nucleophilic attack of the base on the Si atom is facilitated by the electronegative substituents.

Thus in the presence of triethylamine, diphenylsilanediol, which is generally regarded as a very stable silanol, is less stable than dimethylsilanediol, which is one of the least stable silanols. The great difference between acidic and basic catalysis is also confirmed by the discovery that trimethylsilanol in methanol is condensed 500 times as rapidly by hydrogen chloride as by KOH (*103*). Whereas orthosilicic acid has a stability maximum at about pH $\sim$2 to 3 (*134*), this maximum moves into the alkaline range on methylation (*164*).

The increase in reactivity in the above sequences on alkaline catalysis reflects the strong influence of the electronegativity of the substituents, which overrides the opposing influence of steric hindrance (*47*).

A large, if not the largest, part among condensation catalysts is played by organic compounds of metals. They are used in silicone resin technology (Section 8.2) and room-temperature vulcanizable silicone rubbers (Section 8.1). Their mode of action has not been investigated in great detail. It appears that they not only promote the condensation, but also act as oxygen carriers, causing cross-linking of the organic groups by oxidation (*56*). Cobalt and zinc salts probably act mainly as condensation catalysts while lead salts are active through the second mechanism as well. It is also conceivable that certain metal salts can extend the silicone skeleton by the incorporation of metaloxane groups. The fact that crystalline $[(CH_3)_3SiO]_2Pb$ can be prepared from $(CH_3)_3SiOH$ and PbO (*213*, *238*) would indicate that the formation of the

Si—O—Pb—O—Si bond is responsible for the activity of lead oxide. An analogous mechanism might well apply to tin compounds.

Metal salts capable of yielding single-phase mixtures with siloxanes or their solutions are generally preferred. The metal naphthenates, long known as driers in the paint and varnish industry, are the most important (*29*, *60*, *181*, *311*). In addition, carboxylic acid salts in general have been proposed by various authors (*3*, *132*, *171*, *194*, *196*, *216*, *268*), and so have dithiocarbamates and xanthates (*198*), carboxylic acids containing organic groups bonded directly to metals (*33*, *222*, *226*), and other organometallics (*58*, *136*, *167*, *196*, *226*); also, alkoxides and phenoxides (*26*, *33*, *34*, *194*, *216*, *222*), chelate complexes (*171*), metal-modified phenolic resins (*243*), and many more.

The action of the catalysts is strongly temperature-dependent and increases with increasing temperature. A metal catalyst can, however, still accelerate polymerization at room temperature. If the charge to be catalyzed is to remain unchanged at room temperature, it is possible to protect the siloxane at lower temperatures against the action of the metals by complexing, especially chelate-forming, additives. Compounds of the nature of acetylacetone, acetoacetic ester, malonic acid, and biacetyl (*152*) are suitable for this purpose. Certain systems of functional polysiloxanes, on the other hand, react with metal catalysts, particularly organic compounds of tin, in such a way that they can be vulcanized at room temperature (*197*, *222*). It is thus possible to convert liquid linear polysiloxanes within a few minutes or hours—depending on the nature and quantity of the catalyst—into highly elastic materials.

The activity of the metal catalysts varies widely. The following arrangement attempts a classification of their effectiveness, based principally on experience with silicone resins.

Highly active metal compounds—those of Pb, Sn, Zr, Al, Ca, alkali metals

Weakly active metal compounds—those of V, Cr, Mn, Fe, Co, Ni, Cu, Zn, Cd, Hg, Ti, Th, Ce, Mg

For problems in paint and varnish technology, it has often proved advantageous to combine metal catalysts from both groups (Section 8.2.2.1.2).

Detailed investigations on condensation kinetics have revealed the following information. The rate of the thermally activated condensation of diarylsilanediols decreases rapidly with the degree of condensation. At 180°C the condensation process is largely completed within 5 to 10 hours, and the elimination of the residual OH groups proceeds only very slowly (*248*).

The acid-catalyzed condensations of dimethylsilanediol and chloro- and bromomethylmethylsilanediols in dioxane solution proceed, at an initial concentration of 0.7 mole/liter, as a second-order reaction. Higher initial concentrations give a higher order of reaction. The reaction is of first order with respect to the catalyst. The rate constants K in (*48*)

$$-\frac{d\left[\begin{array}{l}\diagdown\\ \text{—SiOH}\\ \diagup\end{array}\right]}{dt} = K[\text{HCl}]\left[\begin{array}{l}\diagdown\\ \text{—SiOH}\\ \diagup\end{array}\right]^2$$

were as given in the accompanying tabulation for the silanols mentioned.

Silanol	K (sec^{-1} $mole^{-2}$ $liter^{-2}$)
$(CH_3)_2Si(OH)_2$	0.33
$ClCH_2(CH_3)Si(OH)_2$	0.0094
$BrCH_2(CH_3)Si(OH)_2$	0.0069

These results indicate that the first step, i.e., attachment of the catalyst proton to the silanol oxygen

$$H^{\oplus} + \diagdown\!\!\!\text{—Si—}\overline{\underline{O}}\text{—H} \rightarrow \text{—Si—}\underset{\oplus}{\overset{H}{\overset{|}{\underline{O}}}}\text{—H}$$

takes place very rapidly and that the reaction

$$\text{—Si}\overline{\underline{O}}\text{H} + \text{—Si—}\underset{\oplus}{\overset{H}{\overset{|}{\underline{O}}}}\text{—H} \rightarrow \text{—Si—}\overline{\underline{O}}\text{—Si—} + H^{\oplus} + H_2O$$

is the rate-determining one. This can be considered a nucleophilic attack of the oxygen in one silanol molecule on the silicon atom in another silanol molecule in the oxonium form.

The marked decrease of the rate of condensation with chloro substitution in a single methyl group can be ascribed only partially to a steric effect. The main factor is probably the inductive effect of the chlorine which exerts an electron-attractive force on the oxygen, and thereby reduces its ability to carry out a nucleophilic attack. Contrary to expectations from this theory, the Br substitution product reacts somewhat more slowly than the Cl compound, but this inversion could be due to steric factors.

The result of a *p*-halogen substitution in a phenyl group leads to an increase in the tendency toward condensation; at first sight this is in contradiction to the results on aliphatically substituted silanols. An explanation might be found in the fact that resonance structures of the type

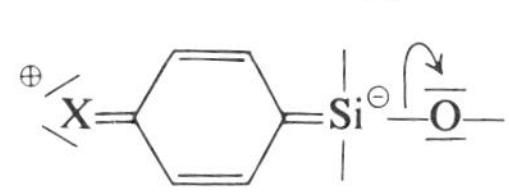

that occur with halogen substitution, tend to oppose the inductive effect of the phenyl radical

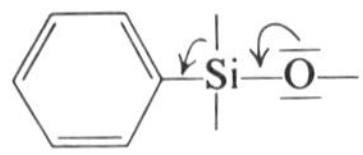

and thereby increase the basicity of the oxygen and its capacity for nucleophilic attack (*248*).

In the base-catalyzed condensation, the initiating step

$$\geq\text{SiOH} + \text{OH}^{\ominus} \rightarrow \geq\text{Si—O}^{\ominus} + \text{H}_2\text{O}$$

is probably followed by nucleophilic attack of the siloxanyl anion on a silanol molecule (*47*):

$$\geq\text{Si—O}^{\ominus} + \geq\text{Si—}\underline{\overline{\text{O}}}\text{—H} \rightarrow \geq\text{Si—}\underline{\overline{\text{O}}}\text{—Si}\leq + \text{OH}^{\ominus}$$

as the rate-determining step. These interpretations of the reaction mechanism are analogous to those of the acid and alkaline polymerization of siloxanes (Sections 5.4.2.2 and 5.4.2.3).

In highly cross-linked siloxanes, the elimination of the last OH groups by means of a simple condensation, with the splitting out of water, is no longer possible. It must, rather, be assumed that in this case the residual OH groups are removed by reactions with organic groups in the following way (*55*):

$$\geq\text{Si—R} + \text{HOSi}\leq \rightarrow \geq\text{Si—O—Si}\leq + \text{RH}$$

Such "silanolysis" of the Si—C bond can be detected in the condensation of phenylsilanols (*250*).

The low reactivity of the silanol groups in highly cross-linked polyorganosiloxanes can be understood if it is considered that the low mobility of the network makes an encounter of two OH groups unlikely. The silanol groups can, however, react with neighboring organic groups, but more sluggishly than in a simple condensation. The removal of such residual OH components can take 150 to 200 hours, even on heating to 240°C.

The influence of the network structure on the condensation is shown in Figs. 3 to 6. Figure 6 also shows how residual silanol groups in a silica glass—and it is known that there are hardly any such glasses free from OH groups—tenaciously resist all condensation at high temperatures, since they no longer have groups capable of reacting with them available in their neighborhood.

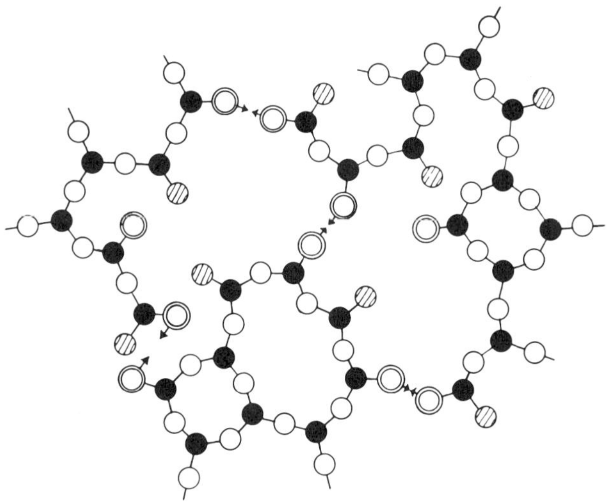

FIG. 3. Polyorganosiloxane network (plane projection) with some OH groups still capable of condensation (●) Si; (○) O; (◎:) OH; (⊘:) C_6H_5.

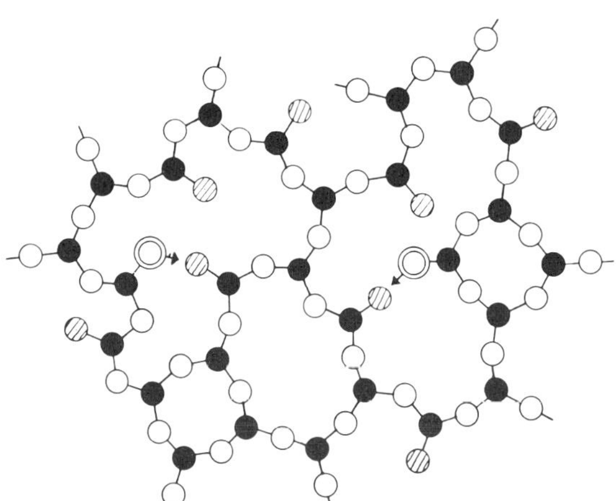

FIG. 4. Network of Fig. 3 after elimination of water is complete.

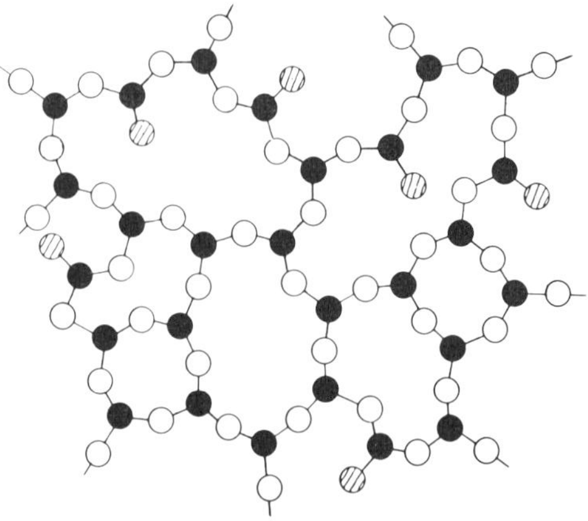

FIG. 5. Network of Fig. 4 with the last OH groups removed by reaction with neighboring C_6H_5 groups.

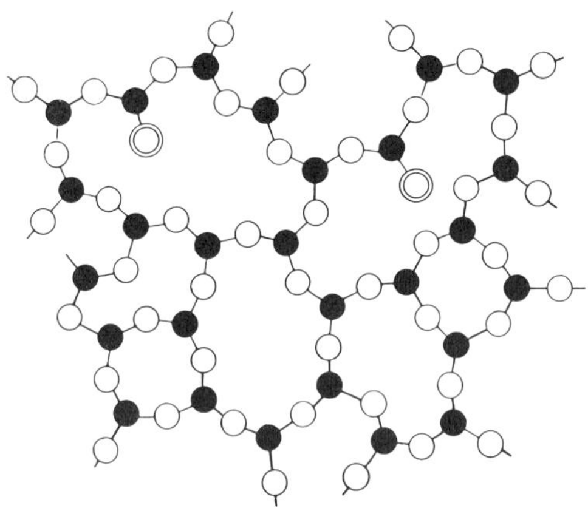

FIG. 6. Isolated OH groups in a siloxane network (silica glass).

5.4.2 POLYMERIZATION

5.4.2.1 *Thermal Polymerization*

Low-molecular-weight polydiorganosiloxanes, particularly cyclic ones, can be converted into high-molecular-weight products by heating them to 250—300°C under their own vapor pressure in a closed system (*199*, *202*, *262*, *319*). The upper temperature limit is determined by the sensitivity of the Si—C bond to thermal cleavage and by the possibility of thermal decomposition of the organic groups. The danger of the formation of isolated cross-linking by thermal degradation also cannot be entirely excluded with this method. The yields of high-molecular-weight products can be increased if pressures of 1000 to 7000 kg/cm^2 are used in conjunction with temperatures of 200 to 350°C (*71*). The pressure is exerted either by the heated fluid itself—being adjusted by the release of part of the fluid—or by a hydraulic or dynamic compressor such as a press or a positive displacement pump.

These processes have not acquired much technical importance, since there are methods based on catalysis that achieve polymerization without pressure at considerably lower temperatures. They are however of considerable theoretical interest, since the reverse effect, i.e., degradation, takes place when the temperature is increased further.

If polydiorganocyclosiloxanes are heated from 250 to 300°C with dimethyldichlorosilane in a closed system, telomerization accompanied by the opening of the siloxane rings takes place. The main products are low molecular α,ω-bis-dimethylchlorosiloxy-polydimethylsiloxane (*14*):

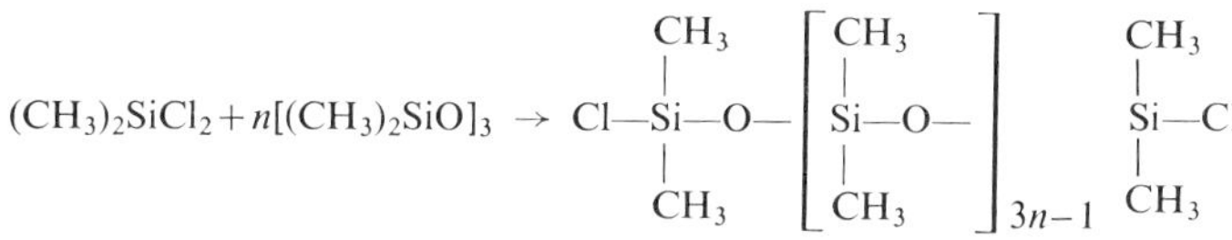

On being heated with an organotrichlorosilane, hexamethylcyclotrisiloxane yields telomerization products of type (I) (*16*) in the formula below, and on being heated with silicon tetrachloride those of type (II) (*15*):

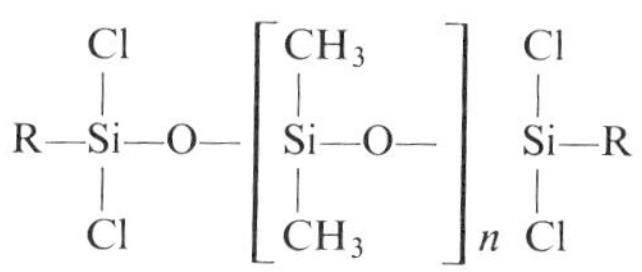
I

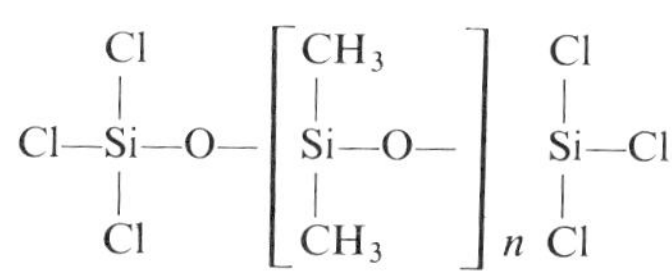
II

5.4.2.2 *Polymerization with Acids*

Proton and Lewis acids are acid catalysts for siloxane polymerization.

The first step when a proton acid reacts with a siloxane bond appears to be the attachment of the proton to the lone electron pair of the siloxane oxygen:

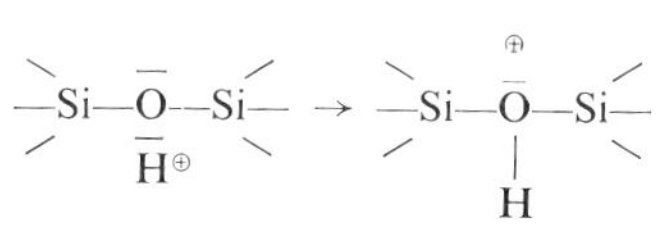
(1)

Since, therefore, the siloxane oxygen acts as electron donor in this case, acid catalysis differs fundamentally from base catalysis: the base functions as an electron donor and adds to the silicon atom, enlarging its valency sphere (*114*) (Section 5.4.2.3).

The subsequent reactions leading to the scission and re-formation of siloxane bonds can be interpreted in two ways. A conceivable mechanism consists in a decomposition of the oxonium complex with the formation of an Si—OH bond and entails the reaction of an Si—X bond with the $X^{\ominus}$ anion:

$$\geq Si-\overset{\oplus}{\underset{H}{O}}-Si\leq \; + \; X^{\ominus} \longrightarrow \; \geq Si-OH \; + \; \geq Si-X \qquad (2)$$

Since there is an energy gradient favoring the formation of a macromolecular polysiloxane chain, condensation reactions follow which reverse the scission of the chain by forming new siloxane bonds (*214*):

$$\geq Si-OH + X-Si\leq \; \rightarrow \; \geq Si-O-Si\leq + HX \qquad (3)$$

$$\geq Si-OH + HO-Si\leq \; \rightarrow \; \geq Si-O-Si\leq + H_2O \qquad (4)$$

$$\geq Si-X + H_2O \; \rightarrow \; \geq Si-OH + HX \qquad (5)$$

While, therefore, intermediate condensation reactions are assumed here, the following mechanism would justify the view that this is a pure polymerization reaction: the siloxane oxygen of a nonprotonized molecule adds nucleophilically to the protonized activated intermediate (1), and this is followed by scission and re-formation of the siloxane bonds:

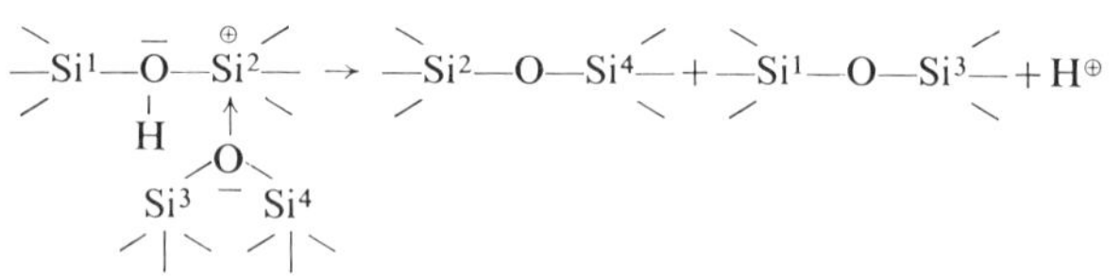
(6)

The first part-reaction [cleavage of the siloxane bond as in (1) and (2)] is a first-order process in which the decomposition of the oxonium complex in (2) is rate-determining (*54*). The over-all process however is of second order (*151*), and one of the condensations (3) and (4) or reaction (6) is therefore the rate-determining step.*

The action of Lewis acids is similar in principle. The oxygen is again the electron donor and makes its lone electron pair available to the catalyst. This

* Russian authors have recently reported that the polymerization of polydimethylcyclosiloxanes in the presence of sulfuric acid can be accelerated by salts of metals of variable valence. From these and other observations it is suggested that redox processes participate in this reaction (*105*).

view is supported by the fact that it has been possible to isolate *adducts* of some Lewis acids, such as that of $SnCl_4$, with the siloxane bond in hexamethyldisiloxane (*257*):

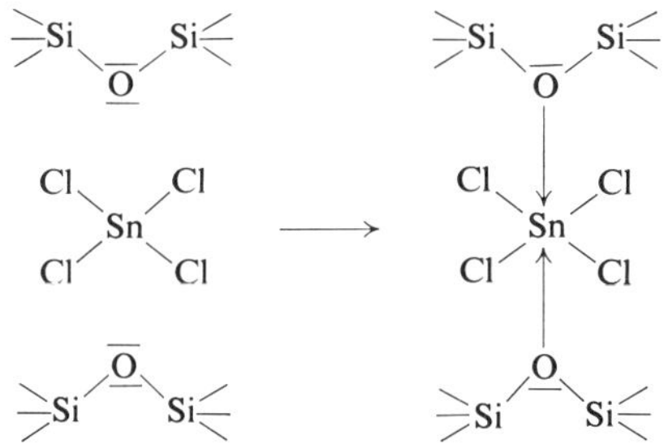

(7)

This reaction is then followed by cleavage of the siloxane bond:

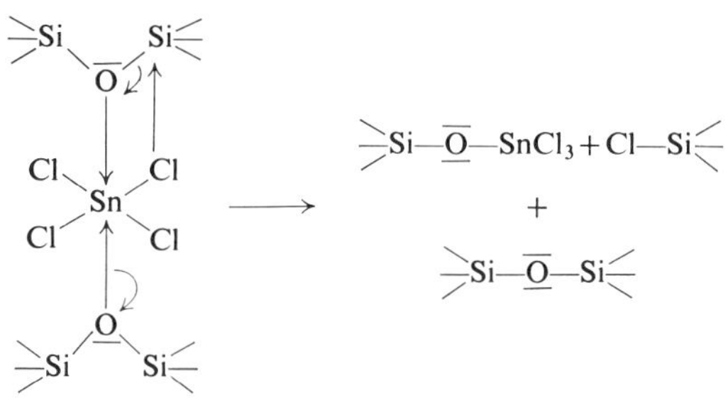

(8)

which results in the formation of a stannosiloxane bond (cf. Section 7.1.4). If the Lewis acid cannot form stable adducts with the siloxane or stable heterosiloxanes, as is the case with ferric chloride, the Lewis acid is again formed by the reverse of reaction (8). It is thus capable of further reaction with siloxane bonds and therefore functions as a genuine catalyst:

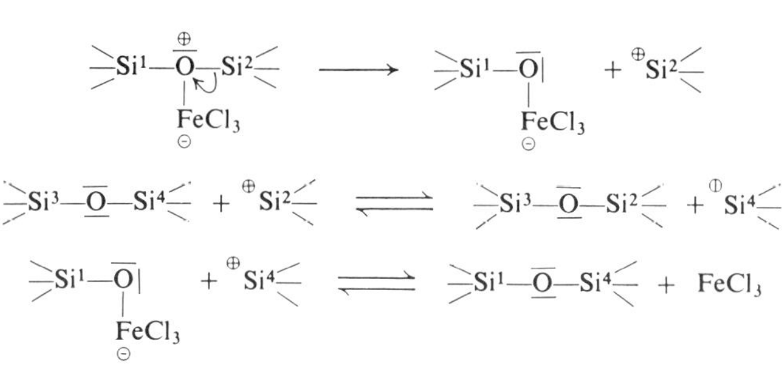

Proton acids useful as *catalysts* are sulfuric acid (*212*, *214*), ethylsulfuric acid (*8*), chlorosulfonic acid (*172*), selenic acid (*225*), nitric acid (*114*), phosphoric acids (*310*), pyrophosphoric acid (*114*), and boric acid (*180*). Cation exchangers, particularly acid-activated fullers' earths, could also come under this heading. They contain exchangeable $H^{\oplus}$ ions and could thus be considered as proton acids with an infinitely large anion represented by the exchanger substrate—montmorillonite in the case of fullers' earths (*37*, *149*).

Whether carboxylic acids acting as proton acids could promote the polymerization of siloxanes is still undecided. Oxalic acid has been patented (*81*) as a catalyst but is not effective with octamethylcyclotetrasiloxane (*114*). Trichloroacetic and perfluorobutyric acids are as ineffective as oxalic (*114*). Trifluoroacetic acid is mentioned as a catalyst in the patent literature (*182*), but it has no effect on octamethylcyclotetrasiloxane in the cold, while at higher temperatures it seems to lead to permanent scission and formation of silanol and silyl ester groupings (*114*).

The following are effective Lewis acid catalysts: iron(III) chloride (*1*), iron(III) chloride hexahydrate (*322*), aluminum chloride mixed with iron(III) oxide (*1*), boron trifluoride (*192*), zinc chloride (*192*), and tin(IV) chloride (*322*).

The result of an acid-catalyzed reaction will of course depend primarily on the nature of the *siloxane components* used.

If cyclic siloxanes consisting only of *D units* are polymerized by means of mineral acid catalysts, linear macromolecular siloxanes with terminal $\equiv$Si—OH and $\equiv$Si—X groups are obtained, the protons and the acid anions acting as chain stoppers.

If, on the other hand, mixtures of mono- and difunctional siloxanes, i.e., oligomers or polymers containing *M and D units* are used, MD_nM chains blocked by triorganosiloxy units are formed. Investigations on the course of the polymerization have led to interesting conclusions (*144*). If one mole of hexamethyldisiloxane [M_2] and one mole of octamethylcyclotetrasiloxane [D_4] are stirred with 4% of sulfuric acid at 25°C, the viscosity changes as shown in curve 1 of Fig. 7. The macromolecular weight distribution reaches equilibrium in about 15 hours and then remains constant. If, however, hexamethyldisiloxane [M_2] and hexamethylcyclotrisiloxane [D_3] in a molecular ratio 1:0.55 are digested with sulfuric acid under otherwise similar conditions, the viscosity is at first found to increase steeply and then falls to a constant value (Fig. 7, curve 2). This effect can be explained by the fact that the D_3 molecule is highly reactive toward the acid. The ring of the trisiloxane is broken more quickly than the hexamethyldisiloxane molecule, so that the catalyst first promotes vigorous chain growth. Only when the M units are increasingly introduced into the reaction are the chains shortened until molecular equilibrium is established. Polymerizations of the type

$$MM + \frac{n}{3}D_3 \rightarrow MD_nM \quad (n \text{ large})$$

preponderate during the initial period when the viscosity increases rapidly. Equilibration reactions of the type

$$MD_nM + MM \rightarrow MD_{n-1}M + MDM$$
$$MD_{n-1}M + MDM \rightarrow MD_{n-2}M + MD_2M, \text{ and so on}$$

follow in the phase of decreasing viscosity and establishment of the final value. The activity of different fully methylated polysiloxanes with acid catalysts

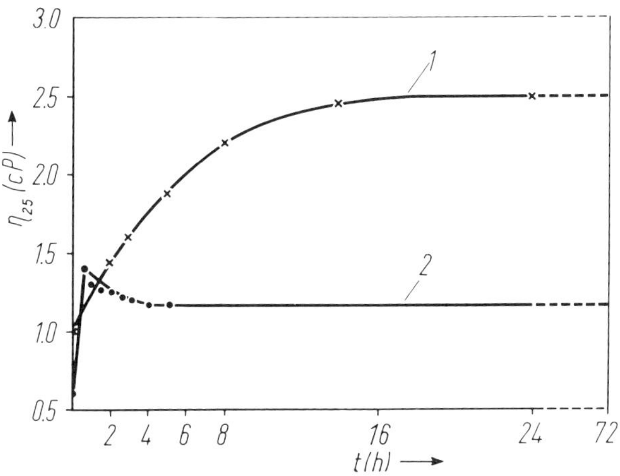

FIG. 7. Change of viscosity during equilibration of polymethylsiloxanes with sulfuric acid [values from Kantor *et al.* (*144*)]. Curve 1: Equilibration of M_2 and D_4 in a molar ratio of 1:1. Curve 2: Equilibration of M_2 and D_3 in a molar ratio of 1:0.55.

decreases in a sequence beginning with hexamethylcyclotrisiloxane and terminating with octamethylcyclotetrasiloxane.

$$D_3 > MM > MDM > MD_2M > D_4$$

Systems with *T and D units* can also be polymerized so as to form cross-linked resinous siloxanes, and be equilibrated with acid catalysts. On the other hand, it is not possible to obtain appreciable yields of defined low-molecular-weight siloxanes of the type M_3T, M_4T_2, M_5T_3, etc., from the hydrolysis products of trimethylchlorosilane and phenyltrichlorosilane by means of proton and Lewis acids. In this case only a "reversed" hydrolysis of the chlorosilanes, i.e., the addition of a definite amount of water to the bulk of the silanes, gives the desired result (*258*).

The Si—H bonds in *organo-H-siloxanes*, which are extremely sensitive toward alkaline catalysts, largely survive acid-catalyzed polymerization, e.g., with sulfuric acid (*242*, *317*). The best polymerization catalyst for these siloxanes would appear to be an acid-activated fullers' earth, which attacks the Si—O—Si bond but leaves the Si—H bond completely intact (*205*, *289*).

The reactivity of the Si—O—Si bond in acid catalysis is strongly dependent on the *nature of the organic groups* or their *substituents*. Methylsiloxanes are easiest to polymerize and equilibrate. The reactivity decreases rapidly as the size of the aliphatic radical is increased, or if aromatic substituents are present. The ratios of the times required to reach the same relative viscosities when a number of cyclotetrasiloxanes with different aliphatic substituents were polymerized with sulfuric acid were as follows (*267*):

$$[CH_3Si(H)O]_4 : [C_2H_5Si(H)O]_4 = 1:11$$

$$[(CH_3)_2SiO]_4 : [(C_2H_5)_2SiO]_4 = 1:18$$

The sluggishness of the reaction of octaphenylcyclotetrasiloxane with sulfuric acid is in very marked contrast to the ease with which octamethylcyclotetrasiloxane forms linear polymers with the same catalyst.

Substituents in *organofunctional* siloxanes capable of acting as electron acceptors decrease their reactivity toward acid catalysts. Thus, sensitivity to cleavage of the halogen-substituted hexaorganodisiloxanes decreases in the following sequence (*259*):

$$(CH_3)_3Si{-}O{-}Si(CH_3)_3 > BrCH_2(CH_3)_2Si{-}O{-}Si(CH_3)_3$$

$$> BrCH_2(CH_3)_2Si{-}O{-}Si(CH_3)_2CH_2Br > ClCH_2(CH_3)_2Si{-}O{-}Si(CH_3)_2CH_2Cl$$

The copolymerization of cyanomethylheptamethylcyclotetrasiloxane (*223*) and chloromethylheptamethylcyclotetrasiloxane (*73*) with hexamethyldisiloxane can be brought about by digestion with sulfuric acid; the same applies to that of $[C_2H_5OOCCH_2CH_2(CH_3)_2Si]_2O$ with octamethylcyclotetrasiloxane (*279*), while for the copolymerization of tris(3,3,3-trifluoropropyl)-trimethylcyclotrisiloxane with hexamethyldisiloxane, special catalysts of the type of the perfluoroalkanesulfonic acids are required (*217*).

Removal of the *catalysts* after the reaction is in general necessary, because their presence would adversely affect the heat-aging properties of the polymer. The separation of the catalyst becomes more difficult the more viscous is the polymeric product. Acid catalysts are most frequently removed by washing with water. This can be achieved with very viscous polymers by mastication in homogenizers or by simple stirring after dissolving the polymer in organic solvents immiscible with water from which the water can then be removed by simple phase separation.

Another method of destroying the catalysts consists of neutralizing them with basic reagents. However, the siloxane then becomes turbid because the salts formed are insoluble. Low-viscosity polymers can be clarified by filtration, but the treatment of highly viscous products involves troublesome thinning with organic solvents and their subsequent removal. If the polymer is intended to be worked up to rubber mixes which do not require to be transparent, the acid residues can be bound by suitable fillers such as ZnO.

The removal of free or combined acid is unnecessary if an ion exchanger yielding H ions, such as acid-activated fullers' earth, is used. Such ion exchangers are mild polymerization and equilibration catalysts. They have the great advantage of being fully removed by filtration from the product and yield polymers having only silanol terminal groups, if no M groups were present in the original charge (*37*, *149*).

Finally, a process should be mentioned which can be considered an acid *equilibration of polysiloxanes with organohalosilanes*. It can be described by the scheme:

$$-\overset{\diagdown}{\underset{\diagup}{Si^1}}-X + -\overset{\diagdown}{\underset{\diagup}{Si^2}}-O-\overset{\diagup}{\underset{\diagdown}{Si^3}}- \rightleftharpoons -\overset{\diagdown}{\underset{\diagup}{Si^2}}-O-\overset{\diagup}{\underset{\diagdown}{Si^1}}- + -\overset{\diagdown}{\underset{\diagup}{Si^3}}-X$$

which is intended to show how the Si atom of the halosilane is taken into the siloxane chain by an exchange reaction. Equilibration is catalyzed by hydrochloric acid or iron(III) chloride hexahydrate (*241*). From the products of the reaction of dimethyldichlorosilane and polydimethylsiloxane, the following compounds have been isolated and identified:

Tetramethyl-1,3-dichlorodisiloxane

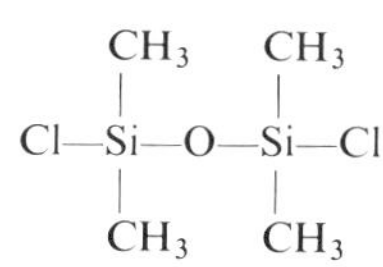

Hexamethyl-1,5-dichlorotrisiloxane

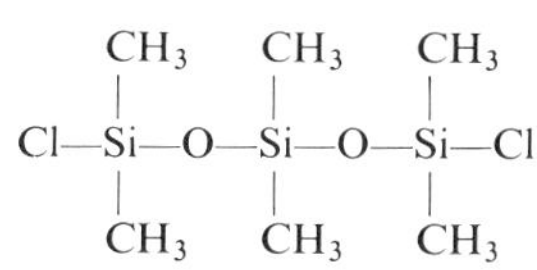

Octamethyl-1,7-dichlorotetrasiloxane

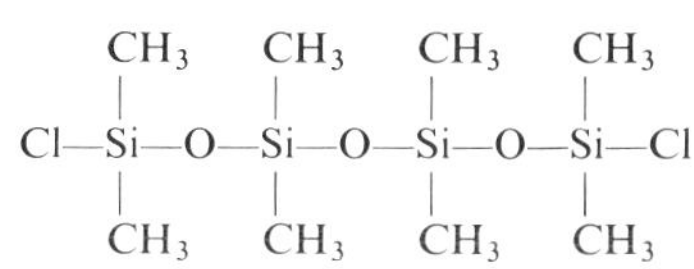

The same method can be applied to the preparation of polymers with siloxane units having different organic substituents. Diphenyldichlorosilane and polydimethylsiloxane can be equilibrated with sulfuric acid to give a polymethylphenylsiloxane in which the diphenylsiloxane units are randomly distributed over the siloxane skeleton (*245*).

An analog to this process is the equilibration of alkylalkoxysilanes with polysiloxanes, which can also be carried out with sulfuric acid as a catalyst (*260*).

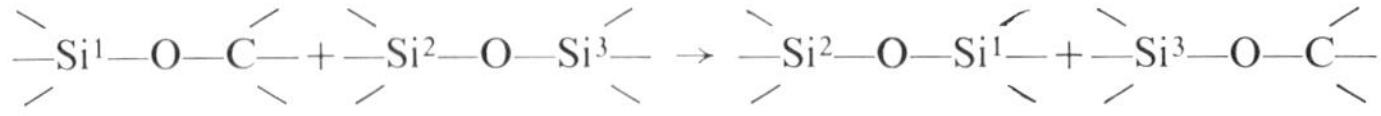

The above reaction can also be catalyzed by bases such as potassium dimethylsiliconate (*19*).

5.4.2.3 *Polymerization with Alkaline Catalysts*

The action of a basic catalyst, like that of an acid catalyst, could be considered to result first in the formation of an activated intermediate at the siloxane bond, which then triggers the second reaction, i.e., cleavage of the siloxane bond; the fragments will then form new Si—O—Si bonds by condensation, in accordance with the energy gradient. The catalyst causes continuous scission and re-formation of siloxane bonds, which leads to further growth of the molecules.

While with acid catalysis the first step is the attachment of the proton or Lewis acid to the siloxane oxygen acting as an electron donor, base catalysis starts with the attachment of the base to the silicon atom, which is thus able to extend its valence sphere to the pentacovalent state (*104*, *116*):

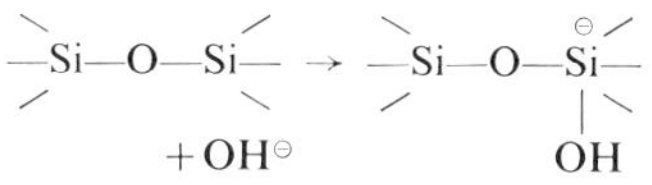
(1)

The subsequent course of the reaction can be explained similarly to the possibilities outlined for acid polymerization, by the assumption that intermediates with Si—OH and Si—OK bonds (assuming KOH as the catalyst) are formed and condense:

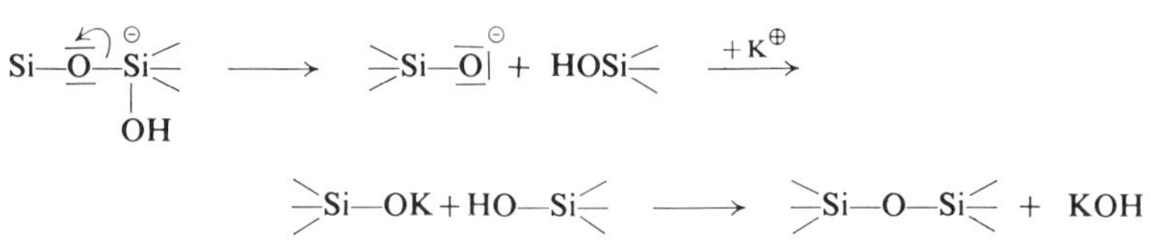
(2)

whereby the base is liberated and reacts again.

The fact that potassium dimethylsiliconate (*130*) also acts as a polymerization catalyst appears to be incapable of a similar explanation, since here only terminal Si—OK bonds are produced and, therefore, the possibility of a "back-reaction" is obviously lacking.

Consequently, special mechanisms have been proposed to explain the catalysis (*104*, *116*). The following mechanisms would be possible for a rigidly anhydrous system:

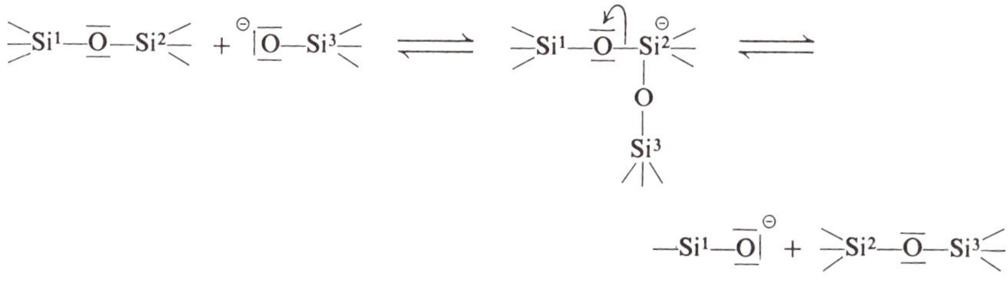

However, the objection must be raised that the presence of water, although only in traces, is a necessary condition for the back-reaction to proceed in the direction given above; also that it is hardly possible to exclude such traces of water from the reaction system completely.

The catalytic activity of the different alkali hydroxides increases in the following sequence:

$$\text{LiOH} < \text{NaOH} < \text{KOH} < \text{CsOH}$$

Lithium hydroxide is practically ineffective. The order of activity shown in this series arises from the fact that the donor action of the oxygen in the base increases with increasing ionic radius of the cation. Compounds with other donor atoms are also good catalysts, e.g., potassium amide, where the nitrogen acts as the electron donor.

Alkaline catalysts have been described in detail in numerous patent applications. The following list surveys the substances proposed and contains also, as far as we know, the oldest literature references:

Alkali-metal hydroxides (*121*), particularly cesium hydroxide (*115*), alkali-metal silanolates (*140*), potassium dimethylsiliconate (*130*), alkali-metal alkoxides (*305*), quaternary ammonium hydroxides (*59*), tetramethylammonium and tetra-*n*-butylphosphonium hydroxides (*143*), tetramethylammonium silanolate (*91*), tetrabutylphosphonium silanolate (*98*), benzyltrimethylammonium butoxide (*173*), β-hydroxyethyltrimethylammonium 2-ethylhexoate (*50*), potassium amide (*116*), amines, and mixtures of amines (*138*), etc.

Some of these either decompose or volatilize above the optimum temperature of polymerization (tetramethylammonium hydroxide, tetrabutylphosphonium hydroxide, tetrabutylphosphonium silanolate). Their reactivity can thus be destroyed very simply by vigorous heating after the polymerization is completed. These catalysts make it possible to carry out polymerization as a continuous process (*148*).

Alkali polymerization can be carried out in the presence of solvents. A high degree of polymerization can be attained if polar solvents such as acetonitrile and dimethylformamide are used (*130*). The rate of polymerization increases with increasing polarity of the solvent, e.g., in the sequence decalin, *p*-chlorotoluene, *o*-dichlorobenzene (*190*). Tetrahydrofuran has a particularly favorable influence on the rate of polymerization, possibly because it raises the concentration of Si—$\overline{\text{O}}|^{\ominus}$ anions in the reacting system (*189*).

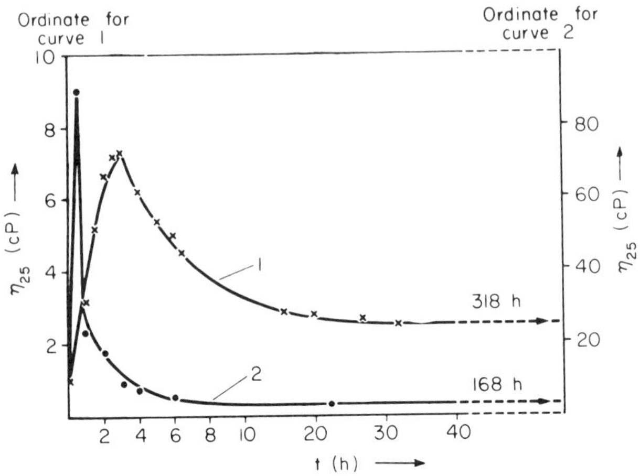

FIG. 8. Change in viscosity during the equilibration of polymethylsiloxanes with tetramethylammonium hydroxide [figures from Kantor *et al.* (*144*)]. Curve 1: Equilibration of M_2 and D_4 in a molar ratio of 1:1. Curve 2: Equilibration of M_2 and D_3 in a molar ratio of 1:1.33.

Investigations in which the rate of change of the viscosity, the mean molecular weight, the density and the refractive index were followed have yielded much information on the kinetics of the polymerization reactions (*144*). If one mole of hexamethyldisiloxane and one mole of octamethylcyclotetrasiloxane are treated with 0.1% of $[(CH_3)_4N]OH$ at 80°C the viscosity changes with time as shown in curve 1 of Fig. 8. This indicates that chain growth

$$\text{MM} + n/4\ \text{D}_4 \rightarrow \text{MD}_n\text{M}$$

predominates at first, and that equilibration reactions of the type

$$\text{MD}_n\text{M} + \text{MM} \rightarrow \text{MD}_{n-1}\text{M} + \text{MDM}$$

$$\text{MD}_{n-1}\text{M} + \text{MDM} \rightarrow \text{MD}_{n-2}\text{M} + \text{MD}_2\text{M}, \text{ and so on}$$

follow. This chain polymerization of D units is still more marked in the system MM/D_3 (1 mole of MM and 1.33 moles of D_3 with 0.1% of $[(CH_3)_4N]OH$ at 80°C), where the viscosity increases abruptly at the beginning, then decreases equally abruptly for a time, and afterwards more gradually as shown in Fig. 8, curve 2.

Systematic investigations of this kind have shown that the reactivity of different low-molecular-weight methylsiloxanes toward bases decreases in the following sequence (*144*):

$$D_3 > D_4 > MD_2M > MDM > MM$$

The behavior of the organosiloxanes on alkaline polymerization is influenced by structural factors and by the nature of the substituents—although their effect is less marked than with acid catalysis. Octaphenylcyclotetrasiloxane, which does not undergo ring cleavage with sulfuric acid or fullers' earth, can be polymerized in admixture with octamethylcyclotetrasiloxane by KOH or CsOH. With siloxanes containing 3,3,3-trifluoropropyl groups it is necessary to go to the more reactive cyclotrisiloxane, e.g., tris(3,3,3-trifluoropropyl)-trimethylcyclotrisiloxane (*185*). The choice of catalysts with *organofunctional* silanes will of course also depend on the reactivity of the substituents on the hydrocarbon radical. Thus, the use of acid catalysts is excluded with amino-alkylsiloxanes (*221*).

If monofunctional and difunctional siloxane units are polymerized, then, depending on the proportions of the two components, longer or shorter linear siloxanes with blocking triorganosiloxy terminal groups are obtained. If, on the other hand, only diorganosiloxanes are polymerized with bases MOH, polymers are formed containing terminal SiOH or SiOM groups. The SiOM groups can be removed by neutralization, and this is necessary in all cases where a "stabilization" of the polymer is to be achieved. The catalyst can be removed by washing with water as described for acid catalysis. It is, however, quicker and simpler to neutralize with an acidic additive soluble in the polymer. Chlorosilanes, such as trimethylchlorosilane (*314*), triaryl phosphites (tri-phenyl or tricresyl phosphite) (*97*), aryl or alkyl phosphates (*166*), and certain bromoalkanes, e.g., 1,2,3-tribromopropane, 1,2,3-tribromobutane, 1,2,3-tribromo-2-methylpropane (*31*), or CO_2 can be used for this purpose. Iodine (*102*) will also remove alkali, with the formation of the iodide and iodate. If thc polymer is to be worked up to rubber compounds the catalyst can also be neutralized by weakly acidic fillers.

5.4.3 The Cross-linking of Polyorganosiloxanes

The vulcanization of silicone rubber depends on the appropriate cross-linking reactions. These can be initiated by peroxide at elevated temperatures or by high-energy radiation at normal temperatures. So-called room-tempera-ture vulcanization with chemical additives is also of considerable interest. This process depends on the catalysis of certain reactions between silicon-functional siloxanes and silanes at room temperature. Appropriate reaction mechanisms are indicated in the preceding sections. The technological aspects of room-temperature-vulcanizable silicone rubber are described in Section 8.1.2.3. Consequently, of the cross-linking reactions that have not so far been described, only cross-linking with peroxide catalysts and high-energy radiations will be discussed in detail below.

5.4.3.1 *Cross-linking with Peroxides*

Linear macromolecular polyorganosiloxanes are converted into rubber-elastic polymers by heating with the addition of small amounts of organic peroxides. The following types of peroxides can be used: diaroyl peroxides, such as dibenzoyl peroxide (*39*, *323*), bis-*p*-chlorobenzoyl peroxide (*93*), and bis-2,4-dichlorobenzoyl peroxide (*324*); dialkyl peroxides, such as di-*t*-butyl peroxide (*303*); diaralkyl peroxides, such as dicumyl peroxide (*318*); alkylaryl peroxides; alkyl aroyl and alkyl acyl peroxides, such as *t*-butyl perbenzoate (*70*, *304*), or *t*-butyl peracetate (*306*); or, finally, mixtures containing representatives of different types like dibenzoyl peroxide and *t*-butyl perbenzoate (*308*). Peroxysiloxanes (*30*) and zinc peroxide (*307*) are also suitable catalysts.

The catalysts from the various groups differ in their optimum working temperatures as well as in their rates of cross-linking. This is discussed in greater detail in connection with the vulcanization of silicone rubbers in Section 8.1.

The action of the peroxides on polymethylsiloxanes would appear to depend on the fact that the free radicals arising from their decomposition at elevated temperatures are able to abstract hydrogen atoms from methyl groups. Ethylene links can thus be formed between the siloxane chains (*70*):

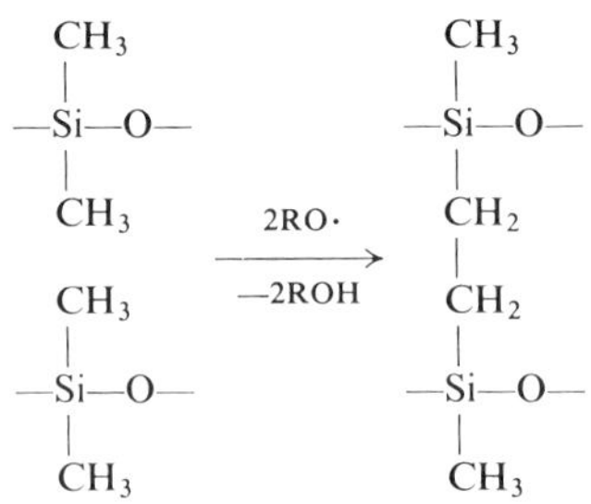

Proof that such reactions can take place has been obtained from a model experiment in which octamethylcyclotetrasiloxane was heated with *t*-butyl perbenzoate, when bis-(heptamethylcyclotetrasiloxanyl)-ethane was formed (*139*):

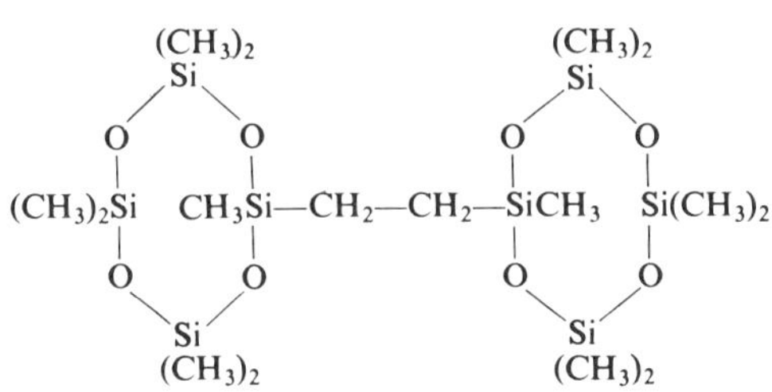

Vinyl groups in polyvinylmethylsiloxanes are particularly reactive with peroxides. This is shown by the ease with which they can be worked up with weakly cross-linking peroxides such as di-*t*-butyl peroxide. It must be assumed that the links formed here are not between vinyl groups, but between vinyl and methyl groups. The mechanisms involved should perhaps be formulated as follows (*70*):

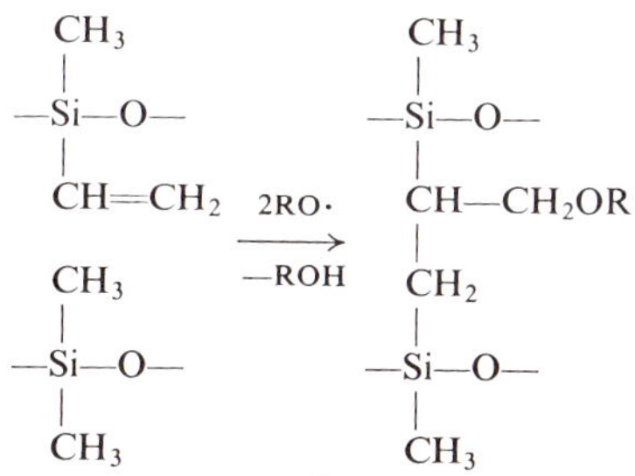

I

or

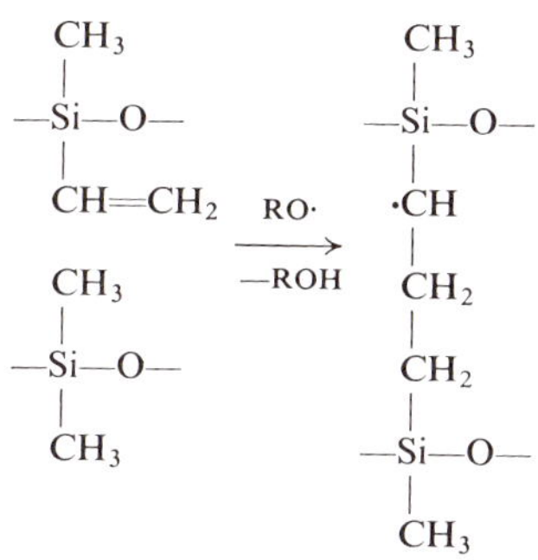

II

If vinyl groups are incorporated into polydimethylsiloxane chains in this way they could be considered in a sense "labeled" cross-linking points and the degree of cross-linking would then depend on the concentration of the vinyl groups. Allyl groups could function similarly.

In addition to peroxides, the following substances—certainly acting differently—have been mentioned as catalysts promoting the conversion of liquid linear polyorganosiloxanes into elastic products: boron hydrides (*236*, *237*), aliphatic azo compounds (*63*), triphenylantimony, dimethylmercury, tributyltin acetate, tetraethyllead, lead tetraacetate, mercuric acetate (*304*), boron trifluoride/piperidine (*170*), also CrO_3, mixtures of CrO_3 and CuO and of Co_2O_3 and CuO (*304*), sulfuryl chloride (*235*), and benzothiazolyl disulfide (*94*).

5.4.3.2 *Cross-linking with High-Energy Radiation*

Linear polydimethyldiloxanes can be cross-linked by irradiation with high-energy radiation. If, as with peroxide cross-linking, the starting materials are high-molecular-weight linear polymers, it is possible to obtain a silicone rubber free from residual catalyst by this method. Since catalyst residues could remain effective after the completion of the cross-linking process, irradiated rubber has a much better heat resistance than peroxide-cured types of rubber (*40*, *44*, *45*, *46*, *158*, *162*, *309*) (cf. Section 8.1.2.4).

Many types of radiation have been found suitable such as electron beams, X-rays, and neutrons. Irradiation leads to formation of radicals such as

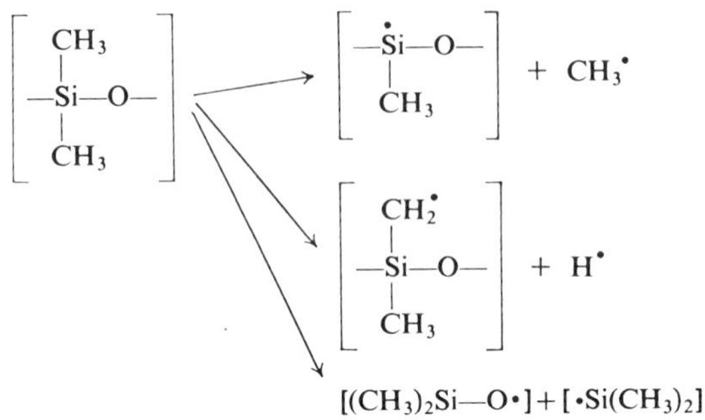

which can be detected by mass spectrometry or electron paramagnetic resonance (*62*, *325*). The chains can therefore be either degraded or cross-linked. Cross-linking by far exceeds degradation. It leads principally to the formation of Si—Si, Si—CH_2—Si, and Si—CH_2—CH_2—Si links (*61*, *143*, *186*, *309*). Bi-(heptamethylcyclotetrasiloxanyl) (I) and bis-(heptamethylcyclotetrasiloxanyl)-methane (II) have been isolated from the products of the irradiation of octamethylcyclotetrasiloxane by distillation:

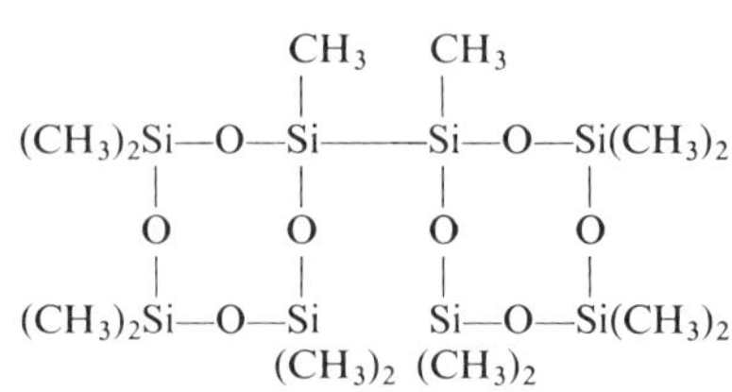

I

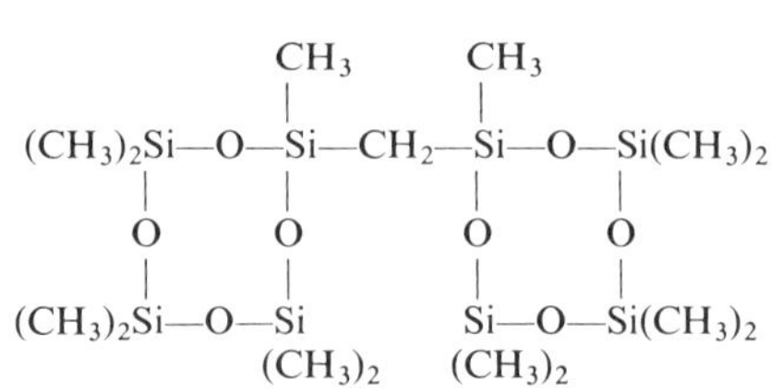

II

Hydrogen, methane, ethane, and (probably) propane were detected as by-products of irradiation (*45*, *75*, *186*). Good agreement between the total gas yield and the polymerization yield is found at temperatures between -40 and $+100°C$; they increase with the temperature and the intensity of irradiation. Divergence occurs above 100°C (*186*).

Oxygen decreases the cross-linking yield. Radical-forming-aromatic compounds such as di-*t*-butyl-*p*-cresol and, to an even greater extent, proton-transfer agents such as *n*-butyl mercaptan retard the cross-linking process. The addition of 10% of mercaptan enables the cleavage of $\rangle$Si—CH_3 groups to be measured directly. Since Si—OH groups are formed in this case it must be concluded that homolytic cleavage of the siloxane backbone can be ignored in irradiation processes (*187*).

Other substituents on the Si atom alter the behavior of the siloxanes toward irradiation. Phenyl groups increase their ability to resist irradiation. On the other hand, siloxanes containing Si—H bonds are particularly sensitive (*325*).

Compounds (I) and (II) are subject to cleavage by polymerization and equilibration catalysts like other polysiloxanes; accordingly, they can also be incorporated into linear polysiloxanes. If the two rings enter different chains, the result is the linking of these two chains through the —CH_2— or Si—Si bridges between the rings. This provides an unusual way of cross-linking linear molecules: Octamethylcyclotetrasiloxane is irradiated in such a way that small portions of (I) and (II) are formed. The reaction mixture is then polymerized with a basic catalyst (KOH, NaOH, $[(CH_3)_4N]OH$ or similar compounds). Not only are chains produced, but they are simultaneously coupled with one another through any available bridge. The end product is similar to that obtained from the subsequent cross-linking of originally linear polymers.

5.5 Depolymerization and Cleavage

The methods employed for the degradation of polysiloxanes are basically similar to those used for their preparation and differ only in their special operating conditions, such as pressure, temperature, or catalyst concentration. This is quite logical if it is remembered that the first step in each case will be a scission of the siloxane bond. Whether the fragments formed recombine to give polymers or remain separate will depend on the stability relationships between the initial and final products, and these again are a function of the pressure, temperature, and chemical composition of the system. Degradation, like polymerization, can therefore be achieved by purely thermal means or by acid or alkaline catalysts.

The addition of agents able to attack siloxane bonds in stoichiometric rather than catalytic concentrations, will result in degradation down to the monomeric silicon compounds, and this can be achieved by acids, alkalis, and a number of special agents.

5.5.1 Thermal Depolymerization

While low-molecular-weight siloxanes can be polymerized at around 300°C and elevated pressures (cf. Section 5.4.2.1), high-molecular-weight siloxanes can be depolymerized at even higher temperatures and reduced pressures (*113, 120, 210*).

Mixtures of cyclic polysiloxanes $[(CH_3)_2SiO]_n$ with $n=3$ to 8 have been obtained from polydimethylsiloxanes as products of thermal degradation at 350 to 400°C in vacuum (*113*).

Low-molecular-weight spirocyclic siloxanes of the empirical compositions given below have been obtained by heating cohydrolysates of dimethyldichlorosilane, methyltrichlorosilane, and silicon tetrachloride to between 350 and 600°C: $(CH_3)_8Si_5O_6$, $(CH_3)_{10}Si_6O_7$, $(CH_3)_{12}Si_7O_8$, $(CH_3)_{14}Si_8O_9$, as well as the polycyclic siloxanes $(CH_3)_8Si_5O_6$, $(CH_3)_{10}Si_6O_7$, $(CH_3)_8Si_6O_8$, $(CH_3)_{10}Si_7O_9$, and $(CH_3)_{10}Si_8O_{11}$ (*252*) (compare Section 6.2.3 and Table 43).

The thermal depolymerization of methylsilsesquioxane polymers $(CH_3SiO_{3/2})_n$ in the region of 250 to 600°C and at pressures above 100 mm Hg gave considerable amounts of tetramethylsilane, besides the hexamethyldisiloxane and hexamethylcyclotrisiloxane, methane, carbon monoxide, and carbon dioxide in the volatile degradation products (*96*).

However, in general, thermal processes have not become important for preparative purposes, since degradation can normally be achieved with alkaline catalysts under considerably milder conditions.

5.5.2 Degradation by Steam and Alcohols

The siloxane bond is cleaved slowly by steam at high temperatures with the formation of siloxanols:

$$\geq Si-O-Si \leq + H_2O \rightleftharpoons \geq SiOH + HOSi \leq$$

Thus, the molecular weight of a polydimethylsiloxane, initially 1,166,000, fell as follows when the siloxane was heated in steam at 220°C (*92*): after 2 hours, 155,000; after 3 hours, 133,000; after 4 hours, 87,000; and after 10.5 hours, 78,000. The variation of the molecular weight with the temperature at a constant time of heating (4 hours) was as follows (initial molecular weight, 1,050,000) (*92*): 200°C, 338,000; 220°C, 70,300; 230°C, 33,000; and 240°C, 18,300. This reaction—especially with the simultaneous use of alkaline catalysts—opens up another route to the production of siloxanols which is of interest, in particular, because of the good reproducibility of the molecular weights. On the other hand, the possibility of depolymerization by steam in the use of silicone rubber in practice must be borne in mind (Section 9.6.1).

By heating with high-boiling primary or secondary alcohols in the presence of an acidic (*22*) or alkaline (*301*) catalyst, siloxanes are converted into alkoxysiloxanes:

$$\equiv Si{-}O{-}Si\equiv + 2ROH \rightleftharpoons \equiv SiOR + ROSi\equiv + H_2O$$

By continuous removal of the water evolved during the reaction, the degradation can be carried as far as the corresponding alkoxysilanes (*22*, *301*).

5.5.3 Depolymerization and Cleavage with Acids

The fact that high-molecular-weight polysiloxanes, as represented by silicone rubber polymers, can be degraded to low-molecular-weight siloxanes by the action of hydrogen chloride can be considered an example of acid-catalyzed depolymerization (*111*, *133*). Most "acid" processes, however, lead to complete depolymerization down to monomers.

The Si—O—Si bond can be broken by stoichiometric amounts of anhydrous hydrogen fluoride (*32*) (anhydrous copper sulfate is used as a catalyst) by adding potassium or ammonium hydrogen fluoride to a sulfuric acid solution of the siloxanes cooled to 0°C (*300*), or by heating with ammonium fluoride (*215*). The corresponding organofluorosilanes are formed from the reaction of the various siloxane units:

$$R_3SiO_{1/2} \rightarrow R_3SiF$$
$$R_2SiO_{2/2} \rightarrow R_2SiF_2$$
$$RSiO_{3/2} \rightarrow RSiF_3$$

If a mixture or a copolymer of different siloxanes is used as the starting material, the organofluorosilanes corresponding to the silicone functionalities are obtained by this method. The products can easily be separated by distillation, and this offers a method of *analysis* of the polymer mixture (*32*).

More highly alkylated siloxanes, particularly hexamethyldisiloxane, are also decomposed by thionyl chloride in the presence of Friedel-Crafts catalysts and converted into the corresponding chlorosilanes in good yield (*161*). If hexamethyldisiloxane is heated for 20 minutes with thionyl chloride in a molar ratio of 1:1.5 in the presence of about 0.3% of anhydrous iron(III) chloride, trimethylchlorosilane is formed in almost quantitative yield:

$$(CH_3)_3Si{-}O{-}Si(CH_3)_3 + SOCl_2 \xrightarrow{FeCl_3} 2(CH_3)_3SiCl + SO_2$$

The Si—O—Si bond in hexamethyldisiloxane is broken by stoichiometric quantities of sulfuric acid with the formation of trimethylsilyl sulfate (*213*):

$$[(CH_3)_3Si]_2O + H_2SO_4 \rightarrow [(CH_3)_3Si]_2SO_4 + H_2O$$

Hexaethyldisiloxane reacts similarly to form triethylsilyl sulfate (*278*).

Trialkylsilyl phosphates are formed by the reaction of phosphorus pentoxide in equimolar ratio with hexamethyldisiloxane and hexaethyldisiloxane (*78*, *240*):

$$3[R_3Si]_2O + P_2O_5 \rightarrow 2[R_3Si]_3PO_4$$

On the other hand, with an excess of P_2O_5, solid trialkylsilyl polyphosphate is the main product (*78*):

$$n[R_3Si]_2O + nP_2O_5 \rightarrow 2\left[\begin{array}{c} O \\ \| \\ -P-O- \\ | \\ OSiR_3 \end{array}\right]_n$$

In the presence of zinc chloride (*299*) or sulfuric acid (*21*), acetic anhydride decomposes siloxanes with the formation of

$$\equiv Si-O-Si\equiv + (CH_3CO)_2O \rightarrow \equiv SiOOCCH_3 + CH_3COOSi\equiv$$

5.5.4 Depolymerization and Cleavage with Alkalis

While alkali-metal hydroxides promote polymerization below about 250°C, they catalyze depolymerization at higher temperatures. This situation is similar to purely thermal polymerization and depolymerization. Even small amounts of NaOH, of the order of 0.1 %, have a considerable effect. The cyclic trimers and tetramers $[R_2SiO]_3$ and $[R_2SiO]_4$ (*120*, *123*) are produced preferentially from polydiorganosiloxanes.

If the initial products contain triorganosiloxy groups, they appear as hexaorganodisiloxanes in the reaction products.

The alkaline depolymerization of cross-linked polymers containing $RSiO_{3/2}$ units gives products from which low-molecular-weight *spherocyclic* siloxanes $[RSiO_{3/2}]_n$ (cf. Section 6.2.3) can be isolated in low yield by distillation under reduced pressure (*27*).

There is a smooth transition from these degradation processes to the alkaline cleavage reactions as the concentration of the alkali-metal hydroxides is increased. If stoichiometric ratios are used, alkali-metal silanolates (cf. Section 3.7) result from the reaction. Alkali-metal salts of tri-, di-, or monoorganosilanols are formed, depending on the functionalities of the siloxane units present. Since these products have different solubilities, it is possible to use alkaline hydrolysis for the separation of polymers based on different siloxane units (*125*). The salts of monoorganosilanols are water soluble, while the salts of di- and triorganosilanols hydrolyze in water to give polysiloxanes soluble in organic solvents. The former can thus be extracted with water, and the latter hydrolyzed and then taken up in organic solvents.

5.5.5 Cleavage by Lewis Acids

Lewis acids have already been mentioned as catalysts for polymerization reactions (Section 5.4.3). It is therefore not surprising to find that they can also be used in stoichiometric amounts for cleaving the siloxane bond. Organohalosilanes and heterosiloxanes are formed as a result of this scission.

Like polymerization, cleavage of a siloxane bond by Lewis acids starts with the attachment of the acid with its electron deficiency to the siloxane oxygen which acts as an electron donor. Examples of such scission of the siloxane bond that have been thoroughly investigated are the reactions with boron trichloride and tribromide.

If an equimolecular mixture of *boron tribromide* and hexamethyldisiloxane is frozen and then thawed, an unstable solid addition compound is formed at about $-40°C$ when the two compounds are melted. This decomposes on further warming to room temperature, with separation of trimethylbromosilane and the formation of a borosiloxane. The reaction can be written as follows:

$$(CH_3)_3Si{-}\overline{\underline{O}}{-}Si(CH_3)_3 + BBr_3 \rightarrow (CH_3)_3Si{-}\overset{\oplus}{\overline{O}}(\ominus BBr_3){-}Si(CH_3)_3 \rightarrow$$

$$(CH_3)_3Si{-}Br + (CH_3)_3Si{-}O{-}BBr_2$$

If the reaction mixture is now heated at 80°C for half an hour, the borosiloxane formed first decomposes into trimethylbromosilane and boron oxybromide:

$$(CH_3)_3Si{-}O{-}BBr_2 \rightarrow (CH_3)_3Si{-}Br + BOBr$$

which finally disproportionates to boron trioxide and boron tribromide:

$$3BOBr \rightarrow B_2O_3 + BBr_3$$

One-third of the BBr_3 originally used is thus recovered at the end of the reaction.

If the reaction is carried out by slowly adding a benzene solution of boron tribromide to a benzene solution of a siloxane so that an excess of siloxane is always available, the borosiloxane formed as an intermediate compound reacts further with the excess of siloxane, and trimethylbromosilane and bis-trimethylsiloxy-boron bromide are formed:

$$(CH_3)_3Si{-}O{-}BBr_2 + (CH_3)_3Si{-}O{-}Si(CH_3)_3 \rightarrow (CH_3)_3Si{-}Br + [(CH_3)_3SiO]_2BBr$$

Cleavage of hexamethyldisiloxane and methyl-*H*-disiloxanes by *boron trifluoride* or *boron trichloride* (*74*, *177*) as well as that of cyclosiloxanes, leads through similar intermediates with Si—O—B bonds to the corresponding organohalosilanes and organo-*H*-silanes, respectively, and B_2O_3 and boron halide.

In the cleavage of the siloxane bond with *aluminum halides*, alkylsiloxyaluminum halides are formed (*127, 306*). Hexamethyldisiloxane and aluminum chloride give trimethylsiloxyaluminum dichloride (*206*):

$$2(CH_3)_3Si—O—Si(CH_3)_3 + 2AlCl_3 \rightarrow 2(CH_3)_3SiCl + [(CH_3)_3Si—O—AlCl_2]_2$$

which is dimeric by coordination (*244*) (Section 7.1.2); $[H_3Si—O—Al(CH_3)_2]_2$ has been prepared from dimethylaluminum bromide and disiloxane (*154*):

$$2H_3Si—O—SiH_3 + 2(CH_3)_2AlBr \rightarrow [H_3Si—O—Al(CH_3)_2]_2 + 2H_3SiBr$$

Chlorine-stopped polymers (*326*) and an alumosiloxane of the empirical formula $[(CH_3)_2Si]_4Al_3Cl_5O_6$ (*244*) whose structure is still obscure have been found as cleavage products of dimethylcyclosiloxanes and linear polydimethylsiloxanes with aluminum chloride at 120°C.

At higher temperatures, the siloxane bond is also ruptured by *titanium tetrachloride*. Hexaalkyldisiloxanes give trialkylsiloxytitanium chlorides, small amounts of aluminum chloride having a catalytic effect (*10, 67*):

$$R_3Si—O—SiR_3 + TiCl_4 \rightarrow R_3SiOTiCl_3 + R_3SiCl$$

Octamethylcyclotetrasiloxane reacts with opening of the ring as follows (*18*):

$$[(CH_3)_2SiO]_4 + TiCl_4 \rightarrow Cl[(CH_3)_2SiO]_4TiCl_3$$

It is interesting that *alkyl titanates*, especially *butyl titanate*, can behave as Lewis acids with respect to the Si—O—Si bond. Octamethylcyclotetrasiloxane (*11, 12*) reacts with butyl titanate after 10 hours of heating at 180°C probably giving titanosiloxanes as intermediates in accordance with the equation

$$[(CH_3)_2SiO]_4 + (C_4H_9O)_4Ti \rightarrow C_4H_9O[(CH_3)_2SiO]_4Ti(OC_4H_9)_3$$

These then disproportionate into dimethylolibutoxysilane and low-molecular-weight α,ω-dibutoxypolydimethylsiloxanes. These compounds can be distilled off from a residue which contains silicon-titanium polymers of unknown structure (*11, 12*). Copolymers are also said to be produced by heating butyl titanate with alkyl- and arylsilsesquioxanes (*234*).

5.5.6 Cleavage with Organometallic Compounds

The fact that organometallic compounds can break the siloxane bond was originally demonstrated by means of the reaction of methylmagnesium iodide with polydimethylsiloxanes (*147*). A siloxanylmagnesium iodide is first formed and this can be converted into the silanol with water (*240*):

$$[(CH_3)_2SiO]_n + nCH_3MgI \rightarrow n(CH_3)_3SiOMgI$$

$$(CH_3)_3SiOMgI + H_2O \rightarrow (CH_3)_3SiOH + Mg(I)OH$$

The silanol formed is more highly alkylated than the original siloxane.

The siloxane bonds in silicic acids or quartz can also be broken by Grignard reagents with the formation of polyorganosiloxanes, but, of course, with correspondingly greater difficulty than with siloxanes already containing organic substituents (*145*). Activation of the quartz surface with HF, OH^-, or OCH_3^- is said to have a beneficial effect on the degree of conversion.

Analogously, polyorganosiloxanes are decomposed by alkylaluminum compounds into lower molecular weight (but more highly alkylated) compounds (*135*). Organolithium compounds also cause scission of hexaorganodisiloxanes (*99*):

$$R_3Si{-}O{-}SiR_3 + R'Li \rightarrow R_3R'Si + R_3Si{-}O{-}Li$$

5.5.7. Cleavage with Lithium Tetrahydroaluminate

If a linear polydimethylsiloxane is added dropwise to a solution of lithium tetrahydroaluminate in ether, gentle warming initiates a reaction which can be detected by the evolution of gas. A white precipitate gradually separates. The gas has been identified as $(CH_3)_2SiH_2$. The reaction probably proceeds in such a way that the siloxane bond is converted into an alumosiloxane bond and silane is evolved:

$$\diagdown\!\!\!-Si{-}O{-}Si\!\!-\!\!\diagup + \diagdown AlH \rightarrow -Si{-}H + \diagdown Al{-}O{-}Si-$$

which shows the analogy with cleavage of the siloxane bond by aluminum halides (*88*) (see Section 5.5.5).

The cleavage reaction proceeds considerably faster with silicic acid esters, which react spontaneously on the dropwise addition of an ethereal solution of lithium tetrahydroaluminate, giving SiH_4.

REFERENCES

(1) M. Agens, General Electric, U.S. Pat. 2448756 [14.3.1944]. (2) —, Thomson-Houston, French Pat. 1124302 [U.S. Prior. 21.5.1954]. (3) —, General Electric, U.S. Pat. 2714585 [26.5.1954]. (4) A. D. Allen and G. Modena, *J. Chem. Soc.* [London] **1957**, 3671. (5) H. H. Anderson, *J. Amer. Chem. Soc.* **72**, 196 [1950]. (6) D. N. Andreev, B. N. Dolgov and S. V. Butz, *J. Gen. Chem.* (*Engl. transl.*) **32**, 1247 [1962]. (7) K. A. Andrianov, A. K. Dabagova and Zh. S. Syrtsova, *Bull. Acad. Sci. USSR, Div. Chem. Sci.* (*Engl. transl.*) **1962**, 1487. (8) —, S. I. Dzhenchelskaya and Yu. K. Petrashkov, *Plast. Massy*, **1960**, No. 3, 20. (9) —, T. N. Ganina and N. N. Sokolov, *J. Gen. Chem.* (*Engl. transl.*) **26**, 1897 [1956].

(10) — and N. A. Kurasheva, *Proc. Acad. Sci. USSR* (*Engl. transl.*) **131**, 297 [1960]. (11) —, S. V. Pichkhadze and V. V. Komarova, *Bull. Acad. Sci. USSR, Div. Chem. Sci.* (*Engl. transl.*) **1962**, No. 4, 672. (12) —, —, — and T. N. Vardosanidze, ibid. **1962**, No. 5, 776. (13) —, A. A. Zhdanov and S. A. Pavlov, *Doklady Akad. Nauk. SSSR* **102**, 85 [1955]. (14) — and V. V. Severnyi, *Proc. Acad. Sci. USSR, Sect. Chem.* (*Engl. transl.*) **134**, 1167 [1960]; *Bull. Acad. Sci. USSR, Div. Chem. Sci.* (*Engl. transl.*) **1962**, 1161. (15) —, —, ibid. **1963**, 72. (16) —, — and B. A. Izmailov, ibid. **1963**, 256. (17) —, N. N. Sokolov and E. N.

KHRUSTALEVA, *J. Gen. Chem.* (*Engl. transl.*) **26**, 1249 [1956]. (18) — and co-workers, *Proc. Acad. Sci. USSR, Sect. Chem.* (*Engl. transl.*) **131**, 259 [1960]; *J. Gen. Chem. USSR* (*Engl. transl.*) **32**, 2279 [1962]. (19) D. L. BAILEY, Union Carbide, French Pat. 1116196 [U.S. Prior. 14.12.1953].

(20) — and F. M. O'CONNOR, —, German Appl. (West German) 1022226 [U.S. Prior. 31.10.1955]. (21) —, —, —, U.S. Pat. 2910496 [26.6.1956]. (22) —, —, —, U.S. Pat. 2881199 [14.11.1956]. (23)—, V. B. JEX and W. T. BLACK, French Pat. 1189993 [U.S. Prior. 12.10.1956]. (24) — and A. N. PINES, *Amer. Chem. Soc.* 125th Meeting, Kansas City 1954, Abstr. 19N, No. 36. (25) J. E. BAINES and C. EABORN, *J. Chem. Soc.* [London] **1955**, 4023; **1956**, 1436. (26) A. J. BARRY, Dow Chemical, U.S. Pat. 2460799 [21.2.1947]. (27) — and J. W. GILKEY, Dow Corning, U.S. Pat. 2465188 [27.3.1948]. (28) — and E. E. MERRILL, —, U.S. Pat. 2483963 [24.11.1948]. (29) L. F. BERHENKE and L. W. BYERS, Dow Chemical, U.S. Pat. 2521672 [29.4.1946].

(30) B. A. BLUESTEIN, General Electric, U.S. Pat. 2970982 [16.9.1957]. (31) R. I. BOOT, —, French Pat. 1307715 [U.S. Prior. 12.12.1960]. (32) H. S. BOOTH and M. L. FREEDMAN, *J. Amer. Chem. Soc.* **72**, 2847 [1950]. (33) O. A. BRALEY, Dow Corning, U.S. Pat. 2504388 [11.6.1949]. (34) —, —, U.S. Pat. 2818906 [10.12.1954]. (35) E. C. BRITTON and L. F. BERHENKE, Dow Chemical, U.S. Pat. 2521673 [8.7.1946]. (36) — and H. C. WHITE, —, U.S. Pat. 2476132 [28.7.1945]. (37) —, — and C. L. MOYLE, —, U.S. Pat. 2460805 [6.12.1945]. (38) L. B. BRUNER, Dow Corning U.S. Pat. 3035016 [16.3.1959, cip. 24.3.1958]. (39) A. M. BUECHE, *J. Polymer Sci.* **15**, 105 [1955].

(40) —, ibid. **19**, 297 [1956]. (41) C. A. BURKHARD, *J. Amer. Chem. Soc.* **67**, 2173 [1945]. (42) L. F. CEYZERIAT and P. DUMONT, Rhône-Poulenc, German Appl. (West German) 1121329 [French Prior. 12.7.1957]. (43) —, —, German Appl. (West German) 1121803 [French Prior. 6.2. and 2.12.1958]. (44) A. CHARLESBY, *Nature* [London] **173**, 679 [1954]. (45) —, *Proc. Roy. Soc.* [London], Series A **230**, 120 [1955]. (46) —, *J. Polymer Sci.* **17**, 379 [1955]. (47) I. CHOJNOWSKI and S. CHRZCZONOWICZ, Internat. Symposium on Organosilicon Chemistry, Scientific Communications, Prague 1965, p. 41. (48) S. CHRZCZONOWICZ and J. CHOJNOWSKI, *Rozn. Chem.* **36**, 1293 [1962]. (49) H. A. CLARK, Dow Corning, German Pat. (West German) 941647 [U.S. Prior. 16.8.1951].

(50) —, —, French Pat. 1085288 [U.S. Prior. 22.10.1952]. (51) H. CLASEN, Metallgesellschaft, German Pat. (West German) 941286 [21.11.1952]. (52) V. S. CHUGUNOV, *Bull. Acad. Sci. USSR, Div. Chem. Sci.* (*Engl. transl.*) **1956**, 1075. (53) G. D. COOPER and M. PROBER, *J. Org. Chem.* **25**, 240 [1960]. (54) K. DAMM, D. GÖLITZ and W. NOLL, *Angew. Chem.* **76**, 273 [1964]; Internat. Edition **3**, 310 [1964]. (55) — and W. NOLL, *Kolloid-Z.* **158**, 97 [1958]. (56) —, —, Farbenfabriken Bayer, unpublished. (57) —, K. SCHNURRBUSCH, W. OETKE, R. ERDMENGER and K. STADE, —, Belg. Pat. 572412 [German Prior. (West German) 31.10. and 25.11.1957]. (58) W. H. DAUDT, Dow Corning, French Pat. 1066163 [U.S. Prior. 14.7.1951]. (59) — and J. F. HYDE, —, German Pat. (West German) 872087 [U.S. Prior. 30.3.1943].

(60) J. B. DECOSTE, Bell Telephone, U.S. Pat. 2516047 [15.11.1945]. (61) H. A. DEWHURST and L. E. S. PIERRE, *J. physic Chem.* **64**, 1063 [1960]. (62) V. H. DIBELER, F. L. MOHLER and R. M. REESE, *J. chem. Physics* **21**, 180 [1953]. (63) P. A. DIGIORGIO and M. M. SAFFORD, General Electric, U.S. Pat. 2613199 [5.4.1951]. (64) —, L. H. SOMMER and F. C. WHITMORE, *J. Amer. Chem. Soc.* **68**, 344 [1946]. (65) —, —, —, ibid. **70**, 3512 [1948]. (66) D. G. DOBAY, Union Carbide, U.S. Pat. 2769829 and 2769830 [29.12.1953]. (67) B. N. DOLGOV, N. F.

ORLOV and M. G. VORONKOV, *Bull. Acad. Sci. USSR, Div. Chem. Sci. (Engl. transl.)* **1959**, 1358. (68) C. D. DOYLE, General Electric, U.S. Pat. 2542641 [7.8.1947]. (69) J. J. DUANE, Union Carbide, U.S. Pat. 2646441 [17.12.1949].

(70) M. L. DUNHAM, D. L. BAILEY and R. Y. MIXER, *Ind. Engng. Chem.* **49**, 1373 [1957]. (71) — and G. H. WAGNER, Union Carbide, U.S. Pat. 2759006 [10.6.1953]. (72) C. EABORN, *J. Chem. Soc.* [London] **1950**, 3077. (73) J. R. ELLIOTT and R. H. KRIEBLE, General Electric, U.S. Pat. 2513924 [5.5.1945]. (74) H. J. EMELÉUS and M. ONYSZCHUK, *J. Chem. Soc.* [London] **1958**, 604. (75) L. M. EPSTEIN and N. S. MARANS, *Rubber Age* (New York) **82**, 825 [1958]. (76) R. A. FALK, Dow Corning, U.S. Pat. 3032530 [13.10.1958, cip. 29.4.1957]. (77) G. O. FANGER, Dow Corning, French Pat. 1277673 [U.S. Prior. 8.2.1960]. (78) F. FEHÉR, G. KUHLBÖRSCH, A. BLÜMCKE, H. KELLER and K. LIPPERT, *Chem. Ber.* **90**, 134 [1957]. (79) S. FELLERMEIER, E. KALB and S. NITZSCHE, Wacker-Chemie, German Pat. (West German) 831694 [22.9.1950].

(80) H. N. FENN and L. A. RAUNER, Dow Corning, German Pat. (West German) 854708 [U.S. Prior. 21.1.1946]. (81) —, —, —, German Pat. (West German) 855163 [U.S. Prior. 21.1.1946]. (82) H. J. FLETCHER and G. L. CONSTANT, —, Canadian Pat. 560981 [U.S. Prior. 7.10.1955]. (83) — and M. J. HUNTER, *J. Amer. Chem. Soc.* **71**, 2918 [1949]. (84) —, —, ibid. **71**, 2922 [1949]. (85) G. S. FORBES and H. H. ANDERSON, ibid. **70**, 1222 [1948]. (86) K. C. FRISCH and P. A. GOODWIN, General Electric, U.S. Pat. 2678938 [9.4.1952]. (87) —, — and R. E. SCOTT, *J. Amer. Chem. Soc.* **74**, 4584 [1952]. (88) G. FRITZ, *Z. Naturforsch.* **10b**, 423 [1955]. (89) L. W. FROST, Amer. Chem. Soc. 125th Meeting, Kansas City 1954, Abstr. 14 N, No. 27.

(90) A. GANCBERG, 27e *Congrès International de Chimie Industrielle, Bruxelles* **3**, 382 [1954]. (91) I. GEBHARDT, B. LENGYEL and F. TÖRÖK, *Mag. Kem. Folyóirat* **67**, 450, [1961]. (92) —, —, —, ibid. **68**, 159 [1962]. (93) General Electric, Belg. Pat. 532704 [U.S. Prior. 22.10.1953]. (94) —, Belg. Pat. 540543 [U.S. Prior. 16.8.1954]. (95) —, Brit. Pat. 818305 [U.S. Prior. 17.3.1955]. (96) P. D. GEORGE and P. D. ZEMANY, Thomson-Houston, French Pat. 1174288 [U.S. Prior. 18.4.1956]. (97) A. R. GILBERT, General Electric, U.S. Pat. 2717902 [24.6.1953]. (98) — and S. W. KANTOR, *J. Polymer Sci.* **40**, 35 [1959]. (99) H. GILMAN. H. N. BENEDICT and H. HARTZFELD, *J. Org. Chem.* **19**, 419 [1954].

(100) — and G. E. DUNN, *J. Amer. Chem. Soc.* **73**, 3404 [1951]; older lit. cited by authors. (101) T. H. GORDON, Allied Chemical and Dye, U.S. Pat. 2832794 [5.11.1952]. (102) W. T. GRUBB, General Electric, U.S. Pat. 2789109 [29.6.1954]. (103) —, *J. Amer. Chem. Soc.* **76**, 3408 [1954]. (104) — and R. C. OSTHOFF, ibid. **77**, 1405 [1955]. (105) W. N. GRUBER and co-workers, *Vysokomol. Soed.* **1**, 1194 [1959]; **3**, 84 [1961]; **3**, 89 [1961]; **3**, 174 [1961]. (106) C. J. GUILLISSEN and A. GANCBERG, Union Chimique Belge, U.S. Pat. 2695307 [Belg. Prior. 20.9.1947]. (107) —, —, —, U.S. Pat. 2732320 [Belg. Prior. 31.1.1951]. (108) W. J. HAGGERTY JR. and L. W. BREED, *J. Org. Chem.* **26**, 2464 [1961]. (109) F. A. HENGLEIN, P. SCHMULDER and H. NIEBERGALL, German Pat. (West German) 947739 [27.11.1953].

(110) —, —, —, German Appl. (West German) H 23937 [26.11.1953]. (111) B. J. HUMPHREY and H. H. WASSERMAN, Connecticut Hard Rubber, U.S. Pat. 2673843 [17.5.1951]. (112) M. J. HUNTER, H. J. FLETCHER and C. C. CURRIE, Dow Chemical, U.S. Pat. 2415389 [31.5.1944]. (113) —, J. F. HYDE, E. L. WARRICK and H. J. FLETCHER, *J. Amer. Chem. Soc.* **68**, 667 [1946]. (114) D. T. HURD, ibid. **77**, 2998 [1955]. (115) — and R. C. OSTHOFF, General Electric, U.S. Pat. 2737506 [9.6.1953]. (116) —, — and M. L. CORRIN, *J. Amer. Chem. Soc.*

76, 249 [1954]. (117) J. F. HYDE, Dow Corning, German Pat. (West German) 833122 [U.S. Prior. 26.2.1942]. (118) —, —, German Pat. (West German) 833255 [U.S. Prior. 26.2.1942]. (119) —, —, German Pat. (West German) 848706 [U.S. Prior. 26.2.1942].

(120) —, —, German Pat. (West German) 875046 [U.S. Prior. 26.2.1942]. (121) —, —, German Pat. (West German) 833126 [U.S. Prior. 30.3.1943]. (122) —, Corning Glass Works, U.S. Pat. 2467976 [30.3.1943]. (123) —, —, U.S. Pat. 2455999 [1.8.1944]. (124) —, Dow Corning, U.S. Pat. 2446135 [13.4.1944]. (125) —, Corning Glass Works, U.S. Pat. 2574265 [15.11.1945]. (126) —, Dow Corning, German Pat. (West German) 851257 [U.S. Prior. 12.7.1947]. (127) —, —, German Pat. (West German) 873085 [U.S. Prior. 17.4.1950]. (128) —, —, U.S. Pat. 2629725 [19.7.1950]. (129) —, —, U.S. Pat. 2629726 [19.7.1950].

(130) —, —, U.S. Pat. 2634284 [29.12.1951]. (131) —, *J. Amer. Chem. Soc.* **75**, 2166 [1953]. (132) — and W. H. DAUDT, Dow Corning, German Pat. (West German) 851258 [U.S. Prior. 27.10.1948]. (133) — and G. J. KOOKOOTSEDES, —, German Pat. (West German) 954455 [U.S. Prior. 1.9.1953]. (134) R. K. ILER, Colloid Chemistry of Silica and Silicates, Cornell Univ. Press 1955, p. 46. (135) H. JENKNER, Kali-Chemie, German Pat. (West German) 954645 [5.5.1955]; *Z. Naturforsch.* **14b**, 133 [1959]. (136) —, —, German Appl. (West German) 1018226 [18.5.1955]. (137) V. B. JEX and D. L. BAILEY, French Pat. 1184096 [U.S. Prior. 12.10.1956]. (138) O. K. JOHANNSEN, Dow Corning, German Pat. (West German) 949604 [U.S. Prior. 3.12.1952]. (139) —, —, U.S. Pat. 2762827 [9.10.1953].

(140) — and J. F. HYDE, —, German Pat. (West German) 886075 [U.S. Prior. 10.1.1947]. (141) S. W. KANTOR, *J. Amer. Chem. Soc.* **75**, 2712 [1953]. (142) —, General Electric, U.S. Pat. 2717258 [25.2.1953]. (143) —, —, U.S. Pat. 2766220 [11.5.1954]. (144) —, W. T. GRUBB and R. C. OSTHOFF, *J. Amer. Chem. Soc.* **76**, 5190 [1954]. (145) H. KAUTSKY, German Appl. (West German) 1028784 [25.6.1955]. (146) M. KIN, Dow Corning, U.S. Pat. 2721855 [16.3.1953]. (147) F. S. KIPPING and J. E. HACKFORD, *J. Chem. Soc.* [London] **99**, 138 [1911]. (148) N. KIRK, *Ind. Engng. Chem.* **51**, 515 [1959]. (149) H. KNOPF, A. BEERWALD and G. BRINKMANN, Farbenfabriken Bayer, German Pat. (West German) 957662 [15.2.1954].

(150) C. F. KOHL, Corning Glass Works, U.S. Pat. 2640063 [27.10.1949]. (151) K. KOJIMA, N. TARUMI and S. WAKATSUKI, *J. Chem. Soc. Japan, pure chem. Sect.* **76**, 1205 [1955]. (152) W. KRAUSS and R. KUBENS, *Dtsch. Farben-Z.* **10**, 1 (1956]. (153) R. H. KRIEBLE and J. R. ELLIOTT, *J. Amer. Chem. Soc.* **67**, 1810 [1945] and **68**, 2291 [1946]. (154) W. A. KRINER, A. G. MACDIARMID and E. C. EVERS, ibid. **80**, 1546 [1958]. (155) A. G. KUZNETSOVA, K. A. ANDRIANOV and D. YA. ZHINKIN, *Plast. massy.* **1962**, No. 11, 15. (156) H. F. LAMOREAUX, General Electric, U.S. Pat. 2470497 [21.11.1945]. (157) E. LARSSON, *Svensk. kem. Tidskr.* **62**, 4 [1950]. (158) E. J. LAWTON, W. T. GRUBB and J. S. BALWIT, *J. Polymer Sci.* **19**, 455 [1956]. (159) R. H. LEITHEISER, Dow Corning, U.S. Pat. 3122579 [17.10.1960].

(160) A. LENGYEL and T. SZÉKELY, Hungarian Pat. 145756 [25.7.1955]. (161) R. N. LEWIS, General Electric, U.S. Pat. 2500761 [19.6.1947]. (162) F. M. LEWIS and E. J. LAWTON, Thomson-Houston, French Pat. 1080171 [U.S. Prior. 3.6.1952]. (163) N. S. LEZNOV, L. A. SABUN and K. A. ANDRIANOV, *J. Gen. Chem. USSR* (*Engl. transl.*) **29**, 1482 [1959]. (164) K. LICHT and H. KRIEGSMANN, *Z. anorg. allg. Chem.* **323**, 252 [1963]; **330**, 151 [1964]. (165) P. LINKE, Th. Goldschmidt, German Pat. (West German) 958253 [4.11.1949]. (166) R. C. LINVILLE, General Electric, U.S. Pat. 2739952 [24.6.1953]. (167) G. R. LUCAS, —, U.S. Pat. 2598402 [3.10.1950]. (168) — and R. W. MARTIN, —, U.S. Pat. 2600307 [15.3.1950]. (169) —, —, *J. Amer. Chem. Soc.* **74**, 5225 [1952].

(170) R. S. LUDINGTON, Westinghouse, U.S. Pat. 2689843 [27.1.1953]. (171) G. P. MACK and E. PARKER, Advance, French Pat. 1101689 [U.S. Prior. 5.3. and 6.8.1953]. (172) J. MARSDEN and G. F. ROEDEL, General Electric, U.S. Pat. 2469883 [11.6.1945]. (173) K. N. MATHES, —, U.S. Pat. 2518160 [18.9.1947]. (174) M. MATSUI, *Sci. Pap. Inst. Phys. Chem. Res. Tokyo* **54**, 141 [1960]. (175) F. R. MAYO, *J. Polymer Sci.* **55**, 65 [1961]. (176) J. J. MCBRIDE JR. and H. C. BEACHELL, *J. Amer. Chem. Soc.* **74**, 5247 [1952]. (177) P. A. MCCUSKER and T. OSTDICK, ibid. **80**, 1103 [1958]. (178) R. R. MCGREGOR and E. L. WARRICK, Dow Corning, German Pat. (West German) 881404 [U.S. Prior. 16.2.1942]. (179) —, —, —, U.S. Pat. 2459387 [5.10.1942].

(180) —, —, —, U.S. Pat. 2431878 [30.3.1943]. (181) H. W. MCNULTY and D. J. KILIAN, National Distillers, U.S. Pat. 2687388 [13.11.1950]. (182) R. L. MERKER, Dow Corning, French Pat. 1124823 [U.S. Prior. 20.5.1954]. (183) — and M. J. SCOTT, *J. Polymer Sci.* [A] **2**, 15 [1964]. (184) —, — and G. G. HABERLAND, ibid. [A] **2**, 31 [1964]. (185) Midland Silicones, Brit. Pat. 832489 [U.S. Prior. 27.6.1956]. (186) A. A. MILLER, *J. Amer. Chem. Soc.* **82**, 3519 [1960]. (187) —, ibid. **83**, 31 [1961]. (188) Monsanto Chemical, Brit. Pat. 657704 [Belg. Prior. 20.9. and 13.12.1947]. (189) M. MORTON and E. E. BOSTICK, *J. Polymer Sci.* [A] **2**, 523 [1964].

(190) —, M. A. DEISZ and E. E. BOSTICK, ibid. [A] **2**, 513 [1964]. (191) Y. NAKAIDO and T. TAKIGUCHI, *J. Org. Chem.* **26**, 4144 [1961]. (192) S. NITZSCHE, Wacker-Chemie, German Pat. (West German) 831098 [25.5.1949]. (193) —, —, German Pat. (West German) 873433 [25.5.1949]. (194) —, —, German Pat. (West German) 831292 [31.12.1950]. (195) — and M. WICK, —, German Pat. (West German) 930481 [13.1.1953]. (196) —, —, —, German Pat. (West German) 954292 [23.5.1953]. (197) —, —, *Kunststoffe* **47**, 431 [1957]. (198) —, — and S. REICHSTALLER, Wacker-Chemie, German Pat. (West German) 935455 [19.11.1953]. (199) W. NOLL and P. SIMONS, Farbenfabriken Bayer, German Pat. (West German) 873142 [Swiss Prior. 24.12.1949].

(200) —, —, —, German Pat. (West German) 864152 [2.1.1951]. (201) —, —, —, German Pat. (West German) 903994 [26.4.1950]. (202) —, —, —, German Pat. (West German) 895651 [23.4.1951]. (203) J. E. NOLL, J. L. SPEIER and B. F. DAUBERT, *J. Amer. Chem. Soc.* **73**, 3867 [1951]. (204) R. OKAWARA, *Bull. Chem. Soc. Japan* **28**, 360 [1955]. (205) — and M. SAKIYAMA, IUPAC-Kongress München 1959, A 114, Verlag Chemie, Weinheim/Bergstr. (206) N. F. ORLOV, *Proc. Acad. Sci. USSR (Engl. transl.)* **114**, 629 [1957]. (207) — and B. N. DOLGOV, ibid. **125**, 266 [1959]. (208) W. I. PATNODE, General Electric, German Pat. (West German) 833647 [U.S. Prior. 2.10.1942]. (209) —, Thomson-Houston, French Pat. 938822 [U.S. Prior. 29.10.1942].

(210) —, —, French Pat. 950305 [U.S. Prior. 29.10.1942]. (211) —, General Electric, U.S. Pat. 2381366 [29.10.1942]. (212) —, —, U.S. Pat. 2469888 [29.10.1942]. (213) — and F. C. SCHMIDT, *J. Amer. Chem. Soc.* **67**, 2272 [1945]. (214) — and D. F. WILCOCK, ibid. **68**, 358 [1946]. (215) D. S. PAYNE, *J. Chem. Soc.* [London] **1954**, 2143. (216) F. F. PETROT, Rhône-Poulenc, German Pat. (West German) 831724 [French Prior. 7.7.1947]. (217) O. R. PIERCE and G. W. HOLBROOK, Dow Corning, German Appl. (West German) 1101768 [U.S. Prior. 7.4.1958]. (218) R. M. PIKE, Union Carbide, French Pat. 1275384 [U.S. Prior. 8.12.1959]. (219) —, —, French Pat. 1278251 [U.S. Prior. 8.12.1959].

(220) —, *J. Polymer Sci.* **50**, 151 [1961]. (221) — and D. C. BAILEY, Union Carbide, French Pat. 1184097 [U.S. Prior. 12.10.1956]. (222) K. E. POLMANTEER, Dow Corning, German Appl.

(West German) 1019462 [U.S. Prior. 18.2.1954]. (223) M. PROBER and G. D. COOPER, Thomson-Houston, French Pat. 1116725 [U.S. Prior. 31.12.1953]. (224) C. E. REED and J. M. TOME, General Electric, U.S. Pat. 2741630 [17.3.1955]. (225) N. REISO, Yushi K. K., Jap. Pat. 3738 [19.1.1955]. (226) W. T. RESSITER and C. C. CURRIE, Dow Corning, French Pat. 1068730 [U.S. Prior. 2.8.1951]. (227) Rhône-Poulenc, French Pat. 950582 [23.7.1947]. (228) —, Brit. Pat. 645389 [French Prior. 8.8.1947]. (229) —, French Pat. 1105714 [4.6.1954].

(230) E. G. ROCHOW, General Electric, U.S. Pat. 2371068 [30.4.1940]. (231) —: *An Introduction to the Chemistry of the Silicones*, 2nd Ed., N.Y., John Wiley & Sons Inc. 1951, p. 25. (232) J. B. RUST, Montclair Research, U.S. Pat. 2698314 [12.3.1951]. (233) — and C. A. MACKENZIE, —, U.S. Pat. 2506616 [29.5.1945]. (234) — and H. H. TAKIMOTO, Hughes Aircraft, U.S. Pat. 3057822 [30.6.1959]. (235) M. M. SAFFORD, General Electric, U.S. Pat. 2454759 [11.6.1946]. (236) —, —, U.S. Pat. 2558560 [12.9.1950]. (237) — and D. T. HURD, —, U.S. Pat. 2558561 [12.9.1950]. (238) — and E. L. MINCHER, —, U.S. Pat. 2455880 [13.5.1946]. (239) R. O. SAUER, —, U.S. Pat. 2398672 [21.8.1942].

(240) —, *J. Amer. Chem. Soc.* **66**, 1707 [1944]. (241) —, General Electric, U.S. Pat. 2421653 [1.7.1944]. (242) —, —, U.S. Pat. 2595890 [21.3.1946]. (243) F. SCHLENKER, Chemische Werke Albert, German Pat. (West German) 951889 [23.4.1954]. (244) H. SCHMIDBAUR, *Angew. Chem.* **77**, 206 [1965]. (245) K. SCHNURRBUSCH, Farbenfabriken Bayer, German Pat. (West German) 1034631 [24.5.1957]. (246) —, —, unpublished. (247) — and W. KNIEGE, —, Belg. Pat. 588831 [German Prior (West German) 20.3.1959]. (248) G. SCHOTT and H. BERGE, *Z. anorg. allg. Chem.* **297**, 32 [1958]. (249) — and co-workers, ibid. **325**, 169, 175 [1963]; other lit. cited by authors.

(250) — and W. D. SPRUNG, ibid. **333**, 76 [1964]. (251) W. A. SCHWENKER, General Electric, U.S. Pat. 2758124 [11.4.1952]. (252) D. W. SCOTT, *J. Amer. Chem. Soc.* **68**, 356 [1946]. (253) P. C. SERVAIS, Dow Chemical, U.S. Pat. 2485928 [8.10.1947]. (254) L. H. SHAFFER and E. M. FLANIGEN, *J. Physic. Chem.* **61**, 1591 [1957]. (255) C. SHAW, W. E. SMITH and H. G. EMBLEM, U.S. Pat. 2580852 [7.1.1947]. (256) M. F. SHOSTAKOVSKY and KH. I. KONDRATYEV, *Bull. Acad. Sci. USSR, Div. Chem. Sci.* (*Engl. transl.*) **1957**, 333. (257) W. SIMMLER, *Diss. Univ. München* 1956. (258) —, Farbenfabriken Bayer, German Appl. (West German) 1046049 [17.8.1957]. (259) —, *Makromol. Chem.* **57**, 12 [1962].

(260) —, W. NOLL and K. SCHNURRBUSCH, Farbenfabriken Bayer, German Appl. (West German) 1060144 [29.11.1957]. (261) P. SIMONS, —, German Pat. (West German) 888851 [18.4.1950]. (262) — and W. NOLL, —, German Pat. (West German) 895649 [27.3.1950]. (263), —, —, —, German Pat. (West German) 895650 [24.3.1951]. (264) B. SMITH, *Svensk. kem. Tidskr.* **65**, 101 [1953]. (265) —, ibid. **67**, 421 [1955]. (266) Soc. Centrale de Recherches et d'Applications techn., French Pat. 1042019 [7.9.1951]. (267) N. N. SOKOLOV, *J. Gen. Chem. USSR* (*Engl. transl.*) **28**, No. 12, 3354 [1958]. (268) M. M. SOLOMON, Thomson-Houston, U.S. Pat. 2803614 [15.3.1954]. (269) L. H. SOMMER, Dow Corning, U.S. Pat. 2589446 [13.4.1950].

(270) —, —, U.S. Pat. 2557803 [13.4.1950]; U.S. Pat. 2662909 [13.12.1951]. (271) —, —, U.S. Pat. 2662910 [13.12.1951]. (272) — and co-workers, Amer. Chem. Soc. 135th Meeting, Boston 1959, Abstr. 22 M/24M, No. 62/63 (cf. *Angew. Chem.* **71**, 532 [1959]). (273) —, W. P. BARIE and J. R. GOULD, *J. Amer. Chem. Soc.* **75**, 3765 [1953]. (274) —, W. D. ENGLISH, C. R. ANSUL and D. N. VIVONA, ibid. **77**, 2485 [1955]. (275) —, G. M. GOLDBERG, G. H. BARNES and L. S. STONE JR., ibid. **76**, 1609 [1954]. (276) —, L. Q. GREEN and F. C. WHITMORE,

ibid. **71**, 3253 [1949]. (277) —, N. S. MARANS, G. M. GOLDBERG, J. ROCKETT and R. P. PIOCH, ibid. **73**, 882 [1951]. (278) —, E. W. PIETRUSZA and F. C. WHITMORE, ibid. **68**, 2282 [1946]. (279) — and R. P. PIOCH, ibid. **75**, 6337 [1953].

(280) —, —, N. S. MARANS, G. M. GOLDBERG, J. ROCKETT and J. KERLIN, ibid. **75**, 2932 [1953]. (281) — and L. J. TYLER, ibid. **76**, 1030 [1954]. (282) F. J. SOWA, U.S. Pat. 2486992 [25.7.1945]. (283) J. L. SPEIER, Dow Corning, U.S. Pat. 2676163 [3.9.1952]. (284) M. M. SPRUNG, General Electric, U.S. Pat. 2472629 [11.6.1946]. (285) — and C. A. BURKHARD, —, U.S. Pat. 2448556 [16.1.1947]. (286) — and F. O. GUENTHER, *J. Amer. Chem. Soc.* **77**, 3990 [1955]. (287) —, —, ibid. **77**, 3996 [1955]. (288) —, —, ibid. **77**, 6045 [1955]. (289) W. SUNDERMEYER and W. NOLL, Farbenfabriken Bayer, Brit. Pat. 922377 [German Prior. (West German) 6.4.1960].

(290) C. G. SWAIN, R. M. ESTEVE and R. H. JONES, *J. Amer. Chem. Soc.* **71**, 965 [1949]. (291) T. TAKIGUCHI, M. SAKURAI, R. KISHI, J. ICHIMURA and Y. IIZUKA, *J. Org. Chem.* **25**, 310 [1960]. (292) P. TARRANT, G. W. DYCKES, R. DUNMIRE and G. B. BUTLER, *J. Amer. Chem. Soc.* **79**, 6536 [1957]. (293) L. J. TYLER, Dow Corning, U.S. Pat. 2605274 [9.1.1950]. (294) —, —, U.S. Pat. 2611774 [9.1.1950]. (295) —, *J. Amer. Chem. Soc.* **77**, 770 [1955]. (296) Union Chimique Belge, French Pat. 1006989 [Belg. Prior. 30.12.1948 and 20.10.1949]. (297) —, French Pat. 1027425 [Belg. Prior. 22.10.1949]. (298) —, French Pat. 1037411 [Belg. Prior. 31.5.1950]. (299) I. VALADE, *C.R.* **246**, 952 [1958].

(300) M. G. VORONKOV, *Bull. Acad. Sci. USSR, Div. Chem. Sci.* (*Engl. transl.*) **1957**, 529, (301) — and Z. I. SCHABAROVA, *J. Gen. Chem. USSR* (*Engl. transl.*) **29**, 1501 [1959]. (302) G. H. WAGNER and C. E. ERICKSON, Union Carbide, U.S. Pat. 2731485 [26.7.1949]. (303) E. L. WARRICK, Dow Corning, German Pat. (West German) 934910 [U.S. Prior. 22.10.1946]. (304) —, —, U.S. Pat. 2480620 [27.12.1946]. (305) —, —, U.S. Pat. 2634252 [25.5.1949]. (306) —, —, German Pat. (West German) 935284 [U.S. Prior. 10.11.1949]. (307) —, —, French Pat. 1061631 [U.S. Prior. 19.4.1950]. (308) —, —, German Pat. (West German) 909995 [U.S. Prior. 12.10.1950]. (309) —, *Ind. Engng. Chem.* **47**, 2388 [1955].

(310) — and R. R. MCGREGOR, Dow Corning, German Pat. (West German) 841506 [U.S. Prior. 5.10.1942 and 30.3.1943]. (311) C. E. WELSH, General Electric, U.S. Pat. 2449572 [11.8.1944]. (312) — and N. G. HOLDSTOCK, —, U.S. Pat. 2661348 [8.2.1951]. (313) R. WEST, *J. Amer. Chem. Soc.* **76**, 6015 [1954]. (314) D. R. WEYENBERG, Dow Corning, U.S. Pat. 2714099 [11.1.1954]. (315) E. WIBERG and P. BUCHHEIT, Süd-Chemie, German Pat. (West German) 960234 [11.6.1951]. (316) — and W. SIMMLER, *Z. anorg. allg. Chem.* **282**, 330 [1955]. (317) D. F. WILCOCK, General Electric, U.S. Pat. 2491843 [21.3.1946]. (318) W. D. WILLIS, Hercules Powder, U.S. Pat. 2816089 [25.3.1955]. (319) W. J. WORMUTH, General Electric, U.S. Pat. 2552247 [8.4.1949].

(320) J. G. E. WRIGHT, —, U.S. Pat. 2452416 [26.4.1944]. (321) J. R. WRIGHT and A. GOLDSCHMIDT, California Research, U.S. Pat. 2780636 [16.6.1954]. (322) J. G. E. WRIGHT and J. MARSDEN, General Electric, U.S. Pat. 2389477 [21.8.1942]. (323) — and C. S. OLIVER, —, U.S. Pat. 2448565 [14.3.1944]. (324) D. C. YOUNGS, Dow Corning, U.S. Pat. 2723966 [19.11.1953]. (325) J. F. ZACK JR., E. L. WARRICK and G. KNOLL, *J. Chem. Engng. Data* **6**, 279 [1961]. (326) A. A. ZHDANOV, K. A. ANDRIANOV and A. A. BOGDANOVA, *Bull. Acad. Sci. USSR, Div. Chem. Sci.* (*Engl. transl.*) **1961**, 1172. (327) W. ZIMMERMANN, *Chem. Ber.* **87**, 887 [1954].

CHAPTER 6

The Polymeric Organosiloxanes

We shall now turn to the molecular properties which determine the bulk characteristics of the polyorganosiloxanes, i.e., those compounds which according to our earlier definition represent the actual silicone polymers. This chapter must necessarily occupy a central position, since a knowledge of their molecular structure, bond strengths, and physicochemical behavior forms the basis for an appreciation of the properties of the industrial silicone products, and at the same time provides the key to the understanding of their uses. The reader interested in silicone chemistry as applied to the material in bulk will, accordingly, find useful the chain of reasoning mentioned earlier, which leads from the presentation of the molecular properties and the physicochemical behavior of the siloxanes (Chapter 6) to the technologically important characteristics of the industrial silicone products (Chapter 9) and then to their applications (Chapter 10). In this way he can systematically follow the indicated path from the molecular characteristics of these products to their technological properties, and from those to their numerous possible applications.

6.1 Systematic Classification of Polyorganosiloxanes based on their Molecular Structure and Chemistry

The general structural principles of the polyorganosiloxanes have already been treated in Chapter 1. We now turn to a classification of the material whose diversity has already been shown. This diversity is due to the possibilities of combination of siloxane units of different functionalities in the synthesis of the polymer, and to the numerous variations possible in the substitution of those valencies on the silicon atoms that are not bound by oxygen.

Different ways of classifying this material are possible, just as in any other attempt at systematization. Silicones are frequently divided phenomenonologically into oils, resins, rubbers, and saltlike compounds. This arrangement is suitable for technological purposes, but from a scientific standpoint it appears clearer, less ambiguous, and more informative to base our classification on structural and chemical characteristics.

In analogy with the systematic classification of organic polymers, we distinguish the following groups of polymers according to their structures:

1. *Linear, or straight-chain polysiloxanes*; these correspond to the structural type

$$MD_nM \quad \text{or} \quad R_3SiO[R_2SiO]_nSiR_3$$

2. *Branched-chain polysiloxanes*: these contain at least one trifunctional (T) or tetrafunctional (Q) siloxane unit as a branching center. The branching unit is incorporated into either a chain or a ring. As opposed to cross-linking, however, branching does not lead to an appreciable spatial extension of the molecule.

3. *Cyclic polysiloxanes*: these are built up from siloxane rings—actually ringlike siloxane polyhedra. *Monocyclic* molecules consist of such rings formed by difunctional siloxane units, and can be characterized by the symbol D_n. Just as in organic chemistry, however, condensed ring systems are also known in silicone chemistry: these may be called *polycyclic*. Such systems no longer consist of D units alone, but also contain T units linking groups between rings. The *spirocyclic* systems are distinguished from polycyclic systems as a separate group; the point of linkage in this case is a Q unit, which thus becomes characteristic of the structure of such molecules together with the D units. As a result of the tetrahedral coordination of the siloxane oxygen atoms around the silicon atom in a Q unit, the planes of alternate rings are at right angles to one another. A fourth subgroup of the cyclic polymers could finally be designated as *spherocyclic*. These are small, not really spherical, but more or less isometrically formed molecules built up from a number of rings. Molecules of this kind contain only T groups and their symbol is thus T_n.

4. *Cross-linked polymers*: This group is characterized by the fact that linear or cyclic molecules are cross-linked by means of T and Q units to form two- or three-dimensional networks. It can easily be seen that a great diversity of structural forms is possible in this case. This also explains why the cross-linked molecules play such an important part in industrial silicone products.

Networks which contain a relatively high proportion of T or Q units can also be considered as high-molecular-weight polycondensed ring systems. Polycyclic, spirocyclic, and spherocyclic ring systems could of course be interpreted as arising from a cross-linking of individual ring systems. It would however, appear desirable to separate the molecular forms characteristic of small, definite, molecules from the three- dimensionally cross-linked macromolecular networks, and thus to distinguish the group of cyclic low-molecular-weight polymers from that of the cross-linked macromolecular ones.

* Page number of chapter references.

Each group of polymers can be further subdivided according to the nature of the substituents bound to the silicon atom. The siloxane skeleton can be substituted by different types of saturated hydrocarbon radicals. It may, in addition, contain silicon-functional or organofunctional groups or indeed both simultaneously. A subdivision of the polymer groups into *nonfunctional* polyorganosiloxanes, polysiloxanes with *silicon-functional or organofunctional* groups, and finally into *silicon- and organofunctional* siloxanes would accordingly appear to be desirable.

Reference will constantly be made to this type of classification. There are, of course, many other ways of classifying the material. The organosiloxane polymers are also frequently classified according to their chemistry, for instance, according to the silicon/oxygen or silicon/carbon ratio. Such a procedure, however, would need to be supplemented by structural data, since any given silicon/oxygen ratio could be associated with a number of different molecular structures.

Finally, another rather formal arrangement is possible, according to the nature of the siloxane units represented in the molecule. The four siloxane units can be combined in the synthesis of the polymers to give combinations of the same units with one another, of two different types of units with one another, or of three or four different types of units with one another; this gives fifteen combinations. Since one of these—the combination of Q units—is not characteristic for silicone chemistry, fourteen combinations of siloxane units remain; Hyde (*74*) classified these as given in the accompanying tabulation.

Class	Siloxane units	Class	Siloxane units
1	Q T	8	Q T M
2	Q D	9	Q D M
3	Q M	10	T D M
4	T D	11	Q T D M
5	T M	12	T T
6	D M	13	D D
7	Q T D	14	M M

6.2 The Various Types of Polysiloxanes

6.2.1 Linear Polysiloxanes

Linear organosiloxane molecules can be synthesized with a wide range of molecular dimensions. The shortest possible chain, consisting of two silicon atoms linked through an oxygen bridge, is represented by a hexaorganodisiloxane:

$$R_3Si—O—SiR_3 \quad [M_2]$$

Low-molecular-weight oligomeric polyorganosiloxanes, which still have definite molecular weights and which can be isolated by distillation or crystallization, can be prepared by the stepwise insertion of difunctional siloxane units:

$$R_3SiO[R_2SiO]_nSiR_3 \quad [MD_nM]$$

Identification of the oligomeric compounds is no longer possible when there are more than about 10 difunctional units in the chain. However, chain growth can be continued to at least 10,000 D units and further extension of the chains seems possible. A prerequisite for the synthesis of such long siloxane chains is, of course, that no branching or cross-linking T or even Q units are present, and that the concentration of M units does not exceed that required for the desired chain length. The presence of other monofunctional groups capable of acting as stoppers and thus terminating chain growth prematurely is, of course, also detrimental.

Examples of a number of short-chain polysiloxanes of definite molecular dimensions are given in Tables 33, 35 and 36. The further one goes into the region of macromolecular organosiloxane chains, the more difficult it becomes to synthesize polymers of sharply defined molecular size. Even "equilibration" processes (see Section 5.4) only reduce the polydispersity of the molecular dimensions without ever producing a uniform molecular weight.

Attempts have been made to determine the *molecular size distribution* in macromolecular linear polyorganosiloxanes by various methods. The earliest investigations of this nature originated with Scott (*158*) who fractionated a macromolecular siloxane by cooling a solution in ethyl acetate and acetone saturated at room temperature. Possibly as a result of the high concentration of the solutions, however, the fractionation was insufficiently sharp (*156*).

In later experiments, solutions of the polymers in benzene, ethyl acetate, or methyl ethyl ketone (*25*, *46*, *85*, *156*, *196*) were repeatedly precipitated in fractions with methanol or acetone, and the molecular weights of the individual fractions were determined by osmotic-pressure measurements and by light-scattering methods. In the investigation of a high-molecular-weight polydimethylsiloxane suitable for the production of silicone rubber the results given in the tabulation were obtained for the mean molecular weight and the molecular weights of the three fractions recovered (*25*).

In polymers produced by different means intended for the same application, a considerably larger distribution of molecular weights, ranging from 1,060,000 down to 33,000, was found (*156*).

Fraction	Mean molecular weight	No. of siloxane units in the chain, approx.
Not fractionated	473,000	6400
1	708,000	9600
2	452,000	6100
3	372,000	5000

Various attempts have been made to formulate the relation between *viscosity* and mean *molecular weight*. Staudinger's law, according to which intrinsic viscosity is proportional to the molecular weight M,

$$[\eta] = K \cdot M$$

is valid for the linear polydimethylsiloxanes of the structural type MD_nM only up to a molecular weight of approximately 2500, which corresponds to about 34 siloxane units in the molecule (*8*). The constant K in this case is 1.43×10^{-5}.

With compounds of higher molecular weight, the relationship between the intrinsic viscosity and the molecular weight can be represented by the exponential function:

$$[\eta] = K \cdot M^{n}$$

The results of a number of authors are available for solutions in toluene. The best curve through all the values available at present from light scattering, sedimentation, and diffusion measurements is represented by the formula (*66*):

$$[\eta] = 8.28 \times 10^{-3} \times \mathrm{M}^{0.72}$$

Measurements in methyl ethyl ketone, where the molecular weights were determined osmotically and by light scattering, gave the relation (*46*):

$$[\eta] = 0.8 \times 10^{-3} \times \mathrm{M}^{0.5}$$

which has also been confirmed by more recent measurements in bromocyclohexane (*66*).

Barry (*8*) found for the relationship between the viscosity and the molecular weight that the results in the region of $M > 2500$ could be represented by the expression

$$\log \eta^{\mathrm{cSt}/25°\mathrm{C}} = 1.00 + 0.0123 \times M^{0.5}$$

and thus correspond to an equation obtained by Flory (*45*) for the bulk viscosity of polymer melts. However, Warrick, *et al.* (*196*) deduced from measurements on a high-molecular-weight polydimethysiloxane on the one hand,

TABLE 33

LOW-MOLECULAR-WEIGHT LINEAR POLYORGANOSILOXANES (NONFUNCTIONAL)[a]

Compound	η (cP) at	t (°C)	Mp (°C)	Bp (°C) at	p (mm Hg)	n_D^t at	t (°C)	d_4^t (gm/cc) at	t (°C)
$(CH_3)_3SiOSi(CH_3)_3$	0.412 (204)	37.8	−68 (71)	100.1 (204)	757	1.3774 (151)	20	0.7619 (151)	20
$(CH_3)_3SiO[(CH_3)_2SiO]Si(CH_3)_3$	0.703 (204)	37.8	−86 (71)	151.7 (204)	747	1.3848 (129)	20	0.8200 (129)	20
$(CH_3)_3SiO[(CH_3)_2SiO]_2Si(CH_3)_3$	1.09 (204)	37.8	−76 (71)	194 (204)	760	1.3895 (129)	20	0.8536 (129)	20
$(CH_3)_3SiO[(CH_3)_2SiO]_3Si(CH_3)_3$	1.50 (204)	37.8	−84 (71)	229 (204)	760	1.3925 (129)	20	0.8755 (129)	20
$(CH_3)_3SiO[(CH_3)_2SiO]_4Si(CH_3)_3$	1.91 (204)	37.8	−59 (71)	184.2 (204)	101	1.3948 (129)	20	0.8910 (129)	20
$(CH_3)_3SiO[(CH_3)_2SiO]_5Si(CH_3)_3$	2.58 (204)	37.8	−78 (71)	184.5 (204)	40	1.3965 (204)	20	0.911 (204)	20
$(CH_3)_3SiO[(CH_3)_2SiO]_6Si(CH_3)_3$	2.87 (204)	37.8	−63 (71)	202.0 (204)	39.3	1.3970 (204)	20	0.913 (204)	20
$(CH_3)_3SiO[(CH_3)_2SiO]_7Si(CH_3)_3$	3.38 (204)	37.8	—	198.8 (204)	16.0	1.3980 (204)	20	0.918 (204)	20
$(CH_3)_3SiO[(CH_3)_2SiO]_8Si(CH_3)_3$	3.94 (204)	37.8	—	202.8 (204)	10.0	1.3988 (204)	20	0.925 (204)	20
$(CH_3)_3SiO[(CH_3)_2SiO]_9Si(CH_3)_3$	4.52 (204)	37.8	—	201.8 (204)	4.7	1.3994 (204)	20	0.930 (204)	20
$(CH_3)_3SiOSi(C_2H_5)_3$	—	—	—	46–47 (168)	7	1.4099 (168)	20	0.8110 (168)	20
$(CH_3)_2(C_2H_5)SiOSi(C_2H_5)(CH_3)_2$	—	—	−120 (10) Solidification	152 (10)	760	1.4000 (10)	20	0.798 (10)	20
$(C_2H_5)_3SiOSi(C_2H_5)_3$	—	—	—	231 (78)	760	1.4340 (32)	20	0.8402 (186)	25
$(CH_3)_3SiO[CH_3(C_2H_5)SiO]Si(CH_3)_3$	—	—	—	173 (36)	760	1.3928 (36)	20	0.830 (36)	20
$(CH_3)_3SiO[CH_3(C_2H_5)SiO]_2Si(CH_3)_3$	—	—	—	227 (36)	760	1,4019 (36)	20	0.867 (36)	20
$(CH_3)_3SiO[CH_3(C_2H_5)SiO]_3Si(CH_3)_3$	—	—	—	263 (36)	760	1.4066 (36)	20	0.891 (36)	20
$(CH_3)_3SiO[CH_3(C_2H_5)SiO]_4Si(CH_3)_3$	—	—	—	311 (36)	760	1.4104 (36)	20	0.906 (36)	20
$(CH_3)_2(C_2H_5)SiO[CH_3(C_2H_5)SiO]Si(CH_3)_2(C_2H_5)$	—	—	—	212 (10)	760	1.4075 (10)	20	0.850 (10)	20
$(CH_3)_2(C_2H_5)SiO[CH_3(C_2H_5)SiO]_2Si(CH_3)_2(C_2H_5)$	—	—	—	259 (10)	760	1.4121 (10)	20	0.880 (10)	20
$(CH_3)_2(C_2H_5)SiO[CH_3(C_2H_5)SiO]_3Si(CH_3)_2(C_2H_5)$	—	—	—	301 (10)	760	1.4150 (10)	20	0.901 (10)	20
$(CH_3)_2(C_2H_5)SiO[CH_3(C_2H_5)SiO]_4Si(CH_3)_2(C_2H_5)$	—	—	—	364 (10)	760	1.4210 (10)	20	—	—
$(CH_3)_3SiO[(C_2H_5)_2SiO]Si(CH_3)_3$	1.441 (174)	20	—	187 (174)	733	1.4005 (174)	20	0.8399 (174)	20
$(CH_3)_3SiO[(C_3H_7)_2SiO]Si(CH_3)_3$	—	—	—	96–98 (91)	17	1.4088 (91)	20	0.8390 (91)	20
$(C_3H_7)_3SiO[(C_3H_7)_2SiO]Si(C_3H_7)_3$	—	—	—	—	—	1.4418 (112)	20	0.8680 (112)	20
$(CH_3)_3SiO[(C_4H_9)_2SiO]Si(CH_3)_3$	—	—	—	120–122 (91)	18	1.4143 (91)	20	0.8406 (91)	20
$(C_6H_5)_3SiOSi(C_6H_5)_3$	—	—	226 (30)	494 (30)	750±10	1.59, 1.68, 1.71 (30)	25	—	—
$(CH_3)_2(C_6H_5)SiOSi(CH_3)_3$	1.5[b] (30)	25	< −80 (30)	206 (30)	750±10	1.4588 (30)	25	0.891 (30)	25
$(CH_3)_2(C_6H_5)SiOSi(CH_3)_2(C_6H_5)$	3.4[b] (30)	25	< −80 (30)	292 (30)	750±10	1.5149 (30)	25	0.971 (30)	25
$(CH_3)(C_6H_5)_2SiOSi(CH_3)_3$	4.9[b] (30)	25	−65 (30)	285 (30)	750±10	1.5166 (30)	25	0.974 (30)	25

(continued on p. 252)

TABLE 33 *(Contd.)*

LOW-MOLECULAR-WEIGHT LINEAR POLYORGANOSILOXANES (NONFUNCTIONAL)[a]

Compound	η (cP) at	t (°C)	Mp (°C)	Bp (°C) at	p (mm Hg)	n_D^t at	t (°C)	d_4^t (gm/cc) at	t (°C)
$(CH_3)(C_6H_5)_2SiOSi(CH_3)_2(C_6H_5)$	13.0[b] *(30)*	25	−45 *(30)*	364 *(30)*	750 ± 10	1.5549 *(30)*	25	1.031 *(30)*	25
$(CH_3)(C_6H_5)_2SiO[(CH_3)_2SiO]Si(CH_3)(C_6H_5)_2$	37.2[b] *(42)*	25	−35 *(42)*	188 *(42)*	5	1.5561 *(42)*	25	1.066 *(42)*	25
$(CH_3)(C_6H_5)_2SiO[(CH_3)(C_6H_5)SiO]Si(CH_3)(C_6H_5)_2$	165[b] *(42)*	25	−15 *(42)*	225 *(42)*	5	1.5785 *(42)*	25	1.0950 *(42)*	25
$(C_6H_5)_3SiOSi(CH_3)_3$	—	—	51 *(30)*	349 *(30)*	750 ± 10	1.5587 *(30)* supercooled	25	1.032 *(30)* supercooled	25
$(CH_3)(C_6H_5)_2SiOSi(CH_3)(C_6H_5)_2$	—	—	50 *(30)*	418 *(30)*	750 ± 10	1.5866 *(30)* supercooled	25	1.076 *(30)* supercooled	25
$(C_6H_5)_3SiOSi(CH_3)_2(C_6H_5)$	—	—	50 *(30)*	415 *(30)*	750 ± 10	1.5843 *(30)* supercooled	25	1.076 *(30)* supercooled	25
$(C_6H_5)_3SiOSi(CH_3)(C_6H_5)_2$	—	—	76 *(30)*	466 *(30)*	750 ± 10	1.6124 *(30)* supercooled	25	1.112 *(30)* supercooled	25
$(C_2H_5)(C_6H_5)_2SiOSi(C_2H_5)(C_6H_5)_2$	—	—	65.5 *(103)*	—	—	—	—	—	—
$(CH_3)_3SiO[CH_3(C_6H_5)SiO]Si(CH_3)_3$	—	—	—	95 *(97)*	5	1.4468 *(97)*	20	0.9080 *(97)*	20
$(CH_3)_3SiO[CH_3(C_6H_5)SiO]_3Si(CH_3)_3$	13[b] *(77)*	25	—	190 *(77)*	0.2	1.4932 *(77)*	25	1.0115 *(77)*	25
$(CH_3)_3SiO[(C_6H_5)_2SiO]Si(CH_3)_3$	7.120 *(72)*	25	−78 *(163)*	110 *(193)*	0.27	1.4981 *(193)*	20	0.984 *(193)*	20
$(CH_3)_3SiO[(C_6H_5)_2SiO]_2Si(CH_3)_3$	32 *(163)*	50	49 *(163)*	208 *(163)*	1	—	—	—	—
$(CH_3)_3SiO[(CH_3)_2SiO]Si(CH_3)_2(C_6H_5)$	2.0[b] *(30)*	25	< −80 *(30)*	233 *(30)*	750 ± 10	1.4450 *(30)*	25	0.911 *(30)*	25
$(CH_3)_2(C_6H_5)SiO[(CH_3)_2SiO]Si(CH_3)_2(C_6H_5)$	3.9[b] *(30)*	25	−75 *(30)*	307 *(30)*	750 ± 10	1.4913 *(30)*	25	0.975 *(30)*	25
$(CH_3)_2(C_6H_5)SiO[(CH_3)(C_6H_5)SiO]Si(CH_3)_2(C_6H_5)$	—	—	—	169 *(97)*	0.7	1.5280 *(97)*	20	1.0227 *(97)*	20
$(CH_3)_2(C_6H_5)SiO[(C_6H_5)_2SiO]Si(CH_3)_2(C_6H_5)$	32[b] *(193)*	25	—	257 *(77)*	10	1.5543 *(77)*	25	1.057 *(77)*	25

[a] Numbers in parentheses indicate reference numbers.
[b] Here, cSt instead of cP.

and from the data for definite low-molecular-weight siloxanes on the other hand (see Table 33), that the intrinsic viscosity measured as an "initial viscosity" depends on the molecular weight M in somewhat different ways above and below $M = 40{,}000$. For $M < 40{,}000$ the empirical relationship

$$\log \eta^{P/40^\circ C} = 1.43 \log M - 5.54$$

(where P stands for poise) is valid; for $M > 40{,}000$ at the same temperature

$$\log \eta^{P/40^\circ C} = 3.64 \log M - 15.44$$

The mean molecular weights calculated from Barry's expression and the number of units in the polymer chain arising from this are given in Table 34

TABLE 34

VISCOSITIES, CALCULATED WEIGHT-AVERAGE MOLECULAR WEIGHTS $\bar{M}$ AND AVERAGE NUMBER OF SILOXANE UNITS $\tilde{N}$ PER MOLECULE IN A SERIES OF LINEAR POLYDIMETHYLSILOXANES

ν_{20}[cSt]	$\bar{M}$	$\tilde{N}$
60	3600	50
140	8000	110
440	17000	230
680	21000	280
1440	30000	400
10000	60000	800
50000	88000	1200
100000	103000	1400
300000	143000	1900

for a series of silicone oils consisting of linear polydimethylsiloxane molecules* having terminal trimethylsiloxy groups. Calculations according to Warrick *et al.* lead to very similar values. In spite of this, care must be taken not to place too great a reliance on accuracy of the data.

Investigations by Weissler (*198*) have shown that the molecular weight can also be evaluated from the velocity of sound. The velocity of sound, like the molecular refraction, is an additive property. The following empirical equation gives the relationship between molecular weight M, density d, velocity of sound v, and refractive index n:

$$M = \frac{b \cdot d}{v^{1/3} - a[(n^2-1)/(n^2+2)]}$$

* Compare Table 78.

In this expression *a* and *b* are constants which have the values $a=40.885$ and $b=48.10$ for a series of linear polysiloxanes. Up to molecular weights of 15,000 the figures evaluated in this way are in good agreement with those obtained from viscosity measurements.

Classical methods such as cryoscopy can, of course, also be used for molecular-weight determinations in the region of low-molecular-weight siloxanes. On the other hand, attempts have been made to determine the molecular dimensions in highly polymeric rubber-elastic vulcanizates by electron microscopy (*143*, *146*). In view of the many possibilities of error inherent in such a procedure, it would be advisable to view the results with some reserve.

We now turn to a discussion of the *variety of possible structures* which linear polysiloxanes can assume. The backbone of the chain structure is determined by the Si—O and Si—C interatomic distances and the valence angles of the Si—O—Si, O—Si—O, and C—Si—C bonds which will be considered in detail when the siloxane and the Si—C bonds are discussed later. We need only say here that the larger volume of silicon as compared with that of carbon is responsible for the relatively large distance of the organic groups from the Si—O—Si axis. If it is further considered that rotation about the siloxane bond is possible because of the steric situation, there arise from this and from the valence angles on the Si and O atoms certain important characteristics of the behavior of siloxane chains. These can be demonstrated by means of Stuart models (in which the atoms are represented by segments of spheres) that include models of Si and O atoms with dimensions adapted for siloxane chemistry (*118*).

The smallest molecule in the series, hexamethyldisiloxane, is represented by the model in Fig. 9.

The number of conformations that the chain can have increases with increasing number of difunctional siloxane units.

The most general form of the linear molecule is a ball; it can coil up in this way owing to the relatively large valence angle of the siloxane oxygen (see Table 52) and the size of the silicon atoms, which results in the methyl groups being pushed so far from the Si—O—Si axis that free rotation around the siloxane bond becomes possible.

Viscosity measurements in solution (*46*) and NMR (nuclear magnetic resonance) investigations (*69*, *93*) demonstrate this great mobility of the polydimethylsiloxane chain. It has also been confirmed by light-scattering measurements in a large number of solvents at 2 to 90°C. These have shown that the flexibility is largely independent of solvent and temperature, and is limited only by the small steric hindrance due to the methyl groups. The "forbidden" angle for movement amounts to only about 20% of the total arc.

Linear polydimethylsiloxanes appear to assume the *randomly coiled conformation* only when dissolved in organic solvents.

Various observations favor the idea that the molecules of the siloxanes under discussion have a *helical* conformation in the absence of solvents or diluents over a wide range of molecular size (Fig. 13). When spread in monomolecular layers on polar surfaces, however, they adopt a characteristic "spread chain" conformation (Fig. 10) (*122*).

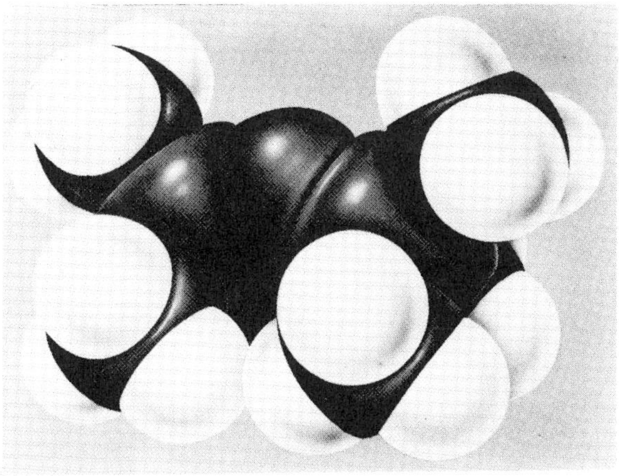

FIG. 9. Stuart model of hexamethyldisiloxane.

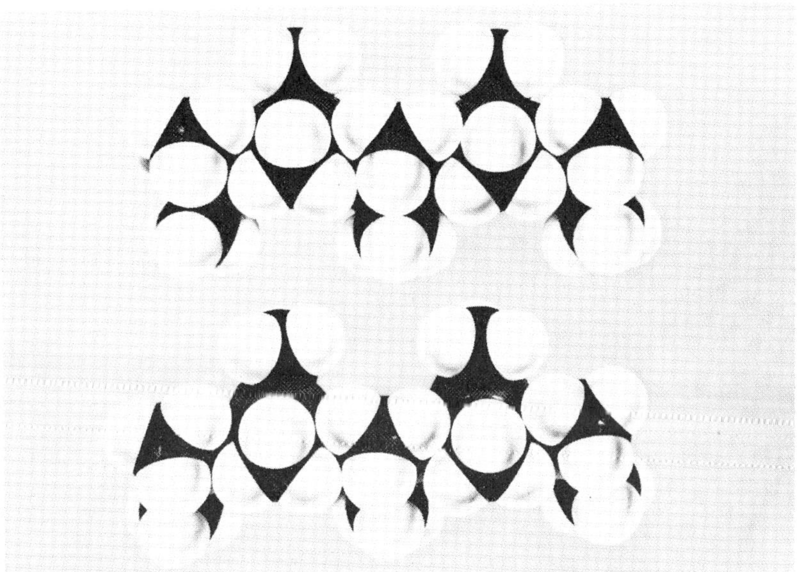

FIG. 10. Sections of polydimethylsiloxane chains. View from the side of the methyl groups.

Characteristic of this type of conformation is a linear structure (linear with respect to the total molecule—in detail, each siloxane chain has a zigzag form) such that the dimethyl groups and the siloxane bonds are oriented in opposite directions. This stretching is obviously only possible with organosiloxanes containing at least one methyl group per silicon atom; with larger and bulkier hydrocarbon radicals, this is limited or even prevented for steric reasons.

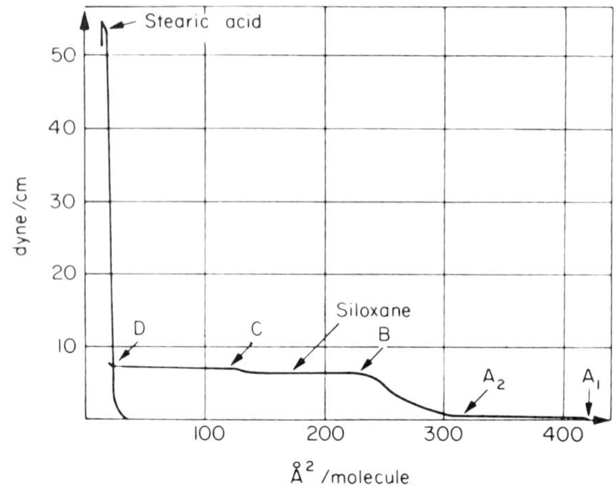

FIG. 11. Force/area diagram of a polymethylsiloxane $MD_{14}M$ and of stearic acid (*122*).

In the *helix* the siloxane bonds are oriented toward the screw axis, while the methyl groups point outward and shield the siloxane bond. The tendency of the chain to assume this form is explained by the fact that the siloxane bond includes an appreciable polar component and thus represents a weak dipole

$$Si^{\delta+}—O^{\delta-}$$

(see Section 6.3.3). Each siloxane dipole in the helix is faced by a compensating dipole of opposite polarity. The tendency of the chains to internal dipole compensation, which is undoubtedly associated with a gain in energy, shows that the helical form is always the stable molecular conformation when no external forces act on the siloxane chain (*119*).

Probably the first indication of the existence of helical siloxane chains (*47*) was derived from measurements of force/area (F/A) isotherms; these provide information on the surface pressure of polysiloxane molecules spread in a monolayer on water, as a function of the area requirements of the molecules.

Figure 11 shows the results of very recent investigations on a linear polydimethylsiloxane $MD_{14}M$ (*122*). Comparison with the F/A isotherms of stearic acid clearly shows that a different type of molecular orientation must take place in the case of polydimethylsiloxanes. In fact, calculation shows that over the entire range of areas, the siloxane chains lie with their axis parallel to

(a)

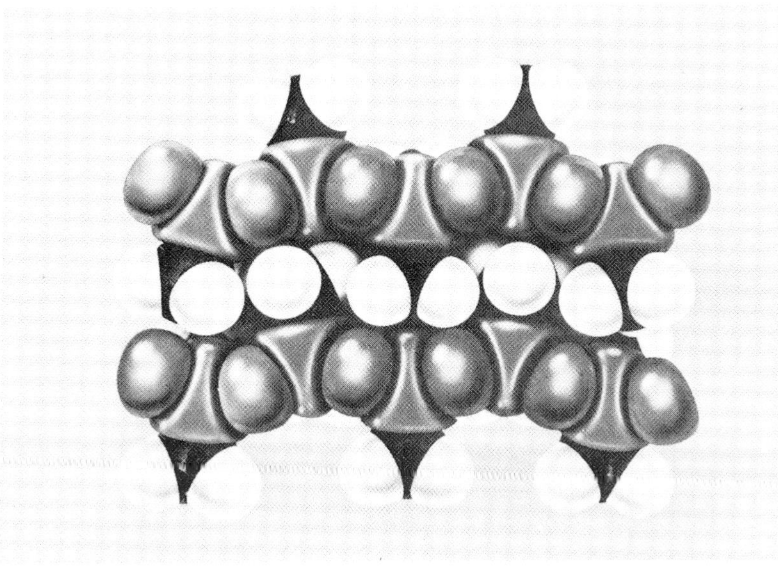

(b)

FIG. 12. Polydimethylsiloxane chains extremely compressed and intermeshed. (a) View from the side of the methyl groups, and (b) view from the side of the siloxane chain.

the water surface, while the carboxylic acid molecules are oriented at right angles to it, as is well known.

With an available surface area of about 400 Å^2 per molecule, the surface pressure is still immeasurably small. In this case there is a quasi-gaseous distribution of the molecules on the water surface, and the conformation and orientation of the siloxane chains would appear to correspond to points A_1 and A_2. Within this range, the siloxane chains exist in the stretched form (analogous to Fig. 10). The area requirements in A_1 and A_2 are very closely satisfied by models in which the chains are hydrated by water molecules between the chain members. Since the siloxane chains are always oriented in such a way that they lie with the partially polar Si—O—Si bonds on the water surface, this hydration could be considered the result of hydrogen bond formation between siloxane oxygen and the water molecules.

On compression of the film beyond A_2, the water molecules are squeezed out of the siloxane-water film, and the surface film is "dehydrated." The siloxane chain molecules—which can intermesh like a zip fastener by means of their methyl groups—are then forced together into a state of densest packing. Since a certain steric hindrance opposes this interlocking, this step is accompanied by a considerable increase of pressure—the only large pressure increase

FIG. 13. Stuart model of a polydimethylsiloxane chain with a helical structure.

of the F/A isotherms. The steric hindrance is demonstrated by the hysteresis taking place on expansion of the film in the region $B \rightarrow A_2$; this means that disengagement of the molecules tends to lag behind the increased surface area available. The processes in all other ranges of the curve are smoothly reversible. The surface at point *B* then consists of a closely packed film of intermeshing chains whose surface turned away from the water surface consists entirely of CH_3 groups (Fig. 12).

Since denser packing of the chains no longer allows the previous characteristic feature (contact of the Si—O—Si bonds with the water surface), further increase in pressure necessarily leads to a new structure with smaller area requirements—a helix. With increasing compression beyond point *B*, chain after chain is raised off the water surface and deposited again in helical form. This process is obviously favored by the mutual compensation of siloxane dipoles in the helix. The gain in energy compensates the energy required for compression to such an extent that only a small increase in the surface pressure is measurable. The turning point *C* of the curve represents a state of the film where all the chains are helical (Fig. 13) and lie densely packed on the surface of the water with their screw axes parallel to it.

With greater compression the system is no longer that of a monomolecular film. Portions of the film are pushed over one another in monomolecular blocks, and it collapses.

Evaluation of the spreading measurements also enables information on the periodicity of the siloxane helixes to be obtained. The area requirements at point *C* (125 Å^2) correspond most closely to a helix in which 6 siloxane units make up one complete turn, as indicated by the accompanying tabulation.

Siloxane units per turn of the helix	Surface required per molecule (Å^2)
3	160
6	125
8	100

The force/area isotherms however show no discontinuity of any kind when siloxanes are spread on nonpolar liquids. Compression of the film proceeds analogously to the compression of a two-dimensional gas. This means that no molecular reorientation, such as that of the stretched chains into helixes, takes place in this case. The molecules appear to lie on the surface of the nonpolar substrate in helical form throughout the process.

The helical structure however is apparently destroyed when a linear polydimethylsiloxane is dissolved. This is indicated by the fact that the entropy of dilution of solutions of the siloxanes in benzene is appreciably greater than could be expected for mixtures of liquids with unordered molecules, and also by the fact that the heat of mixing shows only a slight dependence on the chain length (*113*).

The *physicochemical behavior* of the linear polyorganosiloxanes is to a very large extent determined by the nature of the organic substituents. Constants for selected polyorganosiloxanes with aliphatic and aromatic groups are shown in Table 33. As might be expected from their considerable technological importance, methyl- and phenyl-substituted polymers, as well as those simultaneously containing methyl and phenyl groups, have been investigated with particular thoroughness. Linear polyorganosiloxanes with lower aliphatic organic groups, up to the highest obtainable molecular weights, are highly viscous liquids at room temperature. On the other hand, aromatically substituted siloxanes are solid at room temperature; even the shortest member of the series with exclusively aromatic substituents, hexaphenyldisiloxane, has a melting point of 226°C and a boiling point of almost 500°C. Linear polysiloxanes with mixed aromatic and aliphatic substituents occupy a position intermediate between the two extremes, according to the number and nature of their substituents.

The nature of the linear polymers can be modified not only by variation of the organic substituents but also by the insertion of silicon-functional and/or organofunctional atoms into the chain. Such substituents can be attached to the chain-forming D units or the terminal M units of the siloxane.

An idea of the variety of the polymers characterized by this type of substitution, and which have already been investigated in detail, is given in Tables 35 and 36. They contain only low-molecular-weight, definite oligomers and even these only insofar as they can be characterized by physical constants. Selected as the silicon-functional atoms and groups have been H, OR, Cl, and OH; the organofunctional substituents are unsaturated radicals and halogen-, hydroxy-, amino-, carboxyl-, cyano-, and epoxy-substituted organic groups.

Functional groups substituted in the D units or the terminal M units may differ in reactivity from those substituted in monomers. It must be remembered here, as in the case of reactivity of silanol groups in polysiloxanes (Section 5.4.1), that steric factors, molecular coiling, or inductive effects, possibly through neighboring Si—O bonds, affect the behavior of the functional group. Thus, the CH_2Cl group reacts more readily with KOH or C_2H_5ONa when it is adjacent to the siloxane oxygen in the disiloxane $ClCH_2(CH_3)_2Si—O—Si(CH_3)_3$ than in the monomeric silane $ClCH_2(CH_3)_3Si$.

TABLE 35

LOW-MOLECULAR-WEIGHT LINEAR POLYORGANOSILOXANES WITH SILICON-FUNCTIONAL GROUPS[a]

Compound	η (cP) at	t (°C)	ν (cSt) at	t (°C)	Mp (°C)	Bp (°C) at	p (mm Hg)	n_D^t at	t (°C)	d_4^t (gm/cc) at	t (°C)
(a) With Si—Cl bonds											
$Cl(CH_3)_2SiOSi(CH_3)_2Cl$	—	—	—	—	−37.5 (*129*)	138 (*129*)	760	—	—	1.038 (*129*)	20
$Cl[(CH_3)_2SiO]_2Si(CH_3)_2Cl$	—	—	—	—	−53 (*129*)	184 (*129*)	760	—	—	1.018 (*129*)	20
$Cl[(CH_3)_2SiO]_3Si(CH_3)_2Cl$	—	—	—	—	−62 (*129*)	222 (*129*)	760	—	—	1.011 (*129*)	20
$Cl[(CH_3)_2SiO]_4Si(CH_3)_2Cl$	—	—	—	—	ca. −80 (*129*)	138 (*129*)	20	—	—	1.005 (*129*)	20
$Cl[(CH_3)_2SiO]_5Si(CH_3)_2Cl$	—	—	—	—	ca. −80 (*129*)	161 (*129*)	20	—	—	1.003 (*129*)	20
$Cl[(CH_3)C_6H_5SiO]Si(CH_3)(C_6H_5)Cl$	—	—	—	—	—	154 (*30*)	4	1.5318 (*30*)	25	1.153 (*30*)	25
$Cl[(CH_3)C_6H_5SiO]_2Si(CH_3)(C_6H_5)Cl$	—	—	—	—	—	188–191 (*30*)	1	1.5375 (*30*)	25	1.153 (*30*)	25
$Cl[(CH_3)C_6H_5SiO]_3Si(CH_3)(C_6H_5)Cl$	—	—	—	—	—	231–233 (*30*)	1	1.5411 (*30*)	25	1.155 (*30*)	25
$Cl[(C_6H_5)_2SiO]_2Si(C_6H_5)_2Cl$	—	—	—	—	—	290–303 (*20*)	1	—	—	—	—
(b) With Si—OR bonds											
$CH_3O[CH_3(CH_3O)SiO]_2CH_3$	—	—	—	—	—	32 (*186*)	1	1.3812 (*186*)	25	1.0179 (*186*)	25
$CH_3O[CH_3(CH_3O)SiO]_3CH_3$	—	—	—	—	—	82 (*186*)	2	1.3867 (*186*)	25	1.0488 (*186*)	25
$CH_3O[CH_3(CH_3O)SiO]_4CH_3$	—	—	—	—	—	97 (*186*)	1	1.3910 (*186*)	25	1.0688 (*186*)	25
$C_2H_5O[CH_3(C_2H_5O)SiO]_2C_2H_5$	—	—	1.24 (*44*)	25	—	100 (*44*)	20	1.3895 (*44*)	25	0.9441 (*44*)	25
$C_2H_5O[CH_3(C_2H_5O)SiO]_3C_2H_5$	—	—	1.92 (*44*)	25	—	73 (*44*)	0.5	1.3934 (*44*)	25	0.9744 (*44*)	25
$C_2H_5O[CH_3(C_2H_5O)SiO]_4C_2H_5$	—	—	2.77 (*44*)	25	—	95 (*44*)	0.5	1.3955 (*44*)	25	0.9935 (*44*)	25
$C_2H_5O[CH_3(C_2H_5O)SiO]_5C_2H_5$	—	—	3.48 (*44*)	25	—	120 (*44*)	0.5	1.3974 (*44*)	25	1.0060 (*44*)	25
$CH_3O[(CH_3)_2SiO]_2CH_3$	—	—	0.79 (*95*)	25	—	138 (*95*)	760	1.3836 (*95*)	20	0.9049 (*95*)	20
$CH_3O[(CH_3)_2SiO]_3CH_3$	—	—	1.33 (*95*)	25	—	180 (*95*)	760	1.3890 (*95*)	20	0.9230 (*95*)	20
$CH_3O[(CH_3)_2SiO]_4CH_3$	—	—	1.87 (*95*)	25	—	216 (*95*)	760	1.3923 (*95*)	20	0.9317 (*95*)	20
$CH_3O[(CH_3)_2SiO]_5CH_3$	—	—	2.51 (*95*)	25	—	247 (*95*)	760	1.3949 (*95*)	20	0.9413 (*95*)	20
$CH_3O[(CH_3)_2SiO]_6CH_3$	—	—	3.11 (*95*)	25	—	273 (*95*)	760	1.3963 (*95*)	20	0.9470 (*95*)	20
$CH_3O[(CH_3)_2SiO]_7CH_3$	—	—	3.60 (*95*)	25	—	151 (*95*)	1	1.3975 (*95*)	20	0.9499 (*95*)	20
$CH_3O[(CH_3)_2SiO]_8CH_3$	—	—	4.53 (*95*)	25	—	165 (*95*)	1	1.3989 (*95*)	20	0.9551 (*95*)	20
$CH_3O[(CH_3)_2SiO]_9CH_3$	—	—	5.19 (*95*)	25	—	178 (*95*)	1	1.3998 (*95*)	20	0.9579 (*95*)	20
$CH_3O[(CH_3)_2SiO]_{10}CH_3$	—	—	6.31 (*95*)	25	—	193 (*95*)	1	1.4008 (*95*)	20	0.9603 (*95*)	20
$C_2H_5O[(CH_3)_2SiO]_2C_2H_5$	—	—	0.97 (*43*)	25	−134 (*43*)	161 (*43*)	760	1.3880 (*43*)	20	0.8788 (*43*)	20
$C_2H_5O[(CH_3)_2SiO]_3C_2H_5$	—	—	1.35 (*43*)	25	−126 (*43*)	196 (*43*)	760	1.3922 (*43*)	20	0.9024 (*43*)	20
$C_2H_5O[(CH_3)_2SiO]_4C_2H_5$	—	—	1.78 (*43*)	25	−124 (*43*)	227 (*43*)	760	1.3950 (*43*)	20	0.9157 (*43*)	20
$C_2H_5O[(CH_3)_2SiO]_5C_2H_5$	—	—	2.24 (*43*)	25	−123 (*43*)	253 (*43*)	760	1.3964 (*43*)	20	0.9207 (*43*)	20

(*continued on p. 262*)

TABLE 35 *(Contd.)*

Low-Molecular-Weight Linear Polyorganosiloxanes with Silicon-Functional Groups[a]

Compound	η (cP) at	t (°C)	ν (cSt) at	t (°C)	Mp (°C)	Bp (°C) at	p (mm Hg)	n_D^t at	t (°C)	d_4^t (gm/cc) at	t (°C)
$C_2H_5O[(CH_3)_2SiO]_6C_2H_5$	—	—	2.75 (*43*)	25	−118 (*43*)	274 (*43*)	760	1.3980 (*43*)	20	0.9317 (*43*)	20
$C_2H_5O[(CH_3)_2SiO]_7C_2H_5$	—	—	3.28 (*43*)	25	−112 (*43*)	292 (*43*)	760	1.3990 (*43*)	20	0.9364 (*43*)	20
$C_2H_5O[(CH_3)_2SiO]_8C_2H_5$	—	—	3.86 (*43*)	25	−110 (*43*)	311 (*43*)	760	1.3996 (*43*)	20	0.9409 (*43*)	20
$C_2H_5O[(CH_3)_2SiO]_9C_2H_5$	—	—	4.50 (*43*)	25	−53 (*43*)	—	—	1.4002 (*43*)	20	0.9442 (*43*)	20
$C_2H_5O[(CH_3)_2SiO]_{10}C_2H_5$	—	—	5.17 (*43*)	25	−100 (*43*)	—	—	1.4009 (*43*)	20	0.9471 (*43*)	20
(I)	—	—	—	—	—	187–192 (*163*)	13	1.4910 (*163*)	20	1.025 (*163*)	20
(II)	—	—	—	—	—	210–215 (*163*)	[b]	1.4960 (*163*)	20	—	—
$(C_2H_5O)_2(C_6H_5)SiOSi(C_6H_5)(OC_2H_5)_2$	—	—	—	—	—	200–210 (*163*)	1	1.485 (*163*)	20	1.057 (*163*)	20
(III)	—	—	—	—	—	240–250 (*163*)	1	1,510 (*163*)	20	1.112 (*163*)	20
(IV)	—	—	—	—	—	280–290 (*163*)	1	1.521 (*163*)	20	1.155 (*163*)	20
i-$C_3H_7O[(CH_3)_2SiO]_2i$-C_3H_7	—	—	—	—	—	171 (*125*)	760	1.3914 (*125*)	20	0.8661 (*125*)	20
i-$C_3G_7O[(CH_3)_2SiO]_3i$-C_3H_7	—	—	—	—	—	100.5 (*125*)	15	1.3942 (*125*)	20	0.8872 (*125*)	20
i-$C_3H_7O[(CH_3)_2SiO]_4i$-C_3H_7	—	—	—	—	—	123 (*125*)	14	1.3965 (*125*)	20	0.9095 (*125*)	20
i-$C_3H_7O[(CH_3)_2SiO]_5i$-C_3H_7	—	—	—	—	—	89 (*125*)	0.5	1.3982 (*125*)	20	0.9133 (*125*)	20
$C_4H_9O[(CH_3)_2SiO]_3C_4H_9$	—	—	—	—	—	124–125 (*152*)	10	1.4053 (*152*)	20	0.8932 (*152*)	20
$C_6H_5O[(CH_3)_2SiO]_2C_6H_5$	4.4 (*163*)	20	—	—	—	150 (*163*)	4	1.4970 (*163*)	20	—	—
$C_6H_5O[(CH_3)_2SiO]_3C_6H_5$	5.2 (*163*)	20	—	—	—	185 (*163*)	0.5	1.4851 (*163*)	20	—	—
$C_6H_5O[(CH_3)_2SiO]_4C_6H_5$	6.0 (*163*)	20	—	—	—	150 (*163*)	0.1	1.4697 (*163*)	20	—	—
$C_6H_5O[(CH_3)_2SiO]_5C_6H_5$	7.0 (*163*)	20	—	—	—	210 (*163*)	0.5	1.4613 (*163*)	20	—	—
$C_6H_5O[(CH_3)_2SiO]_6C_6H_5$	8.3 (*163*)	20	—	—	—	180 (*163*)	0.2	1.4573 (*163*)	20	—	—
$C_6H_5O[(CH_3)_2SiO]_7C_6H_5$	9.8 (*163*)	20	—	—	—	265 (*163*)	0.5	1.4525 (*163*)	20	—	—
$C_6H_5O[(CH_3)_2SiO]_8C_6H_5$	11.8 (*163*)	20	—	—	—	280 (*163*)	0.5	1.4473 (*163*)	20	—	—
(V)	—	—	—	—	—	270–280 (*163*)	[b]	1.5720 (*163*)	20	—	—
(c) With Si—H bonds											
$H(CH_3)_2SiOSi(CH_3)_3$	—	—	—	—	—	85.7 (*110*)	760	1.3789 (*110*)	20	0.758 (*110*)	20
$H(CH_3)_2SiOSi(CH_3)_2H$	—	—	—	—	—	71.8 (*14*)	760	1.3694 (*14*)	20	0.7564 (*14*)	20
$H_2(CH_3)SiOSi(CH_3)H_2$	—	—	—	—	—	34.5 (*110*)	760	1.3590 (*110*)	20	0.7487 (*110*)	20
$H(CH_3)_2SiO[CH_3(H)SiO]Si(CH_3)_2H$	—	—	—	—	—	119 (*110*)	760	1.3798 (*110*)	20	0.825 (*110*)	20
$H(CH_3)_2SiO[CH_3(H)SiO]_2Si(CH_3)_2H$	—	—	—	—	—	157 (*110*)	760	1.3831 (*110*)	20	0.867 (*110*)	20
$H(CH_3)_2SiO[CH_3(H)SiO]_3Si(CH_3)_2H$	—	—	—	—	—	188 (*110*)	760	1.3859 (*110*)	20	0.884 (*110*)	20
$H_2(CH_3)SiO[CH_3(H)SiO]Si(CH_3)H_2$	—	—	—	—	—	*ca.* 90 (*110*)	760	1.3761 (*110*)	20	0.8376 (*110*)	20

$H(CH_3)_2SiO[(CH_3)(C_6H_5)SiO]Si(CH_3)_2H$	—	—	—	—	—	88 (*24*)	6	1.4471 (*24*)	25	0.923 (*24*)	25
$H(CH_3)_2SiO[(C_6H_5)_2SiO]Si(CH_3)_2H$	—	—	—	—	—	137 (*24*)	2.8	1.5000 (*24*)	25	—	—
$(CH_3)_3SiO[CH_3(H)SiO]Si(CH_3)_3$	—	—	—	—	—	142 (*17*)	760	1.3818 (*153*)	20	0.8194 (*153*)	20
$(CH_3)_3SiO[CH_3(H)SiO]_2Si(CH_3)_3$	—	—	—	—	—	177 (*17*)	760	1.3854 (*153*)	20	0.8559 (*153*)	20
$(CH_3)_3SiO[CH_3(H)SiO]_3Si(CH_3)_3$	—	—	—	—	—	85.6 (*153*)	10	1.3878 (*153*)	20	0.8806 (*153*)	20
$(CH_3)_3SiO[CH_3(H)SiO]_4Si(CH_3)_3$	—	—	—	—	—	106.5–107.6 (*153*)	10	1.3895 (*153*)	20	0.8991 (*153*)	20
$(CH_3)_3SiO[CH_3(H)SiO]_5Si(CH_3)_3$	—	—	—	—	—	12 (*163*)	1	1.3906 (*163*)	20	—	—
(d) With Si—OH bonds											
$HO[(CH_3)_2SiO]_2H$	—	—	—	—	67–68 (*102*)	—	—	n_1 1.459 ± 0.002 (*75*)	25	1.095 ± 0.008 (*75*)	—
								n_2 1.464 ± 0.002 (*75*)	25	—	—
$HO[(CH_3)_2SiO]_3H$	—	—	—	—	—	79–82 (*3*)	2	1.4089 (*3*)	20	0.9999 (*3*)	20
$HO[(CH_3)_2SiO]_4H$	—	—	—	—	—	97–100 (*3*)	2	1.4054 (*3*)	20	0.9886 (*3*)	20
$HO[(CH_3)_2SiO]_5H$	—	—	—	—	—	104–106 (*3*)	1.5	1.4089 (*3*)	20	0.9914 (*3*)	20
$HO[(CH_3)_2SiO]_6H$	—	—	—	—	—	119–120 (*3*)	2	0.4099 (*3*)	20	0.9916 (*3*)	20
$HO[(CH_3)_2SiO]_7H$	—	—	—	—	—	130–135 (*3*)	2	1.4067 (*3*)	20	0.9891 (*3*)	20
$HO[(CH_3)_2SiO]_8H$	—	—	—	—	—	143–145 (*3*)	2	1.4090 (*3*)	20	0.9912 (*3*)	20
$HO[(CH_3)_2SiO]_9H$	—	—	—	—	—	158–161 (*3*)	2	1.4088 (*3*)	20	0.9921 (*3*)	20
$HO[(CH_3)C_6H_5SiO]_2H$	—	—	—	—	α = 111 (*30*)	—	—	—	—	—	—
					β = 84 (*30*)	—	—	—	—		—
$HO[(C_2H_5)C_6H_5SiO]_2H$	—	—	—	—	87.5 (*140*)	—	—	—	—	—	—
$HO[(C_6H_5)_2SiO]_2H$	—	—	—	—	113–114 (*20,81*)	—	—	—	—	—	—
$HO[(C_6H_5)_2SiO]_3H$	—	—	—	—	111–112 (*20,81*)	—	—	—	—	—	—
$HO[(C_6H_5)_2SiO]_4H$	—	—	—	—	128.5 (*82*)	—	—	—	—	—	—

(I) = $C_2H_5O(CH_3)_2SiO[(C_6H_5)_2SiO(CH_3)_2SiO]C_2H_5$
(II) = $C_2H_5O(CH_3)_2SiO[(C_6H_5)_2SiO(CH_3)_2SiO]_2C_2H_5$
(III) = $(C_2H_5O)_2(C_6H_5)SiO[(C_2H_5O)(C_6H_5)SiO]Si(C_6H_5)(OC_2H_5)_2$
(IV) = $(C_2H_5O)_2(C_6H_5)SiO[(C_2H_5O)(C_6H_5)SiO]_2Si(C_6H_5)(OC_2H_5)_2$
(V) = $C_6H_5O(CH_3)_2SiO(C_6H_5)_2SiO(CH_3)_2SiOC_6H_5$

[a] Numbers in parentheses indicate reference numbers.
[b] High vacuum.

TABLE 36

LOW-MOLECULAR-WEIGHT LINEAR POLYORGANOSILOXANES WITH ORGANOFUNCTIONAL GROUPS[a]

Compound	η (cP)	at t (°C)	Mp (°C)	Bp (°C)	at p (mm Hg)	n_D^t	at t (°C)	d_4^t (gm/cc)	at t (°C)
(a) With alkenyl groups									
$(CH_2{=}CH)(CH_3)_2SiOSi(CH_3)_2(CH{=}CH_2)$	—	—	−99.7 ± 0.1 (*80*)	139 (*80*)	760	1.4123 (*80*)	20	0.811 (*80*)	20
$(CH_2{=}CH)(CH_3)(HO)SiOSi(OH)(CH_3)(CH{=}CH_2)$	—	—	—	95–120 (*53*)	2.5	—	—	—	—
$(CH_2{=}CH)(C_6H_5)(HO)SiOSi(OH)(C_6H_5)(CH{=}CH_2)$	—	—	107 (*53*)	—	—	—	—	—	—
$(CH_2{=}CH)(CH_3O)_2SiOSi(OCH_3)_2(CH{=}CH_2)$	—	—	—	120.0–120.6 (*111*)	46	1.4172 (*111*)	25	1.0627 (*111*)	25
$(CH_2{=}CH)(C_2H_5O)_2SiOSi(OC_2H_5)_2(CH{=}CH_2)$	—	—	—	119 (*111*)	19	1.4100 (*111*)	25	0.9618 (*111*)	25
$(CH_2{=}CH{-}CH_2)(CH_3)_2SiOSi(CH_3)_3$	—	—	—	141.5–142 (*21*)	760	1.4061 (*21*)	20	0.7948 (*21*)	20
$(CH_2{=}CH{-}CH_2)(CH_3)_2SiOSi(CH_3)_2(CH_2{-}CH{=}CH_2)$	—	—	—	179–180 (*21*)	758.3	1.4280 (*21*)	20	0.8208 (*21*)	20
(I)	—	—	—	105–135 (*53*)	2.4	—	—	—	—
$(CH_2{=}CHCH_2)_2(CH_3)SiOSi(CH_3)(CH_2CH{=}CH_2)_2$	—	—	—	70–71 (*159*)	1	1.4591 (*159*)	20	0.8723 (*159*)	22
$(CH_3)_3SiOSi(CH_3)_2(CH_2{-}OOC{-}C(CH_3){=}CH_2)$	—	—	—	86.5 (*105*)	10	1.4202 (*105*)	25	0.903 (*105*)	25
(II)	—	—	—	127 (*104*)	3	1.4472 (*104*)	25	0.996 (*104*)	25
$(CH_3)_3SiO[(CH_2{=}CH)(CH_3)SiO]Si(CH_3)_3$	—	—	—	78 (*162*)	35	1.3962 (*162*)	20	—	—
$(CH_3)_3SiO[(CH_2{=}CH)(CH_3)SiO]_2Si(CH_3)_3$	—	—	—	72 (*162*)	2	1.4146 (*162*)	20	—	—
$(CH_3)_3SiO[(CH_2{=}CH)(CH_3)SiO]_3Si(CH_3)_3$	—	—	—	125–129 (*162*)	11	1.4163 (*162*)	20	—	—
$(CH_3)_3SiO[(CH_2{=}CH)(CH_3)SiO]_4Si(CH_3)_3$	—	—	—	142–147 (*162*)	10	1.4198 (*162*)	20	—	—
$(CH_3)_3SiO[(CH_2{=}CH)(CH_3)SiO]_5Si(CH_3)_3$	—	—	—	167–170 (*162*)	2	1.4228 (*162*)	20	—	—
(b) With haloorganic groups									
$(ClCH_2)(CH_3)_2SiOSi(CH_3)_3$	—	—	—	151.6–151.8 (*86*)	760	1.4106 (*86*)	20	0.9105 (*86*)	20
$(ClCH_2)(CH_3)_2SiOSi(CH_3)_2(CH_2Cl)$	—	—	—	204.5 (*86*)	760	1.4390 (*86*)	20	1.045 (*86*)	20
$(ClCH_2)(CH_3)_2SiO[(CH_3)_2SiO]Si(CH_3)_2(CH_2Cl)$	4.55 (*179*)	0	−79 (*179*)	142 (*179*)	40	1.4283 (*179*)	25	1.020 (*179*)	25
$(ClCH_2)(CH_3)_2SiO[(CH_3)_2SiO]_2Si(CH_3)_2(CH_2Cl)$	5.61 (*179*)	0	−94 (*179*)	168 (*179*)	40	1.4231 (*179*)	25	1.008 (*179*)	25
$(ClCH_2)(CH_3)_2SiO[(CH_3)_2SiO]_3Si(CH_3)_2(CH_2Cl)$	6.80 (*179*)	0	−85 (*179*)	190 (*179*)	40	1,4200 (*179*)	25	1.002 (*179*)	25
$(ClCH_2)(CH_3)_2SiO[(CH_3)_2SiO]_4Si(CH_3)_2(CH_2Cl)$	—	—	−77 (*179*)	223 (*179*)	40	1.4173 (*179*)	25	0.996 (*179*)	25
$(CH_3)_3SiO[(ClCH_2)CH_3SiO]Si(CH_3)_3$	2.46 (*179*)	0	−85 (*179*)	102 (*179*)	40	1.4058 (*179*)	25	0.918 (*179*)	25
$(CH_3)_3SiO[(ClCH_2)CH_3SiO]_2Si(CH_3)_3$	6.02 (*179*)	0	−77 (*179*)	162 (*179*)	40	1.4212 (*179*)	25	1.006 (*179*)	25
$(CH_3)_3SiO[(ClCH_2)CH_3SiO]_3Si(CH_3)_3$	9.83 (*179*)	0	−85 (*179*)	210 (*179*)	40	1.4311 (*179*)	25	1.063 (*179*)	25
$(CH_3)_3SiO[(ClCH_2)CH_3SiO]_4Si(CH_3)_3$	21.17 (*179*)	0	−77 (*179*)	251 (*179*)	40	1.4375 (*179*)	25	1.100 (*179*)	25
$(ClCH_2)(CH_3)_2SiO[(ClCH_2)CH_3SiO]Si(CH_3)_2(CH_2Cl)$	2.75 (*179*)	55.8	—	176 (*179*)	40	1.4465 (*179*)	25	1.112 (*179*)	25
$(ClCH_2)(CH_3)_2SiO[(ClCH_2)CH_3SiO]_2Si(CH_3)_2(CH_2Cl)$	4.97 (*179*)	55.8	—	167 (*179*)	5	1.4520 (*179*)	25	1.157 (*179*)	25
$(ClCH_2)(CH_3)_2SiO[(ClCH_2)CH_3SiO]_3Si(CH_3)_2(CH_2Cl)$	8.00 (*179*)	55.8	—	210 (*179*)	5	1.4553 (*179*)	25	1.182 (*179*)	25

$(Cl_2CH)(CH_3)_2SiOSi(CH_3)_2(CHCl_2)$	—	—	—	149–150 (*87*)	40	1.4660 (*87*)	20	1.2213 (*87*)	20
$(ClCH_2CH_2)(CH_3O)_2SiOSi(OCH_3)_2(CH_2CH_2Cl)$	—	—	—	133 (*13*)	3	1.4333 (*13*)	25	1.189 (*13*)	25
$BrCH_2(CH_3)_2SiOSi(CH_3)_3$	—	—	—	83 (*180*)	49	1.4279 (*180*)	25	1.097 (*180*)	25
$BrCH_2(CH_3)_2SiOSi(CH_3)_2CH_2Br$	—	—	—	232–235 (*180*)	734	1.4719 (*180*)	25	1.3918 (*180*)	25
$(ClCH_2CH_2)(C_2H_5O)_2SiOSi(OC_2H_5)_2(CH_2CH_2Cl)$	—	—	—	134–135 (*13*)	1	1.4280 (*13*)	25	1.091 (*13*)	25
$(CH_3CHCl)(CH_3O)_2SiOSi(OCH_3)_2(CHClCH_3)$	—	—	—	105 (*13*)	2	1.4284 (*13*)	25	1.176 (*13*)	25
$(CH_3CHCl)(C_2H_5O)_2SiOSi(OC_2H_5)_2(CHClCH_3)$	—	—	—	132–133 (*13*)	5.5	1.4243 (*13*)	25	1.080 (*13*)	25
$(p\text{-}ClC_6H_4)(CH_3)_2SiOSi(CH_3)_2(p\text{-}C_6H_4Cl)$	—	—	—	153 (*201*)	4	1.5327 (*201*)	25	1.1175 (*201*)	26
$(p\text{-}ClC_6H_4)_2(CH_3)SiOSi(CH_3)(p\text{-}C_6H_4Cl)_2$	—	—	68–69 (*60*)	174–185 (*60*)	0.1	—	—	—	—
$(p\text{-}ClC_6H_4)_3SiOSi(p\text{-}C_6H_4Cl)_3$	—	—	210 (*62*)	—	—	—	—	—	—
$(p\text{-}FC_6H_4)(CH_3)_2SiOSi(CH_3)_2(p\text{-}C_6H_4F)$	—	—	—	96–98 (*201*)	1–2	1.4950 (*201*)	25	1.0653 (*201*)	23
$[m,m'\text{-}(CF_3)_2C_6H_3]_3SiOSi[m,m'\text{-}C_6H_3(CF_3)_2]_3$	—	—	—	378–382 (*142*)	750	—	—	—	—
(c) With hydroxyorganic groups and their esters									
$(HOCH_2)(CH_3)_2SiOSi(CH_3)_2(CH_2OH)$	32.6 (*181*)	21	−7.5–−5 (*181*)	—	—	1.4358 (*181*)	25	0.979 (*181*)	25
$(CH_3COOCH_2)(CH_3)_2SiO(CH_3)_2SiOSi(CH_3)_2CH_2OOCCH_3$	—	—	—	100 (*114*)	0.7	1.4193 (*114*)	20	—	—
$[HO(CH_2)_4](CH_3)_2SiOSi(CH_3)_2[(CH_2)_4OH]$	—	—	—	Not distillable (*116*)	—	1.4526 (*116*)	20	—	—
(d) With aminoorganic groups									
$(H_2NCH_2)(CH_3)_2SiOSi(CH_3)_2(CH_2NH_2)\cdot 2HCl$	—	—	247–249 (*124*)	—	—	—	—	—	—
$(H_2NCH_2CH_2)(CH_3)_2SiOSi(CH_3)_2(CH_2CH_2NH_2)$	—	—	—	115 (*176*)	13	1.4473 (*176*)	20	0.9075 (*176*)	20
Dihydrochloride	—	—	267–268 (*176*)	—	—	—	—	—	—
$[H_2N(CH_2)_4](CH_3)_2SiOSi(CH_3)_2[(CH_2)_4NH_2]$	—	—	—	120 (*115*)	1	1.4511 (*115*)	20	—	—
(e) With carboxyorganic groups and their esters									
$HOOCCH_2CH_2(CH_3)_2SiOSi(CH_3)_2CH_2CH_2COOH$	—	—	53–54 (*176*)	—	—	—	—	—	—
$HOOC(CH_2)_3(CH_3)_2SiOSi(CH_3)_2(CH_2)_3COOH$	—	—	49–49.5 (*176*)	—	—	—	—	—	—
$p\text{-}HOOCC_6H_4(CH_3)_2SiOSi(CH_3)_2p\text{-}C_6H_4COOH$	—	—	241–242 (*98*)	—	—	—	—	—	—
$[(p\text{-}HOOCC_6H_4)(C_6H_5)_2SiOSi(CH_3)_2]_2O$	—	—	150–152 (*178*)	—	—	—	—	—	—
$C_2H_5OOC(CH_2)_2(CH_3)_2SiOSi(CH_3)_2(CH_2)_2COOC_2H_5$	4.27 (*175*)	25	—	150 (*175*)	7	1.4340 (*175*)	20	0.9788 (*175*)	20
(III)	4.77 (*175*)	25	—	174 (*175*)	7	1.4288 (*175*)	20	0.9787 (*175*)	20
(IV)	4.94 (*175*)	25	—	186 (*175*)	7	1.4257 (*175*)	20	0.9797 (*175*)	20
(V)	6.21 (*175*)	25	—	190 (*175*)	4	1.4255 (*175*)	20	0.9807 (*175*)	20
$(C_2H_5OOC)(CN)CHCH_2(CH_3)_2SiOSi(CH_3)_2CH_2CH(CN)\text{-}COOC_2H_5$	—	—	—	190 (*165*)	0.5	1.4499 (*165*)	20	—	—
$(p\text{-}CH_3OOCC_6H_4)(CH_3)_2SiOSi(CH_3)_2(p\text{-}C_6H_5COOCH_3)$	—	—	57–61 (*178*)	—	—	—	—	—	—
$[(p\text{-}CH_3OOCC_6H_4)(C_6H_5)_2SiOSi(CH_3)_2O]_2$	—	—	134–136 (*178*)	—	—	—	—	—	—
(f) With cyanoorganic groups									
$(NCCH_2)(CH_3)_2SiOSi(CH_3)_3$	—	—	—	82–83 (*136*)	14	1.4118 (*136*)	20	0.875 (*136*)	20
$(p\text{-}NCC_6H_4)(CH_3)_2SiOSi(CH_3)_2(p\text{-}C_6H_4CN)$	—	—	—	190–200 (*98*)	1–2	1.5351 (*98*)	25	1.0460 (*98*)	25

(*continued on p.* 266)

TABLE 36 (*Contd.*)

LOW-MOLECULAR-WEIGHT LINEAR POLYORGANOSILOXANES WITH ORGANOFUNCTIONAL GROUPS[a]

Compound	η (cP)	at t (°C)	Mp (°C)	Bp (°C)	at p (mm Hg)	n_D^t at	t (°C)	d_4^t (gm/cc)	at t (°C)
(g) With epoxyorganic groups									
$\underbrace{CH_2—C}_{O}(CH_3)—CH_2(CH_3)_2Si—O—Si(CH_3)_3$	—	—	—	71–71.5 (*150*)	10	1.4160 (*150*)	20	0.8773 (*150*)	20
$\underbrace{CH_2—CH}_{O}—(CH_2)_2(CH_3)_2Si—O—Si(CH_3)_3$	—	—	—	78–79 (*134*)	8	1.4180 (*134*)	25	0.887 (*134*)	25
$\underbrace{CH_2—CH}_{O}—(CH_2)_2(CH_3)_2Si—O—Si(CH_3)_2(CH_2)_2—\underbrace{CH—CH_2}_{O}$	—	—	—	110–120 (*134*)	1.5	1.4429 (*134*)	25	0.958 (*134*)	25
$\underbrace{CH_2—CH}_{O}—CH_2O(CH_2)_3(CH_3)_2Si—O—Si(CH_3)_3$	—	—	—	122–123 (*134*)	11	1.4267 (*134*)	25	0.915 (*134*)	25
(VI)	—	—	—	—	—	1.5200 (*134*)	25	1.082 (*134*)	25

(I) = $(CH_2{=}CH—CH_2)(CH_3)(HO)SiOSi(OH)(CH_3)(CH_2—CH{=}CH_2)$
(II) = $CH_2{=}C(CH_3)COOCH_2(CH_3)_2SiOSi(CH_3)_2CH_2OOC(CH_3)C{=}CH_2$
(III) = $C_2H_5OOC(CH_2)_2(CH_3)_2SiO[(CH_3)_2SiO]Si(CH_3)_2(CH_2)_2COOC_2H_5$
(IV) = $C_2H_5OOC(CH_2)_2(CH_3)_2SiO[(CH_3)_2SiO]_2Si(CH_3)_2(CH_2)_2COOC_2H_5$
(V) = $C_2H_5OOC(CH_2)_2(CH_3)_2SiO[(CH_3)_2SiO]_3Si(CH_3)_2(CH_2)_2COOC_2H_5$
(VI) = $\underbrace{CH_2—CH}_{O}—CH_2O(CH_2)_3(CH_3)(C_6H_5)Si—O—Si(C_6H_5)(CH_3)(CH_2)_3OCH_2—\underbrace{CH—CH_2}_{O}$

[a] Numbers in parentheses indicate reference numbers.

6.2.2 Branched Polymers

Only a relatively small number of oligomeric organosiloxanes of definite structure is known where the effect of branching and its consequences on the physicochemical behavior pattern can be studied. Table 37 contains a review of methyl- and phenyl-substituted siloxanes of this group.

Branched polymers contain either Q or T units as the branching centers. Definite siloxanes with Q units known at the present time are those of the structural types M_4Q and M_6Q_2:

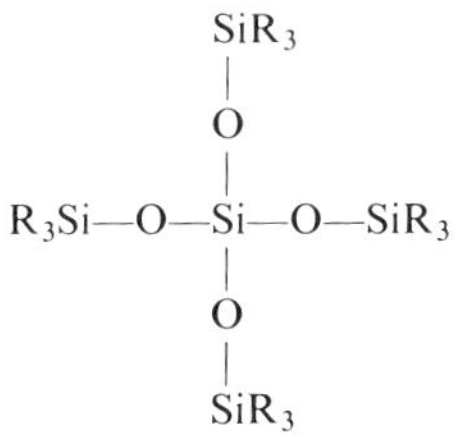

$[M_4Q]$

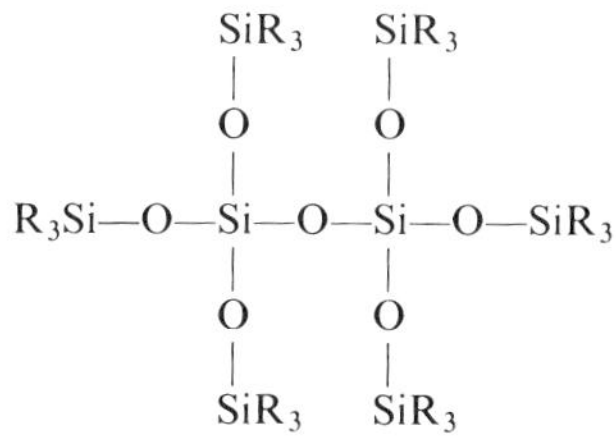

$[M_6Q_2]$

Low-molecular-weight siloxanes containing T and M units have the general composition $M_{n+2}T_n$:

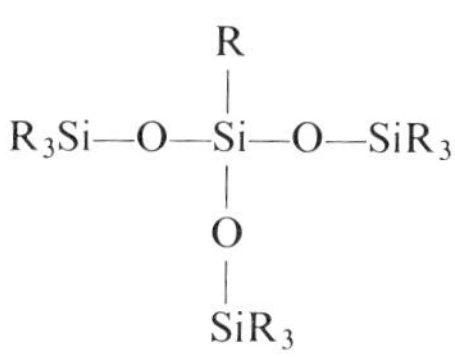

$[M_3T]$

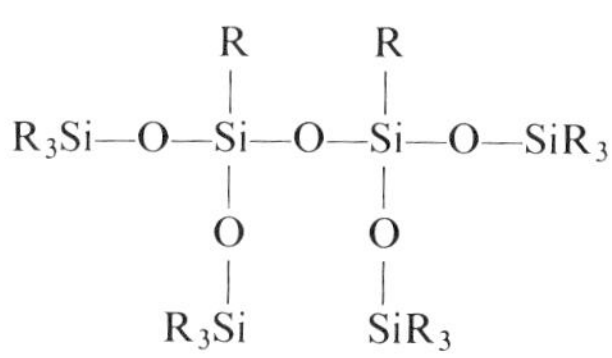

$[M_4T_2]$

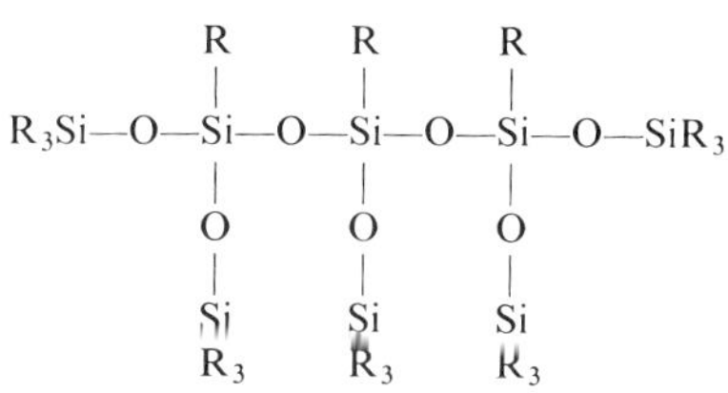

$[M_5T_3]$

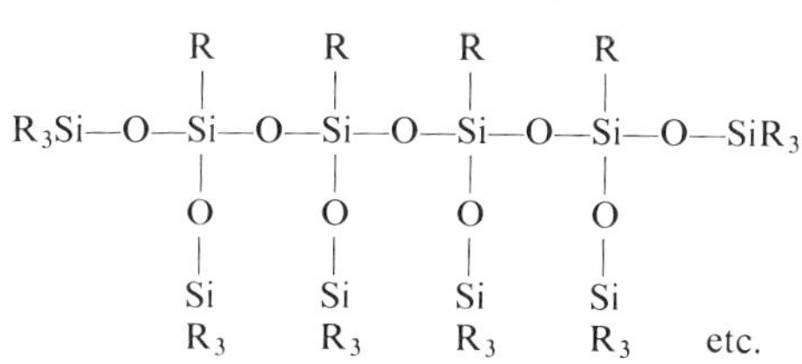

$[M_6T_4]$

etc.

TABLE 37

LOW-MOLECULAR-WEIGHT BRANCHED POLYORGANOSILOXANES (NONFUNCTIONAL)[a]

Type	Compound	η_{20} (cP)	Mp (°C)	Bp (°C)	at p (mm Hg)	n_D^{20}	d_4^{20}(gm/cc)
M_3T	$[(CH_3)_3SiO]_3[CH_3Si]$	1.57[b] (*205*)	−74 (*205*)	190 (*129*)	760	1.3851[d] (*205*)	0.853[e] (*205*)
M_4T_2	$[(CH_3)_3SiO]_4[(CH_3Si)_2O]$	—	—	256 (*109*)	~760	1.3947 (*109*)	0.893 (*109*)
M_4Q	$[(CH_3)_3SiO]_4Si$	2.868 (*174*)	ca. −60 (*174*)	220 (*174*)	733	1.3895 (*174*)	0.8677 (*174*)
M_6Q_2	$[(CH_3)_3SiO]_6[Si_2O]$	—	255 (*108*)	303 (*108*)	~760	—	—
M_3T	$[(CH_3)_3SiO]_3[C_6H_5Si]$	3.8(*160*)	−102 (*160*)	105 (*160*)	1	1.436 (*160*)	0.924 (*160*)
M_3T	$[(CH_3)_2C_6H_5SiO]_3[C_6H_5Si]$	19[c] (*77*)	—	217–225 (*77*)	1	1.528[d] (*77*)	1.060[e] (*77*)
M_4T_2	$[(CH_3)_3SiO]_4[(C_6H_5Si)_2O]$	24 (*160*)	−78 (*160*)	175 (*160*)	1	1.457 (*160*)	0.988 (*160*)
M_5T_3	$[(CH_3)_3SiO_5(C_6H_5Si)_3O_2]$	78 (*160*)	−62 (*160*)	220 (*160*)	1	1.469 (*160*)	1.012 (*160*)
M_6T_4	$[(CH_3)_3SiO]_6[(C_6H_5Si)_4O_3]$	234 (*160*)	—	255 (*160*)	1	1.481 (*160*)	1.035 (*160*)
M_7T_5	$[(CH_3)_3SiO]_7[(C_6H_5Si)_5O_4]$	710 (*163*)	—	270 (*163*)	High vacuum	1.489 (*163*)	1.051 (*163*)
M_4Q	$[(CH_3)_2C_6H_5SiO]_4Si$	—	<-72 (*34*) Solidification	ca. 225 (*34*)	5	1.5178 (*34*)	1.049[f] (*34*)

[a] Numbers in parentheses indicate reference numbers.
[b] η_{25}. [c] η_{25}. Here, cSt instead of cP. [d] n_D^{25}. [e] d_4^{25}. [f] $d_{15.6}^{20}$.

Such siloxanes are related to the linear polymers; they can be derived from them by substituting each of the D units by an [MT] combination which is also difunctional.

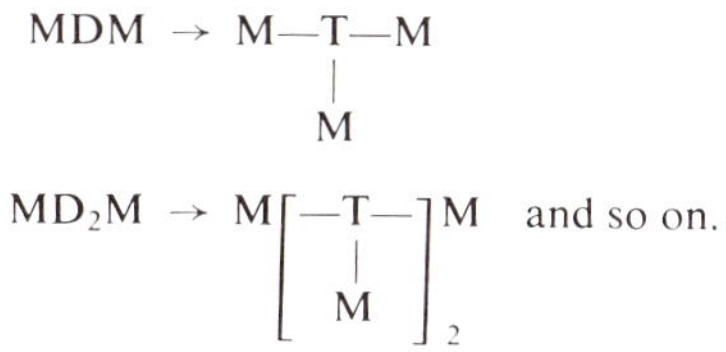

Low-molecular-weight members with phenyl and methyl groups have been isolated (*160*) up to $n = 5$.

These compounds are interesting for a number of different reasons: First, they have a steep vapor pressure curve and are thus easily distillable at low pressures or under high vacuum, and also possess very low vapor pressures at normal temperatures between 10^{-3} and 10^{-9} mm Hg, depending on the number of members of the branched chain. Secondly, by means of these compounds, we can study the influence of the branching of the chain on the viscosity. These relationships are shown in Table 38, in which a number of members of the series MD_nM has been included for comparison with the series $M[TM]_nM$.

TABLE 38

VISCOSITY ν AND MOLECULAR WEIGHT $\bar{M}$ IN THE SERIES $M[TM]_nM$ AND MD_nM (*161*)[a]

$M[TM]_nM$				MD_nM			
n	$\bar{M}$	ν_{20}(cSt)	Siloxane units	n	$\bar{M}$	ν_{20}(cSt)	Siloxane units
1	372	3.8	1+2				
2	583	24	2+2				
3	793	75	3+2	50	3600	60	50+2
4	1003	235	4+2	110	8000	140	110+2
5	1214	700	5+2	280	21000	680	280+2
6	1424	2100	6+2	400	30000	1440	400+2

[a] Here, $M = (CH_3)_3SiO_{1/2}$, $D = (CH_3)_2SiO_{2/2}$, and $T = C_6H_5SiO_{3/2}$

Table 38 shows clearly that in the $M[TM]_nM$ series the stepwise increase in the size of the molecules by one [TM] unit at a time results in quite a considerable increase of the viscosity. Even with five [TM] units in the chain, the viscosity is 700 cSt at 20°C; in the series of the polymethylsiloxanes MD_nM, this viscosity is first found at around 280 D units in the chain. The molecular

TABLE 39

LOW-MOLECULAR-WEIGHT BRANCHED POLYORGANOSILOXANES WITH SILICON-FUNCTIONAL GROUPS[a]

Type	Compound	Mp (°C)	Bp (°C)	at p (mm Hg)	n_D^{20}	d_4^{20} (gm/cc)
M_3T	$[(CH_3)_3SiO]_3SiH$	—	54 (*161*)	4	1.3863 (*161*)	0.849 (*109*)
M_3T	$[H(CH_3)_2SiO]_3SiH$	—	155 (*110*)	760	1.3813 (*110*)	0.8553 (*110*)
M_3T	$[H(CH_3)_2SiO]_3CH_3Si$	−155 (*110*)	164 (*110*)	760	1.3852 (*110*)	0.8613 (*110*)
M_4T_2	$[(CH_3)_3SiO]_4[(SiH)_2O]$	—	95 (*161*)	3	1.3900 (*161*)	0.891 (*109*)
M_4T_2	$[H(CH_3)_2SiO]_4[(SiH)_2O]$	—	208 (*110*)	760	1.3867 (*110*)	—
M_4T_2	$[H(CH_3)_2SiO]_4[(CH_3Si)_2O]$	−140 (*110*)	225 (*110*)	760	1.3908 (*110*)	0.9011 (*110*)
M_5T_3	$[(CH_3)_3SiO]_5[(SiH)_3O_2]$	—	122–126 (*161*)	3	1.3942 (*161*)	—
M_6T_4	$[(CH_3)_3SiO]_6[(SiH)_4O_3]$	—	130–135 (*161*)	3	1.3972 (*161*)	—
M_7T_5	$[(CH_3)_3SiO]_7[(SiH)_5O_4]$	—	155–158 (*161*)	3	1.4000 (*161*)	—
M_4Q	$[H(CH_3)_2SiO]_4Si$	—	189 (*110*)	760	1.3869 (*110*)	0.885 (*110*)
M_4Q	$[H_2(CH_3)SiO]_4Si$	—	159.5 (*110*)	760	1.3833 (*110*)	0.901 (*110*)
M_6Q_2	$[H(CH_3)_2SiO]_6[Si_2O]$	—	261 (*110*)	760	1.3934 (*110*)	0.9298 (*110*)

[a] Numbers in parentheses indicate reference numbers.

weights of the respective compounds are 1214 and 21,000. The influence of branching on the viscosity is thus very pronounced.

Branching in the chain has here been pushed to its extreme. Each chain member has been made into a point of branching. On the other hand, it is also possible to synthesize chains which contain only a few branching units. This type can be formulated as

$$MD_m[TM]_nM$$

for instance with $m=2$ and $n=1$

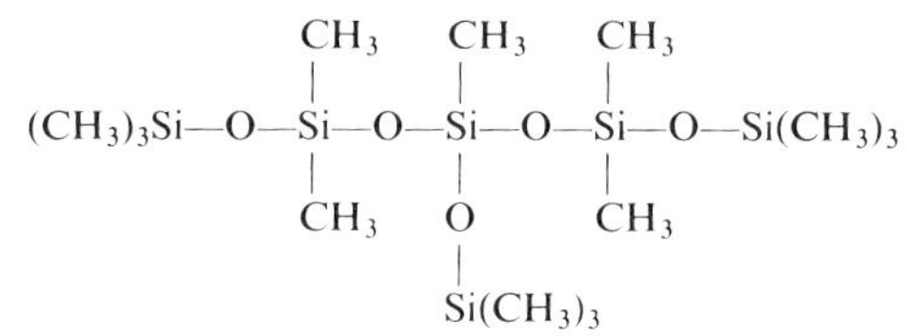

These types of polymers can be found when hydrolysates of dimethylchlorosilane containing methyltrichlorosilane and trimethylchlorosilane as impurities are polymerized. Even with small proportions of branching units in linear polydimethylsiloxanes, a lowering of the pour point of the silicone oils in question is observed. Branching units in long-chain polymers which are made into silicone rubber exert considerable influence on its behavior.

Very little factual material is at present available on low-molecular-weight branched polyorganosiloxanes with silicon-functional or organofunctional groups. A few compounds of this kind, some of which have not been reported in the literature, are given in Tables 39 and 40.

TABLE 40

LOW-MOLECULAR-WEIGHT BRANCHED POLYORGANOSILOXANES WITH ORGANOFUNCTIONAL GROUPS (*161*)

Type	Compound	η_{20}(cP)	Bp(°C)	n_D^{20}	d_4^{20} (gm/cc)
M_3T	$(m\text{-}ClC_6H_4)Si[OSi(CH_3)_3]_3$	6.6	138–140	1.4510	0.995
M_4T_2	$[(m\text{-}ClC_6H_4)Si]_2O[OSi(CH_3)_3]_4$	42.0	205–210	1.4730	1.046
M_5T_3	$[(m\text{-}ClC_6H_4)Si]_3O_2[OSi(CH_3)_3]_5$	—	245–250	1.4895	—
M_3T	$(o,o'\text{-}Br_2C_6H_3)Si[OSi(CH_3)_3]_3$	74.5	155–158	1.4920	1.254

6.2.3 CYCLIC POLYMERS

Of the series of monocyclic siloxanes, i.e., those with molecules consisting of a single ring built up of D units, members containing 3 to 9 D units have been

TABLE 41

LOW-MOLECULAR-WEIGHT MONOCYCLIC POLYORGANOSILOXANES (NONFUNCTIONAL)[a]

Compound	η (cP) at	t (°C)	Mp (°C)	Bp (°C) at	p (mm Hg)	n_D^{20}	d_4^t (gm/cc) at	t (°C)
$[(CH_3)_2SiO]_3$	—	—	64 (*129*)	134 (*129*)	760	—	1.12 (*70*)	[b]
$[(CH_3)_2SiO]_4$	1.69 (*204*)	37.8	17.5 (*129*)	175 (*129*)	760	1.3968 (*129*)	0.9558 (*129*)	20
$[(CH_3)_2SiO]_5$	2.94 (*204*)	37.8	−38 (*129*)	210 (*129*)	760	1.3982 (*129*)	0.9593 (*129*)	20
$[(CH_3)_2SiO]_6$	5.10 (*204*)	37.8	−3 (*129*)	245 (*129*)	760	1.4015 (*129*)	0.9672 (*129*)	20
$[(CH_3)_2SiO]_7$	7.80 (*204*)	37.8	−26 (*129*)	154 (*129*)	20	1.4040 (*129*)	0.9730 (*129*)	20
$[(CH_3)_2SiO]_8$	10.7 (*204*)	37.8	31.5 (*70*)	175 (*129*)	20	1.4060 (*129*)	1.177 (*70*)	[b]
$[(CH_3)_2SiO]_9$	—	—	—	*ca.* 188 (*129*)	20	1.4070 (*129*)	—	—
$[(CH_3)_3SiO(CH_3)SiO][(CH_3)_2SiO]_3$	—	—	—	115–120 (*4*)	18	1.4005 (*4*)	0.9786 (*4*)	20
$[(CH_3)(C_2H_5)SiO]_3$	—	—	−2.97 (*35*) Solidification	198.9 (*35*)	760	1.4110 (*35*)	0.947 (*35*)	20
$[(CH_3)(C_2H_5)SiO]_4$	—	—	37–43 (*35*) Solidification	244.7 (*35*)	760	1.4164 (*35*)	0.955 (*35*)	20
$[(CH_3)(C_2H_5)SiO]_5$	—	—	−122 (*35*) Solidification	287.7 (*35*)	760	1.4208 (*35*)	0.965 (*35*)	20
$[(CH_3)(C_2H_5)SiO]_6$	—	—	115–120 (*35*) Solidification	326.0 (*35*)	760	1.4245 (*35*)	0.976 (*35*)	20
$[(C_2H_5)_2SiO]_3$	3.6 (*208*)	25	9.9 (*208*)	156.7 (*208*)	50	1.4308 (*208*)	0.9549 (*208*)	20
$[(C_2H_5)_2SiO]_4$	11.2 (*208*)	25	−64 (*208*)	127 (*208*)	1	1.4336 (*208*)	0.964 (*208*)	20
$[(C_2H_5)_2SiO]_5$	—	—	38–39 (*73*)	167.5 (*73*)	1	1.4408 (*73*)	0.968 (*73*)	20
$[(CH_3)_2SiO]_2[(t\text{-}C_4H_9)_2SiO]_2$	—	—	244.5–245.7 (*192*)	—	—	—	—	—
$[(C_6H_5)_2SiO]_3$	—	—	188–190 (*76*)	290–300 (*20*)	1	—	—	—
$[(C_6H_5)_2SiO]_4$(I and II)[e]	—	—	200–202 (*20, 76*)	330–340 (*20*)	1	—	1.185 (*23*)	—
$[(CH_3)(C_6H_5)SiO]_3$	—	—	cis: 99.5 (*97*)	—	—	—	—	—
	—	—	trans: 39.5 (*97*)	—	—	—	—	—
$[(CH_3)(C_6H_5)SiO]_4$	—	—	99 (*208*)	—	—	—	—	—
$[(C_6H_5)_2SiO][(CH_3)_2SiO]_2$	—	—	62–64 (*192*)	—	—	—	—	—
$[C_6H_5)_2SiO]_2[(CH_3)_2SiO]_2$	—	—	131–132 (*192*)	—	—	—	—	—
$[(C_6H_5)_2SiO]_3[(CH_3)_2SiO]$	—	—	114 (*161*)	—	—	—	—	—
$[(C_2H_5)(C_6H_5)SiO]_3$	63.6[c] (*208*)	25	—	166 (*208*)	0.025	1.5402[d] (*208*)	1.0932 (*208*)	25
$[(C_2H_5)(C_6H_5)SiO]_4$	220[c] (*208*)	25	106 (*208*)	—	—	1.5430[d] (*208*)	1.1000 (*208*)	25
$[(C_2H_5)(C_6H_5)SiO]_2[(C_2H_5)_2SiO]$	—	—	—	150 (*208*)	0.5	1.5008 (*208*)	1.0561 (*208*)	20
$[(C_2H_5)(C_6H_5)SiO][(C_2H_5)_2SiO]_2$	—	—	—	165 (*208*)	12	1.4734 (*208*)	1.0096 (*208*)	20

[a] Numbers in parentheses indicate reference numbers.

[b] Room temperature.

[c] γ (cSt).

[d] n_D^{25}.

[e] Besides a metastable form (III) with mp = 187–189°C (*76*).

isolated (Table 41). Siloxane rings consisting of 2 D units are unknown in organosiloxane chemistry since they are unstable for reasons of strain.

According to X-ray diffraction (*132*) and electron-diffraction (*1*) investigations, the D_3 ring is almost or completely planar. Electron-diffraction measurements (*1*) in the vapor phase have given an angle of $125 \pm 5°$ for Si—O—Si and $115 \pm 5°$ for O—Si—O. The former is a little smaller and the latter a little larger than the corresponding value in other methylcyclosiloxanes, particularly that in D_4 (Table 52). The ring may thus be slightly strained; the small Si—O—Si valence angle points to a somewhat stronger polar bond than in D_4 (see Section 6.3.3). The high reactivity of hexamethylcyclotrisiloxane toward acidic or alkaline cleavage (Sections 5.4.2.2 and 5.4.2.3 and 6.3.3.3) particularly favors this hypothesis.

Rings with more than three siloxane units are puckered (Fig. 14). The ability of the siloxanes to form rings is very pronounced. Even the D_9 ring, containing

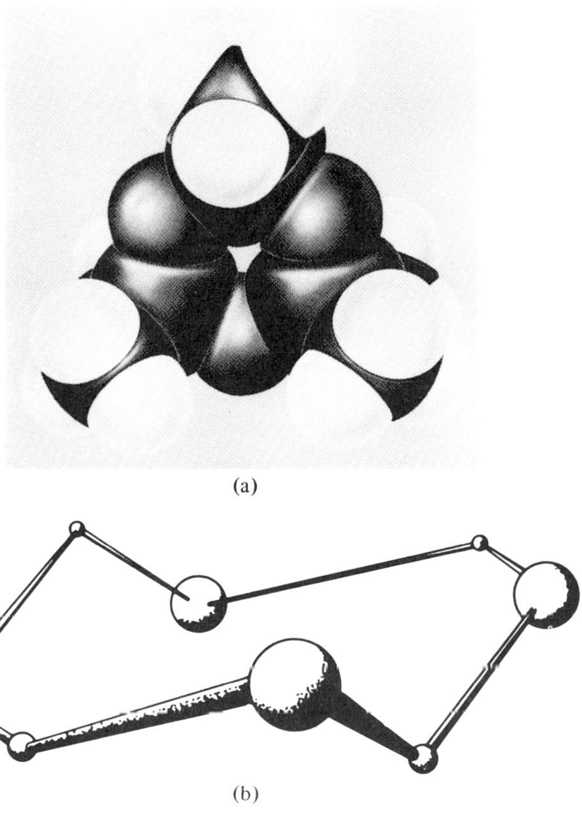

FIG. 14. (a) Planar Si—O—Si ring in hexamethylcyclotrisiloxane. (b) Puckered Si—O—Si ring in octamethylcyclotetrasiloxane.

TABLE 42

CRYSTAL STRUCTURES OF CYCLIC POLYORGANOSILOXANES[a]

Siloxane	Modification	Space group	Unit cell a	b	c (Å)
$[(CH_3)_2SiO]_3$ (*132*)	—	$C^5_{3v}-R\ 3\ m$	4.53		19.23
$[(CH_3)_2SiO]_4$ (*184*)	α (> −16.3°C)	$C^6_{4h}-I\ 4_1/a$	16.10		6.83
	β (< −16.3°C)	$C^4_{4h}-P\ 4_2/n$	16.10 ± 0.02		6.47 ± 0.01
$[(CH_3)_2SiO]_8$ (*52*)	—	$D^4_{2d}-P\ \bar{4}2_1c$	13.95 ± 0.06		8.55 ± 0.04
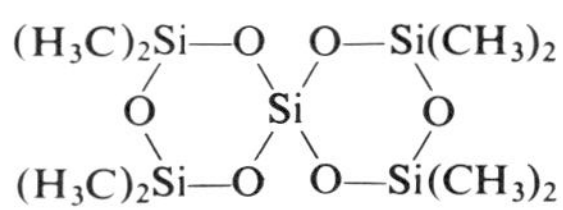 (*149*)	—	$D^{19}_{4h}-I\ 4/amd$	14.09 ± 0.03		10.18 ± 0.03
$[(C_6H_5)_2SiO]_3$ (*76*)	I	D_2-222	15.67 ± 0.1	20.16 ± 0.08	10.03 ± 0.10
	II	$Ci-\bar{1}$	12.89 ± 0.1	14.93 ± 0.15	16.50 ± 0.1[b]
$[(C_6H_5)_2SiO]_4$ (*76*)	I	$Ci-\bar{1}$	10.74 ± 0.08	19.10 ± 0.10	22.00 ± 0.10[c]
	II	$C_{2h}-2/m$	22.93 ± 0.15	10.03 ± 0.05	21.7 ± 0.05[d]
	III[e]	?			

[a] Numbers in parentheses indicate reference numbers.
[b] $\alpha = 96°$; $\beta = 95.3°$; $\gamma = 101°$.
[c] $\alpha = 83.7°$; $\beta = 104°$; $\gamma = 96.5°$.
[d] $\beta = 120°$.
[e] Metastable.

18 atoms (9 silicon and 9 oxygen atoms) has been found. It is possible that even larger rings than D_9 can exist independently, although their isolation has so far failed because of the difficulty of separation.

It is worth noting that rings with three, four, or five siloxane units are usually obtained in relatively large amounts when cyclosiloxanes are formed; this suggests that these molecules are particularly stable.

With respect to their physicochemical behavior, cyclosiloxanes differ in a remarkable manner from the linear polymers. Their solidification points are appreciably higher; hexamethylcyclotrisiloxane is crystalline at room temperature. Octamethylcyclotetrasiloxane melts at 17.5°C. The melting points of the higher cyclic polydimethylsiloxanes vary in a very irregular manner, which is probably connected with the stability of the crystal lattice (see Table 41). Similar phenomena are observed with cyclosiloxanes substituted in different ways.

A few references on the crystal structures of crystalline cyclosiloxanes are given in Table 42. Cyclosiloxanes with two different organic groups can have structural isomers. *Cis* and *trans* forms of $[(CH_3)(C_6H_5)SiO]_3$ have been isolated (see Table 41).

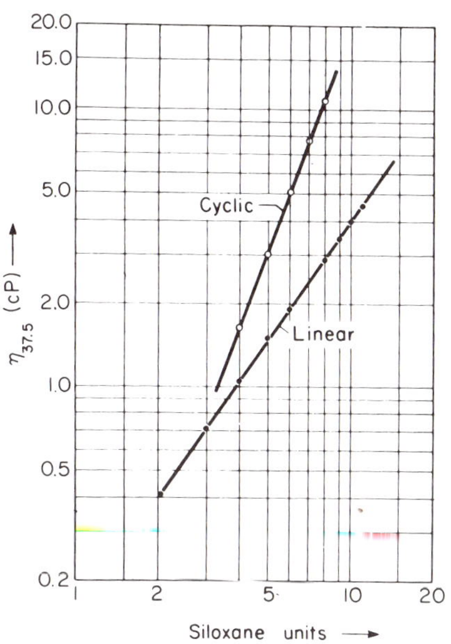

FIG. 15. Viscosities of cyclic and linear polydimethylsiloxanes as functions of the number of siloxane units in the polymer molecule [from Wilcock (*204*) by permission of the American Chemical Society].

TABLE 43

LOW-MOLECULAR-WEIGHT POLYCYCLIC, SPIROCYCLIC, AND SPHEROCYCLIC POLYORGANOSILOXANES (NONFUNCTIONAL)[a, b]

Compound	Mp (°C)	Bp (°C) at	p (mm Hg)	n_D^t at	t (°C)	d_4^t (gm/cc)	at t (°C)
(a) Polycyclic							
$(CH_3)_8Si_5O_6$	118 (*157*)	203 (*157*)	760	—	—	—	—
$(CH_3)_{10}Si_6O_7$	51 (*157*)	232 (*157*)	760	—	—	—	—
$(CH_3)_8Si_6O_8$	119 (*157*)	220 (*157*)	760	—	—	—	—
$(CH_3)_{10}Si_7O_9$	150 (*157*)	248 (*157*)	760	—	—	—	—
$(CH_3)_{10}Si_8O_{11}$	139 (*157*)	257 (*157*)	760	—	—	—	—
(b) Spirocyclic							
$(CH_3)_8Si_5O_6$	121 (*157*)	204 (*157*)	760	—	—	—	—
$(CH_3)_{10}Si_6O_7$	57 (*157*)	230 (*157*)	760	—	—	—	—
$(CH_3)_{12}Si_7O_8$	31 (*157*)	254 (*157*)	760	1.4031 (*157*)	20	1.0352 (*157*)	20
$(CH_3)_{14}Si_8O_9$	−12 (*157*)	157 (*157*)	20	1.4032 (*157*)	20	1.0260 (*157*)	20
(c) Spherocyclic							
$[(CH_3)SiO_{1.5}]_6$	210–211 (*182*)	—	—	—	—	—	—
$[(CH_3)SiO_{1.5}]_8$	365 (sublimes) (*9*)	—	—	—	—	—	—
$[(CH_3)SiO_{1.5}]_{10}$	332.6–334 (*194*)	—	—	—	—	—	—
$[(CH_3)SiO_{1.5}]_{12}$	260.9–261.4 (*194*)	—	—	$\alpha = 1.460$ (*9*) $\gamma = 1.470$ (*9*)	— —	— —	— —
$[(i\text{-}C_3H_7)SiO_{1.5}]_4$	ca. 80 (*202*) softens	—	—	—	—	—	—
$[(t\text{-}C_4H_9)SiO_{1.5}]_4$	ca. 225 (*202*)	—	—	—	—	1.11 ± 0.01 (*202*)	—
$[(C_2H_5)SiO_{1.5}]_8$	280–282 (*9*)	—	—	—	—	1.3079 (*9*)	—
$[(n\text{-}C_3H_7)SiO_{1.5}]_8$	219–220 (*9*)	—	—	—	—	—	—
$[(n\text{-}C_4H_9)SiO_{1.5}]_8$	190–195 (*9*)	—	—	—	—	—	—
$[(C_6H_{11})SiO_{1.5}]_8$	> 400 (*9*)	—	—	—	—	1.174 (*9*)	—
$[(C_6H_5)SiO_{1.5}]_6$	> 400 (*9*)	—	—	$\gamma = 1.580$ (*9*)	—	—	—
$[(C_6H_5)SiO_{1.5}]_8$	> 440 (*18*)	—	—	—	—	—	—
$[(C_6H_5)SiO_{1.5}]_{12}$	385 (*18*)	—	—	—	—	—	—

[a] Numbers in parentheses indicate reference numbers. [b] For structures see text.

Some characteristics of the linear polydimethylsiloxanes, such as the relative independence of the viscosity on the temperature, are not so strongly reproduced in the cyclosiloxanes (*72*). Moreover, in cyclosiloxanes the viscosity increases more rapidly with increasing number of siloxane units than in linear polymers (*204*) (Fig. 15). This and especially the limitations imposed by the ring size on the variation of the molecular weights, is the reason why cyclosiloxanes have attracted no interest as silicone products. On the other hand, they are of considerable importance as intermediates for the production of linear polymers. Tri- and tetracyclosiloxanes are particularly suitable as definite starting materials for the synthesis of chains by reactions catalyzed by alkalis, by acids, or otherwise (see Chapter 5).

The constants important for their characterization are given in Table 43.

The vapor pressure of octamethylcyclotetrasiloxane at any temperature is given by the equation (*128*):

$$\log p_{torr} = -4530.2T - 12.3508 \log T + 45.7216$$

It is of scientific interest that evidence has also been found of a group of *polycyclic* compounds, i.e., those with *condensed ring systems*, among the polymethylsiloxanes. Of the compounds given in Table 43 a structural formula can be assigned only to octamethylbicyclopentasiloxane (*157*):

$(CH_3)_8Si_5O_6$
$[D_3T_2]$

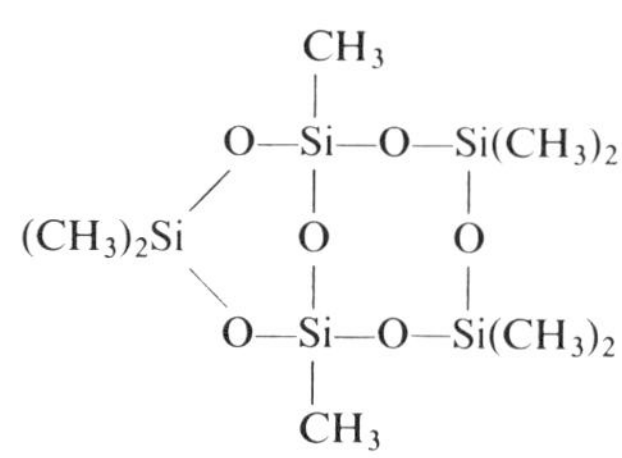

A number of isomeric structures are possible for the other polycyclic compounds but no distinction between them has so far been made.

The third and fourth group of the cyclic siloxanes, the *spirocyclic* and *spherocyclic* compounds, play a part in silicone chemistry similar to that of the polycyclic compounds. They are of considerable fundamental scientific interest but have not so far acquired any practical importance.

In the case of the *spirocyclic* siloxanes, the polysiloxane rings are linked with one another through their *apexes* and not through their *edges*, as in the polycyclic compounds. A central silicon atom occupies the position common to the two rings. It has four oxygen atoms in tetrahedral coordination as the nearest neighbors, each pair of which belongs to a siloxane ring. The two rings bonded through the silicon atom have their planes at *right angles* to one another because of the tetrahedral coordination of the linking *Q* unit. Of the spirocyclic methylsiloxanes mentioned in Table 43, the following two compounds can be assigned a structural formula (*157*):

$(CH_3)_8Si_5O_6$
[D_4Q]

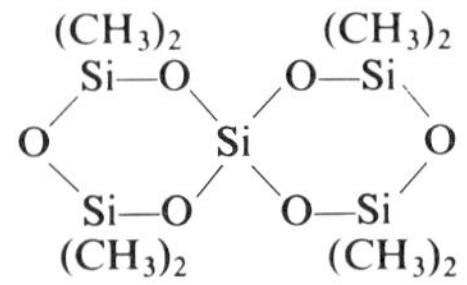

$(CH_3)_{10}Si_6O_7$
[D_5Q]

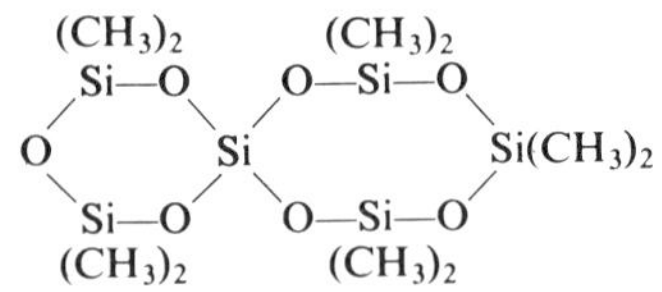

For the other spirocyclic siloxanes, there are again possibilities of isomerism between which no decision has been made.

The group of the *spherocyclic* polysiloxanes is still small. They are low-molecular-weight polysiloxanes built up from T units which can be isolated only with special precautions and at times only in small yields, since the trifunctional siloxane units usually tend to form three-dimensional macromolecules.

The smallest oligomer with four trifunctional siloxane units, T_4, consists of four tetrahedrally constructed $RSiO_{3/2}$ units, linked through the oxygen atoms. The molecule built up in this way has a structure like that of hexamethylenetetramine and can be considered as being made up of four condensed rings each of which contains three tetrahedra (Fig. 16). It has been found (*202*) that this molecule is formed only when the R groups which occupy the three corners of the tetrahedra are relatively *bulky* organic groups, such as isopropyl and *t*-butyl. Only spherocyclic compounds with 6, 8, and 12 T units have been observed with methyl-, ethyl-, *n*-propyl-, *n*-butyl, phenyl-, and cyclohexyl-substituted siloxanes. It is possible to interpret the structure of these molecules by assuming that at any time two rings with three, four, or six trifunctional siloxane units utilize two of the functionalities present for ring formation, and are linked through the third functionality with one another. A small molecule is formed, like that shown in Fig. 17, which can be considered as being built up of two six-membered rings linked with one another and also as six condensed four-membered rings.

Only a small number of definite low-molecular-weight cyclosiloxanes with silicon-functional or organofunctional groups are known, and these are exclusively monocyclic oligomers (Tables 44 and 45).

TABLE 44

LOW-MOLECULAR-WEIGHT CYCLIC POLYORGANOSILOXANES WITH SILICON-FUNCTIONAL GROUPS[a]

Compound	Mp (°C)	Bp (°C) at	p (mm Hg)	n_D^{20}	d_4^{20} (gm/cc)
$[CH_3Si(H)O]_3$	—	93.8–93.9 (*15*)	760	1.3770 (*15*)	0.9677 (*15*)
$[CH_3Si(H)O]_4$	−69 ± 3 (*153*)	134.5–134.9 (*153*)	755	1.3870 (*153*)	0.9912 (*153*)
$[CH_3Si(H)O]_5$	−108 ± 3 (*153*)	168.6–168.9 (*153*)	755	1.3912 (*153*)	0.9985 (*153*)
$[CH_3Si(H)O]_6$	−79 ± 2 (*153*)	92.6–93.0 (*153*)	21	1.3944 (*153*)	1.006 (*153*)
$[CH_3Si(H)O]_7$	−82 (*171*)	114 (*171*)	21	1.3982 (*171*)	1.0142 (*171*)
$[CH_3Si(H)O]_8$	−64 (*171*)	105 (*171*)	2	1.4010 (*171*)	1.0218 (*171*)
$[CH_3Si(H)O]_3[(CH_3)_2SiO]$	−52 (*170*)	69 (*170*)	50	1.3913 (*170*)	0.9718 (*170*)
$[CH_3Si(H)O]_2[(CH_3)_2SiO]_2$	−40 (*170*)	77.5 (*170*)	50	1.3935 (*170*)	0.9674 (*170*)
$[CH_3Si(H)O][(CH_3)_2SiO]_3$	−27 (*170*)	84.5 (*170*)	50	1.3965 (*170*)	0.9583 (*170*)
	—	165 (*127*)	760	1.3950 (*127*)	0.9620 (*127*)
$[C_2H_5Si(H)O]_3$	—	106 (*126*)	20	1.4138 (*126*)	0.9836 (*126*)
$[C_2H_5Si(H)O]_4$	—	135 (*126*)	20	1.4176 (*126*)	0.9920 (*126*)
$[C_2H_5Si(H)O]_5$	—	165 (*126*)	20	1.4200 (*126*)	1.0023 (*126*)
$[CH_3Si(Cl)O]_4$	76 (*171*)	220 (*171*)	745	—	—
$[CH_3Si(Cl)O]_5$	−55 (*171*)	~250 (*171*)	—	—	1.3372 (*171*)
$[CH_3Si(Cl)O]_6$	−10 (*171*)	~276 (*171*)	—	—	1.3391 (*171*)
$[CH_3Si(Cl)O]_7$	−25 (*171*)	~308 (*171*)	—	—	1.3434 (*171*)
$[CH_3Si(Cl)O][(CH_3)_2SiO]_3$	6 (*170*)	85.5–86.5 (*170*)	20	—	1.0231 (*170*)
$[CH_3Si(Cl)O]_2[(CH_3)_2SiO]_2$	24 (*170*)	93–95 (*170*)	20	—	1.1251 (*170*)
$[CH_3Si(Cl)O]_3[(CH_3)_2SiO]$	36 (*170*)	105–106 (*170*)	20	—	—
$[CH_3Si(H)O]_3[CH_3Si(Cl)O]$	−55 (*171*)	150–152 (*171*)	—	—	1.0977 (*171*)
$[CH_3Si(H)O]_2[CH_3Si(Cl)O]_2$	−7.5 (*171*)	88 (*171*)	30	—	1.1786 (*171*)
$[CH_3Si(H)O][CH_3Si(Cl)O]_3$	2 (*171*)	101 (*171*)	30	—	1.2562 (*171*)
$[CH_3Si(OH)O][(CH_3)_2SiO]_3$	−21 (*170*)	100 (*170*)	20	1.4082 (*170*)	1.0478 (*170*)
$[CH_3Si(OH)O]_2[(CH_3)_2SiO]_2$	120 (*170*)	120 (*170*) (sublimes)	20	—	—

(*continued on p.* 280)

TABLE 44 *(Contd.)*

LOW-MOLECULAR-WEIGHT CYCLIC POLYORGANOSILOXANES WITH SILICON-FUNCTIONAL GROUPS[a]

Compound	Mp (°C)	Bp (°C) at	p (mm Hg)	n_D^{20}	d_4^{20} (gm/cc)
$[CH_3Si(OCH_3)O][(CH_3)_2SiO]_3$	—	75–77 (*6*)	1	1.3992 (*6*)	0.9916 (*6*)
$[CH_3Si(OC_2H_5)O][(CH_3)_2SiO]_3$	—	65 (*5*)	1.5	1.3987 (*5*)	0.9833 (*5*)
$[CH_3Si(OC_2H_5)O]_2[(CH_3)_2SiO]_2$	—	162–186 (*6*)	4	1.4029 (*6*)	1.0100 (*6*)
$[CH_3Si(OC_4H_9)O][(CH_3)_2SiO]_3$	—	93–95 (*5*)	1	1.4050 (*5*)	0.9703 (*5*)
$[CH_3Si(OC_4H_9)O]_2[(CH_3)_2SiO]_2$	—	96–114 (*6*)	1	1.4103 (*6*)	0.9804 (*6*)
$[CH_3Si(OC_6H_5)O][(CH_3)_2SiO]_3$	—	119–123 (*6*)	1.5	1.4442 (*6*)	1.0378 (*6*)
$[CH_3Si(NH_2)O][(CH_3)_2SiO]_3$	—	40.5 (*6*)	1	1.4065 (*6*)	1.0208 (*6*)
$[CH_3Si(NHC_6H_5)O][(CH_3)_2SiO]_3$	—	136–139 (*6*)	5	1.4482 (*6*)	0.9711 (*6*)

[a] Numbers in parentheses indicate reference numbers.

TABLE 45

LOW-MOLECULAR-WEIGHT CYCLIC POLYORGANOSILOXANES WITH ORGANOFUNCTIONAL GROUPS[a]

Compound	Mp (°C)	Bp (°C)	at p (mm Hg)	n_D^t	at t (°C)	d_4^t (gm/cc)	at t(°C)
(a) Alkenylcyclosiloxanes							
$[CH_2{=}CH(CH_3)SiO]_3$	—	79.5–80 (*107*)	20	1.4215 (*107*)	25	0.9669 (*107*)	25
$[CH_2{=}CH(CH_3)SiO]_4$	−43.5 ± 0.1 (*80*)	224–224.5 (*80*)	758	1.4342 (*80*)	20	0.9875 (*80*)	20
$[CH_2{=}CH(CH_3)SiO]_5$	−140– −136 (*80*)	261–262 (*80*)	758	1.4373 (*80*)	20	0.9943 (*80*)	20
$[CH_2{=}CH(CH_3)SiO]_6$	−123– −119 (*80*)	296–297 (*80*)	758	1.4400 (*80*)	20	1.0050 (*80*)	20
$[CH_2{=}CHCH_2(CH_3)SiO]_4$	—	91–93 (*159*)	0.5	1.4448 (*159*)	20	0.9622 (*159*)	24
(b) Haloalkylcyclosiloxanes							
$(ClCH_2)(CH_3)SiO[(CH_3)_2SiO]_3$	−1 (*87*)	214 (*87*)	760	1.4158 (*87*)	20	1.0444 (*87*)	20
$[CF_3CH_2CH_2(CH_3)SiO]_3$	—	99 (*106*)	3	1.3658 (*106*)	20	1.2407 (*106*)	25.8
$[CF_3CH_2CH_2(CH_3)SiO]_4$	—	140–145 (*106*)	3	1.3715 (*106*)	20	1.2554 (*106*)	25.7
$[ClC_6H_4(CH_3)SiO]_3$	125–126 (*22*)	—	—	—	—	—	—
$[(CF_3)_2C_6H_3(CH_3)SiO]_3$	—	169 (*84*)	0.4	1.4380 (*84*)	25	1.483 (*84*)	25
(c) Hydroxyalkylcyclosiloxanes and their esters							
$(HOCH_2)(CH_3)SiO[(CH_3)_2SiO]_3$	56–57 (*137*)	55–58 (*137*)	0.5	—	—	—	—
$(HOCH_2CH_2)(CH_3)SiO[(CH_3)_2SiO]_3$	—	112–114 (*137*)	9	1.4163 (*137*)	20	1.020 (*137*)	20
$(CH_3COOCH_2)(CH_3)SiO[(CH_3)_2SiO]_3$	—	110.5 (*57*)	9	1.2127 (*57*)	20	1.033 (*57*)	20
(d) Cyanoalkylcyclosiloxanes							
$(NCCH_2)(CH_3)SiO[(CH_3)_2SiO]_3$	—	92–93 (*136*)	2.0–2.5	1.4155 (*136*)	20	1.021 (*136*)	20
$[NCCH_2CH_2CH_2(CH_3)SiO]_3$	—	242–250 (*79*)	0.2	1.4558 (*79*)	25	—	—
$[NCCH_2CH_2CH_2(CH_3)SiO]_4$	—	175–180 (*79*)	0.025	1.4573 (*79*)	25	—	—
(e) Aminoalkylcyclosiloxanes							
$[NH_2CH_2CH_2CH_2CH_2(CH_3)SiO]_4$	—	240–244 (*138*)	0.5	1.4689 (*138*)	25	—	—

[a] Numbers in parentheses indicate reference numbers.

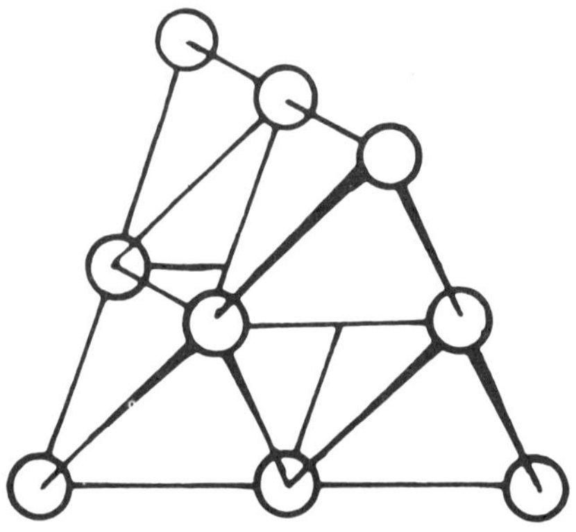

FIG. 16. Combination of four tetrahedra as a model for the structure of the spherocyclic molecule $[RSiO_{3/2}]_4$. The Si atoms occupy the centers of the tetrahedra, the oxygen atoms their linking apexes, and the R groups their free apexes.

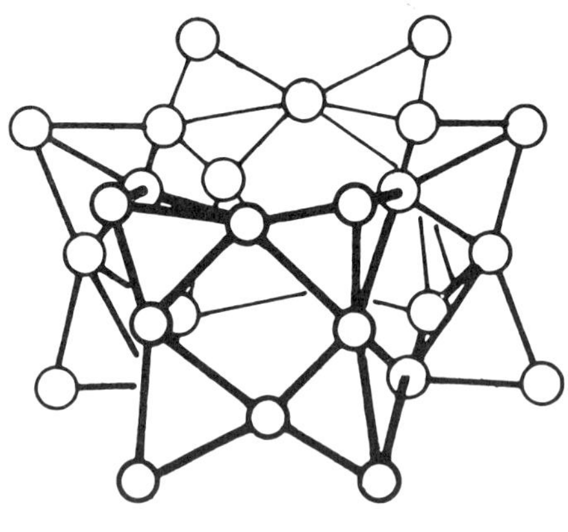

FIG. 17. Combination of twelve tetrahedra as a model for the structure of the spherocyclic molecule $[RSiO_{3/2}]_{12}$. The Si atoms occupy the centers of the tetrahedra, the oxygen atoms their linking apexes, and the R groups their free apexes.

6.2.4 CROSS-LINKED SILOXANES

We have reserved the term "cross-linked" to define the processes involved in the building up of molecular structures leading beyond the field of the low-molecular-weight compounds and yielding in general three-dimensional linkages of siloxane units. The T and Q units are responsible for the cross-linking.

The T and/or Q units can be combined with D, and in certain circumstances also M, units in different ways to synthesize cross-linked polymers; this therefore opens up the path to a variety of polymers, and it is precisely in this group that many technologically important products are found.

It is, of course, immeasurably more difficult to study such polymers and elucidate their structure than to do so with linear or cyclic polymers. A few basic rules of cross-linking can be studied on low-molecular-weight siloxanes, just as the formation of polycyclic, spirocyclic, or spherocyclic siloxanes from monocyclic compounds can be considered the first stage of a cross-linking process. Thus in the polycyclic siloxanes the ring systems are linked with one another by the partial replacement of D by T units; spirocyclic compounds arise from the insertion of Q units, and spherocyclic compounds are formed from monocyclic compounds by the complete replacement of all D by T units.

Starting with linear polysiloxanes, it is possible by the stepwise replacement of D units by T to produce all stages from weakly cross-linked chains (I) linked by only isolated siloxane bridges, up to highly cross-linked systems (II):

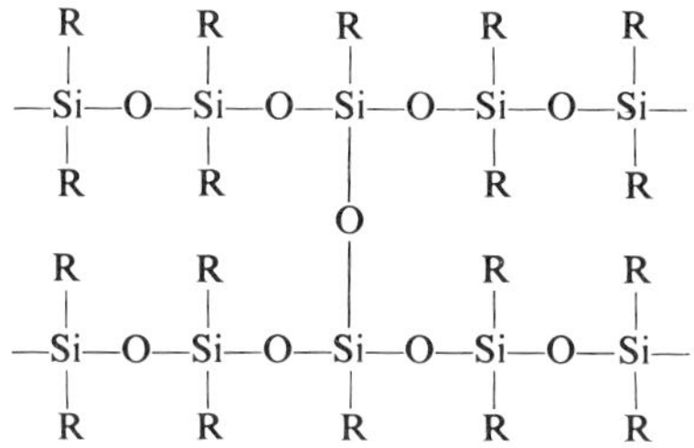

I

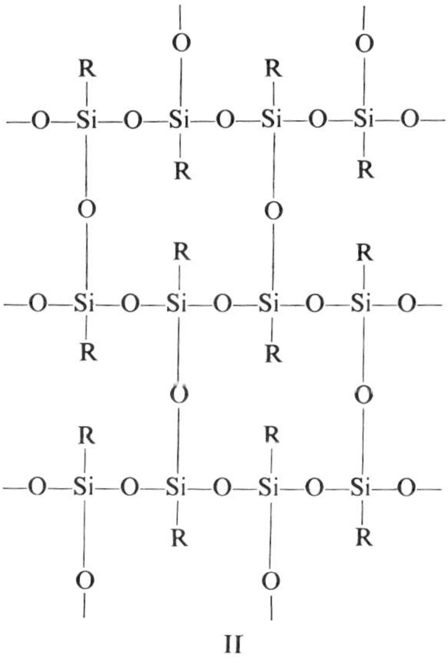

II

A ladderlike polymer, i.e., a one-dimensional infinite linkage (*19*) consisting only of T units is probable for a phenylsilsesquioxane polymer ($R{=}C_6H_5$):

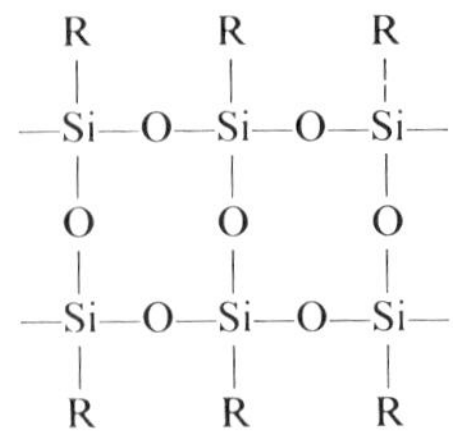

III

In consequence of their high degree of cross-linking, the polymer structures (II) and (III) can be considered as systems of interlinked rings. If the cross-linking is carried on still further by replacing all T units by Q units, the final result of the further development of our polymer series is a structure of the composition $SiO_{4/2}{=}SiO_2$, characterized by complete linking of all Si atoms with one another through bridges consisting of single oxygen atoms:

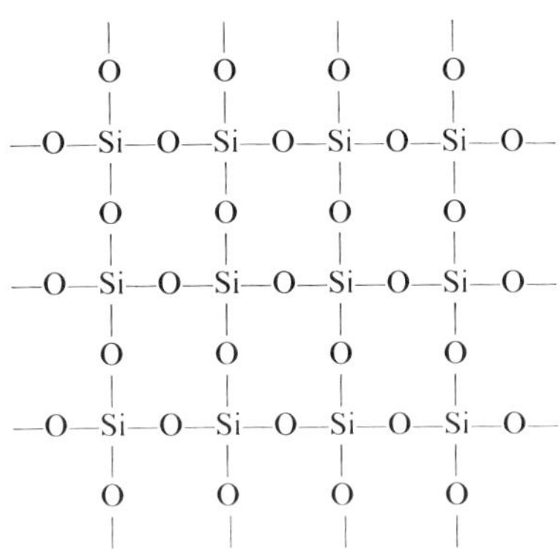

IV

The *degree of cross-linking* also determines the *physical state* of the polymer. To begin with, the viscosity of the liquid linear polysiloxane is increased by weak cross-linking of the polymer chain. As cross-linking progresses, the region of soft resinous silicones is reached and these become harder and more brittle with the increasing degree of cross-linking and, finally, a siloxane consisting only of T units of the type (II) forms a relatively hard and brittle resin. As the configuration of the type (IV) is approached, the products become more gel-like. The series finally ends with silica gel itself.

Experience has shown that strong cross-linking is essential if rapidly condensable and hard silicone resins are to be produced. This can perhaps be understood from the weakness of the intermolecular forces which are specific for the methylsilicones, in particular. The van der Waals forces between the molecules are too weak to restrict the mobility of even large molecules; only by frequent cross-linking of the molecules with one another is it possible to transform the polymer from the liquid to the resinous state.

The transitions between the physical states mentioned occur at different degrees of cross-linking, depending on which organic substituents are bound to the silicon atoms. If one considers for simplicity only the combinations of di- and trifunctional and of tri- and tetrafunctional siloxane units, we have the following situation.

With the *polymethylsiloxanes*, the D polymers remain *liquid* until approximately 30% of T units have been included (ratio $CH_3/Si \approx 1.7$). A region of *resinous* polymers commences with more than about 50% of T units, which extends up to systems with approximately 90% of T and 10% of D units. The products become increasingly glassy and brittle beyond this point and rapidly acquire a gel-like character, particularly if the CH_3/Si ratio falls below unity.

Polyethylsiloxanes are *softer* than polymethylsiloxanes with the same degree of cross-linking. The transitions of the states of the polymers are therefore also displaced when compared with those of the polymethylsiloxanes. *Resinous* products are obtained only in systems having about 50% of D and 50% of T units, i.e., a C_2H_5/Si ratio of about 1.5. On the other hand, here the region of resins extends up to systems with about 50% of T and 50% of Q units, corresponding to a C_2H_5/Si ratio of 0.5 (*141*, *143*). The region of brittle *gel-like* materials is only entered after even stronger cross-linking.

This effect can be explained as due to *internal plasticization* by the larger alkyl groups. However, it is not feasible to utilize this property for most technological purposes, since the heat resistance of the polymers (one of the most important properties of silicone resins) decreases rapidly with the increasing size of the alkyl group.

Similar plasticization can also result from aromatic substituents, particularly *phenyl groups*, in polymethylsiloxanes. The combination of methyl- and phenylsiloxanes in a resin has proved to be particularly favorable; the incorporation of phenyl groups improves not only the mechanical and electrical properties, but also the long-term heat stability of the products (cf. Chapter 9).

The nature of the resinous cross-linked siloxanes can be varied further by furnishing their molecular networks with silicon-functional or organofunctional groups. Si—OH or Si—OR groups are usually present in silicone resins of low molecular weight, since these are normally prepared by hydrolysis and condensation from chlorosilanes or alkoxysilanes. The molecular weight increases with the increasing degree of condensation, and the content of silicon-functional groups is simultaneously decreased. The molecules are still small in initial stages of the condensation, because numerous branching centers on the siloxane network at which siloxane bridges to other Si atoms could be formed are occupied by protons (Si—OH groups) or organic radicals (Si—OR groups). As these groups are progressively removed by condensation, the molecules aggregate and the molecular weight rises, until finally a three-dimensional network is built up. The solubility of the products in organic solvents decreases with increasing degree of condensation, and technical silicone resins are therefore generally condensed to a degree which leaves them only just soluble in organic solvents, so that little further reaction is then required to convert them into a product of higher molecular weight, soluble only with difficulty or not at all.

It is clear from this brief discussion that the question of the possible *molecular weights* of cross-linked silicones can be answered only for certain stages of condensation. In technical methylphenylsiloxane resin solutions with a high content of soluble cross-linked siloxanes, these generally have relatively small molecules with molecular weights of about 1500 to 3000 (*29*). Molecular weights of the order of 6.7×10^4 to 3.4×10^6 have been measured (*135*) on a methylsilicone resin with a CH_3/Si ratio of 1.5 that had been condensed at normal temperature in vacuum; but these values are in no way characteristic of a definite final state and are of interest only when the method of preparation is specified.

A correlation between molecular weight and viscosity in solution, similar to that with linear polymers, has been found for the cross-linked siloxanes. The Flory-Barry equation

$$[\eta] = K \cdot M^n$$

has been shown to hold for two methylsiloxanes with 1.5 and 1.8 CH_3 groups per Si atom; the constants K and n for these products (and for a linear methylsilicone with $2CH_3$ groups per Si, given for comparison) were as shown in the accompanying tabulation (*135*). The constants obviously depend on the method of preparation and on the previous history of the siloxanes (*135*), and it is certain that they will vary from resin to resin.

CH_3/Si	K	n
1.5	0.326	0.21
1.8	0.240	0.28
2.0	0.0805	0.50

Analysis of the molecular structure and conformation of cross-linked siloxanes obviously encounters considerably greater difficulties than in the case of the other polymers discussed in the preceding chapters.

The densities of the dissolved polymer molecules within their sphere of action can be calculated from the viscosities by means of an equation given by Flory for the above-mentioned silicone resins with 1.5 CH_3/Si and 1.8 CH_3/Si; the densities are 0.10 and 0.05 gm/cc, respectively. Comparison with the corresponding value of 0.007 for a linear polysiloxane with 2CH_3/Si shows the relatively dense packing in the resin molecule. This can also be seen from their degree of swelling in solvents, which is given (*135*) for the series $CH_3/Si = 1.5$, 1.8, and 2.0 by factors of 1.15, 2.0, and 3.8, respectively. It is also significant that constant viscosities in solutions of resins are obtained in certain circumstances only after several weeks' standing (*59*). It thus appears that the swelling and dispersion of the molecules by the solvent proceed only very slowly.

There appears to be no indication of a pronounced anisometric structure of the cross-linked molecules. It can be assumed that these have basically isometric dimensions, i.e., a spherical form.

6.3 Siloxane Bonds in Molecules of the Silicones and Anions of Silicates

6.3.1 Anions and Molecular Structures

The preceding discussion has demonstrated the dominant role of the siloxane skeleton in polyorganosiloxane chemistry. Independently of the substituents on the silicon atom or on the organic group, the siloxane configuration tenaciously retains certain characteristic structural forms.

This is of course reminiscent of the situation long recognized in silicate chemistry. The large variety of known silicates is also built up on the basis of a limited number of characteristic frequently recurring arrangements of silicon-oxygen tetrahedra; these offer a relatively simple and clear foundation for a systematic arrangement of this large class of compounds. The classical system due to Bragg, some details of which have now been supplemented, classifies the silicones according to the nature and manner of the combination of the silicon-oxygen complexes with one another (Fig. 18); that is, whether they exist as discrete units in the lattice (*nesosilicates*), whether a small number is combined to form groups (*sorosilicates*) or rings (*cyclosilicates*), or whether the tetrahedra form one-dimensional infinite chains or double chains (*inosilicates*), two-dimensional infinite layer structures (*phyllosilicates*), or, finally, three-dimensional infinite cross-linked networks (*tektosilicates*). When two [SiO_4] tetrahedral complexes unite through a common O atom, one oxygen atom must be split off. Thus each such bond also represents the loss of half an oxygen atom per Si, so that the ratio of O to Si becomes smaller the greater the number of such bonds. The ratio alters from 4:1 for nesosilicates to 2:1 for tektosilicates. Three-dimensional networks of the composition SiO_2 have achieved the highest possible degree of condensation; they have no further valences available for bonding with other ions. All other forms are anionic since they carry negative charges on those free oxygen atoms which do not act as bridge atoms between two silicon atoms. The number of negative charges per Si atom corresponds to the number of free oxygens.

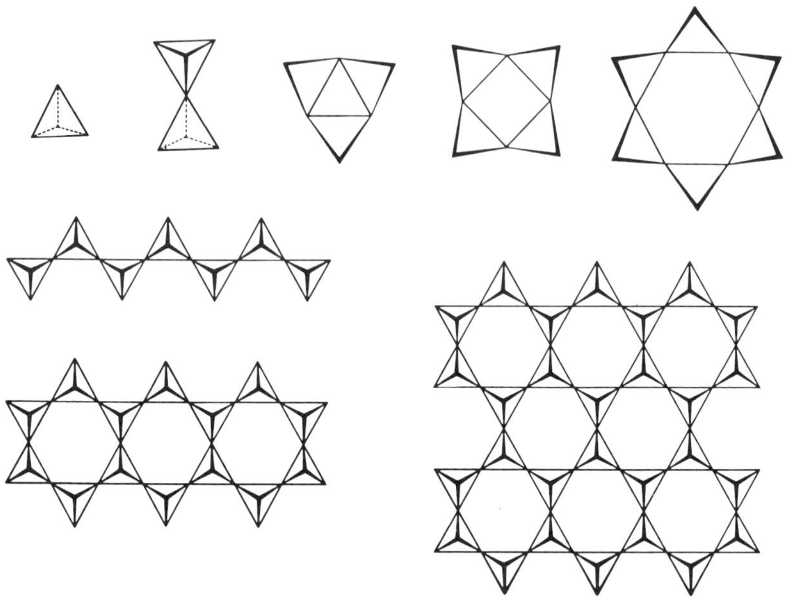

FIG. 18. Polymer structures in silicate anions.

With a single exception, the bonds known up to the present are those which have SiO_4 tetrahedral apexes, but not edges, in common; the exception, however, will be of particular interest in connection with the question of valence angles.

The silicate anions form a system of polymeric anions which can be brought into the closest analogy—at first sight purely formally—with the system of the organosiloxane polymers (*117*). It is possible to pass from the polymeric *anions* of the silicates to the polymeric *molecules* of the organosiloxanes by imagining the free oxygen atoms not engaged in linking silicon atoms to be replaced by organic groups. Analogously—and again formally—substitution by OR groups would lead to organooxysiloxanes (silicic acid and polysilicic acid esters), and substitution by OH groups to the silicic acids themselves; a few silicon oxyhydrides can also be related to the system of the silicate anions by the replacement of free O atoms by H atoms (Table 46).

TABLE 46

POLYMER BONDS IN SILICATE ANIONS AND THE MOLECULES OF THE POLYORGANOSILOXANES, POLYORGANOOXYSILOXANES, SILICIC ACIDS, AND SILICON OXYHYDRIDES

Silicate anion	Structure	Number of Bridging O atoms	Number of Nonbridging oxygen atoms	Polyorgano-siloxanes	Polyorgano-oxysiloxanes	Silicic acids	Silicon oxyhydrides
$[SiO_4]^{4-}$	Single	0	4	R_4Si	$(RO)_4Si$	$Si(OH)_4$	—
$[Si_2O_7]^{6-}$	Group	1	6	R_6Si_2O	$(RO)_6Si_2O$	$Si_2O(OH)_6$	Si_2OH_6
$[Si_3O_9]^{6-}$	Ring	3	6	$R_6Si_3O_3$	$(RO)_6Si_3O_3$	$Si_3O_3(OH)_6$	$[SiOH_2]_n$[a]
$[Si_6O_{18}]^{12-}$	Ring	6	12	$R_{12}Si_6O_6$	$(RO)_{12}Si_6O_6$	$Si_6O_6(OH)_{12}$	
$[SiO_3]_n^{2n-}$	Chain	n	$2n$	$[R_2SiO]_n$	$[(RO)_2SiO]_n$	$[SiO(OH)_2]_n$	—
$[Si_2O_5]_n^{2n-}$	Network	$3n$	$2n$	$[RSiO_{1.5}]_n$	$[ROSiO_{1.5}]_n$	$[SiO_{1.5}OH]_n$	$[SiO_{1.5}H]_n$[b]

[a] Pro-siloxane in the older literature.
[b] Silicoformic acid anhydride in the older literature.

The knowledge obtained from X-ray crystallographic studies of the silicates has been most fruitful for the entire field of silicon compounds with Si—O—Si bonds. Even the problem of the molecular structure of the silicic acids, for instance, has been presented in a new light. It is nowadays generally accepted that in the condensation of silicic acids, the sequence of the various condensation stages shown in Table 46 is in fact followed.

Structural analyses of polymethylsiloxanes can be used (*119*) for a detailed investigation of how far the formally derived relationships do in fact exist. Of all the polyorganosiloxanes, these would appear to be most suitable for the purpose, since the siloxane units in their structure come closest to the tetrahedral SiO_4 complexes of the silicates.

Apart from the question of the valence angles—which will be discussed later —there are direct structural analogies between *group-forming* silicate anions and polymethylsiloxane molecules (Table 47). The expected parallels are also

TABLE 47

GROUP-FORMING ASSOCIATIONS OF TETRAHEDRA (*119*)

As silicate anions	As polymethylsiloxanes
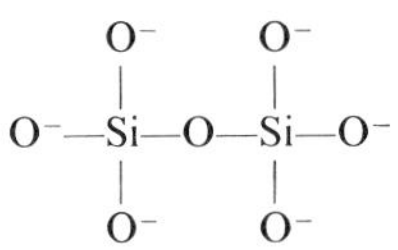	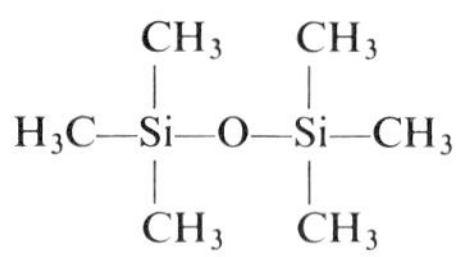
$Si_2O_7^{6-}$	R_6Si_2O
E.g., Thortveitite $Sc_2[Si_2O_7]$ Tilleyite $Ca_5[Si_2O_7\|(CO_3)_2]$	Hexamethyldisiloxane
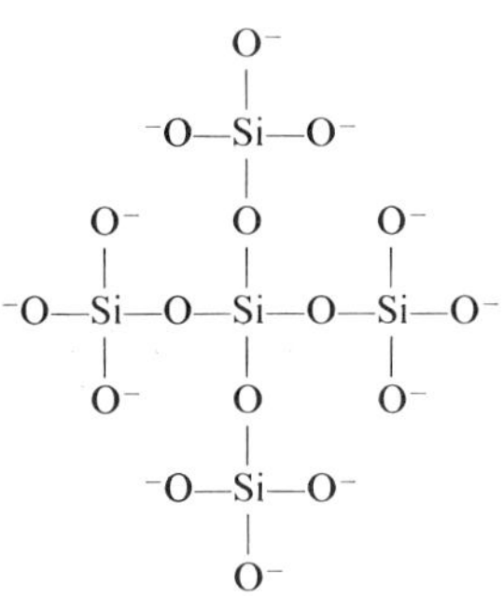	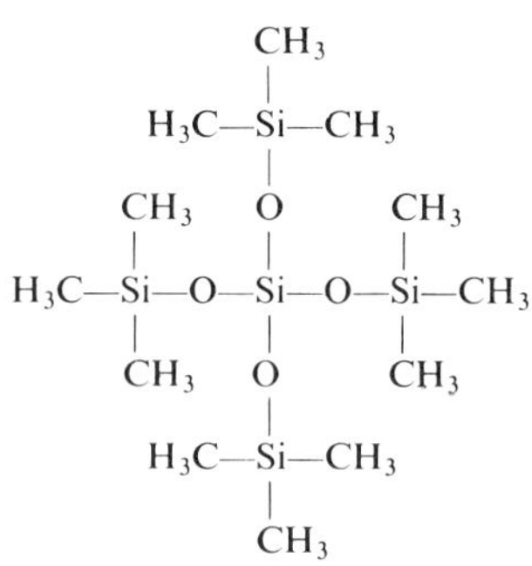
$Si_5O_{16}^{12-}$	$R_{12}Si_5O_4$
Zunyite $Al_{12}[AlO_4\|(OH,F)_{18}Cl\|Si_5O_{16}]$	Dodecamethylpentasiloxane

found with the *monocyclic* structures (Table 48), but the variety of conformations of the polymethylsiloxane rings appears to be greater than that of the monocyclic silicate anions. A whole series of the former with three to nine siloxane units has been prepared; with the latter, however, only monocyclic anions with three or six siloxane units are definitely known (Figs. 18*c* and 18*e*). The existence of four-membered rings (Fig. 18*d*) which earlier formed

TABLE 48

CYCLIC ASSOCIATIONS OF TETRAHEDRA (*119*)

As silicate anions	As polymethylsiloxanes
Monocyclic	
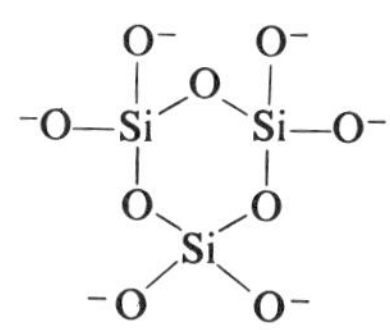 $Si_3O_9^{6-}$ E.g., benitoite $BaTi[Si_3O_9]$	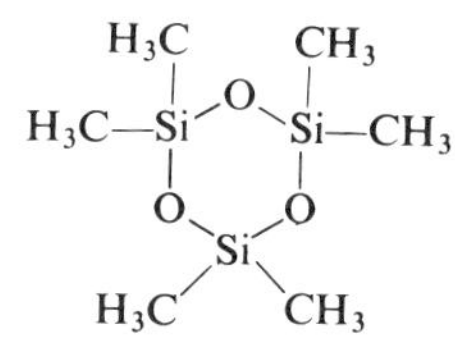 $R_6Si_3O_3$ Hexamethylcyclotrisiloxane
$Si_4O_{12}^{8-}$ Axinite $Ca_2(Fe, Mn)Al_2[OH\vert BO_3\vert Si_4O_{12}]$ (?) Baotite $Ba_4(TiNb)_8ClO_{16}[Si_4O_{12}]$ (?)	$R_8Si_4O_4$ Octamethylcyclotetrasiloxane
$Si_6O_{18}^{12-}$ E.g., beryl $Be_3Al_2[Si_6O_{18}]$ Dioptase $Cu_6[Si_6O_{18}]$ $6H_2O$	$R_{12}Si_6O_6$ Dodecamethylcyclohexasiloxane
$Si_8O_{24}^{16-}$ No example	$R_{16}Si_8O_8$ Hexadecamethylcyclooctasiloxane
Polycyclic	
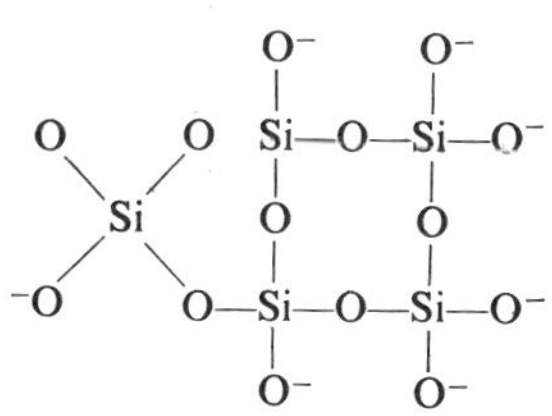 $Si_5O_{14}^{8-}$ No example	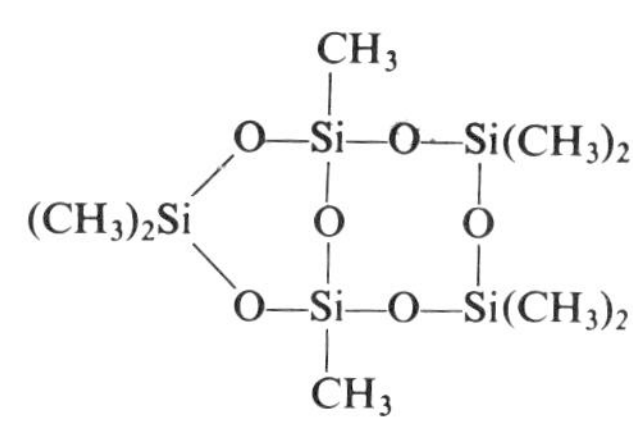 $R_8Si_5O_6$ Octamethyldicyclopentasiloxane

(*continued on p.* 292)

TABLE 48 *(Contd.)*

CYCLIC ASSOCIATIONS OF TETRAHEDRA (*119*)

As silicate anions	As polymethylsiloxanes
Spirocyclic	
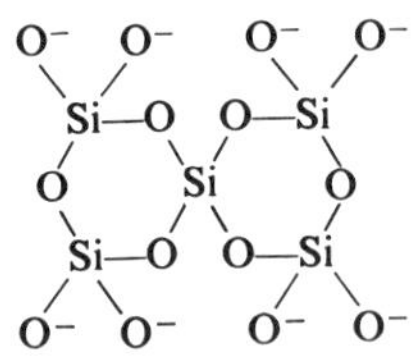	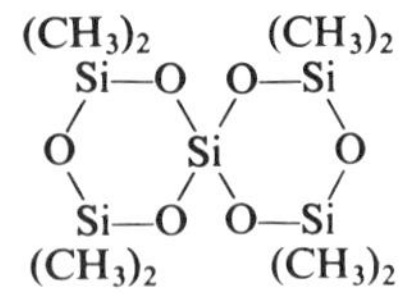
$Si_5O_{14}^{8-}$	$R_8Si_5O_6$
No example	Octamethylspiropentasiloxane
Spherocyclic	
$Si_{12}O_{30}^{12-}$ Milarite $KCa_2AlBe_2[Si_{12}O_{30}]$ $1/2H_2O$	$R_{12}Si_{12}O_{18}$ Dodecamethyldodecasilsesquioxane
$Si_8O_{20}^{8-}$ No example	$R_8Si_8O_{12}$ Octamethyloctasilsesquioxane
$Si_6O_{15}^{6-}$ Possibly dalyite $K_2Zr[Si_6O_{15}]$	$R_6Si_6O_9$ Hexamethylhexasilsequioxane
$Si_4O_{10}^{4-}$ No example	$R_4Si_4O_6$ Tetraisopropyl or tetra-*t*-butyltetrasilsesquioxane

part of the classical conception of systematic silicate classification, has recently been questioned. These structures may perhaps exist in the lattices of axinite and baotite. In any case, their rareness is in striking contrast to the frequent occurrence of the four-membered ring in polymethylsiloxanes; octamethylcyclo-tetrasiloxane is usually the main product in a mixture of siloxanes formed by the hydrolysis of dimethyldichlorosilane and subsequent condensation of the hydrolysis products.*

The variety of structures of the polymethylsiloxanes appears still more markedly in the case of condensed ring systems. *Polycyclic* and *spirocyclic* structures are so far completely unknown as silicate anions. The anion of milarite could be considered a counterpart of the *spherocyclic* molecule of dodecamethyldodecasilsesquioxane (Fig. 17), and the anion of dalyite would

* The relative energy of the tetrahedral rings with n SiO units has been calculated on the basis of a considerably simplified ion potential law (*209*); a minimum was found at $n \approx 5$.

perhaps correspond to hexamethylhexasilsesquioxane; no example among the silicates can be found corresponding to the eight-membered spherocyclic compounds. The smallest structure in this series, i.e., a four-membered spherocyclic compound, is so far only known as a molecule in which the R groups are isopropyl and *t*-butyl (Fig. 16).

If we now look at the entire picture presented by the arrangement of anions and molecules with *linear* associations of tetrahedra, it is possible to distinguish certain characteristic differences between the two series. Linear polydimethylsiloxane molecules do in fact preferentially assume the helical conformation (Fig. 13); the associations of tetrahedra in silicate lattices with chain anions, on the other hand, are always stretched. In this, they mostly have a linear form with a repeating unit of two tetrahedra (pyroxene-type chain, Fig. 20b); more rarely three (β-wollastonite), five (rhodonite), or seven siloxane tetrahedra (pyroxmangite (*100*); or they have a zigzag shape as in the double chains of anions in sodium and lithium metasilicates (Fig. 21) (*64*, *190*). The formation of stretched chainlike compounds with their sometimes remarkably large repeating units is clearly connected with the cation-anion interaction in the crystal lattice, i.e., with the steric constraint on the chain of silicon-oxygen tetrahedra in its condensation with the cation-oxygen polyhedra. This constraint is obviously absent in the case of the polymethylsiloxane molecules. It may be assumed, particularly when considering the weak intermolecular forces between polymethylsiloxane molecules (Section 6.5), that the structure of the molecular chain is determined only by the tendency to compensate its own internal Si—O dipoles and that this is achieved in the helical conformation.

The following special cases can now also be made to fit logically into this picture. As spreading experiments on water show, when a polydimethylsiloxane chain is exposed to the surface forces at an interface, it becomes oriented and assumes conformations which approach the stretched forms of the silicate ions, particularly the double chain of the pyroxene type (Figs. 10 and 20*b*; Section 6.2.1). On the other hand, although the characteristic helical form of the molecule does not appear as a type of linkage of silicon-oxygen tetrahedra in silicate lattices, it does appear in the structures of silica modifications such as β- and α-quartz, and in keatite. It is characteristic that the helical form appears in lattices which are free of foreign cations and which are thus not exposed to the steric constraint which these exert on the siloxane chain. Consequently, all helices consisting of silicon-oxygen tetrahedra are condensed with one another through their free oxygen atoms so that three-dimensional networks of the composition SiO_2 are formed.

A further very interesting difference between the chain structures of the two series of compounds is brought to light by considering the short-chain siloxane polymers. It has been possible to synthesize short-chain molecules with small definite numbers of siloxane units in polymethylsiloxane chemistry (Table 33).

Silicate anions with the corresponding number of small chain members are unknown. Among the silicates, there is a gap not covered by any example between the group anion $[Si_2O_7]^{6-}$, which can also be considered as the first and shortest member of a siloxane chain, and the one-dimensional infinite chain anion:

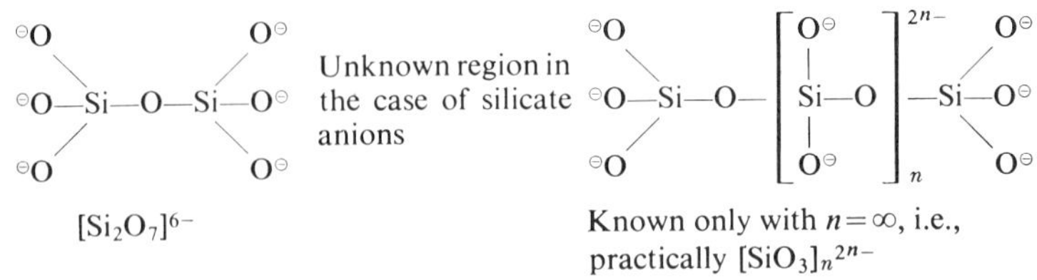

This phenomenon finds its counterpart in the fact that silicate anions corresponding to polycyclic ring systems are unknown. In this case, also, silicate anions form only the first and last members of a series which extends from the simple ring silicate anion to the one-dimensional infinitely condensed polycyclic compounds, usually classified as double-chain anions in the systematic classification (Fig. 18).

Still greater divergencies in structure between molecules and anions are found in the case of siloxanes with high degrees of cross-linking. A wide variety of molecular structures is conceivable in this case: these would in general have the characteristics of glasslike, three-dimensional, cross-linked networks. It is to be expected that it will be possible to prepare siloxanes with structures corresponding to the Si_2O_5 sheets of the phyllosilicates only under special circumstances. *H*-Silsesquioxane $(HSiO_{3/2})_n$ has been produced in micalike platelets (*203*); it is likely that sesquioxanes at least partially substituted with small organic groups could also have a similar structure. A one-dimensional ladder structure, which in principle corresponds to the silicate double chains, perhaps exists in a phenylsilsesquioxane polymer (cf. Section 6.2.4).

This discussion of the relationships between the structures of silicates and silicones cannot be concluded without consideration of the problem from a somewhat different standpoint; this will broach aspects important for the understanding of technological problems. For this purpose we commence with a completely but aperiodically cross-linked system of SiO_4 tetrahedra in the structure of silica glass, taking as a basis the structure of silica glass suggested by Zachariasen. It has been drawn as a simplified plane projection with only free oxygen atoms round each Si atom in Fig. 19*a*. According to this, the glass may be regarded as a network of irregularly formed rings of SiO_4 tetrahedra of different sizes, or as the result of the complete and irregular cross-linkage of chains of SiO_4 tetrahedra.

From such a structure one can pass to the structures of the polyorganosiloxanes by a stepwise substitution of two organic groups per oxygen atom in the framework. This blocks the branching points on the silicon atom and reduces the functionality of the chain elements:

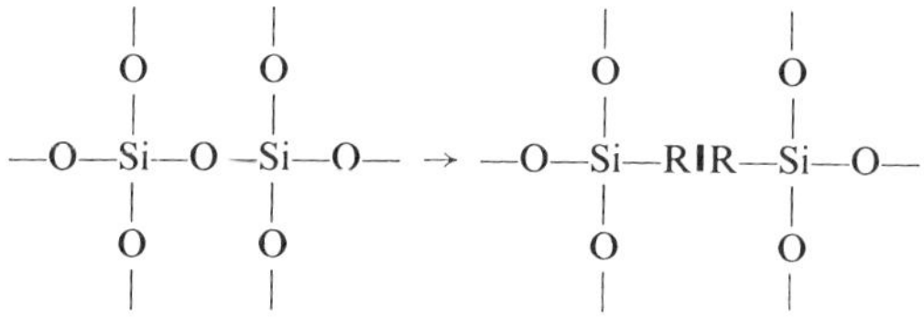

It follows from this that, with the incorporation of increasing numbers of R groups, the degree of cross-linking is decreased and associations are broken up. Two intermediate stages which arise in the course of this "substitutive degradation" of silica glass are shown in Figs. 19*b* and 19*c*.* A resinous polysiloxane with a structure still close to that of silica glass and which could in a sense be considered as a glass plasticized by organic groups is formed as a first stage; incorporation of further organic groups finally leads to linear molecules such as those present in silicone oils. Another kind of degradation may obviously leave ringlike polyorganosiloxane molecules as the final units.

Many properties of silicone resins which are closely related to the network structures of silica glass can be understood in this way. For example, it can easily be seen that phenyl groups must have a stronger plasticizing effect than methyl groups; that with splitting off of organic groups, perhaps by oxidation, the danger arises that the skeleton of the silicone resin can be increasingly cross-linked through oxygen bridges and thus approach the structure of SiO_2 glass; and finally, that danger of embrittlement of a resin is greater, the more glasslike is its structure, the poorer it is in organic groups, and the more easily oxidizable these are. The practical consequences of these observations will be discussed in Chapter 9.

The molecular structures of the polyorganooxysiloxanes can be derived from silica glass in a basically similar manner; the only difference here is that two RO groups are inserted between two Si atoms in place of a bridge oxygen atom

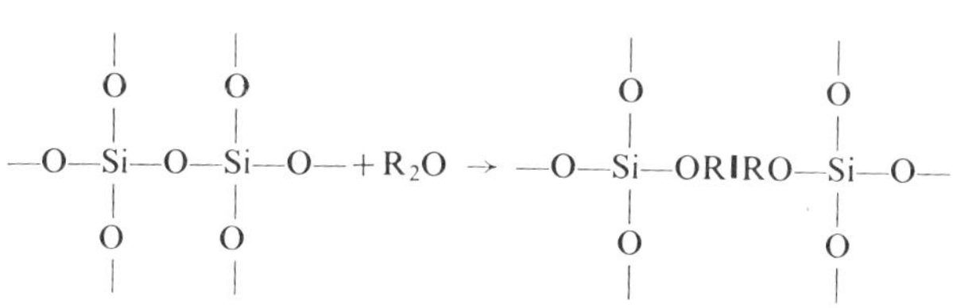

* For the sake of simplicity and clarity, silicon atoms have been taken out of the network, and at the same time O atoms have been replaced by R groups.

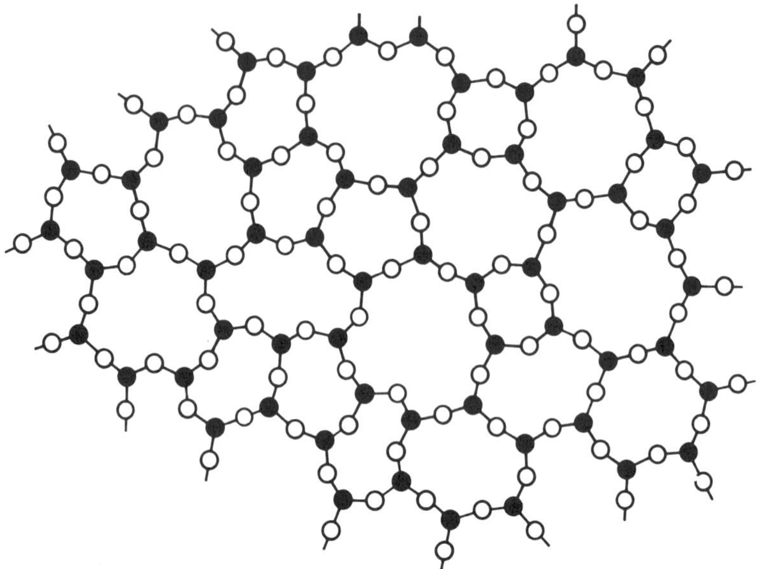

(a)

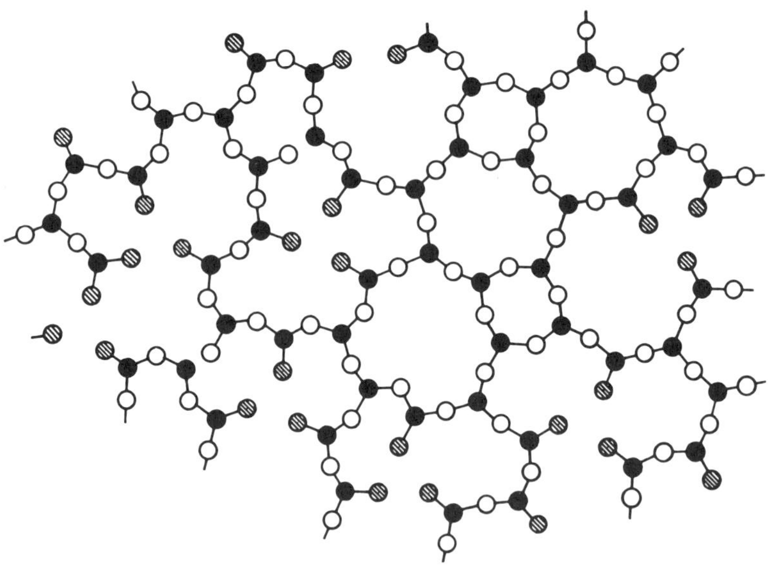

(b)

FIG. 19. Substitutive degradation of silica glass (a) to the structures of silicone resins (b) and silicone oils (c).

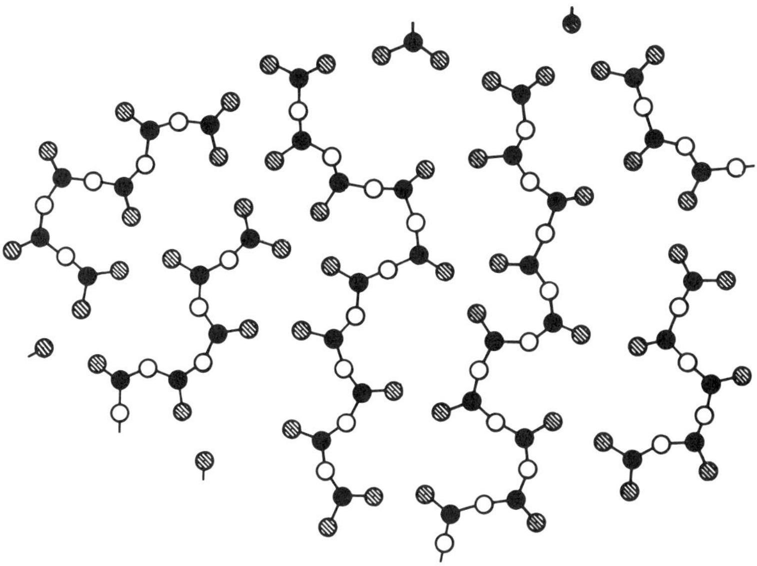

FIG. 19. (c).

so that rupture of the chain occurs at this point. This derivation is in close agreement with the *Trennstellentheorie* (cleavage site theory) of Hofmann *et al.* (*39*) which was developed for silicate glasses and which states that the incorporation of metal oxides causes loss of siloxane bridges:

$$\begin{array}{ccccccccc}
 & | & & | & & | & & | & \\
 & \mathrm{O} & & \mathrm{O} & & \mathrm{O} & & \mathrm{O} & \\
 & | & & | & & | & & | & \\
-\mathrm{O}- & \mathrm{Si} & -\mathrm{O}- & \mathrm{Si} & -\mathrm{O}- + \mathrm{M_2O} \rightarrow -\mathrm{O}- & \mathrm{Si} & -\mathrm{OMIMO}- & \mathrm{Si} & -\mathrm{O}- \\
 & | & & | & & | & & | & \\
 & \mathrm{O} & & \mathrm{O} & & \mathrm{O} & & \mathrm{O} & \\
 & | & & | & & | & & | &
\end{array}$$

Finally, the principle of substitutive network degradation can be applied to networks that also contain heteroatoms other than Si as network formers. In this sense, potassium tetrakis-(trimethylsilyl)-aluminates (Sections 1.2 and 7.1.2) may be regarded as compounds whose monomeric anions have resulted from the degradation of the Al—Si—O linkage of the felspar anions by methyl substituents.

6.3.2 Interatomic Distances, Valence Angles

We shall now extend the comparison of silicates and silicones to a discussion of interatomic distances and valence angles.

The Si—O distances in the $[SiO_4]$ tetrahedra of silicate anions vary considerably, which could be due in particular to the nature and arrangement of the cationic partners in the lattice. Such influences are absent only in lattices of the forms of SiO_2; the Si—O distances in these can therefore be considered representative of the undisturbed Si—O bond. Variations about the average of 1.604 Å are in fact small in this case (Table 49).

TABLE 49

Si—O Distances (Å) in Various Forms of SiO_2 (*120*)

Modification	Individual values	Mean
Low α-quartz	1.61; 1.62; 1.60; 1.62	1.615
	1.605	1.605
High β-quartz	1.61_2	1.61_2
Low cristobalite	1.59; 1.59	1.59
Keatite	1.58; 1.59; 1.61; 1.61	1.59_4
	1.61; 1.61; 1.57; 1.57	
Coesite	1.60; 1.63; 1.61; 1.61;	1.61_3
	1.62; 1.60; 1.62; 1.61	
SiO_2 glass	1.60 ± 0.05	1.60

It is possible to differentiate in silicate lattices between Si—O bonds in which the oxygen acts as a link between Si atoms—that is, as siloxane oxygen—shown here by Si—O(→Si), and those in which the oxygen forms a bond to metal atoms, shown by Si—O(→M). Statistical evaluation of structures which are not too highly cross-linked and are largely free of Al in the anions then shows that the Si—O(→Si) distances are somewhat greater than those of the Si—O(→M) bond (Table 50). This effect and its dependence on the nature of the M atoms can be interpreted from electronic theory (see Section 6.3.3).

The Si—O interatomic distances in polyorganosiloxanes can also be expected to depend on the nature of the bond partners. The data available up to the present, however, are inadequate to establish this point. Nevertheless, it can be stated that the Si—O interatomic distances in polymethylsiloxanes (Table 51) are remarkably constant; the mean value of about 1.645 Å is appreciably higher than the figures for the various forms of SiO_2. The Si—C distance is 1.88 Å and thus about 15% greater than the Si—O distance; the tetrahedral structural units are therefore appreciably distorted when compared with the SiO_4 complexes of the various forms of silica.

TABLE 50

Si—O DISTANCES IN SILICATE LATTICES (Å) (*120*)

		Si—O(→Si)	Mean	Si—O(→M)	Mean
Tilleyite	$Ca_5[Si_2O_7/(CO_3)_2]$	1.65_5; $1.65_5 \pm 0.04$	1.65_5	1.58_0; 1.62_0	
				$1.59_5 \pm 0.02$	1.60_0
Sodium metasilicate	$Na_4[Si_2O_6]$	1.68; 1.67	1.67_5	1.57	1.57
Enstatite	$Mg_2[Si_2O_6]$	1.59; 1.59	1.60	1.54; 1.54	
		1.63; 1.59		1.53; 1.60	1.55
Clinoenstatite	$Mg_2[Si_2O_6]$	1.67; 1.63 ± 0.05	1.65	1.65; 1.61 ± 0.05	1.63
Diopside	$MgCa[Si_2O_6]$	1.60; 1.64	1.62	1.57; 1.62	1.59_5
		1.60; 1.65 ± 0.05	1.62_5	1.56; 1.59 ± 0.05	1.57_5
Pigeonite	$(Mg,Fe,Ca)_2[Si_2O_6]$	1.69; 1.59 ± 0.05	1.64	1.62; 1.55 ± 0.05	1.58_5
Cummingtonite	$(Fe,Mg)_7[Si_4O_{11}/OH]_2$	1.61_9; 1.64_2		1.63_1; 1.62_7	
		1.61_9; 1.64_6	1.63_3	$1.61_7 \pm 0.02$	1.62_5
		$1.63_8 \pm 0.02$			
Crocidolite	$(Na,K,Ca)_{3-4}Mg_6Fe^{2+}$-				
	$(Fe^{3+},Al)_{3-4}[Si_{16}O_{44}/(OH)_4]$	1.65–1.66	1.65_5	1.59–1.60	1.59_5
Actinolite	$Na_2Ca_4(Mg,Fe)_{10}[Si_{16}O_{44}/(OH)_2/O_2]$	1.67	1.67	1.60	1.60
Lithium disilicate	$Li_2[Si_2O_5]$	1.67; 1.66; 1.69	1.65	1.56; 1.57	1.56_5
		1.57; 1.67; 1.66			
β-Sodium disilicate	$Na_2[Si_2O_5]$	1.63; 1.62; 1.65	1.64	1.50; 1.52	1.51
		1.65; 1.64; 1.65			
Sanbornite	$Ba_4[Si_8O_{20}]$	1.68; 1.64; 1.65	1.66	1.60	1.60

TABLE 51

MEASURED INTERATOMIC DISTANCES FOR Si—O AND Si—C[a]

Compound	Interatomic distance (Å) Si—O	Si—C	Method[b]
$(CH_3)_3SiOSi(CH_3)_3$	1.63 ± 0.03 (*207*)	1.88 ± 0.03 (*207*)	*E*(*V*)
$[(CH_3)_2SiO]_3$	1.66 ± 0.04 (*1*)	1.88 ± 0.04 (*1*)	*E*(*V*)
$[(CH_3)_2SiO]_4$	1.65 (*184*)	1.92 (*184*)	*X*(*C*)
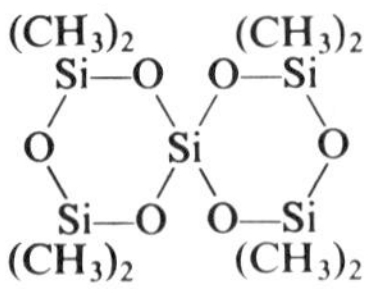	1.64 ± 0.03 (*149*)	1.88 ± 0.03 (*149*)	*X*(*C*)
$H_3Si—O—SiH_3$	1.64 ± 0.01 (*48*)	—	*E*(*V*)
	1.633 ± 0.001 (*26*)	—	*E*(*V*)
$Cl_3SiOSiCl_3$	1.64 ± 0.05 (*207*)	—	*E*(*V*)
	1.61 ± 0.03 (*48*)	—	*X*(*C*)
$Si(OCH_3)_4$	1.64 ± 0.03 (*207*)	—	*E*(*V*)

[a] Numbers in parentheses indicate reference numbers.

[b] Here, *E* indicates electron diffraction; *X*, X-ray diffraction; *D*, measurement of dipole moments; (*V*), vapor; (*L*), liquid; and (*C*), crystal.

TABLE 52

MEASURED VALENCE ANGLES[a]

Compound	Valence angle (deg) Si—O—Si	O—Si—O	C—Si—C	Method[b]
$(CH_3)_3SiOSi(CH_3)_3$	130 ± 10 (*207*)	—	111 ± 4 (*207*)	*E*(*V*)
	131 (*50*)	—	—	*D*(*L*)
	137 ± 7 (*48*)	—	—	*E*(*V*)
$[(CH_3)_2SiO]_3$	125 ± 5 (*1*)	115 ± 5 (*1*)	—	*E*(*V*)
$[(CH_3)_2SiO]_4$	142.5 (*184*)	109 (*184*)	106 (*184*)	*X*(*C*)
$[(CH_3)_2SiO]_8$	130 ± 10 (*51*)	—	—	*X*(*C*)
$(CH_3)_3SiO[(CH_3)_2SiO]_nSi(CH_3)_3$[c]	130 ± 10 (*51*)	—	—	*X*(*C*)
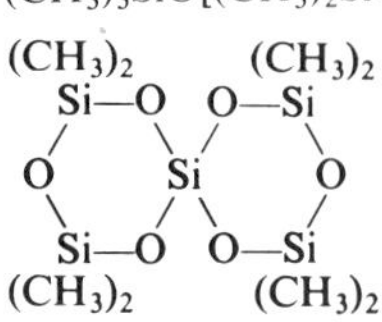	130 ± 4 (*149*)	109 ± 4 (*149*)	109 ± 4 (*149*)	*X*(*C*)
$H_3SiOSiH_3$	154 ± 4 (*27*)	—	—	[d]
	142 ± 4 (*48*)	—	—	*E*(*V*)
	144 ± 1 (*26*)	—	—	*E*(*V*)
$Cl_3SiOSiCl_3$	175 ± 5 (*48*)	—	—	*X*(*C*)

[a] Numbers in parentheses indicate reference numbers.

[b] Same symbol definitions here as given in Table 51.

[c] Measured on stretched highly cooled silicone rubber.

[d] Estimated from infrared spectra.

As regards the valence angles, the data given in Table 52 indicate that the O—Si—O and C—Si—C angles vary only slightly about the tetrahedral angle (109° 28′) which is also, in general, preserved in the SiO_4 complexes of the silicates.

The valence angle of the siloxane oxygen in polymethylsiloxanes shows relatively small variations (Table 52); it is however larger than the tetrahedral angle in the O—Si—O and C—Si—C bonds, amounting to approximately 130°. The measurements obtained for this angle in the hexamethyldisiloxane molecule should be given considerable weight, since they were carried out in the gas phase, when the molecule is exposed to only weak intermolecular forces and the valence angle assumes a value which can be taken as characteristic for the bond.

The more recent data (*119*) presented in Table 53 show that an angle between about 130–140° is also found particularly frequently in silicate anions. No

TABLE 53

VALENCE ANGLE AT SILOXANE OXYGEN IN A NUMBER OF SILICATE LATTICES (*119*)

Anion	Example	Angle (deg.)		
		Mean	Min	Max
Groups	Tilleyite $Ca_5[Si_2O_7\|(CO_3)_2]$	164		
Rings	Benitoite $BaTi[Si_3O_9]$	133		
	Dioptase $Cu_6[Si_6O_{18}]\cdot 6H_2O$	139		
Chains	Diopside $CaMg[Si_2O_6]$	133		
	Sodium metasilicate $Na_4[Si_2O_6]$	137		
	Lithium metasilicate $Li_4[Si_2O_6]$	133		
	Cummingtonite $(Fe,Mg)_7[Si_8O_{22}\|(OH)_2]$	141.5	139.4	142.9
	Actinolite $(Na,Ca)_2(Mg,Fe)_5[(Si,Al)_8O_{22}\|(OH)_2]$	135	132	138
Layers	Amesite $Mg_2Al[(Si,Al)O_5\|(OH)_4]$	133		
	Chlorite $(Mg,Fe,Al)_6[(Si,Al)O_{10}\|(OH)_8]$	134		
	Petalite $LiAl[Si_4O_{10}]$	156	150	166
	Lithium disilicate $Li_2[Si_2O_5]$	137	128	156
	Dickite $Al_2(OH)_4[Si_2O_5]$	134	129	[illegible]
Networks	Orthoclase $K[AlSi_3O_8]$	142	131	153
	Potassium felspar, triclinic $K[AlSi_3O_8]$	142	130	155
	Albite $Na[AlSi_3O_8]$, low-temperature modification	139.5	125	160
	high-temperature modification	139.5	128	155
	Nepheline $KNa_3[Al_4Si_4O_{16}]$	139	133.5	155
	Keatite SiO_2	153	149	156

value below 130° is found, but higher values are possible, which according to data now available may amount to 160–170°. This characterizes the differences between the situations in lattice structures (silicates) and in liquid or gaseous phase (silicones). In the former, the Si—O—Si bonds are subject to symmetry requirements and the influence of lattice forces which may result in a deformation of the valence angle. There is obviously a greater resistance to a decrease than an increase; complete stretching to 180° does not appear to be possible (*7, 101*).

The tendency of the valence angle on the siloxane oxygen to remain close to a value of 130°, or at least not to fall below it, now enables another phenomenon to be understood; we may call it the "rule of structural selectivity." For certain Si/O ratios there is a larger number of geometrically possible arrangements of

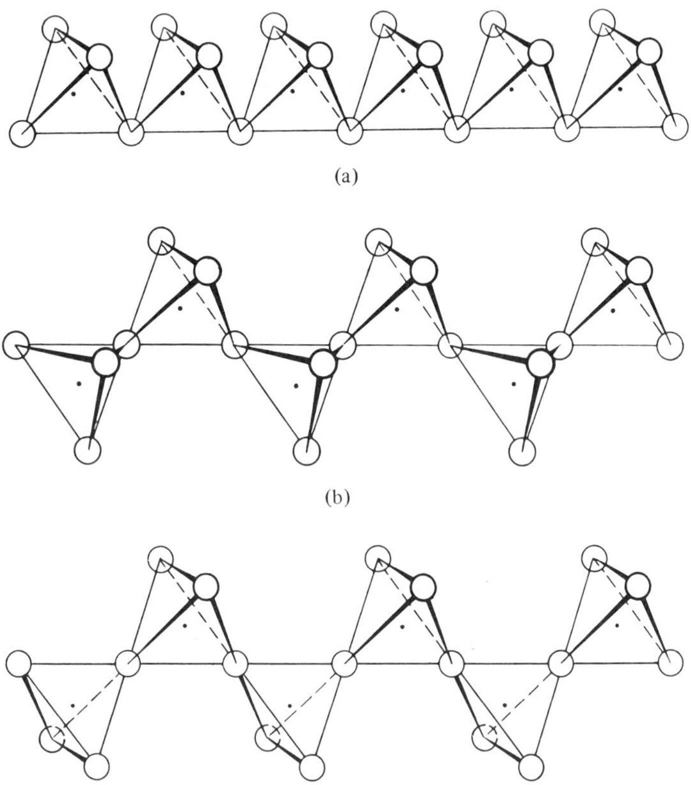

(a)

(b)

(c)

FIG. 20. (a) *Einerkette* consisting of $[SiO_4]$ tetrahedra, composition $[SiO_3]_n^{2n-}$; not found. (b) *Zweierkette* $[Si_2O_6]_n^{4n-}$, e.g., anion of the pyroxenes. (c) *Zweierkette trans*-configuration, not found in silicates.

the silicon-oxygen tetrahedra than is actually found in nature (*96*). It can be shown that nature chooses and constructs only those arrangements from all the possible ones in which the valence angle at the siloxane oxygen (*118*) is not appreciably less than 130°. This explains for instance why the so-called *Einerkette* consisting of $[SiO_4]$ tetrahedra—the repeating unit here is a *single* tetrahedron—is not realized in nature; the Si—O—Si valence angle would have to be deformed to the tetrahedral angle of 109° 28′ in this case (Fig. 20*a*).

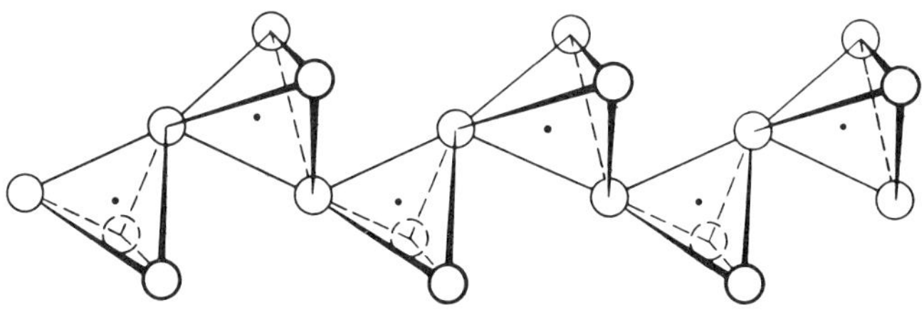

FIG. 21. Zigzag *Zweierkette* derived from the transformation of Fig. 20c by compression (anion of Na and Li metasilicates).

If, however, every second tetrahedron is rotated out of the axis of the bonding oxygen atoms until the Si—O—Si angle is around 130–140° so-called *Zweierketten*, which are frequently found in silicates, are obtained (Fig. 20*b*). Continuation of this rotation finally leads to an Si—O—Si angle of 180° which does not appear to be realized in practice (Fig. 20*c*). Compression of this latter chain into a zigzag form can produce an arrangement (Fig. 21) which again possesses an Si—O—Si valence angle between 130 and 140° and which is found in sodium and lithium metasilicate.

For the same reasons, ribbonlike (Fig. 22) and sheetlike (Fig. 23) configurations formed by the condensation of *Einerketten* and having the composition $[Si_2O_5]_n^{2n-}$ or $[SiO_2]_n$ do not exist. If however aluminum and silicon atoms alternate, the forces stabilizing the valence angle are weakened because of the more highly polar nature of the Al—O—Si bond. The anion of sillimanite $[SiAlO_5]_n^{3n-}$ has the structure given in Fig. 22.

The group of tetrahedra shown schematically in Fig. 16 is also of interest in this connection. It might be thought that a complex of the composition $[Si_2O_5]_2^{4-}$ or $[RSiO_{1.5}]_4$ could have this structure. No anion of this type is, however, found in the silicates. This could again be interpreted as a consequence of the fact that in this case the valence angle on the siloxane oxygen would have to be equal to the tetrahedral angle of 109° 28′. No molecule of this structure has been found in the case of the methyl- and ethylsiloxanes either, but it is formed, as far as can be deduced from investigations available at present, when the organic group has a bulky structure, e.g., when it is an isopropyl or *t*-butyl radical (*202*). The organic groups with their greater space requirements then force the siloxane bond into a configuration in which they have more room than in groups of relatively large numbers of tetrahedra.

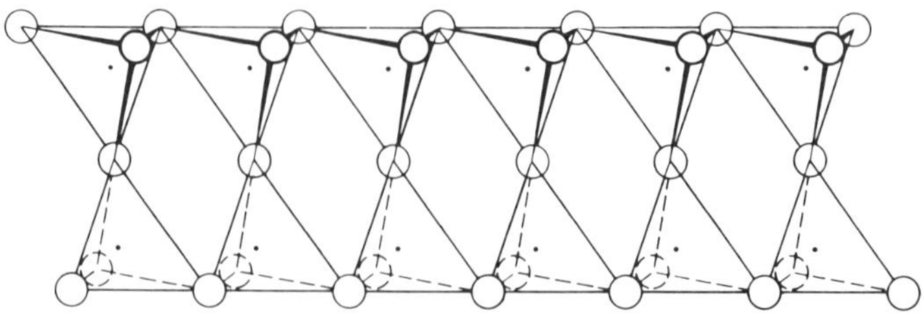

FIG. 22. $[Si_2O_5]_n^{2n-}$ ribbon-shaped condensation of two *Einerketten*; not realized (known only as the $[SiAlO_5]_n^{3n-}$ anion in sillimanite).

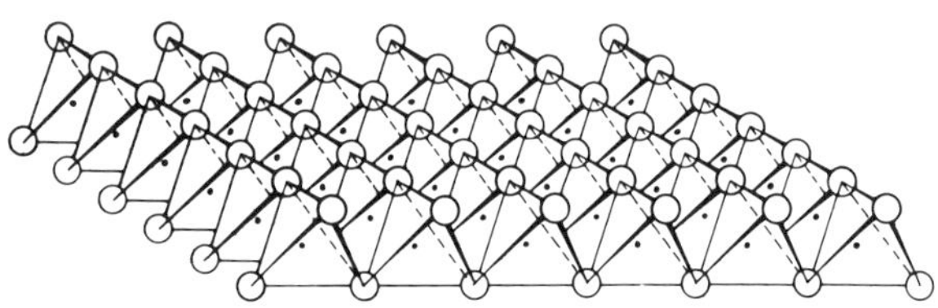

FIG. 23. $[SiO_2]_n$ sheet formed by condensation of *Einerketten*; not realized.

Up to now only a single structure has been found in which the valence angles of the siloxane oxygen is decreased well below 130°; this is the structure of a rhombic modification of silica (*197*). Contrary to the rules valid for the silicate anions and organosiloxane molecules in all other forms of SiO_2, the $[SiO_4]$ tetrahedra are here linked by their edges:

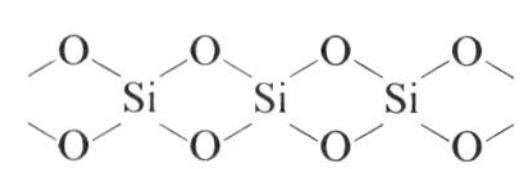

In this case the Si—O—Si valence angle is 88°; it is therefore larger than would be expected from the linking of two undeformed $[SiO_4]$ tetrahedra (theoretically 70° 32′). This indicates that the tetrahedra try to avoid a drastic deformation of the Si—O—Si valence angle by undergoing a slight deformation on being linked through their edges. Such a structure is naturally unstable; the rhombic form of SiO_2 undergoes spontaneous rearrangement to tridymite. It is decomposed by traces of water vapor even at room temperature, being converted with swelling into metasilicic acid (*197*).

6.3.3 Bond Strength, Bond Character, and Reactivity of the Siloxane Bond

6.3.3.1 *Single-Bond Energies*

The average *single-bond energy* is taken as a measure of the strength of a bond A—B; it is defined from the heat of formation of the gas AB_n from its elements at 0°K and $p=0$, and is the nth part of this value; it is measured in kilocalories per mole. Pauling has calculated a number of single-bond energies from thermochemical data (*130*); Skinner has carried out an evaluation of the dissociation energies of diatomic molecules (*167*) on the basis of spectroscopic data. Gilman and Dunn (*61*) have supplemented those figures insofar as they were of interest for a comparison of silicon and carbon chemistry. Taking into account new calculations by Thompson (*191*) from the heat of combustion of siloxanes, by Tannenbaum (*188*) from those of silanes, and by Tanaka and Watase (*187*) from the heats of combustion of organooxysilanes and siloxanes, the present picture is as given in Table 54. It should be noted that the

TABLE 54

SINGLE-BOND ENERGIES[a]

	Data according to Pauling (*130*) (kcal/mole)	Other authors
Si—O	88.2	106 (*61*); 117 (*191*); 101–102 (*187*)
C—O	84.0	80.9 (*61*)
Si—C	69.3	75.0 (*61*); 64 (191); 52–76 (*188*)
Si—Si	42.2	51.3 (*167*)
C—C	83.1	84.9 (*61*)
Si—H	70.4	79.9 (*167*)
C—H	98.8	98.1 (*61*)
Si—F	129.3	147.4 (*167*)
C—F	105.4	102 (*61*)
Si—Cl	85.7	90.3 (*167*)
C—Cl	78.5	77.9 (*61*)

[a] Numbers in parentheses indicate reference numbers.

value for the Si—C bond appears to depend on the nature of the organic group bond to the silicon atom [CH_3—Si, 74 kcal/mole; C_2H_5—Si, 62 kcal/mole; C_4H_9—Si, 54 kcal/mole (*188*)]. Comparison within each column shows that the single-bond energy for the Si—O bond in silicon compounds is exceeded only by that of Si—F bonds; according to Pauling the C—F bond is stronger than the Si—O bond and according to the other authors it is somewhat weaker or very nearly the same. The energy of the siloxane bond is higher in both systems than that of the C—C and Si—C bonds. This fact again quantitatively demonstrates the great affinity of silicon for oxygen; this is a fundamental principle of silicon chemistry.

However, the single-bond energy has only a limited value in the understanding of the chemical behavior of a bond. This is particularly true for all heterolytic decompositions. There is no explanation from thermochemical data of the well-known fact that the Si—O bond is more sensitive to heterolytic attack than the C—C bond, and an understanding of all these phenomena can be gained only from a discussion of the bond from the standpoint of electronic theory.

6.3.3.2 *The Siloxane Bond from the Standpoint of Electronic Theory*

The fact known for decades from the analysis of silicate structures that the Si—O interatomic distance is considerably smaller than the sum of the atomic and the ionic radii is the starting point for the following discussion. This also holds for all the polymethylsiloxanes so far investigated (Table 55).

TABLE 55

MEASURED AND CALCULATED Si—O INTERATOMIC DISTANCES (Å) (*120*)

		Calculated sum of:		As covalent-polar resonance bond	As covalent-polar-double resonance bond
	Measured	ionic radii	atomic radii		
Forms of SiO_2	1.60	1.71	1.83	1.67	1.63
Polymethylsiloxanes	1.64				

Attempts have long been made to explain the shortening of the distance to less than the sum of the atomic radii, by postulating that the siloxane bond has a polar and a covalent bond component* and thus corresponds to a resonance type. This is to be expected in view of the relatively large difference of the electronegativities, which Pauling (*130*) gives as 1.8 for Si and 3.5 for O.

Pauling (*130*) was the first to derive empirically an approximate formula for estimating the contributions of the two types of bonds to a covalent-polar

* From the standpoint of the polar bond the phenomenon can be understood by assuming that Si^{4+} as a strongly polarizing cation will polarize the O^{2-} anion; this reduces the internuclear distance and imparts a covalent-polar character to the bond.

linkage. This was based on measurements of the electric dipole moments of HCl, HBr, and HI. The polar bond component is calculated as a function of the absolute differences in electronegativity $|X_A—X_B|$ of the partners:

$$\text{Polar bond contribution, \%} = 100\,[1 - e^{0.25}(X_A—X_B)^2]$$

From this and the difference of 1.7 between the electronegativities of Si and O, it follows that the covalent and polar contributions to the siloxane bond amount to about 50% each. This value—which must, of course, be considered only as a first approximation—has been extensively quoted in the literature. It should, however, be remembered that it was subsequently much revised. An improvement of this calculation by Hannay (*65*) gave the more accurate equation

$$\text{Polar bond contribution, \%} = 16(X_A—X_B) + 3.5(X_A—X_B)^2$$

which reduces the polarity of the Si—O bond to 38%. Pauling himself recalculated the empirical relation which he had suggested, after the dipole moment of HF had also been measured, and arrived at a polarity of 41% in a bond having a difference in electronegativity of the bond partners (*131*) of 1.7.

The internuclear distance is reduced to less than the sum of the covalent bond radii, in accordance with the difference in the electronegativities. This was formulated by Schomaker and Stevenson (*155*) as follows:

$$d_{AB} = r_A + r_B - \beta(X_A—X_B)$$

where d_{AB} is the measured internuclear distance, r_A and r_B are the covalent bond radii of A and B, $|X_A—X_B|$ is the absolute value of the difference between the electronegativities of the partners, and $\beta = 0.09$. Using this equation we find $d_{Si—O} = 1.67$ (Table 55), i.e., a value just smaller than the measured Si—O distance in silicates and polymethylsiloxanes.

The differences which still exist between measured and calculated values can be reconciled, if it is assumed that additional double-bond components are present in the siloxane bond (*131*). This would make it a resonance bond in which *three* limiting bond types participate; these can be formulated for the SiO_4 complex as follows:

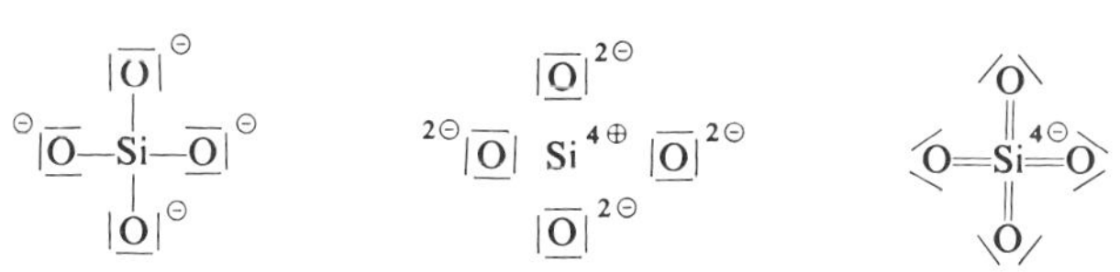

Covalent form Polar form Double-bond form

Further evidence for this resonance bond system can be found in the fact that by its use Pauling's postulate, which demands that in stable molecules or crystals the electric charge on each atom should approach zero, can be satisfied: The charge on the Si atom of the above complex is 4 in the polar form and 0 in the covalent form. Consider a resonance structure which according to the older concepts, had a 1:1 ratio of the two limiting bond types; the formal charge on the Si atom is therefore +2, which cannot be reconciled with the principle of electrical neutrality. In the double-bond form, however, Si has a formal charge of −4. The positive partial charge can be compensated by it, if all three limiting types are considered to constitute the bond in a ratio of approximately 1:1:1. Greater justice is also done to the observed bond moments by the assumption of a resonance bond system of covalent-polar-double bonds than by one consisting only of covalent-polar transition bonds (*92*).

The ability of a silicon atom to develop double-bond components with respect to oxygen is based on a phenomenon which has become increasingly important for the understanding of the chemical behavior of the silicones: its ability and tendency to use unfilled 3*d* levels as shown in the schematic presentation of the orbital arrangement (valence state) below.

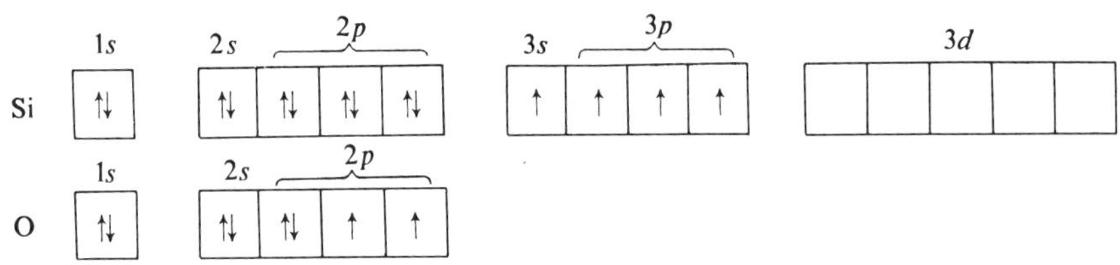

Bonds with a partner such as oxygen, which has lone electron pairs, therefore involve not only the 3*s* and 3*p* electrons of Si as sp^3 hybrids with respect to the *p*-valence electrons of the oxygen, but also the lone pairs of the latter; these can also partially enter the unfilled 3*d* orbitals of the silicon atom:

Of the five *d* orbitals of silicon, the $d_{x^2-y^2}$ and d_{z^2} orbitals participate, since they give the best overlap with the *p* orbitals of the oxygen.

However, we can only be dealing with double-bond *components* of the resonance system in this case. The existence of a pure Si=O double bond has not been experimentally proved in any one case.*

This ability to utilize 3*d* levels is also responsible for the fact that silicon can be penta- or hexacovalent and that it forms hybrid sp^3d and sp^3d^2 bonds. This

* The existence of pro-siloxane ($H_2Si{=}O$) claimed by Stock is not certain; the bond situation in silicon suboxide is still unclear (e.g., ref. *16*).

will occur most easily with those partners which are the strongest electron donors. The longest known example that demonstrates this ability to expand the electron levels is the formation by silicon of the complex $[SiF_6]^{2-}$ with fluorine. Under special circumstances, silicon also has a coordination number of 6 with respect to oxygen, as in the complex cation with acetylacetone (I) (*31*, *33*) and in the complex cation with catechol (II) (*147*), a catechol complex in which the Si atom bears a phenyl substituent containing penta-coordinate silicon (III) (*56*):

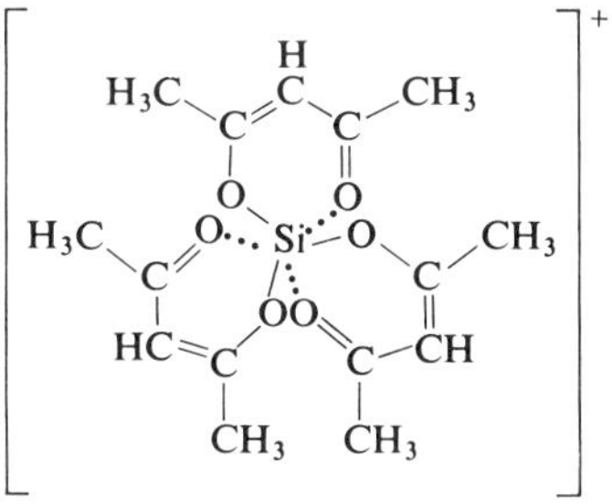

I

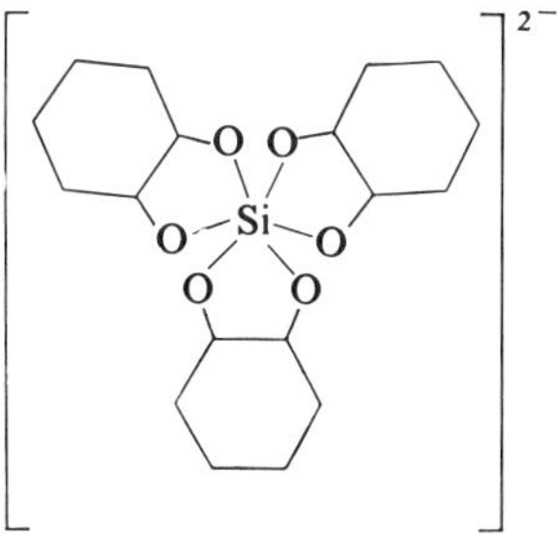

II

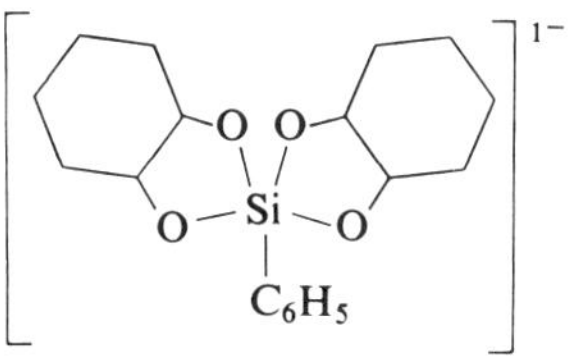

III

Whether more than six ligands can be bound appears doubtful for steric reasons. But it has been suggested that in a nucleophilic attack on the Si-acetylacetonate cation with the coordination number six a sevenfold coordinated state occurs as an intermediate (*131a*).

The idea of polar, covalent, and double-bond contributions to the siloxane bond in a resonance system has become of great importance for an understanding of the physical and chemical properties of the siloxanes as well as those of the silicates. It is not only able to account for the actual internuclear distances, but can also explain the relatively large valence angle at the siloxane oxygen. The nondirectional polar bond component cannot make any contribution to its stabilization. For a pure covalent bond, an angle of 90° might be expected, for pure double bond, an angle of 180°. The double-bond components thus increase the angle to a value of 130–140°, which approximately represents the mean of the two values.

6.3.3.3 *Influence of Substituents*

Substituents alter the ground state of the above "undisturbed" siloxane bond by induction. According to whether the hydrocarbon radicals or other substituents on the polyorganosiloxanes act as electron donors or electron acceptors, the following cases are conceivable:

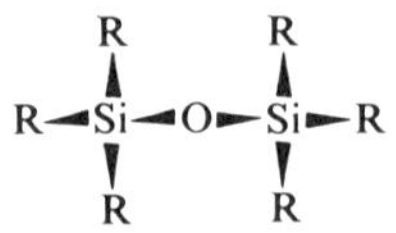

Si—O—Si bond weakened
Si—C bond strengthened

R = electron donor

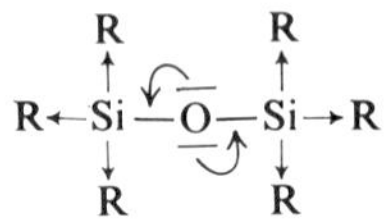

Si—O—Si bond strengthened
Si—C bond weakened

R = electron acceptor

Effects of this type can be demonstrated in many ways in the physical and chemical properties of such compounds.

TABLE 56

INFRARED AND RAMAN FREQUENCIES OF THE Si—O—Si BOND (*120*)

Siloxane	ν_{as}[a,b] Infrared (cm^{-1})	ν_s[a,b] Infrared (cm^{-1})	ν_s[a,b] Raman (cm^{-1})
$(CH_3)_3SiOSi(CH_3)_3$	1050 (S)	523 (U)	
	1055 (U)	522 (U)	519 (U)
$Cl(CH_3)_2SiOSi(CH_3)_2Cl$	1070 (U,S)	578 (U,S)	579 (U)
$Cl_2(CH_3)SiOSi(CH_3)Cl_2$	1085 (S)	638 (S)	634 (U)
$Cl_3SiOSiCl_3$	1110 (S)		722 (U)
$[ClCH_2(CH_3)_2Si]_2O$	1069 (S)	521 (U)	
$ClCH_2(CH_3)_2SiOSi(CH_3)_3$	1064 (S)	521 (U)	
$[BrCH_2(CH_3)_2Si]_2O$	1069 (S)	513 (U)	
$BrCH_2(CH_3)_2SiOSi(CH_3)_3$	1063 (S)	515 (U)	
$[(CH_3)_2(C_6H_5)Si]_2O$	1063 (S,U)	560 (S,U)	562 (U)
$[H(CH_3)_2Si]_2O$	1058 (S,U)	550 (S,U)	550 (U)
$[HO(CH_3)_2Si]_2O$	1073 (S)	555 (S)	554 (S)

[a] Here, subscripts as stands for antisymmetric and s for symmetric.
[b] Here, S stands for in solution and U for undiluted.

If 2, 4, and 6 methyl groups in hexamethyldisiloxane are successively replaced symmetrically by chlorine atoms, it is found that the infrared and Raman spectra clearly show displacement of the symmetric and antisymmetric Si—O—Si valence vibrations toward higher frequencies (*154*) (Table 56). The effect of the strengthening of the bond in the manner shown

is so pronounced that the opposite action of the greater mass of the halogen, as compared with that of the methyl group, is overcompensated.

An alteration of the valence angle cannot be excluded (Table 52), but its influence is overshadowed by the considerable increase of the force constant, since it would have to affect the two vibrations in opposite ways: the frequency of the symmetric vibrations would have to be decreased and that of the antisymmetric increased. Analogous effects have been observed in the infrared spectra of those siloxanes in which the substitution by halogen was made not directly on the Si atom but on the C_α atom (*164*).

Changes in the infrared and Raman frequencies of the Si—O—Si vibrations when methyl groups are replaced by hydrogen, phenyl, or hydroxyl groups are to be understood (*120*) in the same sense (Table 56). The substituent acts in all cases as an electron acceptor and reinforces the Si—O—Si bond, which becomes manifest in the increase of the frequencies ν_s and ν_{as}. It is interesting in this connection that hydrogen also shows this effect as a substituent on silicon; this corresponds to the chemical experience that the Si—H bond has a hydride-like character.

At present only measurements on $H_3Si—O—SiH_3$ and $Cl_3Si—O—SiCl_3$ show how far this substitution alters the *valence angle* of Si—O—Si (Table 52). The trend of these figures compared with those for $(CH_3)_3Si—O—Si(CH_3)_3$ is in the expected sense; the valence angle of Si—O—Si is increased more strongly by Cl than by H, as a result of the increase of the $d\pi—p\pi$ bond components under the influence of the electronegative substituents. Present figures are insufficient to judge the influence of substituents on the *nuclear distances*, since the deviations are obviously too close to the limits of error of the methods of measurement. We can only say that the Si—O nuclear distance of 1.64 Å in polymethylsiloxanes (confirmed by numerous measurements) differs from the value in SiO_2 lattices (1.604 Å), as would be expected from theory. The methyl group, which is a weak electron donor, increases the polar component of the Si—O—Si bond, and this increases the Si—O internuclear distance.

The finer differences due to the dependence of the bond strength on substituents are shown very sensitively, and perhaps most clearly, in the *chemical properties*. Reactions involving electrophilic attack on the free electron pairs of the siloxane oxygen are made more difficult, or are totally suppressed, according to how strongly these electrons are in demand by the electronegative substituents on the Si atoms of the Si—O—Si bond. Such differences become manifest, for instance, in the equilibration of differently substituted siloxanes with *proton acids*, particularly in the presence of a relatively weak catalyst, like acid-activated fullers' earth (*164*).

In the disiloxanes, the sensitivity of the Si—O—Si bond to cleavage decreases in the following sequence:

M—M	$(CH_3)_3Si—O—Si(CH_3)_3$
M′—M	$BrCH_2(CH_3)_2Si—O—Si(CH_3)_3$
M′—M′	$BrCH_2(CH_3)_2Si—O—Si(CH_3)_2CH_2Br$
M″—M″	$ClCH_3(CH_3)_2Si—O—Si(CH_3)_2CH_2Cl$

i.e., with increasing substitution of electron acceptors on the Si atoms (*164*). Octamethylcyclotetrasiloxane responds easily and quickly to acid-activated fullers' earth catalysts. In a mixture of octamethylcyclotetrasiloxane and octaphenylcyclotetrasiloxane, the latter remains unaltered after the same treatment; this is clear evidence of strengthening of the Si—O—Si bond under the influence of the electronegative phenyl substituents. On the other hand, the phenyl—Si bond is split far more easily by acids than the methyl—Si bond, an indication of the greater polarity of the former (compare Section 6.4).

Comparative figures have been obtained for the reactivity of the siloxane bond toward proton attack, since acidic cleavage of the siloxane bond can be followed quantitatively with the Karl Fischer reagent (*28*):

$$\equiv Si—\overline{\underline{O}}—Si\equiv + H^{\oplus} \rightarrow \equiv Si—\underset{H}{\overset{\oplus}{\underset{|}{\overline{O}}}}—Si\equiv \;(I) \rightarrow \equiv Si—\overline{\underline{O}}—H + {}^{\oplus}Si\equiv$$

$$\equiv Si—OH + I_2 + SO_2 + 2HOCH_3 \rightarrow \equiv SiOCH_3 + 2HI + CH_3HSO_4$$

The decomposition of the oxonium complex (I) is the slowest and thus the rate-determining step. The reaction is therefore of first order and can be kinetically characterized by the half-lives and decomposition rate constants given in Table 57.

The differences in reactivity in the sequence:

$$D_3^{Me} \gg D_3^{Ph} > M_2^{Me} > D_4^{Me} > D_5^{Me} > MDM^{Me} > D_6^{Me} \gg D_4^{Ph}$$

are considerable. The effects due to substituents are clearly discernible ($D_3^{Me} \gg D_3^{Ph}$; $D_4^{Me} > D_4^{Ph}$; $M_2^{Me} > MDM^{Me}$). On the other hand, the differences between the reactivities of cyclosiloxanes with the same substituents but rings of different sizes are also evident ($D_3^{Me} > D_4^{Me}$; $D_3^{Ph} > D_4^{Ph}$). They

TABLE 57

HALF-LIVES ($t_{1/2}$) AND DECOMPOSITION RATE CONSTANTS (k) OF THE ACID-CATALYZED CLEAVAGE OF VARIOUS POLYORGANOSILOXANES

Polyorganosiloxanes (silanols for comparison)	Abbreviated formula	$t_{1/2}$ (hr)	$k \times 10^3$ (hr^{-1})
Diphenylsilanediol	—	0.05	—
Diethylsilanediol	—	0.05	—
Hexamethylcyclotrisiloxane	D_3^{Me}	0.44	1575
Hexaphenylcyclotrisiloxane	D_3^{Ph}	21.5	32
Hexamethyldisiloxane	M_2^{Me}	36	19
Octamethylcyclotetrasiloxane	D_4^{Me}	175	3.96
Decamethylcyclopentasiloxane	D_5^{Me}	310	2.2
Octamethyltrisiloxane	MDM^{Me}	780	0.9
Dodecamethylcyclohexasiloxane	D_6^{Me}	1000	0.6
Octaphenylcyclotetrasiloxane	D_4^{Ph}	∞	0

clearly indicate an increased reactivity of the three-membered ring, which is presumably due to a somewhat decreased Si—O—Si valency angle (125°, Table 52) and an increased polar contribution to the resonance bond.

Observations on the reactivity of the Si—O—Si bonds with *Lewis acids* also fit into these considerations. Hexamethyldisiloxane is easily split by boron trifluoride, while hexachlorodisiloxane does not react (*63*). This means that the electron-donor properties of the oxygen in hexachlorodisiloxane are so weakened that attachment to BF_3 is no longer possible:

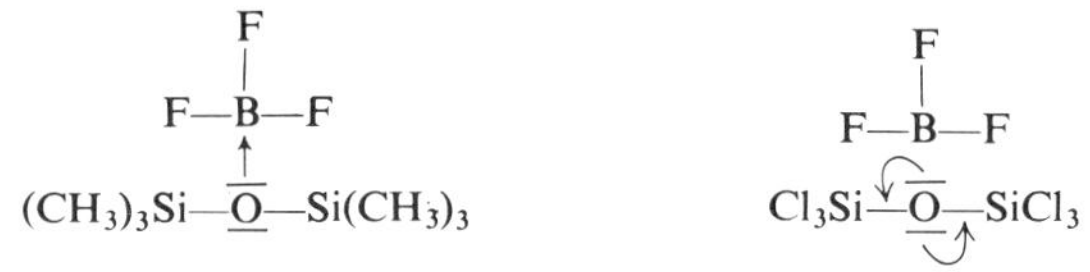

In contrast to the situation with electronegative substitution, an increase of the polar bond component and an increased reactivity to heterolytic attack is to be expected with substituents which donate electrons to the Si—O—Si bond. Particularly striking evidence for this is found in the fact that it has not been possible to make bis-(aminomethyl)-tetramethyldisiloxane from bis-(chloromethyl)-tetramethyldisiloxane and ammonia (*124*). Only the synthesis of the monoamino-substituted disiloxane $NH_2CH_2(CH_3)_2Si—O—Si(CH_3)_3$ was possible by this route (*58*); similarly, four-membered rings containing three dimethylsiloxane units and one aminomethylmethylsiloxane unit (*58*) have been synthesized. These siloxanes, however, tend to undergo molecular rearrangement to an unusual extent even at room temperature; an indication of the strong polarization of the Si—O—Si bond.

The availability of free electron pairs in the siloxane bond is responsible for another of its important properties: its ability to form *hydrogen bonds* with proton donors.

Pure, rigidly dehydrated silica, free of silanol groups, cannot be rehydrated (*121*). This would suggest that the free electron pairs on the Si—$\overline{\underline{O}}$—Si resonance bonds which are not subject to substituent effects are involved so strongly in the $d\pi$—$p\pi$ bond components, that the hydrogen bonds with the hydrogen atoms of the water molecule cannot be formed at all, or only to a very small degree.

Anhydrous SiO_2 would therefore be more properly regarded as hydrophobic than as hydrophilic. The well-known hydrophilic properties of silica glass or silica gel arise from their content of silanol groups. On the other hand, the Si—O bond in polymethylsiloxanes has become so polar that it is capable of forming hydrogen bonds to water molecules with the siloxane oxygen. This can be deduced from the spreading measurements described in Section 6.2.1 which would suggest the formation of "surface hydrates" with the water substrate:

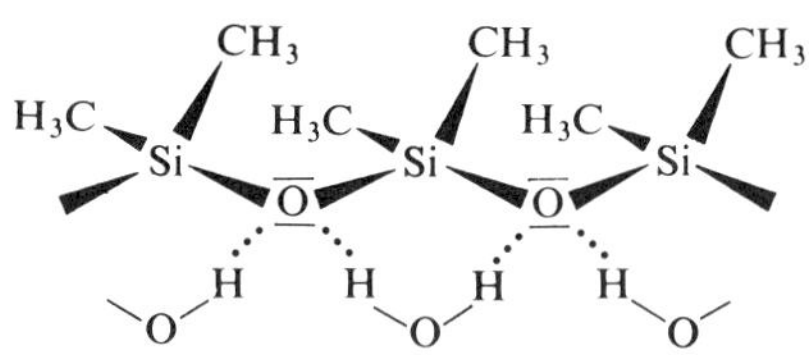

Polyphenylsiloxanes do not spread on water, or at least do so considerably more sluggishly, because they lack the capability of bonding water molecules by utilizing the free electrons on the oxygen for increased $d\pi$—$p\pi$ components of the siloxane bond. Proton magnetic resonance investigations (*68*) and infrared spectroscopic measurements (*200*) have shown that the siloxane bond forms weak hydrogen bonds with stronger organic proton donors. The effect is naturally weaker than that of the Si—O—C bond and considerably weaker than that of the C—O—C bond; it thus follows that contribution of $d\pi$—$p\pi$ bonds decreases in the above sequence.

The preceding examples show how sensitive the chemical behavior of the siloxane bond is to the influence of substituents. Electrophilic reactions and hydrogen bond formation on the siloxane oxygen, in particular, could be considered as indicative of the degree to which the free electron pairs of the siloxane oxygen are involved in the siloxane bond.

Infrared and nuclear resonance spectroscopy is particularly suitable for obtaining quantitative data characterizing the bond which go beyond such merely qualitative discussions. This applies to NMR in particular, since it has been possible to show by means of the chemical shift positions of the ^{29}Si—O bond that it is strongly dependent on the nature of the substituents on the ^{29}Si atom (*67*). In principle, the force constants derived from infrared spectra are suitable as data for the bond order, but because of the participation of mass and angle effects there are still considerable difficulties in the way of their determination. By using the same basic calculations it might be possible to arrive at values which can be compared with one another. Such calculations have been made on the assumption that a >Si—X—Si< bond can be considered as a three-body model and that an effective mass can be assigned to the silicon if, in addition to its own mass, up to one-third of the true masses of its substituents are included. Some valence force constants *f* (*m* dyne Å^{-1}) calculated from the antisymmetric valence vibrations of molecules of the M_2 type are given in the accompanying tabulation (*88–90*):

$(CH_3)_3SiOSi(CH_3)_3$	4.65
$(CH_3)_2C_6H_5Si—O—SiC_6H_5(CH_3)_2$	4.8
$HO(CH_3)_2Si—O—Si(CH_3)_2OH$	4.97
$Cl(CH_3)_2Si—O—Si(CH_3)_2Cl$	5.0
$(CH_3)_3Si—NH—Si(CH_3)_3$	3.84
$(CH_3)_3Si—CH_2—Si(CH_3)_3$	2.77

These show the strengthening of the bond due to electronegative substituents on the silicon atom, but the decreasing participation of $d\pi$—$p\pi$ bond components is very clearly shown in considering the effect due to changes of the bridge atom in the sequence O→NH→C.

To end this section, it may be pointed out that the bond relationships in silicate lattices can be treated from a standpoint analogous to that of the polyorganosiloxanes (*120*). Almost all M atoms bound in the silicates are less electronegative than silicon, they thus act as electron donors in relation to it. An M atom with donor properties polarizes the M—O bond, increases the covalent and double-bond components in the Si—O(→M) bond, and increases the polar components in the Si—O(→Si) bond.

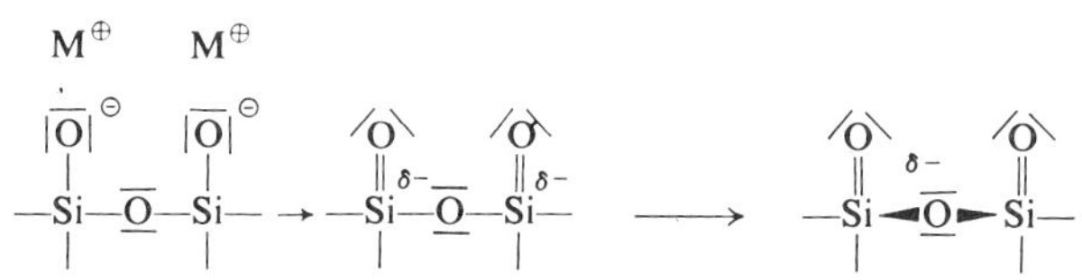

Differences in the internuclear distances of the Si—O bond are accordingly to be expected in the directions →M and →Si, in the sense that the former should be shorter than the latter. This effect has already been shown by the data given in Table 50; the mean values are

Si—O(→Si)	1.66_4 Å
Si—O(→M)	1.58_3 Å

Compared with the value for the Si—O—Si bond in the various forms of SiO_2 (Table 49)

Si—O(→Si)	1.60_4 Å

they make the differences (whose magnitudes obviously depend on the choice of the silicates on which the figures are based) understandable.

The chemical behavior of the Si—O—Si bond in silicates, particularly on acidic cleavage, can also be brought into harmony with the above-mentioned inductive effects (*120*).

6.3.4 Particle Dimensions and Models

It follows from the complex character of the bonds involved here that there are many difficulties in showing particle dimensions and making models. The two viewpoints, one based on the crystal chemistry of the silicates and the other on the chemistry of organosilicon molecules, are here in conflict. In the former, it has become usual to represent silicon and oxygen atoms as spheres of ionic character, whose dimensions are determined by the ionic radius; in the latter approach, on the other hand, Stuart models are used which show that the distance in the bond direction of an atom to a covalently bound neighbor (*covalent radius*) is smaller than the distance in another direction (*van der Waals radius*). Accordingly there are three systems of radii, which characterize the size relationships of silicon and oxygen atoms in different bond states. These are given in the accompanying tabulation. There are no generally recognized symbols for the different systems of radii: those given in the tabulation are merely suggestions. The radius of the metallic bond could appropriately be written as r_m

	Si	0	Values from
Ionic radius r_p(Å)	0.39	1.32	Goldschmidt
Covalent radius r_c (Å)	1.17	0.66	Pauling (*130*)
van der Waals radius R (Å)	1.7	1.3	Stuart (*185*)

The ionic radius r_p describes the size of a spherical ion. The size of a Stuart atomic model is derived from a basic sphere of radius equal to the van der

Waals radius R; segments are cut at a distance from the center of the sphere such that the perpendicular distance from the plane so formed to the center equals the covalent radius r_c. The dimensions of an Si—O atom pair for the spherical ion model and the Stuart model are shown in Fig. 24. In this figure the fact that the Si—O internuclear distances are somewhat smaller than the sum of the ionic and bond radii has also been taken into consideration.

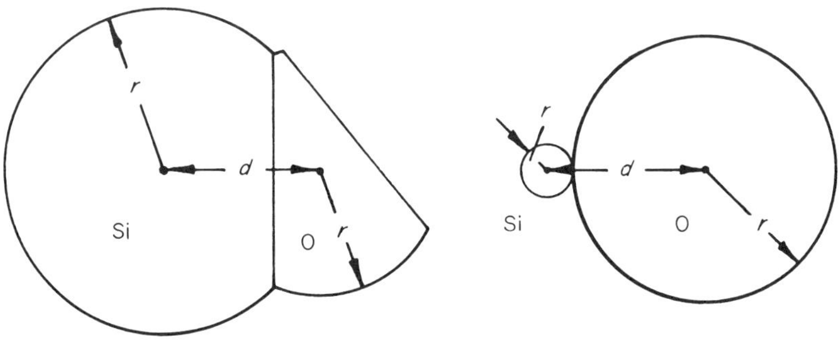

FIG. 24. Size relationships of the silicon and oxygen atoms in a siloxane bond as shown by spherical ion models and Stuart atomic models. Left: Stuart atomic model, $r_{Si}=1.7$ Å; $r_O=1.3$ Å; and $d_{Si\text{-}O}=1.64$ Å. Right: Spherical ion model, $r_{Si}=0.39$ Å; $r_O=1.32$ Å; $d_{Si\text{-}O}=1.71$ Å (corrected to 1.64 Å).

The models can thus be quite different, according to which type of bond is taken as the basis for the representation. In view of the covalent polar character of the Si—O bond, the truth probably lies somewhere in between. As long as an estimation of the bond character in resonance types of bonds depends only on approximate formulas, a compromise between the two viewpoints will continue to have little meaning. Meanwhile, limitations inherent in model representations must be borne in mind when molecular models are used.

6.4 The Silicon—Carbon Bond, Especially as Compared with the Carbon—Carbon Bond

As opposed to the siloxane bond with its "silicatelike" character, the Si—C bond is, as it were, the "organic" bond of the silicones. It suggests a comparison with the C—C bond, with which it seems to have more in common than with the siloxane bond.

The measured values of the internuclear distances in polymethylsiloxanes (Table 51) hardly differ from the sum of the covalent bond radii r_c as given in the accompanying tabulation.

	Si	C	Si—C
r_c	1.17	0.77	Σr_c 1.94 measured, 1.88–1.92

This would suggest that the Si—C bond has a basically covalent character in these compounds. The small difference which has been observed is due mainly to small polar bond components corresponding to the differences in the electronegativities.

$$\text{Si} = 1.8 \qquad \text{C} = 2.5$$

This difference indicates 88% covalent and 12% of polar bond components on the basis of Pauling's approximate formulas (see Section 6.3.3.2). Schomaker and Stevenson's formula gives an internuclear distance of 1.88 Å, which is in good agreement with measured values.

However, a discussion of the chemical behavior of the Si—C bond must also take into consideration the fact that the Si atom will be in a position to form double-bond ($d\pi$—$p\pi$) bond components with carbon just as with the halogens (Section 3.2.3) and oxygen (Section 6.3.3.2); the participation of these components depends on the nature of the substituents on the carbon and silicon atoms. The existence of a pure Si—C double bond has not so far been established with certainty in any single instance. Its presence (*54*) was considered most likely in $(CH_3)_2Si{=}CH{-}Si(CH_3)_3$, since a ring structure

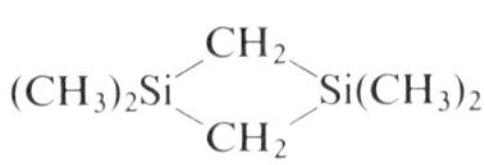

would be unstable because of ring strain. However, spectroscopic investigations have shown that the compound possesses the ring structure and not the double-bond structure, in spite of the strain situation (*55*).

The great differences between silicon and carbon in their readiness to form double bonds with carbon are shown most strikingly in the unsuccessful attempts to prepare a benzene analog from cyclopentamethylenesilane

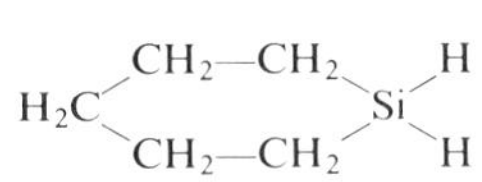

by catalytic dehydrogenation (*199*). The fact that higher $d\pi$—$p\pi$ components are not formed in the Si—C bond would seem to be due to the fact that the larger bond distances, as compared with those of the C—C bond, lead to a smaller overlap of the π-orbitals (*133*).

These special features of the Si—C bond form the basis for an interpretation of its chemical behavior.

As a result of its weakly polar character, it is easily split by *electrophilic* agents, whether they are proton or Lewis acids:

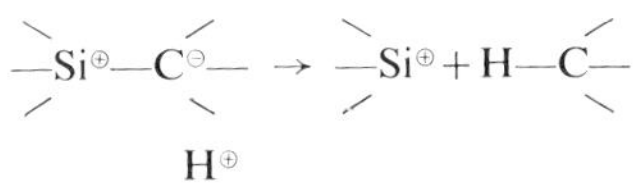

The effect of sulfuric acid on trimethylsilyl groups is an important example, from the preparative point of view, of cleavage by proton acids: by the elimination of a methyl group, methane is liberated and a new siloxane bond is formed (Section 5.3). The ease with which Si—C bonds can be broken by Lewis acids is also important for silicone chemistry, since it enables alkyl groups to be exchanged with one another or with other substituents on the silicon atom (Section 2.5).

As a result of the unfilled $3d$ levels in the silicon atom, *nucleophilic* partners can be added, with the formation of penta- or hexacovalent intermediate complexes. The activated state formed in this way can break down after acceptance of a proton to form a hydrocarbon and a silanol:

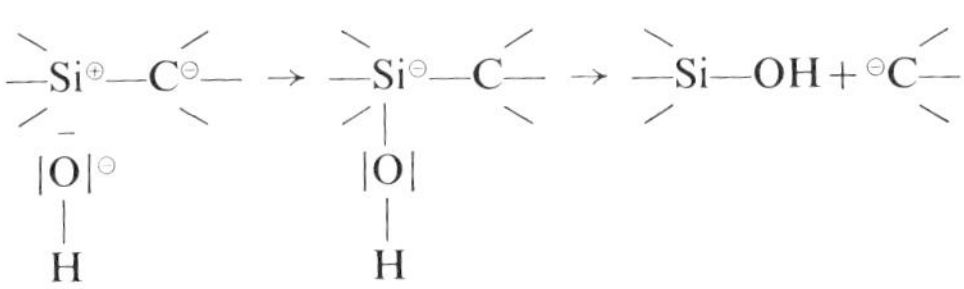

Cleavage of the Si—C bond naturally proceeds with much more difficulty than that of a bond with greater polarity, such as an Si—halogen bond. Hexamethyldisiloxane or polydimethylcyclosiloxanes give off methane only on heating to about 200°C with alkalis under pressure (*189*).

The reactivity of the Si—C bond toward nucleophilic or electrophilic attack can, however, be completely altered if the subsituents in the organic groups or on the Si atom exert an inductive effect due to their donor or acceptor properties.

The methyl group itself behaves toward silicon as a weak electron donor. If it is replaced by the strongly electron-accepting phenyl group, the Si—C bond is polarized and made more sensitive toward acidolytic attack than in methylsilicon compounds.*

* It is however important to note that the effects are reversed with attack by oxygen. Methylphenylsiloxanes are considerably more resistant towards atmospheric oxygen than methylsiloxanes, since the benzene ring is less easily oxidized than the methyl group owing to its resonance stabilization; tetraphenysilane boils without decomposition at 428°C (*99*) and is stable even at temperatures of about 500°C.

The introduction of electronegative substituents such as halogens, OH, carboxyl or cyano groups also has a considerable influence on the reactivity of the Si—C bond. These effects depend on the position of the substituents in the alkyl group and partly on whether the attack is nucleophilic or electrophilic.

Electronegative substitution on the β-carbon always has the greatest effect. The Si—C bond in such compounds is easily accessible to nucleophilic or electrophilic attack:

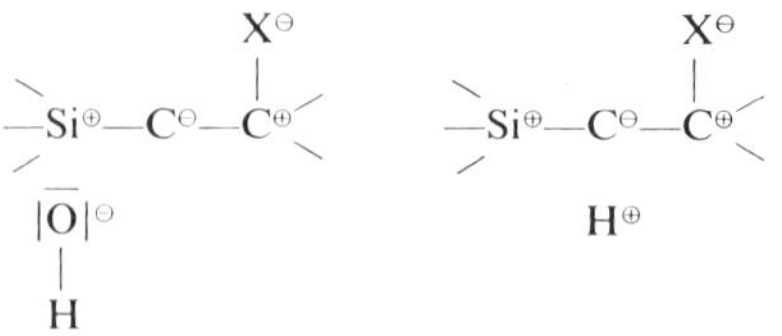

It can also be decomposed by purely thermal treatment (Section 4.5.4).

Nucleophilic attack is still possible with substitution on the α-carbon, since unfilled $3d$ levels of the Si atom are available for filling, as before. Electrophilic attack however is made more difficult than on compounds without X substituents, because of the removal of electrons from the α-carbon.

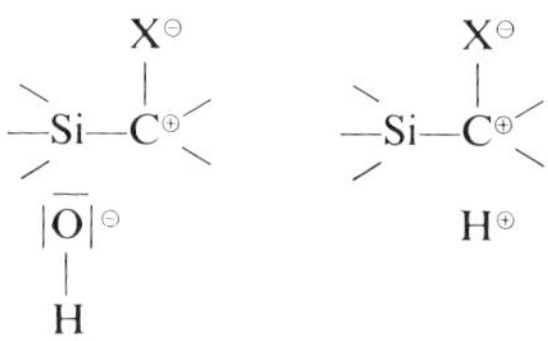

Electronegative substituents on the γ-carbon appear to act like those on the α-carbon; their influence in the δ-position is generally so weak that it is insignificant, at least for preparative purposes. A small effect of fluorosubstitution in the δ-position on silanes containing Si—H bonds is still observable (Section 4.4).

The ease with which cleavage takes place is thus in general:

$$\beta \gg \alpha,\ \gamma > \delta$$

This can be connected, at least formally, with the so-called alternation effect (*38*, *41*, *143*). Its validity has been questioned in organic chemistry (*83*, *183*), but it might have some significance in organometallic chemistry.

The effect of the X substituents on Si—C bonds becomes weaker as the electronegativity of X becomes smaller. The Si—X bond is simultaneously weakened and it is thus possible to observe interesting inversions in chemical behavior in substituent series with decreasing electronegativity of the X atoms. While chloromethylsilicon compounds can still be decomposed by nucleophilic attack on the Si—C bond, it is possible to substitute the halogen in bromo-substituted compounds easily and without appreciable danger of breaking the Si—C bond (*166*):

$$\rangle\!\!-SiCH_2-Cl + OH^{\ominus} \rightarrow \rangle\!\!-Si-OH + {}^{\ominus}CH_2-Cl$$

$$\rangle\!\!-SiCH_2-Br + OH^{\ominus} \rightarrow \rangle\!\!-Si-CH_2OH + {}^{\ominus}Br$$

The influence of substituents on the Si atom itself is also considerable. Measurements of interatomic distances indicate that an increase in the number of electronegative substituents on the Si atom results in an increase of the double bond character of the Si—C bond in methylchlorosilanes and, thus, in a strengthening of the Si—C bond:

$(CH_3)_4Si$	1.93 Å
$(CH_3)_3SiCl$	1.89 Å
$(CH_3)_2SiCl_2$	1.83 Å

This and measurements of dipole moments have led to the conclusion that hyperconjugation effects also play a part (*206*):

$$\begin{array}{c} CH_3 \\ | \\ H^{\oplus}CH_2{=}Si \quad {}^{\ominus}Cl \\ | \\ CH_3 \end{array}$$

The effect of an electronegative substituent on the Si atom on chemical behavior can be seen, for example, in the fact that the Si—C bond in β-chloroethyltriethylsilane is broken relatively easily by heating; whereas that in β-chloroethyltrichlorosilane is only partially subject to β-degradation at high temperatures. It should similarly be clear that the electrophilic attack of $AlCl_3$ on β-chloroethyltrichlorosilane does not cause Si—C cleavage but only removal of halogen chloride (Sections 4.5.4 and 4.3.2.1), while β-chloroethyltriethylsilane displays the opposite behavior,

Nucleophilic attack is facilitated by accumulation of electronegative silicon-functional partners, since electrons are attracted from the silicon atom to the electronegative substituents. The Si—C bond is thus more easily broken in C_α-halo-substituted siloxanes—the siloxane oxygen here acts as electronegative partner—than in silanes which carry besides a halomethyl group only alkyl substituents on the silicon. This demonstrates the contrast in behavior between, for example, the chloromethyl group in chloromethylpentamethyldisiloxane and chloromethyltrimethylsilane (Sections 4.5.4 and 6.2.1).

Organosilicon chemistry is full of examples demonstrating the induction effects described above. The reader is therefore referred to the more detailed descriptions in Chapter 4. Since it is intended to present only a few basic fundamentals at this stage, discussion will be restricted to the influence exerted by the insertion of trimethylsilyl groups into carboxylic acids and amines.

The acid strength of an aliphatic carboxylic acid is decreased by trimethylsilyl substitution (*172*) (Table 58). The same effect is observed with benzoic acid (*37*, *139*), more strongly in the *o*- position (*37*) than in the *m*- or *p*-. This effect is explained by the greater electropositive character of silicon as compared with carbon; the $(CH_3)_3Si$ group thus becomes a stronger electron donor and weakens the dissociability of the proton in the carboxyl group. As was to be expected, this effect falls of rapidly with increasing chain length in aliphatic carboxylic acids.

TABLE 58

INFLUENCE OF THE TRIMETHYLSILYL GROUP ON THE ACID STRENGTH OF CARBOXYLIC ACIDS (*172*, *173*)

Acid	$K \times 10^5$
$(CH_3)_3Si—CH_2—COOH$	0.60
$(CH_3)_3C—CH_2—COOH$	1.00
$CH_3—COOH$	1.75
$(CH_3)_3Si—CH_2—CH_2—COOH$	1.24
$CH_3—CH_2—COOH$	1.32
$(CH_3)_3Si—CH_2—CH_2—CH_2—COOH$	1.30
$(C_6H_5)(CH_3)_2Si—CH_2—COOH$	0.54
CH_3 | $(CH_3)_3Si—O—Si—CH_2—COOH$ | CH_3	0.60

On the other hand, the trimethylsilyl group increases the basicity of aliphatic amines (Table 59).

Thus, silicon does not appear to coordinate with the lone electron pair of the nitrogen in these amines and does not expand its electron shell. Such an effect would of course necessarily result in a considerable decrease of the base strength, as shown by a comparison of the dissociation constants of trimethylamine (10^{-4}) and trimethylamine oxide (10^{-10}). Silicon shows equally little tendency to coordinate with the oxygen of the carboxyl group in carboxylic acids (*177*).

Unlike the trimethylsilyl groups, triorganosilyl groups containing phenyl radicals or oxygen lower the base strength. This is similar to the usual action

TABLE 59

INFLUENCE OF THE TRIMETHYLSILYL GROUP ON THE BASE STRENGTH OF ALIPHATIC AMINES[a]

Amine	$K_B \times 10^4$	
$H_2N—CH_2—Si(CH_3)_3$	9.1 (*177*)	9.6 (*123*)
$H_2N—CH_3$		5.1 (*123*)
$H_2N—CH_2CH_2—Si(CH_3)_3$	9.7 (*177*)	
$H_2N—CH_2CH_3$	5.6 (*94*)	
$H_2N—CH_2CH_2CH_2—Si(CH_3)_3$	5.6 (*177*)	
$H_2N—CH_2CH_2CH_3$	4.9 (*177*)	
$H_2N—CH_2—Si(C_6H_5)(CH_3)_2$		2.8 (*123*)
$[H_2N—CH_2—Si(CH_3)_2]_2O$		2.0 (*123*)

[a] Numbers in parentheses indicate reference numbers.

of phenyl radicals or oxygen as electronegative substituents in amines. These differences between methyl-, phenyl-, and oxygen-containing silyl groups do not appear in carboxylic acids.

These results indicate that the influence of triorganosilyl substituents can be modified by other substituents. This is further confirmed by the fact that a $(CH_3)_3Si$ group can act as an electron acceptor in the position *para* to strongly electron releasing groups such as OH, NH_2, or $(CH_3)_2N$; the base strengths of aniline and dimethylaniline are somewhat decreased by a *p*-$(CH_3)_3Si$ group, while the acidity of phenol is increased (*12*). The effect of the *p*-$(CH_3)_3Si$ group can be interpreted here by the assumption that, owing to its ability to expand its electron shell, the silicon atom forms *resonance structures* of the type

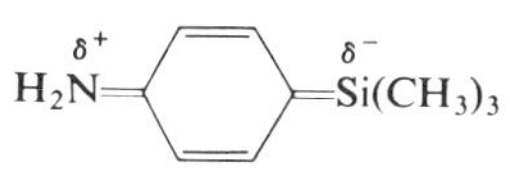

if an electron releasing group is present (*12*, *49*, *169*). The relatively weak influence of a $(CH_3)_3Si$ group in the *m*-position on the dissociation constant of aniline and dimethylaniline would appear to be in agreement with this picture (*12*). The triphenylsilyl group exerts a similar effect in the *p*-position on the dimethylamino group on the ring (*11*).

Finally, it should be pointed out in this context that proton magnetic resonance (PMR) spectroscopy can make a valuable contribution to the study of bond relationships between hydrocarbon radicals and silicon. It has been shown that the electron attraction of a strongly electronegative substituent on the α-C atom is manifest not only in a strengthening of the X—C bond and a weakening of the Si—C bond, but also in acidification of the protons:

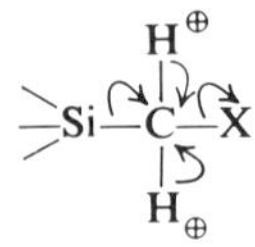

This makes it possible to characterize the inductive effect quantitatively. The PMR spectra of a number of compounds have shown that the δ figures for the methylene protons depend fundamentally on the nature of the substituents X. The influence of further substituents on X or of substituents on Si is small, so that the methylene bridge in different *α-sila compounds* can be characterized by the δ-figures given in the accompanying tabulation (in ppm based on tetramethylsilane=0) (*163*, *165*, *166*):

	$>Si—CH_2—X$
δ (ppm)	X =
0.9–1.3	$>CH(CN)$, $>CCOOH(CN)$
1.6–2.2	S—R
1.7–2.2	N—R
2.3–2.7	Br
2.6–3.0	Cl
2.9–3.0	$S^{\oplus}$—R
3.0–3.5	$N^{\oplus}$—R
3.1–3.6	O—R
3.7–3.85	O—C(=O)R

6.5 Intermolecular Forces

The intermolecular forces in the methylsilicones are unusually weak. This property is most important for an understanding of the technological behavior of these materials.

Indications for this are found in: the relatively low surface tensions of liquid polydimethylsiloxanes, which in general are about 20 dynes cm^{-1}; the small variations of the physical constants, particularly the viscosity of the polydimethylsiloxanes with temperature; the fact that even the highest-molecular-weight linear polydimethylsiloxanes remain liquid; their remarkably low freezing and pour points; their high compressibility; their spreading ability, and many other properties (see Chapter 9). Additional properties of the siloxanes which throw light on these phenomena are given below.

The boiling points of definite low-molecular-weight polyorganosiloxanes are lower than those of corresponding organic compounds of comparable molecular weight. This can also be observed in organosilicon compounds not containing siloxane bonds, and it would thus appear that the Si—R groups are also responsible for the weak intermolecular forces, as is shown in the following examples (*144*).

It is interesting to note that hexamethyldisiloxane and trimethylsilanol have practically the same boiling point, although the molecular weight of the former is almost twice as high (see Table 60). This would indicate, on the one hand, that the intermolecular attraction in disiloxane is weak, and on the other, that the intermolecular bonds are markedly strengthened by the silanol dipole in trimethylsilanol.

TABLE 60

BOILING POINTS OF ORGANOSILICON AND COMPARABLE ORGANIC COMPOUNDS

Mol. wt.	Compound	Bp_{760} (°C)
162	$(CH_3)_3SiOSi(CH_3)_3$	100
158	$(i\text{-}C_5H_{11})O(i\text{-}C_5H_{11})$	173
90	$(CH_3)_3SiOH$	100
88	$(CH_3)_3CCH_2OH$	114
88	$(CH_3)_4Si$	26
86	$(CH_3)_3C(C_2H_5)$	50

Again, in a series of silicone polymers differing only in the nature of the bridge atom between the silicon atoms, those have the weakest intermolecular bonds in which the silicon atoms are linked through oxygen bridges. Table 61 shows that, starting with hexamethyldisiloxane, the boiling points increase (in spite of a decrease in the molecular weights) if oxygen is omitted or replaced by an imino or methylene bridge (*195*).

TABLE 61

BOILING POINTS OF A NUMBER OF SILICON COMPOUNDS HAVING SIMILAR STRUCTURE

Mol. wt.	Compound	Bp_{760} (°C)
162	$(CH_3)_3Si—O—Si(CH_3)_3$	100
146	$(CH_3)_3Si—Si(CH_3)_3$	114
161	$(CH_3)_3Si—NH—Si(CH_3)_3$	126
160	$(CH_3)_3Si—CH_2—Si(CH_3)_3$	134

Further information on the factors of interest here come from investigations on the energy of activation of viscous flow (*71*, *195*, *204*). Linear polymethylsiloxanes are characterized by very low values of this property, which rise from 2.1 for hexamethyldisiloxane to about 4 for polydimethylsiloxanes of high molecular weight. This behavior is specific for polymers with Si—O—Si bonds and methyl substituents, as the following comparison shows (Table 62).

TABLE 62

ENERGY OF ACTIVATION OF VISCOUS FLOW FOR VARIOUS POLYMERS[a]

Compound	Mol. wt.	E_A(kcal·mole^{-1})
Polyisobutylene	3000	14 (*2*)
Polybut-1-ene oxide	2000	12 (*2*)
Polypropylene oxide	2000	9 (*2*)
Polymethylphenylsiloxane	5000	6.5 (*2*)
Polydimethylsiloxane	200000	4 (*2*)
$(CH_3)_3Si[(CH_3)_2Si]_{18}Si(CH_3)_3$	~1200	Infusible (*195*)
$(CH_3)_3SiCH_2[(CH_3)_2SiCH_2]_{14}Si(CH_3)_3$	~1200	6.0 (*195*)
$(CH_3)_3SiO[(CH_3)_2SiO]_{14}Si(CH_3)_3$	~1200	3.8 (*195*)
$(CH_3)_3SiCH_2Si(CH_3)_3$	160	2.6 (*195*)
$(CH_3)_3SiOSi(CH_3)_3$	162	2.1 (*195*)

[a] Numbers in parentheses indicate reference numbers.

Polydimethylsiloxanes also occupy a special position among polymers with respect to the position of the glass temperature T_g, this again being determined not only by the flexibility of the molecules but also by the weakness of the intermolecular bonding forces (see Table 63).

TABLE 63

GLASS TEMPERATURES OF POLYMERS

	T_g(°C)		T_g(°C)
Poly-(methyl methacrylate)	105	Polyisobutylene	−70
Polystyrene	100	Natural rubber	−72
Poly-(vinyl acetate)	20	Polypropylene oxide	−74
Poly-(methyl methacrylate)	0	Poly-1,4-*cis*-butadiene	−85
Poly-(methyl vinyl ether)	−15	Polymethylphenylsiloxane	
Polypropylene	−35	(35% C_6H_5)	−100
Poly-(vinyl fluoride)	−39	Polydimethylsiloxane	−120

The question of what factors determine this important property has remained unanswered to the present day. It might be possible to obtain some insight by using the results of investigations on the mobility of the methyl groups in organosiloxanes. Roth (*148*, *149*) concluded from investigations on octamethylspiropentasiloxane that the methyl group was almost freely rotatable about the Si—O bond. Nuclear magnetic resonance observations by Rochow and Le Clair (*145*) indicate that the hydrogen atoms of the methyl groups possess an unusually high mobility down to very low temperatures. At about 77°K this mobility becomes mainly a rotation around the Si—C bond. It therefore appears that the methyl groups have considerable space requirements because of their mobility and thus keeping the mean distances between the molecules large. This is simply another way of saying that the molecular volumes are large (*72*) and the intermolecular forces small.

REFERENCES

(1) E. H. Aggarwal and S. H. Bauer, *J. Chem. Physics* **18**, 42 [1950]. (2) G. Allen, *J. Appl. Chem.* **14**, 1 [1964]. (3) K. A. Andrianov, V. V. Astakhin and V. K. Pyzhov, *Bull. Acad. Sci. USSR, Div. Chem. Sci.* (*Engl. transl.*) **1962**, 2144. (4) — and S. Ye. Yakushkina, *Vysokomol. Soed.* **3**, 1554 [1961]. (5) — and L. M. Volkova, ibid. **3**, 1580 [1961]. (6) —, —, *Bull. Acad. Sci. USSR, Div. Chem. Sci.* (*Engl. transl.*) **1962**, 241. (7) G. A. Barclay, E. G. Cox and H. Lynton, *Chem. and Ind.* **1956**, 178. (8) A. J. Barry, *J. Appl. Physics* **17**, 1020 [1946]. (9) —, W. H. Daudt, J. J. Domicone and J. W. Gilkey, *J. Amer. Chem. Soc.* **77**, 4248 [1955].

(10) — and J. W. Gilkey, Dow Corning, U.S. Pat. 2495362 [31.7.1948]. (11) R. A. Benkeser, C. E. DeBoer, R. E. Robinson and D. M. Sauve, *J. Amer. Chem. Soc.* **78**, 682 [1956]. (12) — and H. R. Krysiak, ibid. **75**, 2421 [1953]. (13) F. C. Boye and H. W. Post, *J. Org. Chem.* **17**, 1389 [1952]. (14) S. D. Brewer, Thomson-Houston, French Pat. 970424 [U.S. Prior. 13.9.1947]. (15) —, *J. Amer. Chem. Soc.* **70**, 3962 [1948]. (16) L. Brewer and R. K. Edwards, *J. Physic. Chem.* **58**, 351 [1954]. (17) British Thomson-Houston, Brit. Pat. 632954 [U.S. Prior. 21.3.1946]. (18) J. F. Brown jr., General Electric, U.S. Pat. 3000858 [21.1.1959]. (19) —, L. H. Vogt jr., A. Katchman, J. W. Eustance, K. M. Kiser and K. W. Krantz, *J. Amer. Chem. Soc.* **82**, 6194 [1960].

(20) C. A. Burkhard, ibid. **67**, 2173 [1945]. (21) —, ibid. **72**, 1078 [1950]. (22) —, ibid. **74**, 6275 [1952]. (23) —, B. F. Decker and D. Harker, ibid. **67**, 2174 [1945]. (24) H. A. Clark and L. A. Haluska, Dow Corning, U.S. Pat. 2877256 [29.11.1956]. (25) Connecticut Hard Rubber, Brit. Pat. 790370 [U.S. Prior. 16.4.1954]. (26) D. W. J. Cruickshank, *J. Chem. Soc.* [London] **1961**, 5486. (27) R. F. Curl jr., and K. S. Pitzer, *J. Amer. Chem. Soc.* **80**, 2371 [1958]. (28) K. Damm, D. Gölitz and W. Noll, *Angew. Chem.* **76**, 273 [1964]. (29) — and W. Noll, *Kolloid-Z.* **158**, 97 [1958].

(30) W. H. Daudt and J. F. Hyde, *J. Amer. Chem. Soc.* **74**, 386 [1952]. (31) S. K. Dhar, V. Doron and S. Kirschner, Amer. Chem. Soc. 133rd Meeting, San Francisco, Calif. 1958, Abstr. 25 L. (32) P. A. DiGiorgio, W. A. Strong, L. H. Sommer and F. C. Whitmore, *J. Amer. Chem. Soc.* **68**, 1380 [1946]. (33) W. Dilthey, *Chem. Ber.* **36**, 923 [1903]. (34) Dow

Chemical, Brit. Pat. 627136 [U.S. Prior. 18.5.1946]. (35) Dow Corning, Brit. Pat. 659011 [U.S. Prior. 31.7.1948]. (36) —, Brit. Pat. 659012 [U.S. Prior. 31.7.1948]. (37) C. EABORN and S. H. PARKER, *J. Chem. Soc.* [London] **1954**, 939. (38) B. EISTERT: *Tautomerie und Mesomerie*, Enke, Stuttgart, 1938, p. 33. (39) K. ENDELL and H. HELLBRÜGGE, Supplement *Angew. Chemie.* No. **38**, 1 [1940].

(40) R. C. EVANS (*transl. by* E. Thilo): *Einführung in die Kristallchemie*, J. A. Barth, Leipzig, 1954, p. 151. (41) K. FAJANS, *Chem. Engng News* **27**, 900 [1949]. (42) H. J. FLETCHER and G. L. CONSTAN, Dow Corning, U.S. Pat. 2890234 [3.1.1957]. (43) — and M. J. HUNTER, *J. Amer. Chem. Soc.* **71**, 2918 [1949]. (44) —, —, ibid. **71**, 2922 [1949]. (45) P. J. FLORY, ibid. **62**, 1057 [1940]. (46) —, L. MANDELKERN, J. B. KINSINGER and W. B. SHULTZ, ibid. **74**, 3364 [1952]. (47) W. H. FOX, P. W. TAYLOR and W. A. ZISMAN, *Ind. Engng. Chem.* **39**, 1401 [1947]. (48) V. D. FRECHETTE: *Non-Crystalline Solids*, John Wiley, New York, London, **1960**, p. 74. (49) H. FREISER, M. V. EAGLE and J. SPEIER, *J. Amer. Chem. Soc.* **75**, 2821 [1953].

(50) —, —, —, ibid. **75**, 2824 [1953]. (51) L. K. FREVEL, in P. J. Flory, L. Mandelkern, J. B. Kinsinger and W. B. Shultz, ibid. **74**, 3364 [1952]. (52) — and M. J. HUNTER, ibid. **67**, 2275 [1945]. (53) K. C. FRISCH, P. A. GOODWIN and R. E. SCOTT, ibid. **74**, 4584 [1952]. (54) G. FRITZ and J. GROBE, *Z. anorg. allg. Chem.* **311**, 325 [1961]. (55) —, W. KEMMERLING, G. SONNTAG, H. J. BECHER, E. A. V. EBSWORTH and J. GROBE, ibid. **321**, 10 (1963]. (56) C.L. FRYE, *J. Amer. Chem. Soc.* **86**, 3170 [1964]. (57) General Electric, Brit. Pat. 781519 [U.S. Prior. 11.5.1954]. (58) P. D. GEORGE and J. R. ELLIOTT, *J. Amer. Chem. Soc.* **77**, 3493 [1955]. (59) F. GIESEKUS and K. DAMM, Farbenfabriken Bayer, 1958, unpublished.

(60) H. GILMAN, A. G. BROOK and L. S. MILLER, *J. Amer. Chem. Soc.* **75**, 3757 [1953]. (61) — and G. E. DUNN, *Chem. Reviews* **52**, 77 [1953]. (62) — and L. S. MILLER, *J. Amer. Chem. Soc.* **73**, 968 [1951]. (63) H. GROSSE-RUYKEN in H. KRIEGSMANN, *Z. anorg. allg. Chem.* **299**, 149 [1959]. (64) A. GRUND and M. M. PIZY, *Acta crystallogr.* [Copenhagen] **5**, 837 [1952]. (65) N. B. HANNAY and C. P. SMYTH, *J. Amer. Chem. Soc.* **68**, 171 [1946]. (66) A. HAUG and G. MEYERHOFF, *Makromolekulare Chem.* **53**, 91 [1962]. (67) G. R. HOLZMAN, P. C. LAUTERBUR, J. H. ANDERSON and W. KOTH, *J. Chem. Physics* **25**, 172 [1956]. (68) C. M. HUGGINS, *J. Phys. Chem.* **65**, 1881 [1961]. (69) —, L. E. ST. PIERRE and A. M. BUECHE, ibid. **64**, 1304 [1960].

(70) M. J. HUNTER, J. F. HYDE, E. L. WARRICK and H. J. FLETCHER, *J. Amer. Chem. Soc.* **68**, 667 [1946]. (71) —, E. L. WARRICK, J. F. HYDE and C. C. CURRIE, ibid. **68**, 2284 (1946). (72) C. B. HURD, ibid. **68**, 364 [1946]. (73) D. T. HURD and R. C. OSTHOFF, ibid. **75**, 234 [1953]. (74) J. F. HYDE, Corning Glass Works, U.S. Pat. 2486162 [26.2.1942]. (75) —, *J. Amer. Chem. Soc.* **75**, 2166 [1953]. (76) —, L. K. FREVEL, N. S. NUTTING, P. S. PETRIE and M. A. PURCELL, ibid. **69**, 488 [1947]. (77) —, O. K. JOHANNSON, W. H. DAUDT, R. F. FLEMING, H. B. LAUDENSHLAGER and M. P. ROCHE, ibid. **75**, 5615 [1953]. (78) J. W. JENKINS and H. W. POST, *J. Org. Chem.* **15**, 552 [1950]. (79) V. B. JEX and R. Y. MIXER, Union Carbide, French Pat. 1169053 [U.S. Prior. 23.12.1955].

(80) S. W. KANTOR, R. C. OSTHOFF and D. T. HURD, *J. Amer. Chem. Soc.* **77**, 1685 [1955]. (81) F. S. KIPPING, *J. Chem. Soc.* [London] **101**, 2125 [1912]. (82) — and R. ROBISON, ibid. **105**, 484 [1914]. (83) F. KLAGES: *Lehrb. d. organ. Chem.*. II, p. 119; de Gruyter, Berlin 1954. (84) C. F. KOHL, Corning Glass Works, U.S. Pat. 2571090 [27.10.1949]. (85) A. J. KOROLEV, K. A. ANDRIANOV, L. C. UTESEVA and T. E. UVEDENSKAYA, *Doklady Akad. Nauk. SSSR* **89**,

65 [1953]. (86) R. H. KRIEBLE and J. R. ELLIOTT, *J. Amer. Chem. Soc.* **67**, 1810 [1945]. (87) —, —, ibid. **68**, 2291 [1946]. (88) H. KRIEGSMANN, *Z. Elektrochem.* **61**, 1088 [1957]. (89) —, *Z. anorg. allg. Chem.* **299**, 138 [1959].

(90) — and G. ENGELHARDT, ibid. **310**, 320 [1961]. (91) M. KUMADA, *J. Inst. Polytechn. Osaka City Univ. C* **2**, 147 [1952]; *J. chem. Soc. Japan, ind. Chem. Sect.* **55**, 752 [1952]. (92) Y. KURITA and M. KONDO, *Bull. chem. Soc. Japan* **27**, 160 [1954]. (93) H. KUSUMOTO, I. J. LAWRENSON and H. S. GUTOWSKY, *J. Chem. Phys.* **32**, 724 [1960]. (94) LANDOLT-BÖRNSTEIN: *Physikalisch-chemische Tabellen II*, 1923, 5. Aufl., p. 1157. (95) Z. LASOCKI, *Roczniki Chem.* [*Ann. Soc. Chim. Polonorum*] **31**, 837 [1957] Lodz, Politechn., *Kat. Technolog. chem. organ.* (96) F. LAVES, *Z. Kristallogr. Mineralog. Petrogr.*, Abt. A **82**, 1 [1932]. (97) R. N. LEWIS, *J. Amer. Chem. Soc.* **70**, 1115 [1948]. (98) D. W. LEWIS and G. C. GAINER, ibid. **74**, 2931 [1952]. (99) R. N. LEWIS and A. B. NEWKIRK, ibid. **69**, 701 [1947].

(100) F. LIEBAU, *Z. physik. Chem.* **206**, 73 [1956]. (101) —, *Acta crystallogr.* **14**, 1103 [1961]. (102) G. R. LUCAS and R. W. MARTIN, *J. Amer. Chem. Soc.* **74**, 5225 [1952]. (103) H. MARSDEN and F. S. KIPPING, *J. Chem. Soc.* [London] **93**, 198 [1908]. (104) R. L. MERKER, Dow Corning, U.S. Pat. 2833802 [27.2.1956]. (105) — and J. E. NOLL, *J. Org. Chem.* **21**, 1537 [1956]. (106) Midland Silicones, Brit. Pat. 834711 [U.S. Prior. 10.9.1956]. (107) M. MOMONOI and N. SUZUKI, *J. Chem. Soc. Japan, pure Chem. Sect.* [*Nippon Kagaku Zasshi*] **78**, 1324 [1957] through Cbl. **1959**, 100. (108) R. MÜLLER and R. KÖHNE, *Z. anorg. allg. Chem.* **311**, 142 [1961]. (109) —, — and S. SLIWINSKI, *J. prakt. Chem.* [4] **9**, 63 [1959].

(110) —, —, —, ibid. [4] **11**, 336 [1960]. (111) R. NAGEL, C. TAMBORSKI and H. W. POST, *J. Org. Chem.* **16**, 1768 [1951]. (112) N. S. NAMETKIN, A. V. TOPCHIEV and C. I. KARTASEVA, *Doklady Akad. Nauk. SSSR* (2) **93**, 667 [1953]. (113) M. J. NEWING, *Trans. Faraday Soc.* **46**, 613 [1950]. (114) H. NIEDERPRÜM, Farbenfabriken Bayer, 1964, unpublished. (115) —, E. HORN and W. SIMMLER, —, German Appl. (West German) 1195753 [14.3.1964]. (116) —, —, —, —, German Appl. (West German) 1216873 [14.5.1964]. (117) W. NOLL, *Fortschr. Mineralog.* **26**, 93 [1947]. (118) —, ibid. **34**, 63 [1956]. (119) —, *Naturwissenschaften* **49**, 505 [1962].

(120) —, *Angew. Chem.* **75**, 123 [1963]. (121) —, K. DAMM and R. FAUSS, *Kolloid-Z.* **169**, 18 [1960]. (122) —, H. STEINBACH and CHR. SUCKER, *Ber. Bunsenges. Phys. Chem.* **67**, 407 [1963]. (123) J. E. NOLL, B. F. DAUBERT and J. L. SPEIER, *J. Amer. Chem. Soc.* **73**, 3871 [1951]. (124) —, —, —, ibid. **73**, 3867 [1951]. (125) R. OKAWARA, *J. Chem. Soc. Japan, Ind. Chem. Sect.* **61**, 690 [1958], Osaka, Univ. Techn. Fak., through Cbl. **1959**, 8848. (126) —, E. ASADA and T. WATASE, *Bull. Chem. Soc. Japan* **26**, 141 [1953]. (127) —, U. TAKAHASHI and M. SAKIYAMA, ibid. **30**, 608 [1957], through Cbl. **1963**, 13970. (128) R. C. OSTHOFF and W. T. GRUBB, *J. Amer. Chem. Soc.* **76**, 399 [1954]. (129) W. J. PATNODE and D. F. WILCOCK, ibid. **68**, 358 [1946].

(130) L. PAULING: *The Nature of the Chemical Bond*, Cornell University Press, Ithaca (N.Y.), 3rd Ed., 1960. (131) L. PAULING, *J. Physic. Chem.* **56**, 361 [1952]. (131a) R. G. PEARSON, D. N. EDGINGTON and F. BASOLO, *J. Amer. Chem. Soc.* **84**, 3233 [1962]. (132) G. PEYRONEL, *Atti Accad. naz. Lincei, Rend., Cl. Sci. fisiche, mat. natur.*, Serie 8, **15**, 402 [1953]; **16**, 78 [1954]. (133) K. S. PITZER, *J. Amer. Chem. Soc.* **70**, 2140 [1948]. (134) E. P. PLUEDDEMANN and G. FANGER, ibid. **81**, 2632 [1959]. (135) F. P. PRICE, S. G. MARTIN and J. P. BIANCHI, *J. Polymer Sci.* **22**, 41 [1956]. (136) M. PROBER, *J. Amer. Chem. Soc.* **77**, 3224 [1955]. (137) —,

ibid. **77**, 5180 [1955]. (138) V. B. REX and D. L. BAILEY, Union Carbide, Brit. Pat. 882069 [3.10.1957]. (139) J. D. ROBERTS, E. A. MCELHILL and R. ARMSTRONG, *J. Amer. Chem. Soc.* **71**, 2923 [1949].

(140) R. ROBISON and F. S. KIPPING, *J. Chem. Soc.* [London] **101**, 2156 [1912]. (141) E. G. ROCHOW, General Electric, U.S. Pat. 2258220 [27.4.1940]. (142) —, —, French Pat. 982256 [U.S. Prior. 4.3.1948]. (143) —: *An Introduction to the Chemistry of the Silicones*, 2nd Ed., New York, John Wiley & Sons, Inc. 1951. (144) —: ibid., Table p. 110. (145) — and H. G. LECLAIR, *J. Inorg. Nucl. Chem.* **1**, 92 [1955]. (146) T. G. ROCHOW and E. G. ROCHOW, *Science* [Washington] **111**, 271 [1950]. (147) A. ROSENHEIM, B. RAIBMANN and G. SCHENDEL, *Z. anorg. allg. Chem.* **196**, 160 [1931]. (148) W. L. ROTH, *J. Amer. Chem. Soc.* **69**, 474 [1947]. (149) — and D. HARKER, *Acta crystallogr.* [London] **1**, 34 [1948].

(150) S. I. SADYKH-ZADE, L. V. NOZDRINA and A. D. PETROV, *Proc. Acad. Sci. USSR (Engl. transl.)* **118**, 107 [1958]. (151) R. O. SAUER, *J. Amer. Chem. Soc.* **66**, 1707 [1944]. (152) —, ibid. **68**, 138 [1946]. (153) —, W. J. SCHEIBER and S. D. BREWER, ibid. **68**, 962 [1946]. (154) F. K. SCHOLL, *Diss.* T. H. Stuttgart, 1959. (155) V. SCHOMAKER and D. P. STEVENSON, *J. Amer. Chem. Soc.* **63**, 37 [1941]. (156) G. V. SCHULZ and A. HAUG, *Z. Phys. Chem.*, N.F. **34**, 328 [1962]. (157) D. W. SCOTT, *J. Amer. Chem. Soc.* **68**, 356 [1946]. (158) —, ibid **68**, 1877 [1946]. (159) K. SEYFRIED, Farbenfabriken Bayer, 1965, unpublished.

(160) W. SIMMLER, Farbenfabriken Bayer, German Appl. (West German) 1046049 [17.8.1957]. (161) —, —, 1956, unpublished. (162) —, —, 1957, unpublished. (163) —, —, 1958, unpublished. (164) —, *Makromolekulare Chem.* **57**, 12 [1962]. (165) —, H. NIEDERPRÜM and H. WALZ, *Chem. Ber.* **97**, 1047 [1964]. (166) —, —, —, ibid. **96**, 1495 [1963]. (167) H. A. SKINNER, *Trans. Faraday Soc.* **41**, 645 [1945]. (168) B. SMITH, *Svensk. kem. Tidskr.* **65**, 101 [1953]. (169) H. SOFFER and T. DEVRIES, *J. Amer. Chem. Soc.* **73**, 5817 [1951].

(170) N. N. SOKOLOV, *J. Gen. Chem. USSR (Engl. transl.)* **29**, 251 [1959]. (171) —, K. A. ANDRIANOV and S. M. AKIMOVA, ibid. **26**, 1061 [1956]. (172) L. H. SOMMER, J. R. GOLD, G. M. GOLDBERG and N. S. MARANS, *J. Amer. Chem. Soc.* **71**, 1509 [1949]. (173) —, G. M. GOLDBERG, G. H. BARNES and L. S. STONE JR., ibid. **76**, 1609 [1954]. (174) —, L. Q. GREEN and F. C. WHITMORE, ibid. **71**, 3253 [1949]. (175) — and R. P. PIOCH, ibid. **75**, 6337 [1953]. (176) —, —, N. S. MARANS, G. M. GOLDBERG, J. ROCKETT and J. KERLIN, ibid. **75**, 2932 [1953]. (177) — and J. ROCKETT, ibid. **73**, 5130 [1951]. (178) S. B. SPECK, *J. Org. Chem.* **18**, 1689 [1953]. (179) J. L. SPEIER, *J. Amer. Chem. Soc.* **71**, 273 [1949].

(180) —, ibid. **73**, 826 [1951]. (181) —, B. F. DAUBERT and R. R. MCGREGOR, ibid. **71**, 1474 [1949]. (182) M. M. SPRUNG and F. O. GUENTHER, ibid. **77**, 3990 [1955]. (183) H. A. STAAB: *Einführung in die theoretische organische Chemie*, p. 559, Verlag Chemie, Weinheim/Bergstr. 1959. (184) H. STEINFINK, B. POST and I. FANKUCHEN, *Acta crystallogr.* [Copenhagen] **8**, 420 [1955]. (185) H. A. STUART: *Die Physik der Hochpolymeren, Bd 1, Die Struktur des freien Moleküls*, Springer-Verlag, Berlin-Göttingen-Heidelberg 1952. (186) C. TAMBORSKI and H. W. POST, *J. Org. Chem.* **17**, 1400 [1952]. (187) T. TANAKA and T. WATASE, *J. Chem. Physics* **22**, 1268 [1954]. (188) S. TANNENBAUM, *J. Amer. Chem. Soc.* **76**, 1027 [1954]. (189) W. S. TATLOCK and E. G. ROCHOW, ibid. **72**, 528 [1950].

(190) E. THILO, *Forsch. u. Fortschr.* **29**, 161 [1955]. (191) R. THOMPSON, *J. Chem. Soc.* [London] **1953**, 1908. (192) L. J. TYLER, Dow Corning, U.S. Pat. 2605274 [9.1.1950]. (193) —, —, U.S. Pat. 2611774 [9.1.1950]. (194) L. H. VOGT JR. and J. F. BROWN JR., *Inorg. Chem.*

2, 189 [1963]. (195) E. L. WARRICK, M. J. HUNTER and A. J. BARRY, *Ind. Engng. Chem.* **44**, 2196 [1952]. (196) —, W. A. PICCOLI and F. O. STARK, *J. Amer. Chem. Soc.* **77**, 5017 [1955]. (197) A. WEISS and A. WEISS, IUPAC-Colloquium Münster 1954, Verlag Chemie, Weinheim/Bergstr. (198) A. WEISSLER, *J. Amer. Chem. Soc.* **71**, 93 [1949]. (199) R. WEST, ibid. **76**, 6012 [1954],

(200) —, L. S. WHATLEY and K. J. LATTE, ibid. **83**, 761 [1961]. (201) Westinghouse, Brit. Pat. 706703 [U.S. Prior. 24.3.1950]. (202) E. WIBERG and W. SIMMLER, *Z. anorg. allg. Chem.* **282**, 330 [1955]. (203) —, —, ibid. **283**, 401 [1956]. (204) D. F. WILCOCK, *J. Amer. Chem. Soc.* **68**, 691 [1946]. (205) N. WRIGHT and M. J. HUNTER, ibid. **69**, 803 [1947]. (206) A. Y. YAKUBOVICH and V. A. GINSBERG, quoted by H. Gilman and G. E. Dunn, *Chem. Reviews* **52**, 77 [1953]. (207) K. YAMASAKI, A. KOTERA, M. YOKOI and Y. UEDA, *J. Chem. Physics* **18**, 1414 [1950]. (208) C. W. YOUNG, P. C. SERVAIS, C. C. CURRIE and M. J. HUNTER, *J. Amer. Chem. Soc.* **70**, 3758 [1948]. (209) T. ZOLTAI and M. J. BUERGER, *Z. Krist* **114**, 1 [1960].

CHAPTER 7

Other Organosilicon Polymers

7.1 Heterosiloxanes

Under this heading will be classed all polymers derived from siloxanes in which some of the silicon atoms in the siloxane skeleton have been replaced by heteroatoms (M). If every second Si atom is replaced by a divalent heteroatom, the resulting linear molecule would conform to the structural type

$$-\overset{|}{\underset{|}{\mathrm{Si}}}-\mathrm{O}-\mathrm{M}-\mathrm{O}-\overset{|}{\underset{|}{\mathrm{Si}}}-\mathrm{O}-\mathrm{M}-\mathrm{O}-\overset{|}{\underset{|}{\mathrm{Si}}}-$$

The lowest-molecular-weight heterosiloxanes containing at least one Si—O—M bond are thus built up of the linkages

$$\equiv\mathrm{Si}-\mathrm{O}-\mathrm{M}, \qquad \equiv\mathrm{Si}-\mathrm{O}-\mathrm{M}-, \qquad \equiv\mathrm{Si}-\mathrm{O}-\mathrm{M}\lessdot, \text{ and so on,}$$

depending on the valence of the M atoms. The alkali-metal silanolates already mentioned are formally heterosiloxanes. Heterosiloxanes containing boron, aluminum, titanium, tin, lead, phosphorus, arsenic, and other elements have also been described; these will be the subject of the present discussion. Only carbon itself is not considered as a heteroatom in this context.

The introduction of heteroatoms into the siloxane skeleton extends the structural and bonding principles of silicate chemistry to organosilicon chemistry. In exchange, the results obtained from organically substituted heterosiloxanes are of interest in clarifying features of the "heterosiloxane bond" in silicates.

The replacement of Si atoms by heteroatoms in general produces bonds which are *more strongly polar* than the Si—O—Si bond. This increases the tendency to ionic reactions, and some heterosiloxane bonds are therefore

TABLE 64

HETEROSILOXANES[a]

Compound	Mp (°C)	Bp (°C)	at p (mm Hg)	n_D^{20}	d_4^{20} (gm/cc)
(a) Borosiloxanes					
$[(CH_3)_3SiO]_3B$	−35 (*257*)	34 sublimes (*257*)	1	1.3859 (*171*)	0.8299 (*301*)
		184.5 (*301*)	776	1.3860 (*301*)	0.8285 (*301*)
$[(C_2H_5)_3SiO]_3B$	—	134–136 (*174, 310*)	1–2	1.4379 (*301*)	0.8918 (*301*)
		178–179 (*301*)	13	1.4382 (*301*)	—
$[(C_2H_5)_3SiO]_2B(CH_3)$	—	115 (*174, 310*)	~2.5	—	—
(b) Alumosiloxanes					
$[[(CH_3)_3SiO]_3Al]_2$	238 (*253*)	155 sublimes (*253*)	1	—	—
$[(CH_3)_3SiOAlCl_2]_2$	88–89 (*253*)	102 (*253*)	4	—	—
$[(CH_3)_3SiOAlH_2]_2$	80–82 (*253*)	40 sublimes (*253*)	1	—	—
$[(CH_3)_3SiOAl(CH_3)_2]_2$	45.5 (*253*)	81.5 (*253*)	10	—	—
(c) Titanosiloxanes					
$[(CH_3)_3SiO]_4Ti$	—	106 (*90*)	7	1.4278 (*90*)	0.9078 (*90*)
$[(CH_3)_3SiO]_2Ti(OC_3H_7)_2$	—	120 (*75*)	14	1.4408[b] (*75*)	—
$[(CH_3)_3SiO]Ti(OC_3H_7)_3$	—	114 (*75*)	13	1.4490[b] (*75*)	—
$[(CH_3)_3SiO]_3Ti(OC_4H_9)$	—	76–77 (*9*)	2	1.4317 (*9*)	0.9070 (*9*)
$[(CH_3)_3SiO]_3TiCl$	—	103–105 (*9*)	10	—	—
$[(CH_3)_3SiO]_2TiCl_2$	—	72–73 (*9*)	3	—	—
$[(CH_3)_3SiO]TiCl_3$	—	67 (*9*)	9	—	—
$[(C_2H_5)_3SiO]_4Ti$	95 (*20*)	204 (*20*)	4	—	—
$[(C_6H_5)_3SiO]_4Ti$	501–505 (*313*)	370 sublimes (*313*)	1	1.6488[c] (*313*)	1.215[d] (*313*)
$[(C_6H_5)_8Si_4O_5]_2Ti$	313–314.5 (*314*)	—	—	—	—

(*continued on p.* 334)

* Page number of chapter references.

TABLE 64 *(Contd.)*

HETEROSILOXANES[a]

Compound	Mp (°C)	Bp (°C)	at p (mm Hg)	n_D^{20}	d_4^{20} (gm/cc)
(d) Stannosiloxanes					
$(CH_3)_3SiOSn(CH_3)_3$	−59 (*253*)	141 (*253*)	720	—	—
$(CH_3)_3SiOSn(n\text{-}C_3H_7)_3$	—	126 (*215*)	20	1.4575 (*215*)	1.093 (*215*)
$(CH_3)_3SiOSn(n\text{-}C_4H_9)_3$	—	142 (*215*)	2	1.4582 (*215*)	1.059 (*215*)
$[(CH_3)_3SiO]_2Sn(CH_3)_2$	48 (*253*)	75 (*253*)	11	—	—
$[(CH_3)_3SnO]_2Si(CH_3)_2$	—	77 (*253*)	1	—	—
$[(CH_3)_3SiO]_3SnCH_3$	34 (*253*)	49 (*253*)	1	—	—
$[(CH_3)_3SiO]_4Sn$	64 (*253*)	60 (*253*)	1	—	—
$(C_6H_5)_3SiOSn(C_6H_5)_3$	139–140 (*253*)	—	—	—	—
$[(C_6H_5)_3SiO]_2Sn(C_6H_5)_2$	148.5–149.5 (*253*)	—	—	—	—
$[(C_6H_5)_3SnO]_2Si(C_6H_5)_2$	94.5–95.5 (*253*)	—	—	—	—
(e) Plumbosiloxanes					
$(CH_3)_3SiOPb(CH_3)_3$	—	172 (*253*)	720	—	—
(f) Phosphorosiloxanes					
$[(CH_3)_3SiO]\overset{O}{\overset{\|}{P}}(OCH_3)_2$	—	87–88 (*92*)	12	1.4054 (*92*)	1.0705 (*92*)
$[(CH_3)_3SiO]\overset{O}{\overset{\|}{P}}(OC_2H_5)_2$	—	97–98 (*92*)	12	1.4070 (*92*)	1.018 (*92*)
$[(C_2H_5)_3SiO]\overset{O}{\overset{\|}{P}}(OC_2H_5)_2$	—	134–139 (*92*)	12	1.4273 (*92*)	1.004 (*92*)

$[(CH_3)_3SiO]_2P(O)OCH_3$	—	98–99 (*92*)	12	1.4081 (*92*)	1.001 (*92*)
$[(CH_3)_3SiO]_2P(O)OC_2H_5$	—	102–103 (*92*)	12	1.4062 (*92*)	0.981 (*92*)
$[(C_2H_5)_3SiO]_2P(O)OC_2H_5$	—	85–87 (*92*)	10^{-3}	1.4372 (*92*)	0.972 (*92*)
$[(CH_3)_3SiO]_3P{=}O$	—	85–87 (*248*)	4	1.4090 (*248*)	0.9650 (*302*)
	—	108–109 (*92*)	12	1.4092 (*92*)	—
$[(C_2H_5)_3SiO]_3P{=}O$	—	106–108 (*92*)	10^{-3}	1.4475 (*92*)	—
	—	183–186 (*302*)	3	1.4455 (*302*)	0.9668 (*302*)
$(C_2H_5O)_2P(O)[(CH_3)_2SiO]P(O)(OC_2H_5)_2$	—	101–104 (*92*)	10^{-3}	1.4180 (*92*)	1.1536 (*92*)
$(C_2H_5O)_2P(O)[(CH_3)_2SiO]_2P(O)(OC_2H_5)_2$	—	122 (*92*)	10^{-3}	1.4158 (*92*)	1.1197 (*92*)
$(C_2H_5O)_2P(O)[(CH_3)_2SiO]_3P(O)(OC_2H_5)_2$	—	~130 (*92*)	10^{-3}	1.4145 (*92*)	1.1012 (*92*)
$(C_2H_5O)_2P(O)[(CH_3)_2SiO]_4P(O)(OC_2H_5)_2$	—	~140 (*92*)	10^{-3}	1.4140 (*92*)	1.0853 (*92*)
$(C_2H_5O)_2P(O)[(CH_3)_2SiO]_5P(O)(OC_2H_5)_2$	—	~152 (*92*)	10^{-3}	1.4127 (*92*)	1.0691 (*92*)

[a] Numbers in parentheses indicate reference numbers.
[b] n_D^{22}.
[c] n_D^{114}.
[d] d_4^{29}.

highly sensitive toward hydrolytic attack. The relationship of the heterosiloxanes to the siloxanes is the same as that of the silicates to the various forms of silicic acid. Silicates are also more easily broken down by hydrolytic attack—such as weathering in nature—than, for example, quartz. However, the sensitivity to hydrolysis of the heterosiloxane bond is not so marked in the crystals of the silicates as in the molecularly disperse siloxanes; this difference is certainly due, to some extent, to differences in surface development, but the latter can undergo within minutes reactions which go to completion only within geological periods on the surfaces of crystalline solids, where there is an additional retardation by the formation of layers of reaction products inhibiting reaction and diffusion.*

These factors explain why heterosiloxanes based on Al—O—Si have found application mainly in the production of resin moldings, i.e., of compact masses in which, because of the small surface development, the increased sensitivity to hydrolysis of the heterosiloxane bond becomes unimportant.

The practical importance of the heterosiloxanes cannot at present be fully evaluated. They have acquired little interest as materials of construction and are inferior rather than superior to the pure siloxanes. In a few cases it has been possible to obtain special properties in siloxane molecules by inserting heteroatoms. The part played by heterosiloxanes in chemical processes such as curing and vulcanization would appear to be of greater interest, since they form intermediate compounds during catalysis with compounds of tin, titanium, aluminum, or boron.

The methods used to prepare Si—O—M bonds are almost exclusively reactions corresponding to the "nonhydrolytic" processes in the formation of Si—O—Si bonds (Section 5.2). Simple cohydrolysis of hydrolyzable compounds containing M and Si atoms is successful only if the difference in the rates of hydrolysis of the reaction partners is not too great.

7.1.1 Borosiloxanes

The preparation of B—O—Si bonds by the *cleavage of siloxanes with borohalides* has been mentioned previously (Section 5.5.5). This proceeds as follows:

$$>\mathrm{B{-}X} + -\mathrm{Si{-}O{-}Si}- \rightarrow\ >\mathrm{B{-}O{-}Si}- + -\mathrm{Si{-}X}$$

* Comparative experiments on the resistance to hydrolysis of the heterosiloxane bond in the molecule and the silicate crystal would certainly be very interesting. Work by Andrianov and Ganina (*11*) on the stability of kaolinite and alumosilicates toward aqueous hydrochloric acid has not yielded a useful basis for comparison, since the "kaolinite" used was almost completely decomposed in 10% acid within about an hour, contrary to all other known experiments.

Numerous other methods are available for the formation of these bonds such as the following.

(a) The reaction of boric acid with alkoxysilanes (*171*, *301*):

$$>B{-}OH + R{-}O{-}Si{\equiv} \rightarrow\ >B{-}O{-}Si{\equiv} + ROH$$

The first clearly defined borosiloxane, tris-trimethylsilyl borate, was prepared by this method (*171*).

(b) The reaction of boric acid with silanols (*2*, *133*, *217*, *297*):

$$>B{-}OH + HO{-}Si{\equiv} \rightarrow\ >B{-}O{-}Si{\equiv} + H_2O$$

(c) The reaction of boric acid with halosilanes (*221*):

$$>B{-}OH + X{-}Si{\equiv} \rightarrow\ >B{-}O{-}Si{\equiv} + HX$$

The analogous reaction with the functional groups interchanged, i.e., that of boron halides with silanols, does not give borosiloxanes directly. In this case an exchange between the OH and halogen on the two reactants takes place (*310*):

$$>B{-}X + HO{-}Si{\equiv} \rightarrow\ >B{-}OH + X{-}Si{\equiv}$$

analogously to the OR-halogen exchange in the reaction of boron halides with alkoxysilanes (*106*, *309*):

$$>B{-}X + RO{-}Si{\equiv} \rightarrow\ >B{-}OR + X{-}Si{\equiv}$$

after which the formation of the borosiloxane bond can be made to take place at higher temperatures.

(d) The reaction of boric acid with trialkylsilanes in the presence of colloidal nickel (*217*), nickel chloride, or cobalt chloride (*79*):

$$>B{-}OH + H{-}Si{\equiv} \xrightarrow{Ni}\ >B{-}O{-}Si{\equiv} + H_2$$

(e) The reaction of boron halides with alkali-metal silanolates (*2*, *310*):

$$>B{-}X + NaO{-}Si{\equiv} \rightarrow\ >B{-}O{-}Si{\equiv} + NaX$$

If, for example, a benzene solution of sodium triethylsilanolate is added slowly, with stirring, to a benzene solution of boron tribromide in a molar ratio of 3:1 at room temperature, NaBr separates instantaneously in practically quantitative yield.

(f) The reaction of boric esters with acetoxysilanes (*18*, *143*):

$$>B{-}OR + AcO{-}Si{\equiv} \rightarrow\ >B{-}O{-}Si{\equiv} + AcOR$$

and, inversely, of acetoxyboron compounds with alkoxysilanes:

$$>B{-}OAc + RO{-}Si{\equiv} \rightarrow\ >B{-}O{-}Si{\equiv} + AcOR$$

which can be catalyzed by acids and bases (*143*). So far, it has not been possible to prepare definite borosiloxanes by this reaction. The tendency to undergo exchange reactions appears to cause particular difficulty in this case. These reactions cause the attachment of acyloxy or alkoxy groups to the same boron or silicon atom, so that intramolecular reactions can take place that lead to the formation of boric acid or siloxanes as by-products.

The reaction of boric esters with silanols:

$$>B{-}OR + HO{-}Si{\equiv} \rightarrow\ >B{-}O{-}Si{\equiv} + HOR$$

gives only moderate yields (*2*).

(g) Reaction of aminoboron compounds with silanols (*2*):

$$>BNR_2 + HOSi{\equiv} \rightarrow\ >B{-}O{-}Si{\equiv} + R_2NH$$

Tris-triethylsilyl borate has been prepared in excellent yield (98%) by this process (*2*).

Investigations of such syntheses have shown, or made it appear probable, that borosiloxanes can be prepared as different types of polymers. Besides the simplest borosiloxane, tris-trimethylsilyl borate (I), branched borosiloxanes (II) have been prepared. Using phenylboronic acid as difunctional boron component, cyclic (III), linear (IV), and cross-linked, presumably "ladder-like", polymeric condensation products (V), have also been obtained.

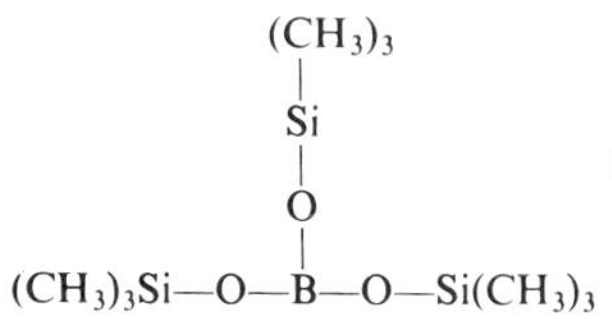

(*171*)

I

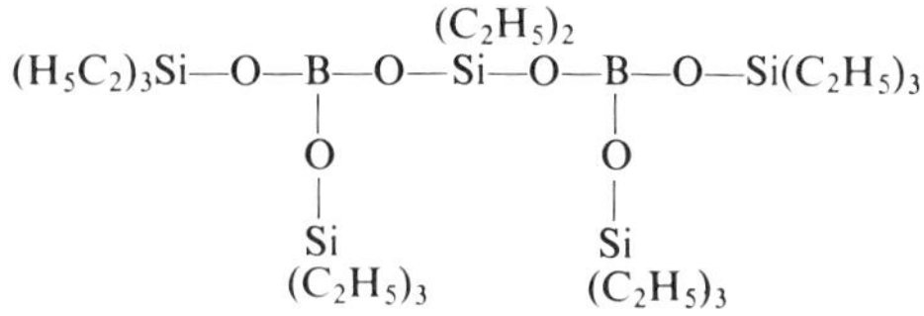

(*10*)

II

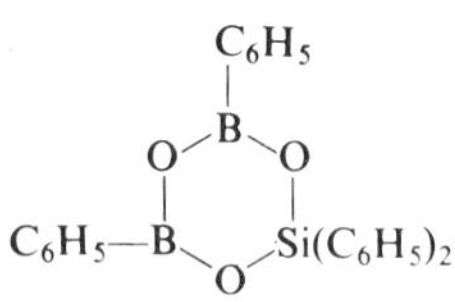

(*133*)

III

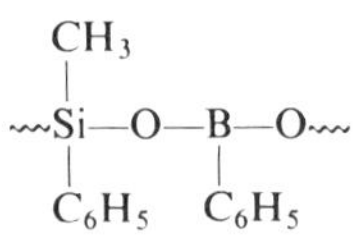

(*16*)

IV

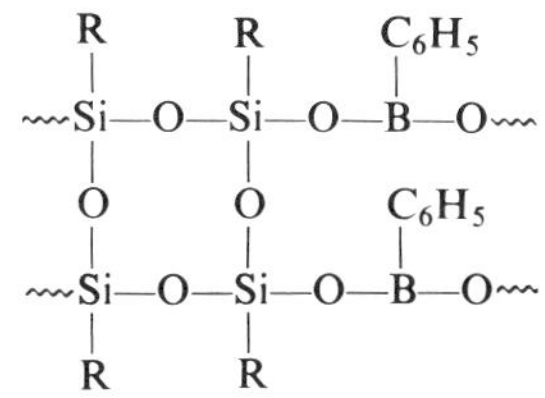

(279)

V

The B—O—Si bond is generally far more sensitive to hydrolysis than the siloxane bond. Trialkylsilyl borates, however, decompose only slowly on shaking with cold water (*2*). The sensitivity to hydrolysis of linear borosiloxane polymers can be decreased by substituting at least one phenylaminomethyl group on the Si (*16*). Cyclic borosiloxanes of type (III) have considerable resistance to hydrolysis (*133*). These compounds are also recommended as plasticizers for polyvinyl chloride (*133*, *134*).

Interesting effects can be achieved by the inclusion of B—O—Si bonds in polydimethylsilicone rubber, since the boron atom with its electron vacancy can act as acceptor for the lone electron pairs of the siloxane oxygen, and electron bridges are formed between neighboring molecules, causing weak intermolecular bonding:

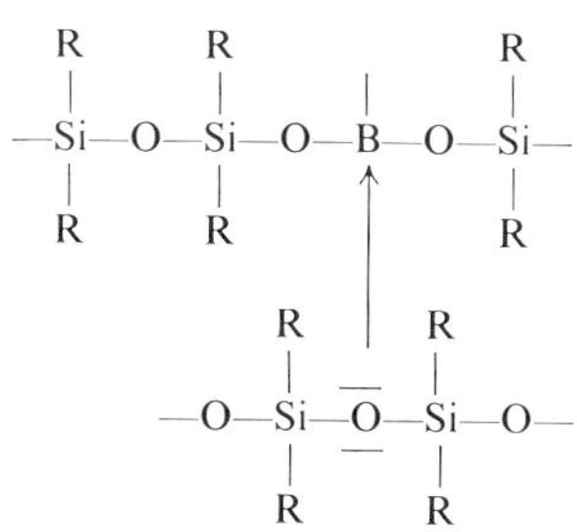

Depending on the amount of boron incorporated, the polymer acquires properties which cannot be achieved by cross-linking through primary valences, siloxane, or, e.g., alkylene bridges. Products having one Si—O—B bond for every three to every hundred Si atoms in the molecule already show the behavior of bouncing putty at relatively low molecular weights. They possess extremely high resilience when subjected to rapid shocklike stresses but react like a plastic putty under slow deformation. If the boron content is reduced to one boron atom per 200 to 500 Si atoms, materials are obtained at molecular weights of about 350,000 to 500,000 that are sticky after a pre-cure and weld together after final vulcanization (*311*) (see Section 8.1.1).

7.1.2 Alumosiloxanes

In addition to cleavage reactions of the Si—O—Si bond with aluminum halides (Section 5.5.5) the following routes are available for synthesis of alumosiloxanes.

(a) The reaction of aluminum chlorides (*19*, *252*) or organosiloxyaluminum chlorides (*156*) with alkali-metal silanolates or, conversely, of chlorosilanes with sodium aluminate (*27*, *281*) proceeds very smoothly:

$$\gt \mathrm{Al{-}Cl} + \mathrm{NaO{-}Si} \lessgtr \;\rightarrow\; \gt \mathrm{Al{-}O{-}Si} \lessgtr + \mathrm{NaCl}$$

$$\gtrless \mathrm{Si{-}Cl} + \mathrm{NaO{-}Al} \lt \;\rightarrow\; \gtrless \mathrm{Si{-}O{-}Al} \lt + \mathrm{NaCl}$$

(b) The reaction of aluminum alkoxides with acyloxysilanes (*168*, *186*, *244*):

$$\gt \mathrm{Al{-}OR} + \mathrm{AcO{-}Si} \lessgtr \;\rightarrow\; \gt \mathrm{Al{-}O{-}Si} \lessgtr + \mathrm{AcOR}$$

(c) The reaction of aluminum alkoxides with silanols, e.g., of aluminum butoxide with triethylsilanol (*213*):

$$\gt \mathrm{Al{-}OR} + \mathrm{HO{-}Si} \lessgtr \;\rightarrow\; \gt \mathrm{Al{-}O{-}Si} \lessgtr + \mathrm{HOR}$$

(d) Finally, the reaction of aluminum with silanols (*11*, *308*):

$$2\mathrm{Al} + 6\mathrm{HOSi} \lessgtr \;\rightarrow\; 2\mathrm{Al}(\mathrm{{-}O{-}Si} \lessgtr)_3 + 3\mathrm{H_2}$$

Attempts have been made to obtain high-molecular-weight products not only directly by reactions of this type (i.e., using polyfunctional starting materials) but also by a stepwise preparation of definite low-molecular-weight alumosiloxanes and subsequent partial hydrolysis (*20*, *151*, *315*):

$$n\mathrm{R_3Si{-}O{-}Al(OR')_2} \xrightarrow{n\mathrm{H_2O}} {-}\left[\begin{array}{l}\mathrm{SiR_3}\\ \;|\\ \mathrm{O}\\ \;|\\ \mathrm{Al{-}O{-}}\end{array}\right]_n + 2n\mathrm{R'OH}$$

$$n[\mathrm{R_3Si{-}O{-}}]_3\mathrm{Al} \xrightarrow{n\mathrm{H_2O}} {-}\left[\begin{array}{l}\mathrm{SiR_3}\\ \;|\\ \mathrm{O}\\ \;|\\ \mathrm{Al{-}O{-}}\end{array}\right]_n + 2n\mathrm{R_3SiOH}$$

Whether cohydrolysis of aluminum chloride with organochlorosilanes (*11*) leads to polymers with a statistically homogeneous distribution of Al—O and Si—O units must remain very questionable, since the differences in the sensitivity of the aluminum and silicon halides to hydrolysis appears to be very great.

A number of definite low-molecular-weight alumosiloxanes have been prepared and thoroughly investigated (*253*). The organosiloxyaluminum halides obtained by the cleavage of hexaorganodisiloxanes with aluminum halides are dimeric by coordination and contain the almost planar skeleton of structure (I). By alkylation with alkyllithiums these have been converted into the corresponding pentaalkylalumosiloxane compounds, which are also dimeric (II). These are thermally very stable and decompose only above 180°C, with the liberation of tetraalkylsilanes. They are sensitive to hydrolysis and oxidation but are less reactive than alkylaluminum compounds (*253*):

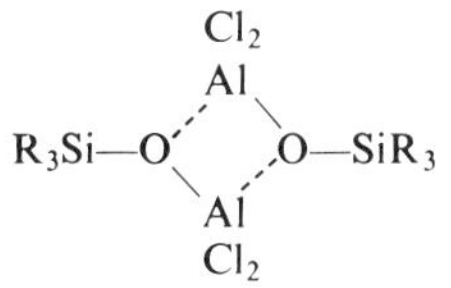

I

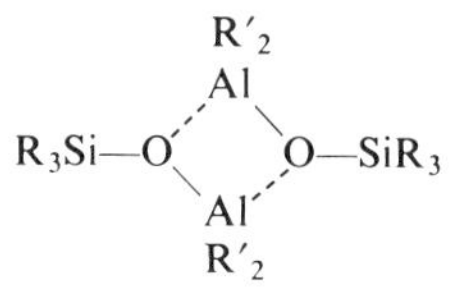

II

The hydrides obtained from (I) by means of lithium tetrahydroaluminate tend to polymerize and are stable in their dimeric form for only short periods (*253*).

The tris-triorganosiloxy compounds of aluminum have also been shown to form coordination dimers (*253*):

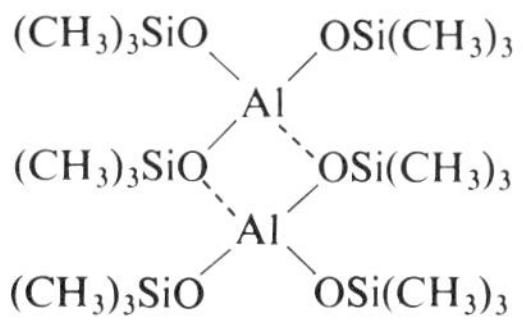

They are remarkably stable compounds which decompose only slowly above 280°C to siloxanes and alumina (*253*). They are relatively insensitive to water alone but are rapidly hydrolyzed in the presence of organic solvents (*253*). With alkali-metal silanolates, e.g., $(CH_3)_3SiOK$, they form alkali-metal tetrakis-trimethylsilyl-aluminates smoothly and completely even at room temperature (*253*):

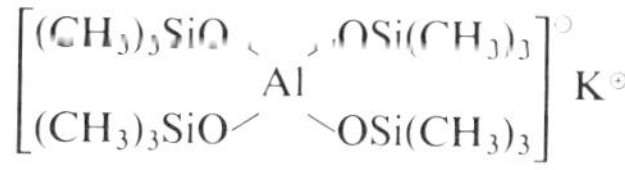

Linear, branched, and cross-linked polymers can be formed in the synthesis of high-molecular-weight alumosiloxanes, depending on the nature of the starting material and on the reaction mechanism, but their structures must still be considered uncertain.

Cross-linked polymers form resinous or glasslike materials, which are said to be useful for the preparation of laminates (*316*) or for bonding glass fibers and phenol-formaldehyde resins (*15*). There is, however, no convincing evidence that alumosiloxanes show decisive advantages over the simple siloxanes, if the complete pattern of their properties is evaluated. It has not been possible to obtain products analogous to silicone rubber (*281*), presumably because linearity cannot be achieved at high molecular weights because of unavoidable cross-linking through coordinative or primary valences.

7.1.3 Titanosiloxanes

Compounds with Ti—O—Si bonds are formed when siloxanes are cleaved with titanium tetrachloride or butyl titanate (Section 5.5.5).

However, it is preferred to use the following processes to prepare titanosiloxanes:

(a) The reaction of titanium tetrachloride with silanols in the presence of ammonia as an acid acceptor (*90*):

$$\equiv Ti{-}Cl + HO{-}Si\equiv \xrightarrow{NH_3} \equiv Ti{-}O{-}Si\equiv + NH_4Cl$$

(b) The reaction of titanium tetrachloride with alkali-metal silanolates (*9*, *21*, *107*):

$$\equiv Ti{-}Cl + MO{-}Si\equiv \rightarrow \equiv Ti{-}O{-}Si\equiv + MCl$$

(c) The reaction of alkyl titanates with silanols (*6*, *38*, *75*, *80*, *137*):

$$\equiv Ti{-}OR + HO{-}Si\equiv \rightarrow \equiv Ti{-}O{-}Si\equiv + ROH$$

This type of transesterification with alkyl titanates is catalyzed by sodium (*80*).

(d) The reaction of titanic acid esters and acetoxysilanes (*108*, *245*):

$$\equiv Ti{-}OR + AcO{-}Si\equiv \rightarrow \equiv Ti{-}O{-}Si\equiv + ROAc$$

As with the analogous reactions of boric acid esters and acetoxysilanes, there is again obviously the possibility of partial exchange reactions of —OR and —OAc groups which result in *intra*molecular rather than *inter*molecular reactions (*13*). Mono-, bis-, tris-, and tetrakis-trimethylsilyl titanates can, however, be obtained in good yield by this reaction if the stoichiometry of the components is carefully controlled (*245*).

Definite trialkylsiloxytitanoxanes are formed if $[(CH_3)_3SiO]_4Ti$ is hydrolyzed at room temperature in dioxane solution (*37*). Polymeric organotitanosiloxanes are claimed to have been obtained by the cohydrolysis of halogen compounds of titanium or alkyl titanates with organohalosilanes (*12*, *232*).

The sensitivity to hydrolysis of the Ti—O—Si bond can be decreased by suitable substitution on the titanium atom. Experiments have tended in the direction of building up polymers containing chelate groups, e.g., with acetylacetone (*41*, *286*, *287*), in which the chelate-substituted titanoxane units are difunctional and form linear polymers with difunctional dialkylsiloxanes (*286*):

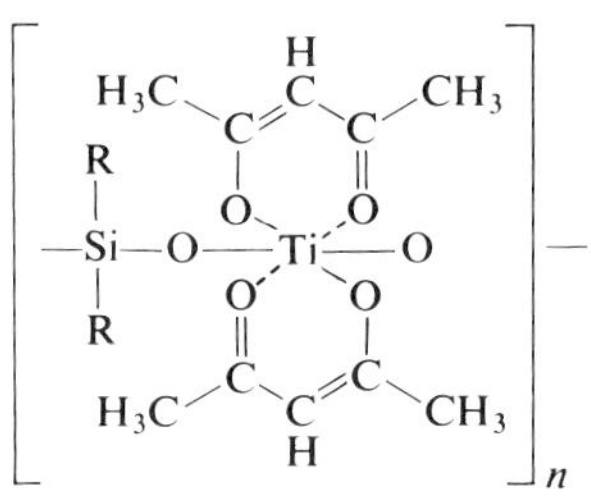

Heterosiloxanes with Ti—O—Si bonds have not found industrial application as such. They could be formed as intermediates during the condensation of siloxanes with the addition of functional titanium compounds, particularly titanium butoxide (Section 8.2.2.1.2).

7.1.4 Stannosiloxanes

The Sn—O—Si bonds are generally obtained by a rapid and smooth reaction between alkali-metal silanolates and tin halides or organotin halides.

$$\equiv Sn{-}Cl + MO{-}Si\equiv \rightarrow \equiv Sn{-}O{-}Si\equiv + MCl$$

A number of definite low-molecular-weight organostannosiloxanes have been prepared by this process (*97*, *150*, *220*, *255*, *256*, *288*, *289*). The converse process is also possible (*254*, *255*):

$$\equiv Sn{-}OLi + Cl{-}Si\equiv \rightarrow \equiv Sn{-}O{-}Si\equiv + LiCl$$

Stannosiloxanes can also be obtained by the reaction between tin diacetate or diisobutyltin diacetate and dimethyldiethoxysilane (*145*):

$$\equiv Sn{-}OAc + RO{-}Si\equiv \rightarrow \equiv Sn{-}O{-}Si\equiv + ROAc$$

or by the reaction of diphenyldimethoxytin with diphenyldichlorosilane (*97*):

$$\geq\!Sn{-}OR + Cl{-}Si\!\leq \rightarrow\ \geq\!Sn{-}O{-}Si\!\leq + RCl$$

With triphenyl-substituted stannanes and silanes, however, the first type of reaction proceeds sluggishly and with poor yields (*289*).

A series of other relatively simple processes also leads to this group of compounds. Dimeric molecules of the type $(CH_3)_3SiO[R_2SnO]_2Si(CH_3)_3$ have been prepared by the cohydrolysis of dialkyltin dichlorides with trimethylchlorosilane (*214*). The condensation of silanediols, e.g., diphenylsilanediol, with diphenyl- or dibutyltin oxide (*97*, *165*, *166*) gives stannosiloxanes:

$$nR_2SnO + mR'_2Si(OH)_2 \xrightarrow{-H_2O} \left[-\overset{R}{\underset{R}{Sn}}-O \right]_n - \left[\overset{R'}{\underset{R'}{Si}}-O- \right]_m$$

Organotin oxides R_2SnO can be copolymerized with dimethylsiloxane rings or chains either by simple heating (*159*, *265*) or in the presence of catalytic amounts of KOH (*97*, *265*). These syntheses clearly indicate the close relationship between stannoxane and siloxane chemistry.

Since the Sn atom in a stannosiloxane chain can act as an electron acceptor for the lone electron pairs on the oxygen of the neighboring chain, the viscosity of a compound is considerably increased by the insertion of an Sn—O unit into a siloxane chain (*265*). Similarly, it is not surprising that linear polymeric stannosiloxanes can behave like bouncing putty, a polymer which is usually prepared with boron acting as an electron acceptor (Section 7.1.1).

Stannosiloxane bonds are presumably formed as intermediates when tin(II) or tin(IV) compounds, as vulcanization or curing catalysts, react with siloxane polymers. The acceptor action of tin with respect to the siloxane oxygen may also be of importance here, since it can initiate cleavage or rearrangement of the siloxane chains. The activity of the Sn—O—Si bond is shown clearly by the fact that, after the addition of ethyl silicate or polymethyl-*H*-siloxanes, stannosiloxanes vulcanize at room temperature in the same way as a mixture of siloxanes with functional Sn compounds (*159*) (Section 8.1.2.3).

By the addition of stannosiloxanes to methylchlorophenylsilicone oils it is possible to improve their lubricant action at elevated temperatures (*149*). However, these effects are generally slight (*267*).

The low-molecular-weight stannosiloxanes investigated so far are toxic and possess an unpleasant odor. Their hydrolytic and thermal stability is low (*253*). The resistance of the triethylsiloxy-metal compounds to hydrolysis decreases in the sequence (*22*):

$$Ti > Al > Sn$$

7.1.5 Plumbosiloxanes

Our knowledge of compounds with Pb—O—Si bonds is scant. Hexamethylplumbosiloxane is formed, even at room temperature, smoothly and in good yield by the reaction of trimethyllead chloride or bromide with alkali-metal trimethylsilanolates (*255*, *256*, *262*):

$$(CH_3)_3PbBr + NaOSi(CH_3)_3 \rightarrow (CH_3)_3SiPb{-}O{-}Si(CH_3)_3 + NaBr$$

Bis-(trimethylsiloxy)-lead has been prepared from trimethylsilanol and lead(II) oxide (*223*, *246*):

$$2(CH_3)_3SiOH + PbO \rightarrow (CH_3)_3Si{-}O{-}Pb{-}O{-}Si(CH_3)_3 + H_2O$$

Hexaalkylplumbosiloxanes are liquids with an unbearable odor and probably high toxicity (*253*). All plumbosiloxanes are extremely reactive and highly sensitive to hydrolytic attack. Plumbosiloxanes have the highest reactivity of all siloxanes with heteroatoms from group 4. Even the Pb—C bond contained in these compounds is labile.

The importance of plumbosiloxanes in silicone chemistry, like that of the titano- and stannosiloxanes, would appear to reside mainly in the fact that they are formed in intermediate stages during the growth of the molecules or the cross-linking of siloxanes with the aid of lead compounds (Section 8.2.2.1.2)

7.1.6 Other Heterosiloxanes

A large number of various other types of heterosiloxanes have been prepared, but none has achieved practical importance. To round off the presentation of this class of substances, therefore, only the following short observations will be made.

Phosphorosiloxanes are formed when siloxanes are cleaved with phosphorus pentoxide or with phosphoric acid (compare Section 3.11). They have also been synthesized by numerous other routes:

$$\equiv Si{-}X + RO{-}P(=O)\!< \;\rightarrow\; \equiv Si{-}O{-}P(=O)\!< + RX \quad (92, 197, 261, 302)$$

$$\equiv Si{-}X + HO{-}P(=O)\!< \;\xrightarrow{\text{pyridine}}\; \equiv Si{-}O{-}P(=O)\!< + \text{pyr.HX} \quad (92, 294, 302)$$

$$\equiv Si{-}X + NaO{-}P(=O)\!< \;\rightarrow\; \equiv Si{-}O{-}P(=O)\!< + NaX \quad (55, 92, 170, 294, 295)$$

$$\equiv Si{-}H + HO{-}P(=O)\!< \;\xrightarrow{\text{Ni}}\; \equiv Si{-}O{-}P(=O)\!< + H_2 \quad (219)$$

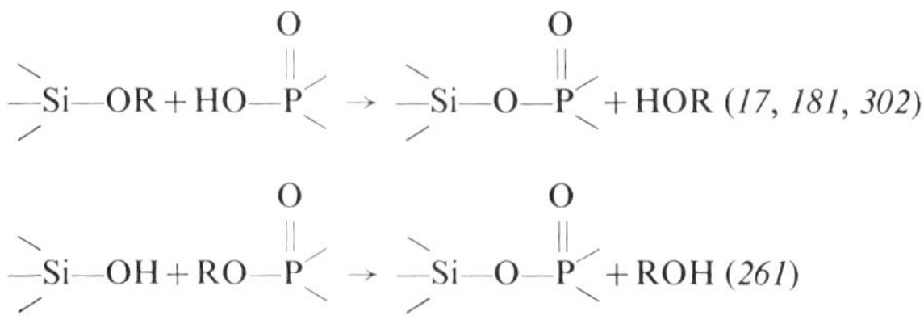

The Si—O—P bond is extremely sensitive to hydrolysis if lower alkyl substituents are present; resistance to hydrolysis can however be considerably increased by bulky substituents (*93*). Certain polymeric phosphorosiloxanes possess good antifoam properties (*295*).

Arsenosiloxanes and *stibiosiloxanes* are obtained by analogous and other processes (cf. ref. *253*):

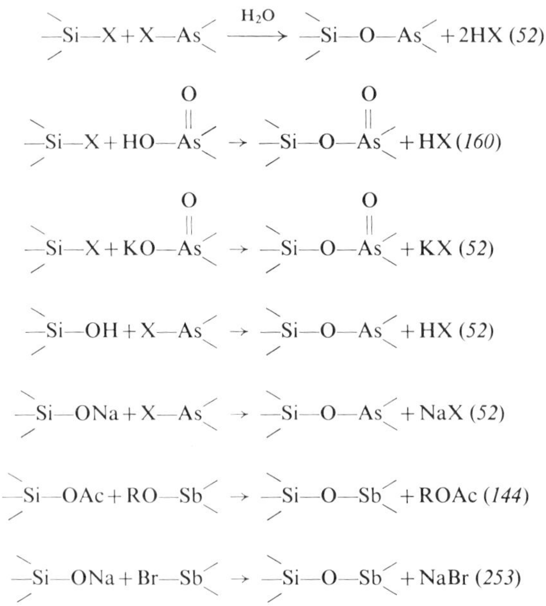

$$6R_3Si\text{—}OH + Sb_2O_3 \rightarrow 2(R_3SiO)_3Sb + 3H_2O \ (\mathit{218})$$

Among others, siloxanes with the following heteroatoms are known: gallium (*253*), germanium (*253*), zirconium (*38*), mercury (*288*), vanadium (*51*, *69*, *216*), tantalum (*38*, *290*), chromium (*1*, *259*), and iron (*258*). Many of these and other heterosiloxanes are short-lived because of the labile nature of the Si—OM bond, and can be detected only as intermediates (*150*). Their practical importance cannot as yet be evaluated. Certain heterosiloxanes containing chromium and vanadium are effective in combination with organoaluminum compounds as catalysts for the polymerization of olefins (*70*, *296*). Arsenosiloxanes are said to combine the water-repellent properties of the silicones with the insecticidal and fungicidal properties of arsenic (*3*).

Siloxanes with ferrocene groupings do not strictly belong to the heterosiloxanes as defined earlier (Section 7.1), since they contain iron atoms bound to silicon through carbon bridges (*251*):

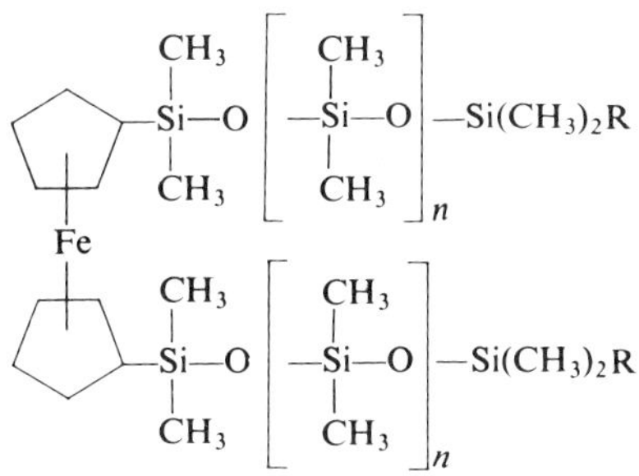

Low-molecular-weight compounds with $n=0$ to 2 and $R=CH_3$, C_6H_5 and $C_6H_3Cl_2$ have been isolated. They are obtained, for example, by the reaction of cyclopentadienyllithium with chlorosilanes or chlorosiloxanes to give the chloro-derivatives and by converting these by means of silanolates or cohydrolysis with chlorosilanes into cyclopentadienylsiloxanes from which the ferrocene derivatives are formed with butyllithium and iron(II)chloride (*251*). Bis-(trimethylsilyl)-ferrocene has been obtained from ferrocene by its conversion with butyllithium into dilithium ferrocene and treatment of this with trimethylchlorosilane (*234a*). Silicone fluid/trimethylsilylferrocene mixtures treated with oxygen at elevated temperature exhibit an improved resistance to oxidation (*23*) (Section 9.2.2).

7.2 Organosilicon Polymers without Siloxane Bridges

7.2.1 Polyorganosilanes

Polyorganosilanes (Table 65) are polymers in which the Si atoms are linked directly to one another without bridge atoms. They are obtained by the reaction of organohalosilanes with alkali metals (*47*, *48*, *65*, *109*, *152*, *158*, *162*, *283*):

$$2\ \mathrm{{\gt}Si{-}X} + 2\mathrm{K} \rightarrow \mathrm{{\gt}Si{-}Si{\lt}} + 2\mathrm{KX}$$

Polymers with Si—Si bonds are of interest as direct analogs of organic C—C polymers. However the results obtained from their synthesis and the investigation of their properties is disappointing. As in Stock's work on polysilanes,

TABLE 65

POLYORGANOSILANES[a]

Compound	Mp (°C)	Bp (°C)	at p (mm Hg)	n_D^{20}	d_4^{20} (gm/cc)
(a) Linear polymers, nonfunctional					
$(CH_3)_3Si—Si(CH_3)_3$	14.0–14.4 (*44*)	112.0–112.5 (*183, 283*)	759.5	1.4229 (*50*)	0.7272 (*50*)
$(C_2H_5)_3Si—Si(C_2H_5)_3$	—	255 (*169*)	760	1.4759 (*111*)	0.8300 (*111*)
$(n\text{-}C_3H_7)_3Si—Si(n\text{-}C_3H_7)_3$	—	114 (*260*)	3	1.4740 (*260*)	0.8693[b] (*260*)
$(C_6H_5)_3Si—Si(C_6H_5)_3$	368–370 (*114*)	—	—	—	—
$(CH_3)(C_6H_5)_2Si—Si(CH_3)(C_6H_5)_2$	145–146 (*111*)	—	—	—	—
$(CH_3)_2(C_6H_5)Si—Si(CH_3)_2(C_6H_5)$	34–35 (*111*)	289–302 (*111*)	760	—	—
$(C_6H_5)_3Si—Si(CH_3)_3$	107–108 (*112*)	—	—	—	—
$(C_6H_5)_3Si[(C_6H_5)_2Si]Si(C_6H_5)_3$	260–262 (*115*)	—	—	—	—
(b) Cyclic polymers, nonfunctional					
$[(C_6H_5)_3Si]_4$	400 (decomp.) (*163*)	—	—	—	—
(c) Linear polymers, functional					
$(C_6H_5)_3Si—Si(C_6H_5)_2H$	128–129 (*110*)	—	—	—	—
$(C_6H_5)_3Si—Si(C_6H_5)_2Cl$	154–155 (*115*)	—	—	—	—
$(C_6H_5)_3Si—Si(C_6H_5)Cl_2$	98–101 (*113*)	—	—	—	—
$(C_6H_5)_3Si—SiCl_3$	143–145 (*113*)	—	—	—	—
$(C_6H_5)_3Si—Si(C_6H_5)_2OC_2H_5$	210.5–212 (*110*)	—	—	—	—
$(C_6H_5)_3Si—Si(C_6H_5)_2OH$	134–134.5 (*110*)	—	—	—	—

[a] Numbers in parentheses indicate reference numbers.

[b] d_4^{26}.

the preparation of macromolecules from organically substituted silanes has been found difficult or impossible. Furthermore, the Si—Si bond is sensitive to hydrolysis, particularly in the presence of alkalis, as well as to oxidation and heating. The difference between carbon and silicon chemistry cannot be shown more clearly than by a comparison of the C—C and the Si—Si bonds.

The low stability of the Si—Si bond also makes it impossible to prepare polyorganosilanes by the alkylation of polyhalosilanes with Grignard compounds or oganic compounds of the alkali metals. The reaction of octachlorotrisilane, Si_3Cl_8, with phenylmagnesium bromide leads to a mixture of tetraphenylsilane and hexaphenyldisilane; i.e., in a chain consisting of 3 Si atoms, one Si—Si bond is cleaved by the Grignard reagent (*260*).

Si—Si bonds have been incorporated into siloxane networks to impart certain properties to silicone resins (Section 8.2.1.2).

Polyorganohalosilanes, which are formed as by-products in the direct synthesis of organohalosilanes, can be converted into these by treatment with hydrogen halides (Section 2.2.8).

7.2.2 Polyorganosilazanes

The Si atoms in polyorganosilazanes are linked through NH or NR bridges to form cyclic or linear polymers (Table 66):

or they are branched or cross-linked through N atoms:

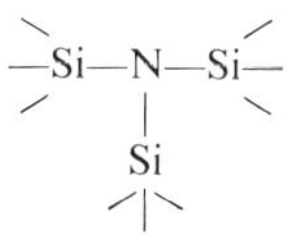

Analogously to the formation of polysiloxanes from silanols, polysilazanes are formed from monomeric aminosilanes by condensation (Section 3.8). Methylaminosilanes condense spontaneously and so rapidly that monomeric compounds having no organic substituent on the nitrogen atom cannot be synthesized. Hexamethyldisilazane is thus obtained directly when trimethylchlorosilane reacts with ammonia (*248*, *250*):

$$2(CH_3)_3SiCl \xrightarrow{NH_3} (CH_3)_3Si{-}NH{-}Si(CH_3)_3$$

Dimethyldichlorosilane reacts on being added in drops to liquid ammonia or when gaseous ammonia is passed through its solution in benzene, diethyl ether, or carbon tetrachloride; hexamethylcyclotrisilazane (I) and octamethylcyclotetrasilazane (II) are formed (*43*, *138*, *180*):

$$\underset{\text{I}}{[(CH_3)_2SiNH]_3} \qquad \underset{\text{II}}{[(CH_3)_2SiNH]_4}$$

TABLE 66

POLYORGANOSILAZANES[a]

Compound	Mp (°C)	Bp (°C)	at p (mm Hg)	n_D^{20}	d_4^{20} (gm/cc)
(a) Linear polymers, nonfunctional					
$(CH_3)_3SiNHSi(CH_3)_3$	—	125.7–126.2 (*250*)	758	1.4078 (*250*)	0.7741 (*250*)
$(CH_3)_3SiN(CH_3)Si(CH_3)_3$	—	148 (*209*)	740	1.4190[b] (*209*)	0.794[c] (*209*)
$(C_6H_5)_3SiNHSi(C_6H_5)_3$	170–175 (*236*)	—	—	—	—
$(CH_3)_3SiNH[(CH_3)_2SiNH]Si(CH_3)_3$	—	192–195 (*43*)	756	1.4281 (*43*)	0.8366 (*43*)
(b) Cyclic polymers, nonfunctional					
$[(CH_3)_2SiNH]_3$	−10 (*43*)	188 (*43*)	756	1.4448 (*43*)	0.9196 (*43*)
$[(CH_3)_2SiNH]_4$	97 (*43*)	225 (*43*)	756	—	—
$[(CH_3)_2SiN(C_6H_5)]_3$	249–250 (*180*)	—	—	—	—
$[(C_6N_5)_2SiNH]_3$	213.5 (*178*)	—	—	—	—
(c) Linear polymers, functional					
$(C_6H_5)(CH_3O)_2SiNHSi(C_6H_5)(OCH_3)_2$	—	92–95 (*242*)	6	—	1.124[d] (*242*)
$(CH_3)_2(CH_3NH)SiN(CH_3)Si(CH_3)_2(NHCH_3)$	—	75–76 (*180*)	10	—	—

[a] Numbers in parentheses indicate reference numbers.
[b] n_D^{25}.
[c] d_4^{25}.
[d] d_4^{15}.

Bicyclic products and products of higher molecular weight are also formed (*14*). On ammonolysis, methyltrihalosilanes give polymers cross-linked through nitrogen atoms (*56–58*).

The third hydrogen on the bridge nitrogen atom does not react with chlorosilane even under the most stringent conditions. The silyl substitution is however successful if metallation is first carried out with an alkyl- or aryllithium, followed by reaction with chlorosilane (*304*):

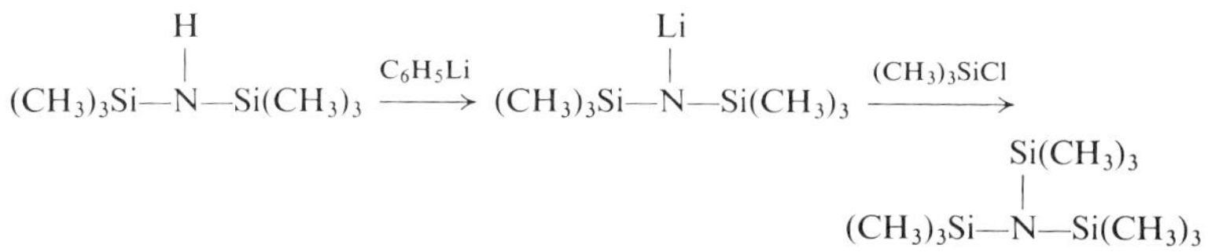

Hexaorganocyclotrisilazanes react with *n*-butyllithium in *n*-hexane to form successively singly, doubly, and triply metallated compounds which can subsequently react with trimethylchlorosilane to give the corresponding *N*-silylated cyclotrisilazanes (*94*):

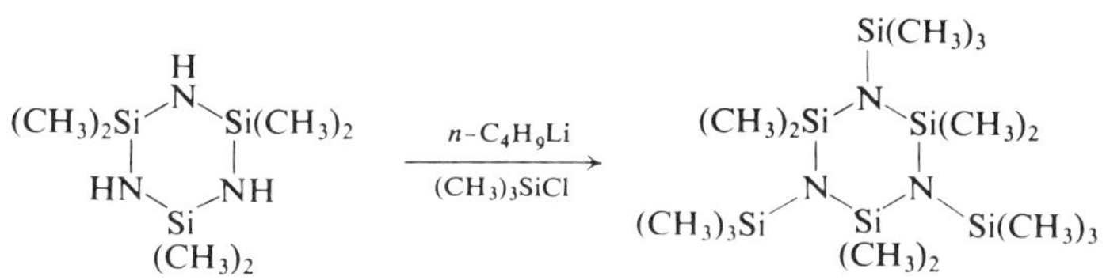

In analogy with the cyclosiloxanes, the cyclosilazanes can be cleaved and converted into linear polymers. Hexaphenylcyclotrisilazane is polymerized even by brief heating at elevated temperatures (*135*). Hexamethylcyclotrisilazane can also be partially converted into polymers by heating with ammonia under pressure; however, part of it is converted into the cyclic tetramer (*235*). Ammonium sulfate and ammonium bromide and iodide are effective *polymerization catalysts*, but they bring about the cleavage of only some of the rings to form linear polymers (I); the remainder undergo cross-linking (II) with elimination of ammonia (*175*, *235*):

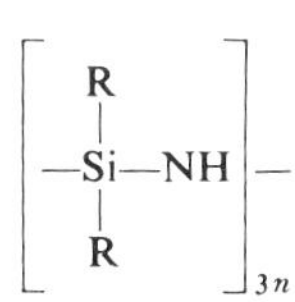

I

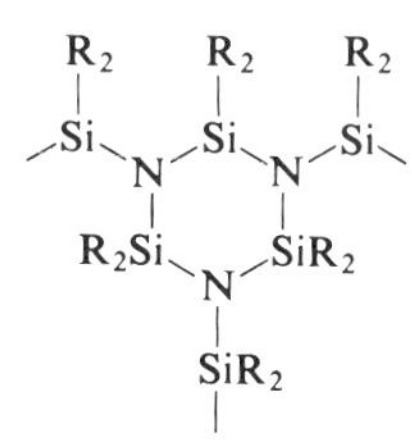

II

By *N*-substitution, with neighboring nitrogen atoms linked by ethylene or propylene bridges, it has been possible to stabilize linear polymers sufficiently for rubberlike products with molecular weights in the region of 8000 to be obtained by polymerization with ammonium sulfate as catalyst (*40*):

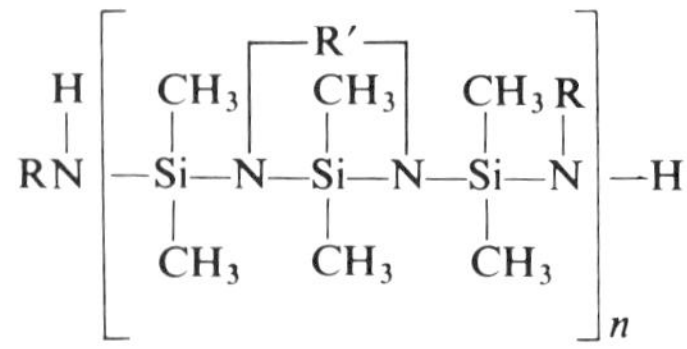

Alkali metals and alkali-metal amides have been recommended as *catalysts* (*36*); cobalt(II) chloride, mercury(II) sulfate, and silver nitrate are also said to be effective (*85*).

Polysilazanes are more sensitive to hydrolysis than polysiloxanes, but sometimes, depending on their substitution, they are remarkably resistant toward water. Only 10% of hexamethyldisilazane is split after four hours of boiling in water. The sensitivity toward acids is greater than toward alkalis (*250*). Hexamethylcyclotrisilazane and octamethylcyclotetrasilazane are relatively stable toward water, and even more stable toward bases. On the other hand, they are easily split by acids, and it is interesting to note that the trisilazane reacts more easily than the tetrasilazane, just as with the corresponding cyclosiloxanes (Section 5.4.2.2). The ethyl-substituted cyclosilazanes (*5*) are still more stable.

Hexamethylcyclotrisilazane can be converted into octamethylcyclotetrasilazane by the addition of catalytic amounts of concentrated sulfuric acid (*263*). The polydimethylcyclosilazanes are split by alcohols and phenols with the liberation of ammonia and the formation of α,ω-bis-OR-substituted polydimethylsilazanes (*196*, *263*) or dimethyldialkoxy- or dimethyldiaryloxysilanes (*7*, *8*). Linear polymers can be obtained by condensation with diols and these have a rubberlike appearance (*39*, *263*).

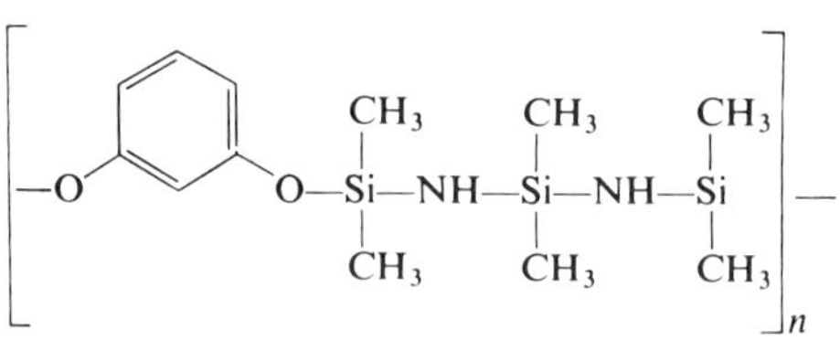

Hexamethyldisilazane reacts with phenyl isocyanate with cleavage of the Si—N bond to form *N,N′*-bis-trimethylsilylphenylurea in 95% yield (*96*):

$$[(CH_3)_3Si]_2NH + C_6H_5NCO \rightarrow (CH_3)_3Si{-}NH{-}CO{-}N\begin{cases} Si(CH_3)_3 \\ C_6H_5 \end{cases}$$

Hexamethylcyclotrisilazane is similarly split at the Si—N bond by phenyl isocyanate, the following ring compound being finally formed after stepwise degradation:

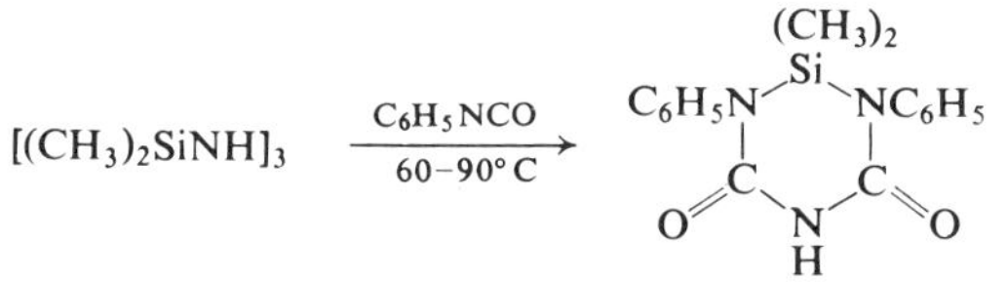

Octamethylcyclotetrasilazane behaves analogously (yields, 62 to 99%) (*95*). With diisocyanates polyadducts are formed (*194*).

Thermodynamic considerations indicate that the Si—N bond has a considerable thermal stability. In fact, polymethylsilazanes, unlike polymethylsiloxanes, are not depolymerized (*239*) even at 400°C. Numerous research groups have therefore attempted to develop types of polymers which could be the basis of technically usable products. The most important question for this is whether, and how far, it is possible to decrease the sensitivity of the Si—N bond to hydrolysis. Both because of the *N*-substitution and because of the cross-linking, a ladderlike polymer presumably of the structure

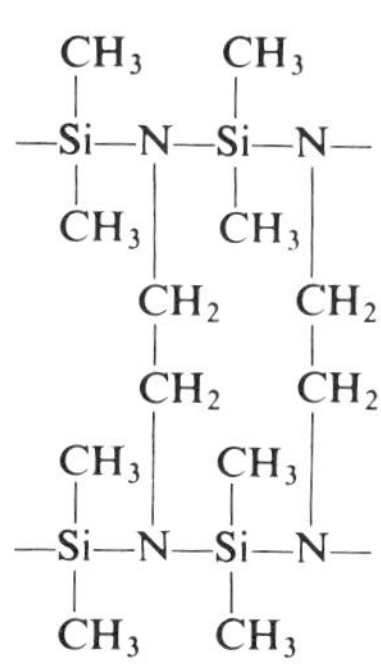

hydrolyzes one-third as fast as the linear polymer (*192*):

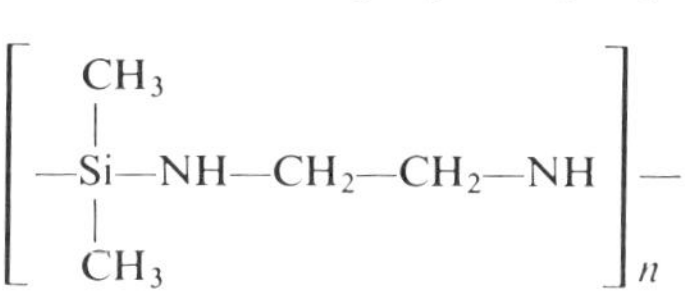

The latter can be stabilized against hydrolysis with metal ions, toward which the nitrogen acts as electron donor evidently with formation of coordinated structures (*192*). Triple silyl substitution on the nitrogen increases the stability toward hydrolysis; however, such compounds tend to undergo migrations of silyl groups, rearrangements, and isomerization reactions (*204*). Therefore,

TABLE 67

POLYORGANOSILTHIANES[a]

Compound	Mp (°C)	Bp (°C)	at p (mm Hg)	n_D^{12}	d_4 (gm/cc)
(a) Linear polymers					
$(CH_3)_3Si—S—Si(CH_3)_3$	—	163–164 (*53*)	760	—	—
$(C_2H_5)_3Si—S—Si(C_2H_5)_3$	—	278–279 (*87*)	750	1.492 (*87*)	~0.9 (*87*)
$(n\text{-}C_3H_7)_3Si—S—Si(n\text{-}C_3H_7)_3$	—	168 (*179*)	7	—	—
(b) Cyclic polymers					
$[(CH_3)_2SiS]_2$	113 (*76, 91*)	172–173 (*76, 212*)	38	—	—
$[(CH_3)_2SiS]_3$	17–18 (*177, 212*)	151.5–153.5 (*177, 212*)	38	—	—
$[(C_2H_5)_2SiS]_2$	20.1 (*53*)	160–162 (*53*)	36	—	—
$[(C_2H_5)_2SiS]_3$	—	111–115 (*76*)	0.2	—	—
$[(C_6H_5)_2SiS]_2$	145–147 (*195*)	—	—	—	—

[a] Numbers in parentheses indicate reference numbers.

even though an increase in the resistance to hydrolysis of the Si—N bond appears possible in principle, it is doubtful whether it will be possible to obtain polymers based on silazanes with an over-all behavior pattern as good as that of the siloxanes. Nevertheless, their incorporation into silicone resins confers certain properties on these resins, precisely because of their sensitivity to hydrolysis (Section 8.2.1.2).

7.2.3 POLYORGANOSILTHIANES

Polyorganosilthianes (Table 67), in which Si atoms are linked through sulfur bridges to give the polymer units

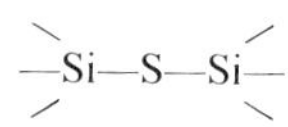

are formed as by-products in larger or smaller amounts during the preparation of the monomeric silanethiols (Section 3.9) since the monomeric silanethiols, like the silanols and the aminosilanes, tend to condense spontaneously. Thus, tetra-*n*-propylcyclodisilthiane (I) is obtained in 65% yield from di-*n*-propyldichlorosilane and hydrogen sulfide in the presence of triethanolamine (*91*):

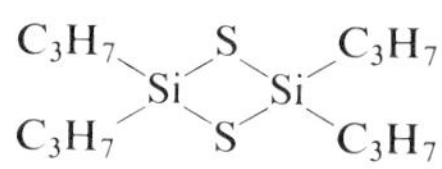

I

With hydrogen sulfide in pyridine *n*-propyltrichlorosilane yields tetra-*n*-propyltricyclotetrasilthiane (II):

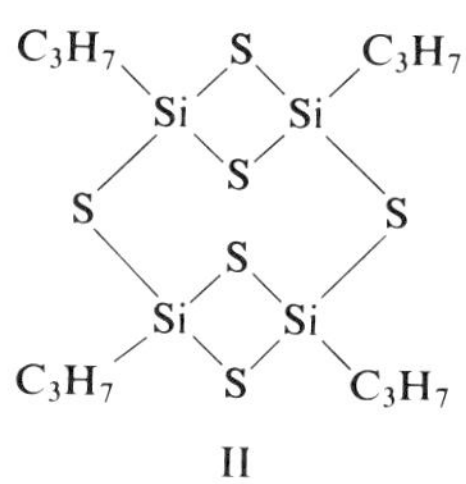

II

as well as small amounts of di-*n*-propyldichlorocyclodisilthiane (III):

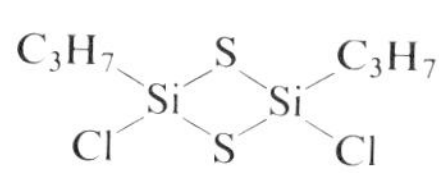

III

Vinyltrichlorosilane reacts with hydrogen sulfide in benzene solution and in the presence of pyridine to give polyvinylsilthianes (*45*):

$$[CH_2{=}CH{-}SiS_{3/2}]_n$$

Polyorganosilthianes differ remarkably in their structure from polysiloxanes, polysilazanes, heterosiloxanes, and polysilcarbanes. Their tetrahedral organosilthiane units can be linked through the *tetrahedron edges* as indicated by the structural formulas (I), (II), and (III) (cf. also ref. *312*). This corresponds to the well-known fact that SiS_2 contains chain molecules in which SiS_4 tetrahedra are also linked through their edges:

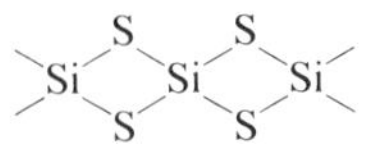

Consequently, the sulfolysis of alkyltrihalosilanes does not give highly polymeric substances, but cyclic tetramers of type (II).

7.2.4 POLYORGANOSILALKYLENES

Finally, let us turn to a number of polymers in which *carbon atoms* act as bridges between Si atoms. Organosilicon polymers in which silicon atoms are linked through $(CH_2)_n$ bridges are termed "polyorganosilalkylenes" (Table 68).

$$-Si-(CH_2)_n-Si-$$

($n=1$ polysilmethylenes; $n=2$ polysilethylenes, etc.)

Since the preparation of silalkylenes involves the formation of Si—C—Si bonds, the processes employed are similar to those used for the preparation of Si—C bonds in monomers.

The first and most important group of processes comprises the condensation of chloroalkylsilanes with chlorosilanes by means of sodium, lithium, or magnesium. The difference between this procedure and that customary in the preparation of Si—C bonds consists therefore solely in the fact that, instead of an organic halide, a haloorganosilicon compound is used:

$$-Si-R-Cl + Cl-Si- \xrightarrow{M} -Si-R-Si-$$

The reaction with sodium appears to be used most often (*61*, *119*, *121*, *122*, *127*). The condensation of haloorganosilanes with one another is also possible (*210*):

$$-Si-CH_2Cl + ClCH_2-Si- \xrightarrow{Na} -Si-CH_2-CH_2-Si-$$

The reaction using magnesium is accomplished either via Grignard compounds (*274*):

$$-Si-CH_2Cl + Mg \xrightarrow{ether} -Si-CH_2MgCl$$

$$-Si-CH_2MgCl + Cl-Si- \rightarrow -Si-CH_2-Si-$$

or is carried out in a single step (*284*); e.g., as follows:

$$2\mathrm{Cl}-\underset{\mathrm{CH_3}}{\overset{\mathrm{CH_3}}{\underset{|}{\overset{|}{\mathrm{Si}}}}}-\mathrm{Cl}+2\mathrm{Mg}+\mathrm{Cl}-(\mathrm{CH_2})_6-\mathrm{Cl} \rightarrow \mathrm{Cl}-\underset{\mathrm{CH_3}}{\overset{\mathrm{CH_3}}{\underset{|}{\overset{|}{\mathrm{Si}}}}}-(\mathrm{CH_2})_6-\underset{\mathrm{CH_3}}{\overset{\mathrm{CH_3}}{\underset{|}{\overset{|}{\mathrm{Si}}}}}-\mathrm{Cl}$$

Metallic lithium is particularly effective. With silicon tetrachloride, $LiCH_2(CH_3)_3Si$ gives $[(CH_3)_3SiCH_2]_2SiCl_2$, $[(CH_3)_3SiCH_2]_3SiCl$, and even $[(CH_3)_3SiCH_2]_4Si$ (*276*). If the reactions of $ClCH_2(CH_3)_3Si$, $(CH_3)_2SiCl_2$, and $ClCH_2(CH_3)_2SiCl$ with lithium are combined, the products are linear compounds (*275*), as in the reaction sequence:

$$\mathrm{ClCH_2(CH_3)_3Si}+\mathrm{Li} \rightarrow \mathrm{LiCH_2(CH_3)_3Si}$$

$$\mathrm{LiCH_2(CH_3)_3Si}+\mathrm{ClCH_2(CH_3)_2SiCl} \rightarrow \mathrm{ClCH_2}-\underset{\mathrm{CH_3}}{\overset{\mathrm{CH_3}}{\underset{|}{\overset{|}{\mathrm{Si}}}}}-\mathrm{CH_2}-\mathrm{Si(CH_3)_3}$$

$$\mathrm{ClCH_2}-\underset{\mathrm{CH_3}}{\overset{\mathrm{CH_3}}{\underset{|}{\overset{|}{\mathrm{Si}}}}}-\mathrm{CH_2}-\mathrm{Si(CH_3)_3}+\mathrm{Li} \rightarrow \mathrm{LiCH_2}-\underset{\mathrm{CH_3}}{\overset{\mathrm{CH_3}}{\underset{|}{\overset{|}{\mathrm{Si}}}}}-\mathrm{CH_2}-\mathrm{Si(CH_3)_3}$$

$$2\mathrm{LiCH_2}-\underset{\mathrm{CH_3}}{\overset{\mathrm{CH_3}}{\underset{|}{\overset{|}{\mathrm{Si}}}}}-\mathrm{CH_2}-\mathrm{Si(CH_3)_3}+\mathrm{(CH_3)_2SiCl_2} \rightarrow$$

$$\mathrm{CH_3}-\underset{\mathrm{CH_3}}{\overset{\mathrm{CH_3}}{\underset{|}{\overset{|}{\mathrm{Si}}}}}-\mathrm{CH_2}-\left[\underset{\mathrm{CH_3}}{\overset{\mathrm{CH_3}}{\underset{|}{\overset{|}{\mathrm{Si}}}}}-\mathrm{CH_2}-\right]_3\underset{\mathrm{CH_3}}{\overset{\mathrm{CH_3}}{\underset{|}{\overset{|}{\mathrm{Si}}}}}-\mathrm{CH_3}$$

By condensation with lithium, moreover, α,ω-dibromoalkanes, $Br(CH_2)_nBr$ with $n=4$, 5, 6 and 10, and trimethylchlorosilane, can be converted into hexamethyldisilalkylenes (*306*). However, the corresponding compounds with $n=2$ and 3 cannot be obtained in this way. Dibromomethane gives only small yields of hexamethyldisilmethylene with trimethylchlorosilane and lithium (*306*).

Besides condensations of this type, the preparation of silalkylenes with two or more C atoms in the alkylene bridge by addition reactions has become of interest. This is shown in the following examples:

$$\equiv\mathrm{Si}-\mathrm{H}+\mathrm{H_2C}=\mathrm{CH}-\mathrm{Si}\equiv \xrightarrow{[\mathrm{C_6H_5CO_2}]_2} \equiv\mathrm{Si}-\mathrm{C_2H_4}-\mathrm{Si}\equiv \quad (49)$$

$$\equiv Si{-}H + CH_2{=}CH{-}Si\equiv \xrightarrow[600^\circ C]{Fe} \equiv Si{-}CH_2{-}CH_2{-}Si\equiv \quad (230)$$

$$n\left[CH_2{=}CH{-}\overset{|}{\underset{|}{Si}}{-}H\right] \xrightarrow{Pt} \left[{-}\overset{|}{\underset{|}{Si}}{-}CH_2{-}CH_2\right]_n{-} \quad (73)$$

$$2\equiv Si{-}H + CH_2{=}CH{-}(CH_2)_2{-}CH{=}CH_2 \xrightarrow{H_2[PtCl_6]} \equiv Si(CH_2)_6Si\equiv \quad (299)$$

On the other hand, processes such as the direct synthesis from dichloromethane and silicon of polychlorosilmethylenes, which can be converted into the alkyl-substituted compounds by alkylation (*103*, *201*), and the pyrolysis of methylchlorosilanes (*99*) have little preparative importance. The latter method has, however, given a most illuminating insight into the chemistry of the polysilalkylenes (*99*).

Summarizing, it has been shown that it is possible to parallel the different molecular forms found in the polyorganosiloxanes with polyorganosilmethylenes as regards low-molecular-weight compounds (see Table 68). The tendency of the silalkylenes to cyclization is very marked.

Numerous cyclic and polycyclic fused ring systems have become known (see particularly refs. *99* and *173*), as can be demonstrated by only a few examples (compare the physical constants in Table 68):

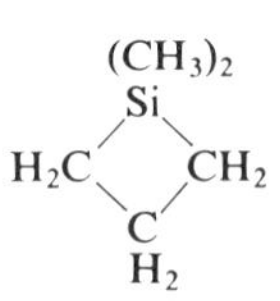

$(CH_3)_2Si(CH_2)_3$ (*202*)

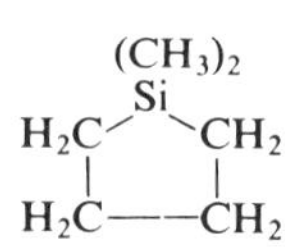

$(CH_3)_2Si(CH_2)_4$ (*203*)

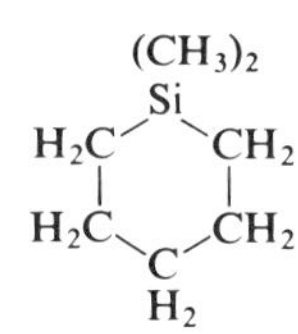

$(CH_3)_2Si(CH_2)_5$ (*305*)

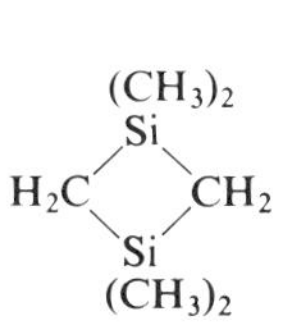

$[(CH_3)_2SiCH_2]_2$ (*173*)

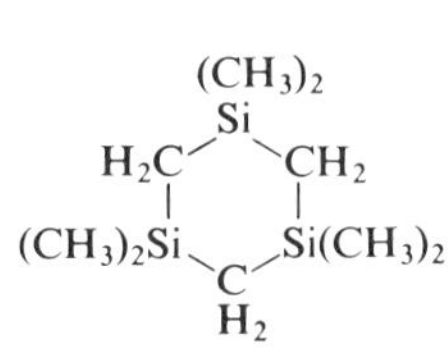

$[(CH_3)_2SiCH_2]_3$ (*81*)

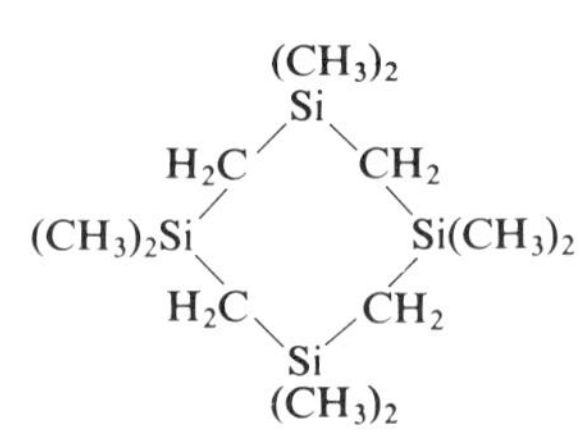

$[(CH_3)_2SiCH_2]_4$ (*173*)

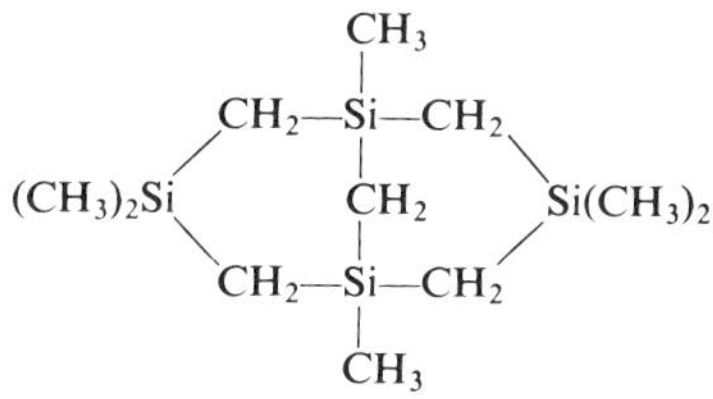

$C_{11}H_{28}Si_4$ *(100, 101)*

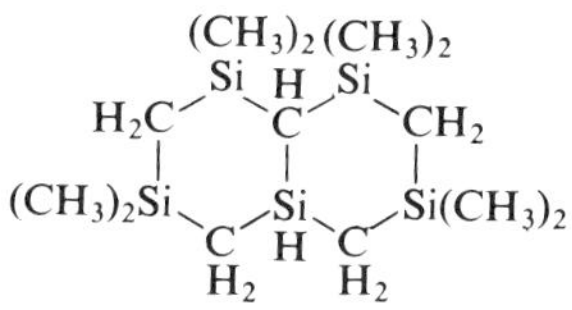

$C_{13}H_{34}Si_5$ *(102)*

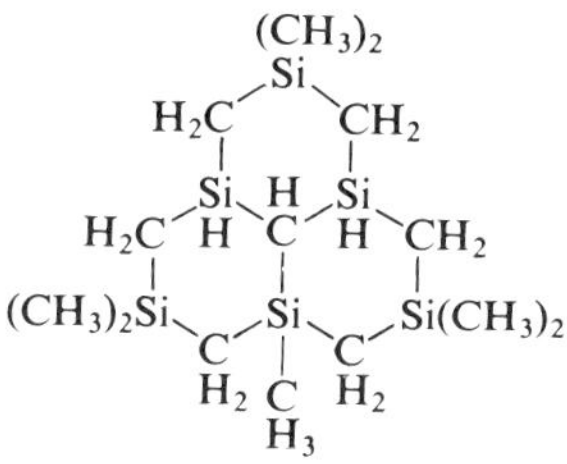

$C_{14}H_{36}Si_6$ *(102)*

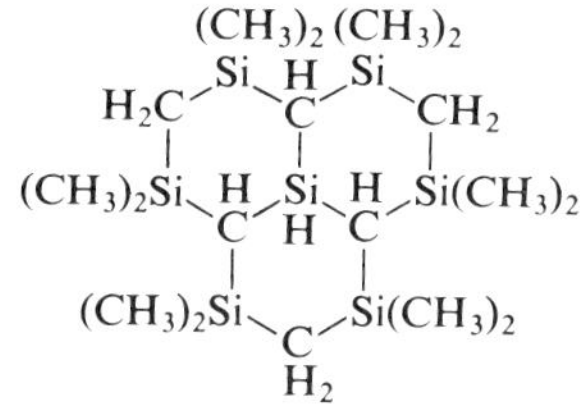

$C_{18}H_{46}Si_7$ *(100, 102)*

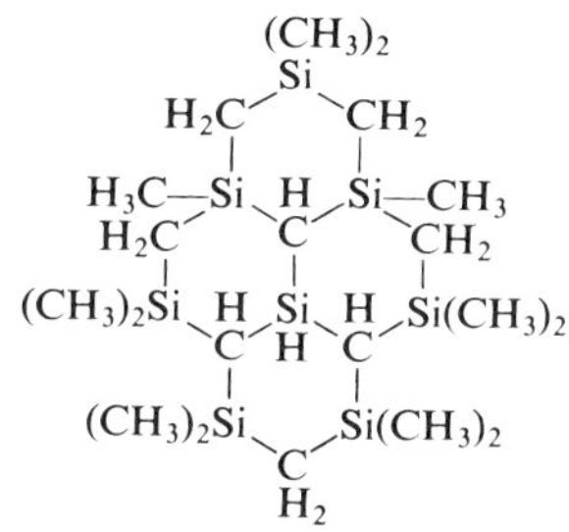

$C_{20}H_{50}Si_8$ *(100, 102)*

The preparation of higher, or macromolecular, products by means of organometallic processes is tedious and difficult; however, it can be achieved by addition reactions. The H_2PtCl_6-catalyzed reaction of methylphenylsilane with hepta-1,6-diyne in the presence of hydroquinone as polymerization inhibitor gave polymers of the structural type:

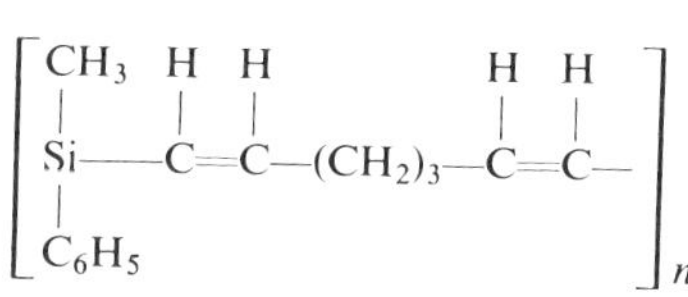

which had molecular weights up to about 100,000 and the physical consistency of oils or rubberlike products, depending on the size of the molecules *(98)*.

In analogy to the ring opening and polymerization processes in siloxanes, similar reactions have been recently investigated with the cyclosilalkylenes.

TABLE 68

POLYORGANOSILALKYLENES AND POLYORGANOSILARYLENES[a]

Compound	η_{20} (cP)	Mp (°C)	Bp (°C)	at p (mm Hg)	n_D^t	at t (°C)	d_4^t (gm/cc)	at t (°C)
(a) Linear polymers, nonfunctional								
$(CH_3)_3SiCH_2Si(CH_3)_3$	0.736 (*275*)	—	134 (*275*)	760	1.4172 (*275*)	20	0.7520 (*275*)	20
$(CH_3)_3SiCH_2CH_2Si(CH_3)_3$	—	—	150–151 (*291*)	760	1.4204 (*291*)	20	0.7536 (*291*)	20
$(CH_3)_3SiCH_2CH_2CH_2Si(CH_3)_3$	—	—	172–173 (*176*)	—	—	—	—	—
$(CH_3)_3SiCH_2CH_2CH_2CH_2Si(CH_3)_3$	—	—	199 (*306*)	748	1.4244 (*306*)	25	0.763 (*306*)	25
$(CH_3)_3SiC{\equiv}CSi(CH_3)_3$	—	—	134–136 (*225*)	760	1.4260 (*225*)	20	0.7703 (*225*)	20
o-$(CH_3)_3SiC_6H_4Si(CH_3)_3$	—	−72 (*68*)	135 (*68*)	20	1.5110 (*68*)	25	0.8985 (*68*)	25
m-$(CH_3)_3SiC_6H_4Si(CH_3)_3$	—	−26 (*68*)	112 (*68*)	22	1.4867 (*68*)	25	0.8685 (*68*)	25
p-$(CH_3)_3SiC_6H_4Si(CH_3)_3$	—	88 (*68*)	143 (*68*)	43	—	—	—	—
$(CH_3)_3SiCH_2[Si(CH_3)_2CH_2]Si(CH_3)_3$	1.788 (*275*)	—	206 (*275*)	760	1.4420 (*275*)	20	0.7987 (*275*)	20
$(CH_3)_3SiCH_2[Si(CH_3)_2CH_2]_2Si(CH_3)_3$	3.590 (*275*)	—	259 (*275*)	760	1.4552 (*275*)	20	0.8245 (*275*)	20
$(CH_3)_3SiCH_2[Si(CH_3)_2CH_2]_3Si(CH_3)_3$	6.514 (*275*)	—	309 (*275*)	760	1.4640 (*275*)	20	0.8408 (*275*)	20
(b) Branched polymers, nonfunctional								
$[(CH_3)_3SiCH_2]_4Si$	—	28.5 (*272*)	146–154 (*272*)	23	—	—	—	—
(c) Cyclic polymers, nonfunctional								
Monocyclic								
$(CH_3)_2Si(CH_3)_2$	—	—	80.2–80.5 (*202*)	755	1.4250 (*202*)	20	0.7700 (*202*)	20
$CH_3(C_6H_5)Si(CH_3)_2$	—	—	61.5–62.0 (*202*)	5	1.5349 (*202*)	20	0.9511 (*202*)	20
$CH_3SiCl(CH_2)_3$	—	—	103.5 (*202*)	755	1.4482 (*202*)	20	0.9854 (*202*)	20
$(CH_3)_2Si(CH_2)_4$	—	—	104–104.5 (*203*)	760	1.4335 (*203*)	20	0.7935 (*203*)	20
$(CH_3)_2Si(CH_2)_5$	—	—	133 (*305*)	760	1.4380 (*305*)	25	0.798 (*305*)	25
$[(CH_3)_2SiCH_2]_2$	—	−9 to −8.5 (*173*)	60 (*173*)	100	1.4401 (*173*)	24	—	—
	—	—	117–119 (*164*)	760	—	—	—	—

$[(CH_3)_2SiCH_2]_3$	—	—	200–201 (*81*)	745	1.4606 (*81*)	25	0.846 (*81*)	25
$[(CH_3)_2SiCH_2]_4$	—	79–80.5 (*173*)	104 (*173*)	2	—	—	—	—
$[CH_3SiClCH_2]_2$	—	−7 to −6.5 (*173*)	59 (*173*)	20	1.4725 (*173*)	25	—	—
$[CH_3SiClCH_2]_3$	—	—	117–118 (*173*)	5	—	—	—	—
$[CH_3Si(OC_2H_5)CH_2]_2$	—	—	78–79 (*173*)	16	1.4336 (*173*)	25	—	—
$[CH_3Si(OC_2H_5)CH_2]_3$	—	—	68–69 (*173*)	0.2–0.25	1.4478 (*173*)	25	—	—
$[CH_3Si(H)CH_2]_2$	—	—	96 (*173*)	760	1.4546 (*173*)	25	—	—
$[CH_3Si(H)CH_2]_3$	—	—	68–71 (*173*)	19	1.4705 (*173*)	25	—	—
$[CH_3(C_6H_5)SiCH_2]_2$	—	—	99 (*173*)	1.1	1.5727 (*173*)	25	—	—
$[CH_3(C_6H_5)SiCH_2]_3$	—	—	—	—	1.5873 (*173*)	25	—	—
Polycyclic[b]								
$C_{11}H_{28}Si_4$	—	106–107 (*101*)	113–114 (*102*)	1	—	—	—	—
$C_{13}H_{34}Si_5$	—	—	126–132 (*102*)	1	—	—	—	—
$C_{14}H_{36}Si_6$	—	—	150–153 (*102*)	1	—	—	—	—
$C_{18}H_{46}Si_7$	—	—	170–195 (*102*)	1	—	—	—	—
$C_{20}H_{50}Si_8$	—	—	200–230 (*102*)	1	—	—	—	—
(d) Linear polymers, functional								
$Cl_3SiCH_2SiCl_3$	—	—	184–185 (*222*)	756	—	—	—	—
$Cl_2(CH_3)SiCH_2Si(CH_3)Cl_2$	—	—	192 (*63*)	760	—	—	1.288 (*63*)	25
$Cl(CH_3)_2SiCH_2Si(CH_3)_2Cl$	—	—	95–96 (*34*)	50	1.4480 (*34*)	20	1.016 (*34*)	20
$(C_2H_5O)_3SiCH_2Si(OC_2H_5)_3$	—	—	114–115 (*200*)	3.5	1.4098 (*200*)	20	0.9741 (*200*)	20
$H_3SiCH_2SiH_3$	—	—	14.7 (*224*)	754.3	1.4115 (*224*)	4	0.6979 (*224*)	4
$HO(CH_3)_2SiCH_2Si(CH_3)_2OH$	—	84–86 (*34*)	—	—	—	—	—	—
$C_2H_5O(CH_3)_2SiCH_2Si(CH_3)_2OC_2H_5$	—	—	186.5 (*176*)	760	—	—	—	—
$(C_2H_5O)_2CH_3SiCH_2Si(CH_3)_2OC_2H_5$	—	—	112 (*62*)	25	1.4148 (*62*)	25	0.8914 (*62*)	25
$Cl_2HSiCH_2SiHCl_2$	—	—	57–58 (*292*)	10	—	—	—	—
$[(CH_3)_3SiCH_2]_2SiCl_2$	—	—	92 (*276*)	9	—	—	—	—
$[(CH_3)_3SiCH_2]_2Si(OH)_2$	—	92 (*276*)	—	—	—	—	—	—
$Cl(CH_3)_2SiC_6H_4Si(CH_3)_2Cl$	—	85–88 (*211*)	116–122 (*211*)	0.3	—	—	—	—
$H(C_6H_5)CH_3SiC_6H_4SiCH_3(C_6H_5)H$	—	—	188 (*32*)	1.4	1.5892 (*32*)	25	1.027 (*32*)	25

(*continued on p.* 362)

TABLE 68 *(Contd.)*

POLYORGANOSILALKYLENES AND POLYORGANOSILARYLENES[a]

Compound	η_{20} (cP)	Mp (°C)	Bp (°C)	at p (mm Hg)	n_D^t	at t (°C)	d_4^t (gm/cc)	at t (°C)
$H(C_6H_5)_2SiC_6H_4Si(C_6H_5)_2H$	—	—	107–108 (*32*)	—	—	—	—	—
$HO(C_6H_5)_2SiC_6H_4Si(C_6H_5)_2OH$	—	—	226–228 (*32*)	—	—	—	—	—
$(C_2H_5O)_2CH_3SiC_6H_4SiCH_3(OC_2H_5)_2$	—	—	128 (*42*)	2	1.4614 (*42*)	25	0.9828 (*42*)	25
$C_2H_5O(CH_3)_2SiC_6H_4Si(CH_3)_2OC_2H_5$	—	—	127–128 (*42*)	3.5	1.4752 (*42*)	25	0.9354 (*42*)	25
$H(CH_3)_2SiC_6H_4Si(CH_3)_2Cl$	—	—	63 (*129*)	0.1	1.5053 (*129*)	20	0.9816 (*129*)	20
$H[(CH_3)_2SiC_6H_4]_2Si(CH_3)_2H$	—	111–112 (*129*)	—	—	—	—	—	—
$H[(CH_3)_2SiC_6H_4]_3Si(CH_3)_2H$	—	118–119 (*129*)	—	—	—	—	—	—
(e) Cyclic polymers, functional								
$[Cl_2SiCH_2]_3$	—	81–82.5 (*222*)	—	—	—	—	—	—
$[H_2SiCH_2]_3$	—	10 (*224*)	134.9 (*224*)	769	1.5059 (*224*)	20	0.9001 (*224*)	20
(f) Branched polymers, functional								
$[(CH_3)_3SiCH_2]_3SiCl$	—	13–14 (*276*)	112 (*276*)	4	1.4600 (*276*)	20	0.8900 (*276*)	20

[a] Numbers in parentheses indicate reference numbers.
[b] Structural formulas, Section 7.2.4.

1,1,3,3-Tetramethyl-1,3-disilacyclobutane is cleaved at 300°C and is converted into linear polysilmethylenes (*164*):

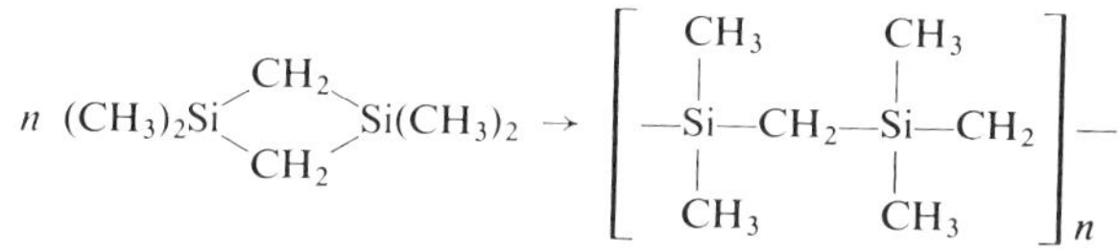

1,1-Dimethylsilacyclobutane has been polymerized to polysilpropylenes by heating to 150°C Molecular weights of the order of 100,000 were achieved (*202*, *300*):

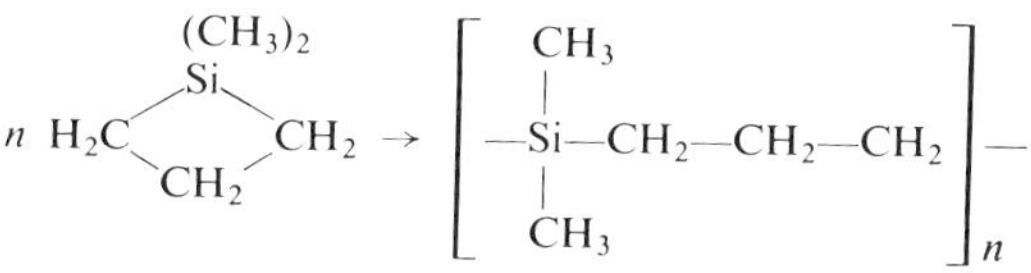

The analogous l-monomethylsilacyclobutane polymerizes (*273*) even at temperatures slightly above 60°C.

The technical expectations attached to the investigations of the polysilalkylenes have not yet been fulfilled. Comparison with the siloxanes shows (*275*, *293*) that liquid silmethylenes invariably have higher boiling points and higher viscosities at the same temperatures than siloxanes of the same molecular weight. Characteristic properties of the siloxanes, such as the unusually low dependence of viscosity on temperature, are not reproduced to the same extent in this field, and one of the most important properties of the siloxanes—their weak intermolecular bonding—is therefore lost on replacing oxygen by methylene groups (compare Section 6.5).

Alkylene bridges between the Si atoms in siloxane networks are formed in many processes already used in industry such as the hot vulcanization of silicone rubber (Sections 5.4.3.1 and 8.1.2.1) and the cross-linking of polyvinylsiloxanes and poly-*H*-siloxanes by means of chloroplatinic acid (Section 8.1.2.3.1).

Attempts to synthesize benzenelike ring systems with conjugated Si=C bonds have been unsuccessful. The failure to prepare double bonds between silicon and carbon demonstrates the limits set to the reproduction of organic chemistry with Si—C compounds (cf. Section 6.4).

7.2.5 POLYORGANOSILARYLENES

The interest in this group of substances has been concentrated mainly on the polysil-*p*-phenylenes, which contain the grouping

$$\geq Si-C_6H_4-Si \leq$$

Processes similar to those used for the polysilalkylenes are employed for their preparation. These are condensation reactions using sodium (*60*, *61*, *68*, *211*):

$$\equiv Si-C_6H_4Cl + Cl-Si\equiv \xrightarrow{Na} \equiv Si-C_6H_4-Si\equiv$$

or Grignard reactions (*32*, *42*, *129*, *136*, *237*, *238*, *241*, *247*), e.g.:

$$2\equiv Si-Cl + BrC_6H_4Br \xrightarrow{Mg} \equiv Si-C_6H_4-Si\equiv$$

Polysilylphenylenes are also formed by a reaction between Si—H and phenyl—Si bonds (*28*, *243*), which corresponds to that between *H*-silanes and benzene (Section 2.4.2) and is catalyzed by boron trichloride:

$$\equiv Si-H + C_6H_5-Si\equiv \xrightarrow{BCl_3} \equiv Si-C_6H_4-Si\equiv + H_2$$

Polysilphenylenes are produced as by-products of the direct synthesis of phenylchlorosilanes, and when phenylchlorosilanes are heated in the presence of aluminum chloride (*59*):

$$\equiv Si-C_6H_5 + C_6H_5-Si\equiv \xrightarrow{AlCl_3} \equiv Si-C_6H_4-Si\equiv + C_6H_6$$

Polysil-*p*-phenylenes are very resistant to heat. This property is due at least in part to their inability to undergo cyclization. Hexaphenoxydisilphenylene is therefore suitable as a heat-transfer fluid, especially when mixed with aryloxysilanes (*161*). The polysil-*p*-phenylenes show a considerably greater tendency to crystallinity than the polysiloxanes because of the more rigid bond through the phenylene bridge. Si—C_6H_4—Si linkages incorporated as copolymeric components into siloxane chains are therefore responsible for interesting effects (Section 8.1.1). Sil-*p*-phenylenes with silicon-functional groups which can be cocondensed with silicon-functional siloxanes, can be used for the incorporation of such linkages [see (d) in Table 68].

The viscosity of such compounds and its temperature coefficient are both greater than in the siloxanes. Sil-*p*-phenylenes are therefore relatively uninteresting for modifying the properties of silicone oils. In resins they have a favorable effect on the elastic properties (Section 8.2.1.2).

7.3 Organosilicon Copolymers (Polysilalkylenesiloxanes, Polysilarylenesiloxanes, Polysilalkylenesilanes, Polysilylenesiloxanes)

This picture of the organosilicon polymers becomes even more varied if it is considered that different types of structural units can be incorporated into one and the same macromolecule. The number of possible polymers is very large. The combination of linkages here is not confined to simple regular alternation of two polymer units each in a one-to-one ratio. Any blocks of a short-chain prepolymer can also be combined as desired with single units or blocks of a different prepolymer. A few selected types will be discussed below (cf. Table 69):

Polysilalkylenesiloxanes, e.g., $-\overset{|}{\underset{|}{\mathrm{Si}}}-\mathrm{CH_2}-\overset{|}{\underset{|}{\mathrm{Si}}}-\mathrm{O}-\overset{|}{\underset{|}{\mathrm{Si}}}-$

Polysilarylenesiloxanes, e.g., $-\overset{|}{\underset{|}{\mathrm{Si}}}-\mathrm{C_6H_4}-\overset{|}{\underset{|}{\mathrm{Si}}}-\mathrm{O}-\overset{|}{\underset{|}{\mathrm{Si}}}-$

Polysilalkylenesilanes, e.g., $-\overset{|}{\underset{|}{\mathrm{Si}}}-\mathrm{CH_2}-\overset{|}{\underset{|}{\mathrm{Si}}}-\overset{|}{\underset{|}{\mathrm{Si}}}-$

Polysilylenesiloxanes, e.g., $-\overset{|}{\underset{|}{\mathrm{Si}}}-\overset{|}{\underset{|}{\mathrm{Si}}}-\mathrm{O}-\overset{|}{\underset{|}{\mathrm{Si}}}-$

For the preparation of *polysilalkylenesiloxanes* and *polysilarylenesiloxanes* it is usual to proceed by stages. Monomeric or polymeric preliminary stages are first built up in which one type of bond is established, but in which the Si atoms still carry functional groups which are utilized in the second stage of the process for preparing the second type of bond. Organosilicon compounds with the skeleton

$$\mathrm{X}-\overset{|}{\underset{|}{\mathrm{Si}}}-\mathrm{CH_2}-\overset{|}{\underset{|}{\mathrm{Si}}}-\mathrm{X} \quad \text{or} \quad \mathrm{X}-\overset{|}{\underset{|}{\mathrm{Si}}}-\mathrm{C_6H_4}-\overset{|}{\underset{|}{\mathrm{Si}}}-\mathrm{X}$$

(where X = halogen, hydrogen, or an alkoxy group) have generally been prepared in this way. This is achieved by the processes already described for the preparation of Si—R—Si bonds (Sections 7.2.4 and 7.2.5), i.e., by the condensation of haloorganosilanes, halosilanes, and/or haloalkenes or haloaromatics with alkali metals or magnesium (*105*, *188*, *208*), as well as by addition reactions (*227*, *270*). The silicon-functional groups are then converted into Si—O—Si bonds in a second stage, e.g.:

$$n\ -\overset{|}{\underset{|}{\mathrm{Si}}}-\mathrm{R}-\overset{|}{\underset{|}{\mathrm{Si}}}-\mathrm{X} \xrightarrow{\mathrm{H_2O}} \left[-\overset{|}{\underset{|}{\mathrm{Si}}}-\mathrm{R}-\overset{|}{\underset{|}{\mathrm{Si}}}-\mathrm{O}\right]_n-\mathrm{H}\ [\mathrm{X} = \mathrm{Cl}, \mathrm{OC_2H_5}]\ (\mathit{105, 120, 227, 270})$$

$$n\ -\overset{|}{\underset{|}{\mathrm{Si}}}-\mathrm{R}-\overset{|}{\underset{|}{\mathrm{Si}}}-\mathrm{H} \xrightarrow{\mathrm{H_2O}} \left[-\overset{|}{\underset{|}{\mathrm{Si}}}-\mathrm{R}-\overset{|}{\underset{|}{\mathrm{Si}}}-\mathrm{O}\right]_n-\mathrm{H}\ (\mathit{188, 199, 208})$$

$$n\ -\overset{|}{\underset{|}{\mathrm{Si}}}-\mathrm{R}-\overset{|}{\underset{|}{\mathrm{Si}}}-\mathrm{OH} \xrightarrow[-\mathrm{H_2O}]{\mathrm{NaOH}} \left[-\overset{|}{\underset{|}{\mathrm{Si}}}-\mathrm{R}-\overset{|}{\underset{|}{\mathrm{Si}}}-\mathrm{O}\right]_n-\mathrm{H}\ (\mathit{188, 285})$$

TABLE 69

POLYSILALKYLENESILOXANES AND POLYSILARYLENESILOXANES[a]

Compound	ν_{25} (cSt)	Mp (°C)	Bp (°C)	at p (mm Hg)	n_D^{25}	d_4^{25} (gm/cc)
(a) Linear polymers						
$(CH_3)_3SiCH_2[(CH_3)_2SiO]Si(CH_3)_3$	—	—	99–101 (*34*)	57	1.4137[b] (*34*)	0.808[c] (*34*)
$(CH_3)_3Si[CH_2Si(CH_3)_2]O[(CH_3)_2SiCH_2]Si(CH_3)_3$	—	−115 (*123*)	126 (*123*)	20	1.4319 (*123*)	0.8311 (*123*)
$(CH_3)_3Si[CH_2Si(CH_3)_2]_2O[(CH_3)_2SiCH_2]_2Si(CH_3)_3$	—	−115 (*123*)	200 (*123*)	19	1.4528 (*123*)	0.8580 (*123*)
$(CH_3)_3Si[CH_2Si(CH_3)_2]_3O[(CH_3)_2SiCH_2]_3Si(CH_3)_3$	—	−115 (*123*)	280 (*123*)	26	1.4645 (*123*)	0.8705 (*123*)
$(CH_3)_3Si[OSi(CH_3)_2]CH_2[(CH_3)_2SiO]Si(CH_3)_3$	—	—	132–133 (*34*)	55	1.4121[b] (*34*)	0.845 (*34*)
(I)	—	—	184–185 (*34*)	34	1.4243[b] (*34*)	0.872[c] (*34*)
(b) Cyclic polymers						
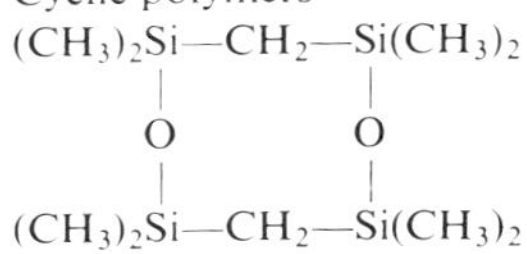	3.12 (*120*)	28–29 (*34*)	103–104 (*34*)	20	1.4328 (*120*)	0.9042 (*120*)
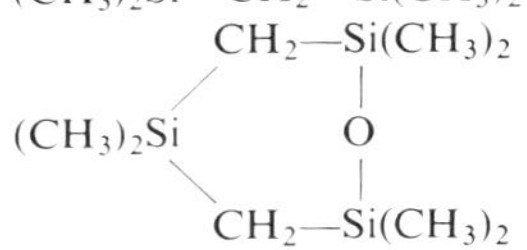	—	—	77 (*120*)	27	1.4371 (*120*)	0.872 (*120*)
$[CH_3[(CH_3)_3SiCH_2]SiO]_3$ (*cis-trans*-isomers)	—	—	177 (*125*)	25	1.4355 (*125*)	0.9111 (*125*)
			183 (*125*)	25	1.4393 (*125*)	0.9180 (*125*)
$[CH_3[(CH_3)_3SiCH_2]SiO]_4$	—	—	223 (*125*)	24	1.4452 (*125*)	0.9390 (*125*)
$[CH_3[(CH_3)_3SiCH_2]SiO]_5$	—	—	270 (*125*)	25	1.4493 (*125*)	0.9303 (*125*)

(I) = $(CH_3)_3SiOSi(CH_3)_2CH_2Si(CH_3)_2O(CH_3)_2SiCH_2(CH_3)_2SiOSi(CH_3)_3$

[a] Numbers in parentheses indicate reference numbers.
[b] n_D^{20}. [c] d_4^2.

Hydrolysis of precursors with two silicon-functional groups is said to lead to ladderlike polymers (*4*):

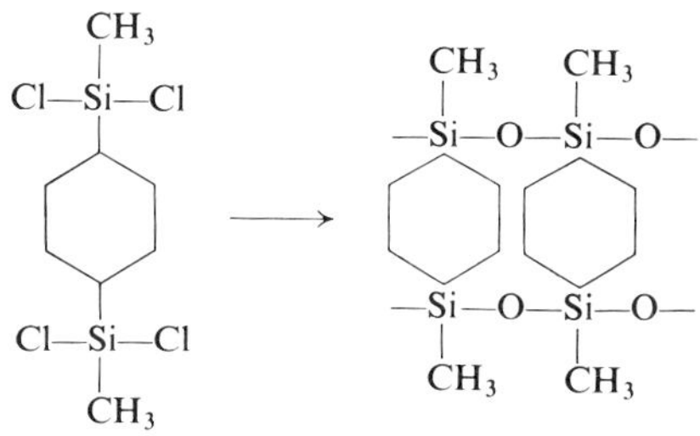

Conversely, the organosiloxane can be formed first and then linked by means of Grignard or addition reactions through alkylene bridges:

$$n\ \text{—Cl—}\overset{|}{\underset{|}{\text{Si}}}\text{—O—}\overset{|}{\underset{|}{\text{Si}}}\text{—CH}_2\text{Cl} \xrightarrow[-\,\text{MgCl}_2]{\text{Mg}} \left[\text{—}\overset{|}{\underset{|}{\text{Si}}}\text{—O—}\overset{|}{\underset{|}{\text{Si}}}\text{—CH}_2\right]_n\text{—} \quad (34)$$

$$\text{—}\overset{|}{\underset{|}{\text{Si}}}\text{—O—}\overset{|}{\underset{|}{\text{Si}}}\text{—H} + \text{CH}_2\text{=CH—CH}_2\text{—}\overset{|}{\underset{|}{\text{Si}}}\text{—} \rightarrow \text{—}\overset{|}{\underset{|}{\text{Si}}}\text{—O—}\overset{|}{\underset{|}{\text{Si}}}\text{—(CH}_2)_3\text{—}\overset{|}{\underset{|}{\text{Si}}}\text{—} \quad (155, 226, 271)$$

The formation of polymers is also possible by the di-addition of di-*H*-disiloxanes to acetylene (*154*, *229*):

$$n\ \text{H—}\overset{|}{\underset{|}{\text{Si}}}\text{—O—}\overset{|}{\underset{|}{\text{Si}}}\text{—H} + n\ \text{CH}\equiv\text{CH} \rightarrow \left[\text{—}\overset{|}{\underset{|}{\text{Si}}}\text{—O—}\overset{|}{\underset{|}{\text{Si}}}\text{—CH}_2\text{—CH}_2\right]_n\text{—}$$

Another route to the copolymers discussed here is indicated by the fact that it is possible in certain cases to cleave cyclic compounds containing Si—O—Si and Si—R—Si bonds, which is done with catalysts used in the cleavage of the cyclosiloxanes. The products may subsequently be subjected to polymerization or copolymerization with cyclosiloxanes:

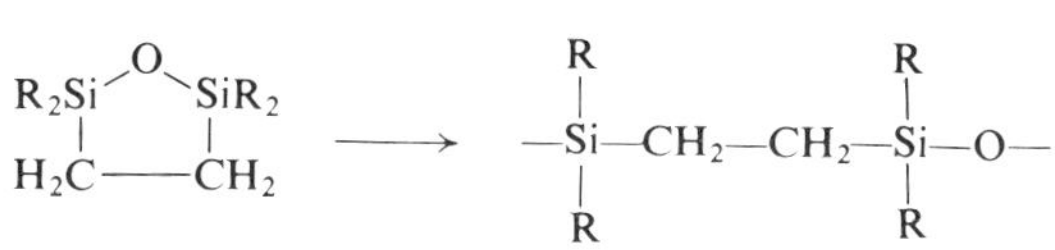

(*189*)

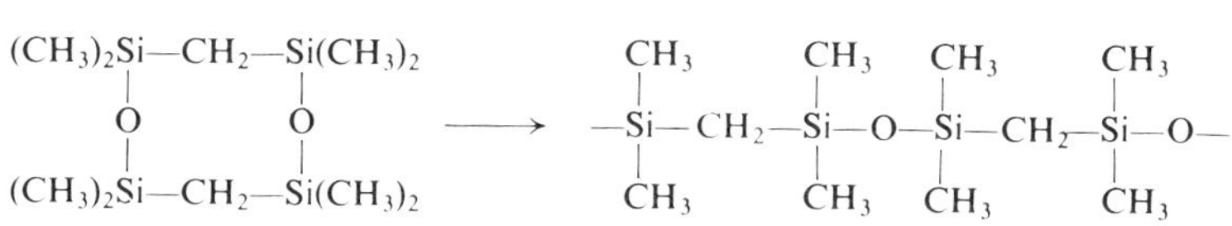

(*124*, *249*)

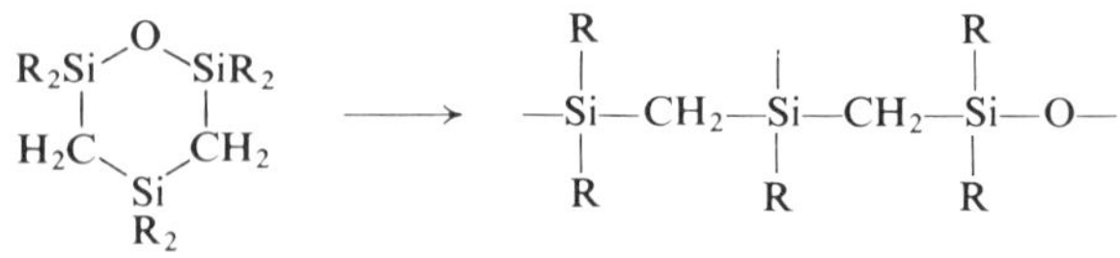 (267)

This enables the techniques which lead to macromolecular siloxanes to be applied also to these compounds and thus opens up further possibilities for copolymerization.

Of all the compounds investigated up to now, polysilphenylenesiloxanes appear the most interesting. They obviously tend less to depolymerization than pure siloxanes and have a better heat resistance (*148*). Introduction of Si—C_6H_4—Si groups into the siloxane chain increases the degree of crystallinity of linear polymers (see Section 8.1.1).

Polysilalkylenesilanes have been obtained by first preparing dichlorotetramethyldisilmethylene and treating this with metallic sodium (*64*); a cyclic dimer is formed, together with compounds of higher molecular weight:

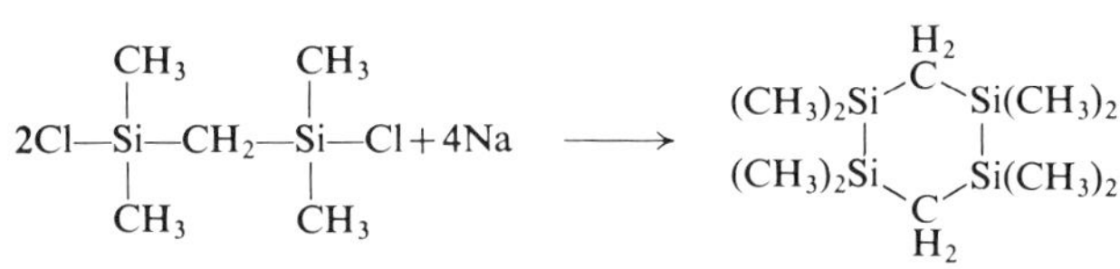 Mp 42°C; Bp_3 85°C

Finally, *polysilylenesiloxanes* have been obtained in two ways: either by causing a part of the halogen atoms of mono- or diorganohalosilanes to react with alkali metals in the first stage, thus obtaining prepolymers with Si—Si bonds and converting these in a second stage into silylenesiloxanes by hydrolytic cleavage of the residual halogen (*66*), or by first partially hydrolyzing the chlorosilane and then condensing the resulting intermediate with an alkali metal (*67*).

7.4 Organosilicon-Organic Copolymers

7.4.1 General

The question of how far the properties of the silicone products can be modified by combination with purely organic components or, conversely, how far the behavior of organic polymers can be influenced by the incorporation of siloxanes has always been of considerable interest. In most cases no effect can be achieved by mixing organosilicon and organic components physically, simply because of the low compatibility of silicone polymers with macromolecular organic compounds. In the vast majority of cases it has proved to be necessary to cause the two to react chemically. Such reactions may comprise *copolymerization*, *cocondensation*, and *coaddition*.

The reactions under consideration are well known from the chemistry of monomers and can take place through silicon-functional or organofunctional substituents (Chapters 3 and 4). They can also be applied to components which are already prepolymers. Since the molecular size, chemical constitution, and functionality of the reactants can be widely varied, there is in principle a very large number of possibilities.

It is essential for the control of copolymerization and cocondensation reactions that the homopolymerization or homocondensation of each reactant with molecules of the same kind is reduced or, if possible, completely eliminated. This is particularly difficult with cocondensation reactions involving silanols or siloxanols, since, as is well known, these possess a pronounced tendency to self-condensation (Chapter 5). This problem is generally solved by controlling the reaction, or by a suitable choice of catalysts preferentially promoting the cocondensation reaction but not the self-condensation of the silanols.

The structure of the copolymers produced, particularly the sequence of the structural units, depends on the type of the reaction, the conditions and course of the process, and the molecular size of the starting components which may be monomers or prepolymers. The polymers therefore contain either structural units in periodic sequence, such as in certain polyalkeneoxysiloxanes (polysilylacetals):

$$\mathrm{HO}-\left[\begin{array}{l}\mathrm{R}\\ |\\ \mathrm{Si}-\mathrm{CH_2}-\mathrm{O}-\\ |\\ \mathrm{R}\end{array}\right]_n \mathrm{H}$$

or they are block polymers:

$$\mathrm{HO}-\left[\begin{array}{l}\mathrm{R}\\ |\\ \mathrm{Si}-\mathrm{O}-\\ |\\ \mathrm{R}\end{array}\right]_m \left[\begin{array}{l}\mathrm{H}\\ |\\ \mathrm{C}-\mathrm{O}-\\ |\\ \mathrm{H}\end{array}\right]_n \left[\begin{array}{l}\mathrm{R}\\ |\\ \mathrm{Si}-\mathrm{O}-\\ |\\ \mathrm{R}\end{array}\right]_q \left[\begin{array}{l}\mathrm{H}\\ |\\ \mathrm{C}-\mathrm{O}-\\ |\\ \mathrm{H}\end{array}\right]_r \mathrm{H}$$

in which blocks of the siloxane and the organic components of different and varying sizes alternate. It is the latter case that is frequently found in technical products.

Copious information (particularly in patents) is available on processes and formulations for such "organic modifications" of silicones. The work is frequently directed toward technical goals, and details of the reaction mechanisms are not made clear. The following discussions give only a selective presentation, which is intended to outline the possibilities rather than to treat them exhaustively.

For the sake of simplicity the concept "copolymer" will be given a wider meaning in this context and will be taken to denote the products without reference to the route by which they have been prepared, such as polymerization, condensation, or addition. The organic components will be used for the classification of these compounds.

7.4.2 Copolymers with Aldehydes

Dimethyldichlorosilane reacts with methylene glycol (formaldehyde hydrate) in the presence of pyridine to give polymethyleneoxysiloxanes (polysilformals) (*146*):

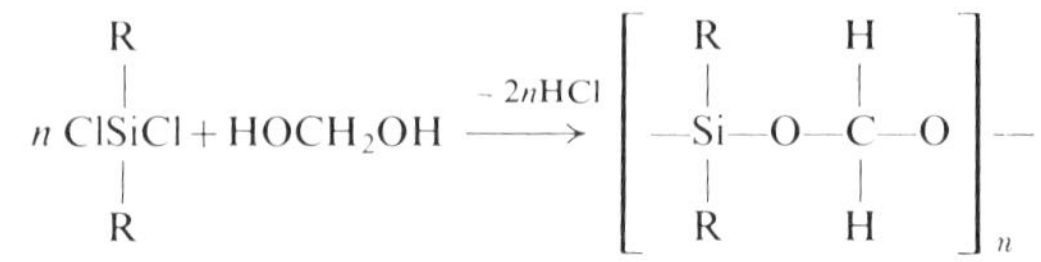

Definite polymers of this type have been obtained by the reaction of methylene diacetate with dimethyldiethoxysilane in the presence of sodium alkoxide (*146*).

Linear polymers of analogous structure but of relatively low molecular weight have also been prepared by the reaction of diphenyldichlorosilane with paraformaldehyde in the presence of water (*89*). Depending on the conditions of working, cyclic compounds containing monosilatrioxane rings (I) or disilatrioxane rings (II) are formed with methylchlorosilanes:

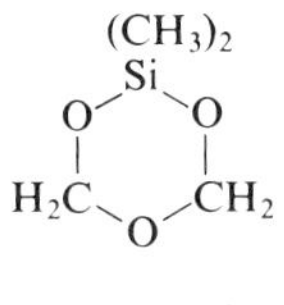

I

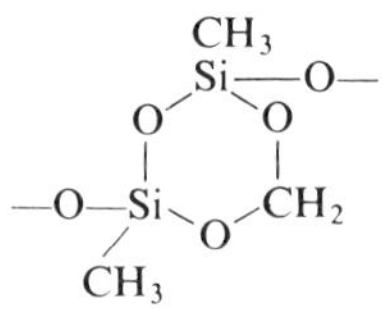

II

These are interesting because of their reactivity. On heating, they split off formaldehyde; they undergo reactions with proteins even at room temperature (Section 10.9.2).

Rings containing not Si—O—C but Si—C—O—C bridges are considerably more stable:

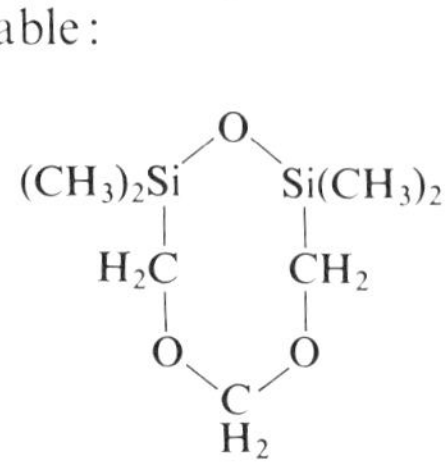

I

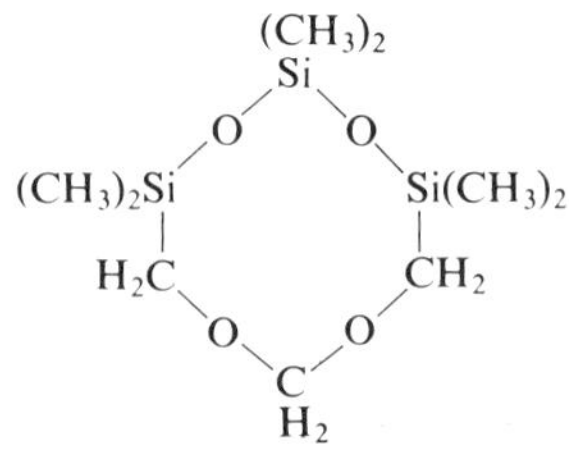

II

Type (I) is formed by the reaction of bis-hydroxymethyl-tetramethyldisiloxane with formaldehyde in the presence of *p*-toluenesulfonic acid. Linear polymers of high molecular weight are first formed and these are depolymerized and cyclized on being heated to 180°C in the presence of the same catalyst (*269*). Rings of this type can be copolymerized with trioxane. The catalysts normally employed for the polymerization of trioxane, in particular BF_3, can be used (*269*):

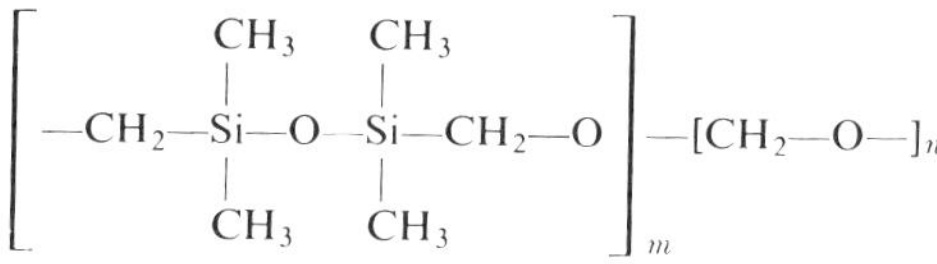

7.4.3 Copolymers with Polyalcohols and Polyphenols

Condensation reactions between alcohols or phenols and silicon-functional compounds can be carried out according to the following reaction schemes, among others (cf. also Chapters 3 and 11), working in the presence of catalysts or acceptors, according to the nature of the reactants:

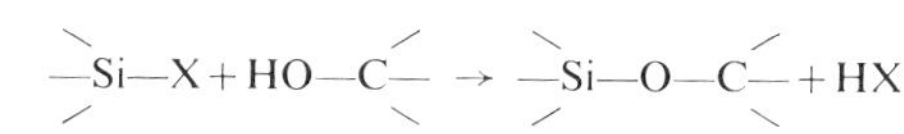
(1)

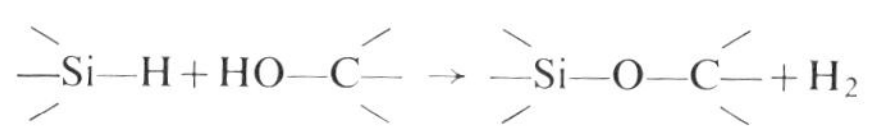
(2)

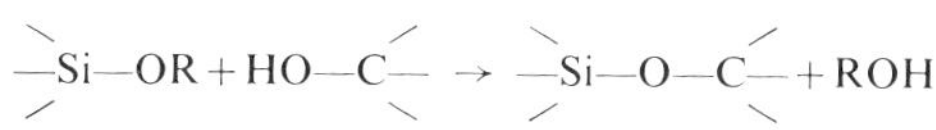
(3)

(where X = halogen).

If these are applied to starting components with at least two functionalities, polycondensation products are formed which have linkages such as:

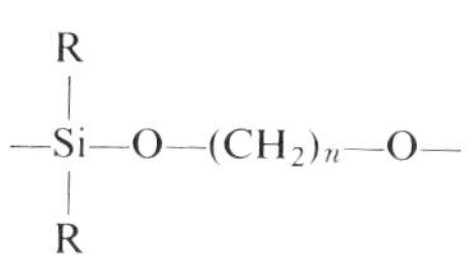

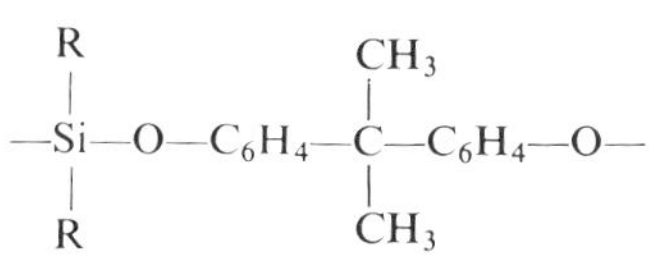

Many syntheses of such copolymers have been carried out using as alcoholic or phenolic components glycols (*141*, *303*), glycerol (*303*), 4,4′-dihydroxybenzophenone (*280*), 4,4′-dihydroxydiphenylmethane (*282*), and 4,4′-dihydroxydiphenylpropane (*117*, *282*, *298*). The silicon component may be a silane or siloxane. Si—OR bonds, particularly those with $R = CH_3$ or C_2H_5, are preferred as reactive groups. Si—halogen bonds can easily give rise to interfering side reactions by elimination of hydrogen halide.

The types of compounds in this group are, to a certain extent, also accessible through copolymerization. A two-phase mixture of polyols with siloxanes can be converted into a single phase by heating in the presence of alkali-metal hydroxides, which indicates molecular rearrangement and formation of siloxane-polyol copolymers (*104*). A ring compound obtained by reaction of neopentyl glycol with dimethyldichlorosilane has been cleaved by means of metallic sodium, potassium silanolate, or naphthylsodium and polymerized like a cyclosiloxane (*54*):

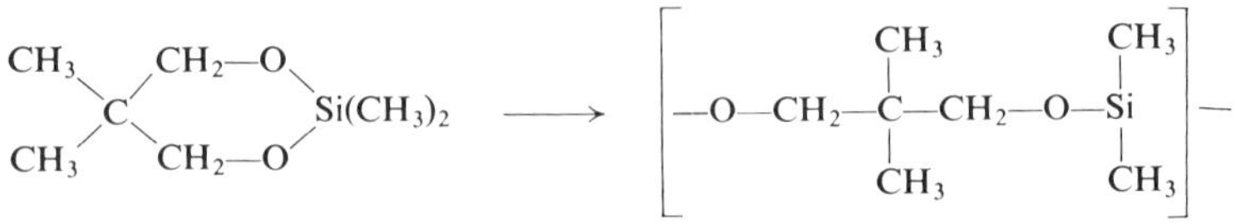

Cyclic esters of structural type (I) which have been obtained by reaction of dimethyldichlorosilanes or dialkoxysilanes with difunctional phenols, particularly 4,4′-dihydroxydiphenylpropane can be cleaved and converted into linear polymers (II) (*172*) using certain catalysts such as zinc stearate, lead naphthenate, and toluenesulfonic acid $[A = C_6H_4C(CH_3)_2C_6H_4]$.

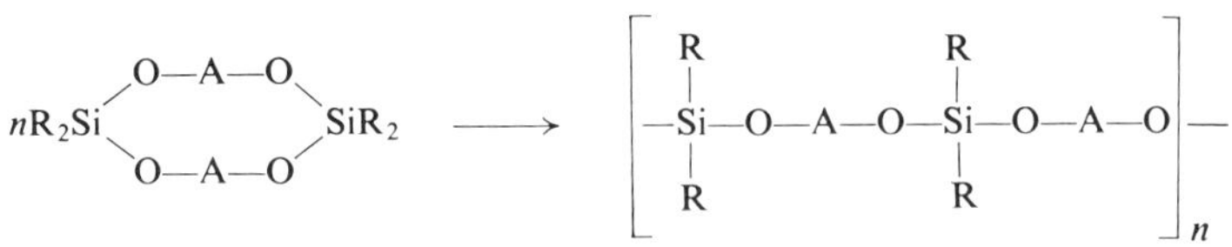

I II

A special feature of all cocondensation products with Si—O—C bonds is that they are more or less sensitive to hydrolysis. Siloxane glycol copolymers are water soluble, but can be kept in aqueous solution for only limited periods (*72*). All synthetic resinlike materials based on cocondensation products with phenols would seem to suffer from the disadvantage of low water resistance. This disappears if the linkage between the silicon and the polyol components takes place directly through Si—C bonds, for example:

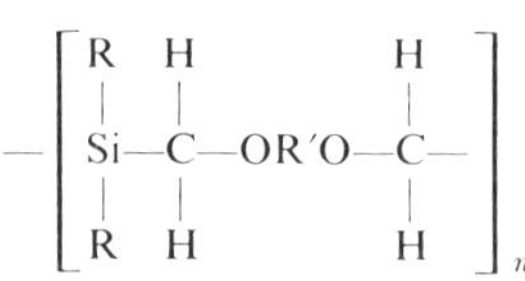

Cocondensation reactions are again available for the preparation of such products, and for these the Si—CH_2—Br grouping is a particularly attractive partner because of its reactivity. Bis-bromomethyl-tetramethyldisiloxane reacts with polyols in the presence of tertiary amines as hydrogen bromide acceptors, to give high-molecular-weight polycondensates (*269*). With the sodium salt of 4,4′-dihydroxydiphenylpropane it can easily be converted into poly-condensation products (*205*):

$$\left[-\overset{\displaystyle R}{\underset{\displaystyle R}{\overset{|}{\underset{|}{Si}}}}-O-\overset{\displaystyle R}{\underset{\displaystyle R}{\overset{|}{\underset{|}{Si}}}}-CH_2-O-C_6H_4-\overset{\displaystyle CH_3}{\underset{\displaystyle CH_3}{\overset{|}{\underset{|}{C}}}}-C_6H_4-O-CH_2 \right]_n -$$

7.4.4 Copolymers with Polyethers

Compounds with extremely interesting properties are found in this group, and preparative methods for these copolymers have therefore received particular attention.

The copolymers can be divided into two large groups, those with Si—O—C and those with Si—C bridges between the organosilicon and the organic parts of the molecule. In this second case this bridge generally does not lead directly to the polyether, but the partners are joined through another linkage, which may be a thioether, urethane, or ester bond, depending on the process used. From these points of view, the most important processes can be arranged as follows (where PE = polyether).

1. Siloxane-polyether copolymers with Si—O—C bridges.

The hydroxyl-containing polyether is caused to react with silanes and siloxanes containing Si—O—C_2H_5 (*25*, *26*), Si—H (*26*, *157*), or Si—N groupings (*206*).

$$\geq Si-OR + HO-PE \rightarrow \geq Si-O-PE + ROH \qquad (1)$$

$$\geq Si-H + HO-PE \rightarrow \geq Si-O-PE + H_2 \qquad (2)$$

$$\geq Si-NH_2 + HO-PE \rightarrow \geq Si-O-PE + NH_3 \qquad (3)$$

Reaction (1), which amounts to a transesterification, is catalyzed especially by trifluoroacetic acid, and reaction (2) by zinc or lead salts of carboxylic acids. Condensation (3) requires no catalyst.

2. Siloxane-polyether copolymers with Si—C bridges.

Direct linkage of the siloxane with the polyether is achieved by the addition of Si—H bonds to polyethers containing alkenyl groups (*139*, *140*, *307*).

$$\geq Si-H + CH_2{=}CH-PE \rightarrow \geq Si-CH_2-CH_2-PE$$

The reaction is catalyzed by chloroplatinic acid.

The linkage between the two parts can also be achieved by reactions on the organofunctional substituents of the silicon component. Chlorine on the α-C atom of chloromethylsiloxanes can be brought into reaction only with difficulty and requires reaction conditions which could make the reproducible synthesis of definite molecules more difficult. On the other hand, the considerably more reactive bromine on the α-C atom is a very suitable partner. Siloxanes containing bromomethyl groups can be condensed with the OH groups of polyethers in the presence of tertiary amines (*266*):

$$\rangle\!\!-Si-CH_2-Br + HO-PE \rightarrow \rangle\!\!-Si-CH_2-O-PE$$

The following routes also give the desired result without difficulty, the bridges between the silicon component and the polyether component being lengthened. The bromomethyl derivative is converted into the hydroxymethyl derivative and this is linked with the OH group of the polyether through a urethane bridge by the addition of a diisocyanate (*268*):

$$\rangle\!\!-Si-CH_2-S-CH_2-CH_2-OH + OCN-R'-NCO + HO-PE \rightarrow$$

$$\rangle\!\!-Si-CH_2-S-CH_2-CH_2-O-\underset{\underset{O}{\|}}{C}-\underset{\underset{H}{|}}{N}-R'-\underset{\underset{H}{|}}{N}-\underset{\underset{O}{\|}}{C}-O-PE$$

or both the hydroxymethyl group and the polyether are esterified with the chloride of a dicarboxylic acid in the presence of a tertiary amine (*207*):

$$\rangle\!\!-Si-CH_2-OH + ClOC-R-COCl + HO-PE \xrightarrow{NR_3}$$

$$\rangle\!\!-Si-CH_2-OOC-R-COO-PE$$

A bridge extended in the same way can be obtained if epoxy-functional silicon compounds are used as the starting materials—those, for instance, that are readily accessible by the addition of the Si—H bond to allyl glycidyl ether and which are attached to the OH group of the polyether (*74*):

$$\rangle\!\!-Si-(CH_2)_3-O-CH_2-\underbrace{CH-CH_2}_{O} + HO-PE \rightarrow$$

$$\rangle\!\!-Si-(CH_2)_3-O-CH_2-CH(OH)-CH_2-O-PE$$

By means of these processes it has been possible to synthesize copolymers of various structural types. They most frequently contain a polyether or a linear block polymer consisting of ethylene oxide and propylene oxide units and often (II and III) an $O-C_4H_9$ group at the ends. Some of the molecules are linear (I) (*140*) and some are branched (II), while others have the nature of graft polymers (III).

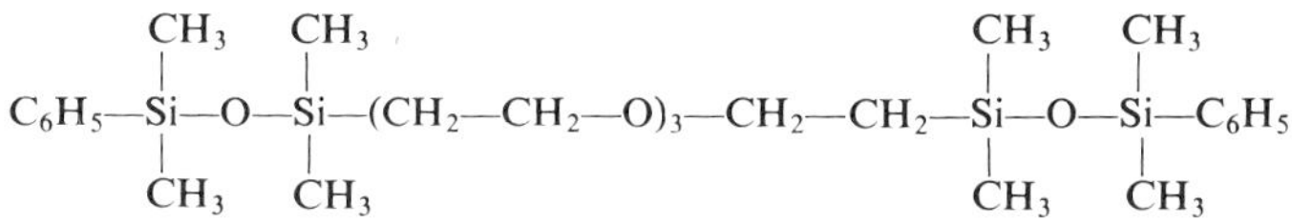

I

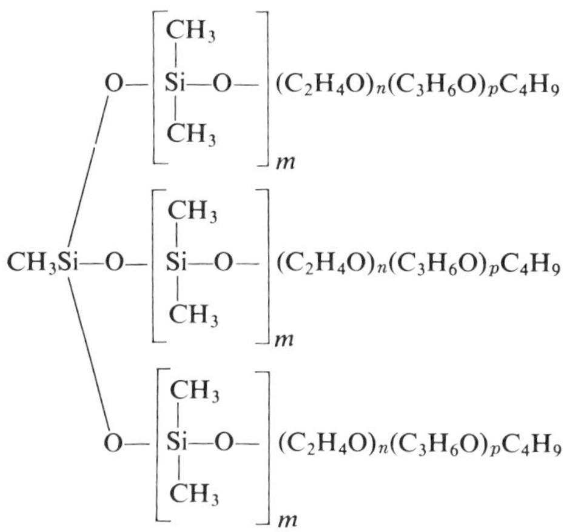

II

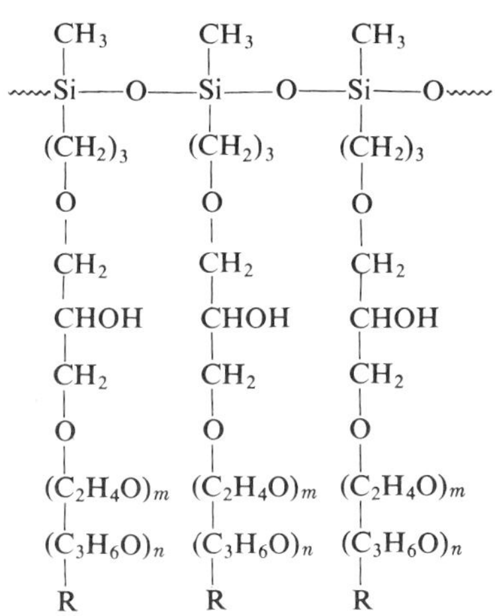

III

A common feature of all these copolymers is that they contain a hydrophobic (siloxane) and a hydrophilic (polyether) portion. Their remarkable properties arise from this structure. Depending on the ratio of the siloxane and the polyether portions, the solubility in water or more weakly polar solvents, such as alcohols, can vary within wide limits. Aqueous solutions of polyether-siloxane copolymers with Si—C bonds are stable; if Si—O—C linkages are present, the solution decomposes in the course of time with separation of a silicone oil phase. All polyether-siloxane copolymers are more or less surface-active, depending on their structural type. The addition of even small amounts of certain polyether-siloxane copolymers to water considerably lowers the surface tension. Concentrations of 10^{-6} gm/liter can reduce σ to 21 dynes cm^{-1}. The change of viscosity of aqueous solutions with concentration is also very interesting with certain polyether-siloxane copolymers; it exhibits a maximum in the region of 40 to 60% of polyether-siloxane copolymer (referred to the solution). Regardless of the structural type and concentration of the copolymer, aqueous solutions display an inverse temperature solubility pattern, i.e., the compounds flocculate out when their solutions are heated to about 30 to 40°C.

Consequently, polyether-siloxane copolymers have also found very interesting applications. They include products which are very effective stabilizers for foamed materials (Section 10.6); other copolymers act—evidently because of their inverse temperature-solubility behaviour—as demulsifiers of latex emulsions (Section 10.4.3); other types again are suitable as leveling (flow-promoting) agents for paints and varnishes (Section 10.5.1). They have been recommended as antistatic agents for synthetic fibers (*77*). Polyether-siloxane copolymers in which the polyether component acts as a solubility promoter may also be of interest for the cosmetic industry because of their solubility in alcohol or water.

7.4.5 Copolymers with Polyesters

This group of materials has also enriched the technology of silicones, particularly that of silicone resins.

A number of routes are available for the preparation of copolymers from siloxanes and polyesters. As in the preceding section, two main groups of processes can be distinguished, depending on whether the linkage of the silicon and polyester components is achieved through an Si—O—C or an Si—C bridge.

In the first case polyesters still containing free OH groups are chosen as starting materials. These are condensed with alkoxysiloxanes by the processes already described for cocondensation with polyalcohols (Section 7.4.3) (*86*, *128*, *167*):

$$\equiv Si—OR + HO—C\equiv \rightarrow \equiv Si—O—C\equiv + HOR$$

In addition, there is a condensation process which starts from silanols or siloxanols (*35*, *116*, *191*, *193*):

$$\equiv\mathrm{Si{-}OH} + \mathrm{HO{-}C}\equiv \;\rightarrow\; \equiv\mathrm{Si{-}O{-}C}\equiv + \mathrm{H_2O}$$

This is particularly attractive because it is based on starting materials which are easily accessible by the hydrolysis of chlorosilanes. A condition for the success of such reactions, however, is that the siloxanol and polyester components have high OH contents and sufficiently high molecular weights to allow reaction temperatures of 200 to 250°C to be reached *in vacuo* (*191*).

In principle, the sequence of operations in the synthesis of the siloxane-polyester copolymers can be modified by first condensing a functional silane or siloxane with an excess of a polyalcohol, and subsequently esterifying the product with a polycarboxylic acid so that the network is completed to form a macromolecular structure (*153*, *303*).

Many such products, mostly highly cross-linked and with a resinlike character, have been prepared. Ethylene glycol (*191*), glycerol (*116*, *128*, *167*, *303*), tris-hydroxymethyl-propane (*191*), and pentaerythritol (*167*, *303*) have been used as alcoholic components; phthalic acid (*116*, *128*, *191*, *303*), isophthalic acid (*88*, *193*), and terephthalic acid (*167*, *191*) and their anhydrides or methyl esters, stilbenedicarboxylic acid (*167*), dibenzyldicarboxylic acid (*167*), and many others have been used as carboxylic acid components. Commercially available alkyd resins which have been partly oil-modified have also been frequently used as starting materials (*35*, *46*).

The second group of copolymers, containing Si—C bonds, is not easily accessible and perhaps for this reason has not achieved similar practical importance. These compounds are obtained by reactions of carboxyorganosilanes, carboxyorganosiloxanes or their esters with polyalcohols (*29*, *33*, *277*):

$$\equiv\mathrm{Si{-}R{-}COOH} + \mathrm{HO{-}C}\equiv \;\rightarrow\; \equiv\mathrm{Si{-}R{-}COOC}\equiv$$

and in principle they make it possible to prepare siloxane-modified alkyd resins.

7.4.6 Copolymers with Olefinic Components

Under certain circumstances, silanes with unsaturated groups can be not only homopolymerized (Section 4.3) but also copolymerized with olefins and other unsaturated purely organic monomers, Ziegler catalysts being used. It has been shown, for example, that ethylene can be copolymerized with vinyl- and allyltrimethylsilanes (*71*), and propylene with diallyldimethyl- and diallylmethylphenylsilanes (*198*). Numerous other organic components such as acrylic acid, methacrylic acid (*118*), maleic anhydride, styrene (*131*), acrylonitrile, and *N*-vinylpyrrolidone (*24*) have been worked up to copolymers with vinyl- and allylsilanes or vinyl- and allylsiloxanes. These are of some interest as water-soluble emulsifiers (*24*) and also, because of their remarkable swelling properties, as thickeners (*118*). Copolymerization, like homopolymerization, appears to be easier with allyl- than with vinylsilicon compounds.

Numerous variants with other polymerizable silicon components are conceivable. *p*-Styrylsiloxanes have been copolymerized with styrene by means of azodiisobutyronitrile (*132*), and acryloxymethylsiloxanes with acrylonitrile, alkyl methacrylates, glycol esters of maleic acid, etc.; the same catalyst or benzoyl peroxide (*187*) was used.

7.4.7 Various Copolymers

The following is a selection from the large number of experiments designed to combine siloxanes with other organic polymers.

Siloxane-urethane copolymers are obtained when siloxane-polyol copolymers containing free alcoholic OH groups react with diisocyanates (*231*). The same principle can be applied to siloxane-polyester copolymers with free OH groups. It enables the resinous products based on these materials to be cured at room temperature (compare Sections 8.2.1.3 and 8.2.2.1.4). Conversely, the siloxane can be added to polyols through its own isocyanate functional groups (*147*). Copolymers with Si—C linkages are obtained by the reaction of aminoalkyl- (*228*) or hydroxyalkylsilanes and aminoalkyl- or hydroxyalkylsiloxanes (*84*, *130*, *278*) with diisocyanates.

Siloxane-polycarbonate copolymers are formed when condensation products of siloxanes with bisphenol A (Section 7.4.3), which contain free phenolic OH groups, are condensed with phosgene (*117*, *233*, *298*).

It is mainly practical problems in resin chemistry that have given rise to further combinations. Organic *epoxy-hydroxy* compounds have been worked up to copolymers with organosilane or siloxane components (*30*, *31*, *78*, *142*). The processes in these frequently rather complex systems are difficult to follow but coupling of the two components presumably again basically involves condensation processes between the Si—X, Si—OR and Si—OH groups of the Si components on the one hand and the functional groups of the organic partner on the other. The latter may, for example, be one of the group of condensation products from dihydric phenols and epichlorohydrin (*30*, *31*, *78*, *142*) or from sulfonamides and epichlorohydrin (*142*), which may be modified by fatty acids (*142*). The silicon components may also contain Si—C_6H_4—Si bonds instead of Si—O—Si bonds (*240*). By the reaction of a bis-ethoxy-functional disilarylene with an epoxyalcohol, definite precursors for copolymers have been obtained:

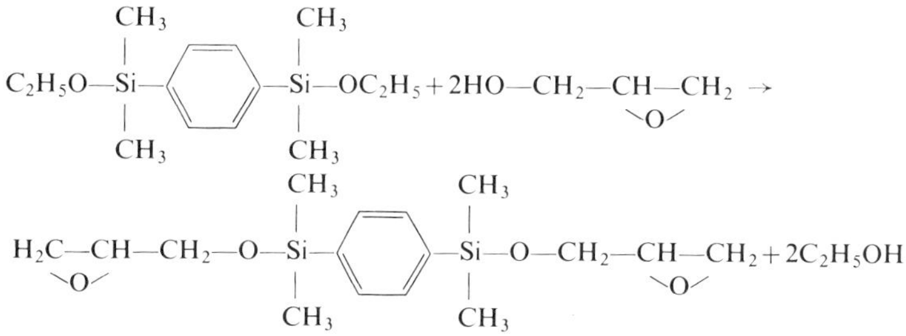

These can be copolymerized with epoxide resins (*240*).

Silicone-polyester condensation products can be blended with epoxy resins so that, when certain conditions are fulfilled, further interesting possibilities of combination arise (*82*, *184*). Besides this, the reaction products of amino-organosiloxanes with epoxy resins (*264*) provide examples of copolymers in which both components are linked through Si—C bonds.

Under certain conditions, siloxanes can also be combined with phenol-formaldehyde (*190*, *234*), melamine-formaldehyde (*83*, *126*, *185*), urea-formaldehyde (*83*, *182*, *185*), and triazine-formaldehyde (*185*) resins. The components are mixed in the presence of organic solvents such as toluene, xylene, ethers, and esters and then hardened to form resins by heating *in situ*. The extent and nature of the chemical reactions taking place between the two components are not always clear. The siloxane components are preferably copolymers of siloxanes with polyesters (*83*, *126*, *190*, *234*), but unmodified siloxanes (185) and silanes (*182*) have also been used. For the special formulations, mixing ratios, and other conditions to be observed, reference must be made to the original literature.

REFERENCES

(1) E. W. Abel, *Z. Naturforsch.* **15b**, 57 [1960]. (2) — and A. Singh, *J. Chem. Soc.* [London] **1959**, 690. (3) American Smelting and Refining's Central Research, *Chem. Engng. News* **34**, 3608 [1956]. (4) K. A. Andrianov et al., *Vysokomol. Soed.* **2**, 1099 [1960]; *Bull. Acad. Sci. USSR, Div. Chem. Sci.* (*Engl. transl.*) **1961**, 407 and 1177. (5) — et al., *J. Gen. Chem. USSR* (*Engl. transl.*) **31**, 3768 [1961]; *Bull. Acad. Sci. USSR, Div. Chem. Sci.* (*Engl. transl.*) **1963**, 1832. (6) — and V. V. Astakhin, *Proc. Acad. Sci. USSR* (*Engl. transl.*) **127**, 633 [1959]. (7) —, —, *J. Gen. Chem. USSR* (*Engl. transl.*) **32**, 2284 [1962]. (8) —, — and V. B. Losev, *Bull. Acad. Sci. USSR, Div. Chem. Sci.* (*Engl. transl.*) **1963**, 863. (9) — and V. G. Dulova, ibid. **1958**, 628.

(10) — and M. N. Ermakova, *J. Gen. Chem. USSR* (*Engl. transl.*) **31**, 1213 [1961]. (11) — and T. N. Ganina, *Bull. Acad. Sci. USSR, Div. Chem. Sci.* (*Engl. transl.*) **1956**, 71. (12) —, — and E. N. Khrustaleva, ibid. **1956**, 817. (13) —, —, *J. Gen. Chem. USSR* (*Engl. transl.*) **29**,

601 [1959]. (14) —, I. KHAIDUK and L. M. KHANANASHVILI, ibid. **34**, 905 [1964]. (15) —, YA. L. SHUGAL and E. Z. ASNOVICH, *Soviet Plastics* (*Engl. transl.*) **1964**, No. 2, 45. (16) —, V. S. TIKHONOV, L. M. KHANANASHVILI, HAN-EN-TSE and HAN-SHU-YU, ibid. **1962**, No. 12, 23. (17) —, T. V. VASIL'EVA and L. M. KHANANASHVILI, *Bull. Acad. Sci. USSR, Div. Chem. Sci* (*Engl. transl.*) **1961**, 952. (18) — and L. M. VOLKOVA, ibid. **1957**, 317. (19) — and A. A. ZHDANOV, *Proc. Acad. Sci. USSR* (*Engl. transl.*) **114**, 597 [1957].

(20) —, —, *J. Polymer Sci.* **30**, 513 [1958]. (21) —, —, *Bull. Acad. Sci. USSR, Div. Chem. Sci.* (*Engl. transl.*) **1958**, 756. (22) —, —, *Vysokomol. Soed.* **1**, No. 6, 894 [1959]. (23) R. W. AWE and H. M. SCHIEFER, Dow Corning, U.S. Pat. 3002989 [9.3.1959]. (24) D. L. BAILEY, Union Carbide, U.S. Pat. 3002987 [17.12.1957]. (25) — and F. M. O'CONNOR, —, U.S. Pat. 2834748 [22.3.1954]. (26) —, —, —, Brit. Pat. 892819 [U.S. Prior. 23.5.1957]. (27) W. R. BAMFORD, ICI, French Pat. 1292189 [Brit. Prior. 15.6.1960 and 23.5.1961]. (28) A. BARRY, Dow Corning, U.S. Pat. 2557931 [3.2.1950]. (29) — and J. W. GILKEY, —, German Pat. (West German) 865804 [U.S. Prior. 5.7.1949].

(30) N. V. de Bataafsche Petroleum, French Pat. 1130032 [U.S. Prior. 30.4.1954]. (31) —, French Pat. 1130033 [U.S. Prior. 30.4.1954]. (32) H. N. BECK and R. G. CHAFFEE, *J. Chem. Engng. Data* **8**, 602 [1963]. (33) W. T. BLACK, D. L. BAILEY and V. B. JEX, Union Carbide, German Appl. (West German) 1113455 [U.S. Prior. 12.10.1956]. (34) B. A. BLUESTEIN, *J. Amer. Chem. Soc.* **70**, 3068 [1948]. (35) A. BOWMAN and E. M. EVANS, ICI, Brit. Pat. 583754 [29.11.1944]. (36) W. A. BOYER, Dow Corning, U.S. Pat. 3143514 [15.6.1960]. (37) D. C. BRADLEY and C. PREVEDOROU-DEMAS, *Canad. J. Chem.* **41**, 629 [1963]. (38) — and I. M. THOMAS, *Chem. and Ind.* **1958**, 17 and 1231. (39) L. W. BREED and R. C. ELLIOTT, AD 286219 (U.S. Department of Commerce, Office of Technical Service).

(40) —, R. C. ELLIOTT and A. F. FERRIS, *J. Org. Chem.* **27**, 1114 [1962]. (41) — and W. J. HAGGERTY JR., ibid. **27**, 257 [1962]. (42) —, — and F. BAIOCCHI, ibid. **25**, 1633 [1960]. (43) S. D. BREWER and C. P. HABER, *J. Amer. Chem. Soc.* **70**, 3888 [1948]. (44) L. O. BROCKWAY and N. R. DAVIDSON, ibid. **63**, 3287 [1941]. (45) M. C. BROOKS and R. K. KUHNS, United States Rubber, U.S. Pat. 2763674 [27.7.1953]. (46) R. H. BUNNELL, Libbey-Owens, U.S. Pat. 2607755 [4.12.1948]. (47) C. A. BURKHARD, General Electric, U.S. Pat. 2554976 [1.9.1948]. (48) —, *J. Amer. Chem. Soc.* **71**, 963 [1949]. (49) — and R. H. KRIEBLE, ibid. **69**, 2687 [1947].

(50) C. C. CERATO, J. L. LAUER and H. C. BEACHELL, *J. chem. Physics* **22**, 1 [1954]. (51) M. M. CHAMBERLAIN, G. A. JABS and B. B. WAYLAND, *J. Org. Chem.* **27**, 3321 [1962]. (52) B. L. CHAMBERLAND and A. G. MACDIARMID, *J. Amer. Chem. Soc.* **82**, 4542 [1960]. (53) G. CHAMPETIER, Y. ÉTIENNE and R. KULLMANN, *C. R. Hebd. Séances Acad. Sci.* **234**, 1985 [1952]. (54) —, N. SPASSKY and P. SIGWALT, *Rev. Chim. Bucarest* **7**, 743 [1962]. (55) E. A. CHERNYSHEV and E. F. BUGERENKO, *Bull. Acad. Sci. USSR, Div. Chem. Sci.* (*Engl. transl.*) **1963**, 696. (56) N. D. CHERONIS, U.S. Pat. 2579416 [25.1.1946]. (57) —, U.S. Pat. 2564674 [7.8.1947]. (58) —, U.S. Pat. 2579418 [24.1.1949]. (59) H. A. CLARK, Dow Corning, German Pat. (West German) 853350 [U.S. Prior. 23.2.1949].

(60) —, —, German Pat. (West German) 845513 [U.S. Prior. 23.2.1949]. (61) —, —, German Pat. (West German) 845514 [U.S. Prior. 23.2.1949]. (62) —, —, U.S. Pat. 2507517 [19.3.1949]. (63) —, —, U.S. Pat. 2557942 [19.3.1949]. (64) —, —, U.S. Pat. 2563004 [15.6.1949]. (65) —, —, German Pat. (West German) 872270 [U.S. Prior. 15.6.1949]. (66) —, —, German Pat. (West German) 917275 [U.S. Prior. 2.9.1949]. (67) —, —, German

Pat. (West German) 917276 [U.S. Prior. 2.9.1949]. (68) —, A. F. GORDON, C. W. YOUNG and M. J. HUNTER, *J. Amer. Chem. Soc.* **73**, 3798 [1951]. (69) H. J. COHEN, National Distillers, U.S. Pat. 2994711 [10.8.1959].

(70) — and J. M. HOYT, —, U.S. Pat. 3046267 [22.9.1959]. (71) G. D. COOPER and A. R. GILBERT, Thomson-Houston, French Pat. 1240983 [U.S. Prior. 25.11.1958]. (72) C. C. CURRIE and R. C. GERGLE, *Amer. Perfumer Aromatics* **68**, 19 [1956]. (73) J. W. CURRY, *J. Amer. Chem. Soc.* **78**, 1686 [1956]. (74) K. DAMM, K. OTT and H. H. STEINBACH, Farbenfabriken Bayer, Belg. Pat. 627281 [German Prior. (West German) 19.1.1962]. (75) J. DANFORTH, *J. Amer. Chem. Soc.* **80**, 2585 [1958]. (76) C. DATHE, Diploma thesis, Munich 1955. (77) F. DAWSON, E. J. VICKERS and D. R. SHEARD, ICI, Brit. Pat. 981814 [1.6.1961]. (78) H. DETERING, Th. Goldschmidt, German Pat. (West German) 954456 [8.1.1954]. (79) B. N. DOLGOV, I. I. KHUDOBIN and N. P. KHARITONOV, *Proc. Acad. Sci. USSR (Engl. transl.)* **122**, 717 [1958].

(80) — and N. F. ORLOV, ibid. **117**, 1037 [1957]. (81) Dow Corning, Brit. Pat. 667435 [U.S. Prior. 19.3.1949]. (82) —, Brit. Pat. 737515 [U.S. Prior. 14.6.1952]. (83) —, Brit. Pat. 726694 [U.S. Prior. 16.4.1952]. (84) —, Brit. Pat. 631536 [26.8.1947]. (85) —, Brit. Pat. 968110 [U.S. Prior. 1.12.1960]. (86) C. D. DOYLE and H. C. NELSON, General Electric, U.S. Pat. 2587295 [22.11.1947]. (87) C. EABORN, *J. Chem. Soc.* [London] **1950**, 3077; *Nature* [London] **165**, 685 [1950]. (88) L. E. EDELMAN and S. ROBBINS, Westinghouse Electric, U.S. Pat. 2821518 [25.5.1954]. (89) H. H. ENDER, *J. Appl. Polym. Sci.* **7**, 1133 [1963].

(90) W. D. ENGLISH and L. H. SOMMER, *J. Amer. Chem. Soc.* **77**, 170 [1955]. (91) Y. ÉTIENNE, *C. R. hebd. Séances Acad. Sci.* **235**, 966 [1952]. (92) F. FEHÉR, G. KUHLBÖRSCH, A. BLÜMCKE, H. KELLER and K. LIPPERT, *Chem. Ber.* **90**, 134 [1957]. (93) — and K. LIPPERT, ibid. **92**, 2998 [1959]; **94**, 2437 [1961]. (94) W. FINK, *Angew. Chem.* **73**, 467 [1961]; *Helv. Chim. Acta* **45**, 1081 [1962]. (95) —, *Chem. Ber.* **97**, 1424 [1964]. (96) —, ibid. **97**, 1433 [1964]. (97) W. E. FOSTER and P. E. KOENIG, Ethyl Corporation, U.S. Pat. 2998440 [31.12.1956]. (98) R. R. FREEMAN, *Polym. Preprints, Amer. Chem. Soc., Philadelphia Meeting 1964*; **5**, 291 [1964]. (99) G. FRITZ, *Fortschr. Chem. Forschg.* **4**, 459 [1964].

(100) — and J. GROBE, *Z. anorg. allg. Chem.* **299**, 302 [1959]. (101) — and B. RAABE, ibid. **286**, 149 [1956]. (102) —, —, ibid. **299**, 232 [1959]. (103) — and H. THIELKING, ibid. **306**, 39 [1960]. (104) C. L. FRYE, Dow Corning, French Pat. 1181151 [U.S. Prior. 5.9.1956]. (105) General Electric, Brit. Pat. 870024 [U.S. Prior. 15.8.1958]. (106) W. GERRARD and J. A. STRICKSON, *Chem. and Ind.* **58**, 860 [1958.] (107) S. A. GIDDINGS, American Cyanamid, U.S. Pat. 3030394 [24.8.1961]. (108) J. W. GILKEY, Dow Corning, French Pat. 1127223 [U.S. Prior. 2.8.1954]. (109) H. GILMAN and G. E. DUNN, *J. Amer. Chem. Soc.* **73**, 5077 [1951].

(110) — and J. J. GOODMAN, ibid. **75**, 1250 [1953]. (111) —, R. K. INGHAM and A. G. SMITH, *J. Org. Chem.* **18**, 1743 [1953]. (112) — and T. C. WU, *J. Amer. Chem. Soc.* **73**, 4031 [1951]. (113) —, —, ibid. **75**, 3762 [1953]. (114) —, —, *J. Org. Chem.* **18**, 753 [1953]. (115) —, —, H. A. HARTZFELD, G. A. GUTER, A. G. SMITH, J. J. GOODMAN and S. H. EIDT, *J. Amer. Chem. Soc.* **74**, 561 [1952]. (116) Glidden, Brit. Pat. 682348 [U.S. Prior. 1.9.1949]. (117) E. P. GOLDBERG and E. J. POWERS, *J. Polymer Sci.* [B] *Pol. Letters* **2**, 835 [1964]. (118) B. F. GOODRICH, German Appl. (West German) 1088718 [U.S. Prior. 9.12.1957]. (119) J. T. GOODWIN JR., Dow Corning, German Pat. (West German) 844901 [Brit. Prior. 1.7.1947].

(120) —, —, German Pat. (West German) 863943 [Brit. Prior. 1.7.1947]. (121) —, —, German Pat. (West German) 847594 [Brit. Prior. 1.7.1947]. (122) —, —, German Pat. (West German) 877306 [Brit. Prior. 1.7.1947]. (123) —, —, U.S. Pat. 2511056 [20.4.1948]. (124) —, —, German Pat. (West German) 863416 [U.S. Prior. 19.3.1949]. (125) —, —, U.S. Pat. 2527808 [19.3.1949]. (126) —, —, Brit. Pat. 701013 [U.S. Prior. 21.8.1950]. (127) —, W. E. BALDWIN and R. R. McGREGOR, *J. Amer. Chem. Soc.* **69**, 2247 [1947]. (128) — and M. J. HUNTER, Dow Corning, U.S. Pat. 2584340 to 2584344 [7.1.1950]. (129) G. GREBER and A. BALCIUNAS, *Makromol. Chem.* **71**, 62 [1964].

(130) — and S. JÄGER, ibid. **57**, 150 [1962]. (131) — and L. METZINGER, ibid. **39**, 217 [1960]. (132) — and J. TÖLLE, ibid. **67**, 98 [1963]. (133) S. J. GROSZOS, American Cyanamid, U.S. Pat. 2957900 [16.9.1957]. (134) —, —, U.S. Pat. 3029210 [30.3.1960]. (135) — and I. A. HALL, —, U.S. Pat. 2885370 [3.10.1957]. (136) G. GRÜTTNER and M. CAUER, *Ber. dtsch. chem. Ges.* **51**, 1283 [1918]. (137) H. C. GULLEDGE, DuPont, U.S. Pat. 2512058 [5.4.1948]. (138) C. P. HABER, General Electric, U.S. Pat. 2462635 [22.10.1946]. (139) L. A. HALUSKA, Dow Corning, U.S. Pat. 2868824 [9.8.1956].

(140) —, —, French Pat. 1179743 [U.S. Prior. 9.8.1956]. (141) W. E. HANFORD, DuPont, U.S. Pat. 2386793 [6.6.1940]; U.S. Pat. 2441066 [23.7.1945]. (142) D. B. HATCHER, Allied Chemical and Dye, U.S. Pat. 2695276 [11.7.1951]. (143) F. A. HENGLEIN, R. LANG and K. SCHEINOST, *Makromol. Chem.* **15**, 177 [1955]. (144) —, —, —, ibid. **18/19**, 102 [1956]. (145) —, — and L. SCHMACK, ibid. **22**, 103 [1957]. (146) — and P. SCHMULDER, ibid. **13**, 53 [1954]. (147) H. HOLTSCHMIDT and O. BAYER, Farbenfabriken Bayer, German Appl. (West German) 1066582 [31.10.1958]. (148) N. G. HOLDSTOCK, General Electric, U.S. Pat. 3117149 [15.12.1958]. (149) — and E. D. BROWN JR., —, German Appl. (West German) 1083007 [U.S. Prior. 11.12.1957].

(150) E. D. HORNBAKER and F. CONRAD, *J. Org. Chem.* **24**, 1858 [1959]. (151) Hughes Aircraft, Brit. Pat. 915714 [U.S. Prior. 31.5.1960]. (152) M. J. HUNTER, Dow Corning, U.S. Pat. 2554193 [15.12.1948]. (153) — and L. A. RAUNER, —, U.S. Pat. 2584351 [10.11.1948]. (154) E. HUSEMANN, G. GREBER and L. METZINGER, German Pat. (West German) 1167031 [27.10.1959]. (155) —, —, —, German Appl. (West German) 1107941 [5.3.1959]. (156) J. F. HYDE, Dow Corning, U.S. Pat. 3002986 [23.10.1959]. (157) A. IBBOTSON and E. J. VICKERS, ICI, Brit. Pat. 954041 [19.10.1960 and 6.7.1961]. (158) R. K. INGHAM and H. GILMAN in F. G. A. STONE and W. A. G. GRAHAM, *Inorganic Polymers*, p. 344, Academic Press 1962. (159) T. ISHIZUKA et al., Shin-Etsu, U.S. Pat. 3096304 [Jap. Prior. 2.12.1958].

(160) R. M. KARY and K. C. FRISCH, *J. Amer. Chem. Soc.* **79**, 2140 [1957]. (161) H. C. KAUFMANN, Pierce Foundation, U.S. Pat. 2834736 [5.3.1956]; U.S. Pat. 2934502 [16.10.1956]. (162) F. S. KIPPING and co-workers, *J. Chem. Soc.* [London] **1921**, 830, 848; **1923**, 2590, 2598; **1928**, 1431; **1929**, 1176, 2545; **1930**, 1029; **1931**, 1290. (163) — and J. E. SANDS, ibid. **119**, 830 [1921]. (164) W. H. KNOTH JR., DuPont, U.S. Pat. 2850514 [25.10.1955]. (165) P. E. KOENIG, Conf. on high temperature polymer and fluid synthesis, WADC Techn. report 57-657, 1958, ed. by United States Dep. Commerce, Office of Technical Services, p. 50. (166) — and W. E. FOSTER, Ethyl Corporation, U.S. Pat. 2998407 [17.12.1956]. (167) C. F. KOHL, Dow Corning, U.S. Pat. 2686739 [31.5.1951]. (168) N. L. KOPNOVA, V. S. CHUGUNOV and A. L. KLEBANSKII, *J. Gen. Chem. USSR* (*Engl. transl.*) **34**, 1357 [1964]. (169) C. A. KRAUS and W. K. NELSON, *J. Amer. Chem. Soc.* **56**, 195 [1934].

(170) A. P. Kreshkov, D. A. Karateev, V. Fyurst and E. N. Pavlova, *J. Gen. Chem. USSR (Engl. transl.)* **33**, 253 [1963]. (171) R. H. Krieble, General Electric, U.S. Pat. 2440101 [7.5.1947]. (172) H. Krimm and H. Schnell, Farbenfabriken Bayer, German Pat. (West German) 1136114 [16.7.1960]. (173) W. A. Kriner, *J. Org. Chem.* **29**, 1601 [1964]. (174) U. Krüerke, Dissertation Munich 1953. (175) C. R. Krüger and E. G. Rochow, *Angew. Chem.* **74**, 588 [1962]; *J. Polymer Sci.* [A] **1964**, 3179. (176) M. Kumada and A. Habuchi, *J. Inst. Polytechn. Osaka Univ.* C3, 65 [1952]. (177) Y. Kurita and M. Kondo, *Bull. Chem. Soc. Japan* **27**, 160 [1954]; (*C.A.* **1955**, 9988). (178) E. Larsson and L. Bjellerup, *J. Amer. Chem. Soc.* **75**, 995 [1953]. (179) — and R. Mårin, *Acta Chem. scand.* **5**, 964 [1951].

(180) — and B. Smith, ibid. **3**, 487 [1949]. (181) N. S. Leznov, L. A. Sabun and K. A. Andrianov, *J. Gen. Chem. USSR (Engl. transl.)* **29**, 1247 [1959]. (182) C. A. MacKenzie and J. B. Rust, Ellis Foster and Montclair Research, U.S. Pat. 2500843 [10.3.1944]. (183) G. Martin, *Ber. dtsch. chem. Ges.* **46**, 3289 [1913]. (184) W. M. McLean, Dow Corning, French Pat. 1081000 [U.S. Prior. 14.7.1952]. (185) —, —, French Pat. 1101368 [U.S. Prior. 27.5.1953]. (186) R. C. Mehrotra and R. C. Pant, *Indian J. Appl. Chem.* **26**, 109 [1963]. (187) R. L. Merker, Midland Silicones, Brit. Pat. 803479 [U.S. Prior. 6.8.1956]. (188) —, Dow Corning, Belg. Pat. 606357 [U.S. Prior. 22.7.1960] and French Pat. 1199434 [U.S. Prior. 20.3.1957]. (189) — and W. A. Piccoli, —, French Pat. 1261082 [U.S. Prior. 14.8.1959].

(190) Midland Silicones, Brit. Pat. 746306 [U.S. Prior. 13.4.1953]. (191) K. H. Mielke and W. Noll, Farbenfabriken Bayer, German Appl. (West German) 1178540 [27.9.1952]. (192) R. Minne and E. G. Rochow, *J. Amer. Chem. Soc.* **82**, 5625, 5628 [1960]. (193) F. J. Modic and J. E. Saville, General Electric, French Pat. 1192426 [U.S. Prior. 26.12.1956]. (194) Monsanto Chemicals, French Pat. 1346317 [Swiss. Prior. 17.10.1961 and 11.5.1962]. (195) L. S. Moody, General Electric, U.S. Pat. 2567724 [29.11.1950]. (196) M. M. Morgunova, D. Ya. Zhinkin and M. V. Sobolevskii, *Soviet Plastics (Engl. transl.)* **1963**, No. 4, 23. (197) W. Moschel, H. Jonas and W. Noll, Farbenfabriken Bayer, German Pat. (West German) 832499 [30.3.1950]. (198) N. S. Nametkin, A. W. Topchiev, S. D. Durgaryan and I. M. Tolchinskii, *Vysokomol. Soed.* **1**, 1739 [1959]. (199) —, — and L. S. Porazov, *Doklady Adkad. Nauk. SSSR* **99**, 403 [1954].

(200) —, — and V. I. Zetkin, ibid. **87**, 601 [1952]. (201) —, —, —, ibid. **93**, 1045 [1953]. (202) — and V. M. Vdovin, *J. Polymer Sci.* [C] No. 4, Part 2, 1043 [1963]. (203) —, — and K. S. Pushchevaya, *Proc. Acad. Sci. USSR (Engl. transl.)* **150**, 439 [1963]. (204) H. Niederprüm, *Angew. Chem.* **75**, 165 [1963]. (205) —, Farbenfabriken Bayer, 1962, unpublished. (206) — and W. Simmler, —, Brit. Pat. 107156 [German prior. 9.4.1965]. (207) —, —, —, Brit. Pat. 1073368 [German Prior. 17.4.1965]. (208) J. M. Nielsen, Thomson-Houston, French Pat. 1334850 [U.S. Prior. 2.10.1961]. (209) J. E. Noll, J. L. Speier and B. F. Daubert, *J. Amer. Chem. Soc.* **73**, 3867 [1951].

(210) D. C. Noller and H. W. Post, ibid. **74**, 1361 [1952]. (211) I. G. Noltes and G. J. M. v.d. Kerk, R. **81**, 565 [1962]. (212) T. Nomura, M. Yokoi and K. Yamasaki, *Proc. Japan Acad.* **29**, 342 [1954] (*C.A.* **1955**, 12274). (213) Z. N. Nudelman, K. A. Andrianov and G. B. Kudryavitskaya, *Vysokomol. Soed.* **4**, 440 [1962]. (214) R. Okawara and coworkers, *Proc. Chem. Soc.* [London] **1961**, 383; *J. Amer. Chem. Soc.* **83**, 1342 [1961]. (215) — and K. Sugita, ibid. **83**, 4480 [1961]. (216) N. F. Orlov, B. N. Dolgov and M. G. Voronkov, *Proc. Acad. Sci. USSR (Engl. transl.)* **122**, 681 [1958]. (217) —, —, —, *Bull. Acad.*

Sci. USSR, Div. Chem. Sci. (*Engl. transl.*) **1960**, 1493. (218) — and M. G. VORONKOV, ibid. **1959**, 1459. (219) —, —, *J. Gen. Chem. USSR* (*Engl. transl.*) **32**, 597 [1962].

(220) S. PAPETTI and H. W. POST, *J. Org. Chem.* **22**, 526 [1957]. (221) W. J. PATNODE, General Electric, U.S. Pat. 2434953 [1.7.1944]. (222) — and R. W. SCHIESSLER, —, U.S. Pat. 2381000 and 2381002 [4.3.1942]. (223) — and F. C. SCHMIDT, *J. Amer. Chem. Soc.* **67**, 2272 [1945]. (224) A. D. PETROV and V. A. PONOMARENKO, *Doklady Adak. Nauk. SSSR* **90**, 387 [1953]. (225) — and L. L. SHCHUKOVSKAYA, ibid. **86**, 551 [1952]. (226) — and V. M. VDOVIN, *Bull. Acad. Sci. USSR, Div. Chem. Sci.* (*Engl. transl.*) **1959**, 905. (227) —, —, ibid. **1959**, 1105. (228) R. M. PIKE, Union Carbide, U.S. Pat. 2907782 [12.10.1956]. (229) A. M. POLYAKOVA and co-workers, *Bull. Acad. Sci. USSR, Div. Chem. Sci.* (*Engl. transl.*) **1959**, 2159.

(230) G. S. POPELEVA, K. A. ANDRIANOV, S. A. GOLUBTSOV and K. K. POPKOV, ibid. **1963**, 1881. (231) W. J. PROBST, Dow Corning, French Pat. 1225229 [U.S. Prior. 2.6.1958]. (232) S. R. RAFIKOV, K. A. ANDRIANOV, S. A. PAVLOVA, I. I. TVERDOKHLEBOVA and SH. V. PICHKHADZE, *Bull. Acad. Sci. USSR, Div. Chem. Sci* (*Engl. transl.*) **1962**, 1496. (233) H. RAUBACH, Dissertation, Humboldt-Universität Berlin 1962. (234) L. A. RAUNER, Dow Corning, U.S. Pat. 2718507 [19.4.1954]. (234a) M. RAUSCH, M. VOGEL and H. ROSENBERG, *J. Org. Chem.* **22**, 900 [1957]. (235) G. REDL and E. G. ROCHOW, *Angew. Chem.* **76**, 650 [1964]. (236) H. H. REYNOLDS, L. A. BIGELOW and C. A. KRAUS, *J. Amer. Chem. Soc.* **51**, 3067 [1929]. (237) E. G. ROCHOW, General Electric, U.S. Pat. 2352974 [30.4.1940]. (238) —, —, U.S. Pat. 2383817 [30.4.1940]. (239) —, *Monatshefte für Chemie* **95**, 750 [1964].

(240) D. A. ROGERS JR. and D. W. LEWIS, Westinghouse, Belg. Pat. 560835 [U.S. Prior. 14.9.1956]. (241) S.D. ROSENBERG and E. G. ROCHOW, *J. Amer. Chem. Soc.* **77**, 2907 [1955]. (242) L. ROSNATI, *Gazz. chim. ital.* **78**, 516 [1948]. (243) J. B. RUST, Montclair Research, U.S. Pat. 2689860 [8.8.1950]. (244) — and H. H. TAKIMOTO, Hughes Aircraft, U.S. Pat. 3061587 [3.5.1960]. (245) —, — and G. C. DENAULT, *J. Org. Chem.* **25**, 2040 [1960]. (246) M. M. SAFFORD and E. L. MINCHER, General Electric, U.S. Pat. 2455880 [13.5.1946]. (247) P. S. SANIN, *J. Gen. Chem. USSR* (*Engl. transl.*) **23**, 1029 [1953]. (248) R. O. SAUER, *J. Amer. Chem. Soc.* **66**, 1707 [1944]. (249) —, General Electric, U.S. Pat. 2582799 [22.1.1948].

(250) — and R. H. HASEK, *J. Amer. Chem. Soc.* **68**, 241 [1946]. (251) R. L. SCHAAF, P. T. KAN, C. T. LENK and E. P. DECK, *J. Org. Chem.* **25**, 1986 [1960]; cf. *J. Org. Chem.* **26**, 1790 and 4038 [1961]; U.S. Pat. 3010982 [7.4.1960]. (252) H. SCHMIDBAUR, *Chem. Ber.* **96**, 2696 [1963]. (253) —, *Angew. Chem.* **77**, 206 [1965]. (254) — and H. HUSSEK, ibid. **75**, 575 [1963]. (255) —, —, *J. Organomet. Chem.* **1**, 244 and 257 [1964]. (256) — and M. SCHMIDT, *J. Amer. Chem. Soc.* **83**, 2963 [1961]. (257) —, —, *Angew. Chem.* **74**, 328 [1962]. (258) —, —, *J. Amer. Chem. Soc.* **84**, 3600 [1962]; *Chem. Ber.* **97**, 836 [1964]. (259) —, —, *Angew. Chem.* **70**, 704 [1958].

(260) W. C. SCHUMB and C. M. SAFFER JR., *J. Amer. Chem. Soc.* **61**, 363 [1939]. (261) R. SCHWARZ and K. SCHOELLER, *Chem. Ber.* **91**, 2103 [1958]. (262) D. SEYFERTH and D. L. ALLESTON, *Inorg. Chem.* **2**, 418 [1963]. (263) D. J. SHINKIN et al. *Plaste und Kautschuk* **10**, 387 [1963]. (264) L. M. SHORR, Dow Corning, French Pat. 1146882 [U.S. Prior. 7.3.1955]. (265) W. SIMMLER, Farbenfabriken Bayer, German Appl. (West German) 1104705 [6.7.1959]. (266) —, —, Belg. Pat. 603832 [German Prior. (West German) 24.5.1960]. (267) —, —, unpublished. (268) —, H. NIEDERPRÜM, W. NOLL and K. SCHNURRBUSCH, —, Brit. Pat. 1038348 [German Prior. 22.7.1963]. (269) —, — and H. SATTLEGGER, *Chem. Ber.* **99**, 1368 [1966].

(270) D. D. SMITH, Dow Corning, U.S. Pat. 2884433 [7.3.1957]. (271) D. P. SMITH, —, French Pat. 1198034 [U.S. Prior. 7.3.1957]. (272) L. H. SOMMER, —, U.S. Pat. 2507551 [19.3.1949]. (273) —, —, U.S. Pat. 3046291 [5.5.1958]. (274) —, J. M. GOLDBERG, J. GOLD and F. C. WHITMORE, *J. Amer. Chem. Soc.* **69**, 980 [1947]. (275) —, F. A. MITCH and G. M. GOLDBERG, ibid. **71**, 2746 [1949]. (276) —, R. M. MURCH and F. A. MITCH, ibid. **76**, 1619 [1954]. (277) J. L. SPEIER JR., Dow Corning, German Appl. (West German) 1105183 [U.S. Prior. 17.12.1953]. (278) —, —, U.S. Pat. 2527590 [Brit. Prior. 26.8.1957]. (279) M. M. SPRUNG, General Electric, French Pat. 1322812 [U.S. Prior. 12.5.1960].

(280) H. STAMM, M. SANDER and G. CRAMER, Th. Goldschmidt, German Pat. (West German) 1133133 [16.1.1958]. (281) I. K. STAVITSKII and S. N. BORISOV, *Vysokomol. Soed.* **1**, 1496 [1959]. (282) S. STERMAN, Union Carbide, Brit. Pat. 797598 [U.S. Prior. 25.5.1954]. (283) W. SUNDERMEYER, *Z. anorg. allg. Chem.* **310**, 50 [1961]. (284) M. SVEDA, DuPont, U.S. Pat. 2561429 [28.10.1946]. (285) —, —, U.S. Pat. 2562000 [28.10.1946]. (286) H. H. TAKIMOTO and J. B. RUST, Hughes Aircraft, U.S. Pat. 3153000 [28.9.1960]. (287) —, —, *J. Org. Chem.* **26**, 2467 [1961]. (288) W. S. TATLOCK and E. G. ROCHOW, ibid. **17**, 1555 [1952]. (289) C. THIES and I. B. KINSINGER, *Inorg. Chem.* **3**, 551 [1964].

(290) I. M. THOMAS, *Canad. J. Chem.* **39**, 1386 [1961]. (291) A. V. TOPCHIEV, N. S. NAMETKIN and A. A., SHCHERBAKOVA, *Doklady Akad. Nauk. SSSR* **86**, 559 [1952]. (292) —, — and O. P. SVLOVOVA, ibid. **86**, 965 [1952]. (293) —, — and V. I. ZETKIN, ibid. **99**, 551 [1954]. (294) C. E. TRAUTMAN, Gulf Research, U.S. Pat. 2488449 [11.9.1946]. (295) —, —, U.S. Pat. 2515024 [11.9.1946]. (296) Union Carbide, Belg. Pat. 637740 [U.S. Prior. 23.9.1963]. (297) R. W. UPSON, DuPont, U.S. Pat. 2517945 [19.3.1947]. (298) H. A. VAUGHN JR., General Electric, French Pat. 1321457 [U.S. Prior. 23.1.1961]. (299) V. M. VDOVIN and A. D. PETROV, *J. Gen. Chem. USSR* (*Engl. transl.*) **30**, 852 [1960].

(300) —, K. S. PUSHCHEVAYA and A. D. PETROV, *Proc. Acad. Sci. USSR* (*Engl. transl.*) **141**, 1203 [1961]. (301) M. G. VORONKOV and V. N. ZGONNIK, *J. Gen, Chem. USSR* (*Engl. transl.*) **27**, 1550 [1957]. (302) —, —, ibid. **27**, 1557 [1957]. (303) Wacker-Chemie, Austrian Pat. 176020 [German Prior. (West German) 20.6.1951]. (304) U. WANNAGAT and H. NIEDERPRÜM, *Angew. Chem.* **71**, 574 [1959]. (305) R. WEST, *J. Amer. Chem. Soc.* **76**, 6013 [1954]. (306) — and E. G. ROCHOW, *J. Org. Chem.* **18**, 1739 [1953]. (307) C. L. WHIPPLE, Dow Corning, Belg. Pat. 603552 [U.S. Prior. 13.5.1960]. (308) E. WIBERG and K. H. KANZLER, German Pat. (West German) 937557 [19.6.1953]. (309) — and U. KRÜERKE, *Z. Naturforsch.* **8b**, 608 [1953].

(310) —, —, ibid. **8b**, 609 [1953]. (311) M. WICK, *Kunststoffe* **50**, 433 [1960]. (312) M. YOKOI, T. NOMURA and K. YAMASAKI, *J. Amer. Chem. Soc.* **77**, 4484 [1955]. (313) V. A. ZEITLER and C. A. BROWN, ibid. **79**, 4616 [1957]. (314) —, —, ibid. **79**, 4618 [1957]. (315) A. A. ZHDANOV and K. A. ANDRIANOV, *J. Polymer Sci.* **55**, 89 [1961]. (316) World's Fair Brussels 1959, Russian Pavilion.

CHAPTER 8

Production of Technical Silicone Products from Polyorganosiloxanes

Only a few of the many siloxane polymers arising from the processes described in the previous chapters can be used directly as technical products; these are the liquid polymers or special monomeric products such as those used for the preparation of impregnating agents or protective films. They include, among others, sodium methylsiliconate and dimethyldiethoxysilane.

Another large, if not actually the largest, part of the polymerizates must however undergo further treatment and be "formulated" to give a technical product ready for use. This applies especially to the siloxane polymers from which rubber vulcanizates, resins, paints and varnishes, and greases or emulsions are obtained in their various forms. If silicone oils, silicone rubber gums (i.e., the vulcanizable raw material for the preparation of rubber articles), silicone resins, and saltlike metal silanolates are regarded as the "basic materials" of silicone chemistry, the evolution of the technical products from the starting materials—the organochlorosilanes—via these basic materials can be represented as a family tree as on p. 387.

Some important process steps leading from the basic materials to the technical products will be outlined below.

Since the problem is to develop products having particular properties, it cannot be solved without a discussion of the complex of properties concerned. The reader should therefore consider this and the following chapter as closely interconnected and should evaluate them accordingly.

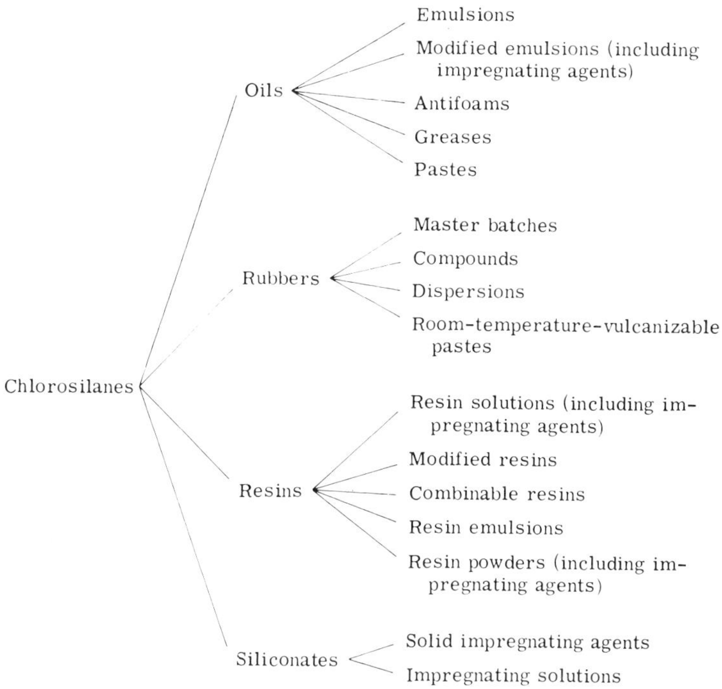

8.1 Technology of Silicone Rubber †

On the basis of the terminology customary in the rubber industry, the following terms have become accepted in the technology of silicone rubber‡:

1. *Silicone rubber gum*, *silicone gum*, or *gum* denotes high-molecular-weight linear polydiorganosiloxanes that can be converted from the highly viscous plastic state into the predominantly elastic state by cross-linking. The term "silicone elastomer" is also found.
2. *Silicone rubber* means, in general, the product prepared from the silicone gum by cross-linking, i.e., the vulcanizate. However, this term is also used as a comprehensive term for 1 and 2.

Like ordinary organic types of rubber, silicone rubber must be cross-linked. This cross-linking, for which the terms "vulcanization" and "curing" are customary in technical terminology, can be achieved in various ways with silicone rubber. The most important technical methods at the present time are

* Page number of chapter references.

† Section 8.1 was written with the cooperation of W. Kniege, Leverkusen, West Germany.

‡ Compare British Standard 3558: 1962 *Glossary of Terms used in the Rubber Industry*, British Standards Institution, 2 Park St., London, England.

cross-linking with organic peroxides at elevated temperatures, and cross-linking with organometallic and organic silicon compounds at room temperature or at only slightly elevated temperatures.

8.1.1 Types of Silicone Rubber Gum

Depending on their chemical and physical structure, it is possible to produce silicone rubbers which have similar basic properties (such as, for example, heat-stability, low-temperature flexibility, electrical insulation properties and abhesive properties) but which differ from one another in certain special characteristics.

The basic skeleton of the types of silicone rubber now on the market is a polydimethylsiloxane chain, generally modified by special substituents. The heat-vulcanizable types are characterized by extremely high molecular weights (300,000–700,000), while the room-temperature-vulcanizable types, since they must generally be suitable for dipping and spreading processes, are adjusted to a lower molecular weight (10,000–100,000) and to a lower viscosity. Since the mechanical properties of a vulcanizate are affected by the molecular weight of the gum—the strength improves with increasing molecular weight—the heat-vulcanizable types of silicone rubber have, at least in most cases, better strengths than room-temperature-vulcanizable types for this reason alone. The optimum molecular weight has been given as 600,000 (*102*).

The molecular structure may also affect other physical properties of the heat-vulcanizable silicone rubbers. Polydimethylsiloxanes having a certain amount of branching exhibit improved low-temperature flexibility. If polydimethylsiloxanes with different proportions of the chain links denoted by (I) and (II) in the formula below are made, at a ratio of I to II equal to 1:10 there is a fall in the stiffening temperature from -76°C for I-free polymers to -116°C. A minimum of -118°C is reached with 22% of I units, while higher proportions of I units lead again to decreased low-temperature flexibility (*214*).

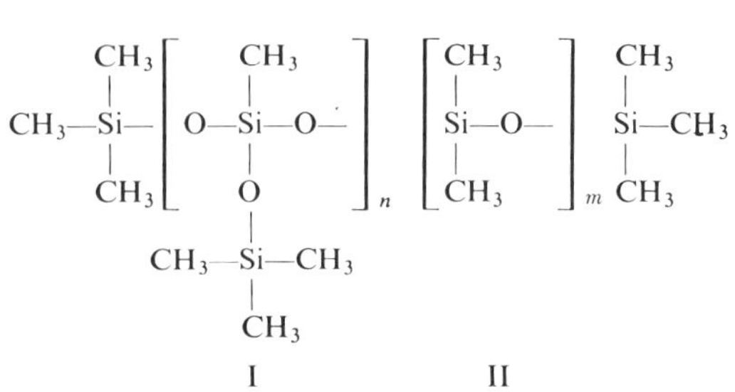

I II

If a proportion of the methyl groups in polydimethylsiloxanes is replaced by phenyl groups, polymers characterized by an improved low-temperature behavior are again obtained. Vulcanizates from polydimethylsiloxanes lose their rubberlike elasticity and become brittle at about −50°C. If the dimethylsiloxane units are replaced by increasing amounts of methylphenylsiloxane units, the stiffening temperature first falls, reaching a minimum of about −115°C at a content of 7.5 mole-% $[CH_3(C_6H_5)SiO]$ units, but it rises again with increasing degree of substitution (*186*) (Fig. 61). At the same time, the tendency toward crystallization changes. While polydimethylsiloxane rubber tends to undergo rapid crystallization without supercooling at −50°C, the addition of 7.5 to 15 mole-% of $[CH_3(C_6H_5)SiO]$ units inhibits crystallization.

However, the introduction of phenyl groups into the molecule decreases the vulcanization yield, as shown particularly by the reduced elasticity, higher tensile strength, and higher tear resistance of the vulcanizates. It may also reduce their compression set and their resistance to swelling in polar or aromatic and naphthenic solvents. The disadvantages attached to phenyl groups can be compensated to a certain extent by suitable means, such as the incorporation of increased amounts of vinyl groups into the polysiloxane molecule. The property of phenyl groups of interfering with cross-linking is used when vulcanizates having an optimum resistance to high-energy radiation are required. Silicone rubbers with high phenyl-group contents are particularly resistant to β- and γ-radiation.

Different properties are obtained according to whether the phenyl groups are incorporated in the molecule in the form of diphenylsiloxy or methylphenylsiloxy units. For a given amount of phenyl groups in the polymerizates, a higher rate of cross-linking is achieved with $[CH_3(C_6H_5)SiO]$ units than with $[(C_6H_5)_2SiO]$ units (*26*).

Other substituents, such as naphthyl, benzyl, and phenylethyl groups may also bring about an improvement in the low-temperature flexibility. In the case of polymethylsiloxanes containing 2.5 mole-% of such substituents, their effectiveness falls in the following sequence: α-naphthyl > benzyl > β-phenylethyl > phenyl > α-phenylethyl (*25*).

The main disadvantage of the classical silicone rubber gum made from nonmodified polydimethylsiloxanes is its marked tendency to swell in many

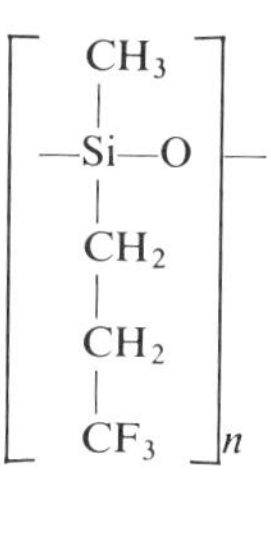

III

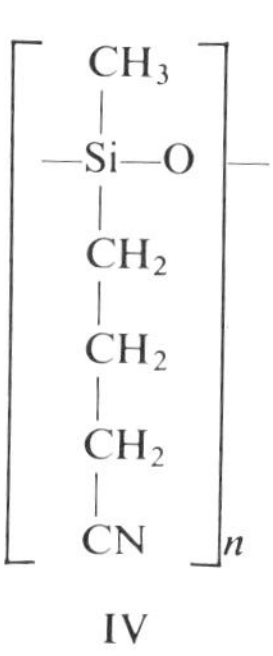

IV

organic solvents such as gasoline and benzene. This defect has been successfully overcome by modifying the structure of the molecule. At the present time, two basic types giving vulcanizates substantially more resistant to swelling are known. These are the so-called fluorosilicone and nitrile-silicone rubbers, the former having acquired industrial importance. Fluorosilicone rubber contains, in addition to methyl groups, γ-trifluoropropyl groups linked directly to silicon (III) (*178*) (cf. Section 4.4), while nitrile-silicone rubber is derived from dimethylsilicone rubber by the partial replacement of CH_3 groups by β-cyanoethyl or γ-cyanopropyl groups (IV) (*259*) (cf. Section 4.11).

Both types give vulcanizates with a good resistance to swelling in common technical lubricants and fuels (*108*, *205*). However, their heat stability is somewhat inferior to that of the pure polydimethylsiloxanes. Fluorosilicone rubber decomposes above 300°C with the liberation of toxic vapors (*142*).

The cross-linking of silicone rubber by means of organic peroxides at elevated temperatures takes place comparatively slowly, and first gives vulcanizates with a low degree of vulcanization. Improvements can be achieved in both respects when part of the methyl groups of the polysiloxanes is replaced by unsaturated organic groups (*57*, *92*, *124*).

In general, about 0.02 to 0.5 mole-% of methylvinylsiloxane units are copolymerized into the rubber molecule, so that in the polymerizate there is one vinyl group for every 500 to 10,000 other organic groups. Although the content of unsaturated groups is thus very low, compounds containing such polymerizates have a higher rate of vulcanization than nonreactive formulations. The vulcanizates are characterized by a substantially higher degree of vulcanization, which is shown, in particular, by increased elasticity and lower compression set (*low compression set gum*). The stability of the silicone rubber in sealed systems (i.e., with the exclusion of atmospheric oxygen) is also markedly improved.

Since, as shown in Section 5.4.3, in the process of cross-linking with peroxides, alkenyl groups react with alkyl groups and form cross-linking sites, suitable mixtures of vinyl-containing and vinyl-free polysiloxanes have also been used industrially. Numerous modifications can be achieved simply by mixing two different types of polymerizates (*56*).

Polyethylmethylsiloxanes have recently come into somewhat greater prominence, since they are said to be distinguished from the polydimethylsiloxanes by the fact that at an appropriately high content of diethylsiloxane units (75 mole-%) they give vulcanizates with the lowest brittle temperature so far found, namely −136°C. Their rate of crystallization is also lower. The presence of ethyl groups has a similar effect to that of vinyl groups: the degree of cross-linking is increased, which reduces the peroxide requirements, and the elasticity and heat stability in closed systems are improved. However, because of the greater susceptibility of ethyl groups to oxidation as compared with methyl groups, their heat stability in air at elevated temperatures is impaired (*27*).

Generally, silicone rubber may contain larger or smaller amounts of low molecular-weight components whose volatilization at the curing temperatures (200°C or more) can cause the vulcanizates to shrink by a relatively large amount. However, the rubber can be freed from these volatile constituents by suitable methods after polymerization of the gum, giving products whose shrinkage is small and of about the same order of magnitude as that of organic types of rubber (*80*). Products treated in this way are called "low-shrinkage rubbers."

Other special properties can be imparted to the silicone rubber by incorporating heteroatoms into the molecular chains. For example, the incorporation of boric acid gives products which, even in the prevulcanized state, weld firmly to one another (*81*, *258*) ("fusible silicone rubber"; cf. Section 7.1.1). Boron-containing polymers are suitable for the manufacture of heat-resistant adhesives (*215*). The incorporation of $BO_{3/2}$ groups into siloxane chains leads to the so-called "bouncing putty" (see Section 7.1.1).

The properties of silicone rubber can be further modified by the incorporation of organosilphenylene units into the chain. Polymerizates of the type

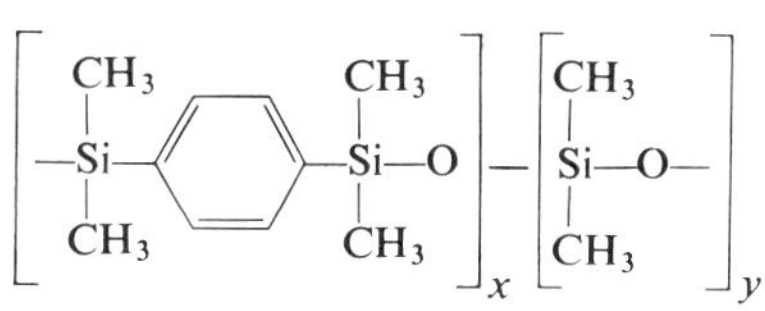

have a higher degree of crystallinity and improve not only the tensile strength but also the radiation and heat resistance of the vulcanizates, though at the expense of the room-temperature resilience (*8*, *135*, *136*, *170*).

The structure of a last group of polymers differs from those mentioned above in that their molecules are not blocked by methyl groups in the form of $[(CH_3)_3SiO_{1/2}]$ units, but still bear reactive groups at the ends. These polymerizates are characterized by the fact that they can be vulcanized by suitable means even at room temperature or only slightly elevated temperatures. In technical terminology, these types of silicone rubber are called RTV (room-temperature-vulcanizing) or LTV (low-temperature-vulcanizing) rubbers. The reactive end groups are generally hydroxyls, but polymerizates with other terminal silicon-functional groups have also found technical application (cf. Section 8.1.2.3).

Other special modifications of the polymer structure are possible. For example, the introduction of hydrogen atoms as substituents of the Si atom gives vulcanizates with particularly good properties of adherence to metals (*107*).

According to ASTM-D 1418-61T, the following abbreviations have been proposed for the various basic types of silicone rubber.

Si: Silicone elastomers having only methyl-containing groups on the polymer chain, such as polydimethylsiloxane.
PSi: Silicone elastomers having both methyl- and phenyl-containing groups on the polymer chain.
VSi: Silicone elastomers having both methyl- and vinyl-containing groups on the polymer chain.
FSi: Silicone elastomers having both methyl- and fluorine-containing groups on the polymer chain.
PVSi: Silicone elastomers having methyl-, phenyl-, and vinyl-containing groups on the polymer chain.

8.1.2 Possibilities of Cross-linking (Vulcanization; Curing)

8.1.2.1 *Heat Vulcanization with Organic Peroxides*

At the present time, the cross-linking of silicone rubber by means of organic peroxides at elevated temperatures (see Section 5.4.3.1) is the most important method of vulcanization (*80*). The peroxides are dispersed homogeneously in the silicone rubber compound at temperatures at which they do not decompose. The compounds obtained in this way are introduced into molds and are subsequently exposed to temperatures above 100°C, whereupon cross-linking takes place. Besides the nature of the peroxide used, its amount also has an essential influence on the processing conditions of the compounds and on the properties of the vulcanizates. Certain peroxides can be used for vulcanization without the application of pressure to obtain nonporous vulcanizates, while others must always be used under pressure (vulcanizing presses, steam, compressed air) to avoid porosity in the vulcanizates. The various peroxides differ from one another also by the temperature at which their decomposition (and therefore the vulcanization process) sets in, and by their rate of decomposition at a given temperature. In particular, the following are of technical importance: diaroyl, dialkyl, diaralkyl, alkyl aralkyl, and alkyl aroyl peroxides. The division of the peroxides into two classes, recommended by certain authors, namely, those that are capable of vulcanizing both saturated and unsaturated polysiloxanes and those which are capable only of polymerizing unsaturated polysiloxanes, is unsound, since it does not accord with practical experience.

Diaroyl peroxides, for example, dibenzoyl peroxide, di-*p*-chlorobenzoyl peroxide, and bis-2,4-dichlorobenzoyl peroxide, are the three classical cross-linking agents for silicone rubber. Of these substances, bis-2,4-dichlorobenzoyl peroxide has the highest rate of decomposition and the lowest decomposition temperature; in this series, di-*p*-chlorobenzoyl peroxide has the lowest rate of decomposition and the highest decomposition temperature. With respect to these properties, dibenzoyl peroxide is between the other two diaroyl peroxides but is closer to di-*p*-chlorobenzoyl peroxide. The low decomposition temperature of bis-2,4-dichlorobenzoyl peroxide and the fact that its decomposition gives rise only to products with low vapor pressures enable this substance to be used for vulcanization without applying pressure. Dibenzoyl peroxide and di-*p*-chlorobenzoyl peroxide are now unimportant, since, in addition to possessing unsatisfactory scorch behavior, they markedly impair the heat stability of the silicone rubber in closed systems and, moreover, may have an adverse effect on the compression set in formulations with diatomaceous earths as fillers.

Of this class of peroxides, therefore, only bis-2,4-dichlorobenzoyl peroxide is still used to any extent for vulcanization without pressure at temperatures above 200°C, and, for the reasons mentioned, its concentration must be kept as low as possible. The latter requirement frequently leads to the use of mixtures of peroxides: This generally enables the amount of diaroyl peroxide to be reduced from 0.1 to 0.2%, the remainder being replaced by another peroxide. Diaroyl peroxides cannot be used in formulations containing carbon black.

Dialkyl peroxides are much less liable to cause scorching than the products mentioned above and also require higher curing temperatures (140 to 180°C). In any case, curing must be carried out under pressure to avoid porosity of the vulcanizates. The members of this class of peroxides that are of the greatest interest from the technical point of view are di-*t*-butyl peroxide and 2,5-dimethyl-2,5-di-(*t*-butylperoxy)-hexane. The latter can, of course, also be included in the class of the alkylperoxyalkanes. Di-*t*-butyl peroxide has the disadvantage of a high vapor pressure and, therefore, of high volatility. Attempts have been made to circumvent this by absorption of the peroxide on a zeolitic molecular sieve (*239*). Since a maximum of only 15% of peroxide can be taken up, undesirably large amounts of zeolite must be incorporated into the silicone rubber. More interesting from the technical point of view is 2,5-dimethyl-2,5-di-(*t*-butylperoxy)-hexane, which has much the same basic properties, but is free from the disadvantage of high volatility. 2,5-Dimethyl-2,5-di-(*t*-butylperoxy)-hex-3-yne has a similar action to 2,5-dimethyl-2,5-di-(*t*-butylperoxy)-hexane but, with otherwise similar properties, it has a much smaller tendency to scorching and a lower rate of decomposition. The hexyne derivative is therefore of interest for the manufacture of thick-walled or very bulky articles.

Diaralkyl peroxides, such as dicumyl peroxide and its halogenated or organo-substituted derivatives, have practically the same properties as dialkyl peroxides. However, they have higher melting points and lower vapor pressures. Since no decomposition products containing carboxyl groups are produced when these peroxides are used, they give vulcanizates with better heat stability in closed systems. They are particularly suitable, like the dialkyl peroxides, for the manufacture of large or metal-bonded articles, and they can be used even in the presence of carbon black. A disadvantage, as compared with the dialkyl peroxides, is their relatively high acid sensitivity, so that when these peroxides are used it is particularly important carefully to check the pH of all the compounding constituents used.

Alkyl aralkyl peroxides, for example, *t*-butyl cumyl peroxide and 1,4-bis-(*t*-butylperoxyisopropyl)-benzene, have an effect intermediate between those of the dialkyl and the diaralkyl peroxides.

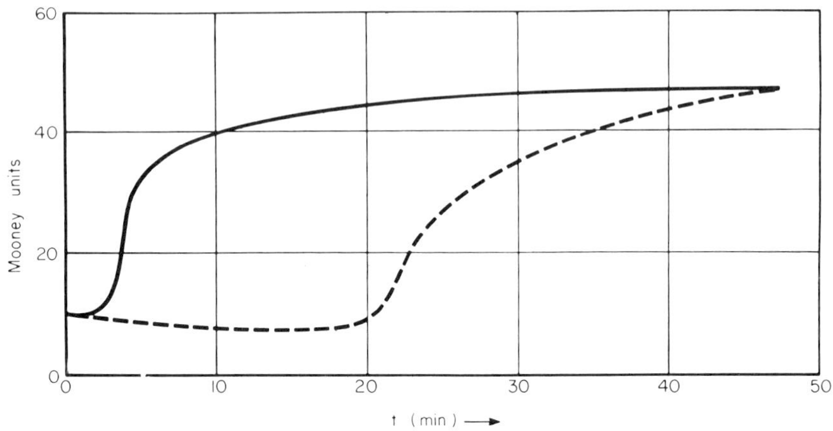

FIG. 25. Mooney scorch of a silicone gum premix at 120°C (Silopren SVA Batch II, Farbenfabriken Bayer AG) by means of peroxides: (——) 0.6% of dibenzoyl peroxide and (---) 0.6% of dicumyl peroxide.

Alkyl aroyl and *alkyl acyl peroxides*, of which *t*-butyl perbenzoate, *t*-butyl peracetate, and *t*-butyl peroctoate have been most frequently used, have the same disadvantages as diaroyl peroxides and are therefore employed only in special cases. *t*-Butyl perbenzoate is still used in the manufacture of porous silicone rubber, while *t*-butyl peroctoate imparts extreme transparency to silicone rubber.

The use of other peroxides has occasionally been recommended. The following, in particular, have been mentioned: peroxysiloxanes (*20*, *105*, *182*) and peroxycarbonates (*105*), such as *t*-butylperoxyisopropyl carbonate, which is roughly comparable, in its technical applications, to bis-2,4-dichlorobenzoyl peroxide without, however, possessing an adverse effect on the depolymerization of silicone rubber.

Figure 25 shows to what extent the vulcanization process can be affected by various peroxides. Figure 26 shows, similarly, the time of decomposition of various types of peroxides in a silicone rubber formulation at 140°C.

As with organic elastomers, the incorporation of certain metal oxides, particularly zinc oxide, and of alkenyl compounds with a functionality of more than 2 (e.g., triallyl cyanurate and triallyl phosphate) can improve the degree of cross-linking in the vulcanization of silicone rubber with peroxides (*14*).

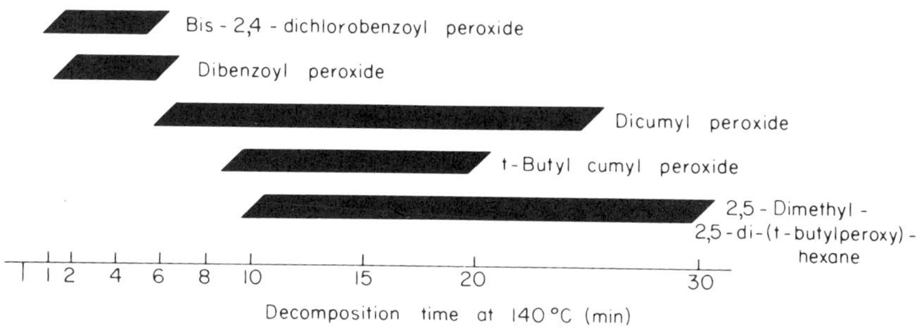

FIG. 26. Decomposition time of various peroxides in a silicone rubber compound at 140°C (Silopren SVA Batch II, Farbenfabriken Bayer AG), with 0.6% by weight of peroxide in each case.

8.1.2.2 *Heat Vulcanization with Sulfur*

Polysiloxanes with relatively large amounts (2 to 5%) of unsaturated groups (e.g., vinyl groups) in the side chains are also capable of vulcanization with sulfur (*91*, *160*, *187*). The curing agents used are either sulfur and the accelerators customary in the rubber industry or products of the tetramethylthiuram disulfide type. The vulcanization is carried out at about 150°C. In general, the vulcanizates have lower heat stabilities than the products vulcanized with peroxides, so that the process has not so far acquired industrial importance. Sulfur enables organic elastomers to be covulcanized with silicone rubber (*160*), but such blends of silicone rubber with, for example, styrene-butadiene, acrylonitrile-butadiene, or polybutadiene rubbers have had no practical success.

8.1.2.3 *Room-Temperature Vulcanization*

8.1.2.3.1 Multicomponent Systems. Linear polymers, especially α,ω-difunctional polydiorganosiloxanes, can be vulcanized even at room temperature by using polyfunctional silanes and siloxanes as cross-linking agents together with certain catalysts. For example, cross-linking may take place as follows for polydimethylsiloxane-α,ω-diols and esters of orthosilicic (I) or polysilicic (II) acids, or for alkyltrialkoxysilanes (III):

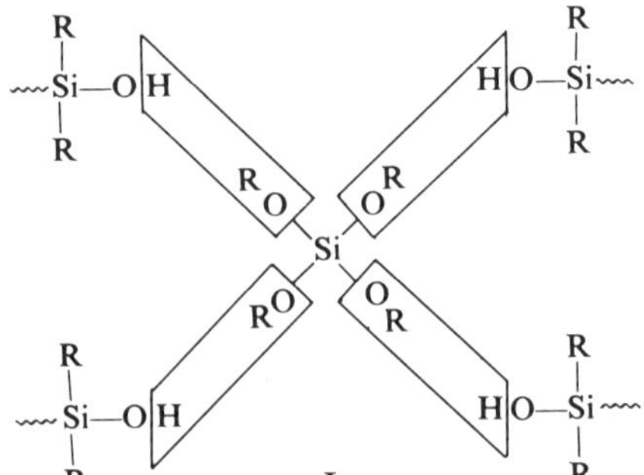

I

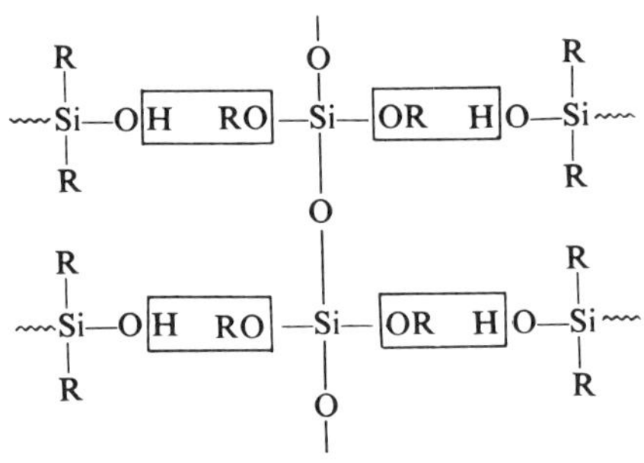

II

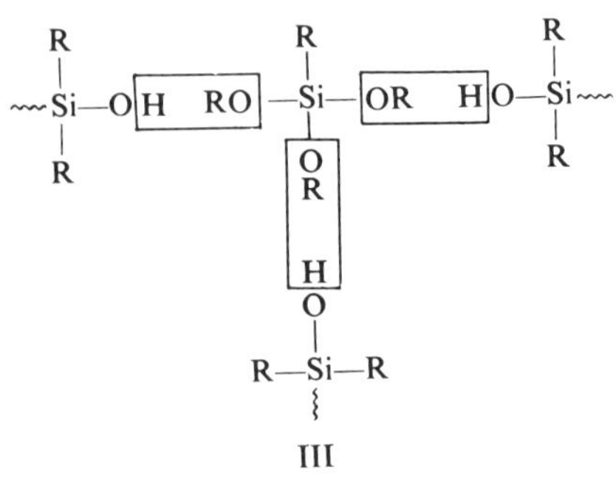

III

All these reactions are therefore catalyzed silanol-alkoxysilane condensations (Section 5.2.12) and take place with the elimination of alcohol. This is the actual reaction mechanism frequently used for room-temperature vulcanization. But there are also other possibilities as will be shown below.

In principle, methods of room-temperature vulcanization have been known for a relatively long time (*53*, *94*, *99*, *101*, *185*) but have only recently been brought to technical maturity (*19*, *44*, *160*, *264*).

The preferred *cross-linking agents* are methoxy- and ethoxysilanes or polysiloxanes, because of their reactivity. Alkoxysilanes simultaneously containing hydrogen bound to the silicon atom, such as methyldiethoxysilane or triethoxysilane are particularly effective (*264*). Suitable *catalysts* are amines and carboxylic acid salts of many metals, such as Pb, Zn, Zr, Sb, Fe, Cd, Sn, Ba, Ca, and Mn, particularly the naphthenates, octoates, hexoates, laurates, and acetates. Tin(II) octoate and dibutyltin dilaurate have proved particularly satisfactory. The addition of a chloroacetic acid to these tin salts considerably accelerates the vulcanization process (*42*).

Although, therefore, at least *three* components have to react, for practical purposes these are usually premixed in such a way that the processer is supplied with only *two* components, which must be compounded shortly before use. Either the siloxanol is mixed with the cross-linking agent or with the catalyst, or the cross-linking agent is mixed with the catalyst. Particularly active combinations of cross-linking agent and catalyst with good storage properties can be obtained from carboxylic acid salts and silicic acid esters by prolonged heating at 50 to 200°C (*119*, *207*).

The cure rate can be varied within wide limits by appropriately selecting the OH content of the siloxanol and the amounts of the cross-linking agent and the catalyst. It can also be affected to a considerable extent by the temperature, the moisture content of the mixture (*159*), and by its pH. In absolutely anhydrous systems, the reaction does not occur, or occurs only very slowly. In the acid pH region, the vulcanization may be retarded or even completely suppressed. Consequently, the addition of acetic acid or acetic anhydride is suitable for stabilization (*67*).

In addition to room-temperature-vulcanizing systems, the peroxides used for heat vulcanization can also be incorporated in the mixtures. This gives vulcanizates with an improved heat stability (*141*, *167*).

Again, siloxanes with Si—H bonds are suitable as cross-linking agents in place of the alkoxy compounds of silicon. The former react with the polysiloxanols, liberating hydrogen. Here again, metal salts of carboxylic acids are used as catalysts (*140*, *156*).

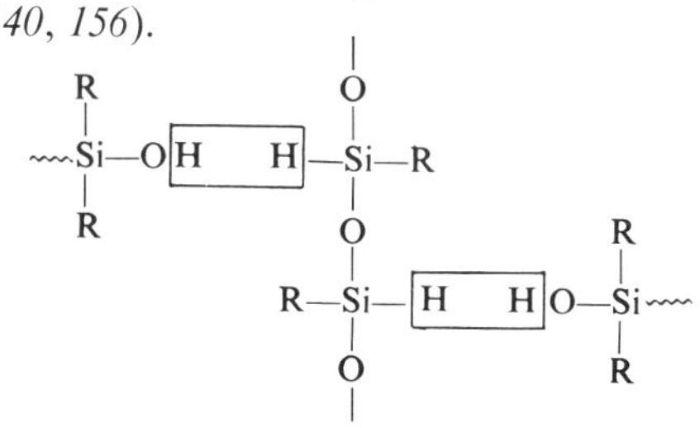

The hydrogen acts as a blowing agent, so that it is possible to obtain foam rubbers with such vulcanizing systems (*33*). However, the appearance of a porous structure can be prevented if compounds reducible by nascent hydrogen such as peroxides, sulfenamides, and dithiocarbamates are added to the mixes before vulcanization (*157*).

Finally, mixtures of polydimethylsiloxanes with unsaturated silicic acid esters also vulcanize rapidly to elastomers at room temperature when amines are added as catalysts (*29*).

While the systems mentioned above react completely even at room temperature (*RTV types*), similar combinations can be adjusted to vulcanize at only slightly elevated temperatures (*LTV types*). This can be achieved not only by the choice of suitable combinations of cross-linking agent and catalyst but also by absorbing the catalyst in a zeolite ("molecular sieve") and adding it to the mix in this form. It is then inactive at room temperature and is activated by heating (*158*). Systems of polyorganosiloxanes containing alkenyl groups and those containing Si—H bonds also belong to the class of LTV rather than RTV mixes; in the presence of platinum or platinum compounds they are vulcanized by heating to 50–80°C (*223*).

Low-temperature-vulcanizing types have the advantage of longer pot life at room temperature in the catalyzed state than RTV types and possess good cure rates at only moderately elevated temperatures.

The technology of the RTV and LTV types differs considerably from that of the heat-vulcanizable types of silicone rubber. It is generally possible to stir in the cross-linking agent and disperse it homogeneously without scorching only if the silicone rubber compounds have a relatively low viscosity. For this reason, the main field of application of RTV and LTV types of rubber is where pourable or spreadable formulations can be used or are required. Although the low vulcanization temperatures are typical for products of the class described here, it has occasionally proved desirable in practice to carry out a heat treatment even in this case. Where the finished articles are exposed to high temperatures in use, RTV or LTV types of rubber should be slowly heated to these temperatures. This applies particularly to bulky articles which otherwise readily tend to undergo depolymerization in their interior.

8.1.2.3.2 Single-Component Systems. Single-component systems based on room-temperature-vulcanizable silicone rubber are compounds which, in contrast to the multicomponent systems described above, already contain all the components necessary for vulcanization, these reacting only by contact with the atmosphere and, in particular, with the atmospheric water vapor. The polysiloxane or the cross-linking agent must therefore contain easily hydrolyzable groups.

The first single-component systems of this type consisted of polydimethylsiloxane-α,ω-diols to which methyltriacetoxysilane was added as the cross-linking agent (*199*), or they contained α,ω-bis-(methyldiacetoxysiloxy)-polydimethylsiloxane (*34*):

$$(CH_3COO)_2\overset{\displaystyle R}{\overset{|}{Si}}-O\left[-\underset{\displaystyle R}{\underset{|}{\overset{\displaystyle R}{\overset{|}{Si}}}}-O\right]_n-\overset{\displaystyle R}{\overset{|}{Si}}(OOCCH_3)_2$$

Mixtures of these compounds with suitable fillers can be stored in the absence of moisture; in contact with atmospheric humidity, vulcanization takes place by hydrolytic cleavage of the acetoxy groups from silicon atoms, a condensation of silanol and acetoxy groups taking place with continuous displacement of the equilibrium by the evaporation of the acetic acid formed (Section 5.2.14). Vulcanization begins at the surface and proceeds toward the interior as water vapor diffuses in and acetic acid diffuses out; consequently, thick layers are slow to vulcanize completely. If, however, magnesium oxide is added to the mixes, this acts as an acid-binding agent and makes the vulcanization of even relatively thick layers possible (*201*). Such products give a silicone rubber which adheres well to silicone rubber itself and to metals, wood, and many other materials, and has therefore become of interest as an adhesive and sealing material in many fields of application.

There are several other methods of preparing such single-component systems. It is possible to combine polydimethylsiloxanes bearing terminal diacetoxy groups with alkoxysilanes, metal catalysts, and/or amines (*161*), or to produce systems consisting of mixtures of polydimethylsiloxane-α,ω-diols, a silicic acid ester, and an organic titanate, particularly tetrabutyl titanate (*43*). It is also possible to make use of the sensitivity to hydrolysis of the Si—O—N═C grouping, which can be obtained by the condensation of Si—Cl or Si—OH bonds with ketoximes, especially acetone oxime, and to attach this group to the ends of a polydimethylsiloxane chain (*233*):

$$[(CH_3)_2C{=}NO]_2\overset{\displaystyle CH_3}{\overset{|}{Si}}-O\left[-\underset{\displaystyle CH_3}{\underset{|}{\overset{\displaystyle CH_3}{\overset{|}{Si}}}}-O\right]_n-\overset{\displaystyle CH_3}{\overset{|}{Si}}[ON{=}C(CH_3)_2]_2$$

Finally, the increased sensitivity to hydrolysis of the Si—N bond in the neighborhood of a carbonyl group enables *N*-silyl-substituted carboxylic amides such as $(CH_3)Si[NCH_3COCH_3]_3$ to be used as cross-linking agents (*74*).

8.1.2.4 *Cross-linking Using High-Energy Radiation*

Vulcanization can also be carried out by means of high-energy radiation. The sources of energy most frequently used industrially are radioactive cobalt (^{60}Co), Van de Graaff generators, and resonance transformers (*66*). However, cascades and linear accelerators or betatrons are also basically suitable

(*145*). Industrially, Van de Graaff generators are generally preferred to a ^{60}Co source, since they provide an accurately focused and uniform radiation. In contrast, a ^{60}Co source gives a diffuse radiation whose energy of 60 to 140 W falls during its use in irradiation. No vulcanization auxiliaries of any kind and, in general, no elevated temperatures are necessary for the radiation cross-linking of silicone rubber.

As cryoscopic molecular-weight determinations on polydimethylsiloxanes irradiated with electrons have shown, 1.44 cross-linking sites appear per ion pair; 22 eV is necessary per cross-linking site (*35*). The average amount of radiation required for cross-linking is about 10 Mrep. However, the necessary dose may deviate considerably from this figure for other types of silicone rubber. In general, polymerizates not containing vinyl groups require a somewhat higher dose than those containing vinyl groups. The energy required also becomes considerably greater as the amount of phenyl groups is increased.

The properties of vulcanizates cross-linked by high-energy radiation scarcely differ, in the unaged state, from products cross-linked with peroxides and with the same degree of cross-linking. The most important advantage of a silicone rubber cross-linked by radiation is its better resistance to hydrolytic depolymerization on aging in humid atmospheres at elevated temperatures. The main reason for this is the absence of peroxide decomposition products which could act as depolymerization catalysts (*243*). However, it is possible that high-energy radiation gives a more stable type of cross-linkage (*184*).

If radiation vulcanization is carried out in the presence of polyvalent metal compounds or if halogen-containing polymers are added to the silicone rubber, the cross-linkage stability and the heat resistance of the products are improved still further (*11*).

Vulcanization by high-energy radiation is also affected by the nature of the filler, besides other factors. For example, compounds containing carbon black can be vulcanized more easily than those containing highly disperse silica (*169*).

8.1.3 Manufacture of Vulcanized Silicone Rubber Articles

8.1.3.1 *Heat-Vulcanizable Silicone Rubber*

The processing of heat-vulcanizable silicone rubber corresponds fundamentally to normal rubber technology and is therefore generally performed by the rubber industry.

8.1.3.1.1 Formulation of compounds. Silicone rubber (gum) compounds have simple formulations as compared with those made up from organic elastomers. They consist essentially of *gum*, *fillers*, and *curing agents* (vulcanization auxiliaries). *Pigments* and special additives are used to achieve special effects. The basic formulation is somewhat as follows: 100 parts by weight of silicone rubber gum, 20.0–200 parts by weight of fillers, 0.5–3.0 parts by weight of the curing agent, 0.0–5.0 parts by weight of pigments, and varying amounts of special additives.

The type of gum to be used in any particular case is selected from the materials described in Section 8.1.1, according to the properties required of the raw material or the vulcanizate. In many cases reclaimed silicone rubber can be used to replace part of the silicone rubber gum.

The properties of both the unvulcanized compounds and vulcanizates can be varied within wide limits as functions of the nature and amount of the *fillers* (*36*, *244*). The mechanical properties of unfilled silicone rubber are unsatisfactory. As with other noncrystallizing types of synthetic rubbers, the addition of fillers, preferably reinforcing fillers, is necessary to obtain technically useful vulcanizates.

The highest reinforcing effect is obtained with *fumed silicas*, which have BET surface areas of 150 to 400 m^2/gm. Unvulcanized silicone rubber compounds containing such fillers exclusively or predominantly are generally characterized by a relatively high Mooney viscosity and undergo a pronounced, but reversible, hardening during storage. The vulcanizates are characterized by high tensile strength. Fumed silicas impart only moderate hot-air stability to the vulcanizates; in so-called closed systems, on the other hand, these fillers are second only to products made hydrophobic with organosilanes.

Precipitated hydrated silicas are somewhat less reinforcing than the fumed silicas mentioned above and, in comparable amounts, therefore give compounds with a lower Mooney viscosity, lower green strength, and lower tendency to hardening during storage. The tensile strength of the vulcanizates containing such fillers is lower than for fumed silicas. The hot-air stability of the vulcanizates is substantially improved by precipitated silicas, while their stability in closed systems is definitely impaired. These fillers are also generally less suitable for the production of optimum dielectric properties and for articles of high transparency than the fumed silicas.

Fillers with a weak reinforcing effect (semi-reinforcing fillers) are, in particular, various specially treated *natural products*. Diatomaceous earths (diatomaceous mineral fillers) are the main products used. Because of their low reinforcing effect, it is possible to use large amounts of fillers while retaining good processing properties. As compared with the highly reinforcing fillers, these semi-reinforcing fillers give, at the same Shore hardness, a smaller increase in volume on swelling in oils and better compression set properties.

The examples of Table 70 show how the mechanical properties depend on the type and amount of filler added. All compounds contained a silicone rubber of type VSi; the figures given under "compounding formulation" are parts by weight.

TABLE 70

PROPERTIES OF VULCANIZATES BASED ON METHYLVINYLSILICONE RUBBER

Formulation of the compound				
Silicone rubber[a]	100.0	100.0	100.0	100.0
Highly active silica[b]	40.0	—	—	—
Medium-active silica[c]	—	43.0	—	—
Kieselguhr[d]	—	—	50.0	100.0
Benzoyl peroxide	1.25	1.25	1.25	1.25
Mechanical characteristics after post-cure for 15 hr at 200°C				
Tensile strength (kg/cm^2)	65–75	55–65	25–35	40–50
Elongation at break (%)	250–300	300–350	200–250	150–200
Hardness (Shore *A*)	60 ± 5	55 ± 5	40 ± 5	60 ± 5
Resilience	35–40	30–35	35–40	30–35
Compression set 22 hr/175°C (%)	30–40	20–30	35–45	40–50
Density (gm/cm^3)	~1.17	~1.18	~1.23	~1.40
Shrinkage (%)	~2.5	~2.5	~1.8	~1.5
After hot-air aging (14 days/200°C)				
Tensile strength (kg/cm^2)	55–65	55–65	25–35	40–50
Elongation at break (%)	200–250	250–300	150–200	100–150
Hardness (Shore *A*)	65 ± 5	60 ± 5	45 ± 5	65 ± 5
Electrical properties (measured on plates 1 mm thick)				
Dielectric strength (kV/cm)	200–220	200–220	220–240	250–270
Dielectric loss factor at 800 cps	$\sim 5 \times 10^{-3}$	9×10^{-2}	2×10^{-2}	3×10^{-2}
Rel. dielectric constant at 800 cps	~2.5	4.5	3.0	3.5
Specific resistance (Ω cm, minimum)	5×10^{16}	4×10^{14}	3×10^{14}	2×10^{14}

[a] Silopren RS, Farbenfabriken Bayer, Leverkusen, West Germany.
[b] Aerosil 130, Degussa, Frankfurt/Main, West Germany.
[c] Vulkasil S, Farbenfabriken Bayer, Leverkusen, West Germany.
[d] Celite Super Floss, Johns-Manville Corporation, New York.

Although fillers based on silica form the bulk of those used in silicone rubber, other products are used such as titanium dioxide, iron oxide, calcium carbonate, alumina, zinc oxide, zirconium silicate, barium silicate, and carbon black. The products mentioned in Section 8.1.2 are used as *curing agents*.

Since silicone rubber is colorless, *pigmentation*, in itself, causes no difficulties. However, only pigments having an adequate heat stability and no adverse side effects can be used. Consequently, inorganic pigments such as iron oxide, titanium dioxide, and cadmium pigments are the ones used most commonly. The majority of metal oxides and numerous other metal compounds have, in addition to their pigment action, the property of imparting an improved hot-air stability to silicone rubber. Certain types of titanium dioxide and iron oxide, in particular, improve the stability of the vulcanizates at elevated temperatures. The addition of iron salts of carboxylic acids also has a favorable action in this respect (*235*).

While nonreinforcing fillers give comparatively soft raw compounds which scarcely harden on prolonged storage, mixtures exclusively or predominantly containing reinforcing fillers tend to undergo hardening to an increased extent on storage, although the mechanical properties of the vulcanizates improve with increasing activity of the fillers. For this reason, silicone rubber compounds containing relatively large amounts of highly reinforcing fillers almost always require the addition of special substances, so-called "*structure-control additives*," which have the object of reducing hardening to a minimum.

At the present time it is impossible to make precise statements on the molecular features that a structure-control additive must have to be active in a mix. The only general characteristic of compounds having good properties appears to be that they have a polar structure (*174*). Examples of useful compounds are diphenylsilanediol (*174*) or the reaction product of pinacol with dimethyldichlorosilane (*7*, *79*):

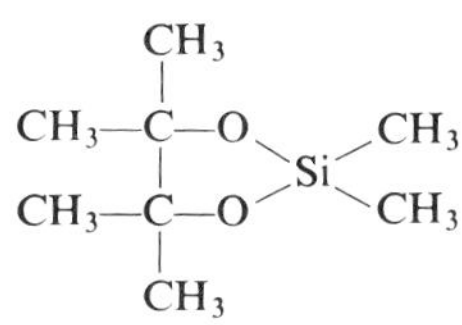

It has been possible to work out certain rules for the relationship between constitution and activity in the case of compounds containing OH groups (*105*). The OH groups should be adjacent and be located as symmetrically as possible within the molecule. The effect of hexanediols decreases in the following sequence of the positions of the OH groups:

$$3,4 > 2,5 > 1,6 > 2,3$$

Tertiary groups are better than secondary or primary groups. For practical use, it is also important as a principle of selection that the dispersants should have a high boiling point so that they cannot volatilize during the high-temperature processing of the rubber compound.

To improve the dispersibility of the fillers in the rubber, they have been treated with organochlorosilanes or organoethoxysilanes after or even during their manufacture (*32*, *37*, *236*). This increases the affinity of the surface of the filler for the polyorganosiloxane and thus makes it more easily wetted by the latter. More highly loaded rubber compounds can be made with fillers pretreated in this way. Instead of treating the fillers with the silicon-functional organosilanes, it is simpler to treat them with cyclic polyorganosiloxanes, particularly octamethylcyclotetrasiloxane (*122*), which can easily be applied in the vapor phase. In this case, there are no by-products (hydrogen halide, alcohol) to be removed during or after the treatment of the filler.

Carbon black is used only in the form of special grades for the manufacture of electrically conducting silicone rubber vulcanizates.

Plasticizers are used in silicone rubber only in rare cases. The majority of organic plasticizers are insufficiently heat stable, so that silicone oils are almost the only plasticizers that can be used. However, because of the lower degree of polymerization of silicone oils as compared with silicone rubber, they reduce its mechanical strength. In general, therefore, instead of adding a silicone oil, the proportion of silicone rubber in the mix is increased. In special cases (for example, for the manufacture of vulcanizates with a high dynamic damping) methylphenylsilicone oils are used as plasticizers.

Porous vulcanizates of the type of sponge or foam rubber can be manufactured from heat-vulcanizable silicone rubber by the incorporation of suitable *blowing agents*, such as *N*,*N'*-dinitrosopentamethylenetetramine, diphenyl oxide 4,4'-disulfonhydrazide, and *N*,*N'*-dimethyl-*N*,*N'*-dinitrosoterephthalamide (*113*, *146*).

The incorporation of polytetrafluoroethylene to improve the *tear resistance* has been described (*203*). However, such additives give compounds with greatly *impaired flow properties* (*51*). Even small amounts (0.2 to 0.5%) of finely subdivided polytetrafluoroethylene considerably improve the calendering properties of silicone rubber.

The use of certain boron compounds, e.g. boron triacetate, as constituents of the compound, like the incorporation of boroxane groups in the molecule (Section 7.1.1), gives the vulcanizates *adhesive properties* with respect to similar material. The incorporation of organoisocyanatosilanes (*123*), of organic silicates (*212*), of certain peroxy compounds such as dipersuccinic acid, *t*-butyl hydroperoxide, di-*t*-butyl diperphthalate, di-*t*-butyl dipersuccinate, or di-*t*-butyl diperadipate (*213*), and of alkyl-*H*-siloxanes (*105*) has been recommended in order to improve the adhesion of silicone rubber to metal surfaces or other materials.

8.1.3.1.2 Stock forms. Silicone gum was originally marketed almost exclusively in the form of ready-for-curing compounds. Later, the gum itself was made available, so that the processer could prepare suitable compounds according to his own formulations. Each of the two possibilities has its advantages and disadvantages.

Ready-for-curing compounds are the simplest to work with, but the possibilities of variation are limited. The shelf life of the compounds may be limited by the peroxides used as curing agents to an extent depending on their nature and amount. The great advantage to the processer is that the compounds can generally be prepared by the silicone rubber manufacturer under better technical conditions than the processer can use.

On the other hand, the possibility of variation is greater with the processing of *unfilled gums*; furthermore, their shelf life is practically unlimited. However, the incorporation of the highly reinforcing fillers, in particular, is often associated with certain technical difficulties.

In order to combine the possibility of variation provided by the gum with simple handling of the compound, master batches consisting essentially of rubber and highly reinforcing fillers have been increasingly marketed for some time. These master batches, also known as reinforced gums or stocks, have good processing properties, permit almost all variations of formulation, and have an excellent shelf life. Ready-for-curing compounds can be prepared simply by incorporating peroxide and any other desired additives.

The most recent development consists of the so-called "nonmilling compounds," which are available for special fields of application, such as, for example, cable sheathing, in the form of direct-feed strips, granulate, or pelletized compounds (*15*). These involve compounds that do not require plastication prior to processing and that can be molded or extruded in a precured state.

For other uses, e.g., for the coating of fabrics, compounds already *dispersed in organic solvents* are available.

8.1.3.1.3 Production and processing of compounds. Silicone rubber compounds are processed on the conventional machines of the rubber industry. Utmost cleanliness is essential for the production and processing of silicone rubber mixes, since contamination may adversely affect the properties of the final products.

Silicone rubber compounds are usually made on conventional rubber mills. In many cases, processing in internal mixers of the dough mixer type presents certain advantages, particularly with respect to uniformity (*121*). The production and processing temperatures should be below 50°C to prevent premature vulcanization. Since compounds with highly reinforcing fillers harden to a greater or lesser extent during storage, they must be replasticized before further processing, i.e., resoftened in a mill. In this case somewhat higher mill speeds and higher friction ratios than for other types of synthetic rubber have been found desirable. While, however, unfilled silicone rubber undergoes no change on milling, the milling of filled compounds has a considerable influence on the silicone rubber. As Table 71 shows, with relatively intensive processing conditions there is a decrease in the molecular weight of silicone rubber (*13*), which is shown by greater softness and tackiness and imparied extrudability of the compounds.

The compounds are then converted into the desired form by molding, extruding, calendering, etc., and cured. Details of the production and processing of silicone rubber compounds can be found in special literature (*55*, *102*, *176*, *177*, *224*).

TABLE 71

EFFECT OF MILLING ON THE PROPERTIES OF SILICA-FILLED POLYSILOXANES

$(Me_2SiO)_x$ polymer (parts by weight)	Aerosil (parts by weight)	Milling time (min)	Polymer (mol. wt.)
5	—	None	458,000
5	2	6.5	232,000
5	2	36.5	108,000
5	2	120.0	69,000

8.1.3.1.4 Curing (vulcanization). Silicone rubber compounds are generally cured in two separate stages. First there is a pre-cure, which has the primary object of imparting dimensional stability to the external form of the article. This is done with silicone rubbers at relatively high temperatures, especially with organic peroxides, and possibly under pressure. Usually, therefore, pre-curing is carried out either in a vulcanizing press, in open steam, or even in hot air.

Heat curing in air is performed continuously and without pressure. Continuous curing in hot liquids such as polyglycols has also been described (*162*). This allows considerably higher cure rates.

The next process to which the pre-cured article is subjected is *post-curing.* In this step, which may be regarded as a purification and short-time aging process, the polymerizate is additionally cross-linked under the influence of atmospheric oxygen without pressure. The post-curing process affects the mechanical, electrical and chemical properties of the product. The quality of a vulcanizate can be greatly affected by the ratio of oxygen to silicone rubber formed in the post-curing process. During post-curing, about 125 liters of fresh air per kilogram of silicone rubber per minute should always be supplied to give vulcanizates with optimum properties.

In the manufacture of molded articles, pre-curing is normally carried out for 5 to 20 minutes. Steam curing of larger articles requires about 20 to 120 minutes with 5 to 7 atm gage steam, while hot-air curing with adequate pre-curing can be achieved in about 2 to 60 seconds at temperatures between 200°C and about 800°C. Post-curing is always carried out at 200 to 250°C for 2 to 24 hours. Special silicone rubber compounds allow the omission of post-curing in certain cases (*4*).

8.1.3.1.5 Reclaiming. As with most organic elastomers, previously cured silicone rubber can be reincorporated into compounds after certain pre-treatments. The material can be reclaimed by various methods. If it is waste material from a silicone rubber compound exclusively or predominantly containing nonreinforcing fillers, it may simply be plasticized mechanically by remilling in a conventional rubber mill or refiner. Wastes from compounds with good mechanical properties, particularly those that are relatively resistant to hydrolytic depolymerization, require special techniques for reclaiming. Depolymerization is achieved by remilling in the presence of strongly acidic compounds such as titanium tetrachloride, dimethyldichlorosilane, or thionyl chloride (*88*, *262*).

Suitable degradation can also be performed in solution (*95*). Reclamation with steam (*127*, *175*) is of particular technical interest. The reclaimed depolymerized material is then generally refined and strained and can be recombined with a silicone rubber compound (*195*, *246*).

8.1.3.2 *Processing of Room-Temperature-Vulcanizable Silicone Rubber* (*Multicomponent Systems*)

While heat-vulcanizable silicone rubber is stable on storage after addition of the curing agent, RTV and LTV types begin to vulcanize immediately after addition of the curing agent. Consequently, in these cases only a limited time is available for processing. For RTV types, this pot life may be between a few seconds and several hours; for LTV types it may be up to several days.

The types of *silicone rubber* used have the functional groups, particularly hydroxyl groups, described in Section 8.1.2.3. The viscosity of these rubbers is generally 1000 to 50,000 cSt. The *fillers* used are almost all inert materials such as diatomaceous earths, barium sulfate, ground quartz, iron oxide, titanium dioxide, calcium carbonate, and carbon black in amounts which may greatly exceed 100% by weight (referred to the polysiloxane). Reinforcing fillers would also improve the mechanical properties of these products but, on the other hand, they would adversely affect the casting or spreading properties of the mixes. Because of the relatively low molecular weight of the rubbers used, and the absence of reinforcing fillers, the mechanical properties of the vulcanizates are therefore mediocre and are inferior to those of heat vulcanizates. The *curing agents* used are of the types described in Section 8.1.2.3.

Catalyst-free compounds can be made on all conventional mixing devices giving good dispersion of the fillers. Besides the ordinary mills and internal mixers, simple stirring devices and dough mixers have been found useful for the softer formulations.

The mixture of rubber, fillers, and other possible additives is generally treated with curing agents shortly before use and is then thoroughly masticated. This is done manually with relatively small batches, and a dough mixer is used with larger batches. Vulcanization begins after a short time, as shown by an increase in the viscosity. The completion time depends on the amount of curing agent, the temperature, the humidity, and the pH. The rate of cure can be adjusted to give the rubber-elastic state after a few minutes, or only after a few hours or days. The batch is brought to the final desired form by casting, spreading, or other methods within the period after the addition of the curing agent during which there is little or no change in the viscosity. Gentle heating is required for LTV types after shaping.

8.1.3.3 *Processing of Dispersions in Organic Solvents*

Both heat-vulcanizable and room-temperature vulcanizable silicone rubbers can be processed as solutions or dispersions in organic solvents, and are preferably used in this form for coating purposes. The solvent must be removed before curing.

The formulation and production of the compounds are generally similar to those described for the processing of gum.

In selecting fillers, one should remember that the solubility of the filler-containing compounds in organic solvents is reduced by reinforcing fillers, particularly when these are used in relatively high concentration. Moreover, the aging of such fillers can lead to a pronounced thickening of the solutions. It is thus generally preferred to use nonreinforcing fillers. Esters of polysilicic acid have been described as processing auxiliaries, particularly for fabric coating (*139*).

The dispersions are prepared in the usual dough mixers. It is preferable to add the solvent gradually and to carry out complete homogenization before the final dilution. Preliminary swelling in solvents is also desirable.

Silicone rubber compounds can be dispersed in most organic solvents, aliphatic and aromatic hydrocarbons and esters being particularly suitable. Chlorinated hydrocarbons are also suitable in principle, but under certain conditions they may promote the decomposition of peroxidic curing agents, so that they are appropriate only in the dispersion of room-temperature vulcanizable types of silicone rubber. Contamination of the solvents by sulfur-containing or nitrogen-containing compounds adversely affects vulcanization and leads to tackiness of the vulcanizate surfaces.

The incorporation of nonsolvents is often used to lower the viscosity of dispersions. Lower alcohols such as butanol are used most frequently. Their addition can also considerably reduce the thickening of the dispersions of mixes containing reinforcing fillers. In the case of room-temperature vulcanizable dispersions, the incorporation of alcohol prolongs the pot life.

Spreading and dipping are the two most important methods of processing. In the *spreading* process the dispersions are spread with the usual devices on fabrics, paper, and similar materials. Textiles made from organic fibers must be cleaned and smoothed as much as possible (by desizing, singeing, calendering) before coating. Glass fiber fabrics are generally freed from size by heating to about 300 to 350°C.

A very thin coating of 5 to 20 gm/m^2 is adequate for paper. Textiles require thicker coatings of ~ 10 to 500 gm/m^2, the thickness depending on the nature and purpose of the fabric.

Dispersions suitable for *dipping* must be more highly diluted. Dipping is used, for example, for impregnating glass fiber fabrics and for coating metallic and other articles with thin films of silicone rubber.

After evaporation of the solvent, the films are cured by heating or (with room-temperature-vulcanizing rubbers) by storage at room temperature, depending on the nature of the formulation. If heating is carried out, the solvent must be evaporated at a temperature which does not decompose the peroxides, as otherwise not only will the product become porous but the peroxide may even react preferentially with the solvent, thus reducing the amount of peroxide for the vulcanization proper and resulting in a tacky surface and under-vulcanization. This is particularly true in the case of aromatic solvents. Oxygen has a similar effect, inhibiting the vulcanization of thin layers. Solutions of room-temperature-vulcanizable silicone rubbers can also be processed by spraying.

8.2 Technology of Silicone Resins*

8.2.1 Types of Silicone Resins

8.2.1.1 *Methylphenylsilicone and Methylsilicone Resins*

Of the silicone resins, the methylsilicones have the lowest carbon content. It is often believed that because of this they must also be the most heat-resistant resins. That this is not the case has been shown by various authors (*71*, *166*, *168*). However, the low organic content and the small space requirement of the methyl groups are responsible for the hardness and relatively low thermoplasticity of these resins. The elasticity is lower than that of methylphenylsilicone resins, although it can be increased by using only M and T units (*196*). Pure methylsilicone resins are only rarely used for heat-resistant lacquers because of their low compatibility with pigments, but they are very satisfactory for the production of abhesive and hydrophobic surface coatings.

The introduction of phenyl groups into the organosiloxane network improves the heat resistance, the elastic properties, and the pigment compatibility of the products. Their compatibility with organic types of resins and their adhesion to various substrates is also improved. On the other hand, resins containing phenyl groups have a greater thermoplasticity; this, however, can be countered by suitable measures.

* Section 8.2 was written by Dr. W. Krauss, Leverkusen, West Germany.

Resins with other organic groups, particularly those with higher aliphatic groups, have not become technically important. It is true that higher alkyl groups act like phenyl groups as internal plasticizers, so that the number of groups required to achieve a certain elasticity is smaller for longer alkyl groups, but the heat stability falls off with the increasing number of carbon atoms in the alkyl radical. Resins containing only phenyl residues as organic groups are highly thermoplastic; on the other hand, the thermoplasticity can be opposed by using only T units (*30*).

All technical silicone resins that are not dissolved or soluble in organic solvents contain small proportions of hydroxyl groups bound to the Si atoms in addition to the organic groups. They are of great importance for the industrial processing of the resins, since they are the reactive centers where the condensation reactions leading to the hardening of the resin occur. Alkoxy groups may also be present instead of hydroxyl groups (*153*).

8.2.1.2 *Resins with Other Substituents*

Even without the use of catalysts, resins with *vinyl* or *allyl* groups harden more rapidly than resins without alkenyl groups. In addition, the hardness of the products is increased, their solvent resistance is improved, and, if a certain proportion of unsaturated groups is not exceeded, their heat resistance is little impaired (*92*, *234*).

Vinyl groups respond relatively readily to peroxide catalysts, so that they give polysiloxanes which can be cured at relatively low temperatures (e.g., 80°C) (*263*). Since little or no volatile reaction products are produced, resins containing vinyl groups are particularly suitable for dipping varnishes (*117*, *248*). They generally have a comparatively low viscosity, so that for many purposes they can be used without solvents (*149*).

With these products, peroxidic cross-linking through alkenyl groups can be combined with condensation of silanol groups to give pre-curing at relatively low temperatures and completion of the network with, for example, quaternary ammonium hydroxides as condensation catalysts (*249*).

The use of *acrylic esters* of substituted polyorganosiloxanes has been proposed for low-temperature curing (*134*). Water-soluble products (*183*) may also be obtained if the acrylic ester has a suitable structure.

Rapid-drying varnish films having good surface hardness after treatment with sulfonic acid esters are obtained from methylsiloxanes containing *ethyleneimino* groups as substituents on the silicon atom (*206*). However, they have the disadvantage of yellowing rapidly at high temperatures.

Polysiloxanes with *halogenated phenyl radicals* generally have a better heat stability than halogen-free products; nevertheless, they are of little interest in the manufacture of lacquers and electric insulating materials, since they tend to liberate hydrogen halides at high temperatures. *Fluorine derivatives* are of interest because of their resistance to chemicals (*62*) and to solvents and fuels (*193*).

Incorporation of *hydroxyorganic* groups allows reactions with isocyanates (*50*); *carboxyorganic* groups permit cocondensation with polyalcohols, alkyd resins containing hydroxyl groups, etc.; aminoalkyl groups permit reactions with epoxy resins and also improve adhesion to various substrates such as metals and glass (*179*, *180*).

Incorporation of *Si—H bonds* in the molecular network (*101*, *202*) can produce centers reacting with, for example, alcoholic OH groups (*150–152*).

The properties of silicone resins may also be modified by partial replacement of Si—O—Si by Si—N—Si, Si—Si, or Si—C_6H_4—Si bonds. Siloxane-*silazane* copolymers (*220*) have improved adhesion properties, in particular to glass and ceramic surfaces (*171*). Curing to form insoluble polysiloxanes takes place more rapidly when both *polysilane* and siloxane bonds are present; furthermore, the viscosity of such products is low, so that they are suitable as solvent-free dipping lacquers (*46*). Resins with improved elastic properties are produced (*135*, *136*) if some of the silicon atoms are linked to one another through *phenylene* groups rather than through oxygen atoms (*118*).

8.2.1.3 *Cocondensation Products with Organic Resins*

The number of products described under this name in the literature and available commercially is very large. All intermediate steps are found, from resins containing 5% of the siloxane component to those with more than 90% of siloxane. These cocondensation products are intended to combine the good properties of the polysiloxanes and those of the organic components. This combination improves stoving conditions, surface hardness, adhesion (particularly to copper), solvent resistance, and compatibility with pigments and organic resins over those of pure silicone resins; it gives a reduction in the thermoplasticity (which can occasionally be troublesome) and also a reduction in cost. Compared with purely organic resin equivalents the heat stability, weather resistance, water-repellency, and incompatibility with other organic materials (which is shown by the excellent release action of fully condensed coatings) are improved. In general, the properties of the predominant partner in the product are found in the cocondensate.

The processes used to produce the various cocondensates have been treated in detail in Section 7.4, so we now may restrict ourselves to the modifications of the properties that occur with particular types of organic partners. The products of reactions with *polyalcohols* and *dicarboxylic acids* and *polyesters containing hydroxyl groups* obtained from them have received the most attention. The advantage of these products is, in addition to lower cost, an improvement of the pigment compatibility, an increase of the surface hardness, and a decrease of the curing time or the curing temperature. Surprisingly, when the components are suitably chosen, the heat resistance is little affected even by the addition of as much as 50% of the nonsiloxane component.

Of the three phenylenedicarboxylic acids, isophthalic acid gives the most heat-stable cocondensation products (*1*, *2*, *64*, *238*). Incorporation of long-chain monocarboxylic acids (*230*) improves the compatibility of the product with other lacquer raw materials; if the monocarboxylic acids are unsaturated (*222*), resins can be obtained that dry at room temperature like the oil-modified alkyd resins. The condensation of the silanol groups takes place only at higher temperatures, so that while such products dry rapidly at lower temperatures, they require elevated temperatures for thorough hardening. In these circumstances, one property or another is predominant according to the amounts of the individual components. A minimum content in the cocondensate of 65% of an alkyd resin (containing at least 30 to 45% of unsaturated fatty acids of the type of linseed oil, dehydrated castor oil, or oiticica oil) appears to be necessary to produce cocondensates with satisfactory drying properties at room temperature (*59*, *192*).

Cocondensation with maleic acid esters (*60*, *247*) gives products that can be copolymerized with reactive monomers (such as styrene and diallyl phthalate) at room temperature and can therefore be processed without a solvent.

If a long-chain aliphatic or cycloaliphatic alcohol is used as the alcoholic component for the cocondensation process, softer products with higher elasticity (*133*) or oily liquids that can be used as plasticizers for polysiloxanes are produced (*90*, *125*). Moreover, such long-chain polyalcohols are said to prevent blistering during the stoving of dipping lacquers (*130*).

Most cocondensation resins obtained from polysiloxanes and polyesters still contain free hydroxyl groups bound to carbon, which can undergo various reactions, e.g., with isocyanates or with urea-formaldehyde and melamine-formaldehyde resins (*3*). This frequently gives an increase in the resistance to chemicals and solvents and an improvement in the surface hardness (*100*, *148*). When the hydroxyl content is sufficiently high, air-drying two-component lacquers or adhesives can be prepared in this way. However, the incorporation of isocyanates (*18*) reduces the heat stability of the products, particularly their fastness to yellowing.

Combination with epoxy resins can also improve resistance to chemicals and solvents. The epoxy resins can be incorporated in the polysiloxane network either as such (*70*, *132*, *191*, *237*), or together with polyesters. Certain combinations of this type show surprisingly good adhesion to copper. With careful condensation, the epoxy groups remain unchanged so that room-temperature curing two-component lacquers can be made.

Cocondensation products with polycarbonates, characterized by particularly high water resistance, have recently been disclosed (*111*).

The cocondensates described here contain Si—O—C bonds, which have a certain susceptibility to saponifying agents. If the siloxane and the organic component of the cocondensate are linked via Si—C bonds by means of organofunctional silanes or siloxanes, resins more resistant to saponification should be generally expected. Hydroxyalkylpolysiloxanes have become of particular interest in this respect, as the starting materials for cocondensation and addition reactions (*50*, *211*, *221*).

8.2.2 Compounding the Raw Materials

8.2.2.1 *Silicone Resin Lacquers*

8.2.2.1.1 Solubility and dilutability. Silicone resins are generally marketed in the form of solutions in aromatic solvents. To obtain clear and not too viscous solutions, more highly precondensed products are dissolved in cyclohexanone. For use in the lacquer industry, these precondensates are diluted and processed in the usual way with pigments, catalysts, and, sometimes, extenders. The selection of the diluent raises no difficulties in the case of the polysiloxanes. In general, esters, chlorinated aliphatic and aromatic hydrocarbons, and ketones, in addition to aromatics, can be used directly. Glycols, glycol ethers, and alcohols are not true solvents for polysiloxanes and can therefore be used only in association with more slowly evaporating active solvents, so that no separation takes place during evaporation. A small percentage of alcohols, especially butanol, eliminates turbidities which may arise through the separation of small amounts of water of condensation when the solutions are allowed to stand for a relatively long time.

Monohydric aliphatic alcohols containing at least four carbon atoms also improve the storage stability of polysiloxane solutions (*47*). Aliphatic hydrocarbons should not be used as solvents or diluents for polysiloxanes; they cause turbidity even in small amounts. The reduction in viscosity caused by addition of gasoline is of interest in only a few cases, since most commercial products already have a relatively low viscosity.

The viscosity of many silicone resin solutions does not remain constant on prolonged storage, since even at room temperature further condensation of the silanol groups or rearrangements of the siloxane skeleton may take place. Besides the alcohols, etc., mentioned, cerium salts (*54*), cyclohexanone, and chelating agents also have a stabilizing action.

The solubility of the polysiloxanes may vary according to the type of polysiloxane used. This applies particularly to cocondensation products as described in Section 8.2.1.3.

Table 72 gives the dilutability of a number of technically important polysiloxane and polysiloxane-polyester resins.

TABLE 72

DILUTABILITY OF SILICONE RESIN SOLUTIONS

	Methylsilicone resin	Methylphenyl-silicone resin	Polysiloxane-polyester resin	
			25% polyester	50% polyester
Methanol	1:0.4	1:0.5	1:0.8	1:1.4
Butanol	In all proportions	1:1.5	1:4.6	1:6.0
Acetone	In all proportions	In all proportions	In all proportions	In all proportions
Ethyl acetate	In all proportions	In all proportions	In all proportions	In all proportions
Methyl glycol acetate	1:1.6	In all proportions	In all proportions	In all proportions
Toluene	In all proportions	In all proportions	In all proportions	In all proportions
White spirit	In all proportions	1:1.3	1:0.1	1:0.1
Solvent naphtha	In all proportions	In all proportions	In all proportions	In all proportions
Cyclohexane	In all proportions	In all proportions	1:0.2	1:0.2

8.2.2.1.2 Curing to give insoluble end-products. One of the greatest difficulties in the use of silicone resins for surface coating was the fact that for many years long times and high temperatures were necessary to complete the condensation. Initially, condensation takes place relatively rapidly, but the reaction then slows down because of steric hindrance (*166*), and high temperatures or very active catalysts are required to accelerate it. The curing temperatures are limited by the construction of the ovens available in the lacquering industry and by the fact that above 250°C, in addition to condensation, oxidation of the organic radicals and splitting of the Si—C bonds take place to an increasing extent (*166*). The most important catalysts are soluble salts of Pb, Fe, Co, Zr, Sn (*41*), Ti (*154*), and Mn (*144*), together with amines and quaternary ammonium hydroxides (*48*) (cf. Section 5.4.1).*

Many attempts have been made to lower the stoving temperature. Amines are the most effective additives, but they are not suitable for lacquers because they tend to cause yellowing under the action of heat. Very low curing temperatures are achieved by using polysiloxanes containing acyloxy groups (*49*). By analogy with the manner in which silicone rubber systems can be cross-

* An exhaustive patent review is given in *Farbe und Lack* **62**, 374 (1956).

linked at room temperature (Section 8.1.2.3.2), the action of atmospheric moisture leads to hydrolysis followed immediately by condensation of the resulting silanols to cross-linked polysiloxanes.

All catalysts initiate the condensation at room temperature and therefore generally bring about gelation very rapidly, thus making the resin solutions unusable. It is consequently recommended to add the catalyst just before the material is used, and to prepare only enough material to last a few days or even hours. Heavy-metal salts that may act as oxygen carriers catalyze not only the silanol condensation but also, to some extent, the oxidation of organic groups, giving additional cross-linking sites (*84*). This is confirmed by lead salts, which bring about a complete hardening of all silicone resins in a very short time, giving, in fact, products harder than are obtained without catalysts; this shows that cross-linking sites additional to the cross-linkages through siloxane bridges must have been formed. Almost all heavy-metal compounds therefore undesirably affect the heat stability of silicone resin films cured with them. The most active catalysts show this effect to the highest degree. In the case of lead compounds, the catalytic effect is so pronounced that silicone resin films hardened with them become completely brittle after a few days at 200 to 250°C, while films hardened without catalysts can be exposed to these temperatures for several months. Since the corresponding resin solutions gel after only a few days, warnings have been given against the use of lead salts as hardening catalysts (*103*, *181*). Salts of cobalt, zinc, and iron have the smallest influence on the heat stability of the products, and these substances are therefore preferred as curing catalysts for silicone resins, even though their catalytic effect is relatively low (*168*).

The observation that the pot life of resin solutions catalyzed with lead salts can be substantially improved by the addition of complex-forming agents such as acetylacetone, acetoacetic ester, malonic acid, and biacetyl (*112*) was important for the use of silicone resins as raw materials for lacquers. However, these complex-forming agents do not improve the heat stability of the products, since they volatilize during the curing process.

The observation (*112*) that the combination of two metal-containing hardeners gives mixes with good pot life, short drying times, and lacquer films with excellent heat stability, is very much more surprising. It is only necessary to combine a highly effective curing accelerator (a salt of lead, aluminum, or calcium) with an excess of a weak accelerator (a salt of iron, zinc, cobalt, manganese, or titanium). A combined hardener of this type exhibits only the "good" properties of each component: it gives drying at least as rapid as that given by a strong hardener alone and, on the other hand, its effect on heat aging is as good as that of a weak hardener. The pot life of the catalyzed silicone resin solution is substantially better than when a highly active catalyst is used alone.

Figure 27 shows the influence of curing catalysts on the viscosity of a silicone resin solution. Curve 2 shows the steep increase in viscosity brought about by a lead salt; the solution has gelled after only one day. A cobalt salt has hardly any effect on the viscosity. In spite of the much higher amount of catalyst when both lead and cobalt salts are used, the viscosity rises much later than in the case of curve 2. The fact that it is actually the cobalt salt that is responsible for

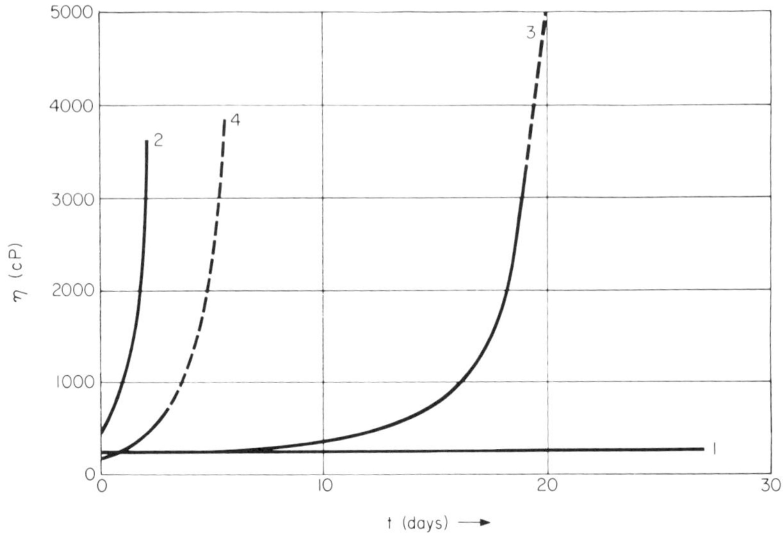

FIG. 27. Influence of curing catalysts on the viscosity of a silicone resin solution with the addition of: (1) 2.66% of cobalt naphthenate; (2) 0.66% of lead naphthenate; (3) 0.66% of lead naphthenate and 2.66% of cobalt naphthenate; and (4) 0.66% of lead naphthenate, 2.66% of cobalt naphthenate, and 3.33% of di-(hydroxyethyl)-*n*-butylamine.

the lower tendency to gel can be shown by adding a complex-forming agent such as dihydroxyethyl-*n*-butylamine. The bulk of the cobalt is masked, so that it is no longer available for stabilization, and the advantage of the combination of catalysts is lost (curve 4).

Table 73 shows, in the last three columns, the most interesting properties of catalyzed silicone resin solutions; namely, drying time, the pot life of the resin solutions, and the heat stability of the completely cured forms. The superiority of the combined hardeners is clearly shown.

In practice, combinations of lead naphthenate and polymeric butyl titanate are mainly used today, since butyl titanate appears to be the most active of the weak catalysts (*110*).

TABLE 73

INFLUENCE OF CURING CATALYSTS ON THE CURING TIME AND HEAT STABILITY OF SILICONE RESINS AND ON THE POT LIFE OF SILICONE RESIN SOLUTIONS

Catalyst	Curing time at 230°C	Pot life of the formulation	Heat stability[a] (hr)
	Hard Methylphenylsilicone Resin		
—	~3 hr	Unlimited	>2000
0.66% of lead naphthenate	<5 min	1 day	24
2.0% of cobalt naphthenate or 2.0% of zinc naphthenate	~60 min	Several months	>2000
0.66% of lead naphthenate and 2.66% of cobalt naphthenate	<5 min	14 days	>2000
0.05% of lead naphthenate and 2.0% of zinc naphthenate	10 min	>3 months	>2000
	Soft Methylphenylsilicone Resin		
—	~30 hr	Unlimited	>2000
0.66% of lead naphthenate	10 min	4 days	100
2.0% of cobalt naphthenate or 2.0% of zinc naphthenate	~10 hr	Several months	>2000
0.66% of lead naphthenate and 2.66% of cobalt naphthenate	10 min	18 days	>2000
0.1% of lead naphthenate and 2.0% of zinc naphthenate	45 min	3 months	>2000

[a] Tested at 240°C with films ~80 μ thick on sheet iron.

8.2.2.1.3 Pigmentability. The pigment compatibility of the polysiloxane resins depends very largely on their structure, improving with the increasing ratio of phenyl to methyl groups. A further increase results from the incorporation of organic compounds, particularly polyesters. The number of pigments available is limited because their heat stability must be at least as good as that of the silicone resins, and they must have no adverse effects on the properties of the polysiloxane solutions or the cured films.

Because of their poor heat stability, therefore, all organic pigments are unusable or can be used only to a very limited extent. Phthalocyanine pigments have been found useful for some special purposes. The necessity for good heat stability also limits the number of suitable inorganic pigments. All hydrated iron oxide pigments are converted at temperatures above 200°C (generally in a very short time) into red oxide (colcothar) which is stable up to high temperatures. Carbon black may be oxidized to carbon dioxide within a few days, and it is thus used only where the film must have a high electrical conductivity (*65*). Inorganic oxides and cadmium sulfides and selenides are very suitable and so also, in general, are the pigments used in the vitreous enamel industry for the coloring of pastes. By adding alkali-metal borosilicate glasses in powdered form, it is possible to make surface coatings in which the action of sufficiently high temperatures destroys the polysiloxanes but fuses the frits to a homogeneous enamel-like coating (*109*, *129*). The color of cadmium sulfide pigments changes at elevated temperatures; this process is reversible, but such differences in color may be objectionable in the case of large surfaces having different temperatures. Surprisingly, fillers such as talc, highly disperse silica, and silicon carbide may have an unfavorable effect on the mechanical properties of the material (particularly its elasticity) so that these substances cannot be used in excessive amounts. Asbestos seems to be an exception, since the addition of about 10% of this material substantially raises the strength of the film (*9*).

The interaction of pigments with silicone resins is much more complicated. This applies particularly to lead- and zinc-containing pigments, especially red lead and zinc chromate, but also to titanium dioxide and purely metallic pigments such as aluminum bronze and zinc dust. From what has been said in Section 8.2.2.1.2, these compounds should have a marked catalytic effect which, in the case of red lead (because of its high concentration of lead), should seriously damage the film; with other pigments it may improve the properties of the film. In fact, silicone resin films pigmented with red lead have an antirust effect but become brittle after a short period of aging at temperatures around 200°C. The pot life of such lacquers is very limited. However, by combining red lead with zinc chromate—as with the combined catalysts described in the preceding section—storage and heat stability are greatly improved, although they are still not completely satisfactory. Thus, the manufacture of anticorrosive primers based on silicone resins is still problematic. Strontium chromate gives the best results, as can be seen from Table 74.

TABLE 74

CORROSION-INHIBITING PIGMENTS FOR SILICONE RESIN LACQUERS

Pigment	Pot life of the lacquers	Heat stability at 260°C (hr)	Applicability
Zinc chromate	> 10 months	45–50	–
Basic Zn chromate	> 10 months	45	–
Sr chromate	> 10 months	550	+
Pb chromate	4 weeks	210	–
Ca plumbate	6 weeks	2–3	–
Red lead	6 weeks	24	–

Silicone resins pigmented with titanium dioxide, red iron oxide, and zinc oxide, and catalyzed with lead naphthenate, possess all the properties that can be obtained with combined catalysts. This point is illustrated in Fig. 28. While a coating pigmented with titanium dioxide and hardened without a catalyst had spalled away from the metal almost completely after aging 500 hours at 230°C, the film made with the addition of lead naphthenate showed no fissures after the same treatment. In general, the combination effect is more pronounced for higher pigment contents. At a given pigment content, the extent

FIG. 28. Influence of lead naphthenate on the heat stability of pigmented silicone resin films. Cocondensate of a methylphenylsilicone resin with 25% of polyester pigmented with titanium dioxide (rutile). Left: cured with 1% of lead naphthenate, after 500 hours of aging at 230°C; right: cured without a catalyst, after 500 hours of aging at 230°C.

of the interaction with lead catalysts depends on the type of pigment. It must therefore be assumed that all pigments consisting of inorganic metal oxides so far used for silicone resins contained small amounts of soluble constituents. In the case of titanium dioxide, this influence on polysiloxanes is so great that surprising improvements can also be obtained in the heat stability of the products. If polysiloxanes are combined with very large amounts of titanium dioxide—the optimum is between 270 and 350 parts of pigment to 100 parts of binder—lacquer coatings with a very good silky gloss are obtained that can be aged at temperatures of 400°C for more than 100 hours with no decrease in the gloss and no chalking or any other appreciable film damage. If the pigment-binder ratio is considerably smaller, chalking and crazing occur very rapidly on heating, and when the ratio is higher the film strength rapidly falls off.

Aluminum bronze and zinc dust occupy a special position among the pigments used for silicone resins, since they, too, can be used to make silicone resin films whose heat stability is far greater than that of unpigmented silicone resins. The amount of pigment required is $1\frac{1}{2}$ or, even better 2, times as much as that of the silicone resin. Because of the high content of catalytically active pigment, such lacquers dry very rapidly and possess a heat stability which is limited only by the melting point of the metal used (cf. Section 9.4.1). Moreover, when their adhesion and elasticity is adequate, they impart to the metal substrate very effective protection against corrosion and the effects of weather. One or two coatings of lacquer pigmented with zinc dust are applied, and these are covered with one or two layers pigmented with aluminum bronze for protection against oxidation (*61*) at high temperatures. Remarkably enough, even heating above the melting point of the zinc does not affect the stability of the coating if the zinc dust primer is protected by an aluminum-bronze layer. At the present time, such combinations of silicone resins and metal powders can be regarded as the lacquers with the highest possible long-term heat stability combined with good adhesion and weather resistance.

8.2.2.1.4 Blends of silicone resins with other resins. Attempts were made very early to extend the field of application of silicone resin lacquers by blending them with cheaper purely organic binders. In most cases, these attempts failed because of the incompatibility of the silicones with the organic components. It is impossible to make any general statements about the organic resins that are at least to some extent compatible with silicone resins, since the compatibility depends very markedly on the chemistry, constitution, and manner of preparation of the silicone resin. Products most likely to be compatible are the following (*71*): polymethacrylates, urea-formaldehyde resins, melamine-formaldehyde resins, chlorinated di- and terphenyls, and some polyesters and alkyd resins. The compatibility of such products with silicone resins modified with organic resins is better. Since these materials often contain relatively large amounts of hydroxyl groups, their compatibility with isocyanates is important, as this makes it possible to prepare air-drying silicone resin combinations with a relatively high silicon content (*110*). In dealing with all these combinations, two groups must be distinguished. The first comprises mixtures in which the silicone resin forms the main constituent and the organic product is added in small amounts to obtain special properties. Here, the added components (e.g., polymethacrylates) have the primary object of shortening the drying time or, indeed, as in the case of isocyanates, of permitting the production of air-drying films. The addition of polyureas improves the surface hardness and reduces the thermoplasticity of the products. Provided that they are compatible, phenol resins or epoxy resins may increase the solvent resistance. Chlorinated biphenyls are used as plasticizers. However, it must not be forgotten that the addition of organic binders decreases the heat stability of the whole system and reduces the resistance to yellowing quite markedly. No heat-resistant white or light-colored lacquers can be produced wich such blends. This is probably the main reason why *cocondensates*, in which the organic constituent is chemically bonded to the silicone, are preferred at the present time (see Section 8.2.1.3).

The second group comprises blends in which the organic binder forms the main constituent and the silicone resin is incorporated in smaller amounts to impart special properties. A relatively large number of silicone resins are marketed specially for this purpose which, in contrast to the "conventional" silicone resins, are characterized by excellent compatibility with various types

TABLE 75

HEAT STABILITIES OF A SILICONE RESIN-POLYESTER COCONDENSATE[a] AND A BLEND OF A HEAT-STABLE ALKYD RESIN AND A METHYLPHENYLSILICONE RESIN[b], BOTH CONTAINING 50% OF THE SILICONE COMPONENT

Time of aging at 180°C (hr)	Gloss[a]		Elasticity[b]		Yellowing[c]	
	a	*b*	*a*	*b*	*a*	*b*
1	100	99	5.2	5.7	0.3	0.3
12	100	97	5.1	2.8	0.3	0.3
24	100	95	5.0	2.8	0.3	0.4
50	100	95	4.6	2.8	0.3	0.7
100	97	90	4.7	2.6	0.4	2.5
500	90	50	4.1	2.4	0.5	4.0

[a] Measured with the Lange gloss meter.
[b] Measured by the Erichsen method in mm depression.
[c] Pure white = 0 → intense brownish discoloration = 4.

of lacquer binders. These are mostly low-polymer polysiloxanes rich in hydroxyl groups (*75*). A high content of phenyl groups increases compatibility (*73*), and so does the presence of alkoxy groups in the polysiloxane (*77*). The general aim of increasing the heat stability of organic binders is not achieved to a completely satisfactory extent. There is still a tendency to yellowing and embrittlement, so that in this case too, the cocondensation products are preferred. This is shown by Table 75, in which a cocondensate containing 50% of silicone has been compared in respect of gloss retention, resistance to yellowing, and elasticity at a temperature of 180°C with a blended product from one part of heat-stable alkyd resin and one part of silicone resin.

The addition of small amounts of silicone resins markedly increases the weather resistance and water repellency, and frequently also the resistance of purely organic binders to chemicals. In this case, a simple blend may be superior to a chemical combination (*39*). This will be discussed more fully in Section 10.5.2.

8.2.2.2 *Foamed Silicone Resins*

Foamed silicone resins have been known for only a few years but, because of a number of good properties—particularly their flame resistance and their outstanding heat stability—they have become of interest as heat-insulating materials.

They are unsuitable as core materials for sandwich elements since, because of their brittleness, they are not capable of withstanding the vibrations occurring in practical use. However, they can be used satisfactorily as components in aircraft construction and as insulating materials for apparatus and instruments (*240*).

The manufacture of foamed silicone resins takes place in five stages: fusion of the silicone resin; compounding with catalysts, blowing agents, and any fillers required; foaming; pre-curing; and post-tempering.

Little information is to be found in the literature (*251*) on the chemical structure of silicone resins suitable for this purpose. They are mostly brittle, solvent-free resins with a softening range between 48 and 60°C (*256*). The usual solvent-containing products can also be used for special processes but, to avoid trouble during foaming, these must first be freed from the solvent by careful heating *in vacuo*.

The mechanical properties are improved by combination with novolac resins (*210*).

The catalysts and any fillers are incorporated by intensive stirring into the resin previously fused by careful heating to 120–140°C.

In addition to the usual naphthenates and stearates of heavy metals, generally lead, cobalt, and zinc, the recommended catalysts are dibutyltin diacetate (*257*) and, in particular, amines such as mono-, di-, and triethanolamines and quaternary ammonium salts of aliphatic carboxylic acids (*254*). The amount of catalyst added generally varies between 0.1 and 2% for the heavy metal compounds and is up to 5% with alkaline substances. Frequently, the alkaline decomposition products of the blowing agent also have a catalytic effect (*251*, *257*) so that it is not always necessary to add catalysts separately.

The blowing agents used are the usual nitrogen-liberating compounds such as *N,N'*-dinitrosopentamethylenetetramine (*251*, *256*) or sulfonhydrazides (*251*). The amount of blowing agent used is governed by the pore structure and the required density. In general, it amounts to about 3% of the silicone resin component. Higher amounts permit lower densities to be obtained. Combinations of silicone resins with polysiloxanes having at least one hydrogen atom on each silicon atom, and compounds containing hydroxyl groups, can be used as self-foaming blends (*252*, *254*). With these mixtures, heating gives rise to the following reaction:

$$\geq C{-}OH + H{-}Si\leq \;\rightarrow\; H_2 + \geq Si{-}O{-}C\leq$$

Fillers are rarely used in foamed silicone resins. Some proposals have been made to add alumina (*253*) or metal powders (*5*)—particularly aluminum powder—to improve the mechanical strength of the products, or to add quartz powder, diatomaceous earths (*250*), titanium dioxide, clay (*166*), etc., as fillers.

General statements on the curing and foaming temperatures are impossible, since these depend on the resin used, on the blowing agent, on the amount of catalyst added, and on the required density. Foaming and curing must be adapted to one another in such a way that an adequate degree of solidification takes place even as the foam is produced. Uniform pore structures are obtained when the heating to the curing temperature is carried out rapidly as soon as foaming begins, and when this temperature is maintained for 3 to 6 hours, depending on the formulation of the blend. After this, a longer period—24 to 80 hours—of tempering at about 250°C had previously been necessary (*257*), but recent developments have made it possible, by stepwise raising of the temperature, to decrease the total curing time to about 15 hours.

Silicone resins have been reported (*255*, *256*) which can be foamed and cured at room temperature by the addition of certain catalysts. Post-tempering at elevated temperatures is no longer considered necessary in these cases.

8.2.2.3 *Silicone Resin Moldings*

Silicone resin moldings are made from mixtures of silicone resins, inorganic fillers, and catalysts at elevated temperatures under pressure. As compared with the silicone resin laminates (see the following section) they have the advantage that even relatively complicated molds can be filled because of the use of finely divided fillers. They are inferior to the laminates with respect to flexural and tensile strength.

Silicone resin moldings contain a very high percentage—generally about 50%—of inorganic fillers. In addition to silica in many forms, comminuted glass fibers are widely used (*84*).

Generally, a 50 to 75% solution of the silicone resin is intimately mixed with the fillers and catalysts, and the solvent is then removed under vacuum. Subsequent heating to 100°C leads to pre-curing, which permits the material to be ground after cooling to 20°C. The pre-condensate so obtained can now be charged into the molds and cured in the press. To obtain good chemical properties, post-curing must finally be carried out outside the press by heating for an extended period.

Unfortunately, it has not yet been possible to substantially shorten the times required for complete curing and, particularly, the long post-curing periods. Catalysts—ethanolamine (*68*, *173*), carboxylic acid salts of quaternary ammonium hydroxides (*48*), triphenoxysilane and tributoxysilane (*147*), and the usual metal curing agents have been recommended—decrease these times only slightly, although they frequently improve the mechanical properties of the moldings. In general, the molding material is kept in the press for 15 to 30 minutes at a temperature of 175 to 200°C and a pressure of up to 750 kg/cm^2 (*128*). Post-tempering then is carried out without pressure, at a temperature slowly rising to 250°C. For optimum properties, this temperature must sometimes be maintained for up to 75 hours. Solvent resistance of the product, particularly its resistance to boiling toluene, is regarded as a good criterion of a sufficient degree of post-curing (*84*).

8.2.2.4 *Silicone Resin Laminates*

Because of their good heat stability, low combustibility, and their outstanding electrical characteristics even under conditions of high humidity, silicone resin laminates have come into use particularly in the electrical industry. The initial difficulties due to very long molding times, high molding pressures, and very long post-curing times have been partially eliminated by the manufacture of special silicone resins. Methylphenylsilicone resins with high contents of tri- or even tetrafunctional siloxane units are particularly useful. In addition, these resins have the advantage that gelation takes place fairly rapidly on heating, so that an excessive flow of resin in the press is avoided. On the other hand, the cure rate cannot be increased excessively since the bonding of the individual layers to one another is adversely affected on too rapid curing. For this reason, too, the glass fiber fabrics used in addition to asbestos fiber fabrics as reinforcements must be completely free from soluble metal compounds that could have a catalytic effect on the curing process (*143*).

Selection of the fabric and its pretreatment are of decisive importance for the properties of the finished laminate. Fine-meshed glass fabric is mainly used and so are the much more loosely woven so-called "Bigelow mats" whose strands of fabric form small loops and so bring about an excellent bonding between the individual layers. The materials involved must be free from alkali, which has a marked adverse effect on the electrical properties (*93*). Careful desizing of the fabric is absolutely essential, since organic materials are used as sizes; this is most safely carried out by prolonged heating (24 to 48 hours) to temperatures between 300 and 325°C. If this temperature is exceeded, the strength of the material may be markedly decreased. A test for measuring the degree of desizing consists in storing a section of the fabric for 1 hour at 150°C and then for 30 minutes at 680°C. The loss in weight after this treatment should not amount to more than 0.2% with a well-desized fabric. Today, some manufacturers supply (under the name Finish 112) fabrics which have been adequately desized by thermal treatment. Glass fibers desized by thermal treatment contain acid groups which, under certain circumstances, may have an effect on the manufacture of the laminates since they change the gelling times of the silicone resins. Desizing with solvents has less effect on the fabric but is frequently not so effective. No separate treatment is required for special fabrics bearing a size prepared from vinylchlorosilanes (*98*). Materials prepared in this way give laminates with outstanding electrical and mechanical properties.

The catalysts have two tasks. On the one hand they should shorten the long pressing times, while on the other hand they should also control the viscosity of the resin to prevent not only excessive flow of resin out of the fabric but also bubble formation as the result of very rapid gelation. Lead salts (*116*), ethanolamine (*173*), choline 2-ethylhexoate, *t*-butylphosphonium acetate, and mixtures of dibutyltin diacetate and lead 2-ethylhexoate (*69*), as well as quaternary ammonium compounds (*22*), have an accelerating effect, and cerium salts soluble in organic solvents (*48*) have been recommended as stabilizers. In general, the catalyst is added to the resins shortly before impregnation, but with the correct amount and combination of catalysts (see Section 8.2.2.1.2) much longer pot lives are possible, so that impregnation of the fabric may also be carried out continuously in immersion baths. With silicone resin mixtures that are to be treated by the low-pressure process, the amount of catalyst is reduced to one-third (*82*). The consistency of the silicone resin solutions permits an average uptake of 30 to 50% of resin on a single immersion of the fabric. Subsequent air-drying removes the bulk of the solvent, and precuring is then carried out at 110°C for several minutes, depending on the type of resin used, until the impregnated layers of fabric are tack-free after cooling.

After drying, the fabric is rolled up; it should now be flexible and dry. Otherwise, more highly cross-linked components are present that will not flow during molding and will have an unfavorable effect on the bonding of the laminate layers. The impregnated fabrics can generally be stored under cold conditions for some months. The pressing process proper is carried out by the high-pressure or the low-pressure process.

In the high-pressure process, pressures between 60 and 100 kg/cm^2 and temperatures of 150 to 175°C are necessary with molding times of 30 minutes to 2 hours. The best results are obtained for a molding cycle in which the pressure and the temperature are simultaneously increased in stages. An excessive initial pressure expels the resin from the fabric.

With low-pressure laminates, the curing time depends on the thickness of the layer; it is 15 minutes for layers 5 mm thick and 60 minutes for layers 50 mm thick. Under these conditions, the pressure may be 2 kg/cm^2. After molding, the laminate is slowly cooled to about 100°C under pressure and is only then removed from the mold. In both processes, prolonged post tempering is necessary to achieve the optimum physical and chemical properties. The laminates must not be allowed to cool at any stage of the post-tempering process (*128*).

8.3 Processing of Silicone Oils

8.3.1 Manufacture of Silicone Greases

The term "silicone greases" is of phenomenological origin. It denotes the consistency of these products, which may range from that of a soft petroleum jelly to stiff compounds. There are very many methods of formulating and manufacturing these products; the only principle common to all of them seems to be that they are made by the incorporation into silicone oils of different viscosities of fatty materials that are already highly dispersed or are brought into the disperse state during incorporation (*23*, *208*). Since the type and amount of the additives determine the consistency, these have been called "consistency regulators" or "thickeners."

In addition to the term "silicone greases," the expression "silicone pastes" is also used; it has been proposed to apply the term "greases" only to products prepared by the swelling of metal soaps in silicone oils, and to call the products obtained by the dispersion of highly disperse fillers in silicone oils "pastes" (*72*). This corresponds to a differentiation between "greases" and "compounds" found occasionally in the English-language terminology (*128*). Since, however, the usage is not uniform and the terminology has no strict basis, all the products concerned may be included below in the term "silicone greases."

A silica aerosol is generally used as an already highly disperse additive for silicone oils; because of the smallness of the primary particles, the most suitable products are those obtained by the high-temperature hydrolysis of silicon tetrachloride (*16*). Nevertheless, other materials have been recommended: silica precipitated from silicate solutions with acids from which the water has been expelled by means of acetone (*12*), and silicas with a surface coating of OR groups (*97*). The dispersion of silica (with its hydrophilic surface) into a silicone oil is facilitated if it is previously coated by a polyorganosiloxane obtained by the use of organohalosilanes, organoalkoxysilanes (*6*, *197*), or low-molecular-weight polyorganosiloxanes such as octamethylcyclotetrasiloxane (*28*); this gives it a greater "silicone affinity." In another process, the greases are prepared by hydrolyzing organohalosilanes, particularly dimethyldichlorosilane, in the presence of highly disperse silica (*198*).

Other suitable thickeners are graphite (carbon black) (*114*, *172*) and copper phthalocyanine (*23*, *137*), but the following have also been mentioned and patented: calcium silicate (*63*), talc (*114*), various organic oxides in colloidal dispersion (*17*), molybdenum disulfide (*115*), alkali-metal, alkaline-earth-metal, and zinc salts of cyanuric acid (*45*), arylureas (*231*), indanthrene compounds (*10*), pteridine derivatives (*232*), and metal salts of polycarboxyalkylsiloxanes (*21*).

The incorporation and dispersal of metal soaps has become still more important for the manufacture of silicone greases. The most suitable soap components, because of their heat stability, are lithium soaps, e.g., lithium stearate (*78*, *138*) and lithium 2-ethylhexoate (*83*), and also the alkali-metal and alkaline-earth-metal salts of cycloaliphatically substituted fatty acids (*52*) and aluminum soaps (*85*). Calcium stearate disperses in polydimethylsiloxane oil readily and rapidly to give relatively stable systems (*163*).

Polydimethylsiloxane oils of various molecular weights, liquid polymethylphenylsiloxanes, and polysiloxanes with other types of substituents can be used as the silicone oil components. Polymethylphenylsiloxanes, and in particular polymethylphenylsiloxanes halogenated in the nucleus, give a better lubricant effect than polymethylsiloxanes; with metal soaps they thus give lubricating greases suitable even for relatively high bearing pressures (*89*). The silicone oil components have been modified by mixing with polyether oils (*87*), mineral oils (*260*), and polyfluoroethylenes (*155*, *245*) or by using condensation products of diols and diorganodihalosilanes (*86*).

Other additives are recommended to improve the properties of the greases in one respect or another. Greases thickened with hydrophilic inorganic fillers are even more resistant to the action of water when lactams are added to them (*189*). The loading of fillers with alkylchlorosilanes or alkoxysilanes or low-molecular-weight siloxanes, which has been mentioned above, has a similar effect. The combination of oil and silica as inorganic thickening agents becomes stronger and more resistant to mechanical action even at high temperatures when small amounts of certain boron compounds are added to the compositions (*261*). The addition of benzotriazole to the greases inhibits the corrosion of copper or steel (*190*).

The manufacture of a satisfactory grease is largely a matter of experience, and therefore the business of laboratories and factories experienced in this field. It is known that the soap is preferably allowed to swell in a hydrocarbon at an elevated temperature, after which the silicone oil is added and the solvent is finally distilled off (*58*, *78*); it is also possible first to swell the soap in an organic oil, then to extract the oil with a solvent such as hexane, to treat the residue with a solution of silicone oil in the same solvent, and finally to evaporate off the solvent (*31*).

8.3.2 Production of Emulsions and Dispersions

For many purposes, silicone oils are required in the form of aqueous emulsions that can be further diluted with water. These products are preferably used for the production of protective and release films, impregnations, and antifoaming agents.

Because of the high interfacial tension at a water/silicone oil boundary, silicone oils are more difficult to emulsify than, for example, paraffin oils. Even more so than with the silicone greases, emulsification is a matter of special experience. The processes vary according to the nature of the silicones and, in particular, the intended application, and have not generally been published by the manufacturers. In addition, silicone oil emulsions are further modified in many ways by the addition of organic compounds, fillers, etc., in order to increase their activity as mold-release agents, water-repellency agents, cleansing and polishing agents, and antifoaming agents. The following discussion will be limited to a few remarks dealing with the basic problems of the emulsification of silicone oils.

The lower the viscosity of silicone oils, particularly the technical polydimethylsiloxane oils, the easier they are to emulsify. The preparation of a stable emulsion, even from a polydimethylsiloxane oil with a viscosity of about 1000 cSt/20°C, presents considerable difficulties, and these increase markedly with rising viscosity. Emulsification can be greatly facilitated by dilution of the silicone oil phase with organic solvents, simply because of the reduction in viscosity. Dilution is particularly effective when the density of the mixture of solvent and siloxane is made equal to that of the external aqueous phase by the incorporation of specifically heavier components, particularly halogenated hydrocarbons. The matching of the density of the methylsilicone oils to that of water can be achieved by the incorporation of small proportions of phenylsiloxanes (*106*). In general, polymethylphenylsiloxanes are easier to emulsify than polydimethylsiloxanes.

The emulsification of polymethyl-*H*-siloxanes is also easier than that of polydimethylsiloxane oils. However, the stability of such emulsions is limited because of the susceptibility of the Si—H bond to hydrolysis. On the other hand, the stability can be improved by special measures. Cation-active emulsions of polymethyl-*H*-siloxanes have an optimum stability at a pH of about 3.5 (Fig. 29) (*226*). The nature of the acid used to adjust the pH has an

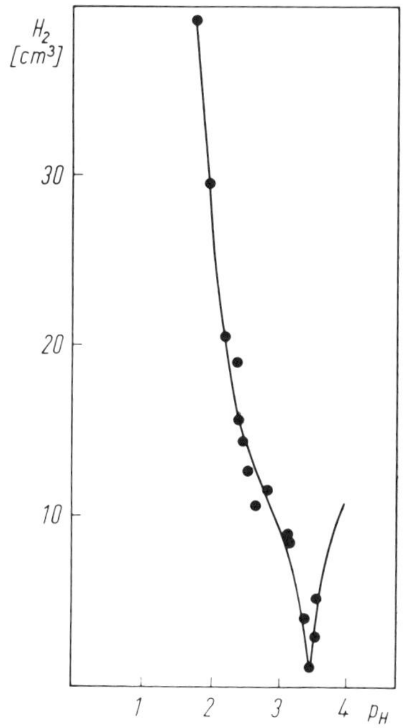

FIG. 29. Stability of an emulsion of polymethyl-*H*-siloxane oil as a function of pH.

influence: hydrochloric, maleic, malonic, monochloroacetic, and nitric acids are suitable, while the addition of acetic, oxalic, sulfuric, malic, or citric acid destabilizes the emulsions (*226*). Furthermore, the addition of aminocarboxylic acids (particularly glycine) to give a pH of about 3.5, stabilizes the emulsions to such an extent that a measurable liberation of hydrogen takes place only on heating and vigorous stirring (*227*). Aldehydes, particularly acetaldehyde, have a stabilizing effect on emulsions with nonionic emulsifying agents (*242*).

Practically all technical silicone oil emulsions are of the "oil-in-water" type. They are obtained with emulsifying agents of the cationic, nonionic, or even anionic type. Under certain circumstances, a synergistic effect can be achieved by combining emulsifying agents of two different classes (*241*).

Emulsifying agents containing a hydrophilic and a "siliconophilic" component in the molecule have been developed for the emulsification of organosiloxanes. These may be either organosilyl- or organosiloxanylcarboxylic acids (*38*, *40*, *209*, *216*) or alcohols (*219*). Another type is represented by alkylsiloxy-substituted polyethylene glycol ethers (*104*, *126*, *229*). Cationic emulsifiers are quaternary ammonium compounds of silicon (*218*), and quaternary salts based on triorganosilylalkylamines (*217*). Interesting possibilities stem from the fact that organofunctional organosilicon esters of sulfuric acid, or their reaction products with alcohols and amines, can be incorporated into the polydimethylsiloxane chain itself. Such molecules therefore contain their "own" emulsifier and, depending on the content of ester groups, give water-insoluble but emulsifiable or spontaneously emulsifying or water-soluble oils (*200*).

High-viscosity silicone oils are converted into very stable emulsions by first pre-emulsifying the oil in a monohydric alcohol which precipitates polyorganosiloxanes (particularly methanol, ethanol, or isopropanol) in the presence of a nonionic emulsifier, and then immediately dispersing this pre-emulsion in water (*164*). Another method is to make an aqueous dispersion of a low-molecular-weight oil and then to treat this with an acid or alkaline catalyst until the oil has been polymerized to the desired molecular size. The catalyst is finally inactivated by neutralization (*96*).

The difficulties of emulsification in an aqueous phase are partly avoided if emulsions are made in glycerol as the external phase. Such emulsions satisfy modest stability demands and can be diluted with water before use (*76*). Stable, spontaneously emulsifying solutions of polydimethyl- or polymethyl-*H*-siloxane oils of not too high molecular weight contain the siloxane dissolved in anhydrous ethanol or isopropanol, together with an emulsifying agent of the class of sorbitol and sorbitan esters, which may be ethoxylated (*165*).

If emulsions of silicone oils are required to make surfaces water-repellent and must thus develop an oil film by evaporation of the aqueous phase at the lowest possible temperature, alkylamines and their reaction products with oleic acid can be used as the emulsifying agents. Large amounts of these emulsifying agents are necessary because of their low emulsifying action, and they are volatile below 100°C (*225*).

The disperse phase of a good silicone oil emulsion has a particle-size distribution whose maximum lies at about 0.2 to 2 μ. The ζ-potentials depend not only on the type of emulsifying agent but also on the other additives. Cationic emulsions may have a ζ-potential of +100 mV or even more; figures of −30 to −50 mV have been measured in anionic emulsions and −4 to +4 mV in nonionic emulsions. The ζ-potential is greatly affected by the addition of alcohol; the addition of 10% of alcohol changes the ζ-potential of a nonionic emulsion from +4 to −25 mV (*228*).

Finally, it may be mentioned that it is also possible to convert solid silanols or siloxanols into aqueous dispersions. This is done, for example, by removing the alkali-metal ions from the water-soluble alkali-metal salts of silanolates by means of ion exchangers and bringing the dispersion to a pH between 3 and 7 (*204*), or by bringing such solutions into this region by acidification so rapidly that the silanol formed does not precipitate from the solution (*120*).

REFERENCES

(1) A. Abdul-Karim and J. Gabrielides, *Official Digest* **34**, 407 [1962]. (2) —, B. Golding and R. A. Morgan, *J. Chem. and Engng. Data* **5**, No. 1, 117 [1960]. (3) S. Abe, Fujisawa, Jap A. p. 448/1959 [3.9.1955]. (4) S. d' Adolf, *Rubber Wld.* **149**, No. 3, 37 (Dec.) [1963]. (5) G. F. Albers and J. F. Kulp jr., General Electric, Brit. Pat. 852887 [24.12.1956]. (6) G. B. Alexander and R. K. Iler, DuPont, U.S. Pat. 2818385 [14.9.1954]. (7) J. Ames and G. B. Ingram, ICI, Brit. Pat. 867485 [23.5.1958]. (8) K. Andrianov, *Proc. Acad. Sci. USSR, Sect. Chem.* (*Engl. transl.*) **151**, 616 [1963]. (9) —, O. Y. Gribanova, N. N. Sokolov and V. S. Tichonov, *Lakokrasochnie materialy i ich primeneniya* **1960**, No. 4, p. 10 (cited in: *Farbe und Lack* **67**, No. 10, 645 [1961]).

(10) J. W. Armstrong and H. A. Woods, N. V. De Bataafsche, German Appl. (West German) 1006996 [U.S. Prior. 15.8.1955]. (11) S. V. Averyanov, I. A. Poddubnyi, Yu. V. Trenke and L. A. Averyanova, *Soviet Rubber Technology* (*Engl. transl.*) **20**, No. 12, 2 [1961]. (12) A. J. Barry, Dow Corning, U.S. Pat. 2645588 [9.3.1950]. (13) — and H. N. Beck in F. G. A. Stone and W. A. G. Graham: *Inorganic Polymers*, p. 233 [1962]. (14) H. Bartl, J. Peter and O. Bayer, Farbenfabriken Bayer, German Appl. (West German) 1136104 [2.6.1960]. (15) H. H. Bashore, E. G. Schwarz and C. E. Simpson: New extrusion techniques with silicone rubber; *Wire and Wire Products* **34**, No. 8, 620 and 960 [1959]. (16) S. L. Bass, Dow Chemical, U.S. Pat. 2428608 [2.11.1942]. (17) —, Dow Corning, German Pat. (West German) 869982 [U.S. Prior. 2.11.1942]. (18) O. Bayer, *Angew. Chem.* **A 59**, 257 [1947]. (19) C. A. Berridge, General Electric, U.S. Pat. 2843555 [1.10.1956].

(20) B. A. Bluestein, —, U.S. Pat. 2970982 [16.9.1957]. (21) —, —, U.S. Pat. 3047497 [15.8.1960]. (22) J. R. Bond, Dow Corning, Belg. Pat. 560547 [U.S. Prior. 5.9.1956]. (23) C. J. Boner, *Manufacture and Applications of Lubricating Greases*, Reinhold Publishers, New York 1954. (24) R. J. Boot, General Electric, German Appl. (West German) 1129690 [U.S. Prior. 26.8.1958]. (25) S. N. Borisov, *Plaste und Kautschuk* **10**, 400 [1963]. (26) — and A. V. Karlin, *Soviet Rubber Technology* (*Engl. transl.*) **21**, No. 12, 4 [1962]. (27) —, N. G. Sviridova and L. A. Averyanova, ibid. **21**, No. 9, 2 [1962]. (28) K. J. Brimley and W. D. Gorden, ICI, Brit. Pat. 854960 [28.4.1958]. (29) H. G. Brod and O. Schweitzer, Degussa, German Appl. (West German) 1035358 [10.8.1955].

(30) J. F. Brown jr., General Electric, Belg. Pat. 586782 [U.S. Prior. 21.1.1959]. (31) G. V. Browning, Standard Oil, U.S. Pat. 2616850 [22.11.1949]. (32) H. Brünner, J. Diether and D. Schutte, Degussa, German Appl. (West German) 1163784 [30.3.1962]. (33) L. B. Bruner jr., Dow Corning, French Pat. 1174146 [U.S. Prior. 30.4.1956]. (34) —, —, Belg. Pat. 577012 [U.S. Prior. 24.3.1958]. (35) A. M. Bueche, *J. Polymer Sci.* **19**, 297 [1956]. (36) —, ibid. **25**, 139 [1957]. (37) — and C. S. Oliver, General Electric, U.S. Pat. 2993809 [23.10.1959]. (38) C. A. Burkhard, —, U.S. Pat. 2583322 [18.8.1949]. (39) H. L. Cahn, *Paint Industry* (Philadelphia) **76**, No. 4, 8 [1961].

(40) R. Calas and N. Duffaut, *Oléagineux* **8**, 21 [1953]. (41) Carlisle Chemical Works, Brit. Pat. 787175 [U.S. Prior. 5.3. and 6.8.1953]. (42) L. F. Ceyzeriat, Rhône-Poulenc,

Belg. Pat. 599356 [French Prior. 25.1.1960]. (43) — and G. L. PAGNI, —, French Pat. 1266528 [30.5.1960]. (44) A. D. CHIPMAN, Dow Corning, U.S. Pat. 2902467 [21.10.1957]. (45) L. J. CHRISTMANN and W. G. DEICHERT, American Cyanamid, U.S. Pat. 2830020 [1.10.1956]. (46) H. A. CLARK, Dow Corning, Brit. Pat. 675912 and 689648 [U.S. Prior. 2.9.1949]. (47) —, —, U.S. Pat. 2755261 [4.5.1953]. (48) —, —, U.S. Pat. 2830968 [27.5.1955]. (49) —, —, U.S. Pat. 3032529 [30.6.1958].

(50) — and J. H. THOMAS, —, U.S. Pat. 2931786 [18.4.1957]. (51) W. H. CRANDELL, *Rubber Wld.* **133**, 236 [1955]. (52) C. C. CURRIE, Dow Corning, U.S. Pat. 2551931 [9.9.1948]. (53) — and J. W. KEIL, —, German Pat. (West German) 960570 [U.S. Prior. 29.4.1954]. (54) K. DAMM, Farbenfabriken Bayer, German Appl. (West German) 1131009 [2.6.1959]. (55) J. J. DEASY and B. B. WHITE, *Rubber Age* **80**, 1001 [1957]. (56) N. G. DICKMANN, Dow Corning, U.S. Pat. 2819236 [23.12.1954]. (57) —, —, U.S. Pat. 2803619 [23.12.1954]. (58) Dow Corning, Brit. Pat. 652754 [U.S. Prior. 21.6.1945] (*C.A.* **1951**, 9260). (59) —, Brit. Pat. 722484 [U.S. Prior. 21.12.1951].

(60) —, French Pat. 1153664 [15.5.1956]. (61) A. E. DURKIN and A. H. HORNER, *Materials and Methods* **38**, 3, 114 [1953]. (62) J. W. DYCKES, Dow Corning, German Appl. (West German) 1056836 [U.S. Prior. 10.9.1956]. (63) G. W. ECKERT, Texas, U.S. Pat. 2656315 [27.2.1952]. (64) L. E. EDELMAN and S. ROLLINS, Westinghouse Electric, U.S. Pat. 2821518 [25.5.1954]. (65) Ekco Products, Brit. Pat. 884208 [U.S. Prior. 6.8.1957]. (66) L. M. EPSTEIN and N. S. MARANS, *Rubber Age* **82**, 825 [1958]. (67) Farbenfabriken Bayer, Trade Brochure: Silopren-Pasten, 1.3.1960, p. 10. (68) H. N. FENN, L. A. RAUNER and W. W. PEDERSEN, Dow Corning, French Pat. 941203 [U.S. Prior. 21.1.1946, 16.4. and 16.8.1946]. (69) S. FORDHAM: *Silicones*, London: George Newnes Ltd. 1960.

(70) C. L. FRYE and W. M. MCLEAN, Dow Corning, U.S. Pat. 3055858 [23.5.1957]. (71) M. A. GLASER, *Ind. Engng. Chem.* **46**, 2334 [1954]. (72) *Gmelins Handbuch der anorg. Chemie*, System Nr. 15, Teil C, p. 455, Verlag Chemie, Weinheim/Bergstr. 1958, 8th Ed. (73) D. R. GODDARD and E. A. MASON, *Paint Manufact.* **29**, No. 5, 155 [1959]. (74) D. GÖLITZ, K. DAMM, R. MÜLLER and W. NOLL, Farbenfabriken Bayer, Belg. Pat. 659254 [German Prior. (West German) 6.2.1964]. (75) T. H. GORDON, Allied Chemical and Dye, U.S. Pat. 2832794 [5.11.1952]. (76) L. Q. GREEN, DuPont, U.S. Pat. 2702276 [10.4.1951]. (77) P. A. GRIFFIN and F. COTON, Midland Silicones, Brit. Pat. 904241 [10.2.1960]. (78) G. M. HAIN and W. A. ZISMAN, U.S. Pat. 2446177 [21.6.1945]. (79) R. A. HALL, ICI, Brit. Pat. 791169 [19.8.1955].

(80) J. R. HARPER, A. D. CHIPMAN and G. M. KONKLE, *Rubber Wld.* **137**, 711 [1958]. (81) — and J. W. KEIL, Dow Corning, Brit. Pat. 859284 [U.S. Prior. 15.11. 1957]. (82) K. R. HOFFMAN, *Mod. Plastics* **30**, No. 3, 146, 219 [1952]. (83) N. G. HOLDSTOCK, General Electric, U.S. Pat. 2606153 [25.4.1951]. (84) H. N. HOMEYER, J. H. PRESTON, S. CASAPULLA and E. M. BEEKMAN, *Ind. Engng. Chem.* **46**, 2349 [1954]. (85) B. W. HOTTEN, California Research, U.S. Pat. 2654710 [22.6.1951]. (86) —, —, U.S. Pat. 2756212 [30.12.1952]. (87) — and R. J. HOUSTON, —, U.S. Pat. 2680095 [15.9.1951]. (88) B. J. HUMPHREY and H. H. WASSERMAN, The Connecticut Hard Rubber, U.S. Pat. 2673843 [17.5.1951]. (89) M. J. HUNTER and H. J. FLETCHER, Dow Corning, German Pat. (West German) 856684 [U.S. Prior. 29.9.1948, 23.2. and 14.9.1949].

(90) — and L. A. RAUNER, —, German Pat. (West German) 851854 [U.S. Prior. 28.12.1949]. (91) D. T. HURD and R. C. OSTHOFF, General Electric, German Appl. (West German)

1011147 [U.S. Prior. 16.8.1954]. (92) — and G. F. ROEDEL, *Ind. Engng. Chem.* **40**, 2078 [1948]. (93) J. K. HYDE, *Brit. Plastics* **26**, No. 288, 174 [1953]. (94) J. F. HYDE, Dow Corning, U.S. Pat. 2571039 [12.4.1950]. (95) — and G. S. KOOKOOTSEDES, —, Brit. Pat. 752860 [U.S. Prior. 1.9.1953]. (96) — and J. R. WEHRLY, —, German Appl. (West German) 1037707 [U.S. Prior. 26.1.1955]. (97) R. K. ILER, DuPont, U.S. Pat. 2705700 [1.4.1952]. (98) M. H. JELLINEK, *Mod. Plastics* **30**, No. 3, 150 [1952]. (99) O. K. JOHANNSON, Dow Corning, German Pat. (West German) 949604 [U.S. Prior. 3.12.1952].

(100) A. JORZIG and A. NUSSPICKEL, Siemens-Schuckert-Werke, German Pat. (West German) 960488 [18.12.1951]. (101) T. A. KAUPPI, Dow Corning, U.S. Pat. 2588393 [27.3.1950]. (102) F. L. KILBOURNE JR., C. M. DOEDE and K. J. STASIUNAS, *Rubber Wld.* **132**, 193 [1955]. (103) M. KIN: Silicone Resins in Heat Resistant Paints, in *Organic Protective Coatings*, Reinhold Publishers, New York, 1953, p. 361. (104) D. X. KLEIN and D. E. KVALNES, DuPont, U.S. Pat. 2476307 [2.10.1946]. (105) W. KNIEGE, Farbenfabriken Bayer, 1962, unpublished. (106) — and K. SCHNURRBUSCH, —, Belg. Pat. 623408 [German Prior. (West German) 13.10.1961]. (107) —, —, —, German Appl. (West German) 1183678 [24.7.1962]. (108) G. M. KONKLE, *Rubber Age* **84**, 975 [1959]. (109) M. KORNBLUTH, USA Secretary of Army, U.S. Pat. 2843559 [2.12.1957].

(110) W. KRAUSS, *Farbe und Lack* **64**, 309 [1958]. (111) — and R. HEBERMEHL, Farbenfabriken Bayer, German Appl. (West German) 1136490 [10.6.1960]. (112) — and R. KUBENS, *Dtsch. Farben-Z.* **10**, 1 [1956]. (113) A. LANGLADE, *Rev. Gén. Caoutchouc* **35**, 423 [1958]. (114) R. G. LARSEN, Shell Development, U.S. Pat. 2466642 [23.1.1946]. (115) —, N. V. De Bataafsche, Brit. Pat. 654533 [U.S. Prior. 28.10.1947]. (116) L. V. LARSEN, J. J. WHELTEN and J. J. PYLE, *Mod. Plastics* **23**, No. 7, 160 [1946]. (117) D. W. LEWIS, Electr. Manufact. (March) 141,360 [1957]. (118) —, Westinghouse Electric, U.S. Pat. 2828279 [21.6.1954]. (119) J. T. LEWIS and N. PAGE, Midland Silicones, Brit. Pat. 936408 [5.9.1960].

(120) Libbey-Owens-Ford Glass, German Appl. (West German) 1057746 [U.S. Prior. 31.3.1952]. (121) A. O. LIERMANN and D. E. MILLER, *Ind. Engng. Chem.* **53**, 702 [1961]. (122) G. R. LUCAS, General Electric, U.S. Pat. 2938009 [11.4.1956]. (123) H. MANAMI and S. NISHIZAKI, *J. Chem. Soc. Japan, Ind. Chem. Sect.* **61**, 218 [1958]. (124) J. MARSDEN, General Electric, U.S. Pat. 2445794 [11.6.1945]. (125) R. W. MARTIN and K. C. FRISCH, —, U.S. Pat. 2707191 [9.8.1952]. (126) W. C. MAY, Dow Corning, German Appl. (West German) 1103022 [U.S. Prior. 1.8.1955]. (127) W. A. MCCOWAN, Dunlop Rubber, Brit. Pat. 716024 [26.10.1950]. (128) R. R. MCGREGOR: *Silicones and their Uses*, McGraw-Hill, New York 1954. (129) J. R. MCKINNIE, *Federat. Soc. Paint Technology Off. Digest* **32**, No. 423, 472 [1960].

(130) W. M. MCLEAN, Dow Corning, French Pat. 1089862 [U.S. Prior. 27.5.1953]. (131) J. R. MCLOUGHLIN, General Electric, German Appl. (West German) 1087350 [U.S. Prior. 26.3.1956]. (132) W. F. MCWHORTER, Devoe and Raynolds, Austral. Pat. 212546 [U.S. Prior. 16.9.1955]. (133) H. MECKBACH and W. NOLL, Farbenfabriken Bayer, German Appl. (West German) F 1328 [26.4.1950]. (134) R. L. MERKER, Dow Corning, French Pat. 1197925 [U.S. Prior. 1.2.1957]. (135) — and M. J. SCOTT, *J. Polymer Sci.* [A] **2**, 15 [1964]. (136) —, — and G. G. HABERLAND, ibid. [A] **2**, 31 [1964]. (137) — and C. R. SINGLETERRY, U.S. Pat. 2597018 [18.5.1948]. (138) — and W. A. ZISMAN, *Ind. Engng. Chem.* **41**, 2546 [1949]. (139) Midland Silicones, Brit. Pat. 765385 [U.S. Prior. 25.1. and 11.6.1954].

(140) —, Brit. Pat. 804199 [U.S. Prior. 22.12.1955]. (141) —, Brit. Pat. 867511 [2.5.1957]. (142) — Data sheet K2/LS-63, Oct. 1959. (143) *Mod. Plastics* **29**, No. 8, 106 [1952]. (144) A. F.

MOISEEV, *Plast. Massy* **1960**, No. 2, 24. (145) J. MOLL, *Chem. Ing. Techn.* **29**, 165 [1957]. (146) TH. S. MORONEY, *Rubber Wld.* **148**, No. 3, 34 [1963]. (147) R. MÜLLER and H. REUTHER, German Pat. (East German) 33 [19.4.1951]. (148) National Research Development, French Pat. 1077714 [17.4.1953]. (149) M. E. NELSON, Soc. of Plastics Engineers, 14th Nat. Techn. Conf., Techn. Papers, Vol. IV, 1958, p. 567.

(150) S. NITZSCHE, *Angew. Chem.* **63**, 490 [1951]. (151) — and E. PIRSON, Wacker-Chemie, German Pat. (West German) 940146 [20.6.1951]. (152) —, —, —, German Pat. (West German) 940855 [20.6.1951]. (153) —, —, —, German Appl. (West German) 1068841 [21.6.1956]. (154) —, —, —, German Appl. (West German) 1077355 [25.9.1956]. (155) — and R. RIEDLE, —, German Appl. (West German) 1069381 [28.2.1957]. (156) — and M. WICK, —, German Appl. (West German) 1058254 [29.9.1955]. (157) —, —, —, German Pat. (West German) 1097133 [27.7.1957]. (158) —, —, —, German Appl. (West German) 1127586 [12.9.1958]. (159) —, —, —, German Appl. (West German) 1118454 [16.7.1959].

(160) —, —, *Kunstoffe* **47**, 431 [1957]. (161) —, — and E. WOHLFARTH, Wacker-Chemie, German Appl. (West German) 1167020 [14.6.1962]. (162) S. J. NIZINSKI: Hot liquid vulcanization of silicone rubber insulated wire and cable; *Wire and Wire Products*, Feb. 1963, p. 227, 274. (163) W. NOLL, Farbenfabriken Bayer, German Appl. (West German) 1000570 [15.9.1953]. (164) —, —, German Appl. (West German) 1033894 [3.8.1955]. (165) —, —, German Appl. (West German) 1106070 [13.11.1959]. (166) —, K. DAMM and W. KRAUSS, *Farbe und Lack* **65**, 17 [1959]. (167) A. S. NOVIKOV, K. F. KALUZHENINA and Z. N. NUDELMAN, *Soviet Rubber Technology* (*Engl. transl.*) **18**, No. 5, 16 [1959]. (168) P. NOWAK and E. RICKLING, *Kunststoffe* **44**, 191 [1954]. (169) L. A. OKSENTEVICH, T. S. NIKITINA and A. S. KUZMINSKI, *Soviet Rubber Technology* (*Engl. transl.*) **18**, No. 5, 21 [1959].

(170) G. M. OMIETANSKI and W. G. REID, Sixth Joint Army-Navy-Air-Force Conference on Elastomer Research and Development, Boston, Oct. 18–20, 1960. (171) W. I. PATNODE, General Electric, U.S. Pat. 2503919 [11.1.1947]. (172) W. W. PEDERSEN, Dow Chemical, U.S. Pat. 2486674 [18.7.1945]. (173) —, Dow Corning, U.S. Pat. 2492498 and 2528606 [16.4.1946]. (174) J. PETER, H. KNOPF and W. NOLL, Farbenfabriken Bayer, German Pat. (West German) 1003441 [12.11.1953]. (175) C. W. PFEIFER, General Electric, U.S. Pat. 2666035 [15.11.1951]. (176) —, *Ind. Engng. Chem.* **46**, 2342 [1954]. (177) —, R. M. SAVAGE and B. B. WHITE, *India Rubber Wld.* **129**, 481 [1953]. (178) O. R. PIERCE, G. W. HOLBROOK, O. K. JOHANNSON, J. C. SAYLOR and E. D. BROWN, *Ind. Engng. Chem.* **52**, 783 [1960]. (179) R. M. PIKE and E. L. MOREHOUSE, Union Carbide, German Appl. (West German) 1114326 [U.S. Prior. 12.10.1956].

(180) A. N. PINES, —, U.S. Pat. 3088847 [26.5.1959; cip 25.7.1957]. (181) E. PIRSON and M. WICK, *Fette, Seifen, Anstrichmittel* **55**, 792 [1953]. (182) E. P. PLUEDDEMANN, Dow Corning, U.S. Pat. 2963501 [6.4.1959]. (183) — and H. A. CLARK, —, Belg. Pat. 613466 [U.S. Prior. 6.2. and 19.5.1961]. (184) I. A. PODDUBNII, S. V. AVERYANOV and L. A. AVERYANOVA, *Doklady Akad. Nauk. SSSR* (*Physical Chem. Sect.*) **139**, 651 [1961]. (185) K. E. POLMANTEER, Dow Corning, U.S. Pat. 2927907 [18.2.1954]. (186) — and M. J. HUNTER, *J. Appl. Polymer Sci.* **1**, 3 [1959]. (187) — and R. J. KOCH, *Ind. Engng. Chem.* **49**, 49 [1957]. (188) W. POSTELNEK, Air Force Development Program, *Rubber Wld.* **136**, 543 and 597 [1957]. (189) D. M. PREISS, N. V. de Bataafsche Petroleum, German Appl. (West German) 1052043 [U.S. Prior. 27.8.1956].

(190) W. H. RAGBORG, Dow Corning, U.S. Pat. 2890170 [6.9.1956]. (191) L. A. RAUNER, Dow Corning, German Pat. (West German) 937554 [U.S. Prior. 14.7.1952]. (192) —, —, U.S.

Pat. 2718508 [27.10.1952]. (193) J. REMOND, *Rev. des Produits Chim.* **65**, No. 1299, 370 [1962]. (194) W. T. RESSITER and C. C. CURRIE, Dow Corning, French Pat. 1068730 [U.S. Prior. 2.8.1951]. (195) R. RIOLFO, *Rev. Gen. du Caoutchouc* **34**, 352 [1957]. (196) Rhône-Poulenc, Brit. Pat. 621883 [French Prior. 27.7.1946]. (197) —, Brit. Pat. 640162 [French Prior. 30.11.1946] (C.A. **1950**, 10378). (198) —, French Pat. 943915 [10.3.1947]. (199) —, French Pat. 1188495 [12.7.1957].

(200) G. ROSSMY, *Chimia* **16**, 226 [1962]. (201) J. R. RUSSELL, Dow Corning, U.S. Pat. 3061575 [1.8.1960]. (202) J. B. RUST, Montclair Research and Ellis-Forster, U.S. Pat. 2628213 [16.4.1948]. (203) M. M. SAFFORD and A. M. BUECHE, General Electric, U.S. Pat. 2710290 [3.4.1953]. (204) TH. R. SANTELLI, Libbey-Owens-Ford Glass, Canad. Pat. 556017 [U.S. Prior. 29.3.1952]. (205) R. M. SAVAGE, *Rubber Age* **84**, 972 [1959]. (206) W. SCHMITT, R. PURRMANN, P. JOCHUM and W. D. ZAHLER, Espe, German Pat. (West German) 1166471 [26.10.1960]. (207) K. SCHNURRBUSCH, Farbenfabriken Bayer, German Appl. (West German) 1167527 [6.12.1961]. (208) G. R. SCHULTZE: Schmierfette, in C. Zerbe: *Mineralöle und verwandte Produkte*, Springer-Verlag, Berlin 1952. (209) A. M. SCHWARTZ, J. W. PERRY and J. BERCH: *Surface Active Agents and Detergents*, Vol. II, Interscience Publishers, New York 1958, p. 148.

(210) R. F. SHANNON and P. W. SULLIVAN, Owens-Corning Fiberglas, U.S. Pat. 3081269 [24.7.1961]. (211) L. M. SHORR, Dow Corning, U.S. Pat. 2924587 [7.3.1955]. (212) R. SMITH-JOHANNSEN, General Electric, U.S. Pat. 2744878 [27.2.1953]. (213) — and C. S. OLIVER, —, U.S. Pat. 2639276 [21.11.1950]. (214) L. V. SOBOLEVSKAYA, T. A. KRASOVSKAYA and M. V. SOBOLEVSKII, *Plast. massy* **1963**, No. 8, 22. (215) V. D. SOLODOVNIK, A. B. DAVYDOV, Z. G. IVANOVA, J. I. MINDLIN and N. S. LEZNOV, ibid. **1963**, No. 3, 39. (216) L. H. SOMMER, Dow Corning, U.S. Pat. 2589445 [3.6.1949]. (217) —, —, U.S. Pat. 2557802 [13.4.1950]. (218) F. J. SOWA and E. I. KENNEY, U.S. Pat. 2580473 [22.3.1946]. (219) J. L. SPEIER, Dow Corning, U.S. Pat. 2584751 [8.1.1951].

(220) —, —, U.S. Pat. 2676163 [3.9.1952]. (221) —, —, German Appl. (West German) 1080782 [U.S. Prior. 17.12.1953]. (222) —, —, German Appl. (West German) 1105183 [U.S. Prior. 17.12.1953]. (223) — and D. E. HOOK, —, U.S. Pat. 2823218 [5.12.1955]. (224) W. B. SPENCER JR., W. B. DAVIS, F. L. KILBOURNE JR. and J. C. MONTERMOSO, *Ind. Engng. Chem.* **45**, 1297 [1953]. (225) H. STAMM and H. ARENS, Th. Goldschmidt, German Pat. (West German) 894771 [13.6.1952]; German Pat. (West German) 900018 [7.8.1952]. (226) H. STEINBACH in O. GLENZ, *Melliand Textilberichte* **41**, 1125 [1960]. (227) —, Farbenfabriken Bayer, German Pat. (West German) 1161419 [23.12.1961]. (228) —, —, 1963, unpublished. (229) —, K. OTT and K. DAMM, —, Belg. Pat. 627282 [German Prior. (West German) 19.1.1962].

(230) R. S. STUART, ICI, Brit. Pat. 874242 [8.10.1958]. (231) E. A. SWAKON and C. G. BRANNEN, Standard Oil, U.S. Pat. 2710840 [18.11.1953]. (232) —, D. KREHBIEL and J. C. GOOSSENS, —, U.S. Pat. 2822334 [17.12.1956]. (233) E. SWEET, Dow Corning, Belg. Pat. 614394 [U.S. Prior. 27.2.1961]. (234) J. SWISS and C. E. ARNTZEN, Westinghouse, U.S. Pat. 2595727, 2595728, 2595729, 2595730 [9.3.1945]. (235) T.D. TALCOTT, Dow Corning, U.S. Pat. 2759904[25.2.1955]. (236) L. J. TYLER, —, U.S. Pat. 2863846 [4.5.1950]. (237) —, —, French Pat. 1145323 [U.S. Prior. 14.1.1955]. (238) —, —, German Appl. (West German) 1165867 [U.S. Prior. 11.1.1960]. (239) Union Carbide, Brit. Pat. 827402 [U.S. Prior. 2.4.1956].

(240) H. VINCENT and K. R. HOFFMAN, *Mod. Plast. Encycl.* **1962**, 406. (241) R. J. VOLKMANN and H. J. BAECKER, Dow Corning, U.S. Pat. 2755194 [7.2.1955]. (242) Wacker-Chemie, Belg. Pat. 607563 [German Prior. (West German) 26.8.1960]. (243) E. L. WARRICK,

Ind. Engng. Chem. **47**, 2388 [1955]. (244) — and P. C. LAUTERBUR, ibid. **47**, 486 [1955]. (245) K. WEISSERMEL and W. JOOS, Farbwerke Hoechst, German Appl. (West German) 1067598 [30.11.1957]. (246) B. R. WENDROW and D. P. SPALDING, *Rubber Wld.* **138**, 738 [1958]. (247) Westinghouse Electric, French Pat. 1153703 [28.5.1956]. (248) D. R. WEYENBERG, Dow Corning, U.S. Pat. 2714099 [11.1.1954]. (249) —, —, French Pat. 1153556 [U.S. Prior. 2.6.1955].

(250) D. E. WEYER, —, U.S. Pat. 2803606 [12.3.1954]. (251) —, —, French Pat. 1120340 [U.S. Prior. 12.3.1954]. (252) —, —, French Pat. 1135903 [U.S. Prior. 25.8.1954]. (253) —, —, French Pat. 1130307 [U.S. Prior. 4.10.1954]. (254) —, —, German Appl. (West German) 1017366 [U.S. Prior. 13.4. and 20.6.1955]. (255) —, Soc. of Plastics Engineers, 14th Nat. Techn. Conf., Techn. Papers, Vol. IV, 1958, p. 855. (256) —, *Mod. Plastics*, Encyclopedia Issue **1959**, p. 351. (257) — and K. R. HOFFMAN, *Mod. Plastics* **31**, No. 9, 133 [1954]. (258) M. WICK, *Kunststoffe* **50**, 433 [1960]. (259) T. C. WILLIAMS, R. A. PIKE and F. FEKETE, *Ind. Engng. Chem.* **51**, 939 [1959].

(260) H. A. WOODS and L. C. BOLLINGER, Shell Development, U.S. Pat. 2652364 [29.12.1951]. (261) J. H. WRIGHT, General Electric, French Pat. 1304847 [U.S. Prior. 31.10.1960]. (262) J. G. E. WRIGHT and C. S. OLIVER, General Electric, U.S. Pat. 2392713 [1.7.1944]. (263) M. YUSEM, USA Secretary of the Navy, U.S. Pat. 2870120 [31.5.1956]. (264) A. ZAPPEL, Farbenfabriken Bayer, German Appl. (West German) 1081663 [26.9.1957]; German Appl. (West German) 1089166 [21.12.1957].

CHAPTER 9

Properties of Technical Products

The silicones have sometimes been described as "miracle substances," and this scientifically inadequate expression has given rise to the danger that the usefulness of this group of materials will be overestimated. Correctly understood, however, the term applied to the silicones represents the special position that they occupy by virtue of their unique properties. This has been responsible for the relatively rapid acceptance of the silicones in practice and the approbation that they have found there.

9.1 General Picture of the Properties of the Silicones and Its Relationship to Molecular Structure and Bond Forces

9.1.1 Review of the Complex of Properties

The properties of silicone products depend to a considerable extent on the type of the organic substituents in the polyorganosiloxane skeleton and on the molecular structure of the skeleton. For polymethylsiloxanes and polymethylphenylsiloxanes, which have become the "classical" products of silicone chemistry and make up the bulk of technical products, the following properties are characteristic and have determined their field of application: heat resistance; low-temperature resistance; resistance to weather, ozone, and corona discharge; low variability of the physical constants with temperature; good dielectric properties; film-forming ability; hydrophobic behavior; release action; surface activity; and physiological inertness.

However heterogeneous and unrelated this juxtaposition of key words may appear at first, it gains in compactness when attempts are made to bring the individual data into relationship with molecular structure and bond forces. This will be done below; for simplicity, the expression "silicones" may here be used only for those polysiloxanes which contain essentially methyl or methyl and phenyl groups as organic substituents. Only when the complex of properties of siloxanes with different substituents varies will the substituents responsible be specially mentioned.

9.1.2 Fundamental Remarks on Heat and Low-Temperature Resistance

There is no question that the heat resistance of the silicones, which usually far exceeds that of all varieties of plastics, rubber, and oils, is an expression of the "silicate" nature of this group of substances. In this respect, silicones occupy an intermediate position between organic polymers and silicate materials such as glasses or enamels; however, even in favorable cases, they are not quite capable of closing the gap between the two extremes. Suitable silicone products can, in general, be heated in air from 180 to 200°C for a year without any appreciable change occurring in their properties; for a short time they can actually be subjected to 350 or 400°C, or even up to 450°C. This puts them in the forefront of all those products that can be processed and used in practice by methods of rubber technology or the technology of resinlike polymers or in the form of liquid phases at room temperature. Obviously, as regards their heat resistance the silicone resins are separated from the silicates by a range of several hundred degrees.

Moreover, it cannot be sufficiently emphasized how difficult it is to give an accurate definition of the "heat resistance" of technical silicone products. These difficulties reside on the one hand in the definition of the term itself, and on the other hand in the wide range of variation of the silicones.

The heat resistance of a material is a function, in the first place, of at least two parameters—the temperature and the time. An exact specification of heat resistance should therefore include information on how long a material can be heated at a given temperature without variation of its properties. It can immediately be seen that there is no sense in talking of a "*permanent* heat resistance" if it is intended to give a definition with accurate time parameters. If the term "permanent heat resistance" is nevertheless used, the limits of its informative capacity must be known. In general, it will be expected that a material to which a "*permanent* heat resistance" of 200°C is ascribed will remain capable of use in practice at this temperature for several years. The experimental evidence even of this, and all the more the evidence for even longer durations of stress, will often have to be dispensed with unless it is possible to carry out extrapolations with a sufficient degree of certainty (see Section 9.4.1).

Again, caution is advised when attempts are made to facilitate extrapolation by shortening the time of the test through raising the temperature, for this may provide the activation energy for reactions which do not take place at lower temperatures.

Information on heat resistance depends on other factors: the type of atmosphere or, more generally, the medium in which the thermal stress is applied, the extent of the boundary surface in contact with this medium, the nature of the change of temperature (whether the heat is applied uniformly or intermittently), the mechanical stress during the thermal exposure, and many others. It is obviously impossible, and even pointless, to attempt to take all these factors into account simultaneously in the test process. They vary in importance according to the type of product and the nature of the application for which the question of heat resistance comes into consideration.

Thus, even a discussion of the type of thermal stress show the difficulty of the problem, but the picture becomes even more complicated when the effects of shape, molecular structure, and nature of the substitution in the silicones are taken into account.

Silicones are degraded thermally either by the oxidative rupture of the Si—C bond or by depolymerization. If a silicone product is exposed to heat in air or oxygen, the *oxidative degradation* is a function of the extent of surface and also, in particular, the internal mobility of the molecules and their capacity for diffusion to the surface. For example, these factors play a relatively large part when silicone oils are heated in air and a smaller part when molded silicone resin products are subjected to the same treatment. In addition, the chemical structure has a decisive effect on the behavior. The oxidative degradation of polyorganosiloxanes begins with an attack of the oxygen on the organic group. The first degradation products in the thermal decomposition of polymethylsiloxanes in air have been found to be formaldehyde and formic acid (*7*, *99*). The oxidation susceptibility of aliphatic groups increases with increasing chain length, so that the heat resistance even of ethylsilicones and more particularly that of butylsilicones is lower than the heat resistance of methylsilicones. On the other hand, the stability of phenyl-substituted siloxanes in the air at elevated temperatures is superior to that of the methylsiloxanes, since, because of resonance stabilization in the nucleus, the phenyl group is less susceptible to oxidation than an aliphatic group. We shall continually encounter this thermal superiority of phenylsiloxanes in describing the properties of individual groups of substances. Figure 30, introduced at this point merely to illustrate the general phenomenon, shows the behavior of a methylsilsesquioxane and a phenylsilsesquioxane in air at various temperatures. The finely powdered substances, which have the nature of brittle resins, were each heated in air for 5 hours at predetermined temperatures and the reduction in their content of carbon and hydrogen was then investigated [after Zappel (*103*)]. It can be seen that at temperatures below 400°C (i.e., up to the upper limit of the temperature of

* Page number of chapter references.

practical interest for the use of silicone products) the phenylsiloxane is considerably superior to the methylsiloxane (cf. also refs. *5* and *131*). It is an interesting fact, which has not yet been explained, that the situation becomes reversed at higher temperatures.

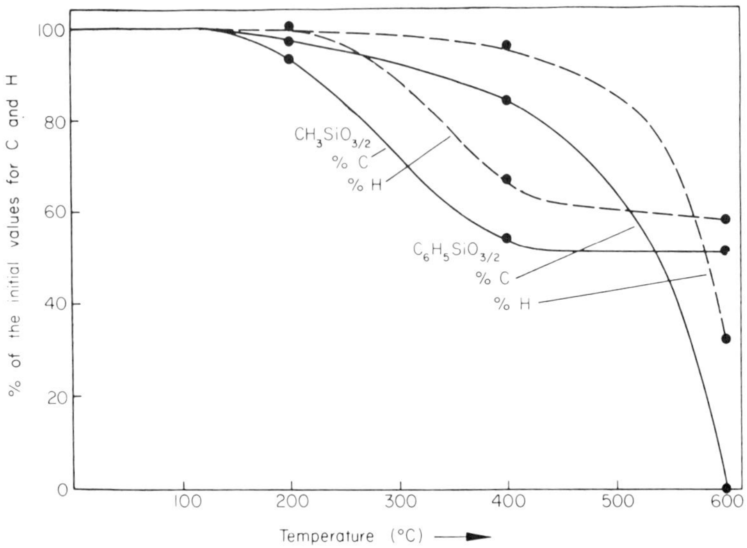

FIG. 30. Thermal behavior of methyl- and phenylsilsesquioxanes $[RSiO_{3/2}]_n$. Time of heating 5 hours in each case [after Zappel (*103*)].

Thermal depolymerization is promoted by residues of catalysts derived from the preparation of the polymers, by metal compounds of the type of Lewis acids, which can form on contact with metals if neutralization has not been complete, and by water vapor. The tendency of polymethylsiloxanes containing D units to undergo depolymerization with the formation of rings of lower molecular weight can be counteracted by the incorporation of phenylsiloxane units. The phenyl group brings about a strengthening of the siloxane bond by increasing $d\pi$—$p\pi$ bond contributions:

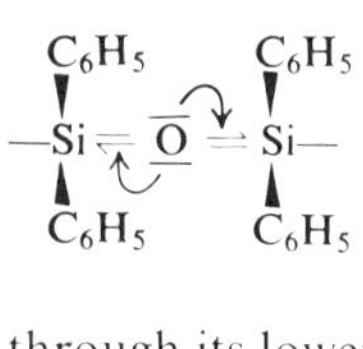

Thus, in this respect as well as through its lower susceptibility to oxidation, it has a favorable influence on the heat resistance of the polymers.

Another chemical factor determining heat resistance is the degree of cross-linkage of the siloxanes. Generalizing, it can be said that with the same type of organo substitution the heat resistance decreases with increasing degree of cross-linkage. This becomes particularly clear within the range of methyl-substituted highly cross-linked silicone resins, the structure of which approaches that of silica glass, as has been shown previously (see Section 6.3). The conversion of these resins into brittle silicalike products by the oxidative removal of the methyl groups and the joining up of siloxane bonds becomes easier and more rapid as they contain fewer organic groups. The heat resistance of such silicone resins is therefore, as experience shows, lower than that of, for instance, polydimethylsiloxane liquids or even of phenyl-substituted siloxanes.

Summarizing, it can be seen from these considerations that it is impossible to give a heat resistance for silicone products as a whole; all data must relate to definite products and can be understood only with a simultaneous definition of the conditions of thermal stress. Reference is made in this sense to the individual data in the following sections.

Corresponding considerations hold for the *low-temperature behavior* of the silicone products. It depends to a pronounced extent on the structure and type of substitution of the siloxane molecules and the nature of the stress.

All silicone products constructed on the basis of linear polydimethylsiloxanes are characterized by an extraordinary resistance to low temperatures. Methylsilicone oils with viscosities in the range of 60 to 300,000 cSt/20°C have pour points between −60 and −40°C (Table 78). The stiffening temperatures of silicone rubber vulcanizates based on polydimethylsiloxane rubber are below about −50°C and depend on the structure of the rubber.

The inclusion of even small proportions of branched siloxane units in silicone oils lowers the pour points considerably at the same viscosity, e.g., from −60 to −90°C for an oil with a viscosity of 60 cSt/20°C. The same measure lowers the stiffening temperatures of silicone rubber (Section 8.1.1). Similarly, the incorporation of small proportions of phenyl groups in linear polydimethylsiloxane molecules leads to a reduction of the pour points of oils (Section 9.2.2) and the stiffening temperature of rubbers (Sections 8.1.1 and 9.6.2). In both cases a minimum is reached at 6 to 8 moles % of phenyl groups.

The low sensitivity of the linear polydimethylsiloxanes to the action of cold must be brought into relationship with the weak intermolecular bond forces, in consequence of which there is only a slight tendency of the molecules to associate even at very low temperatures. With increasing indentation of the structure through branching positions, or indeed by cross-linking, this characteristic is lost. Highly cross-linked silicone resins are therefore not characterized by unusual behavior to cold.

The favorable effect of small contents of phenyl groups on pour points and stiffening temperatures may perhaps be ascribed to a disturbance of the state of order of the molecular chain. With increasing content of phenyl groups, this effect is again compensated or overcompensated, since the strengthening of the intermolecular bond by the phenyl substitution then makes itself increasingly noticeable.

The crystallization behavior of silicone rubber at low temperatures is to be understood in the same sense. Filled silicone rubber vulcanizates based on polydimethylsiloxanes form partially crystalline regions only at temperatures below −50°C (*109*, *162*), while, for example, natural rubber crystallizes spontaneously in a few hours at −20°C (*150*). By incorporating only 3.5 mole-%

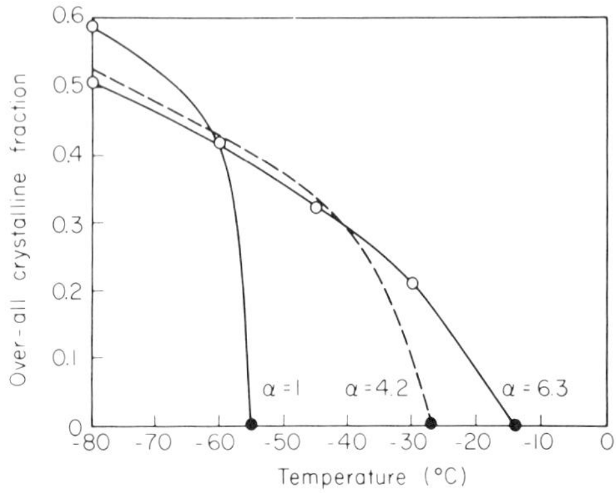

FIG. 31. Over-all crystalline fraction in silicone rubber as a function of temperature for α (degree of elongation) = 1, 4.2, and 6.3 (*109*).

of $[CH_3(C_6H_5)SiO]$ units, the crystallization temperature is reduced to about −70°C. When the content of these units is 7.5 mole-%, crystallization no longer takes place (*119*).

To what extent other parameters must also be taken into account to explain the low-temperature behavior of the substances is shown by an investigation of the crystallization behavior of silicone rubber at various degrees of stretching. With increasing stretch, the crystallization temperature becomes higher, since the stretching favors the orientation of crystallites (Fig. 31) (*109*).

9.1.3 General Questions of Resistance

Silicone possess good-to-outstanding resistance against the effects of weather, against ozone, and against glow discharges. This behavior is explained by the inert nature of the polyorganosiloxanes and the absence of reactive groups and, particularly, of double bonds. The weathering resistance is favored and promoted by the high interfacial tension of all silicones against water.

In spite of these aspects, which are favorable for many practical purposes, it must not be overlooked that the Si—C bond is capable of certain chemical reactions to a higher degree than the C—C bond (see Section 6.4), so that the resistance of the silicones to many chemicals is lower than that of purely organic polymers. Since these questions are of very special nature and again depend particularly on the structure of the silicone product in question, reference must be made to the discussion of the individual products.

9.1.4 Temperature Dependence of Physical Properties

The physical properties of the silicones change with the temperature to such a small extent that the silicones form a completely new class in this respect. We may merely refer here to the low temperature-dependence of the viscosity of silicone oils (Section 9.2.4), the small variation of the dielectric properties of silicone resins with changes in temperature (Section 9.4.6), the damping behavior of silicone rubber (Section 9.6.4), etc.

The low temperature-dependence of the viscosity of silicone oils has been related to the chemical structure of the linear polysiloxane molecules (*41*). According to these ideas, it would be the result of the superposition of two effects: on the one hand, a rise in temperature will increase the mean intermolecular distance while, on the other hand, it will expand the helices and thus diminish this distance. The expansion of the helices opposes the effect of the increase of distance and may compensate it to a certain extent, so that in the final effect the mean intermolecular distance, and therefore the viscosity, appears to be affected only slightly by the temperature. If, however, it is considered that silicone resins with their sometimes highly cross-linked molecules exhibit similar small temperature effects on their physical properties, it appears desirable to consider the phenomenon apart from the special shape of the molecules and to regard it as an expression of the low strength of the intermolecular bond, the temperature coefficient of which may be assumed to be correspondingly low.

9.1.5 Film-Forming Ability

Methylsilicone oils spread rapidly on surfaces of solids and liquids to form *films of molecular dimensions*. Such films impart the typical silicone properties to the surface: they make them water repellent and reduce their adhesion to sticky materials. The combination of film-forming capacity, water repellency, and release action has become one of the most interesting properties of the silicones. On it is based the possibility of using silicones as water-repellency agents, as release agents, and for the purposes of surface finishing.

The spreading behavior of methylsilicone oils has been investigated in detail with respect to films produced on water (*41*, *42*, *104*), and on oleic acid, olive oil, and ethylene glycol (*10*, *34*). In monomolecular surface films, the molecules have different shapes according to the surface pressure and are oriented differently with respect to the surface (see Section 6.2.1). However, it is most remarkable that any replacement of the methyl groups by larger aliphatic or by aromatic hydrocarbon residues as substituents markedly decreases the spreading capacity of the oils and the capacity of the molecules for becoming orientated on polar surfaces (*102*). Even the replacement of one methyl group in polydimethylsiloxanes by an ethyl group has a remarkable effect in this direction. Methylphenylsiloxanes spread on water (*102*) or on the surfaces of solid bodies (*123*) much more slowly than liquid polydimethylsiloxanes. Consequently, methylsilicones play a special role in silicone chemistry as surface-active compounds. The tenacious adherence of industry to methylsilicones in particular is largely explained by the special position which they occupy in virtue of this fact.

The film-forming capacity of methylsilicone oils can be understood as a result of the low strength of the intermolecular binding forces revealed in their low surface tensions. Consequently, in general the work of cohesion $2\sigma_f$ of the liquid phase having a surface tension σ is smaller than the work of adhesion ζ_{sf} to the surface of the substrate, so that the spreading pressure

$$p_{sp} = \zeta_{sf} - 2\sigma_f$$

becomes positive (*166*).

Consequently, the surface film forms the more easily the greater the "affinity" which the film-former and the substrate have for one another; the strength of adhesion of the film increases in the same sense. Skins preventing or impeding access to the surface of a solid body must be removed; this explains why, in order to obtain a closely highly adherent film, it is necessary to remove size from textile or glass fibers or to desorb adsorbed films of water from solid bodies before the silicone treatment.

Under certain conditions, however, a film of water present at the solid/gas phase boundary may also be displaced more or less rapidly by a silicone film. Whether and how far this occurs depends on the affinity of the water and the silicone for the surface of the solid. The adsorbed film of water is displaced by silicone oil from the cleavage surfaces of galena and zinc blende (i.e., from minerals which are in fact somewhat hydrophobic) substantially more rapidly than from the surface of the hydrophilic silicates (*147*). The marked difference in the velocity of the displacement of the water layer, and therefore the imparting of water repellency to various minerals makes possible a selective flotation separation with silicones (cf. Section 10.14.2).

The affinity of the film-forming agent for a crystalline substrate is also increased if there is a possibility of an oriented deposition (*100*), since, on the one hand, a film with ordered molecules forms a state of lower energy than one with an unordered molecular structure and, on the other hand, the ordered molecules give a better saturation of the surface valences. Again, investigation of the flotation phenomena with silicones and systems consisting of galena, zinc blende, and silicate minerals supports this statement (*147*).*

The ratios of the distances of the Pb atoms on the {100} cleavage faces of galena and the O—O distances in the siloxane chain are such that the siloxane chain can adhere to a Pb atom at every other oxygen atom in the manner shown in Fig. 32; the situation with the {110} plane of zinc blende is similar. A discussion of the fluorspar lattice shows that, because of the atomic separations and molecular dimensions, *no* oriented adhesion of the molecules is to be expected here. In harmony with this is the fact that fluorspar cannot be floated with silicones.

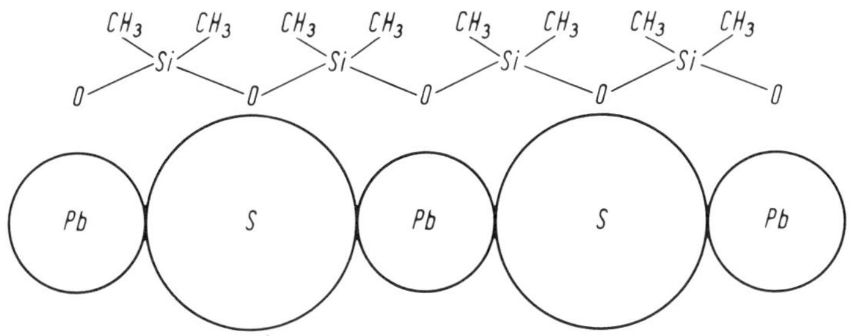

FIG. 32. Orientated adhesion of a polydimethylsiloxane chain to a {100} galena cleavage face [after Steinbach (*147*)].

Because of the semipolar nature of the siloxane bond, the binding forces with which organosiloxane molecules can adhere to surfaces are dipole or hydrogen-bonding forces. But the formation of the film and its adherence can be substantially enhanced by chemical reactions with reactive surface groups. In this way, it is possible for polymethylsiloxane molecules to be anchored to glass surfaces at high baking temperatures by the occasional cleavage of a Si—CH_3 bond through the simultaneous action of oxygen and water followed by condensation with a Si—OH group of the surface to form a primary valence bond:

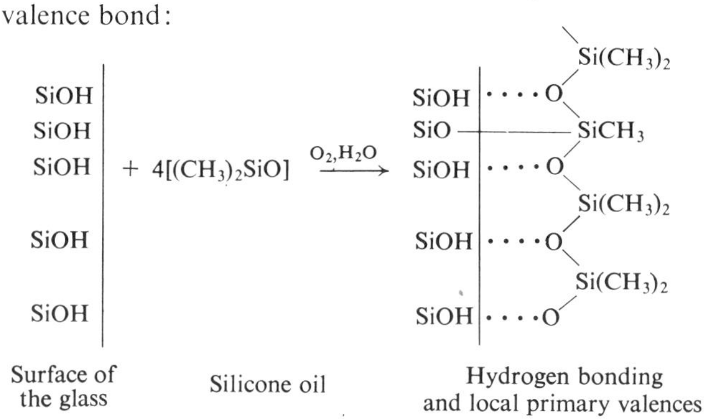

* The same idea has also been discussed repeatedly in the literature for other collectors; for literature see ref. *34*.

However, surface coatings can be obtained through chemical reactions much more easily with silanes or siloxanes which contain chemically reactive bonds such as the Si—Cl, Si—OC_2H_5, Si—OH, and Si—H bonds. When such substances are brought into contact with a surface containing reactive partners, primary valence bonding arises, somewhat in the manner of the reaction of dimethyldichlorosilane with the Si—OH groups of a quartz or glass surface:

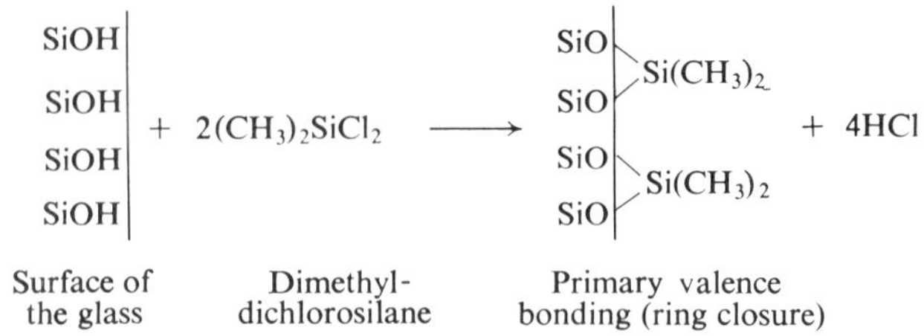

Or, if adsorbed water is present simultaneously and takes part in the reaction, as follows:

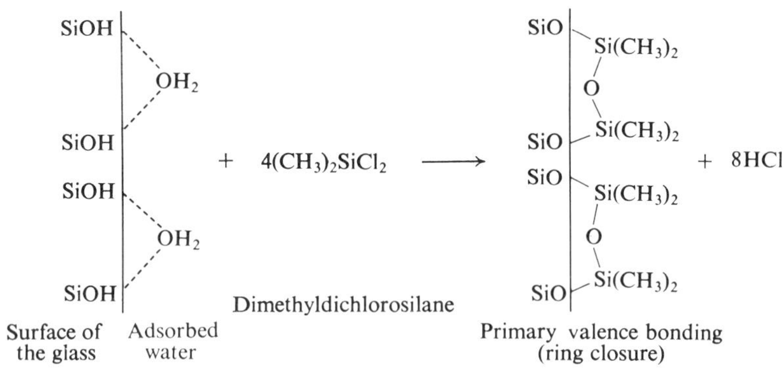

A "surface compound" is produced and the skeleton of the siloxane chain now grows directly without transition into the framework of the structure of the quartz or glass (*105*).

It may be concluded from desorption measurements on quartz that the concentration of OH groups on the surface is about 8 μmole/m^2 (*148*). If trimethylchlorosilane is caused to react with the Si—OH groups of the surface, it can be deduced from a discussion of the stereometric relationships that, because of the space occupancy of the trimethylsilyl group, only every third hydroxyl group can react (Fig. 33); consequently, it may be expected that only slightly more than 2 μmole of trimethylchlorosilane/m^2 can be fixed by chemical reactions. Measurement has, in fact, shown that the chemical adsorption amounts to 2.5 μmole/m^2 (*149*).

Adsorbents whose surfaces contain hydroxyl groups but no absorbed film of water accordingly form a monomolecular surface layer with trimethylchlorosilane. Since completely desorbed substances are never used in practice, treatment with chlorosilane vapors leads, with the participation of the absorbed water, to polymolecular layers whose thickness obviously depends on the particular conditions of working. It has been shown by photobromination with radioactive bromine that a silicone film produced on glass must consist of at least 10 to 30 layers of molecules (*156*).

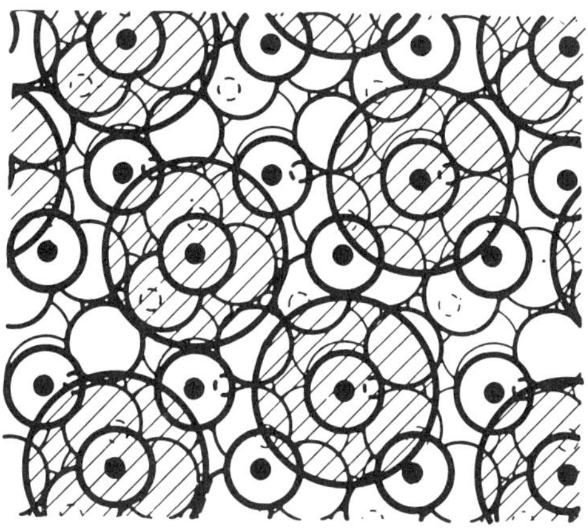

FIG. 33. Maximum possible density of coating of a rhombohedric cleavage face (1011) of quartz—as the most frequent face of cleavage—with trimethylsilyl groups [after Stöber (*149*)]. Hatched circles: $(CH_3)_3Si$ groups; large circles: oxygen ions of the quartz; small circles: silicon ions of the quartz; and black circles: hydrogen ions of the hydroxyl groups on the surface of the quartz.

9.1.6 Hydrophobic Behavior and Release Action

It is a general and characteristic property of all silicones that they are hydrophobic. By "hydrophoby" is understood that water in the liquid phase wets a silicone/air boundary surface to a relatively small extent. This must not be confused with the behavior of water vapor with respect to a silicone: silicones, like all plastics, absorb water vapor and are permeable to it to a certain extent.

A measure of the water repellency of a surface which is important in practice is the magnitude of the contact angle θ which small drops of water placed on the surface make with it (Fig. 34). The contact angle θ, the surface tensions of the solid σ_s and the liquid σ_f, and the interfacial tension γ_{sf} are connected by the relation

$$\sigma_s = \gamma_{sf} + \sigma_f \cos\theta$$

The work of adhesion ζ_{sf} (i.e., the work necessary to remove the liquid from the surface of the solid) is connected with the surface tensions and the interfacial tensions, according to Dupré, by the relation

$$\zeta_{sf} = \sigma_s + \sigma_f - \gamma_{sf}$$

According to Young, the two equations give the work of adhesion as

$$\zeta_{sf} = \sigma_f(1 + \cos\theta)$$

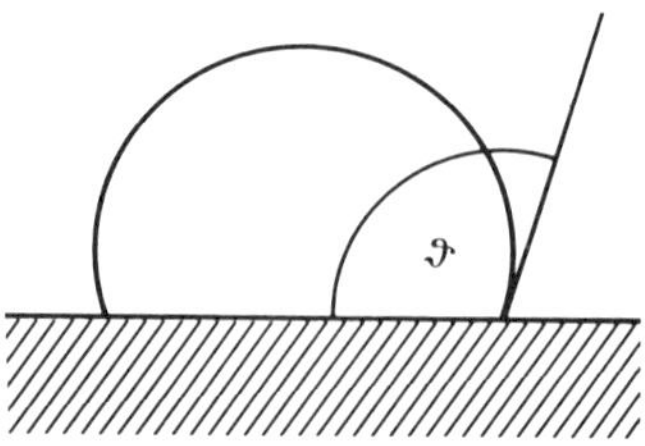

FIG. 34. Illustrating the definition of the contact angle θ, where s stands for solid, f for fluid, and g for gas.

The work of adhesion can therefore easily be derived from measurement of the contact angle, if the surface tension of the liquid phase is known. On the other hand, Young's equation shows that the contact angle depends on the ratio of the work of adhesion between the liquid and the surface of the solid to the work of cohesion of the liquid.

Accurate measurement of the contact angle is subject to considerable difficulties; θ can be strongly affected by traces of surface-active materials dissolved in the liquid phase, by the structure of the surface of the solid, by the phenomenon known as contact angle hysteresis (*166*), etc. Consequently, only an indicative value should be attached to the figures given below, which, moreover, are derived from measurements by various authors using different methods.

The greatest values of the contact angle that have so far been determined between water and smooth surfaces are 100 to 110° (*2*). Contact angles of this magnitude have been measured on paraffin wax. Silicone surfaces, if they are smooth and suitably treated, exhibit similar figures; they are found with glass surfaces on which films of silicone oil have been deposited and fixed by heating.

If a carefully degreased lime-soda glass is immersed in a 0.5% solution of a polydimethylsiloxane oil in chloroform, the solvent is allowed to evaporate, and the contact angles of drops of water that have been deposited are measured, figures of the order of 60° are obtained (see Table 76). If the glass with the silicone film is heated to temperatures of up to 400°C, the contact angle becomes larger, rising to 105°. At still higher temperatures, the angle decreases again, owing to incipient decomposition of the silicone (*105*) (Figs. 35 and 36). With rising temperature, however, not only the water-repelling effect but also the film adhesion are improved. While the silicone film applied at normal temperatures can be emulsified off again by solutions of wetting agents, this is no longer possible with the "baked-on" films.

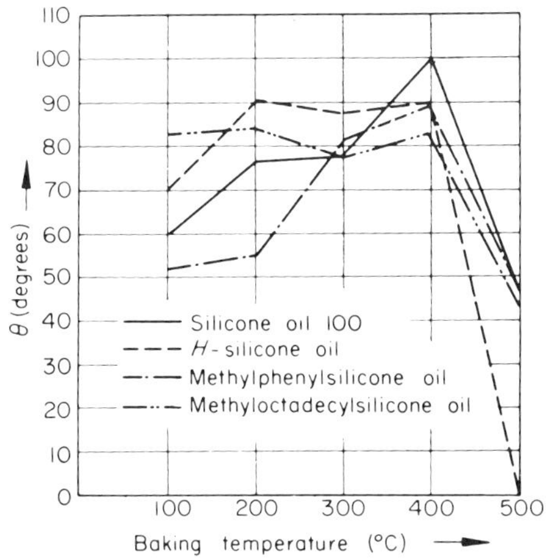

FIG. 35. Hydrophobic effect of films of silicone oil on glass evaluated by the magnitude of the contact angles of drops of water as a function of the baking temperature. All oils applied in the form of 0.5% solutions in chloroform.

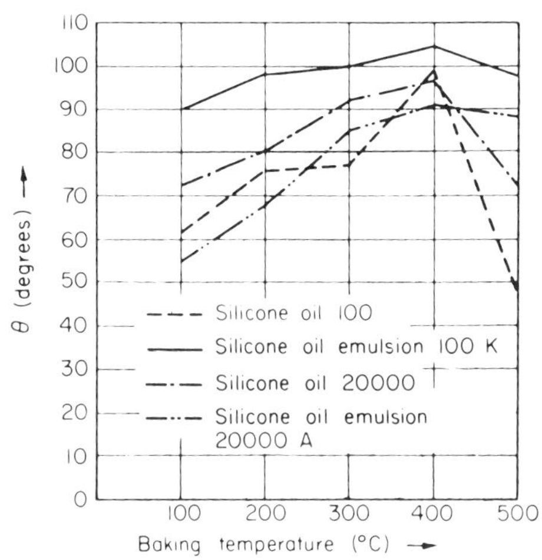

FIG. 36. Hydrophobic effect of films of silicone oil on glass evaluated by the magnitude of the contact angles of drops of water as a function of the baking temperature. Emulsions with 0.5% of silicone oil. Silicone oils in the form of 0.5% solution in chloroform for comparison.

TABLE 76

CONTACT ANGLES AND WORKS OF ADHESION IN SOME WATER/AIR/SOLID SYSTEMS

Solid	Contact angle (deg)	Work of adhesion (erg/cm^2)
Hexatriacontane	110	48
Polytetrafluoroethylene	108	50
Paraffin waxes	~105	~54
Polydimethylsiloxane oil as a film on glass, heated at 400°C	105	54
Polyethylene	94	68
Long-chain alcohols	~90	73
Polydimethylsiloxane oil as a film on glass, heated at 100°C	60	109
Long-chain dicarboxylic acids	45	123
Glass (degreased)	0	>145

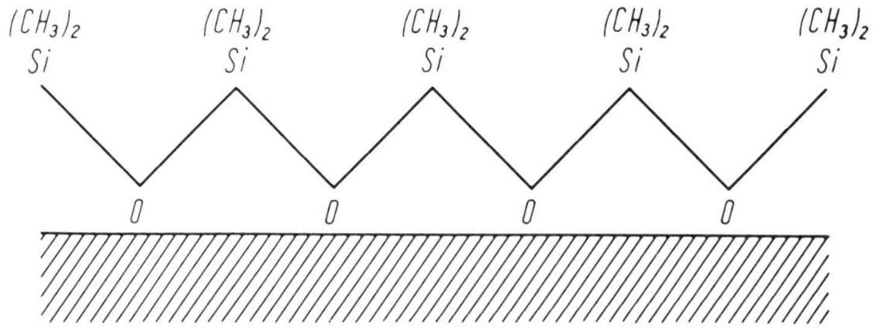

FIG. 37. Polydimethylsiloxane molecules spread on a solid surface after desorption of passivating molecules.

These results must be explained on the assumption that a rise in temperature increases the surface forces of the solids through the desorption of passivating films, the polyorganosiloxane chains being forced preferentially into an orientation such that the siloxane dipoles point to the phase boundary while the free surface is covered by methyl groups in close packing (Fig. 37). An increase in the density of the packing of the methyl groups raises the hydrophobic effect, and, finally, chemical reactions at still higher temperatures increase the adhesion of the film.

This picture, which is intended only to show the significance of an activation of the surface for the success of water-repellency treatment with silicones, may form a guiding line for water-repellency problems with other substances besides the special glass/silicone system. Obviously, the method and procedure for "activation" may be quite different with different substrates, and it is not always possible to achieve this objective by desorption. On the contrary, in many cases, as in the hydrophobing of textiles, the formation of "intermediate layers" which increase the affinity between substrate and silicone films also gives the desired result.

The phenomenon of the dependence of hydrophobic properties on the orientation of molecules in the silicone film has a counterpart in the observation, made on stearic acid, that the contact angle of deposited drops of water can vary between 50 and 105° according to the orientation of the cut surface in the crystals of the acid and according to whether hydrocarbon residues or carboxyl groups are preferentially exposed at the cut surface (*2*).

To a certain extent, the hydrophobic effect that can be obtained with silicones depends also on the nature of the substituents linked to the silicon atom. Silanes or siloxanes containing higher alkyl groups have a stronger hydrophobic effect on glass substrates at low temperatures than the corresponding compounds containing methyl groups; on the other hand, the hydrophobic effect is impaired by substitution with phenyl groups. Oils containing Si—H bonds have remarkably strong water-repellency imparting properties in spite of the relatively low content of organic groups (*63*, *105*). The polydimethylsiloxane oils react most strongly when the temperature is increased (*105*), possibly because their molecules, in contrast to those of siloxanes with bulky organic groups, can most easily direct themselves with respect to the surface in such a way that a phase boundary occupied by methyl groups can arise (Fig. 37).

The possibilities of obtaining water-repellent silicone films on surfaces are enhanced by the use of silanes or siloxanes with reactive groups (see Section 9.1.5).

When the degree of cross-linkage of the siloxanes increases and the number of organic groups per Si atom decreases, a weakening of the hydrophobic effect should be expected, since these changes lead to an increase in the intermolecular forces and a decrease in the film-forming ability. In practice, the effects are not serious provided that one remains in the region of resinlike siloxanes, so that cross-linked resinlike siloxanes are also good water-repellency agents. Because of the high content of siloxane bonds, they have better adhesion than silicone oils to silicate substrates—such as building materials and ceramics, in particular—even at low temperatures, and rapidly make these materials water repellent.

If the efficiency of the silicones is compared with that of the paraffin waxes widely used as water-repellency agents, it will be seen that differences exist less in the values of the contact angle that can be obtained under optimum conditions and more in the behavior with respect to the substrate. Under certain conditions, silicone oils can be anchored particularly firmly to solids; they are capable of spreading to a high degree, and exhibit their typical properties even in molecular films. Consequently, they are outstandingly suitable for water-repellency treatment when the properties of the material must not be changed by the treatment. The porosity of porous materials is not, or not substantially, reduced by silicone treatment; permeability to air and even to water vapor remains unchanged. However, liquid water is prevented from entering the pores because of the increase in the interfacial tension, as Fig. 38 shows with a textile as an example. This bar to the entry of water becomes ineffective only if the water is under a certain static pressure. The breakthrough limit depends on the pressure, the interfacial tension, and the pore size (cf. Section 10.12.3.1).

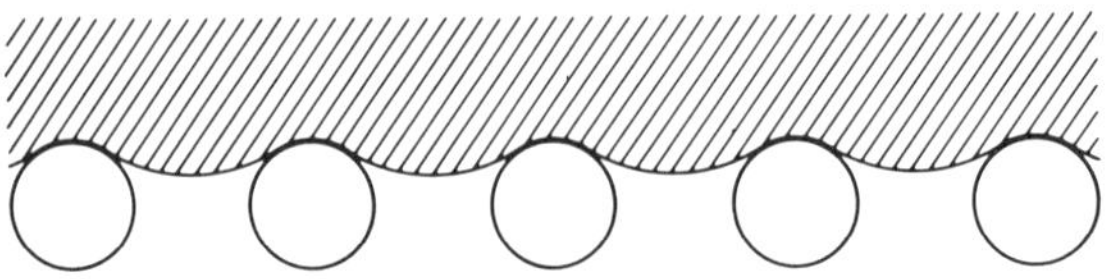

FIG. 38. Pore-retaining hydrophobing of a textile with silicones (schematic). Hatched: water; circles: cross section of the fibers.

Very closely related to the water-repelling behavior of the silicones is the fact that they have an outstanding release action with respect to adhesive materials such as high organic polymers of the nature of rubber or plastics, adhesives, and gel-like or greasy products (including foodstuffs). Like water repellency, this property is an expression of a fundamental chemical characteristic of the polymeric organosiloxane skeleton: its paraffinlike nature, the saturated character of the molecules, and the corresponding smallness of Van der Waals forces which permit the appearance of only slight, if any, affinity for substances of different nature.

9.1.7 Surface Activity

Practical acquaintance with silicones shows that they may act as *foaming agents* and *antifoaming* or *defoaming agents* or *foam suppressors*; they regulate the *pore structure* in foamed materials; they act as *collectors* with respect to mixtures of minerals; they affect the *flow* and *surface structure* of paint and varnish films; they prevent the *segregation* of pigments in paints. We are far from being able to explain all these phenomena in detail, but they can all be regarded as expressions of the *surface activity* of the silicones. The situations in foaming, foam suppression, and flotation seem to be most easily capable of explanation.

The fact that silicones are surface active in the sense that they are enriched in the surface of liquid phases in which they are soluble follows from the lowering of the surface tension produced by silicones in organic solvents

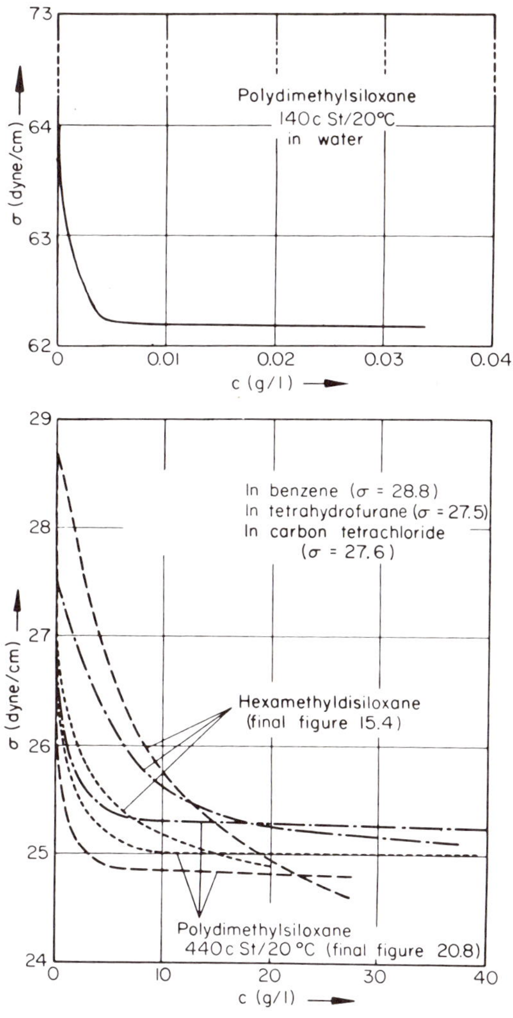

FIG. 39. Variation of the surface tension of organic solvents and water due to the addition of silicone oil. Top: emulsion of a polyorganosiloxane in water [after Steinbach (*147*)]; bottom: solutions of polyorganosiloxanes in organic solvents [after Wolf (*166*)].

(Fig. 39). The emulsification of a silicone oil in water leads to a reduction in the surface tension of the water (Fig. 39, top), which indicates that silicones are capable of forming a surface solution in aqueous systems (*147*). In flotation with silicones, the silicone film is probably transferred to the surface of the mineral particles by means of such surface solutions, which form at the liquid/gas phase boundary (*147*).

The lowering of the surface tension by surface solutions is the reason why silicones can act as foaming agents. Their mode of action depends on the concentration. If the concentration of silicone in the aqueous system is increased, defoaming action becomes increasingly noticeable. In the foam flotation process, the regions in which flotation can be effected with silicones can be delimited from that in which they act as foam-suppressing agents as a function of the silicone concentration. The defoaming action appears to set in when condensed and, particularly, polymolecular surface films are formed (*147*).

These results can be brought into harmony with Gibbs' theory that the stability of a foam increases with increasing surface elasticity of the system (*47*, *138*). Surface-active substances are capable of taking up mechanical and other effects on bubbles elastically; they react to an increase in the surface of a bubble with a reduction in the concentration of the surface-active material, so that the surface tension is raised while, conversely, a reduction in the surface of the bubble brings about an increase in the concentration of the surface-active substance and therefore a lowering of the surface tension. Consequently, the most stable foams are obtained not at those concentrations of surface-active substances at which the surface tension of the system already has the lowest figure but in that range in which the surface tension changes in a particularly marked fashion with the concentration, i.e., where the ratio $d\sigma/dc$ is high. In aqueous systems, this is the case at very low silicone concentrations (Fig. 39). If a condensed polymolecular surface film is formed by increasing the amount of silicone added, the surface elasticity is reduced and the system undergoes defoaming.*

The theory of defoaming also requires *inter alia* (*31*) that the defoaming agent must be as insoluble as possible in the medium to be defoamed, since high surface concentrations are easier to obtain when the amount of substance taken up by the total volume of the liquid is smaller; moreover, the spreading pressure must be positive (*126*) so that a closed film can form rapidly on the surface of the foam lamellas, and the foam-suppressing agent must not form a mixture with the foaming agent in the surface film. All these requirements are probably satisfied particularly well by methylsilicones in aqueous systems, since they have a minimum volume solubility in water, are capable of spreading (Section 9.1.5), and are characterized by very low compatibility with many organic and most inorganic substances.

However, in all discussions of surface activity it must be borne in mind that the substituents of the polysiloxanes have a great influence on their surface-active properties. When the number of carbon atoms in the aliphatic R groups

* There is as yet no general theory of foam suppressors. For the ideas developed specially for silicones cf. refs. *12*, *34*, and *126*.

is increased, the spreading ability falls rapidly since, because of the accumulation of hydrocarbon residues, the polyorganosiloxane becomes increasingly paraffinlike (Fig. 39a). Similarly, electronegative substituents, such as in particular phenyl groups, reduce the spreading ability. This is explained as a consequence of electron attraction that the substituents exert on the lone electron pairs of the siloxane oxygen atoms, which reduces their capacity for forming hydrogen bonds with the protons of water (cf. Section 6.3.3.3). In polymethyl-*H*-siloxanes the hydrogen atom attached to silicon is negative and therefore the Si—O dipole of the siloxane chain is weakened. Consequently, their spreading ability is lower than that of polydimethylsiloxanes.

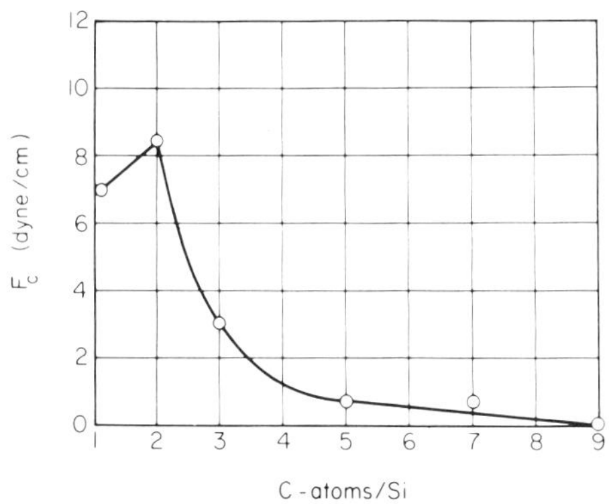

FIG. 39a. Variation of the spreading pressure (F_C) with the C number of the alkyl groups in polyalkylmethylsiloxanes.

If a siloxane is cocondensed with highly hydrophilic organic components, the surface-active properties of the latter may completely overshadow those of the siloxane. In accordance with the classical principles of the structure of tensides, such molecules combine extremely hydrophilic and hydrophobic (siloxane) components; their hydrophilic nature can be adjusted within wide limits as a function of the ratio of the two components.

9.2 Properties of Technical Silicone Oils

9.2.1 The Products

A review of the constitution of the types of silicone oils that have become known as technical products is given in Table 77.

Their technical importance is very variable. The bulk of the technical silicone oils consists of methylsilicone oils (I). Next in importance are methylphenylsilicone oils of various types (II), (III) and (V), and those with chlorine substituents in the phenyl group (VI) (*1*). The oils of types (IV), (VII), (VIII) and

(IX) are required only for special purposes. Silicone oils with Si—H bonds have become of particular interest for impregnation purposes, and will be described in this connection in Section 10.8. The following remarks therefore center essentially on silicone oils with methyl or with methyl and phenyl substituents.

TABLE 77

TECHNICAL TYPES OF SILICONE OILS

Constitution	Symbol	Short name
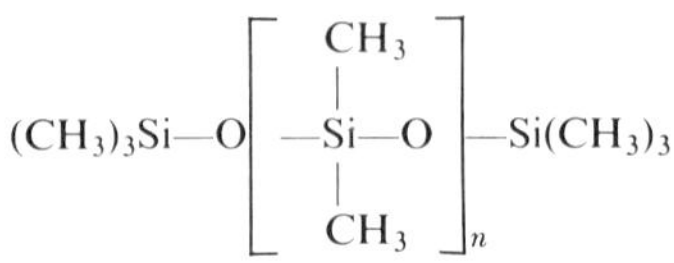 I	MD_nM	Methylsilicone oils
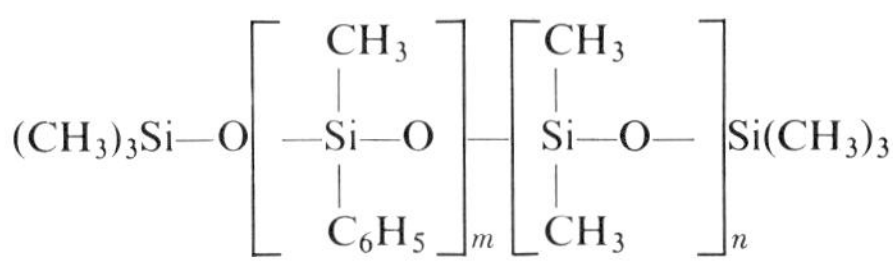 II	MD_nM	Methylphenylsilicone oils
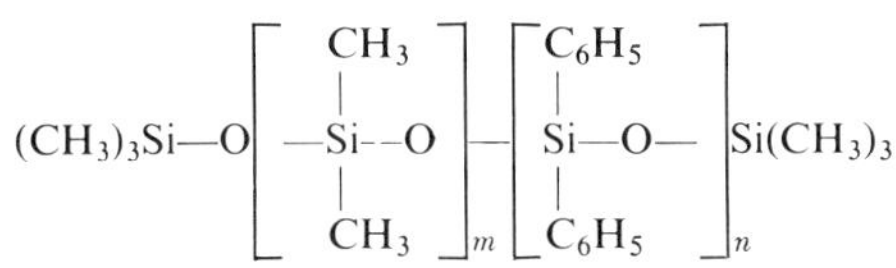 III	MD_nM	Methylphenylsilicone oils
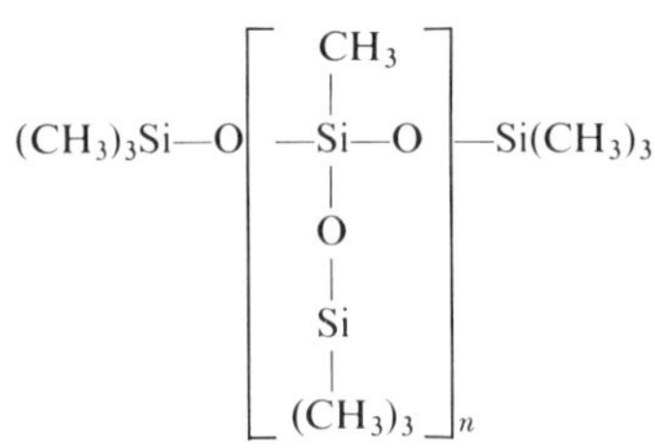 IV	$M\begin{bmatrix}T\\M\end{bmatrix}_n M$	Branched methylsilicone oils
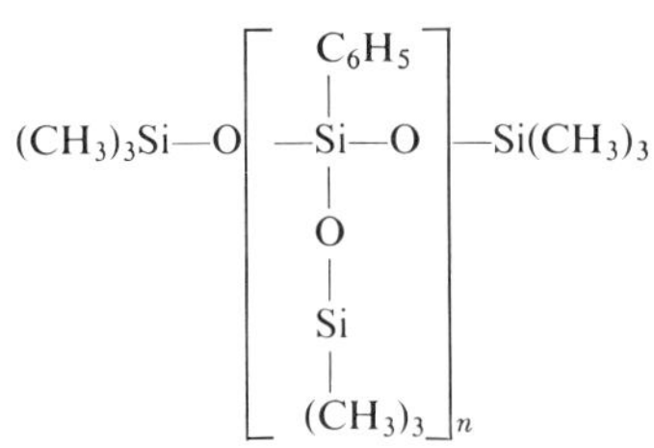 V	$M\begin{bmatrix}T\\M\end{bmatrix}_n M$	Branched methylphenylsilicone oils

TABLE 77 (*contd.*)

TECHNICAL TYPES OF SILICONE OILS

Constitution	Symbol	Short name
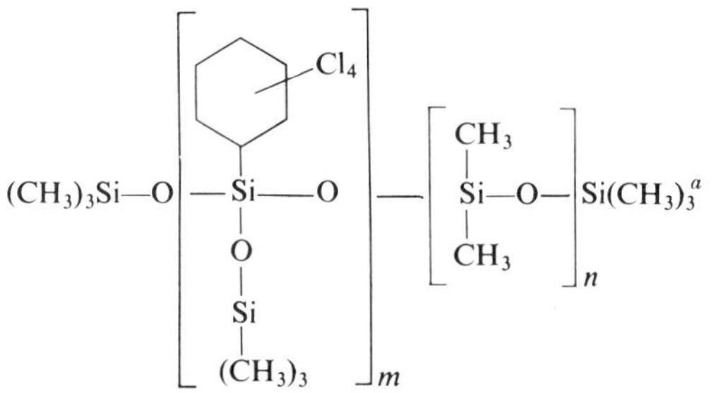	$M\begin{bmatrix}T\\M\end{bmatrix}_m D_n M^a$	Branched chlorophenyl-methylsilicone oils
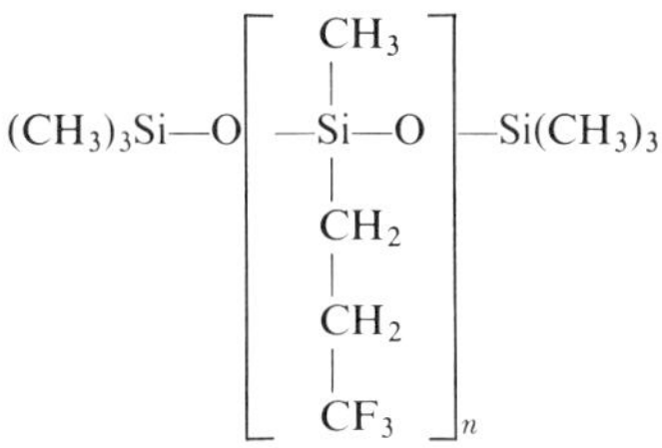 VII	MD_nM	Fluorosilicone oils
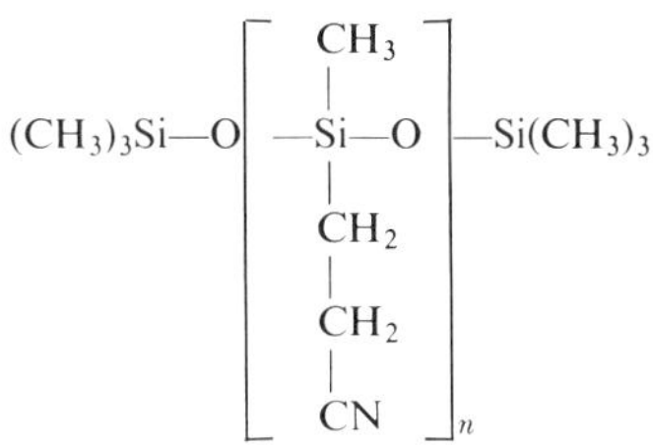 VIII	MD_nM	Nitrilesilicone oils
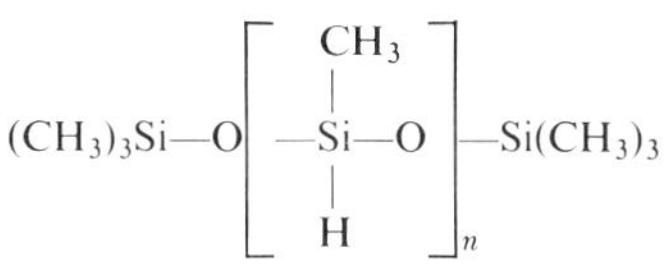 IX	MD_nM	Methyl-*H*-silicone oils

[a] Simplified formulations.

It is one of the most striking properties of the linear *polydimethylsiloxanes* that they form *liquid phases* over an unusually large range of molecular weights: this ranges from $M = 162$ (hexamethyldisiloxane) to the order of magnitude of

500,000, i.e., to molecular sizes at which other polymers are already solid. The range of viscosities is correspondingly wide. It begins at 0.65 cSt/20°C for hexamethyldisiloxane and extends into regions that are no longer accessible to the means used in viscometry. The series of technical silicone oils consist mostly of polydimethylsiloxanes, the viscosity of which extends from 0.65 to about 1,000,000 cSt (Table 78). At still higher viscosities, we reach the silicone rubber polymers from which rubber-elastic vulcanizates can be prepared by cross-linking reactions.

TABLE 78

PHYSICAL PROPERTIES OF SOME TECHNICAL METHYLSILICONE OILS AND METHYLPHENYLSILICONE OILS

ν_{20}°C (cSt)	Pour point (°C)	Flash point (°C)	Flame point (°C)	d_4^{20}	n_D^{20}
		Methylsilicone Oils			
60	< −60	> 300	> 350	0.96	1.4041
140	< −50	> 315	> 380	0.97	1.4045
440	< −50	> 315	> 380	0.97	1.4053
680	< −50	> 315	> 380	0.97	1.4053
1,440	< −50	> 320	> 390	0.97	1.4053
10,000	< −50	> 320	> 390	0.97	1.4058
50,000	< −50	> 320	> 390	0.97	1.4058
100,000	< −50	> 350	> 400	0.97	1.4058
300,000	< −40	> 350	> 400	0.97	1.4058
		Methylphenylsilicone Oils Low Phenyl Content			
200	~ −65	> 300	> 360	1.03	1.465
1000	~ −55	> 315	> 360	1.04	1.475
		High Phenyl Content			
300	~ −40	> 300	> 360	1.06	1.505
1000	~ −30	> 305	> 360	1.09	1.515
		Branched Methylphenylsilicone Oils			
5	~ −102[a]	130	160	0.92	1.436
10	~ −70[b]	145	175	0.98	1.493
25	~ −78[c]	170	200	0.99	1.457
75	~ −62[d]	210	260	1.01	1.469

[a] Boiling point is 105°C/1 torr.
[b] Boiling point is 135°C/1 torr.
[c] Boiling point is 175°C/1 torr.
[d] Boiling point is 220°C/1 torr.

The striking phenomenon of the unusually wide range of existence of liquid phases is specific for methylsilicone oils and is not found to the same extent with other silicone oils, particularly those with stronger polar groups such as the methylphenylsilicone oils. In return, the phenyl groups impart to the oils a series of other valuable properties, including better heat resistance and sometimes also lower pour points and better lubricating properties.

The viscosity behavior of branched silicone oils of the structural type $M\begin{bmatrix}T\\M\end{bmatrix}M$ (i.e., those in which D units have been replaced by [TM] units) differs completely from that of oils with unbranched molecules. The viscosity rises very rapidly with increasing molecular weight (Section 6.2.2) so that a branched silicone oil has much smaller molecules than a chain-polymeric unbranched oil of the same viscosity. Of these types of oil, therefore, only the members of relatively low molecular weight have become of technical interest.

Table 78 gives a review of selected types of technical silicone oils based on polymethylsiloxanes and polymethylphenylsiloxanes. A determination of *density* and *refractive index* makes it possible to distinguish methylsilicone oils and methylphenylsilicone oils. In the methylphenylsilicone oil group, when the constitution remains the same the density and the refractive index increase with the content of phenyl groups.

All these silicone oils are water-clear colorless liquids that are absolutely stable on storage.

9.2.2 Heat and Low-Temperature Resistance

To evaluate heat resistance, the general point of view mentioned in Section 9.1 must be taken into account. Methylsilicone oils are stable up to 150°C for a practically unlimited time, even in thin films and exposed to air. The temperature at which the oil remains usable without decomposition can be increased to about 200°C by reducing the area of contact with air and in closed systems; in vacuum or in inert gas atmospheres the oils can be subjected to temperature peaks of 300°C. The changes that occur in air and in an inert gas are different: in air the influence of the oxidative rupture of the Si—C bond predominates; the cross-linking to which this leads raises the viscosity. In an inert gas, the oils are depolymerized by thermal overstressing, and the viscosity consequently falls off.

The heat resistance of the oils containing phenyl groups exceeds that of the pure methylsilicone oils (*99*, *141*); high-molecular-weight methylphenylsilicone oils can be used in air at 250°C for several hundred hours, and in closed systems temperatures of 250 to 300°C can be tolerated for months, as well as short-term temperature peaks of up to 450°C. In an inert gas, such oils can be used up to 400°C.

With all these data it is assumed that temperature equalization within the oil phase is achieved by a natural (on the thermosiphon principle) or forced (pumping) circulation, i.e., that no appreciable overheating takes place through stagnation, particularly on the heating walls. Such a phenomenon would obviously have a very deleterious effect on the heat resistance.

Of the methylphenylsilicone oils, those of type (V) (Table 77) are particularly heat resistant but at temperatures above about 350°C, of course, they can be used only in closed systems under pressure (*141*).

Since the methyl groups are the components of the silicone oils that are least resistant to oxidation and the atmospheric oxygen usually attacks them first (*7*), attempts have repeatedly been made to increase the resistance of methylsilicone oils to air at elevated temperatures by the addition of antiaging agents. Examples of substances that have been proposed for this purpose are: *p*-aminophenol, phenyl-1-naphthylamine, diphenylamine (*87*); substituted catechols, 2-phenylhydroquinone, and naphthols (*88*); 2-alkyl-4-alkoxyphenols (*25*); metal chelate compounds, particularly the acetylacetonates of numerous heavy metals, which are active even in concentrations of 0.01% (*151*); and metal salts of organic acids (*33*). Among other substances, bis-trimethylsilyl-ferrocene (*8*, *137*), 1,2-benzanthrene, pyrene, fluoranthene, and, of the heavy-metal compounds, iron octoate have been found to be effective anti-oxidants in methylsilicone oils containing chlorophenyl groups [type (VI), Table 77] (*1*).

Data on ignition and flame points, which limit the possibilities of use of the oils in air at elevated temperatures, are given in Table 78.

Besides the data just given, the so-called *gelling time* G_T can provide points of reference for a comparative evaluation of heat resistance. This denotes the time in hours for which an oil must be heated in air at 250°C, in a layer 3 mm thick in a glass dish 35 mm in diameter, before gelling sets in (*141*) (see the accompanying tabulation).

Oils	G_T (hr)
Methylsilicone oils	10
Methylphenylsilicone oils, high molecular weight	
Low phenyl content	400
High phenyl content	1750
Chlorophenylmethylsilicone oils (with the addition of iron octoate)	1000

In evaluating the behavior of methylphenylsilicone oils, however, it must be borne in mind that although the *oxidation resistance* expressed by the magnitude of G_T does indeed rise with the increasing content of phenyl groups, the loss in weight before the onset of gelling simultaneously becomes greater (*141*).

Silicone oils are also characterized in some cases by unusually low pour points, which make it possible to use them at low temperatures. The pour points are a function of the composition and the molecular structure. Even in the region of the low-molecular-weight methylsiloxanes, we have found that the freezing points oscillate with increasing molecular size, in the sense that a chain molecule with an odd number of silicon atoms always has a lower freezing point than one with an even number of members (Table 33). When phenyl groups are incorporated in polymethylsiloxanes, the pour points are first lowered, until a minimum is reached at about 6 mole-% of C_6H_5 groups; a further increase in the phenyl content raises the pour point continuously (*159*) (Fig. 40).

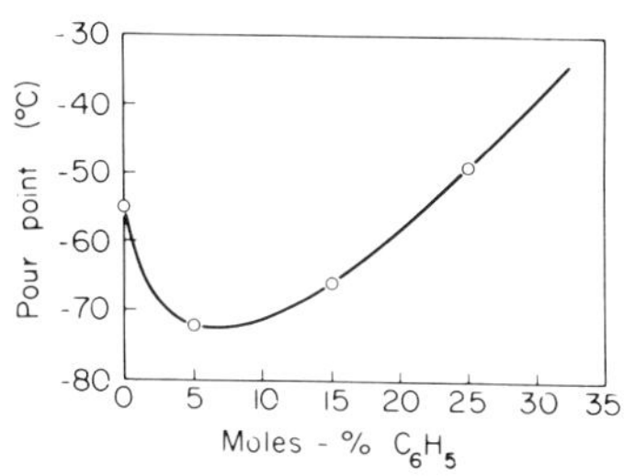

FIG. 40. Pour points of linear polymeric methylphenylsilicone oils as a function of the content of C_6H_5 groups [after Warrick *et al.* (*159*) by permission of Industrial and Engineering Chemistry].

9.2.3 VAPOR PRESSURES

Only the lowest members of the series MD_nM are distillable (cf. Table 33). The vapor pressures of some selected low-molecular-weight siloxanes containing methyl and phenyl groups that are available as technical products* are shown in Fig. 41.

It is worth noting that the products whose molecules consist of branched low-molecular-weight polymers possess vapor pressure curves of unusual steepness and very low vapor pressure figures at normal temperature. They are therefore used preferentially as diffusion pump oils (see also Section 10.14.4).

Temperature	Vapor pressure (torr)
< 140°	$< 10^{-5}$
170°	1×10^{-4}
200°	8×10^{-4}

* PD oils of Farbenfabriken Bayer AG, Leverkusen, West Germany.

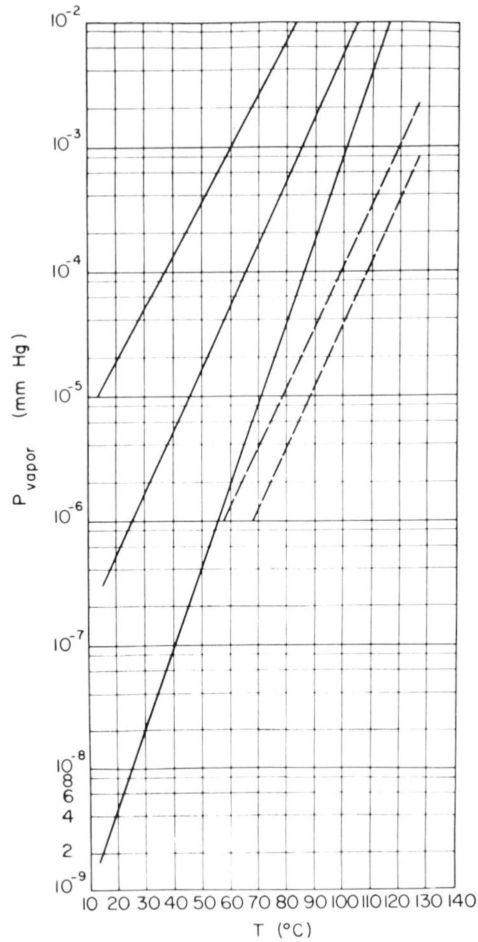

FIG. 41. Vapor pressure curves of some low-molecular-weight polymethylphenylsiloxanes (——) compared with two purely organic diffusion pump oils (– – –) [after Simmler and Bächler (*143*)].

The vapor pressures of high-molecular-weight silicone oils are very low under stationary conditions provided that they have been carefully freed from volatile constituents during their preparation. The figures given in the tabulation on p. 461 were measured (*94*) at various temperatures on a polydimethylsiloxane oil with a viscosity of 140 cSt at 20°C (cf. Table 78).

Other types of oil give very similar figures. A methylsilicone oil with a viscosity of 30,000 cSt/20°C gave 5×10^{-5} torr at 100°C and 3×10^{-4} torr at

220°C (*94*). The methylphenylsilicone oils given in Table 78 have vapor pressures of the order of 10^{-6}/20°C, 10^{-5}/100°C, and 10^{-4}/150°C (*142*).

A measure of the behavior of the oils under nonstationary conditions is provided by the so-called Noack test in which the percentage of volatile constituents arising in the course of a definite heat treatment is measured. The figure obtained includes not only the amounts of siloxanes volatile under the given conditions that were originally present in the oil but also those formed by depolymerization during the heat treatment. In practice, the test is carried out by heating a 65 gm sample of the oil at 250°C under a pressure of 20 torr for one hour. The data in the accompanying tabulation characterize the behavior of various oils (for types, cf. Table 78).

	Viscosity $\nu_{20°C}$ (cSt)	Loss, (max. wt-%)
Methylsilicone oils	60	5
	140	3
	440–1440	0.5
Methylphenylsilicone oils		
Low phenyl content	200	1.0
	1000	0.5
High phenyl content	300	2.0
	1000	1.5

9.2.4 Viscosity Behavior

One of the most striking properties of the methylsilicone oils is the low variation of their viscosity with the temperature. This is shown, in the form of an Ubbelohde diagram, in Fig. 42 for a series of silicone oils with different viscosities, $\log T$ (abs.) being plotted on the abscissa and $\log\log(\nu + 0.8)$ (where ν is the kinematic viscosity in cSt) on the ordinate.* For comparison, the corresponding curves for two technical mineral oils are given.

If the methyl groups are replaced by other groups, the temperature dependence of the viscosity increases. The graph of Fig. 43 shows a comparison of this property of polydimethylsiloxanes and polymethylphenylsiloxanes, both of linear structure. A transition from the methyl group to higher aliphatic groups also raises the temperature dependence of the viscosity. Cyclic dimethylsiloxanes do not exhibit this property to the same extent as the corresponding linear polymers.

* The temperature dependence of the viscosity deviates from linearity at temperatures below −20°C (*160*).

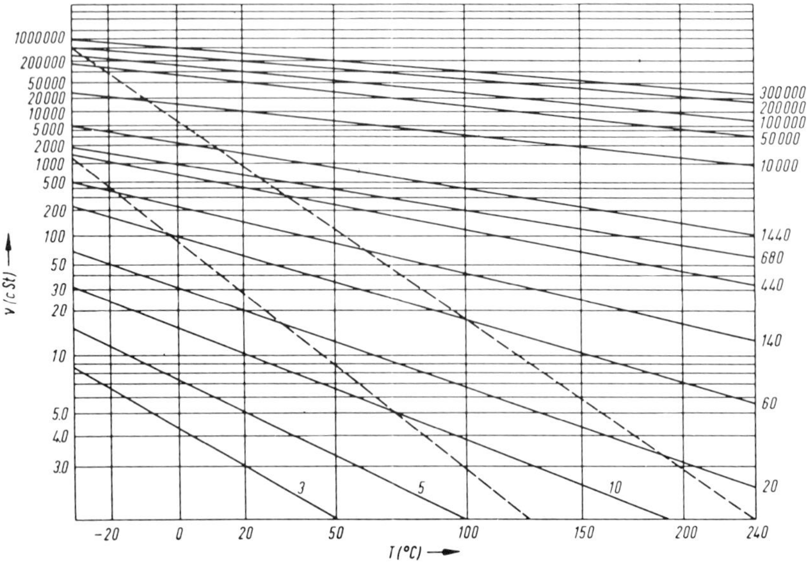

FIG. 42. Temperature dependence of the viscosity ν of methylsilicone oils (——) and two mineral oils (---) for comparison. Data written against the curves of the methylsilicone oils (on the right) denote their viscosity ν_{20} (cSt).

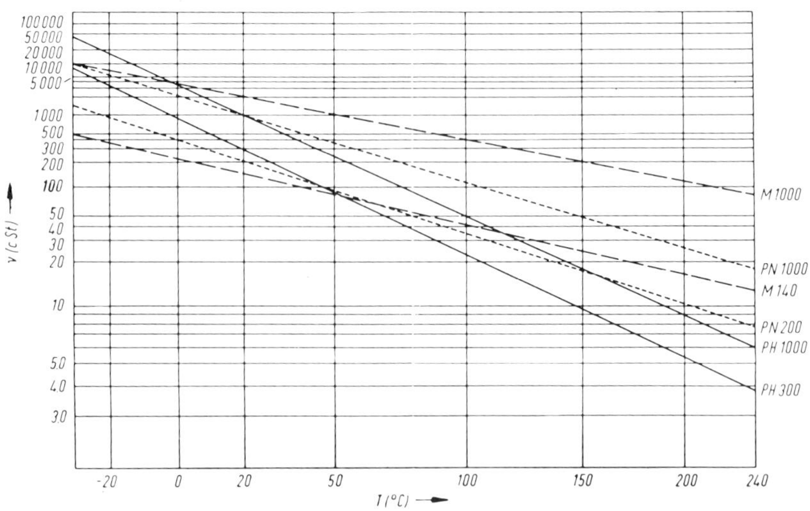

FIG. 43. Temperature dependence of the viscosity ν of methylphenylsilicone oils. The figures written against the curves (on the right) denote the viscosity ν_{20} (cSt). The PH types are methylphenylsilicone oils with a high phenyl content and the PN types ones with a low phenyl content. The two M oils (methylsilicone oils, cf. Fig. 42) are given for comparison.

The viscosity of the methylsilicone oils also depends only slightly on the pressure, as a comparison with technical mineral oils again shows (Fig. 44). While the viscosity of a mineral oil at 25°C is raised by a pressure of 2000 atm about 50-fold to 5000-fold, depending on the nature of the oil, the viscosity of a silicone oil having 140 cSt/20°C is raised by the same treatment only 16-fold.*

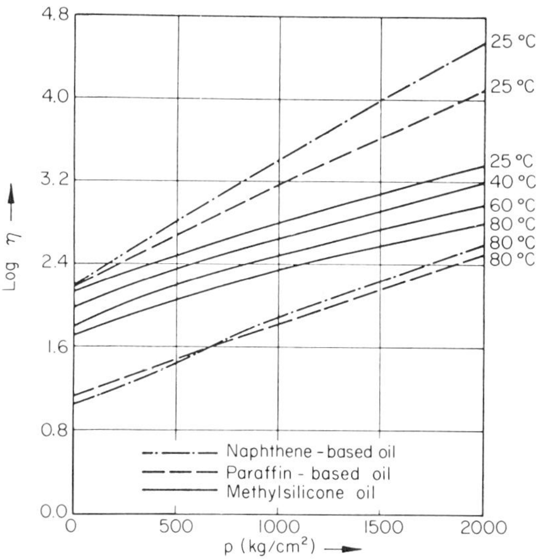

FIG. 44. Pressure dependence of the viscosity η (cP) of a methylsilicone oil in comparison with mineral oils*.

As regards the influence of shear forces on the viscosity, measurements have shown that liquid siloxanes of up to 1000 cSt undergo relatively little change in viscosity even at high velocity gradients; however, in the case of more viscous oils the viscosity decreases with increasing velocity gradient, which is obviously due to the fact that the molecules become ordered in the direction of flow and the frictional resistances are therefore reduced (*29*, *65*) (Fig. 45). This effect becomes increasingly noticeable with increasing molecular weight. The orientation of the molecules falls off as the velocity gradients are reduced; because of the viscosity, a certain time is necessary for deorientation, so that when the velocity gradient is increased and decreased in a rotary viscometer the curves of the apparent viscosity show a hysteresis loop (*29*). The inherent viscosity of an oil is not substantially changed, either, by very long-lasting shear stresses; a polydimethylsiloxane oil of about 70 cSt at 25°C was stressed in a hydraulic

* According to investigations by E. Kuss, Institut für Erdölforschung der Technischen Hochschule Hannover, West Germany (cf. refs. *80* and *101*).

device at 105 kg/cm^2 periodically 105,000 times over 500 hours; after this, its viscosity had changed by less than 2%. A mineral oil lost 50% of its inherent viscosity after being stressed only 18,000 times (*39*).

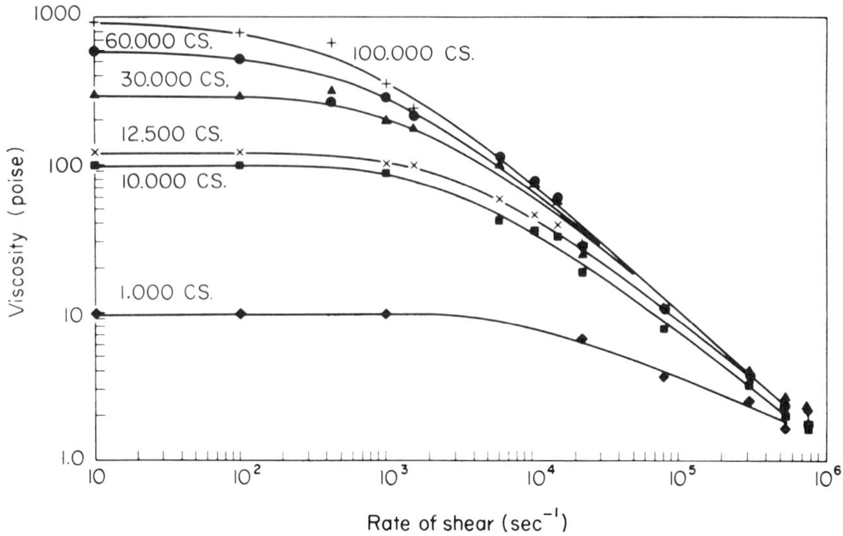

FIG. 45. Viscosity behavior of methylsilicone oils at different rates of shear.

9.2.5 THERMAL CONDUCTIVITY AND SPECIFIC HEAT

The *thermal conductivity* of the silicone oils depends on their molecular weight and viscosity. In the polymethylsiloxane series, it first increases rapidly from the member with the shortest chain, hexamethyldisiloxane, for which it is 2.36×10^{-4} cal sec^{-1} cm^{-1} deg^{-1} at 50°C to 3.7×10^{-4} for an oil with a viscosity of about 100 cSt/25°C, after which it remains practically constant when the viscosity increases further (*13*). The temperature coefficient of the thermal conductivity is low; for practical purposes, it may be taken (*13*) that the coefficient α_{50} in the equation

$$K_t = K_{50}[1 + \alpha_{50}(t - 50)]$$

is about -6×10^{-4}. An empirical relationship can be drawn up between viscosity, temperature, and thermal conductivity (*13*). Methylphenylsilicone oils do not differ from methylsilicone oils with respect to their thermal conductivity.

The specific heat c [cal gm^{-1} °C^{-1}] and the heat content W_{20} [cal gm^{-1}] have been measured in the range of viscosities from 50 to 12,000 cSt/20°C at temperatures of 20 to 200°C (*61*). The results, given in the accompanying tabulation, were found to be practically independent of viscosity (cf. refs. *86* and *122*).

t (°C)	W_{20} (cal gm^{-1})	c (cal gm^{-1} °C^{-1})
20	0	0.37
40	7.3	0.37
60	14.8	0.37
80	22.3	0.37
100	30.0	0.38
120	37.8	0.38
140	46.0	0.39
160	54.8	0.39
180	63.8	0.40
200	73.8	0.41

9.2.6 Thermal Expansion

The mean *coefficient of cubic expansion* γ of methylsilicone oils with linear polysiloxane molecules is 95 to 100×10^{-5} deg^{-1} between 25 and 175°C. Linear methylphenylsilicone oils exhibit somewhat lower figures (78 to 87×10^{-5} deg^{-1} depending on the content of phenyl groups and the viscosity). With increasing C_6H_5 content (*35*), γ falls.

9.2.7 Compressibility

The *compressibility* of the liquid methylsilicones is comparatively high; with oils of between 40 and 1000 cSt/20°C the coefficient of adiabatic compressibility is in the range of 99.8×10^{-12} to 101.1×10^{-12} cm^2 dyne^{-1}. As with liquid hydrocarbons, the compressibility falls off somewhat with increasing molecular weight; in addition, it falls considerably when phenyl groups are incorporated—to 60×10^{-12} cm^2 dyne^{-1} or less, according to the content of phenyl groups. Methylsilicone oils are still liquid at pressures under which hydrocarbon oils have already solidified. While, for example, *n*-hexadecane solidifies at 420 kg/cm^2, hexamethyldisiloxane does so only at a pressure of 4010 kg/cm^2; all liquid linear polymethylsiloxanes with a viscosity of more than 1.0 cSt/20°C remain liquid at pressures as high as 40,000 kg/cm^2, even though they are compressed by about 35% (*22*).

9.2.8 Surface Tension

The *surface tension* of all liquid silicones is surprisingly low. With the mobile silicones of type MD_nM it rises from 15.7 dynes cm^{-1} for hexamethyldisiloxane to about 20 dynes cm^{-1}, and then stays at this figure with increasing viscosity. The surface tension of the methylphenylsilicone oils is somewhat higher than that of the methylsilicone oils; it increases with the increasing content of phenyl groups (*41*) (see the accompanying tabulation).

Approx. C_6H_5/CH_3 ratio[a]	Viscosity $\nu_{25°C}$ (cSt)	Surface tension (dyne cm^{-1}/20°C)
0.14	50	22.2
0.49	102	26.1
0.63	27	28.6

[a] C_6H_5/CH_3 ratio estimated from infrared spectroscopic data

With the low-viscosity methyl-substituted siloxanes of the series MD_nM, the interfacial tension against water varies around 40 dynes cm^{-1}.

9.2.9 Dielectric Properties

The figures for dielectric constant, loss factor, specific resistance, and dielectric strength are good at the normal temperature and vary only slightly with the temperature. Data for a methylsilicone oil are given in Table 79. The dielectric constant and loss factor of methylsilicone oils also depend little on the frequency. The values of the dielectric strength depend markedly on the physical and chemical purity of the oils.

TABLE 79

Dielectric Data for a Polydimethylsiloxane Oil (440 cSt/20°C) at Various Temperatures

	20°C	100°C	200°C
Dielectric constant	2.8	2.5	2.3
Loss factor tan δ/800 cps[a]	1.2×10^{-4}	1.3×10^{-4}	1.5×10^{-4}
Specific resistance (Ω cm)	4×10^{15}	6×10^{14}	10^{14}
Dielectric strength (kV/cm), 50 cps, 6 kV/sec	120	100	95

[a] According to DIN 53483 and VDE 0303, part 4 *tan* δ.

The dielectric properties of special types of silicone oils of different constitution may differ considerably from those of the methylsilicone oils. Silicone oils having polar cyano groups in the β-position of a substituting ethyl group ("nitrilesilicone oils," cf. Section 9.2.1.) have an unusually high dielectric constant which can vary between 4 and 20, according to type (*142*).

9.2.10 Lubricating Properties

It is misleading to call the silicone oils in general lubricating oils. The lubricating properties are a function of the chemistry and the constitution of the oil, on the one hand, and the nature of the system to be lubricated, on the other hand.

The load-bearing properties of a film of methylsilicone oils are low because of the weak intermolecular forces in this phase. It is best when plastics such as polyamide, polystyrene, or phenol resins are used as the bearing materials; in the range of metallic materials, relatively the best results are obtained with combinations of steel against bronze, chromium, zinc, or cadmium, or of chromium against bronze (*28*, *51*). Within the region of boundary friction, methylsilicone oils possess no lubricating properties with the combination of steel against steel.

TABLE 80

Load-Bearing Capacity of Various Films of Silicone Oil on the Almen-Wieland Machine until the Bearing Seizes

Type of oil	Load (kg)
Polydimethylsiloxane	50–150
Polymethylphenylsiloxane with low phenyl content	100–200
Polymethylphenylsiloxane with higher phenyl content	200–350
Polychlorophenylmethylsiloxane	~1000
Uncompounded mineral oils	~400–600
Mineral oils with lubricant additive	~800–2000

By incorporating phenyl groups, the lubricating properties of methylsilicone oils are improved (Table 80) so that such oils are suitable as lubricating agents at bearing pressure that are not too high. A further improvement in lubricating properties can be achieved by halogenating the phenyl groups (*40*) or the aliphatic groups [poly-γ-trifluoropropylmethylsiloxane (*133*)]. In the case of nuclear substitution, the activity rises from fluorine to chlorine and bromine and with the number of halogen atoms. Substitution by the trifluoromethyl group in the *m* position is very effective, and simultaneous substitution of the nucleus by at rifluoromethyl group (CF_3) in position 3 and a chlorine atom in postion 4 is even better (*44*).

With halogen-substituted oils, as well, the lubricating properties depend on the combination of metals in the bearing; the least favorable combinations are those of hard and soft steels, and the best those of hard steel and copper. It appears that the improved lubricating effect is brought about by a splitting off, even if only slight, of hydrogen halide with the formation of a metal halide film; the low coefficients of friction of the steel/copper combination can be explained by the fact that the formation of a metal halide film takes place particularly readily on copper (*19*).

The lubricating properties of silicone oils are improved by various additives; the following have been proposed for this purpose: chlorophenylmethylsilanes, e.g., (p-ClC_6H_4)$(CH_3)_3Si$ and (p-ClC_6H_4)$_2(CH_3)_2Si$ (*23*); Diels-Alder adducts of vinylsilanes and hexachlorocyclopentadiene (*140*); chlorinated biphenyls and organic disulfides (*62*); chlorinated and fluorinated aliphatic hydrocarbons (*3*); chlorinated and brominated fatty acids with 1 to 12 carbon atoms (*84*); esters of lower chlorinated fatty acids and thioglycollic esters (*165*); esters of phosphoric acid (*45*); and copper, magnesium, and lead salts of aliphatic carboxylic acids with 4 to 9 carbon atoms (*27*). However, it must be noted that, while such additives do in fact enhance the lubricant effect at the normal temperature, the effect generally disappears again at higher temperatures because of the inadequate heat stabilities of the components of the additives.

9.2.11 Chemical Behavior

Methylsilicone and methylphenylsilicone oils are, in general, inert substances. At normal temperatures they are stable to water,* air, oxygen, metals, wood, paper, and plastics, and also to solutions of metal salts, liquid ammonia, and 3% hydrogen peroxide; with concentrated hydrogen peroxide, silicone oils give explosive mixtures like other organic substances (*92*). On the other hand, they are decomposed, especially at elevated temperatures, by the stronger mineral acids, particularly hydrofluoric acid (*91*), alkalis, and strong oxidizing agents such as concentrated nitric acid or elementary chlorine. As a relative measure for their behavior with respect to some chemicals, Table 81 gives the change in the viscosity of the oils after various treatments for 12 hours at 20 and 100°C. Methylphenylsilicone oils are more sensitive to superheated steam than methylsilicone oils.

Their behavior in relation to atmospheric oxygen at elevated temperatures determines their heat resistance and has already been described (Section 9.2.2).

Methylsilicone and methylphenylsilicone oils are soluble in a large number of different solvents. Their solubility behavior depends to some extent on viscosity, molecular weight, and constitution.

Generally good solvents are hydrocarbons (gasoline, benzene, toluene), chlorinated hydrocarbons (chloroform, carbon tetrachloride, trichloroethylene), ethers, esters, and higher alcohols from butanol onward. Methylsilicone

* Only with very thin films spread on water, especially in the acid or alkaline regions or at elevated temperatures, have changes been observed (*42*, *104*).

TABLE 81

CHANGE IN VISCOSITY IN PERCENT AFTER 12 HOURS

After treatment with	Polydimethylsiloxane oil 140 cSt/20°C		Polymethylphenylsiloxane oils 200 cSt/20°C low phenyl content		300 cSt/20°C high phenyl content		5 cSt/20°C	
	20°C	100°C	20°C	100°C	20°C	100°C	20°C	100°C
1 *N* NaOH	+0.1	−2.0	+6.7	+5.6	0	+11.4	0	+1.9
1 *N* HCl	+1.2	−5.0	+2.8	+4.8	+0.7	+1.9	+1.9	+1.9
38% HCl	+12.1	−4.2	+16.3	+24.2	+32.4	+17.3	+1.9	−0.4
13% HNO_3	+3.8	−18.8	+4.5	+5.6	+0.8	+1.0	+1.2	+1.2
30% H_2SO_4	—	+6.0	+3.4	+6.7	+0.7	+1.7	+1.5	+1.2

oils with viscosities of 60 to 140 cSt (see Table 78) also dissolve in isopropanol, *n*-propanol, *n*-butanol, acetone, and dioxane; the members of the methylsilicone series with the lowest viscosities ($\nu < 10$ cSt/20°C) are also soluble in ethanol. Low-molecular-weight methylphenylsilicone oils with viscosities of 5, 10, and 25 cSt (Table 78) dissolve in methanol; moreover, in contrast to the other types of oils, they are compatible with paraffins, most vegetable oils, and petroleum jellies.

The miscibility of methylsilicone oils and methylphenylsilicone oils is limited and depends on the phenyl content and structure of the siloxanes.

The solubility of some gases in dimethylpolysiloxane oils at 25°C is shown by the following figures (milliliter of gas dissolved in 1 ml of oil (*86*)): air, 0.168–0.190; N_2, 0.163–0.172; and CO_2, 1.00.

9.2.12 RADIATION RESISTANCE

Under the influence of energy-rich radiation, the molecules of silicone oils are partially cross-linked and partially degraded, in the same way as those of other polymers. Cross-linking outweighs degradation. Consequently, silicone oils first react with an increase in the viscosity which can finally lead to gelling.

The sensitivity of the compounds to radiation depends markedly on the type of substituents. Polymers with methyl groups or with hydrogen or hydroxyl or γ-trifluoropropyl groups undergo change comparatively readily, while those with aromatic groups, such as, in particular, phenyl groups, exhibit a resistance which increases with the content of such groups (*73*, *167*). Table 82 shows some figures obtained in the ^{60}Co irradiation of methylsilicone and methylphenylsilicone oils.

TABLE 82

INFLUENCE OF ENERGY-RICH RADIATION ON METHYLSILICONE AND METHYLPHENYLSILICONE OILS

C_6H_5 (mole-%)	Viscosity (cSt)	*Rg* (rad)[a]	*Rg* × *M*[b] (rad)	*E* (eV)[c]	*G*[d]
	100	41×10^6	2.7×10^{11}	56	1.79
0	1000	9.8×10^6	3.0×10^{11}	62	1.61
	10000	5.9×10^6	3.3×10^{11}	68	1.47
	100	63×10^6	4.3×10^{11}	89	1.12
5	1000	16×10^6	4.7×10^{11}	97	1.03
	10000	9.8×10^6	4.0×10^{11}	83	1.21
25	450	440×10^6	8.0×10^{12}	1600	0.06

[a] *Rg* (rad) is the minimum absorbed dose needed for gel formation.
[b] *M* is the molecular weight.
[c] *E* (eV) is the absorbed energy per cross-linkage.
[d] *G* value is the number of cross-links (*73*) per 100 eV of absorbed energy.

Compared with the *G* values of other authors, the *G* figures given in Table 82 are very low: *G*/cross-linking, 2.99 (*134*) and 2.7 (*98*); *G*/degradation, 1.61 (*134*).

It must be borne in mind that the figure of 2.7 simultaneously includes chain rupture. The differences in *G* are probably connected with traces of impurities in the oils.

9.3 Properties of Technical Silicone Greases

To understand the properties of silicone greases (*51*), it must be realized (cf. Section 8.3.1) that these owe their consistency to the incorporation of solid-disperse *thickening agents* in an *oil phase*; the solid substance forms a *spongelike framework* in which the liquid phase is held and prevented from flowing by the surface forces. The strength of this bond determines the stability of the grease.

Silicone greases, which are prepared by the distribution of highly disperse *silica* in silicone oils, generally do not possess any particularly stable and load-resistant bond. Their penetration figures alter comparatively greatly on milling, and they are thixotropic to a certain extent. Consequently, they are not suitable as lubricating greases; their fields of application lie in a different direction (paint auxiliaries, surface protection, release agents, stopcock greases).

Greases of this type have no measurable drop point. Like the silicone oils themselves, their consistency changes little over a wide range of temperatures and they can be used between about -40 and $+200$°C. They can be subjected briefly to even higher temperatures, the absolute value of which depends on the external conditions. Their resistance to chemicals is generally the same as that of pure silicone oils; their oxidation resistance is good, and their vapor pressure and evaporation loss are low. Silicone greases based on poly-γ-trifluoropropylmethylsiloxanes have a better resistance to solvents than those prepared from the classical polydimethylsiloxanes or polymethylphenylsiloxanes (*133*).

Greases containing *soaps*—generally lithium stearate—as thickening agents can be so formulated that a comparatively stable bond arises. The penetration figures can be adjusted to different values; they vary between about 180 with heavy greases and about 260 with light greases but can be increased still further with "liquid" types. It can be taken as a measure of their stability that an average grease possessing a rest penetration of 220 has a penetration figure of about 280 after milling.* The drop points of soap-thickened silicone greases are measurable and are generally about 200°C.

With a suitable structure, greases are obtained that can be used for the lubrication of bearings; the lubricating properties depend not only on the type of silicone oil used but also on the thickening agent, which can increase the load-bearing capacity of the film of grease.

The heat resistance of soap-thickened greases is determined by the thermal stability of the thickening agent and of the oil in the presence of atmospheric oxygen and by the degree of possible interactions between the two components. The temperatures up to which the greases of this series can be subjected are between 160 and 200°C, according to type. The limiting low temperatures range from -40 to -70°C. Greases consisting of lithium soaps, methylphenylsiloxanes, and diesters such as those of the type of bis-(2-ethylhexyl) adipate, appear to be suitable for still lower temperatures, down to about -80 to -90°C (*95*). The best heat resistance is possessed by greases that have been prepared with carbon black or phthalocyanines as thickening agents.

Normal types of lubricating grease cannot be displaced by silicone greases from all positions in which they work satisfactorily; silicone greases are used where the temperatures are so high—or so low—that normal types of grease based on hydrocarbons fail. Consequently, there is very little point in comparing the lubricating action of a silicone grease with that of other greases at normal temperature; this should rather be done at the desired working temperature. If this is carried out with strict maintenance of similar conditions, as is the case with the data in Table 83, an idea of the efficiency of the silicone products can be obtained.

* Measured in accordance with ASTM D217.

TABLE 83

MEAN LIFE OF GREASES IN OPEN BALL BEARINGS AT 150°C, 3600 rpm, 73 kg RADIAL LOAD (*17*)

Type of grease	Hours
Based on hydrocarbons	
The best of 22 greases	850
The worst of 22 greases	100
Based on silicones	
Methylphenylsilicone oils thickened with lithium soaps, various types	3800; 1900; 1700; 580
Based on esters	
Polyesters thickened with lithium soaps	1000

In their flow behavior, the soap-thickened silicone greases resemble most organic lubricating greases (*51*). It is an interesting fact that the flow behavior of carbon-black-filled greases is similar to that of soap-thickened greases. Measurements of the viscosity as a function of the velocity of flow have shown that the flow behavior of greases based on methylphenylsilicone oils and lithium soaps is determined at low velocity gradients by the concentration of the soap and the degree of phenylation of the oil; at higher velocity gradients, both factors become secondary and the viscosity of the liquid then plays an increasingly decisive part (*29*).

Höppler's rheoviscometer, the principle of measurement of which also takes into account the structural viscosity behavior of the grease, has been proposed for comparative determinations of consistency figures (*144*). In the testing and evaluation of silicone greases, it must be borne in mind that many of the test methods customary for organic types of grease cannot be transferred directly to this field. In contrast to greases based on mineral oils, silicone greases have no saponification and neutralization numbers.

9.4 Properties of Cured Silicone Resin Films*

9.4.1 HEAT RESISTANCE

While in the case of *unpigmented silicone resin films*, *phenomena of cracking* or, as is customary in the electrical industry, *electrical characteristics* are used as criteria of *heat resistance*, in the case of *pigmented lacquers* the factors concerned are *elasticity*, *gloss*, and *resistance to yellowing*. In the paint industry, there is a general tendency to regard the yellowing resistance, in particular, as

* Section 9.4 was written by Dr. W. Krauss, Leverkusen, West Germany.

a very sensitive index of the heat resistance of an enamel. Care must be taken in transferring experience in this direction to silicone resins. Thus, certain white-pigmented silicone enamels based on polymethylphenylsiloxanes can be heated to 350 or even 400°C for several hundred hours without yellowing or loss of gloss. This perhaps explains the frequently encountered opinion that silicone resins are permanently stable at these temperatures. That this is not

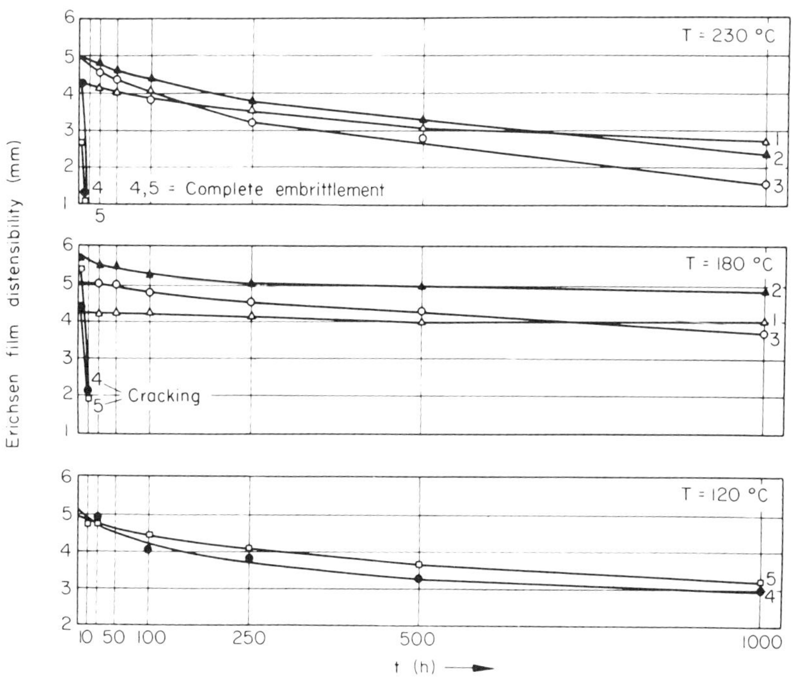

FIG. 46. Elasticity of paint films on heat aging. (1) Straight methylphenylsilicone resin; (2) cocondensation product of silicone and polyester with 75% of silicone component; (3) cocondensation product of silicone and polyester with 50% of silicone component; (4) alkyd resin, silicone-free; (5) epoxide resin, silicone-free. Pigment: titanium dioxide (rutile); pigment/binder ratio 1:2.

correct is shown by a simple elasticity determination of films aged in this way or a testing of the state of the film with a knife. With silicone lacquers of different structures and composition, the change in the gloss is again a good measure of heat resistance. To evaluate heat resistance, therefore, several properties, including at least those mentioned, must be used.

Figures 46 to 48 show the heat resistances of paint films evaluated with respect to the points of view sketched; Fig. 49 summarizes all the results in a manner which shows how long the types of resin investigated can be used at a given temperature. On considering the curves, what has just been said is found to be confirmed. The determination of the elasticity of pure silicone resins gives a sharper criterion of their heat resistance than their tendency to undergo yellowing (curve 1 in Figs. 46 and 47), while with cocondensation products of silicones and polyesters the tendency to yellowing corresponds more sensitively to the heat treatment. It can be seen from Fig. 49 how superior the heat

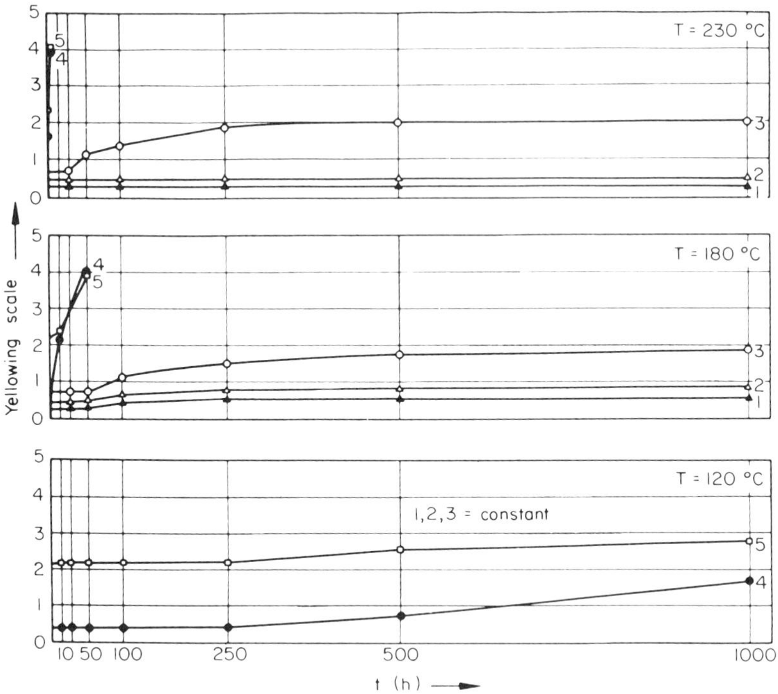

FIG. 47. Yellowing of paint films on heat aging. See Fig. 46 legend for description.

resistance of the silicone resins is to that of resins on a purely organic basis. While the latter can be subjected for 10,000 hours to temperatures up to 130°C at the most, the limiting temperature to which silicone resins can be exposed for the same period is generally 100°C higher. Curve 1 in Fig. 49 may perhaps show the optimum heat resistance that can be achieved today with paints based on silicone resins. In the case of apparently more favorable data that are found occasionally in the literature, there is no reference to the time during which the given temperature was maintained.*

* Better figures for the thermal stability of silicone resins, namely, 1000 hours at 350°C, were also obtained by A. L. Smith *et al.* (*145*) on the basis of infrared spectroscopic investigations, but these are not in direct relationship with the properties of practical importance for the paint and varnish technology.

As already mentioned in Section 8.2.2.1.3, pigments may have a decisive influence on the heat resistance of silicone resins: active pigments accelerate the oxidative heat aging of the resin films, while others exert an inhibiting effect on further cross-linking and therefore embrittlement. These effects can be

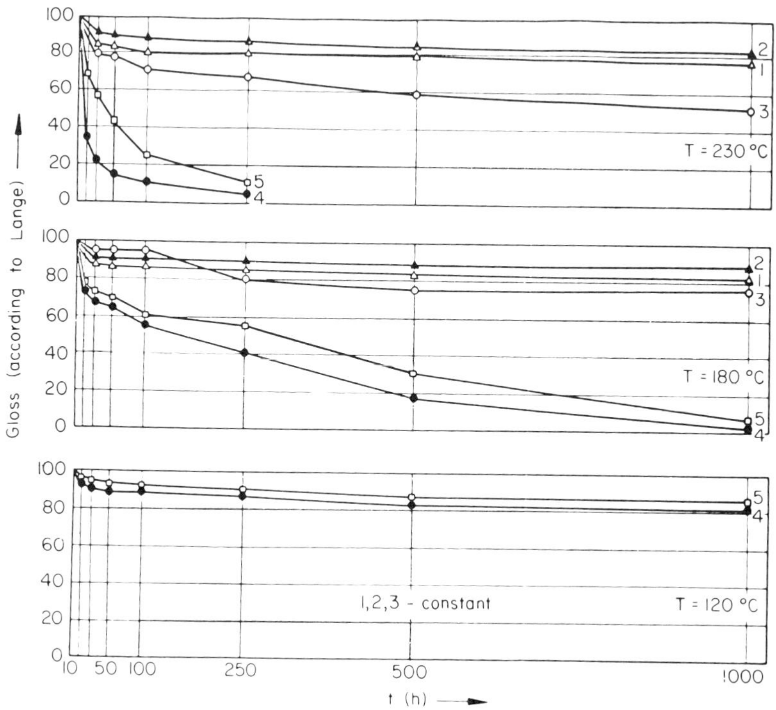

FIG. 48. Gloss retention of paint films on heat aging. See Fig. 46 legend for description.

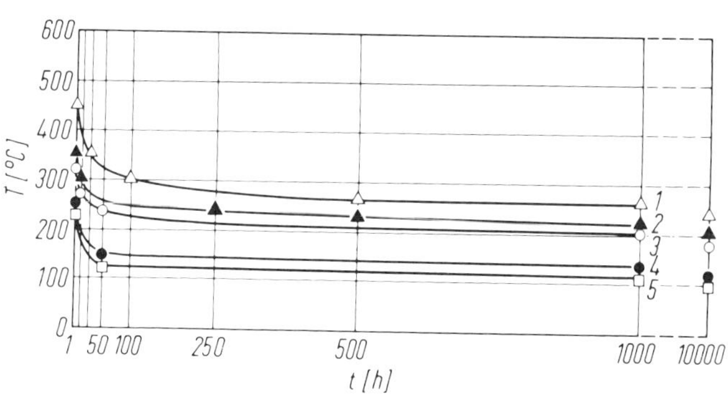

FIG. 49. Heat resistance of various paint films. See Fig. 46 legend for description.

overlaid at higher temperatures by reactions between pigments and the silica arising by the oxidative decomposition of the silicone resins, which lead to enamels with long-term heat resistances of 400 to 600°C. Particularly satisfactory as pigments in this respect are titanium dioxide, aluminum, zinc dust, cadmium, and micaceous iron ore. In fact, these can actually be combined with silicone resins which, in the unpigmented state, give films of only low heat resistance. The nature of the reaction between pigment and binding agent is not known with certainty. However, it appears that the silica reacts with surface oxide skins of the pigment and thereby gives a comparatively coherent protective film which covers the substrate and inhibits corrosion (*48*, *77*).

9.4.2 Mechanical Properties

The mechanical properties of the silicone resins that are of most interest for their use in paints are *hardness*, *elasticity*, and *thermoplasticity*. The hardness and elasticity of silicone resin films can be varied within wide limits by modifying the structure of the resins. Highly cross-linked silicone resins give relatively hard and less elastic films, and those with a low degree of cross-linking give softer and more elastic films. The incorporation of organic substituents with higher space occupancy also gives softer and generally more elastic films. With a structure otherwise similar, methylphenylsilicone resins are therefore softer than the corresponding methylsilicone resins (on this, see also Section 8.2.1). No plasticization of the silicone resins by special plasticizers is necessary; desired degrees of hardness can be obtained by mixing hard and soft types of resin. Methylsilicone resins and methylphenylsilicone resins are, however, compatible with one another only to a limited extent.

The film hardness of pure silicone resins is generally too low for the purposes of the paint industry, while, on the other hand, the thermoplasticity is too high, for which reason cocondensation products of silicones and polyesters are preferred with respect to these properties as well. With a generally better elasticity, they have a higher surface hardness and closely approach the hard two-component lacquers of the type of polyurethane or epoxide lacquers.

Table 84 gives a comparison of the surface hardness and elasticities of some purely organic and silicone-based enamels.

In the purely phenomenological evaluation of the surface hardness of silicone resins such as is customary in the paint industry by scratching the surface with the fingernail, excessive hardnesses are generally found, and these must be corrected by accurate testing with physical methods. This is connected with the high surface smoothness of the silicone films, as a consequence of which the nail slides without scratching.

As has already been mentioned in Sections 8.2.2.1.2 and 8.2.2.1.3, pigments and catalysts also exert a great influence on the hardness and elasticity of silicone resins. They may accelerate the oxidation of the silicone resin film,

TABLE 84

SURFACE HARDNESS AND ELASTICITY OF VARIOUS ENAMELS

Binder	Conditions of curing	Surface hardness[a]	Elasticity[b]
Hard methylsilicone resin	Without catalyst 1 hr/180°C	2B	3.0
Soft methylphenylsilicone resin	With the addition of 0.2% of lead naphthenate and 2% of zinc naphthenate 1 hr/230°C	B	6.0
Silicone-polyester cocondensate with 75% of silicone	With the addition of 1% of lead naphthenate and 1% of butyl titanate 1 hr/230°C	2H	5.0
Silicone-polyester cocondensate with 50% of silicone	With the addition of 1% of lead naphthenate and 1% of butyl titanate 1 hr/230°C	2–3H	3.7
Linseed oil/stand oil cooked with hard resin	Dried for 14 days at room temp. followed by 3 days at 60°C	HB	7.1
Alkyd resin with ~50% of linseed oil	Dried for 14 days at room temp. followed by 3 days at 60°C	H	7.0
Alkyd resin with ~67% of linseed oil	Dried for 14 days at room temp. followed by 3 days at 60°C	H	7.0
Chlorinated rubber with 50% of plasticizer	Dried for 14 days at room temp. followed by 3 days at 60°C	H	7.1
Alkyd resin with 40% of coconut oil and urea-formaldehyde resin	30 min/140°C	H	6.7
Polyurethane with elastic polyester component	Dried for 14 days at room temp. followed by 3 days at 60°C	H–2H	6.1

[a] The hardness determination on the so-called pencil scale (very soft: 6B→B, H→6H very hard) given here is only a relative evaluation.
[b] Distensibility according to Erichsen.

i.e., the transition to silica glass, which gives rise to hard but relatively non-elastic films. Under certain conditions, overpigmentation considerably decreases the elasticity of types of resin with very poor pigment compatibilities. In general, therefore, it is not advisable to add more than one part of pigment per part of silicone resin. Depending on the pigments to be used and the resin to be pigmented, the most favorable mixing ratio is probably 10 to 70 parts of pigment per 100 parts of resin.

The hardness and elasticity of the films are excellent criteria for the effectiveness of catalysts; if the catalyst is of low activity the films produced are soft, and if it is highly active they are hard and brittle. Certain catalysts that can be incorporated in the siloxane lattice—particularly esters of titanic acid—bring about an increase in hardness without substantially decreasing the elasticity.

Of decisive importance for the use of silicone resins as paint and varnish raw materials is their *thermoplastic behavior*, for paint films which become soft at the temperatures of use and are damaged or detached from the substrate by the slightest mechanical stress are unusable, however good their other properties may be.

Hitherto, no great attention has been paid to thermoplasticity in investigating the properties of silicone resins. The thermoplasticity of silicone resins decreases with increasing degree of cross-linking and increases on passing from methylsilicone to methylphenylsilicone resins. It is markedly depressed by condensation of the silicones with certain organic components. Suitable organic components are, in the first place, polyesters, but they also include polycarbonates and epoxide resins containing hydroxyl groups. Table 85 shows how the thermoplasticity of a cocondensation product decreases with increasing content of hydroxyl-group-containing polyesters. The thermoplasticity can easily be estimated almost quantitatively by comparing the surface hardness of a film at 20 and 180°C. An exact determination is given by measuring the adhesive power according to VDE 0362.

Catalysts also have a certain influence on the thermoplastic behavior of the resin. This is clearly detectable in the case of lead and aluminum compounds. Limits are set to their use for the reasons already mentioned in Section 8.2.2.1.2. However, catalysts of the type of esters of titanic acid have proved to be very suitable for improving thermoplastic behavior. The effect of an addition of

TABLE 85

REDUCTION IN THE THERMOPLASTICITY OF SILICONE RESINS BY COCONDENSATION WITH POLYESTERS CONTAINING HYDROXYL GROUPS

Material	Surface hardness[a] at 18°C	Surface hardness[a] at 180°C	Adhesive power according to VDE 0362 at 18°C (kg/cm²)	Adhesive power according to VDE 0362 at 150°C (kg/cm²)
Straight methylphenylsilicone resin	B	6B	40	1
Silicone-polyester cocondensate with 75% of silicone	2H	B	100	2.5
Silicone-polyester cocondensate with 50% of silicone	3H	H–2H	250	15

[a] The hardness determination on the so-called pencil scale (very soft 6B→B, H→6H very hard) given here is only a relative evaluation.

tetrabutyl titanate can be seen from Table 86. Since the titanic ester can react with free hydroxyl groups of the polyestersiloxane resin, it is obvious to assume that the improvement in the thermoplastic behavior depends merely on additional consolidation of the molecular network by tetrafunctional $[TiO_{4/2}]$ complexes. Since, however, orthosilicic esters do not give a similar effect, a specific activity must be ascribed to titanium.

TABLE 86

INFLUENCE OF TETRABUTYL TITANATE ON THE THERMOPLASTICITY OF SILICONE RESINS

Type	Amount of tetrabutyl titanate added	Surface hardness[a] at 20°C	Surface hardness[a] at 180°C
Soft methylphenylsilicone resin	—	B	6B
	5%	B	3B
	10%	HB	2B
Silicone-polyester cocondensate with 75% of silicone	—	H–2H	2B
	5%	2H	B
	10%	2H	HB
Silicone-polyester cocondensate with 50% of silicone	—	2H	HB
	5%	3H	H
	10%	3H–4H	H–2H

[a] The hardness determination on the so-called pencil scale (very soft: 6B→B, H→6H very hard) given here is only a relative evaluation.

Finally, the *adhesion* must be mentioned as another important mechanical property of silicone resin films. Silicone resins adhere in an outstanding manner to metals such as iron, aluminum, silver, and tin* and also to glass and ceramics, so that no pretreatment of these materials is required. The opinion often found that phosphating (*152*) improves the adhesion is correct only to a limited extent, for at the actual temperatures of use of silicone resin paints the phosphate films are destroyed after a short time and then reduce the strength of adhesion quite considerably (*48*, *117*).

Where phosphating is necessary for protection against corrosion, films of 1 μ at the most should be produced, at not too low a pH. All types of priming that do not contain silicones as binders (*112*) likewise substantially reduce the

* The adherence to zinc is sometimes stated to be very poor. However, in this case the constitution of the resin appears to be of decisive influence, since the results cannot be reproduced with any particular silicone resins that may be selected; cf. on this refs. *58* and *146*.

adhesion of films of silicone resin applied on top of them. However, a pre-treatment of the substrate by mechanical cleaning, such as sand blasting, always improves the adhesion of the film to metals, particularly iron.

Only to copper is the adhesion of silicone resin films unsatisfactory, particularly on aging and at elevated temperatures. Adhesion to this metal is improved by modification with organic components; the presence of epoxide groups has a favorable effect. It appears that the oxide films present on copper both impair the anchoring of the film at low temperatures and, by their oxidative action, accelerate the decomposition of the silicone film at high temperatures.

9.4.3 Chemical Behavior

The resistance of a silicone resin film depends on the degree of condensation. Noncondensed hydroxyl groups represent reactive centers, for which reason in describing the resistance of silicone resin films to chemicals in all cases only completely hardened films should be taken into consideration (*58*). Since silicone resins are poorer in Si—C and richer in Si—O bonds than silicone oils and silicone rubber vulcanizates, their chemical resistance differs to some extent from that of the latter.

Fully hardened silicone resin films possess good resistance to aqueous acids (*48*, *90*). Thus, for example, Hedlund (*58*) found for attack lasting more than 100 hours at 25°C complete resistance to 50% sulfuric acid, nitric acid, and even to concentrated hydrochloric acid. Particularly striking is their good resistance to oxidizing acids and also, to some extent, to chlorine and chlorine water (*89*). In this respect, the silicone resins are probably superior to most purely organic raw materials for paints. More aggressive are short-chain organic acids, such as formic and acetic acids, which is probably to be ascribed less to the decomposing than to the dissolving properties of these substances. Fairly strong alkalis such as 10% sodium hydroxide or potassium hydroxide destroy silicone resin films.

The resistance of modified silicone resins to chemicals depends on the organic component incorporated and can be deduced from its chemical structure. However, advantages with respect to resistance to chemicals are provided in individual cases only by combinations with epoxide resins.

Contradictory information is found in the literature about the *water resistance* of films of silicone resins. Outstanding stability is often mentioned; other authors discourage the use of silicone resins for water-resistant paints. The situation becomes clearer when the complex property of water resistance is separated into three different resistances:

1. Resistance to *cold water*, which must be described as outstanding. Silicone enamels undergo no change on contact with distilled water for several years.
2. Resistance to *boiling water*, which depends essentially on the structure of

the silicone resin—generally in the sense that hard pigmented products of low thermoplasticity exhibit a better resistance than soft thermoplastic nonpigmented resins. On prolonged attack (10 to 20 hours), the latter tend to form bubbles.

3. Resistance to *steam*, particularly *superheated steam*, which is generally poor. This is due to the fact that the silicone resin film has a high water vapor permeability, so that within a short time water vapor can penetrate in large amounts to the substrate and may perhaps react there or can at least substantially reduce the adhesion of the film. A typical example of this behavior is provided by sheet-metal strips coated with silicone resin that were half-immersed in boiling water for a long period. After this treatment, the part of the film that was immersed was not attacked while the other was covered with bubbles. Table 87 shows the higher water vapor permeability (*153*) but very much lower water absorption as compared with a purely organic lacquer film (polyurethane); it also shows the permeability-reducing effect of a pigment.

TABLE 87

WATER ABSORPTION AND WATER-VAPOR PERMEABILITY OF A SILICONE RESIN FILM IN COMPARISON WITH A POLYURETHANE FILM

Binder	Pigmentation	Water-vapor permeability (gm/cm·hr·torr)	Water absorption (%)
Silicone resin	—	8.7×10^{-8}	0.36
	Titanium dioxide	4.7×10^{-8}	0.31
Polyurethane resin	—	1.5×10^{-8}	2.3
	Titanium dioxide	1.2×10^{-8}	2.3

The *solvent resistance* of silicone resin films is low. It increases somewhat with increasing consolidation of the siloxane network, i.e., hard silicone resins are more resistant than soft products of a lower degree of condensation, and the resistance is improved by aging the films at high temperatures or by using hardening catalysts. A denser siloxane network is also produced by the exclusive use of methyl substituents on the silicon atom, so that a better solvent resistance can be expected from pure methylsilicone resins than from methylphenylsilicone resins. Pigments and fillers also have a film-consolidating effect.

In considering resistance to the individual solvents, an approximate parallel can be found to the solubility of the silicone resin intermediates, for good solvents for these intermediates generally destroy the hardened films very rapidly. Thus, *aromatic hydrocarbons* such as benzene, toluene, and xylene, *esters* such as ethyl acetate, butyl acetate, and methyl glycol acetate, *ketones* such as acetone and cyclohexanone, and *chlorinated hydrocarbons* such as dichloromethane and chlorobenzene lead to complete degradation within minutes. Silicone resin films exhibit a comparatively good resistance to petroleum hydrocarbons and the lower alcohols. While gasoline brings about only a more or less pronounced and generally reversible softening, depending on the constitution, the lower alcohols have no effect. The resistance to polyhydric alcohols is also good. Thus, for example, silicone resin films are not attacked within 30 days by glycerol at 80°C. Apart from the use of silicone resins as impregnating varnishes, the resistance properties mentioned here are of no great importance technically, since only in the most exceptional cases will subjection to heat and solvents take place simultaneously.

Much more interesting from a technical standpoint is the resistance of silicone resins to *fats* and *oils*. They may come into contact with these substances in coatings of furnaces and ovens, gearboxes, and similar devices. As can be seen from the chemical structure of the fats and oils and their decomposition products (long-chain aliphatic fatty acids and glycerol), these are poor solvents for silicone resins; they are incompatible with them and react neither strongly acid nor strongly alkaline, and consequently, they do not attack silicone resin films. Hardened silicone resin films are not affected even by heating for 40 days in mineral oil, margarine, palmitic acid, etc., at 160°C. They are also resistant to the formation of spots by fruit juice, coffee, and the like, which is an important factor for their use as refrigerator enamels.

In general, the incompatibility of the silicones with a large number of organic materials means that silicone resin lacquers often possess excellent *release properties*. Coating with silicone resins is therefore suitable, for example, for the solution of mold-release problems with plastics or for overcoming release difficulties in the drying of highly adhesive and sticky materials.

9.4.4 Weather Resistance

Together with their heat resistance, the outstanding *weather resistance* of the silicone resin lacquers is probably their most interesting property. Now that it has been possible to collect the experience of weathering experiments lasting over several years (*48*, *112*) it can be stated that the weather resistance of silicone resin lacquers is not even approached by any organic binder. Silicone resins are also resistant to yellowing even under intense irradiation with ultraviolet, so that the hue of lacquer based on them does not change for years when light-fast pigments are used. A silicone resin film pigmented with anatase still showed no chalking phenomena after exposure to the weather for two years.

Silicone resin films are also resistant to attack by molds, since the silicone does not offer fungal cultures a suitable nutrient substrate. On sheets lacquered with silicone resin and placed horizontally in the open air one can see how, after rain, colonies of molds form in the drops of water that have run together because of the hydrophobic nature of the surface, and how these disappear again when the water evaporates. In the case of purely organic lacquers simultaneously subjected to weathering for comparison, these mold formations do not disappear but increase in the course of time to form the well-known dark mold spots.

With various weathering experiments on silicone resin films (*48*), soiling and the formation of water spots have been observed. According to what has been said in Section 9.4.2, this can only be the result of a phenomenon caused by thermoplasticity. In direct insolation, coated iron sheets may easily reach temperatures of 70°C and more, i.e., temperatures at which the thermoplasticity of inadequately cross-linked silicone resins begins to make itself noticeable. If, now, particles of dust fall onto the soft surface, they become anchored in the surface and drops of rain which strike it cause a deformation. To obtain silicone resin films with good gloss retention and a clean surface, therefore, the same processes in principle must be used as those which cause a diminution in thermoplasticity (cf. Section 9.4.2).

Surprisingly, the weather resistance of cocondensation resins does not fall in proportion to increases in their content of organic binder, so that silicone resins containing 50% of the organic binder still have an outstanding weather

TABLE 88

WEATHER RESISTANCE OF SILICONE RESIN FILMS AS COMPARED WITH SOME COMMERCIAL PURELY ORGANIC BINDERS

Binder	Properties after 1 year of weathering		
	Gloss[a]	Yellowing[b]	Chalkiness[c]
Pure silicone resin	0	0	0
Silicone-polyester cocondensate with 75% of silicone	0.1	0.1	0
Silicone-polyester cocondensate with 50% of silicone	0.3	0.3	0
Organic binder *A*	2.5	2.5	2.5
Organic binder *B*	2.5	2.5	2.5
Organic binder *C*	2.0	1.5	2.0
Organic binder *D*	1.5	1.5	0.5

[a] 0 indicates high gloss → 4 indicates absolute matt.
[b] 0 indicates pure white → 4 indicates brown.
[c] 0 indicates no chalking → 4 indicates very marked chalking.

resistance (*79*). It has in fact been found that additions of only 10% of certain types of silicone resins to an alkyd resin or a cocondensation of the two materials significantly increases the resistance of the product to weathering (*21*).

Table 88 gives the results of a series of weathering experiments involving exposure for 1 year at Miami, Florida, showing the superiority of the lacquers containing silicone resin.

Of great practical importance is the weather resistance of silicone lacquers pigmented with metal powders, particularly aluminum bronze and zinc dust, since by means of these it is even possible to protect satisfactorily articles that are exposed alternately to high temperatures and weather influences (*53*, *59*, *78*, *106*).

9.4.5 Incompatibility

The incompatibility of the polymethylsiloxanes and the films prepared from them with respect to many organic high polymers, which has been referred to already in the preceding sections, is often a thoroughly desirable property. It permits problems of the release of molded goods from molds and processing machines in many branches of industry to be solved. Up to the present time, the mold-discharging processes have been studied to the greatest extent in the baking industry, since here difficulties commonly occur when sugar is present. The best releasing effect and resistance to sugar or goods containing much sugar is shown by straight polymethylsiloxanes, while the activity of the polysiloxanes containing phenyl groups can be increased by the addition of polymethylsiloxane oils (*96*, *97*). The silicone coatings in question are only 8 to 10μ thick as a maximum, but can be used for years without renewal (*163*). In addition to silicone resins, combinations of polysiloxanes and polytetrafluoroethylene (*155*) or, indeed, the latter alone (*157*) are used for release lacquers.

The same polysiloxanes also give coatings with extremely high release action for uses in the rubber and plastics industry or in the drying of dyestuffs. They also inhibit, for example, the accretion of ice (*4*, *43*). The releasing effect is greatest with unpigmented systems. If a lubricating effect is desired as well, pigmentation with molybdenum sulfide has proved to be advantageous (*136*).

9.4.6 Electrical Properties

Silicone resins are important raw materials for the provision of electrical machines with silicone insulation (compare Section 10.2). They owe their introduction into the electrical sector not only to their heat resistance but also to their electrical properties, which are good—sometimes excellent—and change little with temperature. The properties of electrically insulating materials made from silicone resins, generally in combination with glass fiber and glass silk fabric, are described later (Section 10.2); here we may give only a few data on the electrical behavior of pure unfilled silicone resins.

To investigate the *loss factor*, *the dielectric constant*, and the *specific resistance*, silicone resin films were fixed by heating on steel wires 2 mm in diameter and were coated with silver as the second electrode. The electrical characteristics were measured from room temperature up to nearly 300°C.* The result (Fig. 50) is very impressive when compared with the figures for other types of organic resin used for enameling of wire. The absolute value of the *loss factor*

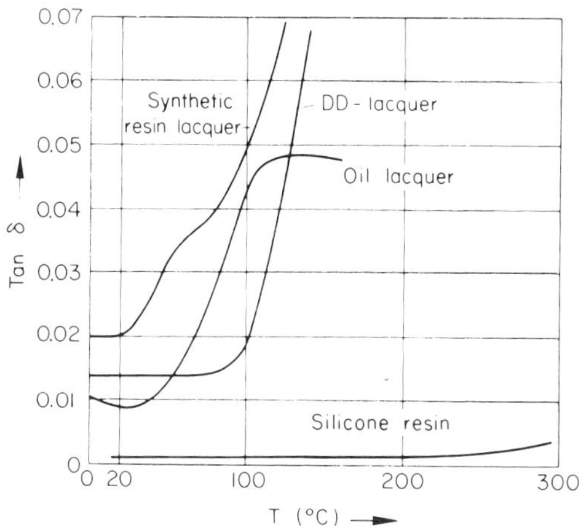

FIG. 50. Loss factor tan δ at 800 cps/50 V of a methylphenylsilicone resin in comparison with organic types of lacquer as a function of the temperature.

at room temperature for silicone resins of about 2×10^{-3} is substantially smaller than for the organic lacquers used for comparison; it remains practically constant up to 200°C and then rises slowly to about 3×10^{-3} on approaching 300°C.† Of the other lacquers used for comparison, only the polyurethane lacquers show an almost constant loss factor up to ~100°C.

No less informative is a comparison of the *dielectric constants* as functions of the temperature, these being shown for some lacquers in Fig. 51. With purely organic insulating lacquers, the dielectric constant rises considerably with increasing temperature, while with silicone resins it falls and is smaller than that of the organic lacquers, particularly in the range of temperatures above 100°C. Since the losses occurring in a dielectric are proportional to the

* According to investigations of K. Mielke and K. Gronbach, Farbenfabriken Bayer AG, Leverkusen, West Germany; see Noll (*101*).

†It cannot be concluded from the fact that the electrical characteristics can be measured up to nearly 300°C that silicone resins can also be used at these temperatures in practice; their limit of application for "long-term" stressing is about 200°C (cf. Section 9.4.1).

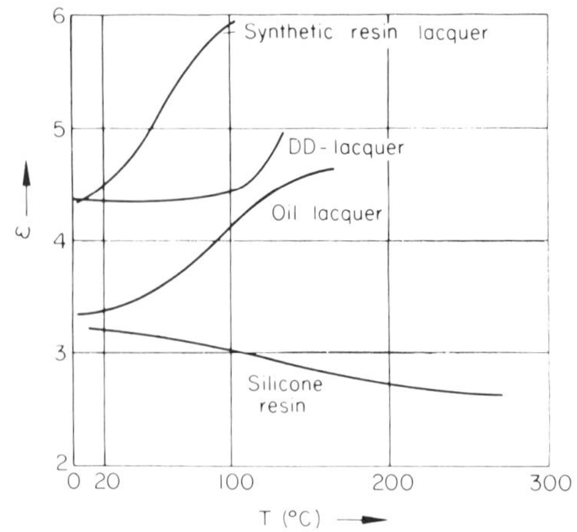

FIG. 51. Dielectric constant ε at 800 cps/50 V of a methylphenylsilicone resin in comparison with organic types of lacquers as a function of the temperature.

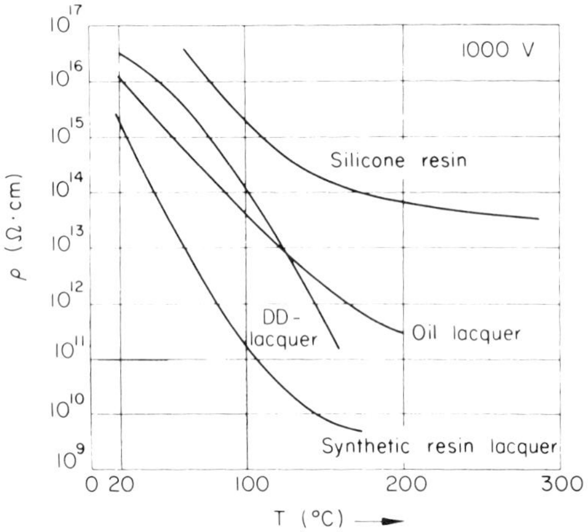

FIG. 52. Specific resistance ρ at 1000 V of a methylphenylsilicone resin in comparison with organic types of lacquers as a function of the temperature.

dielectric constant, among other things, this circumstance is of particular interest in the use of silicone resins for high-tension insulation. If, finally, the *specific resistances* of the various types of lacquers are compared, likewise as functions of the temperature, Fig. 52 again shows the clear superiority of the silicone resin. It is true that the specific resistance of the silicone resin also increases with rising temperature, but substantially more slowly than those of the organic lacquers. Even in the range between 200 and 300°C, i.e., considerably above the highest permissible long-term temperature of 180°C for heat class *H*, the figures measured for silicone lacquers must be described as unusually good. Also of interest is the *dielectric strength* determined at various layer thicknesses, which was measured at room temperature and is shown in the form of a graph in Fig. 53, curve 1.* The shape of the curve shows

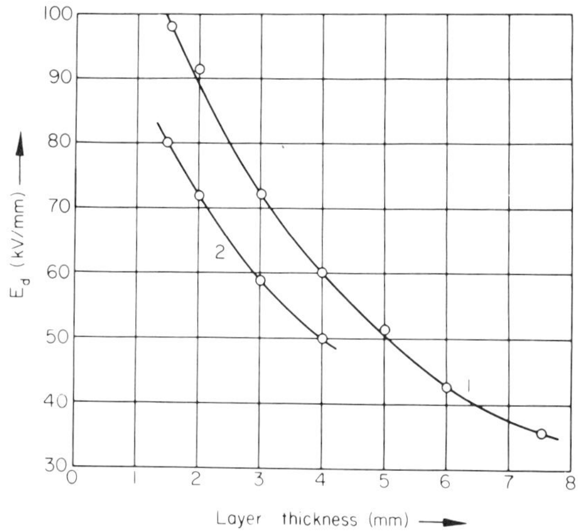

FIG. 53. Dielectric strength E_d of silicone resins as a function of the layer thickness: (1) pure silicone resin, and (2) silicone-polyester cocondensate.

that quite considerable breakdown figures are obtained for small layer thicknesses, which is of particular importance for the evaluation of the insulating power of silicone resin lacquers in electrical machines.

The electrical indices of cocondensation resins depend on the proportion and the type of the organic component. The data given in the accompanying tabulation were measured on cocondensation products of a methylphenylsilicone and a phthalic acid ester component on steel wires in the manner described above after heating for 2 hours at 240°C.

Loss factor tan δ/20°C	0.01
Dielectric constant ϵ/20°C	3.5
Specific resistance Ω cm/20°C	10^{16}

* Testing at room temperature in accordance with the "Leitsätzen für elektrische Prüfungen' von Isolierstoffen" ("Guiding Principles for the Electrical Testing of Insulating Materials") (VDE 0303, part 2/10.55).

9.5 Properties of Silicone Resin Moldings, Laminates, and Foams*

9.5.1 Heat Resistance

Silicone resin moldings, laminates, and foams offer no advantages over products based on organic materials with respect to their mechanical properties

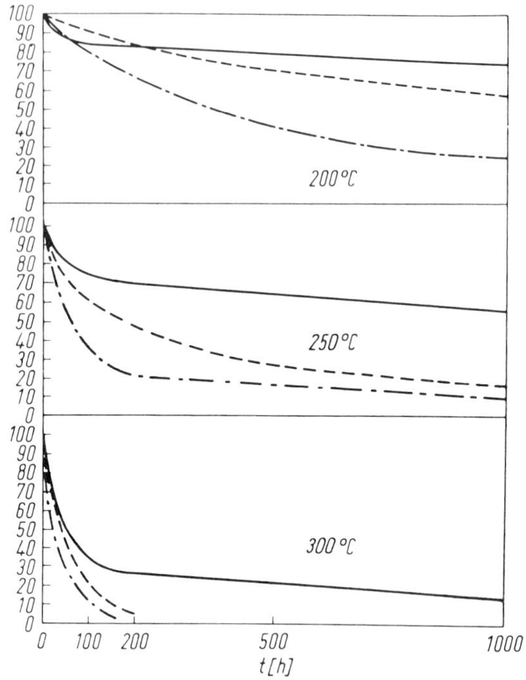

FIG. 54. Fall in the flexural strength of various laminates in percent of the initial amount after thermal aging at 200, 250, and 300°C (K. Andres, Farbenfabriken Bayer AG, unpublished): (——) Silicone resin laminate (initial figure 2050 kg cm^{-2}); (---) epoxide resin laminate (initial figure 4700 kg cm^{-2}); and (-.-) polyester laminate (initial figure 2970 kg cm^{-2}).

and their chemical resistance at the ordinary temperature. However, as soon as these materials are exposed to high temperatures for a long time their superiority becomes appreciable (Fig. 54). For this reason, their use is advantageous only where high temperatures occur for short periods or permanently.

In an attempt to obtain simple criteria for the evaluation of the heat resistance of molding products and laminates, the loss in weight on thermal stressing has been used, among other things, as a basis for comparative figures.

* Section 9.5 was written by Dr. W. Krauss, Leverkusen, West Germany.

Taking the content of filler into account, the term "critical thermal instability temperature" (CTIT) was created, this being defined by the following relation (*139*):

$$\frac{W_{(t+25)} - W_t}{B} \times 100 \leqq C$$

in which t is CTIT in °C, W_t is the loss in weight in per cent at t°C, $W_{(t+25)}$ is the loss in weight in per cent at $(t+25)$°C, B is the content of binder in the molding material in per cent, and C is the constant selected arbitrarily according to the severity of the test conditions; in practice, generally $C=4$. According to this, that temperature is defined as the "critical thermal instability temperature" above which the corresponding material should not be used since it decomposes thermally. Table 89 gives such figures for some laminates (*139*).

TABLE 89

CRITICAL THERMAL INSTABILITY TEMPERATURE OF LAMINATES

Material	Resin content (%)	CTIT (°C)
Paper/phenol resin	50	160
Cotton fabric/phenol resin	50	145
Glass fiber/melamine resin	40	240
Glass fiber/phenol resin (low-pressure laminate)	28	210
Glass fiber/silicone resin	52	350
Glass fiber/silicone resin	40	355

TABLE 90

HEAT RESISTANCE OF RESIN FOAMS

Material	Highest temp. withstood (°C)
Polystyrene	80
Cellulose acetate	95
Phenol resin	180
Silicone resin	360
Polyurethane	180
PVC	60

TABLE 91

MECHANICAL PROPERTIES OF LAMINATES[a]

	Polyester/ glass fiber	Phenol resin/ glass fiber	Epoxide resin/ glass fiber	Melamine resin/glass fiber	Silicone resin/ glass fiber	Silicone resin/ asbestos fiber
Flexural strength (psi)	$30–63 \times 10^3$	$20–35 \times 10^3$	$45–80 \times 10^3$	$40–65 \times 10^3$	$10–38 \times 10^3$	$12–16 \times 10^3$
Notch impact strength (ft lb/in. of notch)	19–35	4.0–10	10–25	5–15	8–15	6–9
Tensile strength (psi)	$20–50 \times 10^3$	$9–30 \times 10^3$	$33–50 \times 10^3$	$25–40 \times 10^3$	$10–30 \times 10^3$	—
Compressive strength (psi)	$30–60 \times 10^3$	$40–65 \times 10^3$	$50–90 \times 10^3$	$30–85 \times 10^3$	$35–46 \times 10^3$	$40–50 \times 10^3$
Hardness (Rockwell)	M 100–M 110	M 105–M 110	M 105–M 120	M 115–M 130	M 100	—

[a] From *Modern Plastics Encyclopedia*, Issue for 1959, **36**, 624.

If the exposure to heat is made short (e.g., for only a few hours or minutes, as may be the case with structural sections of rockets) the critical temperatures of all the types of laminates cited are higher. Silicone resins behave better under such extreme conditions than other resins since, in addition to a high heat absorption capacity, they have a low thermal conductivity. It is true that with respect to strength, phenol resin laminates, particularly those with asbestos fiber reinforcement, must be regarded as better, but if good dielectric properties are required in addition, the silicone resin laminates are superior (*20*, *32*, *125*).

A comparative review of the heat resistance of foamed materials (*164*) can be found in Table 90.

9.5.2 Mechanical Properties

The mechanical properties of silicone resin foams and moldings depend substantially on the filler used, the silicone resin, and also on the hardening conditions such as molding pressure, post-curing time, and temperature, so that a detailed description of these properties is impossible without an accurate statement of the origin of the individual materials. Table 91, which is intended to give an approximate idea of the properties of silicone resin laminates and those with purely organic binders, may serve for general orientation.

Table 92 (*86*) gives some characteristic properties of silicone resin moldings, and Table 93 (*9*) shows the superiority of the silicone resin laminates to aging at elevated temperatures.

9.5.3 Electrical Properties

Certain silicone resin laminates have acquired importance in the electrical industry as insulating materials for the construction of silicone-isolated machines. The properties of these materials will be described later together with those of other special electrically insulating materials based on silicones (Section 10.2.2).

TABLE 92

Characteristic Data for Silicone Resin Moldings (*86*)

Specific gravity	1.6–1.7 gm/cc
Tensile strength	4000–6000 psi
Flexural strength at:	
20°C	8000–10000 psi
150°C	5000–6000 psi
Notch impact strength at:	
25°C	7 ft lb/in.
150°C	5 ft lb/in.
Water absorption in 24 hr (thickness of test piece 1/8 in.)	0.2%

TABLE 93

FLEXURAL STRENGTH OF LAMINATES AFTER HEAT AGING (9)

Test temp. (°C)	Heat aging Time (hr)	Heat aging Temp (°C)	Laminates: Silicone resin/glass fiber (psi)	Laminates: Phenol resin/glass fiber (psi)	Laminates: Melamine resin/glass fiber (psi)
25	—	—	29200	46500	51600
25	200	250	26400	200	17600
150	200	150	14300	39100	30300
250	200	250	13400	550	15600
325	200	325	9800	—	1500

9.6 Properties of Silicone Rubber*

As with vulcanizates from organic rubbers, the special properties of silicone rubber are not affected only by the type of gum used; they also depend on the nature and amount of the components of the formulation, the processing technique, and the method and type of vulcanization. In particular, differences exist in many respects in the complex of properties of room-temperature and heat-vulcanized products. Such differences are expressly noted in each case below. However, the discussion below is intended to give only a picture of the general nature of the vulcanizates which is capable of many special variants in the individual cases.

9.6.1 HEAT RESISTANCE

One of the most remarkable properties of silicone rubber is its resistance to dry heat. In general, on the basis of practical experience, it may be considered that high-quality silicone rubbers still have an elongation of 200% and a tensile strength of 50 kg/cm^2, almost independent of the initial figures, even after exposure to unfavorable conditions in air at 180°C for 10,000 hours. Temperatures of up to about 300°C can even be withstood for a short time without substantial changes in physical properties. Both with respect to this stability in hot air and its low-temperature behavior (cf. Section 9.6.2) silicone

* The first version of Section 9.6 in the first German edition of this book was written by Dr. J. Peter, Leverkusen, West Germany; this has been revised by W. Kniege, Leverkusen, West Germany.

rubber represents an optimum among all elastomers (Fig. 55). However, it must be noted that these statements depend on the formulation, the nature of the mechanical stress (*30*, *76*), and whether hot or room-temperature curing was performed.

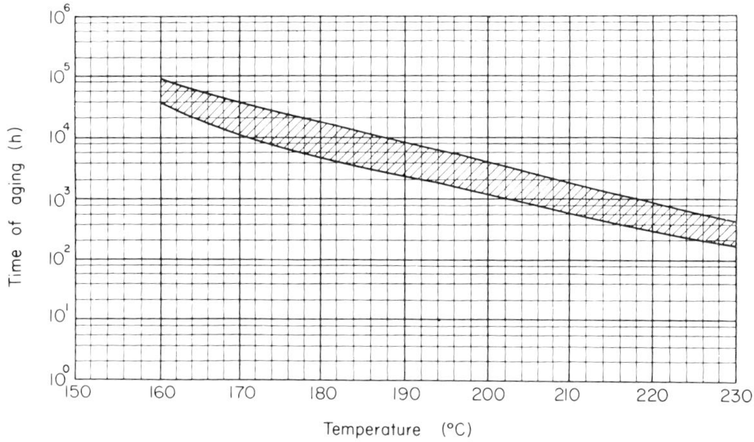

FIG. 55. Life of a silicone rubber according to VDE 0304 (time of a heat treatment causing a fall in the elongation at break to 200% and in the tensile strength to 50 kg/cm²). Compare also VDE 0283. Formulation: Silopren-SVA Batch II, 100 parts by wt; Aerosil 130, 5 parts by wt; dichlorobenzoyl peroxide 40%, 2 parts by wt. Vulcanization: 4 minutes at 250°C.

At temperatures above 150 to 180°C, there are essentially two processes that may differ in their intensities but together determine the course of aging. On the one hand a splitting off of organic groups can take place under the influence of oxygen, which leads to an ever-increasing degree of cross-linking and therefore increasing hardness. On the other hand, however, a depolymerization of the high-molecular-weight linear polysiloxane can take place, whereupon the high-viscosity silicone rubber is degraded to siloxanes of lower molecular weights and therefore the vulcanizate becomes softer.

Experience shows that with hot-air aging for a relatively long time, particularly at temperatures above 200°C, cross-linking always predominates in the end and the vulcanizates become hard (Fig. 56). In the extreme case, cross-linkage proceeds so far that practically all the organic groups are split off and the silicone rubber is degraded to silica.

If atmospheric oxygen, causing cross-linkage, is absent during the action of heat, the influence of depolymerization always predominates: the vulcanizates are degraded to a soft mass, particularly if hydrolyzing media such as water act in the absence of oxygen. In practice, the conditions for this are always present if the silicone rubber is encapsulated all around during its use (silicone-rubber-insulated cable with tight braiding, O-rings).

In order to obtain optimum results even with attack in a closed system, the rubber must be selected from special points of view. In this case, practically only polymers containing vinyl groups come into consideration. However, their resistance to aging depends to a large extent on the type of fillers used (*16*). As Fig. 56 shows, water and fillers rich in hydroxyl groups (for example,

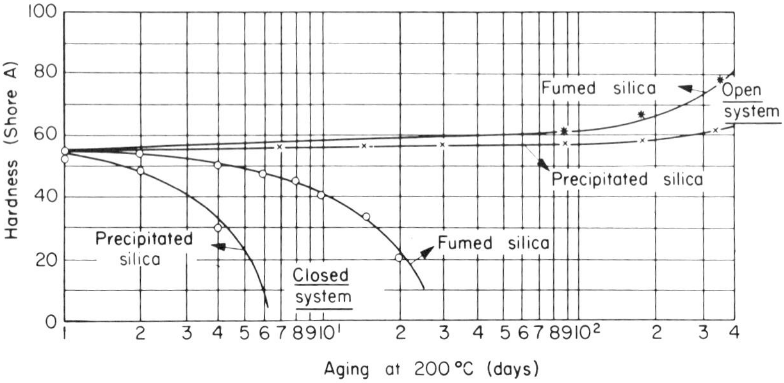

FIG. 56. Influence of the fillers on the hardness of silicone rubber during prolonged exposure to heat. Formulation: 100.0 parts by wt of Silopren RS, 30.0 parts by wt of filler, and 0.8 part by wt of Di-cup 40C. Curing: 15 minutes at 150°C (press) and 15 hours at 200°C (in hot air).

precipitated silicas) are particularly suitable when optimum hot-air stability is required; these fillers are, however, poor in closed systems. Here, conversely, the types of fillers with low contents of water and hydroxyl groups such as, for example, the classical fumed silicas are to be preferred. The reason for this situation is to be seen in the fact that in *closed systems* any water or hydroxyl groups in the fillers would lead to a degradation of the polymers. On the other hand, while the use of water-containing silicas in hot-air aging would lead to some slight degradation of the polymers, here, in contrast to the closed system, this is desirable since it opposes the oxidative cross-linking of the polymer chains.

If the vulcanizate is subjected to stress in closed systems or under similar conditions, peroxides will be used that contain no acyl or aroyl groups. The post-vulcanization is preferably carried out for a short time at high temperatures, e.g. 3 to 5 hours at 250°C. A longer post-vulcanization at such temperatures is to be avoided.

It is also interesting to note that the resistance of silicone rubber in closed systems can be considerably improved by the use with them of low-melting metal alloys containing bismuth or cadmium as constituents of the formulation (*70*).

The fact that many physical (mechanical, electrical) properties of silicone rubber depend on the temperature to a smaller degree than do those of the purely organic elastomers is of considerable importance. The dependence of the tensile strength on the temperature may be mentioned as an example (Fig. 57). Although the tensile strength of silicone rubber measured at room

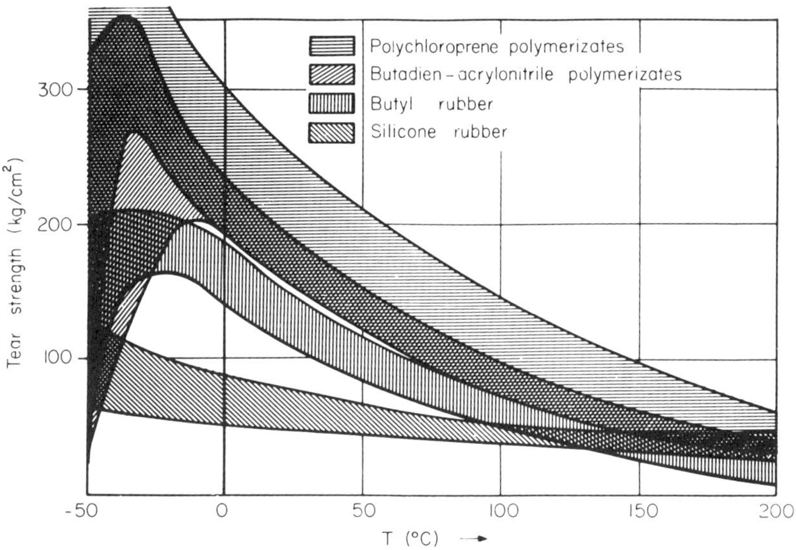

FIG. 57. Tensile strength of vulcanizates of various types of rubber as a function of the test temperature.

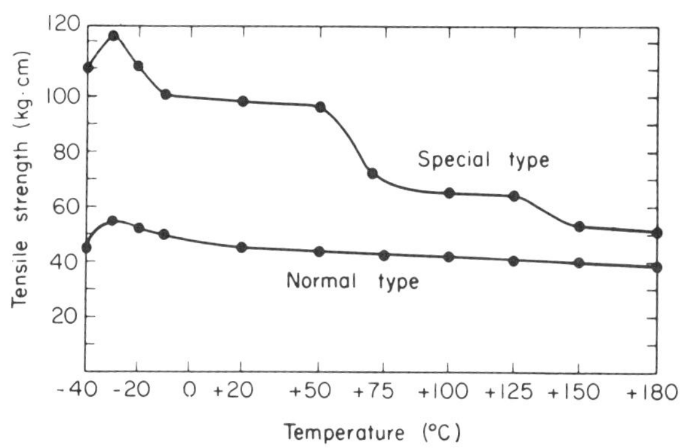

FIG. 58. Tensile strength of silicone rubbers (heat vulcanizates) at various temperatures.

temperature is below that of most organic elastomers, when the temperature is raised the situation changes in favor of the silicone rubber. It is interesting, however, that silicone rubbers with high tensile strength differ from those of a moderate tensile strength in the shape of the strength-temperature curves (Fig. 58) (*71*).

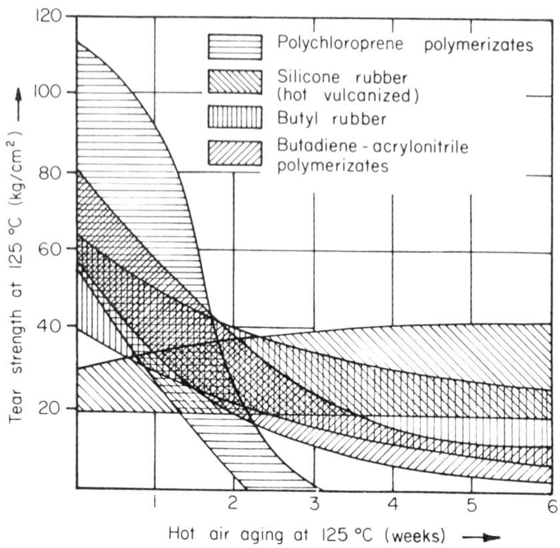

FIG. 59. *Tensile strength* at 125°C as a function of the time of aging in hot air at 125°C.

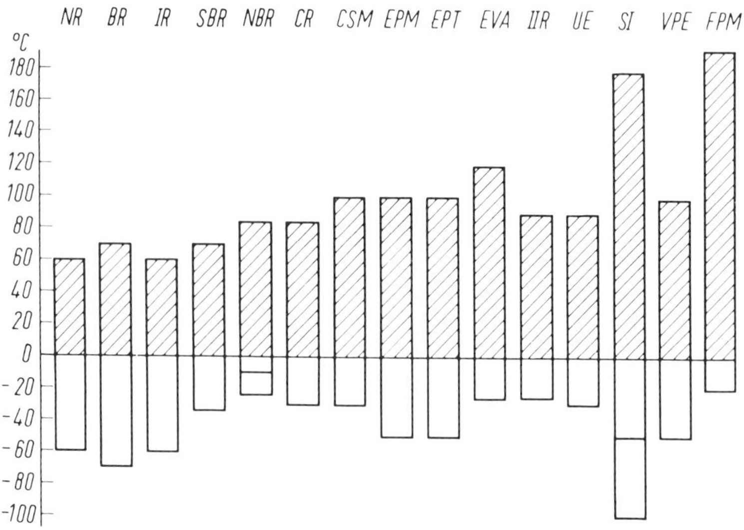

Fig. 60. Limits of applicability of various elastomers. For nomenclature, see p. 499. Lowest values for NBR due to a low nitrile content and those for SI to phenyl groups.

The superiority of silicone rubber becomes still clearer if such measurements are carried out on vulcanizates that have already been subjected to a relatively long heat-aging process. If, for example, aging is carried out at 125°C in hot air and the tensile strength is then determined also at 125°C (i.e., at the temperature of use) the picture that can be seen from Fig. 59 is obtained (*75*, *113*). It shows that silicone rubber, unlike the organic types of rubber, retains its strength even after aging for weeks at 125°C.

From the investigations of the properties of silicone rubber at elevated temperatures, on hot-air aging, and at low temperatures, it is possible to form an approximate picture of its limits of applicability in comparison with other elastomers (Fig. 60). The particular advantages of silicone rubber consist in

NOMENCLATURE ACCORDING TO ASTM D 1418 (WITH ADDITIONS)

Abbreviation	Designation	Type used
NR	Isoprene rubber, natural	Smoked sheets
IR	Isoprene rubbers, synthetic	Solution polymerizate with mixed Ti catalyst with a *cis*-1,4 content of more than 95%
BR	Butadiene rubbers	Solution polymerizate with a mixed Co catalyst with a *cis*-1,4 content of about 95%
SBR	Styrene-butadiene rubbers	23.5% styrene, 76.5% butadiene, not discoloring
NBR	Nitrile-butadiene rubbers	33% acrylonitrile, 67% butadiene, not discoloring
CR	Chloroprene rubbers	Non-sulfur-modified, medium crystalline
CSM	Chloro-sulfonyl-polyethylene	34.5% chlorine, 0.9% sulfur
EPM	Ethylene-propylene copolymer	Ethylene content about 60%, Mooney viscosity (ML-4) about 39
EPT	Ethylene-propylene terpolymer	With about 1% of reactive diene as vulcanizing component
EVA	Ethylene-vinyl acetate	With 40% of vinyl acetate
IIR	Isobutylene-isoprene rubbers	Degree of unsaturation about 0.65 mole-%
UE	Urethane elastomers	Addition product of a mixed adipic acid ester and diphenylmethane diisocyanate
SI	Silicone elastomers	Polydimethylsiloxane with small amounts of reactive groups
VPE	Cross-linked polyethylene	High-pressure polyethylene with a melt index of about 2.0
FPM	Vinylidene fluoride-hexafluoropropylene copolymer	With 65% F

the extremely wide range of temperature within which the material remains rubber-elastic even after long aging at the temperature of use. In every case, however, it must be borne in mind that silicone rubber can develop its typical good resistance to aging only in hot air and that this is lost to a greater or smaller extent in closed systems.

The resistance of silicone rubber to water vapor is practically comparable with its stability in closed systems.

It is true that, because of their thermal stability, vulcanizates of silicone rubber possess a higher ignition temperature than many other elastomers, but they are combustible with the development of the typical white silica fumes. Suitable formulation can impart to the products a certain flame resistance which, for example, satisfies the technical requirements of the aircraft and shipbuilding industries (*46*). With heat-vulcanizable silicone rubber, even small additions of copper or copper compounds, particularly copper(II) oxide, increase the flame resistance (*129*). With cold-vulcanizing silicone rubber, small amounts of chlorinated paraffin wax or similar compounds are sometimes used successfully. In particular, polymethylphenylsiloxanes have proved to be useful for the manufacture of flame-resistant compositions.

9.6.2 Low-Temperature Flexibility

Silicone rubber retains its flexibility down to unusually low temperatures. The glass temperature of dimethylsiloxane rubber is −123°C (*162*); for comparison, natural rubber has a glass temperature of −72°C and butyl rubber

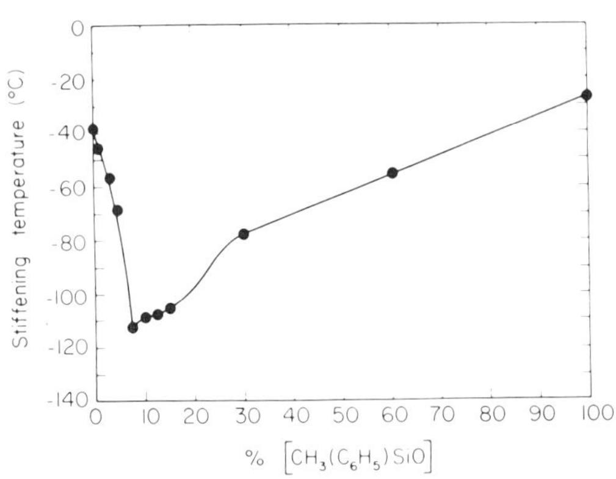

FIG. 61. Stiffening temperature as a function of methylphenylsiloxane molar content.

one of −70°C. The stiffening temperatures range from −50 to −70°C, depending on the formulation (*161*, *162*). A content of 7.5 mole-% of phenyl groups lowers the stiffening temperature to −112°C (Sections 8.1.1 and 9.1.2) (Fig. 61).

Polyethylmethylsiloxanes actually have rubber-elastic properties down to about −135°C. Trifluoropropyl and cyanoethyl groups also lower the stiffening temperature. It is remarkable that the hardness does not change substantially until shortly before the stiffening temperature is reached. The extraordinary behavior of silicone rubber can be seen from Fig. 62, which shows the change in the modulus with the temperature for natural rubber, a polydimethylsiloxane rubber, and a polymethylphenylsiloxane rubber (*120*).

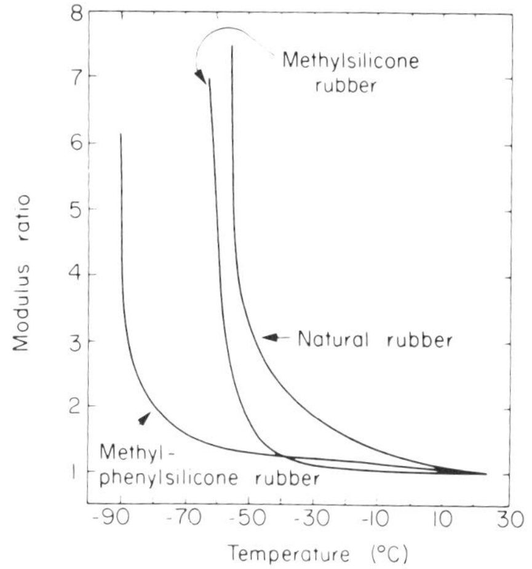

FIG. 62. Modulus ratio vs temperature.

Flexibility at low temperatures is a typical silicone property and is not obtained with most other elastomers even by the addition of plasticizers. Silicone rubber can therefore be subjected alternately to very high and to very low temperatures without any reduction in quality.

9.6.3 RESISTANCE TO OZONE AND WEATHER

Silicone rubber is practically completely resistant to the action of oxygen and ozone on natural aging. Even very high concentrations of ozone have no deleterious effect on the vulcanizates. Silicone rubber may also be regarded as one of the most weather-resistant elastomers, even though there is some fall in the mechanical properties on storage in the open air for several years. Depending on the atmospheric temperature and humidity, a fall of up to 50% in the tensile strength and the elongation at break has been observed within five years (*72*). No changes worth mentioning have been found in silicone rubber under the action of intense ultraviolet irradiation. The transparency qualities of silicone rubbers used, for example, as seals in high-pressure mercury-vapor lamps form an exception; they exhibit an absorption maximum in the region of about 2900 Å and, when the energy of irradiation is sufficiently high, may undergo an increasing degree of cross-linkage. The addition of pigmenting fillers, such as TiO_2, to such rubbers is advantageous in this respect.

9.6.4 Mechanical Properties

The strength properties even of the hitherto best-known grades of silicone rubber are inferior to those of most other types of rubber at room temperatures. In the region of higher and lower temperatures, the picture changes and the superiority of silicone rubber becomes clear, particularly with long-lasting heat stress. While it is true that the strength of the silicone rubber falls to a greater or smaller extent at elevated temperatures of use, an improvement of the properties is frequently associated with increased rate of deformation. This can be seen from Figs. 63 to 65 on the basis of the tensile strength, the elongation at break, and the tear resistance.

In these tests, vulcanizates from a polymethylvinylsiloxane rubber containing 40% by weight of fumed silica and 5% by weight of diphenylsilanediol were used. For vulcanization, 0.72% by weight of dicumyl peroxide, referred to the whole mixture, was added. The vulcanization was carried out for 15 minutes at 151°C in the press and the post-vulcanization for 15 hours at 200°C in hot air (*71*).

Room-temperature vulcanizates have poorer mechanical properties than heat vulcanizates. Some typical data are given in Table 94.

Silicone rubber can be made in the range of hardnesses from 20 to 90 Shore A; vulcanizates of 50 to 70 Shore A generally give the best mechanical properties.

With suitable mixing formulations, very good elastic properties are obtained, and these are preserved in the region of relatively high temperatures. The

TABLE 94

Mechanical Properties of Heat- and Room-Temperature-Vulcanized Silicone Rubber

	Heat vulcanizate	Room-temp. vulcanizate
Tensile strength (kg/cm²)	90–110	30
Elongation at break (%)	300–500	100–150
Hardness (Shore A)	60	60
Elasticity (% resilience)	40–50	50
Structural strength measured on the ring (kg/4 mm)	6–10	1–2

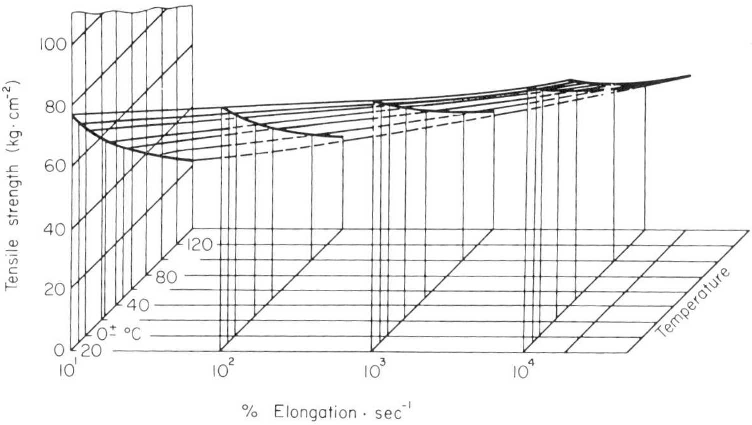

FIG. 63. Tensile strength vs temperature and vs speed of elongation for silicone rubber.

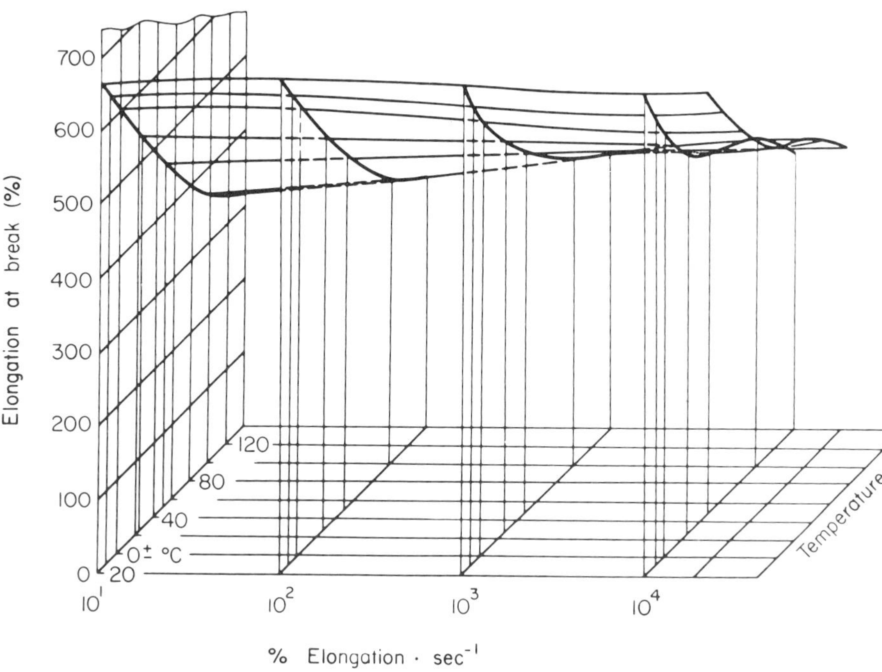

FIG. 64. Elongation at break vs temperature and vs speed of elongation for silicone rubber.

compression set is in fact not very different from that of other organic elastomers at room temperature, but in the region of higher temperatures it is considerably better than with organic types of rubber. Figure 66 shows this behavior.

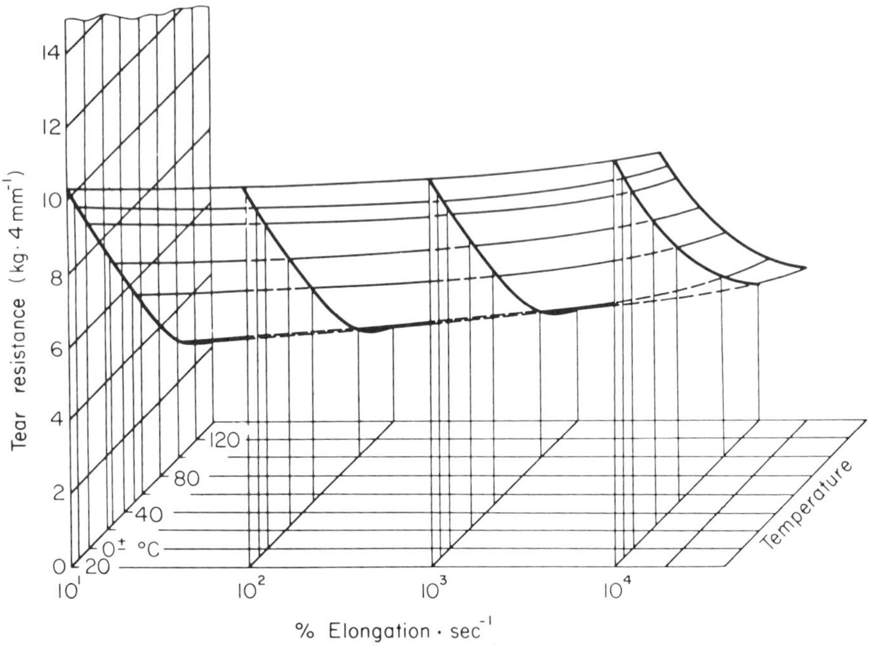

FIG. 65. Tear resistance vs temperature and vs speed of elongation for silicone rubber.

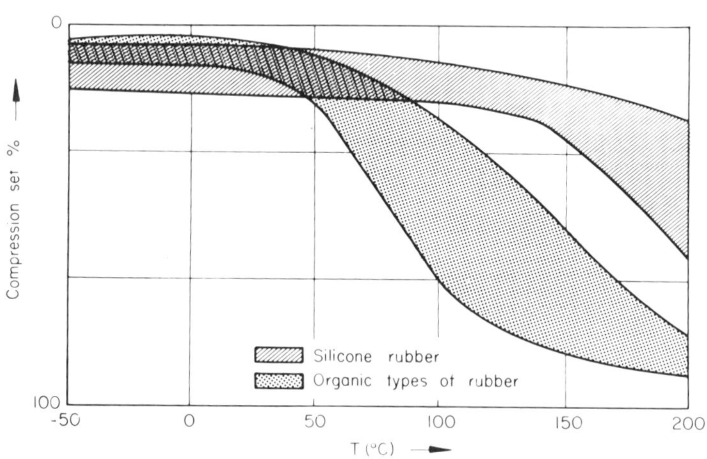

FIG. 66. Compression set of silicone rubber as a function of temperature in comparison with organic types of rubber. Compression and recovery for 24 hours each.

As can be seen, the compression set is substantially smaller with silicone rubber above about 80 to 100°C than with organic types of rubber (*127*). It is lower with polymers containing vinyl groups than with the types free from vinyl groups. With the last-mentioned types of silicone rubber, in particular, it can be improved by additives such as mercury oxide (*66*), cadmium or zinc compounds (oxides, peroxides, or carbonates) (*74*), quinones (*114*), and metal dialkyldithiocarbamates (*115*).

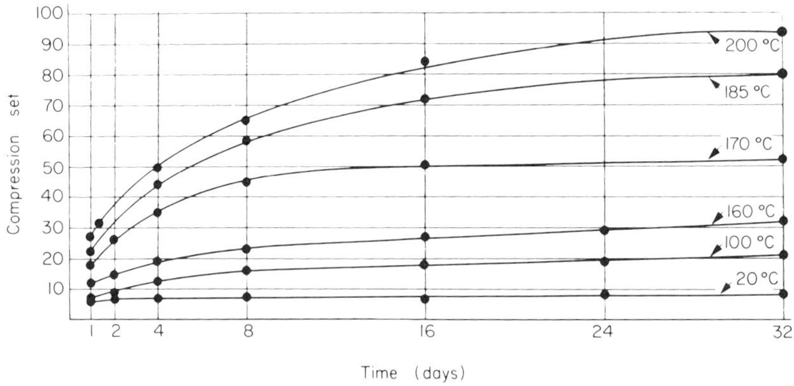

FIG. 67. Compression set of VSi—silicone rubber vs time and temperature.

While the compression set of a silicone rubber is affected to only a subordinate extent by the short-term deformation mentioned, a more pronounced change can be found with longer times of deformation, particularly when the deformation is carried out above 160°C. Figure 67 shows this with a vinyl-groups-containing polymethylsiloxane as an example (*71*).

Another, particularly important, property is the low temperature-dependence of the elasticity constant (Fig. 68) (*111*). Consequently, spring elements can be made from silicone rubber the properties of which are practically independent of the temperature in the usual range of application.

A special position among the rubber-elastic silicone polymers is occupied by a material that, because of its peculiar mechanical behavior, has been called "bouncing putty." It can be obtained by a special kind of polymerization in the presence of boric acid. This product is a curiosity inasmuch as its mechanical behavior depends on the manner and mode in which it is stressed. At rest, it flows like a viscous liquid and therefore spreads out with a smooth surface in a container; with slow and continuous deformation it behaves like a plastic kneadable material that can be drawn out to very thin fibers; with shocklike stresses it reacts as a rubber-elastic material so that balls into which it has been shaped spring back to almost 80% of the height of drop after incidence on a smooth surface; under the impact of a hammer it actually behaves like a brittle substance, exploding into innumerable fragments. Since similar phenomena have also been observed with borosiloxanes, which can be obtained by the reaction of triethyl borate with dimethyldiacetoxysilane (*60*), it may be

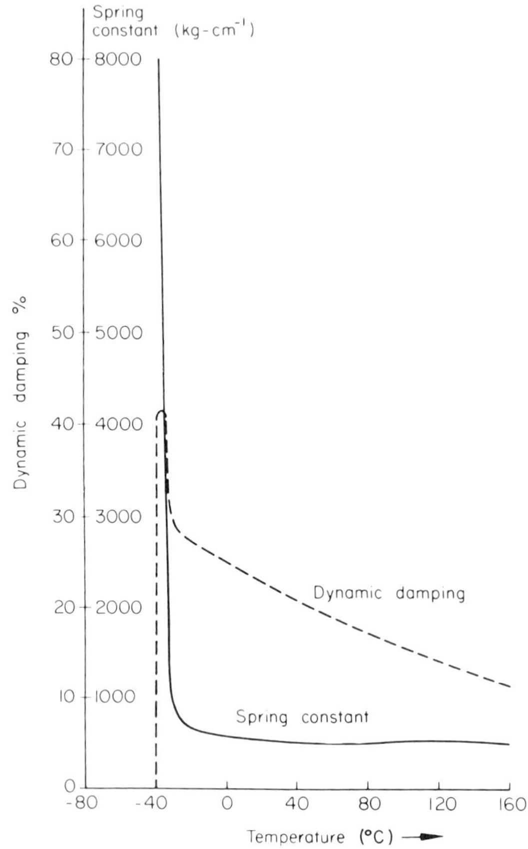

FIG. 68. Elasticity constant and damping of silicone rubber under 20% pre-elongation as a function of the temperature.

assumed that the particular type of cross-linkage of the polymer chains that is brought about by $BO_{3/2}$ groups is responsible for this peculiar behavior.

The abrasion resistance of silicone rubber is in general smaller than that of most other types of rubber.

The heat produced in the dynamic stressing of silicone rubber can—as with other types of rubber—lead to fracture of the material in the course of time.

An idea of its stability on dynamic loading is given by the testing of the material in accordance with ASTM-D 623-52 T. The test results obtained so far permit the conclusion that silicone rubber should not be loaded to a greater extent than 8 to 10 kg/cm^2 under dynamic stress, although within this range of loads other conditions, such as frequency and amplitude, also play a part.

9.6.5 Electrical Properties

Even at room temperature, the electrical properties of silicone rubber are of the order of magnitude of good rubber-elastic insulating materials. At elevated temperatures, however, because of their low temperature-dependence they

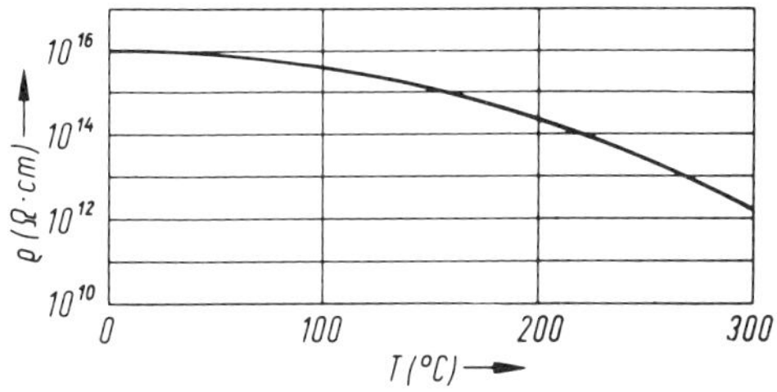

Fig. 69. Specific direct-current resistance ρ of silicone rubber as a function of the temperature.

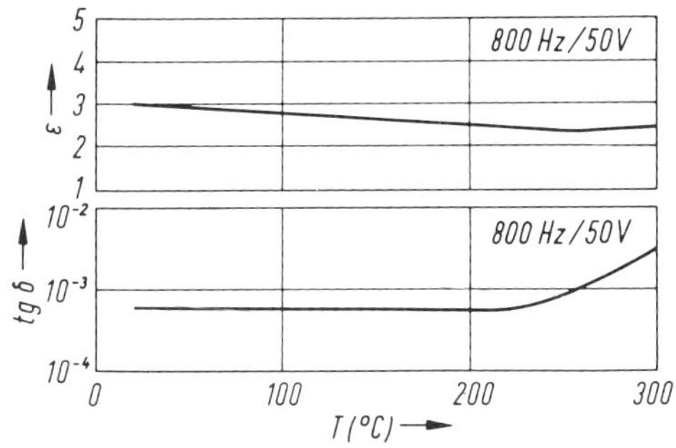

Fig. 70. Loss factor tan δ and dielectric constant ϵ of silicone rubber as a function of the temperature (at 800 cps/50 V).

become even better than the latter. The electrical properties of a high-quality silicone rubber are shown in Table 95 which also gives other data of a non-electrical nature that are of interest for insulation problems. At elevated temperatures at which other insulating materials have already become useless, the vulcanizates still exhibit outstanding electrical characteristics (Figs. 69 and 70).

TABLE 95

PROPERTIES OF A HIGH-QUALITY SILICONE RUBBER VULCANIZATE SUITABLE FOR ELECTRICAL USE

	Test specification	Test specimen	Results
Dielectric strength (kV/cm)			
1. Dry	VDE 0303 part	95 mm diam.,	250
2. 4 days at 80% rel. humidity	2; DIN 53481	1.2 mm thick	250
3. 24 hr storage in water			250
Surface resistance (Ω)			
1. Dry	VDE 0303	Standard rod	3.0×10^{14}
2. 4 days at 80% rel. humidity	part 3;		3.0×10^{13}
3. 24 hr storage in water	DIN 53482		5.0×10^{13}
Resistance between plugs (Ω)			
1. Dry	VDE 0303	Standard rod	4.5×10^{14}
2. 4 days at 80% rel. humidity	part 3;		1.0×10^{13}
3. 24 hr storage in water	DIN 53482		3.0×10^{13}
Specific volume resistance (Ω cm)			
1. Dry	VDE 0303	95 mm diam.	2.1×10^{16}
2. 4 days at 80% rel. humidity	part 3;		1.0×10^{16}
3. 24 hr storage in water	DIN 53482		1.0×10^{16}
Dielectric constant ϵ			
Dry			
1. At 50 cps/50 V	VDE 0303	95 and 30 mm	3.1
2. At 800 cps/50 V	part 4;	diam.	3.0
3. At 10^6 cps/50 V	DIN 53483		2.9
Dielectric loss factor tan δ			
Dry			
1. At 50 cps/50 V	VDE 0303	95 and 30 mm	0.0005[a]
2. At 800 cps/50 V	part 4;	diam.	0.0008
3. At 10^6 cps/50 V	DIN 53484		0.0040
Water-vapor permeability D (gm/hr cm torr) $\times 10^{-8}$	[b]	95 mm diam.	~120
Water absorption (after 7 days) (mg/100 cm^2 of surface)	[b]	80 mm diam.	~35
Thermal conductivity λ (70–80°C) (kcal/m hr deg)	[b]	118 mm diam.	~0.250
Coefficient of linear expansion α (deg^{-1})	[b]	Small standard rod	$\sim 2.1 \times 10^{-4}$

[a] At 50 cps/6 kV tan $\delta = 0.06$.
[b] No standard specifications yet available.

These obviously depend also on the nature of the rubber formulation and the type of vulcanization. By and large, room-temperature vulcanizates have electrical properties inferior to those of heat vulcanizates.

The dielectric strength of silicone rubber scarcely depends upon the temperature in the range from 20 to 200°C, but—as with many other materials—there is a marked dependence on the thickness of the sample in the sense that with increasing thickness the specific dielectric strength decreases (Fig. 71).

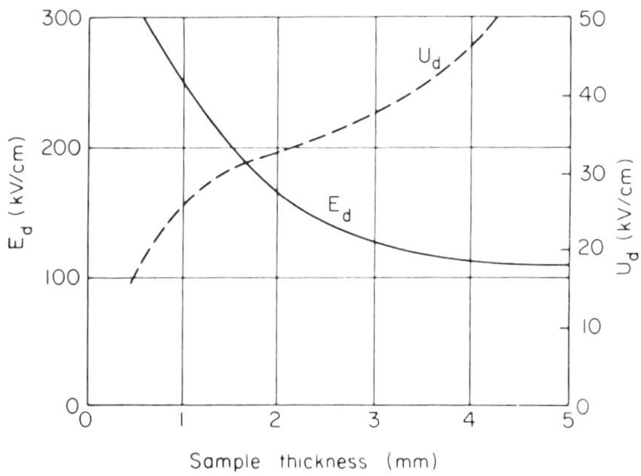

FIG. 71. Dielectric strength E_d and breakdown voltage U_d of silicone rubber as functions of the sample thickness.

Both the dielectric constant and the dielectric loss factor are almost independent of the frequency up to about 10^6 cps. Above 10^6 cps the dielectric loss factor does increase more sharply, but the dielectric constant continues to remain almost unchanged (Fig. 73). With a knowledge of the dielectric properties of silicone rubber, the alternating current resistance can be calculated from the relation

$$\rho_w = \frac{1.8}{f \tan \delta \cdot \epsilon_r} \times 10^{12} \, \Omega \text{ cm}$$

where ρ_w is the alternating current resistance (effective resistance), f is the alternating current frequency in cycles per second, $\tan \delta$ is the dielectric loss factor, and ϵ_r is the dielectric constant. Figure 72 shows the alternating current resistance of two cable mixtures.

Silicone rubber is outstandingly resistant to surface leakage currents and corresponds to the standard resistance quality T 5*; the corona resistance reaches that of mica (*124*).

* According to VDE 0303 or DIN 53 480.

When they burn under the influence of an arc or an electric spark, purely organic types of rubber leave behind electrically conducting carbon. In contrast, silicone rubber leaves behind an insulating silica skeleton.

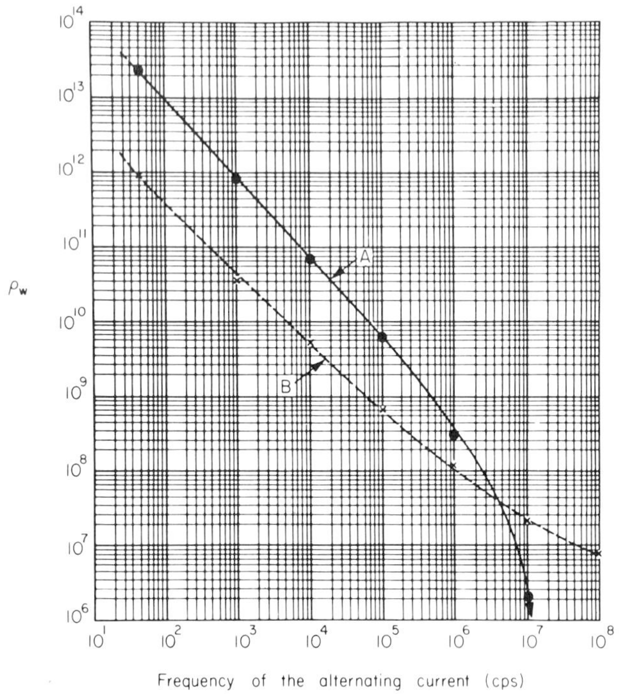

FIG. 72. Specific alternating current resistance ρ_w of silicone rubber mixtures as a function of the frequency: compound A with fumed silica and compound B with diatomaceous earth.

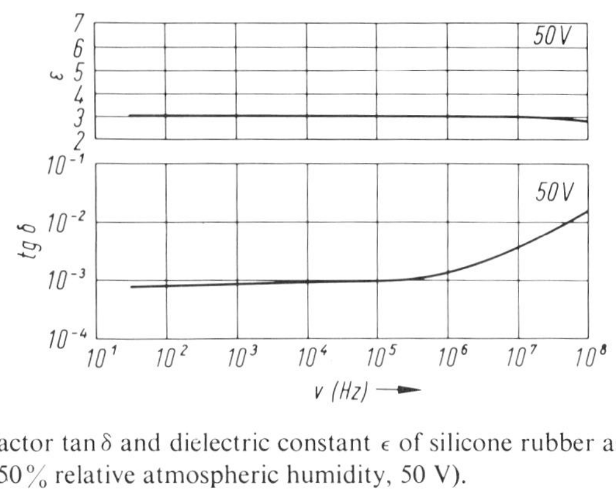

FIG. 73. Loss factor $\tan\delta$ and dielectric constant ϵ of silicone rubber as functions of the frequency (22°C, 50% relative atmospheric humidity, 50 V).

Although silicone rubber is an excellent insulator, there is nevertheless a possibility of obtaining, by the use of suitable types of carbon black as fillers, vulcanizates whose specific resistance can be lowered to about 10 Ω cm (*15*). Figure 74 shows the electrical conductivity of a typical vulcanizate; the complex of properties depends to a marked extent on correct vulcanization to completion in hot air. Since the high conductivity is hardly affected by dynamic pre-stressing of the vulcanizates, the possibility exists here of opening up completely new fields of application for silicone rubber.

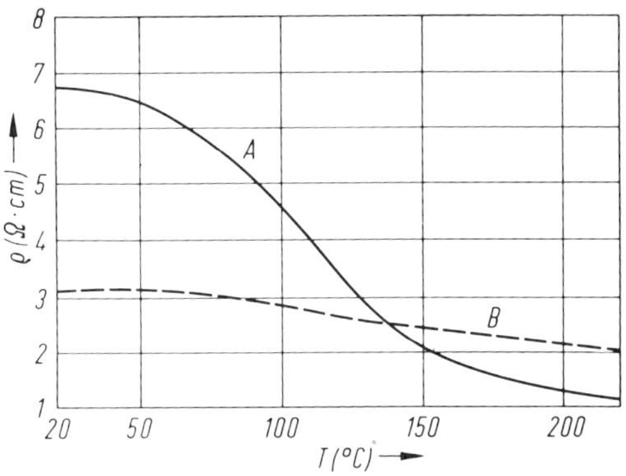

FIG. 74. Specific resistance ρ of a carbon-black–filled silicone rubber as a function of the temperature. Thickness of the plate, 0.25 cm; vulcanization, 15 minutes at 150°C. (A) Without post-vulcanization; (B) post-vulcanized for 15 hours at 200°C.

9.6.6 CHEMICAL BEHAVIOR

Silicone rubber is very resistant to water from 0 to about 100°C, and still possesses adequate stability to low-pressure steam at 3 atm, so that it can, for example, be vulcanized in steam. Steam at higher temperatures destroys the vulcanizate, particularly on prolonged attack.

The resistance of the classical hot-vulcanized types of silicone rubber constructed on the basis of polymethylsiloxanes and polymethylphenylsiloxanes to mineral oils is approximately comparable with that of polychloroprene at room temperature and at moderately high temperatures, but at higher temperatures it is superior. In general, it is good with respect to motor and machinery oils of aliphatic nature. Naphthene-based oils cause somewhat more pronounced swelling. However, in aromatic oils silicone rubber is generally unusable, particularly at temperatures above 150°C. The swelling also depends on the viscosity of the oils. The volume change of the silicone rubber decreases with increasing viscosity. The additives that technical oils contain may also have an effect.

TABLE 96

RESISTANCE OF SILICONE RUBBER TO SOME SWELLING AGENTS IN COMPARISON WITH OTHER TYPES OF RUBBER[a]

Swelling agent	T (°C)	Silicone rubber	Natural rubber	Mixed butadiene-styrene polymerizate	Mixed butadiene-acrylonitrile polymerizate with 33% of nitrile	Polychloro-prene	Butyl rubber	Chloro-sulfonated poly-ethylene
Cleaning gasoline	50	220	250	140	10	55	240	85
ASTM oil I	50	3	60	15	−1.5	4	20	4
ASTM oil I	100	3	320	50	−6	15	170	20
ASTM oil II	50	5	100	40	−0.5	15	50	12
ASTM oil II	100	6	Destroyed	120	7	55	250	75
ASTM oil III	50	20	200	125	3	60	120	65
ASTM oil III	100	26	Destroyed	190	12	100	370	200
Diesel oil	50	110	250	150	12	70	250	120
Diesel oil	100	120	Destroyed	220	13	80	400	160
Shell Oil X-100	50	5	110	40	−3	12	7	12
Shell Oil X-100	100	6	270	80	1	28	250	45
Olive oil	50	1	100	50	−2	22	10	40

Lard	50	1	110	50	1	25	10	45
Formaldehyde	50	2	6	5	10	20	0.5	1.2
Ethanol	50	12	3	0.5	20	6	2	5
Glycol	50	0.5	0.5	−0.5	0.5	2	−0.2	0.5
Diethyl ether	20	230	170	130	30	85	90	85
Ethyl acetate	50	125	90	90	160	100	20	100
Methyl ethyl ketone	50	100	85	80	250	115	15	150
Trichloroethylene	50	250	420	350	230	320	300	600
Carbon tetrachloride	50	250	420	320	75	280	275	350
Benzene	50	200	350	300	200	300	150	430
Xylene	50	220	350	275	100	270	240	480
Aniline	50	3	15	30	380	100	10	70
Phenol	50	5	35	50	470	70	3	80
Cyclohexanol	50	10	55	35	40	35	7	20
Cyclohexanone	50	45	350	240	370	330	55	520
Diphyl	100	15	Destroyed	350	260	280	60	320
Clophen T 64	100	10	Destroyed	310	175	300	260	480
Silicone oil Bayer M 100	50	40	−2	−4	−2	−0.5	−0.5	−0.5
Distilled water	100	3	10	4	11	10	5	4
Sea water	50	0.5	2	1.5	3	4	0.5	0.5

[a] Increase in volume in per cent after 7 days.

The classical types of silicone rubber swell in many organic solvents such as, for example, aliphatic, aromatic, and chlorinated hydrocarbons (*56*). A considerably improved behavior with respect to oils and also with respect to solvents is exhibited by types of silicone rubber based on poly-γ-trifluoropropyl- (*116*), β-cyanoethyl-, and γ-cyanopropylmethylsiloxanes (Section 8.1.1).

Room-temperature-vulcanized types of silicone rubber generally have a lower resistance.

It is remarkable that, because of their microporosity, many vulcanizates absorb oil beyond the amount corresponding to the usual swelling equilibrium even at relatively high temperatures, and sweat out this oil again in the course of time after removal of the liquid phase. This explains why oil-treated silicone rubber parts possess so-called "dry-state properties." For example, shaft seals that take up certain amounts of lubricating oil during use give up part of the oil again as soon as the external supply of lubricating oil ceases; the seals then provide lubrication on their own for a short time.

Silicone rubber is destroyed by strong acids and alkalis, particularly at elevated temperatures. Silicone rubber is resistant to monohydric and polyhydric alcohols, phenol, heat-transfer media of the type of Diphyl, and to Askarels such as Clophen. Details can be seen from Table 96.

9.6.7 Permeability to Gases and Liquids

Many technological properties of silicone rubber point to a microporous structure of this material. Thus, for example, the swelling power and the dry-state properties that have just been mentioned (Section 9.6.6) are explained by this structure. It may also be partially responsible for the high compressibility (cf. Section 9.2.7). The microporous structure of silicone rubber shows itself to a particular degree in its permeability to gases and liquids.

The gas permeability of silicone rubber is substantially greater than that of the other organic elastomers (*52*). Table 97 gives a numerical comparison. The figures given there were measured on vulcanizates which all contained only 25% of fumed silica as filler and a cross-linking system suitable for the rubber concerned, so that it is possible to make a direct comparison.

As Table 97 shows, the gas permeability of the silicone rubber is 10 to 100 times greater than that of natural or butyl rubber. Since the rate of diffusion of different gases is different, it becomes technically possible to separate mixtures of gases by means of silicone rubber foils (*67*).

Silicone rubber is permeable both to gases and liquids, but the permeability coefficients for liquids are substantially lower than those for gases.

9.6.8 Surface Properties

Heat- and room-temperature-vulcanized silicone rubbers exhibit intrinsic releasing properties. They scarcely adhere even to very highly *ad*hesive materials

TABLE 97

GAS PERMEABILITY OF VULCANIZATES IN 10^{-9} cm^2 sec^{-1} atm^{-1} AT 80°C

Rubber composition based on	Air	Nitrogen	Carbon dioxide
Natural rubber	400	360	2100
SBR (Buna-Hüls 152)	260	200	1500
NBR with 28% of acrylonitrile (Perbunan N 2810)	210	70	970
NBR with 33% of acrylonitrile (Perbunan N 3307)	70	55	630
NBR with 38% of acrylonitrile (Perbunan N 3810)	55	25	480
Polychloroprene (Perbunan C 110)	90	80	810
Butyl rubber (Butyl 301)	50	35	290
Chlorosulfonated polyethylene (Hypalon 20)	70	50	450
Urethane rubber (Urepan 600)	70	55	730
Silicone rubber (Silopren RS)	4100	3600	15000

such as, for example, tar, rubber mixtures, resin, asphalt, and molten polyethylene, provided there is no mechanical anchorage to the surface. Silicone rubber owes a series of important uses to this *ab*hesivity. Among the heat vulcanizates, types of rubber based on dimethylpolysiloxanes with small proportions of vinyl groups, which have the highest possible Shore hardnesses with a low content of filler, are particularly suitable.

However, by the use of primers or self-bonding compounds, heat- and room-temperature-vulcanizable silicone rubbers can be adhesively bonded to metals, glass, ceramics, and many other materials during vulcanization. Certain silicon-functional silanes or siloxanes, such as those with alkoxy groups or Si—H bonds, are particularly suitable for this purpose. Such groupings are pre-incorporated into special types of rubber. Of the room-temperature-vulcanizable one-component systems, those with functional acetoxy groups, in particular, develop a good adhesive power, which is evoked by the acetic acid liberated by atmospheric moisture, during culvanization (Section 8.1.2.3.2).

Vulcanizates made from silicone rubber have good slip properties. They can be used to make sealing materials for shafts which permit running at a higher speed than many other types of rubber.

9.6.9 RADIATION RESISTANCE

In view of its technical importance, the question of the radiation resistance of silicone rubbers has already been the object of numerous investigations. Under the action of energy-rich radiation a cross-linkage of the silicone rubber takes place, with, accordingly, an increase in the moduli and the Shore hardness together with a reduction in the elongation at break. These effects make themselves apparent to different extents in accordance with the nature of the polymer, the other constituents of the formulation, and the radiation dose (*18*, *24*, *82*).

Generalizing, it can be said that the radiation resistance of a silicone polymer is better the lower the cross-linking yield on peroxide vulcanization. It also depends to a decisive extent on the nature of the substituents in the polysiloxanes (*38*, *54*, *73*, *158*, *167*) (cf. Section 9.2.12). Since the resistance increases with increasing content of phenyl groups, copolymers with diphenylsiloxane units and methylphenylsiloxane units have proved particularly suitable (*130*). As comparative investigations have shown (*158*), a polymerizate with $[CH_3(C_6H_5)SiO]$ units is cross-linked by 186 Mrep (million roentgen equivalent physical) only to the same extent as a polydimethylsiloxane polymerizate by 10 Mrep. Additions of aromatic compounds also improve the radiation resistance. The combined use of up to 10% by weight of *N*-dihydroanthraquinonacine, phthalocyanine, or their metal complexes or diphenylbenzene as compounding ingredients with selected types of silicone rubber is said to be capable of extending the resistance of the vulcanizates even at 200°C to radiation doses of more than 200 Mrad (*36*, *37*), while ordinary types of silicone rubber generally tolerate no radiation doses higher than 1 to 10 Mrad.

9.7 Physiological Behavior*

Silica is always found in plants and, in low concentrations, in animal and human organisms (*64*); organosilicon compounds with direct bonds between carbon and silicon are however completely unknown in the biochemical field. Surprisingly, silicones have been repeatedly demonstrated to be physiologically and toxicologically inert, so that they can be safely used for technical purposes, and particularly in cosmetics, pharmacy, and medicine. Below will be given the experimental bases that show the physiological behavior of the silicones in various types of application in the animal organism.

A generally interesting reference to the biological inertness of these compounds may be mentioned first: cultures of various tissues of warm-blooded animals, which are well known to be extraordinarily sensitive to foreign influences, show no deviations from the usual growth picture on contact with liquid, semisolid, and rubberlike silicone products (*85*).

9.7.1 Monomeric Silanes

It is very illuminating that a connection must exist between the physiological inertness of the polymeric silicones and their chemical stability in the

* Section 9.7 was written by Prof. G. Hecht, Wuppertal-Elberfeld, West Germany.

biological milieu. This is shown particularly by the different behavior of the starting products for the silicones, the functional monomeric *silanes*. Silanes are by no means physiologically inert, but, in accordance with their reactivity, give rise to caustic effects wherever they come into contact with the body. The action of the following compounds on contact with the skin, with the *ocular conjunctiva*, on *inhalation*, on administration in the *stomach*, or on *intraperitoneal injection* have been investigated on experimental animals (*128*): $SiCl_4$, CH_3SiCl_3, $(CH_3)_2SiCl_2$, $C_2H_5SiCl_3$, $(C_2H_5)_2SiCl_2$, $Si(OC_2H_5)_4$, $CH_3Si(OC_2H_5)_3$, $(CH_3)_2Si(OC_2H_5)_2$, $(CH_3)_3SiOC_2H_5$.

Even the *contact* of the liquid *chlorosilanes* with the shaved abdominal skin of rabbits rapidly leads to tissue disturbances—with $SiCl_4$ in 10 minutes, with CH_3SiCl_3, $C_2H_5SiCl_3$, and $(C_2H_5)_2SiCl_2$ in 2 to 3 minutes, and with $(CH_3)_2SiCl_2$ in about 1 minute. In contrast, none of the *ethoxysilanes* mentioned has irritant properties, apart from the fact that, like organic solvents in general, they exert a "desiccating" effect.

When 0.05 ml of each of the *chlorosilanes* was dripped into the *conjunctival sac* of the rabbit eye, all compounds caused severe erosions of the cornea and the conjunctiva. In each case 0.2 ml of the *ethoxysilanes* caused only transient irritations that died away after 24 hours in the rabbit eye. In contrast to this, however, is the behavior of *tetramethoxysilane* $Si(OCH_3)_4$, which, in the form of the liquid or the vapor, gives rise to a severe necrosis of the cornea that can lead to complete blindness. Consequently, appropriate protective measures must be taken (*168*) when this substance is handled.

Inhalation of air containing 10 to 50 mg/liter of these substances in vapor form gives rise to erosion effects on the mucosa and on the tips of the ears of rats in only 5 to 15 minutes. The fact that specific erosive effects, and not merely those due to the hydrogen chloride liberated by hydrolysis, are present here is shown by experiments on ethoxysilanes: erosive effects on the mucosa, particularly the respiratory tract, were also observed when the vapors of methylethoxysilanes were inhaled. The erosive effect of the vapors caused by breathing 1000 to 4000 ppm for several hours decreases clearly in the sequence (*128*):

$$Si(OC_2H_5)_4 > CH_3Si(OC_2H_5)_3 > (CH_3)_2Si(OC_2H_5)_2 > (CH_3)_3SiOC_2H_5$$

The last-mentioned compound is, however, more toxic than dimethyldiethoxysilane and in fact shows indications of narcotic action. When ethoxysilanes are inhaled repeatedly, resorptive toxic effects appear—in particular in degenerative changes in the kidney and liver (*121*, *128*).

Oral and *intraperitoneal* administration of *chlorosilanes* has a lethal effect on rats after doses of 1.0 or 0.1 gm/kg and occasionally even below, because of the caustic effect on the tissue concerned. There is no information on resorptive toxic effects. The *ethoxysilanes* appear to be somewhat less toxic (*128*). *Sodium methylsiliconate* (Section 3.7) is tolerated by mice, on intravenous administration of the aqueous solution neutralized with HCl immediately before injection, in doses of up to 0.6 gm/kg (*50*).

The toxicological properties of the compounds can be affected to a large extent by substitutions of the organic groups. Thus, δ-aminobutylmethyldiethoxysilane

$$H_2N-CH_2-CH_2-CH_2-CH_2-\underset{\displaystyle CH_3}{\underset{|}{Si}}(OC_2H_5)_2$$

is a compound with an extremely toxic effect on resorption through the skin. In animal experiments, half the rats were killed by the epidermal application of 0.045 ml/kg of body weight. However, in oral administration or on inhalation the compound is obviously much less dangerous. γ-Aminobutylmethyldiethoxysilane exhibits neither the same resorbability through the skin nor the same toxicity, but causes severe erosions on the skin (*154*).

9.7.2 Polymeric Siloxanes

Of the polymeric siloxanes, essentially those with methyl and/or phenyl groups have been tested with respect to their toxicological behavior.

9.7.2.1 *Relationship between the Degree of Polymerization and the Physiological Effect*

The series of *linear polymethylsiloxanes* (cf. Section 6.2.1) permits a comparative consideration of the physiological effect of compounds with different degrees of polymerization. Physiological inertness is inherent only in the higher polymeric siloxanes. The lowest member of the series, hexamethyldisiloxane, is still sufficiently volatile to be able to evoke toxic effects when the vapor is *inhaled*. An atmosphere saturated with it (about 40,000 ppm at 21 to 22°C) leads to the death of guinea pigs after a breathing time of 15 to 20 minutes. Animals removed from this atmosphere after shorter periods generally survive, and so do those that have breathed 25,000 ppm for 30 minutes. Even the repeated inhalation of small concentrations of hexamethyldisiloxane (4400 ppm, 10 rats, 15 times every 7 hours over 18 days; 8 guinea pigs with the same exposure over 26 days) no longer led to recognizable toxic effects; there is therefore no danger of a cumulative poisoning (*128*). The effects of high-molecular-weight technical silicones on inhalation in the form of drops or aerosols will be described later. However, it must be mentioned here that a mixture of relatively low-molecular-weight *cyclic polymethylsiloxanes* led to certain irritation phenomena in the respiratory tract and the ocular conjunctiva when it was inhaled by cats and other experimental animals in a concentration of 5 to 10 ml/liter. Smaller doses of 1 ml/liter administered for 4 hours every day were however without effect, even when repeated 12 times (*57*).

The results of the *oral* administration of various polymeric methylsiloxanes to guinea pigs can be seen from Table 98 (*128*).

With single *intraperitoneal* administrations of the polymethylsiloxanes mentioned in Table 98 to rats in doses of 0.1, 0.3, 1.0, 3.0, and 10 ml/kg, only hexamethyldisiloxane in the highest three doses exhibited a lethal effect. This material has a tissue irritating effect; on autopsy, extensive adhesions of the viscera were found. All other compounds produced only typical uncomplicated

TABLE 98

RESULTS OF ORAL ADMINISTRATIONS OF SILICONE OILS IN ANIMAL EXPERIMENTS

Viscosity and type of polyorganosiloxane (cSt/25°C)				Observed purgative effect in various times after application over			
Polydimethylsiloxanes	Polymethylphenylsiloxanes	Dose (ml/kg)	Mortality rate	2.5 hr	8 hr	24 hr	48 hr
0.65[a]		3.0	0/7	–	–	–	–
		10.0	0/7	–	–	–	–
		30.0	0/7				
		50.0	1/10				
2.0[a]		10.0	0/3	–	+	–	–
		30.0	0/6	–	++	–	–
		50.0	3/3				
50[a]		10.0	0/2	+++	+++	+++	+
		30.0	0/6	+++	+++	+++	+++
		50.0	0/3				
	35[b]	3.0	0/3				
		10.0	0/3	+	++	+++	
		30.0	0/6		++	+++	++
	75[c]	3.0	0/3				
		10.0	0/3	+	+	+	
		30.0	0/6		++	+++	+++
350[a]		5.0	0/2	–	–	+	
		10.0	0/5	–	+	+	–
		30.0	0/6	–	+	+	–
		50.0	0/3	–	–	++	++
12500[a]			Could not be administered				
Liquid paraffin (for comparison)		10.0	0/2	++	++	+++	+
		30.0	0/3	+++	+++	+++	+

[a] Dow Corning, DC 200.
[b] Dow Corning, DC 702.
[c] Dow Corning, DC 550.

foreign-body reactions. Nodules containing liquid and present in the tissue appear. No inflammatory reactions were observed.

On *intracutaneous* application (rabbit) again only hexamethyldisiloxane evoked a visible change in the skin. Inflammations, edema, and necroses appeared at the site of the injection. Other silicone liquids produced weals that were visible for a few days but disappeared after a week and could no longer be observed in the autopsy.

On *subcutaneous* application in rabbits, small skin traumas were exhibited only after 24 hours with all preparations other than hexamethyldisiloxane. However, hexamethyldisiloxane evoked a considerable irritation of the skin with induration. Irritations and necroses were still present on autopsy 13 days after the injection of hexamethyldisiloxane, while with the compounds of higher molecular weight there were no further traces to be seen.

Instillation into the *conjunctival sac* of the rabbit led to the results given in Table 99.

TABLE 99

Action of Silicone Oils on the Conjunctiva of the Eye in Animal Experiments (Rabbits)

Viscosity and type of polyorganosiloxane (cSt/25°C)		Occurrence and duration of irritant effect					
Polydimethylsiloxanes	Polymethylphenylsiloxanes	Immediately	1 hr	4 hr	8 hr	24 hr	48 hr
0.65		+	0	0	0	0	0
2.0		0	0	0	0	0	0
50.0		0	+	+	+	0	0
	35.0	0	0	+	+	0	0
	75.0	0	0	+	+	0	0
350.0		0	+	+	+	+	0
12500.0		0	+	+	+	+	0

All these phenomena disappeared after 48 hours. There was no damage to the cornea.

9.7.2.2 *Inhalation*

Since the more highly polymeric methylsiloxanes are no longer volatile, their physiological action on inhalation cannot be studied on vapors. However, observations on the action of *mists* or *aerosols* on inhalation are of great importance since in numerous applications persons come into contact with mists produced by the spraying of silicone oils. In the treatment of pulmonary edemas by the inhalation of silicone aerosols, which has been proposed medically, the aerosols are actually inhaled directly.

The inhalation of a dense aerosol of a linear dimethylpolysiloxane with an average of 110 siloxane units is tolerated by rats and rabbits without ill effects. Rats were exposed for 4 hours every day; this experiment was repeated 15 times. Immediately after the ending of the experiment, a chemical determination of SiO_2 showed the presence of around 15 mg of SiO_2 in the lungs. Within 3 months after the end of the experiment, this figure had fallen to 5 mg of SiO_2. With rabbits treated in the same way, the SiO_2 figures found by analysis were 70 and 20 mg per lung, respectively. Histological investigations of the lungs of animals that were removed from the experiment at various times provided no surprising results (*69*).

In addition, various types of experimental animals were exposed on 29 test days within one month to mists produced by the spraying of a 40% emulsion of the above-mentioned polydimethylsiloxane with an average of 110 siloxane units per molecule. Neither during nor after the time of action did they exhibit toxic phenomena, which again shows that entry of such products into the respiratory tract in the form of fine droplets has no objectionable consequences (*50*, *118*).

The same results were obtained on repeated inhalation of a silicone defoaming emulsion by rats and guinea pigs (*118*).

No striking consequences occurred when 1 ml of silicone oil was instilled into the trachea of rabbits, and the animals remained unharmed when the treatment was repeated 8 times at intervals of 1 week. One month after the last treatment, their lungs still contained amounts of silicone oil corresponding to around 400 mg of SiO_2. After 4 and 7 months, these figures had fallen to 115 and 40 mg, respectively, and after 14 months silicone oils could no longer be detected. It was also impossible to find residues of the treatment histologically (*69*).

9.7.2.3 *Subcutaneous, Intracutaneous, and Intramuscular Administration*

The results already mentioned in Section 9.7.2.1 on intracutaneous administration of silicone oils must now be supplemented with further observations according to which the subcutaneous and intramuscular administration of silicone products has given rise to no harmful consequences (*11*, *50*, *68*, *83*, *93*). For example, the subcutaneous administration of a 40% aqueous silicone oil emulsion containing a linear polydimethylsiloxane with a viscosity of about 50 cSt/20°C was tolerated well by mice observed for a month (*50*). Moreover, the injection of a suspension of a cross-linked polymethylsiloxane subcutaneously and intramuscularly also had no sequelae. On intracutaneous injection, a slight ingress of macrophages was found on histological examination after 4 days (*68*).

9.7.2.4 *Intraperitoneal and Intravenous Administration*

When silicone oils of the type of polydimethylsiloxanes were introduced into the *free abdominal cavity* of small rodents, they remained unchanged at the point of introduction for a very long time and evoked only slight reactions on the part of the peritoneum. This has been confirmed by a series of observations, of which the following is particularly characteristic (*118*).

Sixteen mice each received 0.2 ml and 24 mice 0.5 ml of silicone oil. On autopsy after 3 weeks, silicone oil was still found in the free abdominal cavity, while on further autopsies after 8 weeks its presence was only doubtful. There were no deposits organized with respect to the connective tissue. In 4 cases, multichambered cysts occurred, which were present either in the peritoneum or in the subcutaneous abdominal tissue and which contained the silicone oil.

Repeated intraperitoneal injection of polydimethylsiloxane oil into rats was carried out (*118*). According to this, no acute or chronic toxicity was shown even with a total dose of 10 to 29 ml in single doses of 2 to 10 ml. In the first 67 days, the rats showed no changes that could have been caused by the silicone oil. The oil itself was found in the abdominal cavity of the animals during the whole duration of the experiments. From the 132nd day onward, it was milkily turbid. From the same day onward, deposits of pinhead size could be found in the liver and spleen, and these increased in number and extent with time and were finally found also in the peritoneum, which had occasionally undergone granulomatous changes. Foci on the diaphragm were rarer and were observed only once in the kidney. In one case, a cyst with residues of the kidney parenchyma was found at the upper pole, which must be regarded as a spontaneous change. The other organs showed no changes. Histological investigation showed that no organ had undergone any specific changes due to the silicone oil. The cysts had a honeycomb structure and exhibited proliferations of connective tissue with ingrown capillaries, which are to be regarded as nonspecific foreign-body reactions. No foreign-body giant cells were found.

In experiments with the intraperitoneal injection into rats of polydimethylsiloxane oils having viscosities of 60 and 140 cSt/20°C, the doses being between 1 and 5 gm per animal, the animals again showed no ill effects. They were either killed after a month or were observed up to their natural death (an average of 496 days). On autopsy, it was seen that the preparations did not remain in the free abdominal cavity. They were obviously distributed in fine drops and probably deposited in the tissue, a reactive multiplication of the reticular tissue taking place. In no case was tumor formation seen, even with a time of observation of more than a year (*50*).

Emulsions of silicone antifoaming agents also proved to be well tolerated on the intraperitoneal administration of quite large doses (3 to 5 ml per animal) to rats. At the most, they exhibited side reactions which must be ascribed to the emulsifier (*128*).

The same thing also applied to silicone resins based on polymethylsiloxanes or polymethylphenylsiloxanes. Such preparations, in some cases suspended in olive oil (*128*) and in some case dissolved in toluene-cyclohexanol (*118*), were given to rats. Again, after times of observations of some months, only nonspecific foreign-body reactions and no proliferative processes were found (*68*, *118*).

It may be mentioned in addition that some experiments with the *intravenous* administration of polymethylsiloxanes to mice have been described (*68*). The suspension was prepared by mixing 1 ml of the methylchlorosilane preparation Dri-Film 9977 with 50 gm of dextrose. After removal in vacuum of the hydrogen chloride formed, the dextrose was made into about a 15% solution in water containing a small amount of gum arabic. Up to 13 doses of 0.5 ml each of the suspensions formed was injected into mice without any recognizable results of the treatment having occurred.

9.7.2.5 *Oral Administration*

The compatibility of the polysiloxane derivatives on oral administration has quite particular importance, since the possibility exists that with their manifold uses they may come into contact with human food and may migrate into them in small amounts. From this point of view it is comforting that single administrations of very considerable amounts are tolerated and have no consequences other than diarrheal phenomena. Considerably more important are the results of investigations with administration of the preparation for a longer time.

The level of the dose of liquid polydimethylsiloxanes tolerated by experimental animals after a *single* administration can be seen from Table 98. In rabbits, cats, and dogs, no harmful effects were seen after single doses of 10 ml/kg or after they had been fed with a powdered methylphenylsilicone resin. Rats were given up to 54 ml/kg of silicone oil with their food without any harmful effects having been observed (*81*). It can hardly be doubted that the preparations remain unresorbed and are excreted again with the feces.

Oral doses of 1.0 to 10.0 gm/kg each were given to rats 20 times over *one month*; here again, there were no harmful effects with respect to growth, blood picture, bone marrow, weights of the individual organs, and histological, findings (*128*). Likewise, a *cyclic* polydimethylsiloxane of relatively low molecular weight was tolerated without harmful effects by rats when 24 individual doses of 5 ml/kg each were given over a month (*49*).

Little material is available on chronic experiments with oral administration over a period of *months*. Groups containing 5 female rats consumed a diet containing 0.1, 0.3, or 1.0% of a silicone antifoaming agent for 3 months. They were just as healthy and increased in weight just as much as control animals. At the end of the experiment, no damage to organs was found (*128*).

The feeding and observation of 25 male and 25 female rats with a diet containing 0.3% of the same silicone antifoaming agent extended over 18 months. Here again, the experimental animals showed the same increase in weight and state of health as control animals not treated with silicones (*26*).

Similar results were obtained with the chronic feeding of a 100% silicone antifoaming agent and a 40% silicone antifoaming emulsion which were given to male rats for more than 2 years so that the individual rats ingested up to 10 gm and more of the product concerned during their lifetime (*49*).

Feeding experiments were also carried out with dogs over 6 months (*26*). The animals received 0.9, 1.0, or 3.0 gm/kg of a silicone antifoaming agent with their food for five days a week. In spite of this massive treatment, during which the silicone could be found in the feces, there were no manifest effects on health and no microscopically detectable changes in the organs investigated, with the possible exception of a deposit of bile pigment in certain cell elements of the liver. This was not explained further but it does not give rise to any objections against the amounts coming into consideration in practice for man.

Methylphenylsilicone resins have also been given in food, although not over such long periods. A lacquer resin used to lacquer baking tins was fed to rats in the diet at the rate of 3% for 50 days and 10% for 34 days, again with no pathological results (*128*). With other investigations, as well, up to 32 individual doses each of 1.0 gm/kg of methylphenylsilicone resin had no ill effects (*49*).

9.7.2.6 *Epidermal Application*

The human *skin* can come into contact with silicone products much more easily and more frequently than any of the organs so far described. This applies not only to the preparation of silicone-containing products but in particular to their daily use. Polydimethylsiloxane oils have proved to be so *inert* to human skin that they have become competitors with the paraffin oils as constituents of *cosmetics* (cf. Section 10.13). The advantages that they exhibit in their behavior on the skin were first recognized by Vonkennel *et al.* (*135*).

The absence of primary skin irritations not only with a single application but also with repeated applications of the polydimethylsiloxanes, which has been shown in Table 98 (Section 9.7.2.1), was recognized very early. Even 20 treatments over a month had no harmful consequences on the shaved abdominal skin of rabbits (*128*). Again, the painting of a rabbit ear lobe with silicone oils on 60 successive days showed no change in comparison with the untreated control side (*50*). The application of a liquid polydimethylsiloxane in the patch test over 24 to 48 hours never led to skin irritation (*135*).

The skin compatibility of silicone oils or greases can be demonstrated particularly clearly by the absence of an epithelium-thickening action of the skin of experimental animals. The experimental arrangement is called the "*acanthosis test*" which expresses the fact that it is particularly the so-called prickle-cell layer of the skin epithelium that contributes substantially to dissemination under the influence of materials damaging the skin. When a paraffin oil or petroleum jelly was allowed to act for 10 days, this dissemination occurred quite clearly, which shows that the acanthosis test responds even to materials which are frequently used in cosmetics. On the other hand, this effect was almost absent when silicone preparations were tested in the same way (*14*, *132* (Figs. 75 to 77).

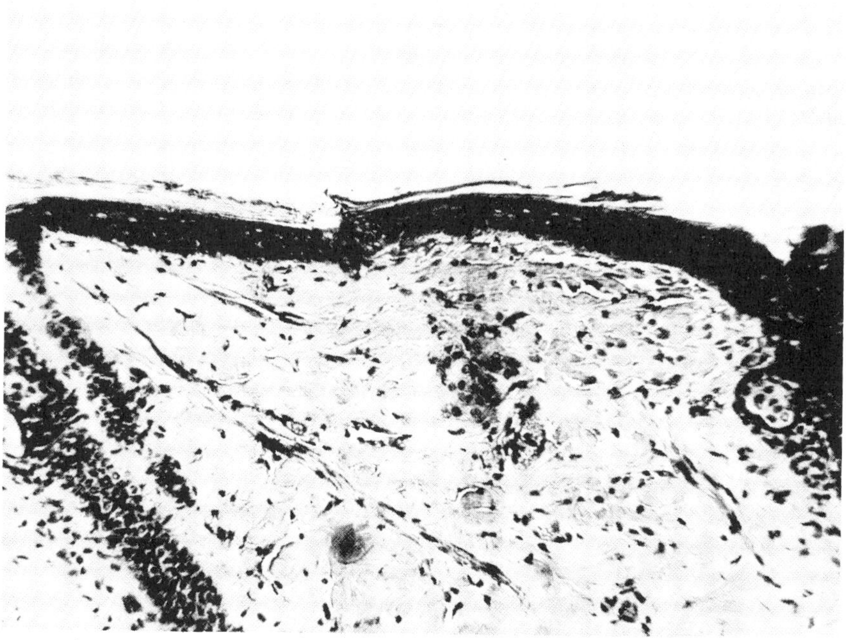

FIG. 75. Normal skin of guinea pig; epidermis, dermis, and hair follicle (after Vonkennel, with kind permission of the author).

9.7.2.7 *Effect on the Eye* (*Conjunctiva, Cornea*)

Peculiarly, there is one exception to the inertness of the silicones, shown repeatedly in the preceding sections, which has long been known . Silicone oils instilled into the eye of rabbits give rise to moderate irritation phenomena, in which, however, the cornea is not involved (*128*) (cf. Table 99). A slight conjunctivitis develops, occasionally accompanied by mild edema of the eyelid, which disappears with 24 or, at most, 48 hours.

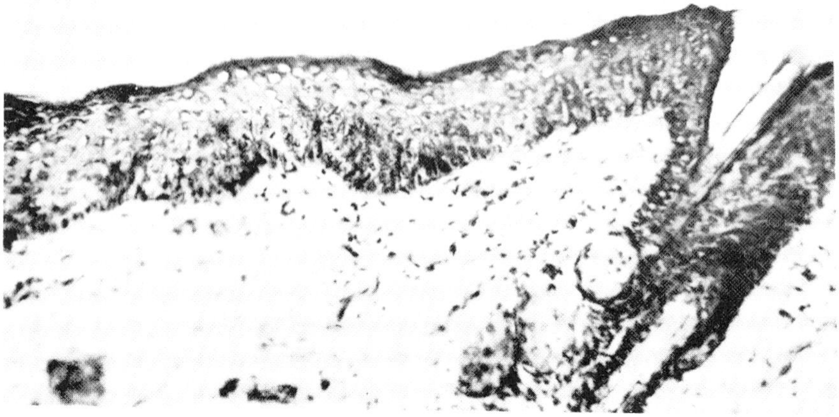

FIG. 76. Skin of a guinea pig after the rubbing in of white petroleum jelly. Marked dissemination of the epidermis; acanthosis test strongly positive (after Vonkennel, with kind permission of the author).

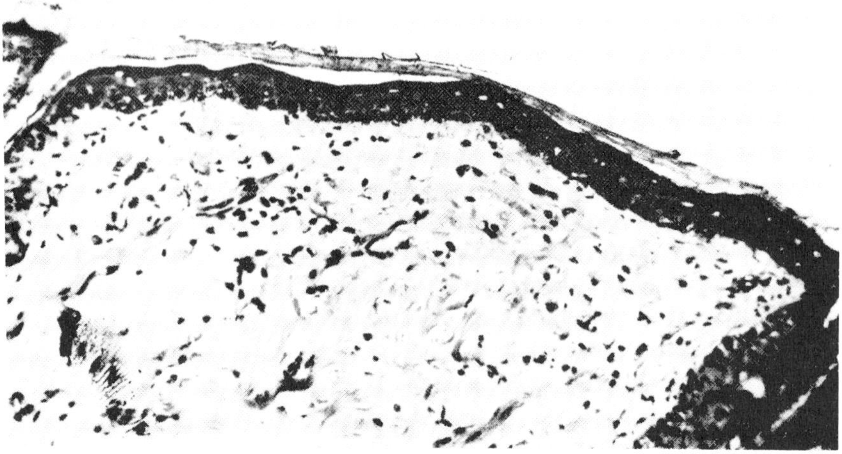

FIG. 77. Skin of a guinea pig after the rubbing in of a silicone grease. No changes in the epidermis; acanthosis test negative (after Vonkennel, with kind permission of the author).

The human eye reacts to direct contact with silicone oil with painful burning, which also affects anyone who touches the eyelids with a finger soiled with silicone oils.

There is still no explanation for this phenomenon. According to our own observations, similar effects can also be obtained with mobile purified fatty oils, such as isopropyl myristate.

In the low concentrations in which silicone oils are present in protective skin creams and cosmetic preparations, the irritant phenomena are either not observed at all or only in cases of particular sensitivity. Direct application of the cream to the eye should however be avoided.

9.7.2.8 *Implantations*

After the *parenteral* introduction of liquid silicone oils, no harmful consequences have been described; the process observed under these conditions have been mentioned above. However, the delayed consequences that occur after the implantation of silicone resin foils are somewhat peculiar. Sarcomas have been observed by various authors under such conditions after the foils have remained for years in tissues of rats and mice (*108*, *110*).

Even before this effect became known for silicone foils, it had been described for other plastic foils; it can also be brought about by metal foils and glass (*107*, *110*). Remarkably, this effect depends on the size of the implantation. The same amount of substance, which in the form of a large disk will lead to a sarcoma with high probability, is harmless when implanted in smaller amounts. The available results permit the conclusion that the chemical effects of the implantations on the surrounding tissue are unconnected with the origin of the sarcomas.

REFERENCES

(1) E. M. Acton, K. E. Moran and R. M. Silverstein, *J. Chem. Engng. Data* **6**, 64 [1961]. (2) N. K. Adam, *Endeavour* **17**, 37 [1958]. (3) M. C. Agens, General Electric, Swedish Pat. 135061 [U.S. Prior. 23.3.1950] (Zbl. **1954**, 10155). (4) K. A. Andrianov, B. V. Deryagin, N. N. Zakhavaeva, M. V. Sobolevskii and M. V. Talaev, *J. Appl. Chem. USSR* (*Engl. transl.*) **32**, No. 12, 2759 [1959]. (5) — and N. N. Sokolov, *Khim. Prom.* **6**, 329 [1955]. (6) D. van Arsdale Brown and G. R. Lucas, Thomson-Houston, French Pat. 1209480 [U. S.Prior. 14.8.1957]. (7) D. C. Atkins, C. M. Murphy and C. E. Saunders, *Ind. Engng. Chem.* **39**, 1395 [1947]. (8) R. W. Awe and H. M. Schiefer, Dow Corning, U.S. Pat. 3002989 [9.3.1959]. (9) B. M. Axilrod and M. A. Sherman, *J. Res. nat. Bur. Standards* **45**, 65 [1950].

(10) W. H. Banks, *Nature* [London] **174**, 365 [1954]. (11) R. de R. Barondes, W. D. Judge, C. G. Towne and M. L. Baxter, *Military Surgeon* **106**, 379 [1950]. (12) R. L. Bass, *Chem. and Ind.* **1959**, 912 (13) O. K. Bates. *Ind. Engng. Chem.* **41**, 1966 [1949]. (14) H. H. Berres, *Arch. f. Dermatologie u. Syph.* **194**, 259 [1952]. (15) C. L. Bitner and M. L. Dunham: *Carbon Black-Filled Silicone Rubber*, Publication of Union Carbide, Silicones Division. (16) W. J. Bobear, *Rubber Age* **84**, 448 [1958]. (17) C. J. Boner: *Manufacture and Application of Lubricating*

Greases, Reinhold Publishers, New York 1954. (18) F. A. Bovey: "The effects of ionizing radiation on natural and synthetic high polymers", Interscience Publishers, New York, 1958 (*Polymer Reviews*, Vol. 1). (19) R. C. Bowers, R. L. Cottington, T. M. Thomas and W. A. Zisman, *Ind. Engng. Chem.* **48**, 943 [1956].

(20) F. Bozzacco and P. Swanson, Proceedings Annual Technical and Management Conference, Reinforced Plastics Division, 13 [1958], Sect 8C. (21) S. A. Brady, J. C. Johnson and J. D. Lyons, Paint Varnish Product. **48**, No. 1, 25 [1958]. (22) P. W. Bridgeman, *Proc. Amer. Acad. Arts Sci.* **77**, 115 [1949]. (23) C. A. Burkhard, General Electric, U.S. Pat. 2561178 [28.7.1949]. (24) A. Charlesby, *Proc. Roy. Soc.* [A] **230**, 120 [1955]. (25) J. A. Chenicek, Universal Oil Prod., U.S. Pat. 2517536 [28.6.1946]. (26) G. P. Child, H. O. Paquin and W. B. Deichmann, *Arch. ind. Hyg. occupat. Med.* **3**, 479 [1951]. (27) C. C. Currie, Dow Corning, U.S. Pat. 2642395 [8.2.1952]. (28) — and M. C. Hommel, *Ind. Engng. Chem.* **42**, 2452 [1950]. (29) — and B. F. Smith, ibid. **42**, 2457 [1950].

(30) A. J. DeFrancesco, R. D. Alling and J. H. Baldrige, *Rubber Wld.* **134**, 866 [1956]. (31) D. G. Dervichian, *Z. Elektrochem.* **59**, 290 [1955]. (32) G. I. Duncan and M. M. Felger, *Electr. Engng.* **77**, No. 4, 318 [1958]. (33) J. R. Elliott, General Electric, U.S. Pat. 2445567 [27.8.1946]. (34) A. H. Ellison and W. A. Zisman, *J. Physic. Chem.* **60**, 416 [1956]. (35) Farbenfabriken Bayer, Company Publications: Siliconöle Bayer M and Siliconöle Bayer P. (36) D. J. Fischer, Dow Corning, Brit. Pat. 893399 [U.S. Prior. 24.9.1959]. (37) —, R. G. Chaffee and V. Flegel, *Rubber Age* **87**, 59 [1960]. (38) —, — and E. L. Warrick, ibid. **88**, 77 [1960]. (39) V. G. Fitzsimmons, D. L. Pickett, R. O. Militz and W. A. Zisman, *Trans. Amer. Soc. Mechan. Engr.* **58**, 365 [1946].

(40) H. J. Fletcher and M. J. Hunter, Dow Corning, U.S. Pat. 2599984 [14.9.1949]. (41) H. W. Fox, P. W. Taylor and W. A. Zisman, *Ind. Engng. Chem.* **39**, 1401 [1947]. (42) —, E. M. Solomon and W. A. Zisman, *J. Physic. Chem.* **54**, 723 [1950]. (43) A. Freiberger and H. Lacks, *Naval Res. Revs.* **1961**, May 1, No. 7; *Brit. Paint Res. Assoc., Rev., Current Lit.* **35**, No. 237, 187 [1962].; cited in *Werkstoffe und Korrosion* **13**, 790 [1962]. (44) G. C. Gainer, *Ind. Engng. Chem.* **46**, 2355 [1954]. (45) W. D. Garden and J. S. McFarlane, ICI, German Pat. (West German) 955539 [Brit. Prior. 15.1. and 25.10.1954]. (46) General Electric, Section B-2, Publication: Silicone Rubber Handbook, January 1956. (47) J. W. Gibbs: *Coll. Works I*, Longmans Green and Co., New York 1928, p. 309. (48) M. A. Glaser, *Ind. Engng. Chem.* **46**, 2334 [1954]. (49) C. Gloxhuber, Farbenfabriken Bayer, Wuppertal-Elberfeld, unpublished 1956.

(50) — and G. Hecht, *Arzneimittelforsch.* **5**, 10 [1955]. (51) G. Grant and C. C. Currie, *Mechan. Engng.* **73**, 311 [1951]. (52) J. B. Gregory, *Rubber Wld.* **133**, 389 [1955]. (53) P. A. Griffin, *Paint Manufacture* **31**, 411 [1961]. (54) R. Harrington, *Rubber Age* **81**, 971 [1957]; **82**, 461 [1957]. **83**, 472 [1958]; **85**, 963 [1959]; **88**, 475 [1960]. (55) H. A. Hartung and R. C. Borchert, *ACS Div. Polym. Chem.* **4**, No. 1, 243 [1963]. (56) R. L. Hauser, C. A. Walker and F. L. Kilbourne jr., *Ind. Engng. Chem.* **48**, 1202 [1956]. (57) G. Hecht, Farbenfabriken Bayer, Wuppertal-Elberfeld, unpublished. (58) R. C. Hedlund, *Jb. Chem. Ind.* **1957/58**, 66. (59) —, *Paint Ind. Mag.* **70**, No. 9, 17 [1955].

(60) F. A. Henglein, R. Lang and K. Scheinost, *Makromolekulare Chem.* **15**, 177 [1955]. (61) G. Hentze, Farbenfabriken Bayer, 1962, unpublished. (62) M. C. Hommel, Dow Corning, U.S. Pat. 2599917 [25.1.1950]. (63) M. J. Hunter, M. S. Gordon, A. J. Barry, J. F. Hyde and R. D. Heidenreich, *Ind. Engng. Chem.* **39**, 1389 [1947]. (64) R. K. Iler: *The Colloid*

Chemistry of Silica and Silicates, Cornell Univ. Press, Ithaca, N.Y. 1955. (65) G. C. JOHNSON, *J. Chem. Engng. Data* **6**, 275 [1961]. (66) H. F. JONES, General Electric, U.S. Pat. 2448530 [30.11.1945]. (67) K. KAMMERMEYER, *Ind. Engng. Chem.* **49**, 1685 [1957]. (68) S. F. KERN, R. C. ANDERSON and P. N. HARRIS, *J. Amer. Pharmac. Ass.* **38**, 575 [1949]. (69) G. KIMMERLE, Farbenfabriken Bayer, 1957, personal communication.

(70) W. KNIEGE, Farbenfabriken Bayer, German Appl. (West German) 1162560 [10.5. 1962]. (71) —, —, unpublished. (72) S. KOGUT, *New York Purchasing Review*, March 1961, p. 4. (73) M. KOIKE and A. DANNO, *J. Phys. Soc. Japan* **15**, 1501 [1960]. (74) G. M. KONKLE, Dow Corning, U.S. Pat. 2684957 [2.5.1952]. (75) —, J. T. MCINTYRE and J. V. FENNER, *Rubber Age* **79**, 445 [1956]. (76) —, R. R. SELFRIDGE and P. C. SERVAIS, *Ind. Engng. Chem.* **39**, 1410 [1947]. (77) W. KRAUSS, *Farbe u. Lack* **64**, 309 [1958]. (78) —, *VDI-Zeitschrift* **101**, 377 [1959]. (79) B. H. KRESS and H. A. HOPPENS, *Paintindia* **2**, No. 9, 35 [1952].

(80) E. KUSS and G. R. SCHULTZE, Proc. 4th World Petrol. Congr., Section V/A, Papers (81) E. LARGENT, M. BLACKSTONE and J. ROTH, US Air Force med. Service **1950**, 1p. (82) E. J. LAWTON, A. M. BUECHE and J. S. BALWIT, *Nature* **172**, 76 [1953]. (83) R. MAGIMEL, *J. Méd. Bordeaux Sud Ouest* **133**, 1196 [1956]. (84) J. J. MAY, Dow Corning, German Pat. (West German) 938739 [28.4.1954]. (85) J. D. B. MCDOUGALL, *Nature* [London] **172**, 124 [1953]. (86) R. R. MCGREGOR: *Silicones and their Uses*, McGraw-Hill, New York 1954. (87) — and E. L. WARRICK, Corning Glass Works, U.S. Pat. 2389802 [22.6.1944]. (88) —, —, —, U.S. Pat. 2389804 to U.S. Pat. 2389806 [22.6.1944]. (89) J. A. MCHARD, *Chem. Engng.* **54**, No. 2, 219 [1947].

(90) —, ibid. **55**, No. 7, 228 [1948]. (91) — and A. F. KOLB, ibid. **56**, No. 10, 227 [1949]. (92) — and L. VAN VOLKINBURG, ibid. **58**, No. 10, 270 [1951]. (93) B. P. MCNAMARA, E. A. MCKAY and M. M. QUILLE, *Fed. Proc. Amer. Soc. Exp. Biol.* **9**, 301 [1950]. (94) G. MENNICKEN, Farbenfabriken Bayer, 1959, unpublished. (95) R. L. MERKER and W .A. ZISMAN, *Ind. Engng. Chem.* **41**, 2546 [1949]. (96) Midland Silicones, Brit. Pat. 823720 [U.S. Prior. 24.1.1957]. (97) —, Brit. Pat. 843074 [U.S. Prior. 25.11.1957]. (98) A. A. MILLER, *J. Amer. Chem. Soc.* **82**, 3519 [1960]. (99) C. M. MURPHY, C. E. SAUNDERS and D. C. SMITH, *Ind. Engng. Chem.* **42**, 2462 [1950].

(100) A. NEUHAUS, *Fortschr. Mineralog.* **29/30**, 136 [1950/51]. (101) W. NOLL, *Angew. Chem.* **66**, 41 [1954]. (102) —, IUPAC-Symposium for Organosilicon Chemistry, Prague, 1965. (103) —, K. DAMM and W. KRAUSS, *Farbe u. Lack* **65**, 17 [1959]. (104) —, H. STEINBACH and CHR. SUCKER, *Ber. Bunsenges. phys. Chem.* **67**, 407 [1963]. (105) — and H. WEISSBACH, *Zement-Kalk-Gips* **9**, 476 [1956]. (106) G. NORMAN, *Corrosion* **17**, No. 9, 12 [1961]. (107) H. NOTHDURFT, *Naturwissenschaften* **45**, 549 [1958]. (108) —, Heidelberg, personal communication. (109) S. M. OHLBERG, L. E. ALEXANDER and E. L. WARRICK, *J. Polymer Sci.* **27**, 1 [1958].

(110) B. S. OPPENHEIMER et al. *Cancer Res.* **16**, 333 [1955]. (111) G. W. PAINTER, *Rubber Age* **74**, 701 [1954]. (112) J. R. Patterson, *Ind. Engng. Chem.* **39**, 1376 [1947]. (113) J. PETER, *Kautschuk u. Gummi* **8**, WT 59 [1955]. (114) C. W. PFEIFER, General Electric, U.S. Pat. 2666041 [3.4. 1952]. (115) — and R. M. SAVAGE, —, U.S. Pat. 2713564 [11.5.1953]. (116) O. R. PIERCE, G. W. HOLBROOK, O. K. JOHANNSON, J. C. SAYLOR and E. D. BROWN, *Ind. Engng. Chem.* **52**, 783 [1960]. (117) E. PIRSON and M. WICK, *Fette, Seifen, Anstrichmittel* **55**, 792 [1953]. (118) G. POLEMANN and G. FROITZHEIM, *Arzneimittelforsch.* **3**, 457 [1953]. (119) K. E. POLMANTEER and M. J. HUNTER, *J. Appl. Polymer Sci.* **1**, 3 [1959].

(120) —, P. C. SERVAIS and G. M. KONKLE, *Ind. Engng. Chem.* **44**, 1576 [1952]. (121) U. C. POZZANI and C. P. CARPENTER, *Arch. ind. Hyg. occupat. Med.* **4**, 465 [1951]. (122) H. REUTHER and C. MAASS, *Chem. Techn.* **6**, 127 [1954]. (123) — and G. REICHEL, *Plaste und Kautschuk* **7**, 171 [1960]. (124) S. I. REYNOLDS: Corona Resistance of Silicone Rubber, General Electric Publication, (Report RL-659) February 1952. (125) D. V. ROSATO, Reinforced Plastics Techn. Conf. of the Brit. Plast. Fed. 1958. (126) S. ROSS, *J. Physic. Chem.* **54**, 429 [1950]. (127) C. W. ROUSH and S. A. BRALEY JR., *Rubber Age* **84**, 75 [1958]. (128) V. K. ROWE, H. C. SPENCER and S. L. BASS, *J. ind. Hyg. Toxicol.* **30**, 332 [1948]. (129) R. M. SAVAGE, General Electric, German Appl. (West German) 1051001 [U.S. Prior. 7.2.1956].

(130) J. C. SAYLOR, Dow Corning, U.S. Pat. 3032531 [1.6.1959]. (131) L. C. SCALA and W. M. HICKAM, *Ind. Engng. Chem.* **50**, 1583 [1958]. (132) F. SCHAAF and F. GROSS, *Dermatologica* [Basel] **106**, 170 and 357 [1953]. (133) H. M. SCHIEFER, R. W. AWE and C. C. WHIPPLE, *J. Chem. Engng. Data* **6**, 155 [1961]. (134) W. SCHNABEL and A. HENGLEIN, *Makromol. Chem.* **44-46**, 611 [1961]. (135) M. SCHOOG, *Arzneimittelforsch.* **1**, 167 [1951]. (136) L. N. SENTYURIKHINA, *Khimiya i Tekhnologiya topliv i masel* **5**, No. 7, 24 [1960]. (137) D. SEYFERTH, Dow Corning, German Appl. (West German) 1116396 [U.S. Prior. 12.3.1956]. (138) L. T. SHEARER and W. W. AKERS, *J. Physic. Chem.* **62**, 1264 [1958]. (139) L. E. SIEFFERT and E. M. SCHOENBORN, *Ind. Engng. Chem.* **42**, 496 [1950].

(140) W. SIMMLER, Farbenfabriken Bayer, German Appl. (West German) 1093506 [3.4.1959]. (141) —, *G-I-T-Fachzeitschrift für das Laboratorium* **6**, 101, 189, 278 [1962]. (142) —, *Dechema-Werkstoff-Tabellen*, DWT-Werkstoffgruppe 92 [1963]. (143) — and W. BÄCHLER, *Vakuumtechnik* **8**, 155 [1959]. (144) S. SLIWINSKI and H. REUTHER, *Chem. Techn.* **7**, 252 [1955]. (145) A. L. SMITH, L. H. BROWN, L. J. TYLER and M. J. HUNTER, *Ind. Engng. Chem.* **49**, 1903 [1957]. (146) L. SPRAGUE, *Precision Metal Molding* **1954** (Sept.), 83. (147) H. STEINBACH, Dissertation, Technische Hochschule Aachen 1959. (148) W. STÖBER, *Kolloid-Z.* **145**, 17 [1956]. (149) —, ibid. **149**, 39 [1956].

(150) H. A. STUART, *Physik der Hochpolymeren III*, p. 254, Springer-Verlag, Berlin-Göttingen-Heidelberg 1955. (151) J. SWISS, Westinghouse, U.S. Pat. 2465296 [20.9.1944]. (152) R. THEUMANN, *Jb. Chem. Ind.* **1957/58**, 70. (153) E. THUN and A. ÖVREGARD, *Färg och Lack (Copenhagen)* **6**, No. 2, 25 [1960]. (154) Union Carbide, Silicones Division: Publications X 1901 and X 1902. (155) Union Chimique Belge, Belg. Pat. 552625 [16.11.1956]. (156) G. L. VANDERVORT and J. E. WILLARD, *J. Amer. Chem. Soc.* **70**, 3148 [1948]. (157) G. WAMANT, Swiss Pat. 338126 [French Prior. 13.12.1954 and 2.12.1955]. (158) E. L. WARRICK, *Ind. Engng. Chem.* **47**, 2388 [1955]. (159) —, M. J. HUNTER and A. J. BARRY, ibid. **44**, 2196 [1952].

(160) W. WEBER, *Rheol. Acta* **1**, 63 [1958]. (161) C. E. WEIR, W. H. LESER and L. A. WOOD, *Rubber Age* **67**, 686 [1950]. (162) —, —, —, *Rubber Chem. Technol.* **24**, 366 [1951]. (163) F. WELCHES and W. BURGMER, *Der Industriebackmeister* **11**, No. 3, 64 [1963]. (164) D. E. WEYER and K. R. HOFFMAN, *Mod. Plastics* **31**, No. 9, 133 [1954]. (165) D. F. WILCOCK and M. M. SPRUNG, General Electric, U.S. Pat. 2597045 [16.1.1947]. (166) K. L. WOLF: *Physik und Chemie der Grenzflächen I*, Springer-Verlag, Berlin-Göttingen-Heidelberg 1957. (167) J. F. ZACK JR., E. L. WARRICK and G. KNOLL, *J. Chem. Engng. Data* **6**, No. 2, 279 [1961]. (168) *Chem. Engng. News* **24**, 1690 [1946].

CHAPTER 10

Applications of Technical Silicone Products in Various Branches of Industry

10.1 Review

Few chemical compounds are capable of so many applications as the silicones. They are used as *engineering materials*, and in many different ways as *auxiliary products*, small concentrations of which enable certain effects to be obtained. As engineering materials, for example, articles of silicone rubber can be used in the construction of machines and apparatus; electrical insulating materials from silicone resins and silicone rubber in the construction of electrical machines; silicone oils as hydraulic liquids or heat-transfer media; and silicone greases as lubricants for ball bearings in machine construction. Silicones play the part of auxiliaries in various branches of industry: as lacquers they are used for the surface protection of metals; as mold-release agents in the rubber and plastics industry; as water-repellency agents in the textile industry, the leather industry, and the polishing and cleansing agent industry; in the protection of masonry; in the glass industry, and in many others; minute additions of silicones affect the surface properties of paints and act as defoaming agents in various systems. Although the concentration of silicones used in these cases is always low, nevertheless, because of their wide distribution, the silicone auxiliaries are no less important than silicone engineering materials.

The astonishing diversity of these applications can easily be understood when we recall the complex of properties of the silicones and relate the applications to these properties. The fact that engineers have shown such interest in silicone products is primarily due to their unique heat resistance and also, to some extent, to their low-temperature properties, the low variation of their physical constants with the temperature, their good electrical behavior, and their weather resistance. On the other hand, the principal characteristics that have made silicones indispensable as auxiliaries in very different branches of industry are: capacity for film formation, water repellency and release effect in extremely thin layers, physiological inertness, and surface activity.

10.2 Silicones in the Electrical Industry*

10.2.1 Reasons for the Introduction of Silicone Insulation

An electric machine must not only withstand mechanical and thermal stresses but must also satisfy stringent requirements resulting from the existence of electric and magnetic fields. Current-carrying parts in the stator and rotor must be insulated up to voltages of about 30,000 V. Difficult problems often arise from the fact that there is in general very little space for the insertion of the insulating materials. The insulation must also stand up to variations in humidity and to the action of any chemically aggressive surroundings; in high-voltage windings it must resist glow discharges. Finally, if we consider that when a rise in temperature causes the expansion of bunched conductors, which in large machines are sometimes several meters long, the insulation must expand with them and not become detached in the process, these remarks will have given some idea of the many demands made upon the insulating material (*87*).

These manifold requirements very soon led to the drawing up of specifications for the use of insulating materials by the electrical engineering associations in individual countries; in particular, on the basis of practical experience, electrical insulation classes were established, the temperature limits of which were not to be exceeded in the hottest parts of the machines.

The highest permissible long-term temperatures of 120 and 130°C, applying to classes *E* and *B* (*355*), first decided the limiting ratings of electrical machines on the usual assumption of a life of 25 to 30 years for large machines and of about 20 years for small and medium-sized machines. A higher degree of utilization necessarily required a considerable increase in the highest permissible long-term temperatures of the insulating materials with no reduction in the level already attained in their other properties. Efforts made in this direction finally led to insulating materials permitting the establishment of classes *F* (155°C) and *H* (180°C). Today, both classes are internationally recognized (*19*). According to the standards, class *F* can be satisfied by organically modified silicone resins—or other equivalent resins—and class *H* by silicones. However, the statements in Section 9.1.2 that the heat resistance of silicone resins depends on their constitution may be recalled to show that by no means all silicone resins can be included in class *H*. Thus, for example,

* Sections 10.2.1 to 10.2.7 were written by Prof. H. Rothert, Berlin, Germany.

highly cross-linked methylsilicone resins do not satisfy the requirements set out for this class.

In evaluating the life of insulation systems, an exponential relation discovered by Montsinger (*209*) has been found to be useful; it was first derived for insulation systems using class *A* materials but was later transferred to other classes (*329*). The life is extrapolated graphically from a diagram on which the ordinates represent the logarithm of the time and the abscissas the logarithm

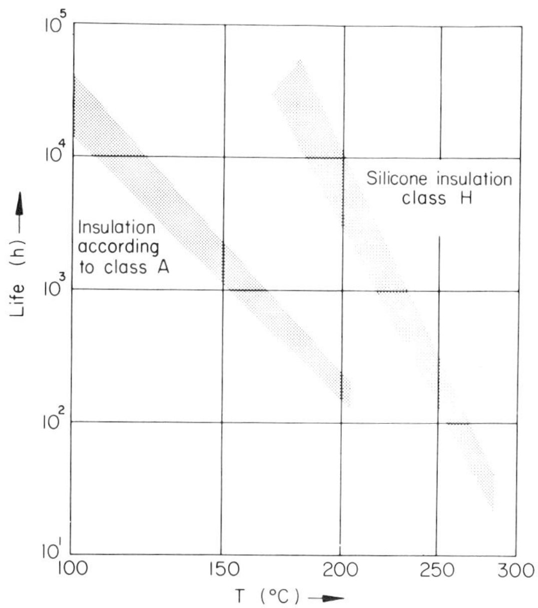

FIG. 78. Life of classes *A* and *H* insulation systems [after Dexter (*75*)].

of the absolute temperature; on this diagram, measured values are plotted showing the duration of heating necessary at various temperatures in order to make a certain property, such as the weight or the dielectric strength, fall by a constant amount, in each case within experimentally acceptable times. If, for example, such measurements are based on the time after which the dielectric strength has fallen to one-half of the initial value, lives are obtained for silicone resins, which are at least equal to those that have been established from long experience for organic resins of classes *A* and *B* (*122*, *123*). The life as a function of the temperature determined on the basis of such tests can be seen for insulation systems of classes *A* and *H* from Fig. 78 (*75*).

* Page number of chapter references.

A generally important factor for the use of silicones as electrical insulating materials is that types of silicone resin can be prepared which satisfy the "Vorschriften für schlagwetter- und explosionsgeschützte elektrische Betriebsmittel" ("Specifications for Firedamp- and Explosion-proof Electrical Machinery") (*353*) with respect to the resistance to mixtures of solvents and steam of the various ignition groups (*234*).

10.2.2 Electrical Insulating Materials

The electrical properties of pure silicone varnishes and lacquers have been treated in an earlier section (Section 9.4.6). Below we shall describe the methods that must be followed in the construction of electrical insulation and give an account of its properties.

10.2.2.1 *Glass-Fiber-Covered Silicone-Enameled Wires*

The use of silicone-enameled wires covered with an insulating layer of pure nonfilled silicone resins has now been abandoned because the mechanical strength of the enamel film is too low and its resistance to aging on copper is poor. After prolonged heat aging at ~180°C, the silicone film turns brittle and decolorized and becomes detached from the copper substrate, the copper oxide present on the copper apparently exerting a catalytically degrading effect. Attempts have therefore been made to remove the oxide film and to improve the adhesion by pretreatment in hydrogen at elevated temperatures (*48*). Other measures are removing the film of copper oxide by means of an aqueous solution of chromic and phosphoric acids (*291*), adding boric acid or boric acid esters to the silicone resin (*218*), coating the copper conductor with a thin homogeneous layer of an inorganic silica-containing powder (*199*), producing an intermediate layer of metals of the iron group, chromium, or a noble metal (*285*), or tinning the copper wire.

Attempts have also been made to produce enamels with high abrasion resistance and increased adhesion to the surface of the copper by means of terephthalate polyesters treated with polyfunctional isocyanates and combined with cocondensates from terephthalates and polysiloxanes (*235*). However, none of these measures has gained acceptance.

In evaluating cocondensates derived from silicone resins and organic types of resin that have been recommended for enameling wire, it must be borne in mind that in some cases these contain only a small proportion of silicone and therefore differ widely from the true silicone resins with their typical properties.

The insulation of copper wire with silicones is today generally carried out by repeatedly impregnating with silicone resins copper wires that have been covered *with glass fiber*, the product being heated to suitable temperatures after each treatment to harden the film of enamel partially or completely. This type of insulation of the conductor is necessarily associated with a greater increase in the diameter than when singly enameled wires are used. For example, with doubly enameled wires having an initial diameter in the range from 0.4 to 1.0 mm, the enamel coating increases the diameter by 0.06 to 0.105 mm or even up to about 0.120 mm; on the other hand, in the case of glass-fiber-covered silicone-enameled wires with a wire diameter of 1.0 mm, and when very thin glass fibers are used, the diameter is increased by about 0.120 mm. The increase rises to 0.160 mm for wire diameters of 2.30 mm. This makes the space factor for the slots of electric machines worse than when enameled wires are used; however, this failing can be partially compensated by the higher load capacity of the silicone-insulated wire. As with singly enameled wires, glass-fiber-covered wires doubly impregnated with silicone varnish are today made with very uniform diameters.

In order to evaluate the behavior of wires with respect to *mechanical stresses* when they are being laid in the slot and the end connections are being formed, use is made of various methods which, however, are not yet generally accepted. If the abrasion number* is used as a means of comparison, this is in fact lower when silicone resins of class *H* are used than for single-enameled wires coated with organic insulating enamels; it is, however, sufficiently high to permit the use of silicone resins in practice. With modified silicone resins of class *F*, the abrasion numbers of singly enameled wires can be reached.

In order to increase the mechanical abrasion resistance of glass-fiber-covered wires impregnated with silicone resins of class *H* in processing, they can be covered with thin foils based on terephthalic acid polyesters "welded" on by heat treatment. It is true that at operating temperatures above 180°C such surface protection is gradually destroyed, but this has no disadvantageous effect on the life of the machine, particularly if the wire has been embedded in a heat-resistant impregnating resin (*157*).

The *electrical properties* of the insulation can be assessed from the behavior of the insulation resistance as a function of elongation (Fig. 79) and from the fact that the average breakdown voltages of silicone-insulated wires [0.8 mm with a 0.12 mm increase in the diameter owing to the insulation (*346*)] amount to about 600 V at up to 25% elongation, practically independently of the elongation. These values are far above the winding voltages usually encountered in practice, which have a maximum of about 30 V. In addition, in order to approach the practical situation, breakdown voltages on transposed wires 0.8 mm in diameter have been determined; the measured values of 1000 to 1400 V must be regarded as completely satisfactory for practical use. The number of defects in well-prepared 0.8 mm wires having glass fiber silicone insulation was found to be 0 in 40 meters when the diameter was increased by

* Determined by abrading a 200 mm stretched length of the insulated wire by a steel wire of prescribed diameter at a prescribed load until destruction of the insulating layers is shown by the completion of an electrical circuit.

0.12 mm by the insulation, and 1 when it was increased by 0.09 mm; for singly enameled wires a maximum of 1 and for doubly enameled wires a maximum of 0.3 defects per meter is permitted. The resistance to caking of the bunched enameled wires completely satisfies all practical requirements even under large forces due to high acceleration (*153*, *154*).

In the selection of the glass fiber used for insulating wire, particular attention must be paid to its quality and treatment (*155*).

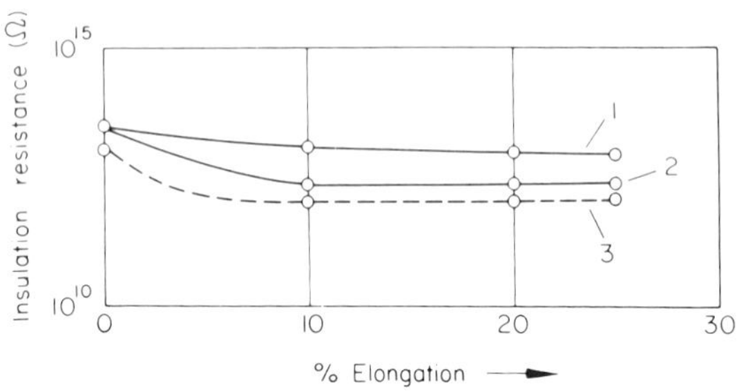

FIG. 79. Insulation resistance of two glass fiber silicone wires (——) and an enameled wire with purely organic insulation (– – –) as a function of the elongation. (1) Glass-fiber-covered silicone-enameled wire, 0.8 mm with a 0.12 mm increase in diameter; (2) glass-fiber-covered silicone-enameled wire, 0.8 mm with a 0.09 mm increase in diameter; and (3) enameled wire, 0.75 mm with a 0.06 mm increase in diameter.

10.2.2.2 *Tapes, Separators, Foils, and Slot Insulations Based on Silicone Resin*

Silicone-bonded glass fabric in the form of tapes, fabric, and foils in combination with mica or processed mica has become very important for slot linings, separators, insulation between coils and core, and for bars, slot wedges, etc.

Glass fiber/silicone enamel tapes have the simplest structure, and are frequently used as retaining tapes for end connections. The thickness varies from about 0.12 to 0.15 mm, occasionally even to 0.20 mm. The proportion of enamel is comparatively high, amounting to about 75%, so that the pores of the glass fiber fabric are completely filled. Roughly speaking, the tapes have the following electrical characteristics:

Power factor, tan δ/20°C/800 cps	$6\text{–}7 \times 10^{-3}$
Dielectric constant ϵ/20°C/800 cps	3.5–4.5
Specific resistance/20°C	$\sim 10^{17}$ Ω cm
Mean dielectric strength	~20 kV/mm

Figure 80 shows the variation in the dielectric constant ϵ and power factor $\tan\delta$ as a function of frequency up to 10^8 cps at room temperature. Within the limits of elongation of up to about 10% necessary for the taping process, the electrical properties change only comparatively slightly. The resistance to leakage currents evaluated by the Nekal drop process* normally corresponds to grade T 2, but it can be raised to grade 3 by selecting special fabrics and by carefully monitoring the resin component. Since such tapes are used predominantly as covering and retaining tapes for end connections, the figures obtained satisfy practical conditions for low-voltage machines.

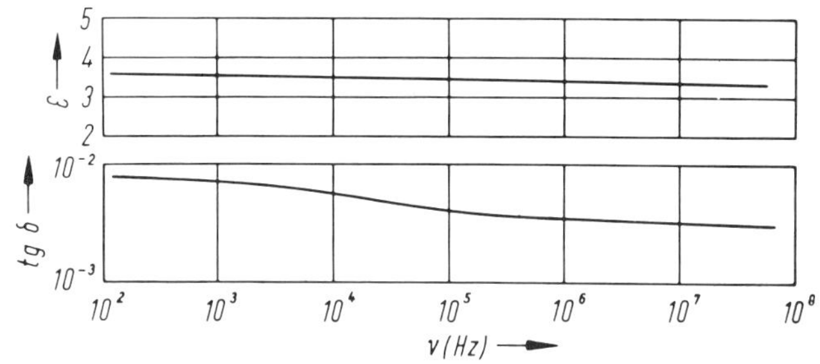

FIG. 80. Dielectric constant ϵ and power factor $\tan\delta$ of a glass fiber/silicone enamel tape as a function of frequency.

Laminated materials consisting of glass fiber, mica, and silicone resins as binders are particularly important for the insulation of flat copper strips and of coils and for lining slots. Such materials frequently have a three-layer structure, the mica layer generally being sandwiched between two outer layers of glass fiber. Attempts are made to keep the proportion of glass fiber low and to select mica having a high dielectric strength. The dielectric strengths measured on materials of this construction with a thickness of about 0.2 mm in a double layer average 25 kV/mm, and with double layers 0.3 mm thick about 29 kV/mm.

If silicone/glass fiber/mica tape is used for high-voltage purposes, the change in the power factor $\tan\delta$ is of particular importance. Figure 81 shows the figures measured for a laminate 0.75 mm thick as a function of the voltage and the temperature. The very flat nature of the curves is worth noting. Only at 180°C do the curves become steeper and the absolute figures rise to about 180×10^{-3} at 10 kV. No organic material or synthetic resin used in the present-day technique of insulating high-voltage coils would be capable of withstanding such a stress. The dielectric strength can be taken as 20 kV/mm, but

* DIN 53 480, October 1955: "Prüfung von Isolierstoffen, Bestimmung der Kriechstromfestigkeit bei Betriebsspannungen unter 1 kV" ("Testing of Insulating Materials, Determination of the Leakage Current Resistance at Operating Voltages below 1 kV").

this can be raised to 30 kV/mm. The leakage current resistance corresponds to grade T 5c; it is better than that of the insulating material used at present; the superiority of the silicones becomes particularly appreciable above 120°C. Silicone/glass fiber/mica tape is not absolutely resistant to glow voltages, but in this respect it is considerably superior to other insulating materials.

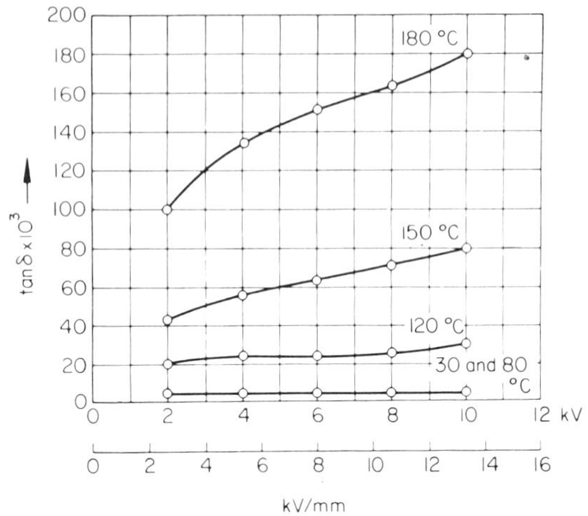

FIG. 81. Power factor of a silicone resin/glass fiber/mica laminate as a function of voltage and temperature (thickness of the laminate, 0.75 mm).

Very recently, materials have been developed, based on processed mica, that are of particular importance for use on high-voltage coils and bars. By suitable hardening, as in the case of shellac-mica foil, and wrapping around the bunched conductors under pressure, this material can give an extremely tough insulation of ringing hardness and a high mechanical strength which also largely satisfies the electrical requirements. For example, Röbel bars only 8 mm in diameter with an insulating coating of glass fiber/silicone/processed mica with a wall thickness of 0.8 mm, are wrapped for a nominal voltage of 3 kV, 500 cps. The nominal field strength of 3.75 kV/mm is above the figures usual today even for plastic insulating materials (*4*, *152*), which generally do not amount to 3 kV/mm.

In combination with insulating materials having such a high heat resistance, the slot wedges must also be made from suitably selected materials, preferably silicone/glass fabric laminates molded under high pressure and high temperature to give hard fabrics (cf. Section 8.2.2.4). In addition to outstanding electrical characteristics, these materials also exhibit good mechanical properties. Since the tensile strengths can be raised to about 2000 kg/cm^2 without the materials becoming too brittle, constructional parts made from them can be used successfully in those positions in which a nonmetallic material with a high long-term heat resistance and a high insulating value is necessary. As can be seen from Fig. 82, the bending strength shows a very satisfactory variation as a function of the time and temperature of aging. The water uptake after storage for 24 hours in water was found to be 0.2 to 0.4%, and no longer changed as a

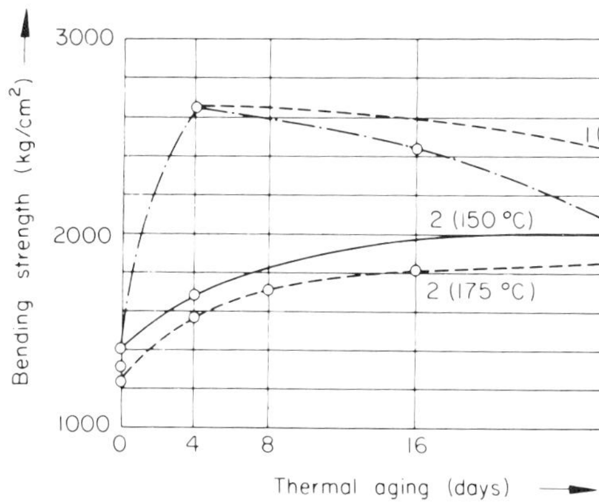

FIG. 82. Bending strength of glass fabric/resin laminates after thermal aging: (1) phenoplastic bonding and (2) bonding with silicone resins.

function of time (*63*). The thermal conductivity was measured in the range of up to 100°C and amounted to about 2×10^{-3} W/cm deg corresponding to 0.478×10^{-3} cal/sec cm deg; it increases still further with rising temperature, so that particularly satisfactory figures are obtained in the range of application of classses *F* and *H* (*330*).

Finally, in many cases glass fiber/silicone resin sheaths are used for the protection of shunts and junctions. However, it is preferable to use glass fiber sheaths built up with silicone rubber for the insulation of the ends of coils and junction wires.

10.2.2.3 *Silicone Rubber/Glass Fiber Laminates*

Heat-vulcanized silicone rubber products as such are used to only a very limited extent in the construction of electrical machinery. Filled silicone rubber is used, for example, for the manufacture of rectangular-section bars as an intermediate layer between the thrust rings and the end-windings of high-voltage coils, where it is particularly suitable because of its mechanical strength, adequate elastic properties, good insulating properties, and substantial insensitivity to moisture (cf. Section 9.6).

However, silicone rubber is acquiring great importance for the construction of electrical machinery as a coating material for glass fiber tapes. Two types of such tapes are in use: one type in which the silicone rubber is used in the prevulcanized state, and the second type in which the silicone rubber is vulcanized only after the coils or bars have been wound with the tape. The tapes provided with the coating of prevulcanized silicone rubber are 0.15 to 0.25 mm thick and have about 35% of glass fiber as the supporting material. On the average the electrical strength, measured with the spherical electrode, is about 15 to 18 kV/mm. This figure remains unchanged even after prolonged aging at 180

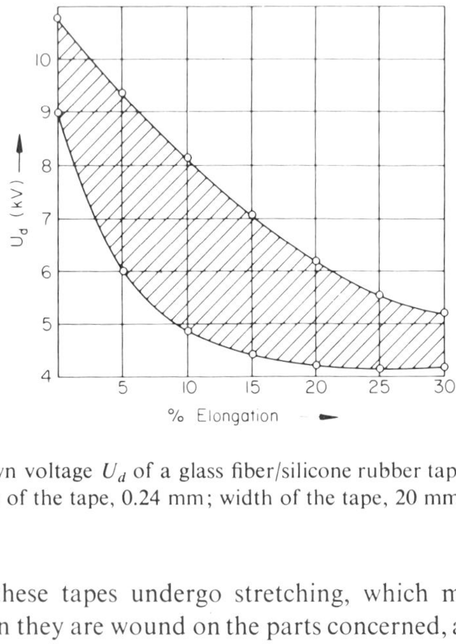

FIG. 83. Breakdown voltage U_d of a glass fiber/silicone rubber tape as a function of the elongation (thickness of the tape, 0.24 mm; width of the tape, 20 mm; spherical electrode, 10 mm diam).

to 200°C. Since these tapes undergo stretching, which may sometimes be considerable, when they are wound on the parts concerned, a knowledge of the minimum and maximum breakdown voltages as functions of the elongation is important (Fig. 83). The result satisfies far-reaching requirements. Measurement of the power factor as a function of the voltage or the field strength with the temperature as the parameter gives the curves shown in Fig. 84. Only in the initial section at 155 and 180°C does the rise reach fairly high values, then it becomes very small and falls almost to 0. The absolute value of 6.5% at 180°C and a field strength of about 11 kV/mm must be regarded as very satisfactory; coils for high-voltage windings can be constructed with the aid of these tapes. Under these conditions, with carefully made coils, maximum figures are obtained for the absolute value of the power factor, in the range between 2 and 16 kV, of about 120×10^{-3} to 135×10^{-3}, and for the rise, of about 3×10^{-3}/kV. The breakdown voltage applied between core and winding for insulation with a wall thickness of about 2.2 to 2.5 mm was found to be about 30 to 35 kV, a value which is greater than the double test voltage of 6 kV machines. The electrical strength of 14 to 16 kV/mm found with multiple layers of tape in this case is by no means as high as with combinations of silicones and mica, but it is sufficiently high to permit the construction of machines which are safe in operation (*156*).

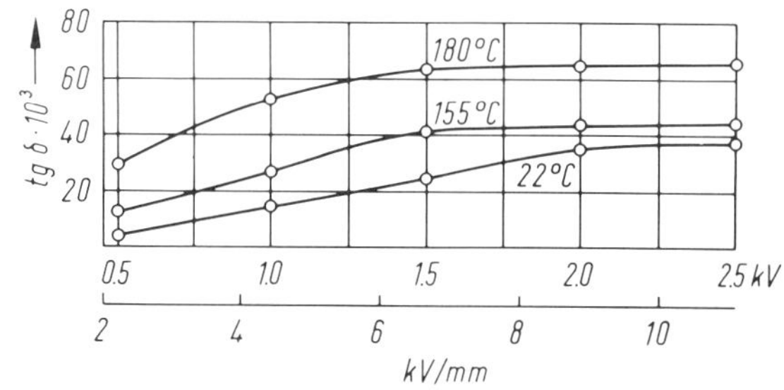

FIG. 84. Power factor of a glass fiber/silicone rubber tape as a function of the voltage and the temperature (fabric thickness, 0.225 mm).

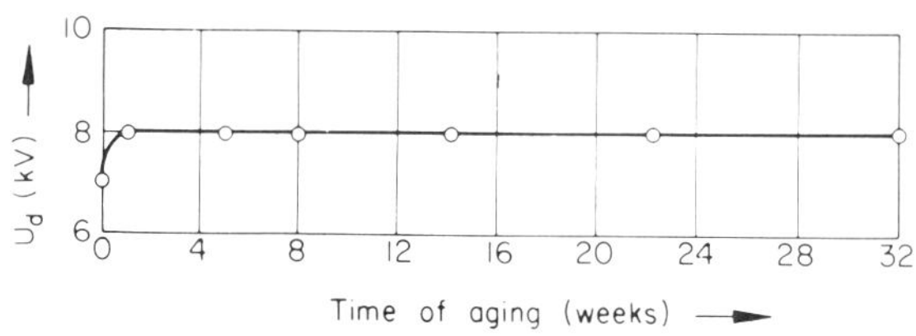

FIG. 85. Constancy of the breakdown voltage U_d of a glass fiber/silicone rubber tape on aging in hot air at 200°C (thickness of the tape, 0.35 mm).

Particular care must be taken in the preparation of tapes bearing only a vulcanized-on layer of silicone rubber, since at normal temperature they can be stored for only a limited time. The breakdown voltage of these tapes remains completely unchanged during heat aging at 200°C lasting more than 32 weeks (Fig. 85). Coils insulated with such tapes are covered, after vulcanization, with a thick moisture-insensitive coating. Such insulations are therefore frequently used very successfully in combination with glass fiber/silicone/mica materials for traction motors (*351*, *352*).

Recently, castable and spreadable silicone rubber pastes that can be converted into silicone rubber at room temperature (cf. Section 8.1) have also become of interest for the electrical industry. They can be used, for example, for embedding coils, for filling cavities, or as auxiliaries in the layer construction of insulation systems (*21*, *64*) (see Section 10.2.10).

10.2.3 Tests on Motorettes and Motors

Tests on individual insulating materials are often inconclusive, since the very complex stresses to which the insulation is subjected in operation are difficult to reproduce. However, a method has recently been introduced which is intended to enable the behavior of the insulating material to be investigated in a simulated electric machine system giving a closer approximation to actual conditions (*49*). This method is being used for low-voltage tests on the so-called "motorettes" (*339*) in the range of 110 to 550 V (Fig. 86).

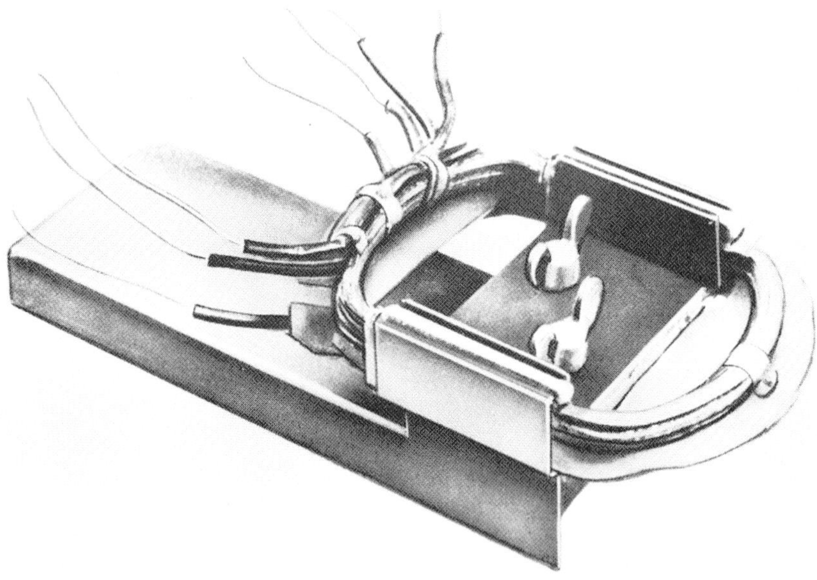

Fig. 86. A motorette.

The motorette represents a stack of armature laminations of an electric machine and contains in the slots that can be seen in the figure a two-layer winding, the upper and lower layers of which consist of a coil with conductors wound in parallel, so that tests from winding to winding are possible, as well as other tests. The slot insulation, separators, slot wedge, and structure of the coils, as well as the insulation of the winding, correspond to those found in a motor. Impregnation is carried out by the same processes and under the same conditions as with normal machines, to give close simulation to the situation occurring in motors when the slots are filled with insulating varnish and heat is transferred from the bunched conductors to the wall of the slot. The tests proposed on motorettes include various conditions that are made substantially more severe than those encountered in practice. Their aim is to give information on the probable life of the insulation system under investigation. Preferably, a batch of ten similarly prepared motorettes is subjected to a predetermined cycle of stresses, and the results are evaluated by a statistical process. In spite of certain objections to lifetime determination by motorette experiments, it provides a basis of discussion and comparison which, in any case, permits valuable information to be gained (*145*).

The proposed test cycle comprises the following:

1. A temperature-time stress; the times and temperatures proposed for classes *F* and *H* are shown in Fig. 87.

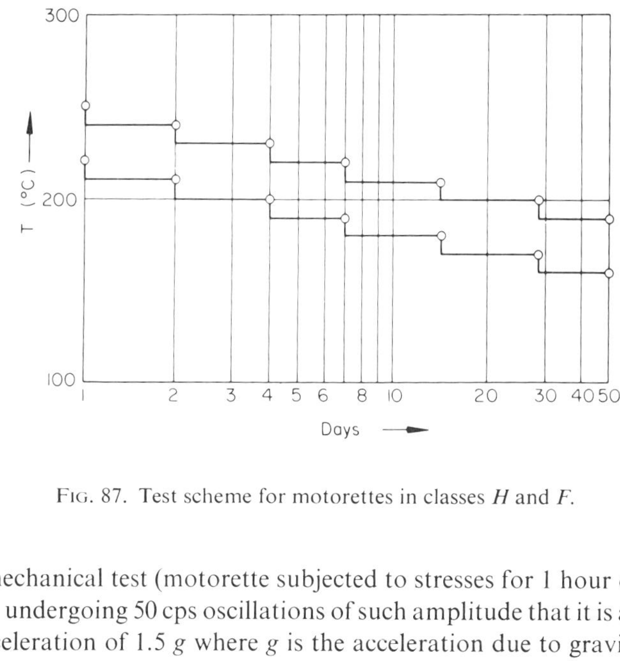

FIG. 87. Test scheme for motorettes in classes *H* and *F*.

2. A mechanical test (motorette subjected to stresses for 1 hour on a vibration table undergoing 50 cps oscillations of such amplitude that it is acted upon by an acceleration of 1.5 *g* where *g* is the acceleration due to gravity).

3. A moisture test, in which an atmosphere of 100% relative humidity is maintained for a period of at least 48 hours; condensation must appear on the winding. As in cycle 2, no voltage is applied during this period.

4. A voltage test, in which the motorettes are still in the humid room and room temperature is approximately maintained. The voltage is applied for a period of 10 minutes, first 50 V between neighboring conductors, then 600 V between the upper and lower layers, and finally 600 V between the layers and earth.

In the investigations on silicone insulating materials discussed here, a change was made from the proposals of the "test code" with respect to the mechanical test, the conditions being more severe in the case of a number of motorettes which were subjected to the action of $2\frac{1}{2}\,g$ for up to 300 hours. Similarly, the moisture test was supplemented by subjecting a motorette in the humid room to changes of temperature. As a further departure from the test cycle, the storage time in the humid room was extended to 5 days (*272*).

If the results evaluated statistically* from the testing of ten similar motorettes are in each case plotted in a coordinate system such as that shown in Fig. 78, it is possible to deduce the expected life. It is found that the mean life of a motor with silicone insulation is the same as that of motors using conventional insulating materials, i.e., about 20 to 25 years. Tests by other workers on individual materials and motorettes have also led to lifetimes such as are customary with present-day systems.

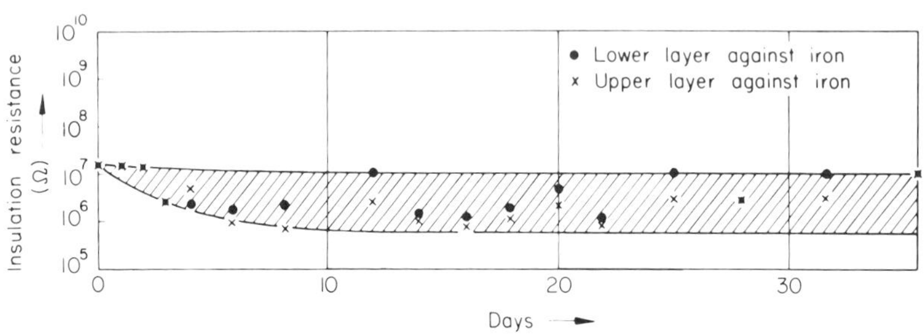

FIG. 88. Insulation resistance of silicone insulation in a motorette test with exposure to heat and moisture.

Finally, in addition to the investigations outlined above, motorettes were subjected to heat and moisture for 6 weeks, the insulation resistance being measured continuously (*272*). The result given in Fig. 88 shows no further appreciable change after an initial fall from 2×10^7 to about $10^6\ \Omega$.

Small motors have actually been run under water without any damage occurring. Similarly, investigations have also been carried out on motorettes half-immersed perpendicularly in water; the resistance to earth determined after several days of exposure was always sufficiently high to indicate the absence of any damage to the insulation in the applied voltage test. After removal of the motorette from the water, the insulation resistance returned to the initial figure in a short time.

Motorette tests similar to those outlined for the low-voltage range have also been proposed in modified form for pre-formed coils up to voltages of 6600 V and machine ratings in the range from 50 to 2000 hp (*340*).

A somewhat modified process of investigation has also been proposed for motors (*339*). An exposure to elevated temperatures follows a mechanical test

* For the method of evaluation cf. ref. *339*.

in which at least 250 high-speed runs and brakings are performed on each day of the temperature cycle for single-phase motors and 5000 reversings each day for polyphase motors. Moisture is then allowed to act again for 2 days, after which an applied-voltage test is carried out, for which, however, no definitive test voltages have yet been laid down. In addition, motors have been tested under particularly severe conditions (*1*).

For this purpose, two totally enclosed explosion-proof motors with a nominal rating of 2.5 kW, a voltage of 500 V, and a speed of 1500 rpm have been used.

One motor was tested in the following way:

1. The stator was exposed for 2 days to a temperature of 300°C and
2. subsequently for 8 days to a temperature of 280°C,
3. the complete motor was then run with no load for 4 hours at 50 cps at an acceleration of 3 *g*;
4. the moisture cycle in an atmosphere corresponding to a moist and humid climate was extended to 15 days in each case;
5. the applied-voltage tests were sometimes carried out with the test voltage.

The second motor was subjected to a moisture test for 2 days when at rest, and was then run with no load for 2 days; these conditions were continued alternately for long periods.

If the expected lifetime is determined from these investigations, figures which correspond to those of the motorette experiments are again obtained.

It is also interesting to investigate the power ratios for a small motor and a medium motor, each with silicone insulation, on the basis of classes *E*, *F*, and *H* with the same active material (windings and laminations). In order to obtain a basis of comparison, the slot space factor was made the same in all three cases and in the evaluation a uniform cooling agent temperature of 40°C was prescribed: the motors were then loaded in each case up to the limiting temperature rise of the heat class in question. The mean rise in temperature, obtained from measurement of the resistance, was regarded as decisive for determining the limiting temperature.

Table 100 gives a comparison of the most important characteristics of the motor for the various classes.

In addition, the motor was operated at the power laid down for class *E* without any ventilation. Under these conditions, the temperature rise of the winding, determined from a measurement of the resistance, was 145°C, i.e., 20°C, higher than the permissible temperature limit for class *H* (125°C). If the latter were to be adhered to, the power would have to be decreased by about 6%. When the motor was operated at the power laid down for class *H* without any ventilation, the temperature of the winding reached a figure of 260°C. In view of the fact that the windings were exposed to a temperature of 300°C for one or two days without any defects occurring, it may be concluded that

TABLE 100

CHARACTERISTICS OF A FOUR-POLE MOTOR WITH SILICONE INSULATION[a]

Characteristic	Classes		
	E	*F*	*H*
Power ratio	1.0	1.11	1.24
Efficiency	0.798	0.780	0.747
Power factor	0.835	0.886	0.900
Slip	0.050	0.068	0.080
Power-to-weight ratio (kW/kg)	0.063	0.070	0.079

[a] Explosion-proof construction, 2.6 kW.

TABLE 101

CHARACTERISTICS OF A SIX-POLE MOTOR WITH SILICONE INSULATION[a]

Characteristic	Classes		
	E	*F*	*H*
Power ratio	1.0	1.22	1.36
Efficiency	0.840	0.840	0.836
Power factor	0.647	0.714	0.752
Slip	0.029	0.041	0.054
Power-to-weight ratio (kW/kg)	0.048	0.058	0.065

[a] Explosion-proof construction, 10.35 kW.

motors provided with silicone insulation and rated as class *E* can operate without protest for short times when ventilation of the machine is prevented for any reason (*240*).

The increase in power achieved for a six-pole motor with a rating of 10.35 kW when the limiting temperature according to class *E* was fully utilized is given in Table 101 together with the most important characteristics.

In evaluating the results, the following points must be taken into account: the slot space factor of the four-pole motor subjected to the conditions of class *E* was somewhat better if singly enameled wires were used instead of glass-fiber-covered enameled wires. Consequently, with this motor the increase in power is somewhat lower than with the six-pole motor for which such differences in the slot space factor do not exist for exposure to the conditions of class *E*.

It follows from the results that a considerable increase in power can be achieved by changing to silicone insulation. These questions are particularly important when, for example, in the repair or rewinding of a motor, a considerable increase in power is to be obtained.

For reversing connections in which the rotating masses have to be accelerated and braked by the motor, there is a possibility of achieving considerably higher switching rates or of accelerating rotating masses many times greater than the inherent moment of inertia of the motor as compared with motors insulated normally in accordance with class *E* or *B*.

In order to be able to evaluate the behavior of silicone insulating materials in high-voltage motors, silicone-insulated coils were made, some by the wrapping process and some by the tape-winding process, and were investigated in particular with respect to their behavior to the application of a substantially increased voltage at temperatures of up to 180°C (Fig. 89). Since the main factor determining the losses occurring in the dielectric is proportional to the square of the electric field strength, a combined thermal and electrical stress corresponds to a particularly high loading by means of which weak points in in the insulation can be found very rapidly. Consequently, coils insulated for a voltage of 6 kV were subjected to a voltage of 15 kV at 180°C for at least 100 hours.

Comparison of the power factors at room temperature before and after exposure shows the good agreement in the series of measurements (Fig. 89). The subsequent increase in voltage up to breakdown likewise showed no change in the electric strength.

High-voltage motors up to 6 kV with silicone insulation are built in large numbers, and have so far proved to be suitable for all operational requirements with respect to temperature and life. It is at present impossible to say how far silicone-insulated high-voltage windings can also be used for voltages of 10 kV and above.

10.2.4 Coatings and Lubricants for Silicone-Insulated Motors

With the increase in the highest permissible operating temperatures of class *H* motors, the question necessarily arises of paints capable of withstanding the more severe operating conditions, and of lubricants suitable for bearing temperatures above 100°C.

Both these problems can again be solved with silicone products; composite resins of polyesters and silicone resins or silicone resin paints pigmented with aluminum bronze are available for heat-resistant coatings (see Sections 8.2 and 9.4). Silicone greases made from methylphenylsiloxanes and lithium soaps have proved to be suitable for bearing lubrication, provided that the bearings are not loaded too highly* (cf. Section 9.3). If the bearing load is high, it is preferable to use organic greases, even at high temperatures, but care must then be taken that the bearings are cooled and, if necessary, that the periods between the lubrications are made very short.

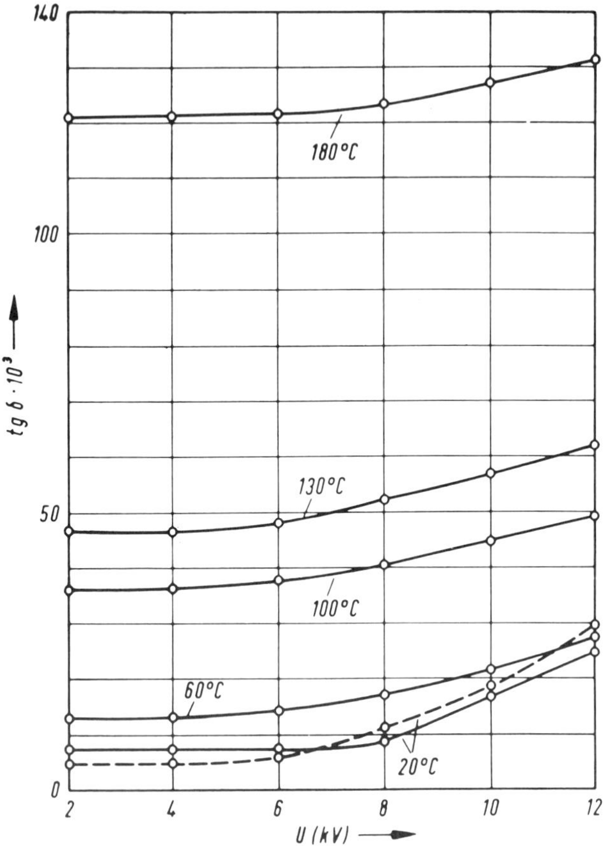

FIG. 89. Power factor of the silicone insulation in a high-voltage coil before and after exposure to thermal and electrical stress (broken line: after exposure).

* Unpublished investigations from the Allgemeine Elektrizitäts-Gesellschaft (AEG) laboratories, 1957. Tests were carried out on 1.1 kW, 1500 rpm Ex motors at bearing temperatures of 120 to 125°C and 140 to 145°C, the ball bearings of which corresponded to type 6204. In general, the experiments were carried out on motors running on no load, with flywheels placed on the shaft extensions. To increase the mechanical stress, the motors were reversed 20 times per hour. The frictional losses of the bearings remained practically unchanged over 6000 hours.

10.2.5 Examples of Silicone-Insulated Motors

Fundamentally, silicone-insulated motors are used in particular under conditions of high *ambient temperatures*, especially if the cooling conditions are difficult; examples of this are found in the chemical industry, in underground mining, in foundry and metallurgical work, in shipbuilding, and in petroleum drilling. Since for a given space, silicone insulation permits the accommodation of a higher electrical power than insulation with organic types

Fig. 90. Scraping centrifuge with a silicone-insulated motor drive.

of resin, such motors are also preferred wherever the power must be obtained with a saving of space. Examples of this are found in the construction of machine tools, in motor vehicles, and in aircraft.

Frequently, silicone-insulated motors are used not for the increase in power permitted by the higher class to which they belong, but rather for their great *safety* in the face of sudden unexpected overloads. From this point of view, they are also of interest in machine tools, as motors for hoists and cranes, as drives for conveying machinery and excavators, in mining, and in the chemical industry (*273*).

Finally, owing to the *water repellency* of the insulation, silicone-insulated motors give particularly good length of service when used in sites affected by condensed water humidity, or even under water.

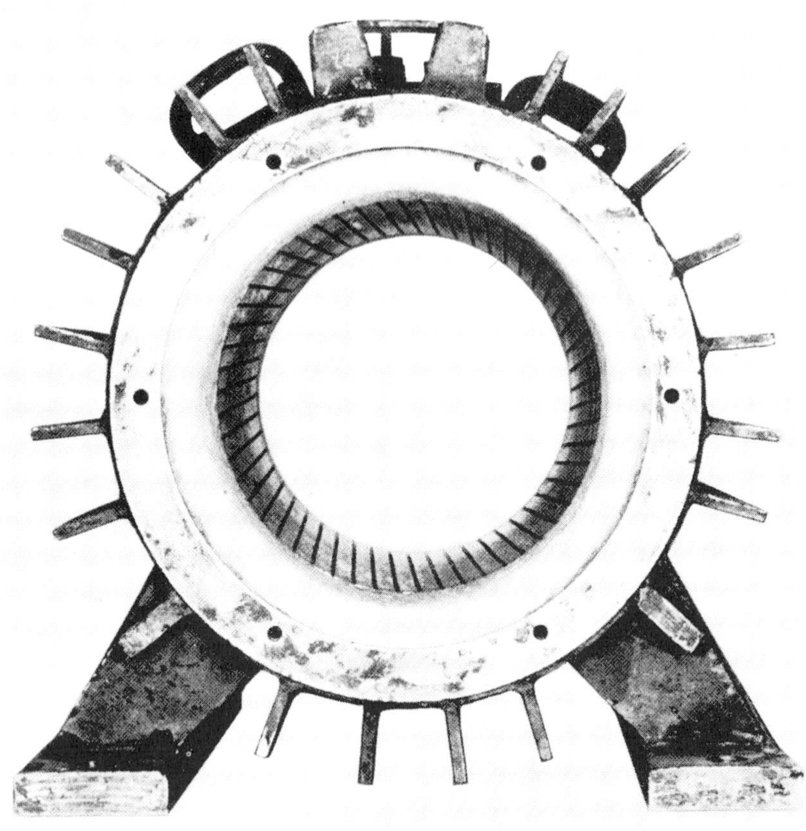

FIG. 91. Stator of a mine motor with silicone insulation.

Figure 90 shows a motor intended for driving a scraping centrifuge of the surface-cooled type (72 kW, 500 V, and 1000 rpm). The moment of inertia of the empty centrifuge calculated from the speed of the motor is 210 times greater, and that of the filled centrifuge 327 times greater, than the moment of inertia of the motor. During the operating time of 14.5 minutes, the motor is mostly taking a current load substantially greater than the rated value, so that it has to work under conditions of severe acceleration and high Joule heat.

Particularly remarkable is the ratio of the size of the motor (on the right in the illustration) to the centrifuge (left).

Another embodiment is the drive motor of a throw-shovel loader, which has a speed of 1500 rpm, and an output of 15 kW in intermittent operation. The motor is encapsulated in a pressure-resistant enclosure and has a completely smooth outer casing and no external fan. With about 245 switchings per hour and braking by "plugging" the motor must dissipate all the heat due to losses through the jacket substantially by means of radiation and convection. In addition, the motor frequently works under short circuit conditions when it stalls and is buried by the material to be loaded. Even under these extraordinarily severe working conditions, silicone-insulated motors have proved to be completely safe in operation.

The type represented by a six-pole 110 kW motor used in potash mining also claims particular interest (Fig. 91). The conditions laid down for the operation of this motor provide for very frequent braking almost to a stop. In addition to the very high thermal stressing of the winding, the end windings had to be effectively protected against the considerable mechanical forces. This was achieved by embedding them in a silicone-asbestos paste which filled all the spaces between the individual coil end windings and the iron so that a completely closed ring (which can be seen in the illustration) was produced, which had a high mechanical strength and was capable of taking up all the applied forces.

Let us take as our final example a fan motor for drying ovens, with an output of 1.4 kW at 950 rpm. This motor works in hot air at an ambient temperature of 150°C; under these conditions, particular demands are made on the lubrication of the roller bearings which can be satisfied only with silicones.

10.2.6 The Use of Silicones in Direct-Current Machines

Silicone insulation is used for direct-current machines just as for three-phase asynchronous motors. Here a distinction must be made between the insulation of the armature coils, to which substantially the same considerations apply as for the stator and rotor windings of three-phase motors, and the field and interpole coils. These coils are generally made of flat strip conductors insulated with glass fiber impregnated with silicone resin and finally wrapped with glass fiber/silicone rubber tapes. To make use of the good thermal conductivity of the silicone materials in these coils, it is necessary to displace all the inclusions of air that unavoidably arise in the wrapping process with silicone resin by a vacuum-pressure-heat treatment. Electrical and mechanical protection of the coils is achieved by wrapping them with post-vulcanized glass fiber/silicone rubber tapes which are vulcanized together under pressure in a pressure mold and give a homogeneous, very compact, and elastic insulating coating. It is very important to adjust the time and temperature in such a way that vulcanization is ensured and that the solvents can escape, since otherwise the mechanical strength and thermal conductivity can be seriously impaired by the formation of bubbles within the insulating coating. For example, many thousands of silicone-insulated dc motors have proved outstandingly satisfactory in diesel-electric locomotives under moist-hot and dry-hot tropical conditions. They have also been used with excellent results as the auxiliary drives in rolling mills in temperate climates and in particularly unfavorable tropical climates (*320*). Finally, it may be mentioned that even large dc ship-propelling machines rated at 2040 kW, 945 V, and 425 rpm have been insulated with silicones and have proved eminently satisfactory in the marine climate.

While the insulation technique presents no particular problems in the case of dc machines, some questions arise that must be taken into account in the heat treatment of the materials for class H motors and in the operation of enclosed machines. In the first place, the heat treatment of the winding requires such high temperatures that soft solder junctions cannot be used for connecting the winding with the winding lugs and the commutator. These joints must be made with a high-melting solder or by brazing or hard soldering. Care is also taken with respect to the commutator since the excess temperature is normally limited to 80 or 90°C (*356*). In totally *enclosed* machines, because of the elevated temperatures, brush wear may increase (*10*, *197*). This phenomenon can be countered not only by careful selection of the carbon but also by separating the collector chamber from the winding chamber as far as possible and specially ventilating it. In *open* machines, which permit efficient ventilation of the commutator, the commutator conditions can be controlled without difficulty.

10.2.7 The Use of Silicones in the Construction of Transformers

Silicone insulation can also be used with particular advantage for transformers. At the present time, it is primarily the explosion-proof dry transformers for the mining industry, encased in pressure-resistant casings, that are furnished with silicones.

As with machines, depending on the output and the coil structure, the windings are made of glass-fiber-covered circular wires or strips impregnated with silicone resins. The coils are treated with a filled silicone resin solution, either during the winding process or afterward, and this is then hardened in accordance with a time-temperature program. As with high-voltage windings, it is very important to avoid cavities and bubble formation to prevent undesirable glow phenomena. The baking process is of great importance as regards the very considerable radial and axial mechanical forces which arise under short-circuit conditions. By selecting suitable silicone pastes in combination with a carefully determined temperature-time program for the hardening process, mechanically highly resistant coils can be made which are equal to all short-circuit stresses. Because of the high temperatures occurring in the winding, the spacers and support pieces between layers used to maintain the shape and to accept the short-circuiting forces must be made of ceramic

FIG. 92. Dry transformer with silicone insulation.

materials or glass fabric/silicone laminates adhesively attached to the coils by means of special silicone pastes. Dry silicone transformers with a primary voltage of up to 6000 V are now made.

In order to investigate the behavior of the windings with respect to surges, a $1/50\mu$ sec full-wave impulse voltage was chosen, the maximum value of which was 75% of the 60 kV full-wave impulse voltage (*2*, *357*). The stresses arising under these conditions were satisfactorily tolerated by the winding.

Figure 92 shows the core built into the tank of a 5000/525 V 315 kVA transformer with an impedance voltage of 3.8% and an efficiency of 98.71%; the high-voltage coils are clearly shown. The dimensions of the windings were chosen in such a way that on full load the limiting temperature remains within the figures prescribed for class *H*. The temperature gradient between the winding and the tank wall is such that the surface temperature of the tank wall satisfies the specifications for fire- and explosion-proof electric machines (*353*). The transformer loses heat to the surroundings only by the convection and radiation.

Only with the introduction of silicones has it become possible to use transformers of this type for such severe operating conditions and to produce a design that is both lighter and mechanically stronger than that of earlier machines (*80*, *192*, *234*).

In addition to the transformers made for the mining industry, distribution transformers for outputs of up to 2000 kVA and voltages of up to 15 kV have been made and put into service (*212*). As dry transformers, they have the advantages of requiring no maintenance at all, and, in contrast to oil-filled transformers, of containing no combustible oil. In special cases, the tank has been filled with nitrogen (*348*).

10.2.8 Wire and Cable Insulation*

Silicone rubber materials have opened up new possibilities for the wire and cable industry. While, for example, with a number of types of electrical leads elastic insulating materials have previously been rejected because of their inadequate stability, and it has been necessary to use rigid to semirigid products, silicone rubber now provides a material which permits electrical leads subjected to high temperatures to be made flexible.

The temperature range in which flexible insulation can be used has been extended by the introduction of silicone rubber up to class *H*, i.e., up to 180°C; these temperatures can even be substantially exceeded for brief periods. Silicone rubber insulation also provides an advance in the direction of low temperatures; while with the usual types of insulation the low-temperature flexibility depends substantially on the formulation and reaches the magnitude

* Sections 10.2.8 to 10.2.10 were written by W. Kniege, Leverkusen, West Germany.

of that of silicone rubber only in extreme cases and at the cost of reduced thermal stability, silicone rubber vulcanizates always ensure flexibility down to about −50°C without any sacrifice of heat resistance. By means of special types of silicone rubbers (Section 9.6.2) vulcanizates can actually be prepared that remain soft and flexible down to temperatures of about −100°C.

The danger of insulation failure due to overloading of the conductor and accumulation of heat has been practically eliminated by the use of silicone rubber and appropriate cable construction. Owing to the high thermal conductivity of silicone rubber, the heat is removed rapidly. This fact and the reserve provided by the heat resistance of the silicone rubber means that such insulation can actually remain fully efficient when the *electrical* overloading of the conductor amounts to 50 to 100%. Cables insulated with silicone rubber can normally be *thermally* stressed 50% more than cables with other rubber-elastic insulating materials.

If silicone rubber is destroyed by fire, a skeleton of silica is formed which still possesses a certain insulating effect. Consequently, after a fire, fixed silicone-insulated conductors generally remain usable for a short time if the insulation is provided with glass fiber brading or tape and is therefore protected mechanically.

Other applications of silicone-rubber-insulated wires and cables are based on their corona, arc, and leakage current resistances and on the resistance of the insulating material to natural aging, ozone, and oxygen. Silicone rubber is even more resistant to γ-rays up to about 1×10^8 r than the fluorine-containing insulating materials (*349*).

The use of silicone rubber for high-voltage cables demands special technical measures in the construction of the cable (*221*) as well as, generally, insulation with a multilayer construction.

Since cables sheathed in silicone rubber are frequently used in closed systems, attention must be paid to the measures described in Section 9.6.1.

Below we shall give some typical examples of the use of silicone-insulated wires and cables.

Because of the continuously growing consumption of current due to increasing industrialization, *underground distribution cables* are frequently overloaded. In addition to general overloading, peak loads often occur periodically as well. In large towns and large industrial plants, in particular, it is often impossible, without considerable expense, to increase the size of the existing cable conduits in such a way that distribution cables with a higher conducting cross section and therefore higher load-carrying capacity can be laid. Since cables insulated with silicone rubber can be very severely overloaded, this offers the possibility of keeping the same conductor diameter and merely changing the quality of the insulation. The special construction of the cable must be adapted to each individual case.

In various branches of industry, e.g., the steel, glass, and cement industries, and in ceramics, *power cables* are necessary for portable electrical appliances. In addition to having the usual insulating properties, these must be particularly heat resistant, very insensitive to moisture, and have the maximum possible resistance to ozone and arcing. Silicone rubber satisfies these requirements and, moreover, permits flexible cables which can be overloaded to be used even where rigid fixed cables were previously required. In the case of power cables with single-core or multicore construction, it will be possible to use tin-covered copper conductors in almost all the cases, since the conductor temperature rarely exceeds 125°C.

Power cables are laid primarily for voltages of 600 to 5000 V. The silicone rubber insulation is preferably protected by means of an asbestos reinforcement (asbestos braid). The steel wire reinforcement occasionally encountered does have the advantage of higher mechanical strength, but any broken steel wires may penetrate through the insulation leading to short circuits or electrical breakthroughs. In any case, with the use of such types of cables the metal armoring must be satisfactorily earthed.

Aircraft and rocket technology, like *shipbuilding*, demands a large number of complicated electrical systems that must remain capable of functioning even under unfavorable conditions of storage and operation, and must in emergencies be able to supply certain reserves of function. It is not only its applicability at relatively high or low temperatures that makes silicone rubber fully capable of use in these fields; its behavior with respect to fire, its resistance to ozone and oxygen, and the small change in its electrical properties caused by moisture makes it an almost ideal insulating material for this purpose. High load-carrying capacity at the same or smaller conductor cross section, and therefore reduced weight and volume with the same load-carrying capacity, are further decisive advantages of such silicone rubber insulation as compared with other insulating materials. While *aircraft wires* have a comparatively simple construction in most cases—generally a multistrand silver-coated copper conductor is sheathed with silicone rubber and covered with glass fiber braiding—*ship's cables* have a somewhat more complicated structure. The typical construction of such cables for general purpose control circuits is described, for example, in MIL-C-2194/5 A.

Heating cables with silicone rubber insulation are used in various embodiments. Heating cables are generally made of nickel-chromium resistance wire, and in the simplest case these are merely insulated with the silicone rubber. If very high requirements are set for the thermal conductivity and mechanical resistance, a braid of tinned copper is applied over a first layer of insulation, and over this a second layer of silicone rubber insulation. Such cables are already being used for heating crude oil in boreholes, liquid baths, etc., for deicing equipment in aircraft and shipbuilding, for the deicing of roofs and windows, and for heating the ground, street surfaces, or playing fields. In addition to its heat resistance and insulating properties, it is particularly its low weight, low moisture uptake, and high flexibility which gives silicone rubber an advantage over other materials. If the heating cables are to be used in contact with swelling oils or similar materials, another braid made of glass fiber or polyester or polyamide fibers is generally applied as an external protection.

In highly stressed internal combustion motors, *ignition cables* are necessary the dielectric strength of which must frequently be greater than 40 kV. These types of cables are frequently exposed to extremely high temperatures and to the action of ozone. A silicone rubber insulation in combination with an appropriate construction of the cable will satisfy all these requirements. Silicone-rubber-insulated high-tension ignition cables have already proved generally useful, particularly with high-compression internal combustion engines such as those used in automobiles and aircraft.

Because of the outstanding electrical properties of silicone rubber, the latter is used to make various *special cables* in which one or other property of the vulcanizate may be the decisive factor for its use. Its outstanding heat resistance is utilized in lamp-holder wires (*354*), in wires and cables for high-wattage lamps, searchlights, operation lamps, etc., in compensating leads for thermocouples, in connection leads for irons, heaters, etc., and in appliance fixtures and hookup wires. In all cases, the use of silicone rubber is also justified by its low moisture sensitivity, its high resistance to ozone and corona discharge, and its leakage current resistance.

In the case of *generator leads*, particularly for high-current and high-frequency air-space cables, the particularly valuable property is the low variation of its electrical characteristics with temperature. The mechanical construction of such wires or cables differs considerably according to the particular field of application. In microwave techniques, silicone-rubber-insulated cables have also proved valuable, for example, in making antenna feeders, etc. (*245*).

Finally, as described in Section 9.6.5, silicone rubber can be used to produce *electrically conducting types of rubber*, the volume resistivity of which is about 10 Ω cm. Silicone rubbers of this kind have found acceptance in practice for the manufacture of special ignition cables, in which the silicone rubber is used as the conductor instead of a metal and the ignition cable therefore does not radiate interference in the ultrashort wave band. It has also been known to be used for resistance wires which can be shaped to suit a particular application. However, such products are used mainly in the manufacture of high-voltage cables, where, when applied in the form of a thin layer to the metal conductor, they contribute to the smoothness of the conductor and therefore produce a homogeneous field of force in the insulation.

10.2.9 Electrical Potting and Encapsulating

By using heat-vulcanizable silicone rubber it has been possible in many cases to improve the conventional construction of certain apparatus, machines, or whole plants, or to remodel them completely. In almost all cases, however, the manufacture of an article from heat-vulcanizable silicone rubber is reserved for the rubber or plastics industry, since certain technical requirements must always be satisfied. With room-temperature-vulcanizing and low-temperature-vulcanizing silicone rubbers, materials have been available for a long time that can be processed without great technical expense, the manipulation of which demands no special knowledge, and which, nevertheless, possess all the typical properties of silicone rubber. Room- and low-temperature-vulcanizing compounds, which can be obtained with viscosities ranging from thin liquids to thick pastes, vulcanize after the addition of the curing agent to a solid permanently elastic silicone rubber. The good electrical properties of these special types of silicone rubber, their elasticity, their low moisture uptake, and their resistance to ozone and heat justify their extensive use as electrical insulating materials. The products have proved equally useful for potting electronic components and subassemblies, and for the impregnation or encapsulation of small transformers or for the encapsulation of random-wound electric motors.

In particular, with the use of room-temperature-vulcanizing compounds it is advantageous that no melting process is necessary and also that no heat is evolved by an exothermic reaction during the vulcanization process. Room-temperature-vulcanizing compounds are therefore preeminently suitable for the potting or encapsulating of electronic components furnished with transistors. Silicone rubber is not thermoplastic, nor does it form rigid bodies; in combination with the heat resistance of these products, we therefore have the possibility of encapsulating sensitive apparatus or components in a manner absorbing shocks and vibrations, protecting these components against corroding influences such as weathering, moisture, ozone, and fungicides, and making them insensitive to variations of temperature. Pressure-sensitive electronic parts that have been embedded in the usual rigid casting resins such as, for example, epoxy resins, are sensitive to the mechanical stresses arising with changes of temperature because of the different thermal expansions of the materials used. Here it has proved to be desirable either to embed the electronic part in the paste completely or else to provide it with a thin intermediate layer of the paste before it is embedded in a rigid resin. The elastic intermediate layer is then capable of absorbing and nullifying any mechanical stresses that may occur.

Room-temperature-vulcanizing silicone rubber can be used with advantage even in those cases where resistance against the influence of high voltage in a high-frequency field is required. With the epoxide resins formerly used, dielectric heating leads to the formation of fissures and to glow discharges and therefore to degradation, while under these conditions the silicone material undergoes practically no change (*61*).

The addition of electrically conducting fillers, e.g., carbon black, to room-temperature-vulcanizing compounds enables an electrically conducting silicone rubber to be obtained from the normally good insulator. Such conducting types are used where electrostatic charges must be avoided or conducted away.

Room-temperature-vulcanizing silicone rubber is used for the production of cable terminations (for example, for the sealing of pygmy connectors) and for encapsulating cable repeaters.

In the construction of electric motors, it has proved satisfactory to impregnate or coat the end connections and windings with room-temperature-vulcanizing compounds. In addition to vibration damping and stabilization against jarring, this gives protection against the entry of dust and moisture.

The above fields of application and examples are typical for room-temperature-vulcanizing compounds, and are capable of many variations in detail. A special version of silicone rubber consists of the so-called low-temperature-vulcanizing compounds. They have the viscosity characteristics of the room-temperature-vulcanizing compounds but, like the heat-vulcanizing types of silicone rubber, for vulcanization they require tempering for a shorter or longer time. In contrast to the pigmented room-temperature-vulcanizing compounds, the low-temperature-vulcanizing products are water-clear and transparent. In all cases where a continuous or periodic visual inspection of an encapsulated apparatus is necessary, it is therefore preferable to use low-temperature-vulcanizing silicone rubber.

With the aid of special primers, both types of rubber can be bonded adhesively to various materials. Repairs can be carried out without difficulty.

10.2.10 Other Applications in the Electrical Field

Electric *plugs* are made of silicone rubber both for industrial requirements and for household use. The highest demands are made on electrical *connectors* of class *H*, since these must be not only stable to heat and resistant to leakage currents but also highly flexible in the cold, as described, for example, in MIL-C-26500. Even where none of the typical properties of silicone rubbers is a decisive factor, special *multiway connectors* are advantageously made with silicone rubber; this is when usual types of rubber vulcanized with sulfur would deleteriously affect the contact resistance of, for example, silver contacts. Plug-in connectors used on many domestic appliances, such as toasters and grills, generally have a porcelain or ceramic insert. It is true that the heat resistance of such plugs is outstanding, but their sensitivity to shock, impact, or dropping is naturally also very high. These disadvantages are avoided when such plugs are made from silicone rubber.

Insulating tapes based on unsupported or glass-reinforced types of silicone rubber are increasingly used for the insulation of transformers, heating pads, and blankets, and also for the insulation of various electrical parts and for cable splicing or patching. Self-adhesive tapes of this type are available. Such so-called fusible silicone rubber tapes are used in the manufacture of special cable designs (*26*).

10.3 Silicones as Materials and Auxiliaries for Machine Construction

10.3.1 Silicone Rubber*

The special position of silicone rubber among the well-known rubber-elastic materials often provides the designer in the machines, apparatus, and instrumentation industries, in the automobile and the aircraft industries, and in shipbuilding, with completely new possibilities. With the exception of the expensive fluoroelastomers, the rubberlike materials available previously failed above 125 to 130°C, particularly on prolonged use. Their low-temperature flexibility, moreover, was not always adequate. When one considers that certain types of silicone rubber are flexible down to about −100°C, that silicone rubber can be exposed to temperatures of 180 to 200°C and even up to about 300°C for a short time, and that in addition its mechanical properties change only slightly with temperature, the interest in this material will be readily appreciated.

Consider some typical examples. *Shaft* and *sleeve seals* are used in various embodiments for sealing the moving parts of machines such as shafts, pistons, and axles. Since the coefficient of friction of silicone rubber is low, higher peripheral speeds or rates of rotation are permissible, and reductions in the size of the whole structure are normally possible. Silicone rubber also permits higher working temperatures; in particular, the surface hardening, dreaded with unsaturated organic elastomers, that may occur after storage in oils with sulfur-containing additives, is safely avoided by using silicone rubber. The somewhat microporous structure of silicone rubber imparts to it a kind of storage capacity for oils. In turn, this means that silicone rubber can take up a certain amount of lubricant during use, which is then available as a reserve in the case of a failure in the external supply of lubricant. For the same reason, spring-loaded transmission oil seals should also be stored in a suitable transmission oil at an elevated temperature for a short time before incorporation in the machine. This improves not only the behavior of the seals in the starting of a shaft but also facilitates mounting. For the above reasons, silicone rubber oil seals exhibit so-called "dry state properties" without lubricants (see Section 9.6.6). Since their hardness does not fall off appreciably at elevated

* Section 10.3.1 was written by W. Kniege, Leverkusen, West Germany.

temperatures, or increase appreciably at low temperatures, effective operation of the silicone rubber seal is ensured over a wide range of temperatures.

The necessity of protecting moving machine parts from external influences and soil is closely associated with the use of collar seals. This problem is solved by the constructive cooperation of protecting elements, particularly *bellows*. Silicone rubber has also proved satisfactory in this field in cases of exceptional heat stress, particularly when the bellows must exhibit no stiffening at low temperatures. Its resistance to aging and oxygen, and its high dynamic strength and flexibility even in the cold, are also in its favor as compared with other elastomers. Under certain circumstances the possibility of making transparent or semitransparent silicone rubbers permits observation of parts of machines and apparatus during use.

In contrast to shaft and collar seals, ring seals such as, for example, O-ring seals are used for sealing predominantly stationary machine parts. With silicone rubber ring seals, sealing can be effected up to about 180 to 200°C. The relatively low permanent deformation and the small change in hardness at elevated temperatures again give silicone rubber a considerable superiority to other materials. With such seals, care should be taken in the design that oxygen can always act on the silicone rubber during the action of heat, since otherwise its heat resistance may suffer (see Section 9.6.1).

While the spring characteristics of organic elastomers change appreciably with the temperature, *spring elements* made from silicone rubber remain fully capable of functioning for a long time at relatively high temperatures. The spring effect also changes only slightly in the cold. Problems of adhesion occurring in the combination of silicone rubber with metals can be solved today by means of suitable adhesives or by the use of self-bonding silicone rubber compounds.

By incorporating heat-resistant silicone rubber *seals* in furnaces and ovens, the construction of the doors can be simplified or made with larger tolerances. Analogous remarks apply to refrigerator cabinets because of the cold resistance of silicone rubber. An additional valuable feature in the latter application is that silicone rubber does not give off any odoriferous components and does not stick to ice or other materials.

The automobile industry uses *spring-loaded transmission oil seals* and similar seals, as well as silicone rubber seals for the valve-tappet protecting tubes. The same applies to the *protective caps* on sparking plugs and sparking plug connectors as well as to dynamos and distributors.

Silicone rubber *tubes* are used, for example, for hot-air, oxygen, and hot-oil conduits, as insulating sheaths, as flexible sealing materials at low temperature, and also, since they are physiologically inert and contain no extractable constituents, for medical purposes (blood transfusion, etc.). *Profiled cords* are preferred for special purposes, not only because of the heat and cold flexibility of the silicone rubber but also because of its resistance to uv and ozone (aircraft construction), to certain chemicals (Askarel-filled transformers), and to weathering (external seals).

Its heat resistance and release properties have made silicone rubber of interest as elastic *coverings for rollers*. In the metal and glass industries, and in ceramics, silicone-rubber rolls are used in the transporting of hot sheet metal, glass articles, or ceramic parts, and as driving and guide rolls for the stripping of hot materials, e.g., for stripping off the excess of tin in the tinning of sheet iron. In the plastics industry, they are used in the *coating* or *lining* of *paper*, *metals*, and *fabrics* with plastics such as polyethylene and in the hot-sealing of plastics. Silicone rubber *conveyor belts* have the advantage not only of allowing sticky materials to fall off easily but also of not giving off any harmful foreign materials. These are properties of interest, in particular, for the transport of taste- and odor-sensitive foodstuffs and luxuries such as tea and coffee.

Since, in contrast to sulfur-containing rubber gums, silicone rubber does not affect photographic materials, it is also widely used in this field.

10.3.2 Silicone Resin Lacquers*

Lacquering with silicone resins has become of interest for the manufacture of machines, apparatus, and instruments predominantly because of their heat resistance, their release action, and their weather resistance.

An idea of the fields of application open to silicone resin lacquers may be given by the list of examples in Table 102. It shows that, because of their heat resistance, silicone enamels may be recommended for uses which were previously closed to purely organic lacquers and were reserved to vitreous enamels. It must by no means be concluded from this that vitreous enamels could be replaced directly by silicone resin lacquering. Vitreous enamels are harder and have a greater resistance to heat and chemicals than silicone lacquers, but the latter are characterized by good elasticity and can be applied by painting techniques and baked on at relatively low temperatures. Because of such differences, the two groups of products are not in competition with one another; on the contrary, they supplement one another and give the designer the possibility of choosing the most suitable material for any particular case. Similar considerations apply to the use of silicones and vitreous enamels for the production of weather-resistant protective coatings.

Of particular importance for the construction of apparatus, machines, and instruments, is the fact that it is possible with silicone lacquers to protect reliably against corrosion and weathering metal parts exposed to temperatures of up to 600°C even when these temperatures are intermittent (*96*, *105*, *113*).

* Section 10.3.2 was written by Dr. W. Krauss, Leverkusen, West Germany.

TABLE 102

RECOMMENDED USES OF SILICONE RESIN LACQUERS

Property determining the application	Type of silicone resin lacquer	Examples of application
Heat resistance		
(a) Up to 230°C	Pure or modified silicone resin mixed with heat-resistant pigments (Stoving enamels)	Smoothing irons Geysers Electrical heating appliances Domestic furnaces High-pressure steam conduits Furnaces and ovens, particularly oil furnaces Chimneys
	Unpigmented coatings or enamels colored with transparent pigments (Stoving enamels)	Incandescent lamps Headlamp glasses Inspection glasses
(b) Up to 400°C	Silicone lacquers with pigment (titanium dioxide)/binder ratio$\approx 3/1$ (Stoving enamels with poor corrosion resistance)	Furnace tubes and doors Ovens
(c) Up to 500°C, in special cases 600°C	Straight silicone resins pigmented with metallic pigments, especially Zn and Al, or with iron, mica, or graphite (Air drying)	Silencers and exhaust pipes Rotary kilns Furnace doors Aircraft parts Blast furnace units Engine parts Rockets Chimneys
Release effect	Clear or pigmented pure or modified silicone resins, including those containing relatively large amounts of silicone oil (Stoving enamels)	Baking sheets and molds Frying pans Dust-removing plants Deicing units Molds for the plastics and rubber industry Drying pans and drums
Weather resistance	Mainly pigmented and modified silicone resins (Stoving enamels)	Lighting units Reflectors Traffic signs Facade cladding sheets

All known lacquers based on organic binders have so far failed when exposed in this way.

Furthermore, all those fields of application in which the release effect of the silicones is utilized remain specific for silicone resins. They permit the production of *antisticking layers* on drying sheets, injection and vulcanizing molds, and the like, which greatly facilitate the removal of sticky articles. In dust-removing plants or tube systems through which pulverulent material is conveyed, films of silicone resins prevent an encrustation of the tube.

10.3.3 Silicone Oils (or Emulsions)

As in many other branches of industry, silicone oils have gained popularity in machine construction for the solution of special release problems, for which they appear to be suitable not only because of their good release effect but also because of their heat resistance.

Thus, they are used in the *precision casting of metals*, particularly in the so-called Croning process. In this, the removal of the shell mold, which consists of a mixture of sand and phenol resin, from a wooden or metal mold is greatly facilitated by the previous spraying of the mold with a silicone oil emulsion; in general, because of the clean release, no subsequent mechanical treatment of the shell is necessary. Combinations of silicone oil emulsions with graphite dispersions have proved to be particularly satisfactory and permit rapid working even in large-scale mass production.

Likewise, silicones have found acceptance as release agents in the precision casting of low-melting metals, particularly zinc and aluminum, because of their resistance to decomposition at high temperatures.

Again, silicones have proved to be satisfactory as release agents in the *welding process* (*259*). If the surroundings of the welded seam are sprayed or painted with silicone oils, silicone emulsions, or, still better, emulsions of certain silicone resins, the beads of the welding materials do not adhere to the metal surface and are easily removed. This measure is being used to an increasing extent in machine welding, where it makes a subsequent removal of the beads by grinding superfluous.

The unusually favorable viscosity behavior, the wide range of viscosities, the heat resistance, and the stability of the silicone oils also make them a valuable group of synthetic oils, the application of which in machine construction is, of course, limited by their cost; they are preferably used in the precision mechanical field, e.g., as damping liquids or hydraulic oils in instrument construction. Owing to their relatively high compressibility, they can be used in liquid springs and springless oscillation dampers of the type, for instance, of the Holset damper (*89*).

If demands are made on their lubricating properties, the differences in the behavior of the various types of silicone oil must be borne in mind (Section

9.2.10). In general, normal polydimethylsiloxane oils are excluded because the bearing capacity of the film is too low, or because of their creep behavior. It is particularly the latter property that makes it impossible to use them as watch and clock lubricants. On the other hand, methylphenylsilicone and, especially, chlorophenylmethylsilicone oils are very valuable for high-temperature lubricating problems (Section 9.2.10).

Finally, silicone oils satisfy all the requirements of a good *heat-transfer medium* for high temperatures. Special types permit the transfer of heat in the range from 200 to 300°C without pressure (Section 9.2.2) and are used increasingly for the operation of high-temperature thermostats (cf. Section 10.14.4).

Their use as diffusion pump oils is discussed in Section 10.14.4.

10.3.4 Silicone Greases

Silicone greases are unsurpassed as *lubricants*, since nothing like their resistance to heat and cold is reached by any lubricating grease based on mineral oil. The load-bearing capacity of the film of silicone grease is not adequate for sliding friction. At the present time, therefore, silicone greases cannot be used for the sliding friction of steel and boundary lubrication but only for *roller bearings*.

In the use of silicone greases, the appropriate type must be selected. The commercial greases, normally thickened with lithium soaps, can be used up to 200°C and for roller bearings running at up to 20,000 rpm. Roller bearings in thermally stressed positions such as in silicone-insulated high-temperature motors, in hot-air fans, and drying-belt units require less attention and suffer fewer breakdowns when lubricated with silicone greases than when less heat-resistant greases are used. Silicone greases are also used where a machine is required to work at both high and low temperatures, as, for instance, in automobile starters and dynamos.

Black greases thickened with carbon black are extremely heat resistant, but they are not suitable for roller bearings with high rates of rotation. They are used on furnace doors, wagons, drying chamber or tunnel cars, etc., when the rate of rotation in the roller bearings is less than $n = 1000$. Blue greases thickened with phthalocyanines, with a similar excellent heat stability, can be used in roller bearings up to $n \approx 3000$. However, they tend to thicken and must be diluted from time to time with lighter types of grease in order to keep their consistency constant.

Greases containing highly disperse silicas as the thickening agents are used only where intermittent stress is expected, e.g., for the greasing of hot-air blowers and motors, regulating valves, stuffing boxes, and hinges when these work at high temperatures. The seizure of screws in thermally exposed positions such as on hot vessels can be prevented by treatment with silicone greases;

rubber seals or rubber sheaths do not adhere to metals at high temperatures when the surfaces of contact are smeared with silicone grease. Rubber is not affected by contact with silicone greases. They are resistant to certain chemicals such as liquid ammonia, acetic acid, phenol, concentrated phosphoric acid, solutions of iron or aluminum chloride, sodium chlorate, chromates, 30% hydrogen peroxide, dilute mineral acids, lubricating oils, and vegetable and animal oils.

Heavy formulations act as *high vacuum greases* which can be used down to pressures of the order of 10^{-6} torr.

10.4 Silicones as Auxiliaries in the Rubber Industry*

10.4.1 Use as Mold Release Agents

10.4.1.1 *Problems*

Efficient manufacture of high-quality rubber articles depends not the least on the use of suitable mold release agents, whose main task is to make possible rapid removal of even complicated rubber articles from the mold without mechanical damage. The particular importance of these auxiliaries is enhanced still further by the fact that the removal of a rubber article from the mold must generally be carried out at the vulcanization temperature, at which the usual types of rubber have only low strengths and can easily be mechanically damaged. Effective mold coating agents therefore contribute to a large extent to keeping the number of defective products low and to rationalizing manufacture both qualitatively and quantitatively. In order to facilitate removal from the mold, vulcanizing molds, mandrels, etc., are treated with a mold release agent before being charged with the raw material.

The quality of a mold release agent is evaluated not only with respect to its release effect but also, in particular, from the following points of view. No deposits must form on the walls of the mold, and the surface of the mold must not corrode. In vulcanization, no flow defects due to the release agent should occur. The article should have a good "finish"; the film remaining on the surface of the molded articles should have no undesirable side effects such as, for examples, a deleterious effect on its gluability and paintability; the diluted dispersions of the mold release agent should be stable; the solvent used should be physiologically inert, evaporate rapidly, and be noninflammable; the mold release agent should be cheap and should give an effect lasting as long as possible with one spraying.

Because of their release properties for sticky materials, and because of their heat stability, silicone oils are particularly suitable for solving the above

* Section 10.4 was written by W. Kniege, Leverkusen, West Germany.

problems. The use of silicone oils as mold release agents was therefore one of the first industrial fields of application of the silicones. They exhibit a good releasing action, do not corrode the molds, permit good ejection, give the surface a good "finish," can be used in aqueous emulsion, and are active in low concentrations. It should be noted that a silicone film adhering to a rubber article can usually adversely affect subsequent lacquering, gluing, or printing of the surface if it is not removed. If a surface treatment is still necessary, special release agents in which liquid polydimethylsiloxanes have been replaced by copolymers with phenylethyl- (*241*) or phenylpropylmethylsiloxane units (*168*) ("paintable release agents") may be used.

The release effect of the silicone oils becomes better with increasing molecular weight. On the other hand, silicone oils of low molecular weight spread over the surface of the mold better, so that as a compromise oils of about 100 to 20,000 cSt/20°C are generally used in the industry.

Because of their thermal stability and their chemical inertness, silicone mold release agents have no corrosive effects. While the residues of organic mold release agents (mostly substances of the wetting agent type) remaining in the mold are gradually cracked at vulcanization temperatures with the separation of carbon, silicone oil residues adhering to the mold walls do not decompose. When mold release agents easily decomposed by heat are used, the encrustation of the mold surface necessitates frequent cleaning; the resulting roughening of the surface of the mold makes the removal of a rubber article increasingly more difficult.

However, the force necessary to remove a rubber article from a mold depends not only on the quality of the mold release agent and the state of the mold, but also on the type of rubber itself. The problems vary considerably according to the type of elastomer and the type and amount of the fillers, plasticizers, and other additives used. It has been shown experimentally that to release various types of vulcanizates from a mold with the use of a silicone mold release agent, natural rubber requires a force of 40 kg, polychloroprene 120 kg, nitrile-butadiene rubber 150 kg, butadiene-styrene rubber 52 kg, and butyl rubber only 18 kg. A similar situation is observed in the use of different amounts of various fillers, where the force necessary for mold removal may be between about 10 and 270 kg (*22*, *23*).

10.4.1.2 *Forms of Application of the Silicone Mold Release Agents*

The silicone oils are generally used in the form of aqueous emulsions; even silicone oils of relatively high molecular weight are available today in the form of stable emulsions (see Section 8.3.2).

The emulsions, which generally contain 30 to 40% of silicone oil in the state in which they are supplied, are diluted with water before use to a concentration of 0.1 to 1% and are generally sprayed on to the hot molds with a spray gun.

TABLE 103

QUALITY COMPARISON OF MOLD RELEASE AGENTS

Agent	Release effect	Mold corrosion	"Finish" of the finished article	Subsequent effects on the surface of the finished article	Over-all evaluation
(a) Aqueous emulsions of silicone oils of low viscosity (100–1000 cSt)	2–3[a]	1	1	Sometimes difficulties in gluing and painting	Release effect not quite so good as (b), (c), and (d); low mold corrosion and otherwise good properties lead to extensive use
(b) Aqueous emulsions of silicone oils of medium viscosity (10,000–20,000 cSt)	1–2[a]	1	1	Sometimes difficulties in gluing and painting	The most productive of all mold release agents
(c) Aqueous solutions of organic mold release agents (wetting agent type)	1	4–5	3	None	Sometimes highly effective, but uneconomic for many purposes because of the high mold corrosion
(d) Combination of aqueous silicone oil emulsions [type (a)] with solutions of organic mold release agents [type (c)]	1–2	2–3	1–2	Sometimes slight difficulties in gluing and painting	Very good compromise, very good results, particularly with highly filled mixtures difficult to release from the mold
(e) Solutions of silicones of high viscosity in organic solvents	1–2[a]	1	1	Sometimes difficulties in gluing and painting	Very good results by treating cold molds and by coating blanks

[a] Danger of the formation of sliding creases with excessive doses and local concentration.

In the spraying process, the finest possible mist must be produced to avoid formation of spots. Aqueous emulsions cannot be used for cold molds or for the painting of compounds. In this case, solutions of silicone oils in organic solvents, such as gasoline or chlorinated hydrocarbons, may be applied. Since it is possible to dissolve silicone oils of even extremely high molecular weights, the emulsification of which in water would present difficulties, in this form of application the more favorable release action of high-molecular-weight silicone oils can be brought to bear. Solutions of silicone rubber have also been employed successfully. Solutions of silicone oils are frequently used in spray cans.

Combinations of organic mold release agents and silicones soil the molds only to a relatively slight extent. The presence of the silicones suffices to prevent the deposition of cracking products on the mold surface (*224*, *244*). On the other hand, the addition of suitable organic mold release agents improves the release effect. Such mold release agents are also preferred frequently for economic reasons, and are employed on a large scale.

For special cases, in which the silicone film remaining on the article has a deleterious effect (for example, on objects that must be painted or glued), the curing of a *silicone resin film* on the mold has occasionally proved satisfactory. However, the surface of the mold must be carefully cleaned by sandblasting to anchor the resin film. For this reason, and because of the greater thickness of the resin film, the process can only be used when the retention of dimensions is not a decisive factor. The introduction of the "paintable silicone release agents" (Section 10.4.1.1) has, however, made this process substantially pointless.

For various reasons, in the manufacture of silicone rubber no materials based on silicones but only organic mold release agents can be used. However, with silicone rubber the use of release agents is necessary only in exceptional cases.

10.4.1.3 *Quality Comparison of Mold Release Agents Based on Silicones and on Organic Wetting Agents*

Table 103 gives a comparison of the two groups of mold release agents: 1 indicates very good and 5 indicates very bad.

10.4.2 Silicone-Rubber-Coated Liner Fabrics

Unvulcanized rubber compounds are generally very sticky and therefore permit rubber articles to be made up in layers. However, this stickiness makes the handling of the raw mixes difficult. Consequently, rubber mixes are, for example, wrapped in liner fabrics immediately after calendering. To avoid adhesion of the mixes to the fabric, the latter are coated with silicone rubber.

Silicone-rubber-coated liner fabrics are characterized by outstanding release properties even from very highly adhesive rubber mixes, by heat resistance, resistance to plasticizers and to aging, and effectiveness even in extremely thin layers.

The fabrics used are generally cotton or artificial silk fabrics.

Both heat-vulcanizable and room-temperature-vulcanizable silicone rubber may be used for the coating. Coatings made with heat vulcanizates have the advantage of being mechanically stronger, while room-temperature coatings are easier to make and offer greater certainty against insufficient vulcanization (*342*). The so-called one-component types are particularly suitable for the purpose described (cf. Section 8.1.2.3.2).

10.4.3 Other Applications

In the manufacture of dispersions or latex mixtures, particularly when synthetic rubber latices are being processed, pronounced foam formation often occurs which has a very harmful effect on the manufacture. This foam formation can be eliminated almost completely by the addition of small amounts of silicone defoaming agents.

The formation of a floating skin which often occurs as an undesirable phenomenon in the manufacture of rubber articles by the dip process can be prevented by the addition of siloxane-polyether copolymers (*290*).

Another very interesting observation is that certain siloxane-polyether copolymers have a sensitizing action in latex emulsions, so that these can be stored at the normal temperature but break at only slightly elevated temperatures—an effect which is particularly valuable for the manufacture of dipped articles (*60*, *290*).

As mixing constituents in solid rubber, silicone oils sometimes show particular effects; for example, they give a low coefficient of friction and better water resistance. Some of these effects are as yet little utilized (*247*). Additions of minute amounts of silicone oils to butadiene-styrene rubber latices lead to less sticky rubber (*110*).

10.5 Silicones as Auxiliaries in the Paint Industry*

10.5.1 Silicone Oils and Silicone Greases

Silicone oils are used in the paint industry for various purposes: even small amounts prevent the floating of pigments and bring about an improvement in the flow; they are also suitable as polishing oils; silicone oils and greases act as defoaming agents and affect the surface structure of paints (hammer finishes).

* Section 10.5 was written by Dr. W. Krauss, Leverkusen, West Germany.

The causes of *floating* are very complex. Generally, the particle size of the pigments, the specific gravity of the pigment and the binder, the film thickness, the rate of drying, the choice of solvents, and many other factors are given as causes. In considering the phenomenon, a distinction must be made between separation in the horizontal direction (floating) and in the vertical direction (flooding). Silicone oils are effective only against horizontal separation (floating). The causes of this type of separation may perhaps be density differences in the paint or, more probably, differences in the surface tension between the surface and the lower layers of a paint. In the case of differences in density, cells of turbulence (so-called Bénard cells) arise (*18*, *283*); with differences of surface tension, which come about by evaporation of solvents from the surface of the wet film, the solvent-richer layers with the lower surface tension break through the surface and spread (*148*). When a mixture of pigments is present, this may cause the various pigment particles to be transported to the surface and distributed there to different extents, which then allows the spreading fronts to become clearly visible when the film dries and hardens. The addition of silicone oil lowers the surface tension to such an extent that even the evaporation of the solvent leads to no further marked changes in it. The breakthrough of inner layers to the surface, causing its destruction, is therefore suppressed.

The concentration of the silicone oil additives at the surface forms a film which can act as a slip layer simulating a higher hardness of the coating. Such a silicone film also appears to have a favorable effect on the weather resistance (*309*).

Flow difficulties of various kinds are known. With nitro lacquers the so-called orange peel effect is frequently observed, this probably being ascribable to the same causes as pigment floating; in air-drying alkyd resins or stoving resins, a spray structure is often seen. All these phenomena can generally be eliminated by the addition of small amounts of silicone oil. Methylphenylsilicone oils appear to be more suitable than methylsilicone oils for improving the flow of unsaturated polyester lacquers, since they exhitit better compatibility with the binder and do not lower the surface tension to such an extent that the film breaks during processing on the flow coating machine (*176*). The subsequent spraying of a second coat of paint is not adversely affected by additions of methylphenylsilicone oils. Similarly active, because of their better compatibility with the various paint raw materials, are those methylsilicone oils that are constructed exclusively of mono- and trifunctional units (*177*).

Of the medium viscosity silicone oils, those with a viscosity between 300 and 500 cSt have become of particular interest as *polishing oils* for nitro polishing lacquers and flat varnishes. Compared with the polishing oils used previously, such as paraffin oil, etc., they have the advantage that because of their incompatibility they do not have a solvent effect on hard resins sensitive to polishing oils and therefore do not give rise to cloudy films. When added to nitro polishing lacquers, they sweat out on the surface almost completely and there form an excellent glide layer. In most cases, as little as 1 % of the amounts of polishing oils previously used is sufficient for an adequate effect.

Silicone oils of medium viscosity or silicone greases are used in the paint industry, particularly in the cooking of resins, as outstandingly active *defoaming agents*. Surprisingly, concentrations as low as 1 in 10^5 to 1 in 10^6 are sufficient for this purpose. Because of the nonvolatility and the stability of silicone oils, their activity is retained even at relatively high temperatures. Now and again, it is observed that silicone oils cause an increased formation of foam and bubbles in many paints (*104*), so that in all cases preliminary trials of the action of added silicones must be recommended.

Finally, silicone greases, and also high-viscosity silicone oils (*268*), are used as auxiliaries for the production of *hammer finishes* (*112*). For this purpose oils are to be preferred since they are more certain and more uniform in their action; moreover, greases lose their activity in the course of time, particularly when shear forces occur, as may be the case with rapid stirring.

TABLE 104

SILICONE OILS AS PAINT AUXILIARIES

Viscosity (cSt)	Structure	Application	Amount added (oil to solid resin material) (%)
~50	Polydimethylsiloxane	Flow improvement, prevention of separation, enhancement of surface smoothness, gloss, and water repellency	0.01
~300	Polydimethylsiloxane	Polishing oil in nitro lacquers and flat varnishes	About 1% of the amount of polishing oil previously used
~100,000 or silicone greases	Polydimethylsiloxane	Imparting the effect in hammer finishes	0.1–0.3
~40	Polymethylphenylsiloxane	Flow improvement especially in one-coat paints	0.1–1
Various	Siloxane-polyether cocondensate	Antisilicone, flow agent, improvement of adhesion	0.01–1

In addition to the polymethylsiloxanes and polymethylphenylsiloxanes, silicone oils forming cocondensation products of siloxanes and polyethers (Section 7.4.4) have also acquired importance in the paint industry. In addition to a siliconophilic part, their molecules contain an organophilic part, as a result of which they are substantially more compatible with organic high polymers than pure methylsilicone or methylphenylsilicone oils. Used as flow-improving agents, they do not adversely affect overcoating (*310*). In many cases, they give valuable service as "*antisilicones*" in the overcoating or recoating of surfaces contaminated with silicone oils, a problem arising to an ever-increasing extent, e.g., in the painting of damaged motor vehicles (*175*).

The possibilities of application of the various silicone oils are summarized in Table 104.

10.5.2 SILICONE RESINS

Silicone resins are used as auxiliaries in the paint industry primarily because of their surface-active properties and their weather resistance. For these properties to have an effect, the binder mixture must in general contain 1 to 10% of silicone resin. Under these conditions, compatibility between organic binder and silicone resin is achieved by the use of special types of silicone resins (see Section 8.2.2.1.4).

An effect on the surface tension is detectable with additions of silicone resin of only 1 to 3%; consequently, silicone resins in such concentrations are outstanding as flow-improving agents. Their flow-improving properties are particularly striking in the case of stoving enamels. The great advantage of the resins as compared with the silicone oils described in Section 10.5.1 is that on overcoating with silicone-resin-containing paints no difficulties of any kind in the flow or adhesion properties are observed.

With additions of about 3 to 5% of silicone resin, the water repellency of the paint surfaces is substantially increased, so that a better water resistance is achieved, as is shown particularly in stability tests against superheated steam or in the salt spray test. A reduction in the spot sensitivity of paints brought about by adding silicone resins has also been reported. Additions of at least 5 to 10% of silicone resin to the organic binder are necessary to improve the weather resistance.

It is frequently discussed (*35*, *179*) whether cocondensation products (so-called silicone alkyd resins) are not better when their properties are considered as a whole than mixed products prepared by the simple addition of silicone resins to alkyd resins at the same silicone content (*125*). With high proportions of silicone this is true, since in mixtures of air-drying paints the unsatisfactory drying properties of the silicone resins may make themselves felt. However, the addition of large amounts of silicone does not help to improve the weather resistance of air-drying lacquers, which scarcely changes further at silicone contents greater than 30% (*179*, *282*), nor is it economically acceptable for wide utilization. Consequently, in practice the additions usually amount to 10%, which roughly doubles the weather resistance and does not adversely affect the other properties of the binder as a paint component.

It has been proposed that the paint industry should itself manufacture silicone alkyd resins by the reaction of alkyd resins with certain silicone intermediates containing reactive alkoxy groups (*124*). In principle such siloxanes can react with all binders containing free hydroxyl groups (*27*, *251*). Although cocondensates prepared in this way have already been investigated in great detail, they do not appear to have been introduced widely in practice (*24*), probably because it has not so far been possible to impart the weather resistance of the silicone resins to any given organic binder.

Silicone resin emulsions improve the weather resistance of emulsion paints based on polyvinyl acetate or styrene-butadiene latices very considerably, even in amounts as small as 3 to 5%. In some cases, this makes it possible to increase the proportion of pigment in these products, thus enabling the water vapor permeability of the paint film to be increased. The use of *sodium methylsiliconate*, described in more detail in Section 10.12.2.1, sometimes gives even better results (*159*).

Table 105 summarizes the most important possibilities of the use of silicone resins as auxiliaries in the paint industry.

TABLE 105

SILICONE RESINS AS PAINT AUXILIARIES

Amount added (%)	Silicone product	Improvement achieved
1	Combinable silicone resin	Improved flow, particularly with stoving enamels
1–3	Combinable silicone resin	Improved water resistance, enhanced spotting resistance in washing machine and refrigerator enamels
5–10	Combinable silicone resin	Improved weather resistance with air-drying lacquers
10	Polysiloxanol products, cocondensed	Improved weather resistance, water resistance, heat resistance, depending on the nature of the organic binder
3–5	Silicone resin emulsions Sodium methylsiliconate	Improved resistance to chalking with aqueous emulsion paints; improved resistance to wiping and washing

10.6 Silicones as Auxiliaries in the Plastics and Synthetic Fiber Industry

As in the rubber industry, the silicones act as outstanding release and slip agents because of their incompatibility with most organic plastics. The spreading power of silicones, as a result of which they can rapidly and uniformly coat complicated mold surfaces, their capacity of exerting a releasing effect even in extremely thin films, their resistance to elevated temperatures, and the fact that they impart a good "finish" and a pleasant handle to the article concerned make them particularly valuable. Thus, silicone oils are also widely used in the plastics industry as *mold release agents*. The releasing effect naturally depends on the type of silicone and the type of plastic. Polydimethylsiloxane oils give the best release effects with PVC, polyethylenes, polyurethanes, polyamides, cellulose acetate, and fluoroplastics; the results with phenoplasts and polystyrene are somewhat inferior, but with polystyrene they are still better than when other, conventional, mold release agents are used.

Since the molding temperatures are generally not so high as in the vulcanization of rubber, there is no preference for aqueous emulsions in this case. Solutions of silicone oils in organic solvents are frequently used, which can be handled particularly easily in aerosol spray cans. Porous molds made from wood, plaster, or ceramic are best treated with silicone greases (or pastes), which are not drawn in from the surface as readily as oils.

The transfer of the mold release film to the surface of the article (which is favorable for the "finish") may have an undesirable effect in that printing inks and paints can no longer be accepted; in that case either the mold release agents known as "paintable release agents" (Section 10.4.1.1) are used, or the mold can be furnished with a film of silicone resin, which is cured onto the well-cleaned surface of the mold. Such films possess great permanence and last for several thousand releases.

Recently, room-temperature-vulcanizing silicone rubber has gained interest for the preparation of shell molds for casting resins. The shells prepared for the casting reproduce all features of the mold surface and can easily be detached from it and from the cast resin article because of the rubber elasticity of the materials and its releasing properties. It must merely be borne in mind that the presence of monostyrene in polyester casting resins swells the silicone rubber and may thus hinder discharge from the mold.

In connection with the releasing properties of silicone oils for plastics and the fact that they are relatively good lubricating agents for them (*200*, *314*) (Section 9.2.10), they are also used successfully as *slip agents* in the processing of plastics. They facilitate the movement of the plastic against metal in injection-molding machines, extrusion presses, rolling mills, and calenders. Because of their incompatibility with the organic polymers, they do not penetrate into them and have no plasticising effect.* The slip properties of certain plastics are also improved by the addition of small amounts of silicone oil (e.g., 0.05% in the case of soft PVC) on processing by the injection-molding or the extrusion process (*341*). Plastic sheets made from PVC and copolymers containing plasticizers lose a certain stickiness and acquire a low coefficient of friction when they are treated with silicone-containing preparations (*213*).

Small amounts of silicone resins may be added to synthetic resins that are to bind slag wool, glass wool, or asbestos. The silicone forms a *water-repelling film* on the binder resin and reduces the water uptake of the fibrous mats and fleeces. Silicone resin emulsions compatible in the mixture are today available for combinations with the synthetic resins, which are generally used in the form of dispersions.

Further, silicones have become indispensable aids in the preparation of *foamed materials* based on polyethers. The addition of certain silicones stabilizes the foam and modifies the pore structure.

If the polyether foams are made from prepolymers, a normal polydimethylsiloxane oil with a viscosity of from about 50 to 200 cSt/20°C is effective (*91*, *239*). If a one-shot process is used, in which all the starting materials—polyether, polyisocyanate, water, *t*-amine activator, and silicone stabilizer—are caused to react in a *single* stage, the addition of polydimethylsiloxane oil has no effect. On the other hand, the use of special silicone/alkylene oxide copolymers has proved to be successful in this case (*194*, *276*), particularly when they are combined with organic Sn(IV) (*134*) or Sn(II) compounds (*98*).

The copolymers may have various types of structure; in particular, the bridge through which the siloxane component of the molecule is linked to the polyether component is variable (*12*, *57*, *58*, *134*, *289*, *328*) (cf. Section 7.4.4).

The pore structure of foamed materials obtained in this way can be seen from Fig. 93, bottom left.

If small amounts of a pastelike dispersion of SiO_2 fillers in a polydimethylsiloxane oil are also added to such combinations before foaming, the pore structure is changed so that the foamed material resembles a natural sponge (Fig. 93, bottom right).

Polyester foamed materials with a uniform pore distribution can generally be prepared without the addition of silicone oil (Fig. 93, top left). However, here again the size and distribution of the pores can be affected by the addition of small amounts of polydimethylsiloxane oils (*30*, *315*) (Fig. 93, top right), while the system reacts in a sensitive manner to an overdose, which leads to collapse of the foam. More recent developments, aimed at using polyester foamed materials as linings in the textile field, make it necessary to impart a

* Only linear polymethylphenylsiloxanes with terminal *t*-C_4H_9O and *t*-$C_5H_{11}O$ groups can act as true plasticizers for polyvinyl chloride (*138*).

particularly fine-pored structure to these materials. Again, special copolymerization products still undergoing development appear to be suitable and necessary for this purpose.

In the synthetic fiber industry, silicone oils are used in the *spinning* of polyamides, particularly polycaprolactam, usually being sprayed in diluted form onto the spinning nozzles from time to time. The silicone film keeps the

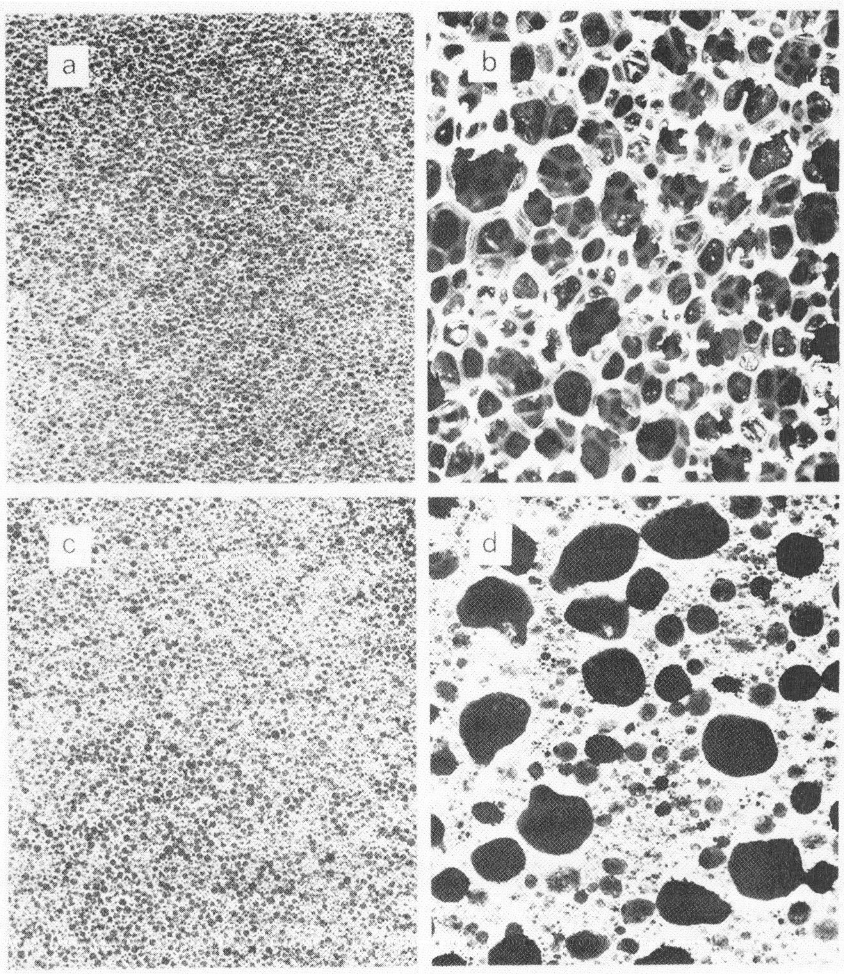

FIG. 93. Action of silicone additives in polyether and polyester foamed materials. (a) Polyester without silicone, (b) polyester with 0.2% of polydimethylsiloxane oil, (c) polyether with 1% of siloxane-polyether copolymers, and (d) polyether with 1% of siloxane-polyether copolymers and 0.05% of silicone oil paste. (H. W. Patzelt, Farbenfabriken Bayer AG.)

nozzles clean, reduces the bending or flattening of the fibers, improves their slip properties, and thereby reduces the failure of the nozzles due to blockage and encrustation (*136*, *236*).* Since the heat stress at the nozzles is high, special types of oils distinguished by high heat resistance and a release effect are used.

Silicone oils also prove satisfactory as slip agents when thermoplastic synthetic fibers have to be passed at high velocity through nozzles, lugs, and other guide members in which they become hot.

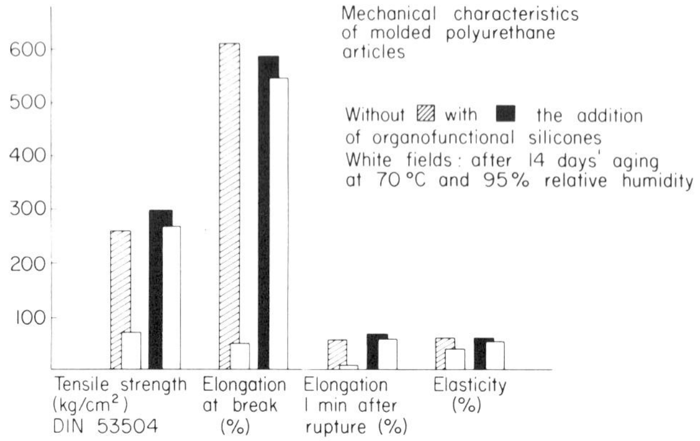

FIG. 94. Influence of modification with siloxanes on the aging resistance of molded polyurethane articles (*226*).

Some very important possibilities of development have recently been found in quite another direction. By means of organofunctional groups, it is possible with the aid of classical types of organic reactions (see Section 7.4) to introduce the siloxane configuration into plastics and synthetic fibers. In contrast to all the applications of the silicones mentioned previously, therefore, the organofunctional siloxane component here becomes part of the molecular skeleton. The possibilities so offered are probably manifold and have not yet been investigated exhaustively by any means. In view of the questions of economy, particular effects that can be evoked by only relatively small amounts of organofunctional siloxanes are naturally of interest. Thus, surprisingly, it has been found that the addition of only 0.5% of 1,3-bis-(β,β'-dihydroxyethylaminomethyl)-tetramethyldisiloxane in the preparation of a polyurethane elastomer improved quite considerably the aging resistance of the molded article, tested at 70°C and 95% relative humidity (*226*, *262*) (Fig. 94). In combination with carbodiimides, the siloxane has a synergistic effect (*263*).

* Compare also the use of resins for the same purpose (*167*).

The following effect is of a different type: if hydroxyalkylsiloxanes and polyalcohols are treated with diisocyanates in such a way that part of the isocyanate groups remain unchanged, and polyvinyl polymers such as poly(vinyl acetate), poly(vinyl chloride), and poly(vinylidene chloride) or their copolymers with ethylene are incorporated in these products, it is found that they exert a kind of slip effect. The stickiness of the article is reduced and the extruder output is increased. This effect is appreciable even with the addition of as little as 0.01%, calculated on the silicone. The amounts preferred in practice vary from 0.1 to 5% (*264*).

The obvious expectation that cocondensation or copolymerization with organofunctional siloxanes would also lead to an increase in the heat resistance of an organic plastic appears to be fulfilled only with relatively high siloxane contents. In this connection, we may recall siloxane-polyester cocondensates. Those containing 75% of siloxane differed only slightly in heat resistance from the pure siloxane, while those containing 50% of siloxane already exhibited a fall in heat resistance (Section 9.4.1). It has been possible to improve the heat-aging properties of epoxide resins, as judged from the loss in weight, by the addition of 25% of polyepoxyorganosiloxanes (*253*).

10.7 Silicones in the Glass Industry and Ceramics

When used in a suitable manner, silicones adhere very strongly to glass. One of the first developments of technical silicone chemistry was laminating resins with an affinity for glass for glass fiber tissues for use in the electrical industry (see Section 10.2). Attempts were also made very early to modify the surface properties of glass and ceramic articles by coating them with thin films of silicones, partly to make them water repellent, partly to improve their electrical or mechanical properties, and partly to improve the adhesion between glass and plastics by the production of an intermediate layer. Furthermore, certain release problems in glass production are solved by the use of silicones.

10.7.1 The Siliconization of Hollow Glassware and Insulators

Organochlorosilanes, particularly methylchlorosilanes, were first used to produce silicone films on glass; in the vapor form they react partly with the film of water adhering to the glass and partly with surface Si—OH groups of the glass and thereby form silicone films bound to the substrate by means of primary valences (*233*, *242*). It was later found that the water repellency of a silicone film prepared in this way, as judged by measurements of the contact angle of drops of water placed on it (Section 9.1.6) is increased if organochlorosilanes with relatively large aliphatic groups are used (*292*). However, this process could not be widely used in practice since the handling of the hydrogen-chloride-liberating silanes is too difficult. Organosilanes with other functional groups, such as ethoxy (*142*) or acetoxy groups, give similar effects without the production of aggressive hydrolysis products by reaction with water and OH groups on the glass surface. Since it has been learned how to anchor even silicone oils relatively firmly to glass surfaces, interest has been directed predominantly to these products and to processes for their application.

For a satisfactory and permanent anchorage of films of silicone oil to glass surfaces, it is essential that the films are fixed by heating to a sufficiently high temperature for a sufficiently long time (see Section 9.1.6). Optimum values for water repellency are obtained, for example, by heating for half an hour to 300 to 400°C. Films baked on in this way can then be removed only with difficulty by wetting agents or organic solvents. It is interesting that better water-repelling effects are obtained, even at relatively low temperatures, when the silicone oils are used not as solutions in organic solvents but in the form of aqueous emulsions (*222*) (Fig. 36). In agreement with this observation, films of silicone oils taken up by immersing glass in silicone oil emulsions are thicker than those obtained from solutions of the oils in organic solvents at the same concentration of the silicone (*326*).

Working with aqueous emulsions has been adopted predominantly in practice, probably because they can be diluted with water. The glass articles are immersed in, or sprayed with, a dilute aqueous emulsion generally containing between about 0.5 and 1% of silicone oil. Depending on the size of the objects and the number to be produced, they are then heated, either intermittently or continuously (e.g., in tunnel ovens). For a good effect, it is essential that the surface of the glass is clean. Films of grease, oil, or soil must, if necessary, be removed by immersion in hot solutions of wetting agents with thorough subsequent rinsing. Glass articles cleaned in this way can be treated with the silicone oil emulsions without intermediate drying. It has been recommended that after the emulsion has been taken up on the glass the latter should be rinsed with water again before the film is baked on (*295*) in order to achieve a particularly uniform distribution of the oil.

Probably without exception, polydimethylsiloxane oils are today used as the silicone components. Siloxanes with higher aliphatic groups do have a better effect when they are fixed at low temperatures, but at higher baking temperatures they are inferior to the polydimethylsiloxanes (cf. Section 9.1.6).

Liquids drain from bottles that have been coated with thin films of silicone of this type without leaving behind any residue worth mentioning. Moreover, there are no losses caused by liquids running down the outer wall. These effects are particularly valuable when expensive or accurately measured substances are to be removed completely from their containers. This problem frequently arises in the pouring out of pharmaceutical preparations, and therefore siliconization of bottles and ampuls was first used to a large extent in this field (*108*). It is advantageous that the silicones simultaneously have a foam-destroying action; to a certain extent, the silicone improves the resistance of the glass to hydrolysis and decreases the catalytic decomposing action that the glass wall exerts on blood, sperms, or sensitive drugs. Consequently, siliconized glass vessels are suitable, for example, for the storage of preserved blood. Of course, it must be borne in mind that the water-repellent effect of siliconized glass depends also on the nature of the liquid phase and that silicone films may in time be destroyed in alkaline media.

Silicone films can also affect the *mechanical* behavior of hollow glassware. Siliconization of the glass, which can be carried out during the cooling process by spraying diluted silicone emulsions onto the bottles still at 200°C, reduces the breakage rate on filling in automatic filling machines and in transport; at the same time the glass bottles become less sensitive to scratching (*31*, *114*, *135*, *257*). The bursting pressure is increased by at least 10% (*281*).

The phenomena may be ascribed partly to a healing up and closure of the surface by the silicone film and partly also to increased slipperiness of the surface. The significance of the process and the limits of its application are still uncertain (*269*); the slipperiness of the glass surface reduces their stackability, so that bottles must be transported in cartons or baskets with individual compartments; in addition, the effectiveness of the silicone film falls off on repeated rinsing in solutions of wetting agents. Consequently, the use of siliconized bottles will probably be useful only with one-trip packing (*210*).

Another very interesting observation in this connection is that the quenching of glass preheated to the softening temperature in a silicone oil bath the temperature of which was first adjusted to 200°C and then lowered in steps of 20°C, together with subsequent drying at 200°C, substantially increased the bending strength of the glass. This effect too is probably based on a healing up of the glass surface and, in part, on an equalization of internal stresses (*286*).

It can be regarded as a necessary consequence of water-repellency treatment that the *surface electrical resistance* of glass and ceramic parts can be increased quite considerably by siliconization. Figure 95 shows the surface resistance of siliconized and untreated glass as a function of the relative atmospheric humidity. Silicone treatment gives glass, ceramic, or steatite insulators a rain breakdown voltage approximating to that of the dry breakdown voltage. Moreover, the self-cleaning action of the insulators by wind and rain is improved. However, the silicone film is destroyed by electrical breakdowns wherever the arc finds its path, so that its use is limited predominantly to low-tension techniques (*215*).

In very dirty regions, the insulators of open-air cables must be cleaned from time to time in order to prevent flashovers. If an unsoiled or cleansed insulator is provided with a coating of silicone grease, it will still become soiled but will not exhibit the usual formation of electrolyte bridges which finally lead to the flashover. This peculiar effect is due to the fact that the application of silicone grease containing silicone oil capable of spreading creates a reserve of silicone oil on the insulator, so that each particle of soil depositing on it is immediately coated with a hydrophobic film capable of repelling wet soil. This prevents the formation of conducting bridges or channels and is also capable of regeneration after mechanical or electrical damage (*303*).

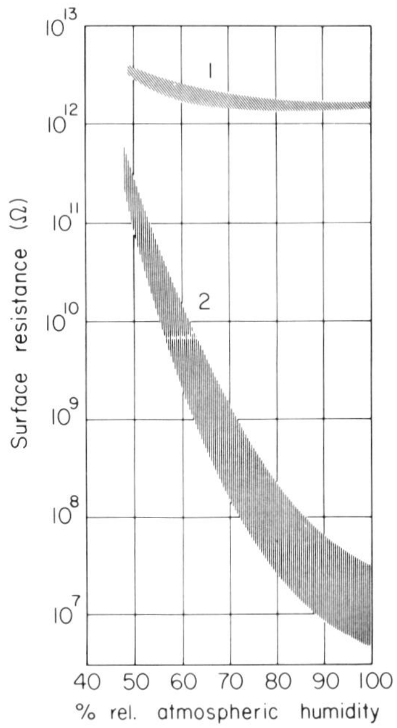

FIG. 95. Surface resistance of (1) siliconized and (2) untreated glass at various atmospheric humidities.

Ceramic materials can be siliconized not only with liquid polymethylsiloxanes or their aqueous emulsions or solutions but also with solutions or emulsions of silicone resins or with solutions of alkali-metal methylsiliconates. Particularly resistant films are said to be obtained by a combined treatment, first with the siliconate and then with the silicone oil (*144*).

10.7.2 SILICONES AS BONDING AGENTS FOR THE BONDING OF GLASS FIBERS TO SYNTHETIC RESINS AND OF PIGMENTS TO GLASS FIBERS

If the surface of a silicate glass is coated with suitable silanes or siloxanes, the behavior of the silicate material, particularly its affinity for synthetic resins or dyestuffs, can be modified. Use is made of these possibilities in order to improve the wet strength of glass fiber laminates and molding masses or the coloring of glass fibers.

When silanes or siloxanes are to undertake such a role of "intermediary," they must contain both silicon-functional groups and organofunctional groups. The silane becomes anchored to the silicate surface by means of the former, while the latter take over bonding to the organic substrate (Fig. 96).

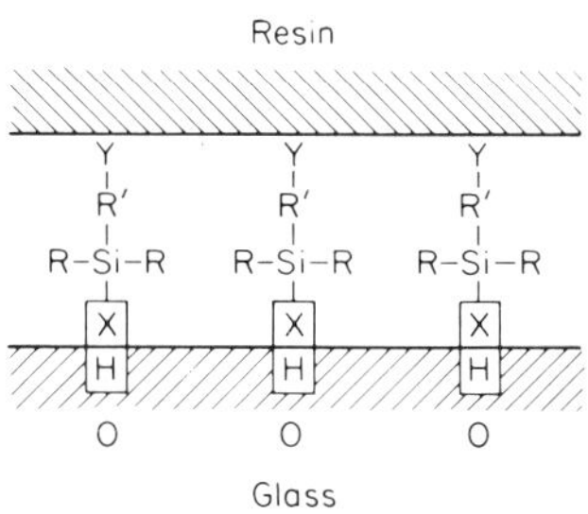

FIG. 96. Silicon-functional and organofunctional silanes as adhesion-imparting agents for the bonding of glass and resin (*226*).

R′Y = organofunctional group,
e.g., $-CH_2CH_2CH_2NH_2$
$-CH{=}CH_2$;
X = silicon-functional group, e.g., $-OC_2H_5$.

The silicon-functional group, and in particular the organofunctional group, can be varied in many ways. A large number of patent specifications deals with the discovery of such reactive groups giving optimum properties for the various synthetic resin bonds. In this respect, the improvement of the wet strength of laminates or molding masses is in the focus of interest, but the same principle can be used for other types of filled resin systems (*302*).

This is done either by precoating the glass fibers with the silicon compounds or by adding these compounds to the resin (*302*). Silicones or siloxanes with vinyl groups are the longest-known compounds suitable for this purpose. Vinyltrichlorosilanes or, better, vinyltriethoxysilanes taken up onto the glass fiber from partially hydrolyzed solution in the dipping process (*25*, *301*) and vinylalkoxysilanes solubilized by condensation with polyglycols (*151*) are used; other embodiments are dispersions containing vinylsilanol (*275*), vinyltriacetoxysilane (*81*), quaternary ammonium salts of vinylsilanol (*43*), alkali-metal polyvinylsiloxanolates (*238*), water-soluble salts of the reaction products of vinyltrichlorosilane with diethanolamine (*92*), or, finally, polyvinylsiloxanes themselves (*149*).

More recently, methacryloxypropyltrimethoxysilane

$$\begin{array}{l} CH_2{=}C{-}COO(CH_2)_3Si(OCH_3)_3 \\ \quad\;\; | \\ \quad\; CH_3 \end{array}$$

and the corresponding ethoxysilane have been found to be particularly good bonding agents for unsaturated polyesters, styrene or acrylic ester resins (*44*, *256*, *302*).

Pretreatment of the glass fibers with aminoalkylalkoxysilanes results in increased adhesion to resins such as aldehyde condensation products, epoxides, or urethanes (*249*, *316*). Organoalkoxysilanes containing the epoxide group as a functional group in the organic residue are suitable for strengthening epoxide resins and phenol resins (*44*, *254*, *255*).

Finally, silanes or siloxanes with carboxyl groups in the hydrocarbon residue are suitable for increasing the adhesion between glass fibers and various resins (*46*, *204*).

In view of the increasing use of glass fibers for textile fabrics and the known difficulties of an abrasion-fast, light-fast, and wash-fast coloration of the glass fibers, attempts to use silicones for such problems are of interest. In analogy with the principle of reactive dyes, processes have been developed which permit coloring materials to be fixed reactively to glass fibers by means of an intermediary containing organic and silicon functions. Suitable substances for this purpose are arylaminoalkylalkoxysilanes, for example, the compounds (*37*):

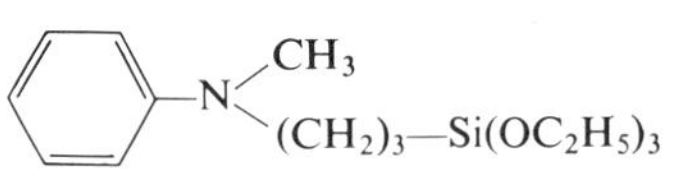

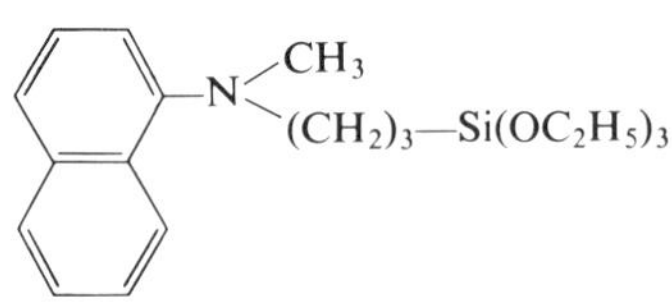

The glass fibers to be colored are treated in a boiling aqueous bath with these compounds and then with solutions of diazotized arylamines or diazotized fast bases (*37*).

In pigment-binding processes, similar fastness effects can be achieved more simply by binding the pigments in a known manner with a binder that may be based on acrylic polymerizates and may possibly be combined with urea-formaldehyde condensation products, applying them nonreactively to the glass fiber and treating the fiber or fabric in a second stage of the process with emulsions of polydimethylsiloxane-α,ω-diols (*132*).

10.7.3 Silicones as Release Agents in the Production of Hollow Glassware

Good results have been obtained with silicone oils as release agents in glass-blowing machines (*33*, *296*). Combinations of silicone oil emulsions with dispersions of graphite proved to be particularly effective release agents for this purpose. As compared with the previous release agents, based on mineral oils, they have the advantage of not smoking, or smoking relatively little, and giving cleaner articles; the size accuracy is improved, and the concentration used is substantially lower than in the case of the mineral oils, so that despite the higher cost of silicones, an over-all reduction in the blowing costs can be achieved (*202*).

10.8 Silicones in the Textile Industry*

10.8.1 Water-Repellency Treatment of Textiles

The use of silicones as water-repellency agents has found ever-increasing acceptance in the textile industry in the last few years. As compared with other highly hydrophobic substances, e.g., paraffin and other waxes, silicones exhibit the following advantages: They do not possess a fatlike nature, which has a good effect on the finish of the fabric; they are fixed on the fiber in such a way that they withstand fairly well washing or treatment with solvents; they are substantially resistant to chemicals and weathering; and, finally, they have a very considerable softening and smoothing effect on the textile.

10.8.1.1 *General Remarks on the Water Repellency of Textiles*

The water-repelling effect of impregnated textiles is evaluated from the following three points of view: water permeability, water uptake, and the behavior in the spray test.

The best-known methods for testing the water repellency effect on textiles are those of Schopper (DIN 53 886), the spray test according to AATCC 22-25, and the test with Bundesmann's apparatus. The last of these is used most widely in Germany, while in the United States evaluation by means of the spray test is preferred. The uptake of water is estimated visually from the spreading of the wetted parts on the cloth, in comparison with a scale of pieces wetted to different extents. The Schopper method corresponds approximately to the AATCC hydrostatic pressure test.

10.8.1.2 *Formation of Polysiloxane on the Fiber*

At the very beginning of industrial silicone chemistry, attempts were made to render fibers and fabrics water repellent by treatment with methylchlorosilanes (*242*). In order to neutralize the hydrogen chloride liberated by the hydrolysis of the chlorosilanes, the process was carried out in the presence of ammonia (*242*, *294*). Hydrolysis may occur by the action not only of adhering films of water but also of hydroxyl groups of the cellulose. The existence of

* Section 10.8 was written by Dr. O. Glenz, Leverkusen, West Germany.

compounds derived from trimethylchlorosilanes and cellulose has been demonstrated (*280*); however, a trimethylsilylcellulose is stable only in an absolutely dry atmosphere and is rapidly hydrolyzed by moisture.

A relatively large number of patents has been granted for processes which describe the condensation of silane derivatives to form polysiloxanes on the fiber without the liberation of products having a harmful effect on the fiber. In place of the organochlorosilanes, the corresponding organoacetoxysilanes (*13*), organoisocyanatosilanes (*164*), organoaminosilanes (*29*), silylcarbamic acids (*324*), and alkoxysilanes (*51*, *308*) have been proposed.

Another group of compounds that are condensed on the fiber are the siliconates, especially sodium methylsiliconate (*82*). The fabric is impregnated with an alkaline solution of the siliconate and is treated with a weak acid. On subsequent heating from 110 to 120°C, the polysiloxane is formed and gives a good water-repellency effect. Although this process appears to be very attractive from the point of view of application technique, it has not become established since the high alkalinity adversely affects the impregnating baths (*143*, *169*).

10.8.1.3 *Use of Substantially Precondensed Polysiloxanes*

None of the processes described for the formation of polysiloxanes from polymers or oligomers on the fiber has acquired practical importance. Their use has been opposed either by the corrosiveness of the materials involved, the laborious mode of application, or the fact that the condensation conditions were too severe for the textile. The success of silicones in the textile industry became possible only through the application of pre-formed siloxane polymers to the textile. This can be done from solution in organic solvents or from aqueous emulsions. Suitable polyorganosiloxanes were found to be, in the first place, those containing Si—H bonds (*68*, *71*, *72*, *232*); excellent results were also obtained later with polydimethylsiloxane-α,ω-diols, sometimes in admixture with α,ω-bis-(trimethylsiloxy)-polydimethylsiloxanes (*73*). Mixtures of polyorgano-*H*-siloxanes with polymethylsiloxanes can be used with advantage (*69*, *70*).

Special measures are necessary to ensure an orientation of the polysiloxane chains such that the siloxane oxygen is turned toward the substrate and the methyl groups are forced outwards (*321*) (cf. Section 9.1). In this way, the density of hydrophobic methyl groups at the surface and the adhesion to the substrate are enhanced. This probably explains why mixtures of silicones and paraffin waxes generally lead to an impairment of the water repellency since a reversal of the siloxane molecules then takes place, the methyl groups being turned toward the paraffin wax and the Si—O groups being directed outward,

A further condition for good water repellency appears to be complete enclosure of the fiber by a film of silicone. It can be imagined that the water-repellent properties are impaired as soon as this coating becomes defective, since wetting of the fiber at the defective sites causes swelling. These swelling processes finally cause bursting of the film and perhaps even its detachment.

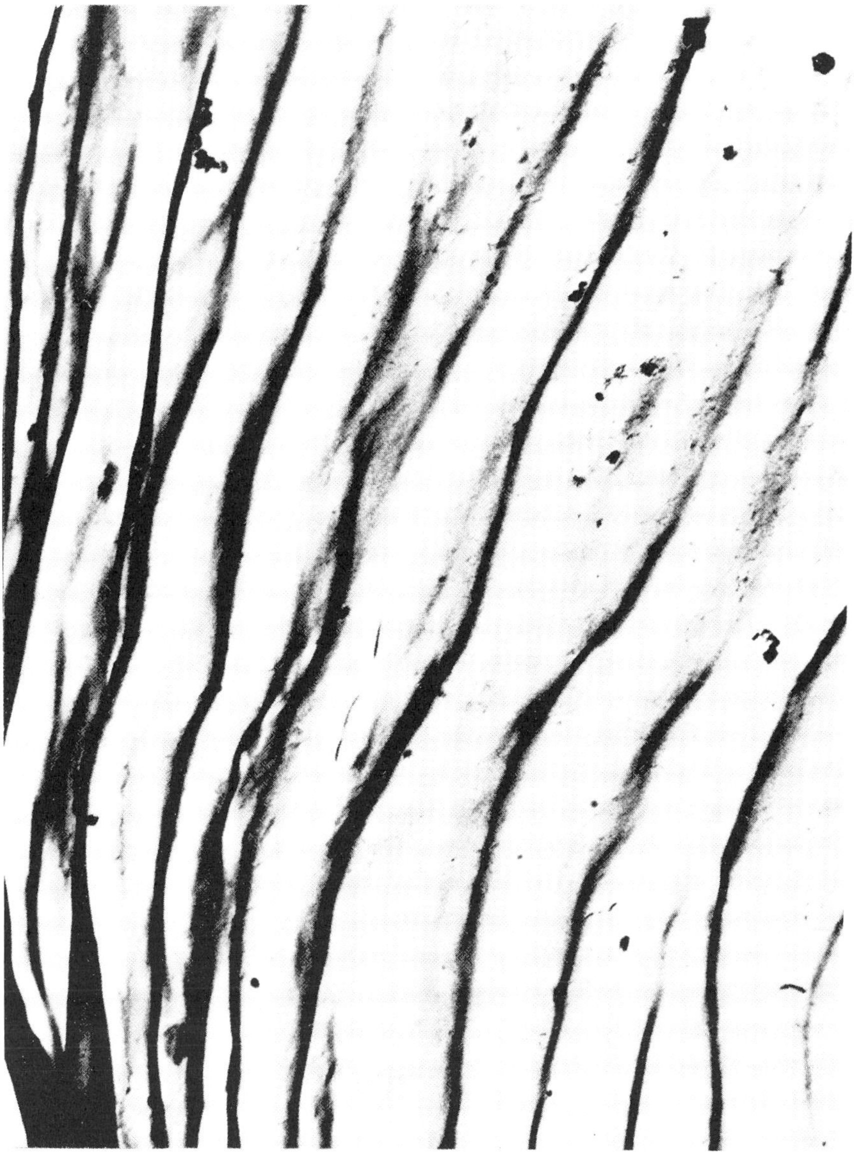

FIG. 97. Electron micrograph of a Plexiglas replica of a cotton fiber. The replica was vapor-coated obliquely with carbon while rotating; 14,000 ×.

Investigations with the electron microscope (Figs. 97 and 98) show that silicone treatment surrounds cotton fibers with a skinlike sheath. This gives an impressive illustration of the protection conferred by silicone treatment.

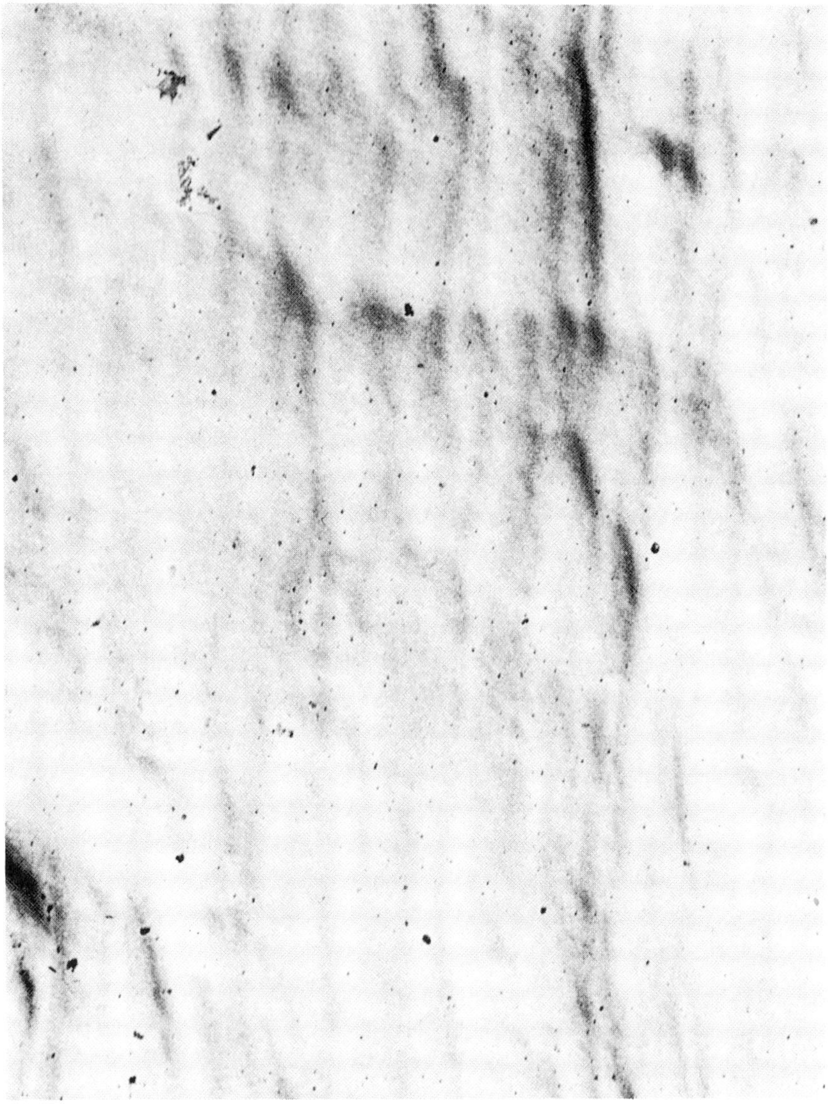

FIG. 98. Replica similar to that shown in Fig. 97 of a cotton fiber coated with a film of silicone.

10.8.1.3.1 Polymethyl-H-siloxanes. When they are condensed on the fiber, polymethyl-*H*- siloxanes give a hard nonsticky film. Regarded chemically, this is a case of cross-linking through the formation of Si—O—Si bonds in place of the Si—H bond:

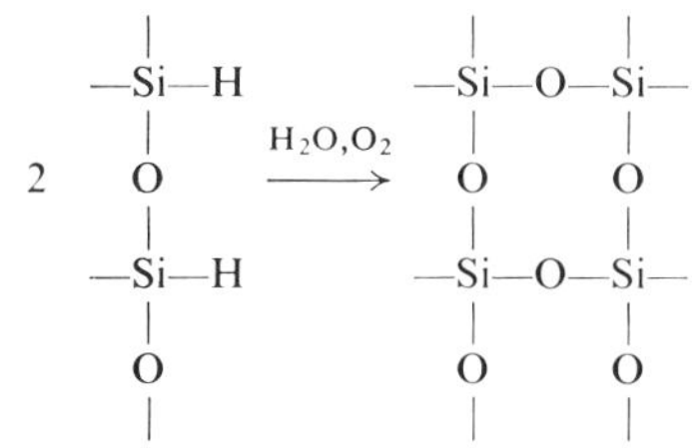

The cross-linking reaction can be initiated by water, more quickly in an alkaline medium; the former gives rise to hydrolysis of the Si—H bonds to silanol groups, which subsequently condense to give siloxanes. This type of reaction also takes place, of course, in aqueous emulsion and leads to eventual breakdown. It can be counteracted to a certain extent by adjusting the emulsion to a pH of about 3–4 and incorporating certain additives which experience has shown to stabilize the Si—H bond in the presence of water—by mechanisms which are obscure. Substances having this effect are, for example, poly(vinyl alcohol) (*84*), glycine (*298*), and certain aldehydes (*319*). However, even such emulsions liberate some hydrogen on storage, although to a greatly reduced extent.

Another possible way of solving the question of stability consists in preparing the aqueous emulsions immediately before use. To allow this to be done in a simple manner and without the expense of machines, self-emulsifying solutions of the siloxane in nonaqueous water-soluble solvents have been developed. When poured into water, they spontaneously give a siloxane emulsion. They are obtained by dissolving the polymethyl-*H*-siloxane in anhydrous isopropanol together with emulsifiers based on esters of fatty acids with polyglycols or polyhydroxy compounds containing more than three OH groups or their etherification products with polyglycols (*225*, *229*). Such solutions can be stored without decomposition over long periods, particularly when small amounts of glycine are added (*299*); the emulsifier has no harmful effect on the water repellency.

The cross-linking reaction of the Si—H bond can be brought about not only by hydrolytic rupture but also by oxidation at elevated temperatures, possibly under the action of a catalyst. Attempts have been made to initiate this reaction on the fiber. The fact that the silicone formed a film on the fiber was shown as follows (*90*). A cellulose acetate fiber was impregnated with polyorgano-*H*-siloxanes which were then cured. When the fiber was dissolved out, a thin sheathlike residue of the silicone was left behind. It was also shown that the film did in fact consist of a cross-linked polysiloxane and that it no longer contained hydrogen bound to silicon (*119*). The condensation takes place only under the action of oxygen, the reaction rate depending on the thickness of the layer. No cross-linking takes place in an atmosphere of nitrogen (*119*).

Temperatures of ~200°C are necessary for curing without a catalyst. Certain catalysts—particularly the compounds of lead, zinc, tin, zirconium, and titanium—accelerate the process, so that the latter can be carried out at 140 to 150°C. Organic titanates, especially butyl titanate, are powerful catalysts and permit a lowering of the condensation temperature or a shortening of the condensation time; because of the sensitivity of butyl titanate to hydrolysis, one is restricted to application in organic solvents (*45*, *115*, *116*).

However, suitable auxiliaries facilitating and improving the water-repellency treatment have also been found for the use of the polymethyl-*H*-siloxanes in aqueous emulsion. The combination with stearatochromium(III) chloride gives a synergistic enhancement of the hydrophobic effect (*171*), i.e., the latter is stronger than would be expected from a purely additive action of the silicones and the complex chromium salts.

Products of the reaction of titanium tetrachloride with polyols, particularly butanediol, propanediol, or glycerol (*317*), or alkanolamine titanates, such as triethanolamine titanate (*85*), can be added to the aqueous phase of the emulsion as catalysts.

Excellent water-repellency agents based on silicones can be developed without metallic catalysts but with certain reactive organic components. Suitable for this purpose are the reaction products of epoxide resins with di- or polyamines in such a ratio that unchanged epoxide groups remain in the product (*185*). This combination is particularly suitable for the water-repellency treatment of fabrics made entirely from synthetic fibers or from mixtures of synthetic and cellulose fibers. Thus, mixed fabrics of polyesters and cotton can be impregnated with a silicone formulation free of zirconium salts to give good abrasion resistance almost without reducing the dye fastness. This is not always possible with zirconium-containing impregnating agents (*331*).

10.8.1.3.2 Polydimethylsiloxanes. Most of the phenomena reported for the polymethyl-*H*-siloxanes also apply to polydimethylsiloxanes. Again, the use of "catalysts" is necessary for the development of optimum water-repellency effects. Certain inorganic and organic salts, certain alkoxides, and certain complex compounds of zirconium, titanium, tin, zinc, and lead are suitable for this purpose. The mode of action of these compounds has not been fully elucidated and may be different in different formulations. It has been assumed that some hydrolyzable zirconium and titanium compounds owe their effect to the fact that on hydrolysis they deposit on the fiber as sparingly soluble hydrated oxides which enable the siloxane to adhere. This assumption is thought to be justified by the fact that even siloxanes without any reactive groups give enhanced water repellency when used together with hydrolyzable titanium and zirconium compounds.

Polyorganosiloxanes containing silanol groups also have a cross-linking action on polymethyl-*H*-siloxanes. Many water-repellency agents are therefore based on a combination of these two types of siloxanes.

For the development of the water-repelling effect, a heat treatment is necessary to promote the cross-linking, the orientation, and the adhesion of the polysiloxane molecules. In the ideal case, a monomolecular layer of silicone deposited on the substrate would be sufficient, i.e., 0.003% by weight of silicone on a cylindrical fiber with a diameter of 20 μ. Under practical conditions, the full effect cannot in fact be obtained with such small amounts of silicone, but 0.05 to 0.1% of silicone is sufficient in the presence of zirconium. Nevertheless, the process is in fact carried out with higher amounts of silicone since this gives better reproducibility of the effects and better fastness to washing.

10.8.1.4 *Commercial Forms of Silicone Water-Repellency Agents*

In the textile industry, the silicones are used either as solutions in organic solvents or as aqueous emulsions. Water-repellency agents in the form of *solutions* generally contain chlorinated hydrocarbons as solvents together with the compounds necessary for curing and fixing.

With *aqueous emulsions*, considerable difficulties had to be overcome because the highly hydrophobic nature of the silicones makes them difficult to emulsify. On the other hand, effective emulsifiers are also wetting agents which partially mask the water-repellency effect. As in the case of paraffin wax products, emulsifiers were sought which lose their activity during the heat treatment, either through reaction with another component or through thermal decomposition.

Anionic emulsifiers of the fatty acid salt type tend to flocculate with the zirconium salt and a polynuclear zirconium complex is deposited on the fiber during the condensation process. When zirconium salts are added to an emulsion containing such emulsifiers there is some danger that the emulsion will break. Techniques have been found for adding the zirconium salt right at the beginning, so that ionic charge adjustments occur before the emulsifying and homogenizing process is completed. Subsequent heat conditioning at 40 to 50°C stabilizes such systems (*279*). Nonionic emulsifiers permit the preparation of excellent emulsions compatible with zirconium salts, but it is impossible to obtain good water-repellency effects with such emulsifiers, since the commercial nonionics are strong wetting agents and exhibit such marked affinity for textile fibers that they can be washed out only with considerable difficulty.

The water-repellency effect depends not only on the emulsifier but also on the type of silicone oil used. It appears that polymethyl-*H*-siloxanes are less sensitive to the wetting effect of the emulsifier than polydimethylsiloxanes.

The charge on the emulsion particles is very important. Highly cationic emulsifiers exhibit affinity for the textile material; emulsions made with them are eminently suitable for batch application, particularly for wool. However, during a continous padding operation preferential adsorption over a relatively long running time reduces the amount of cationic emulsifier in the bath and the emulsion breaks.

10.8.1.5 *Technique of Silicone Impregnation*

10.8.1.5.1 Pretreatment of the fabrics. For the silicone impregnation process to have a good outcome, the fabrics must be thoroughly cleaned before impregnation. Residues of wetting agents, detergents, sizes, padding auxiliaries, etc., may completely nullify the effect of impregnation—particularly in the spray test. In this respect, nonionic auxiliaries are particularly harmful, since they not only adhere firmly to the fabric and cannot be removed completely by any method of treatment, but they have a particularly strong wetting effect. It is therefore essential to avoid nonionic auxiliaries in the preparation of fabrics which are to be silicone-proofed.

10.8.1.5.2 Impregnation of the fabrics. Padding process. The impregnation of piece goods made of cotton, regenerated cellulose, or synthetic fibers, is carried out on a two- or three-roll padder. The concentration of the initial bath and the pickup must be chosen in such a way that about 1.5 to 2% of silicone is taken up by the fabric; for a 25% silicone emulsion, this means a bath concentration of 75 gm/liter when the pickup is 80%. A hardening component must be added to the initial bath (see Section 10.8.1.3).

Commercial silicone oil emulsions are formulated in such a way that they can be used together with synthetic resins for crease-resistant finishing (*198*). Other additives such as optical brighteners, nonslip agents, and softening agents may be added to the silicone baths, but in each case it is recommended to test the stability of the bath and check any modifications of the effects. After impregnation the fabrics are preferably dried first at 80 to 100°C before curing. The time and temperature of curing are of great importance for the water-repellency effect. In general, the impregnated fabric must be cured for about 5 minutes at 140 to 150°C.

A good repellency effect also develops (*170*) when the fabrics are cured at lower temperatures and are then stored for a relatively long period. This so-called post-ripening process depends on the duration and intensity of the preceding drying process.

Batch application. This is used primarily for wool. The wool is impregnated on the winch beck or rope-washing machine. This is a "long bath" process in which exhaustion of the silicone onto the fiber takes place. Emulsions of polyorgano-*H*-siloxanes are particularly suitable, and exhaustion is assured by the use of cationic emulsifiers which give the oily phase a positive charge. For normal woollen cloth *ca.* 4% on the weight of the goods of a 25% silicone oil emulsion is used, and the impregnation is begun at pH 5. After 15 to 20 minutes, ammonium acetate is added and the bath temperature is simultaneously raised to from 30 to 40°C. The lowering of the pH accelerates the exhaustion. Completely exhausted baths are seldom obtained, but silicone wastage is very small.

10.8.1.5.3 Removal of silicones. It is obvious that a finish which largely resists washing and solvent treatment is difficult to remove from the fiber. However, removal may be necessary to correct defects. The removal of a silicone is easier before it has been cured but if curing has already been performed it is feasible to carry out an acid decomposition of the silicone film in the hot, possibly in combination with solvent-containing detergents.

10.8.1.6 *Properties of the Treated Fabric*

10.8.1.6.1 Water-repellency effect and its permanence. The water-repellency figures that can be obtained on various fibers with a commercial silicone water-repellency agent are shown graphically in Fig. 99. From this can be seen the

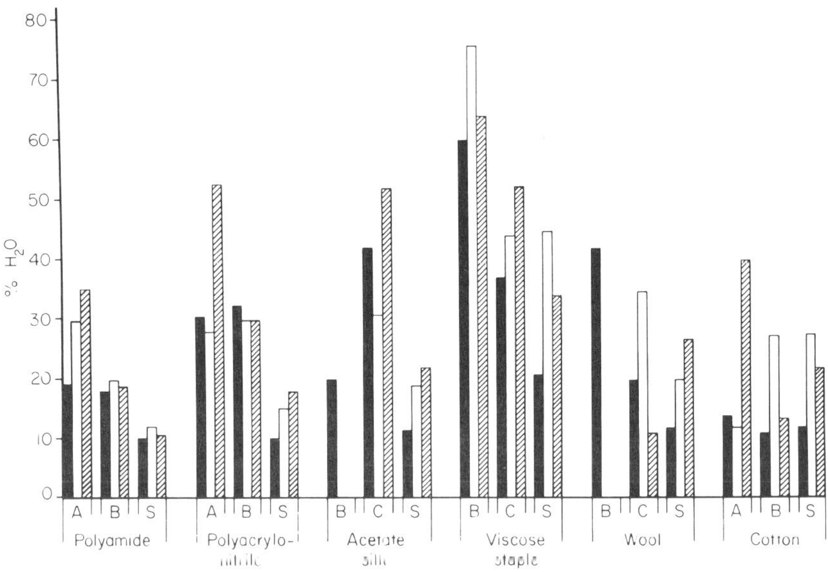

FIG. 99. Water-repellency effects on various fibers. The percentage uptake of water measured by Bundesmann's method has been plotted. A, B, C: finishes with organic water repellents. S: finish with a silicone water repellent. Black columns: initial percentage weight increase. White columns: after a single soap wash. Shaded columns: after a single gasoline wash.

particular importance of the silicone water-repellency treatment for synthetic fibers. Obviously, these figures depend not only on the impregnating agent used but also on the method of application and many other factors. The silicones are relatively fast to washing (Table 106).

It must not be overlooked that these figures depend to a marked extent on the type and amount of detergent used and on the rinsing process. Residues of detergents with a pronounced wetting action mask the water-repellency effect, even when very little silicone has been washed from the fiber. Still more pronounced is the effect of dry cleaning. Here again, hardly any silicone is dissolved off by the solvent, but the effect is modified to a very considerable extent since the detergent additives always present in the cleaning baths are highly hydrophilic compounds which adhere tenaciously to the fiber and mask the water-repellency effect. If fabrics impregnated with silicone water-repellency agents

TABLE 106

WASHING RESISTANCE OF SILICONE WATER-REPELLENCY TREATMENT[a]

Fabric	Amount of water taken up	
	Cotton (%)	Perlon (%)
Untreated	85	55
Water repellent	15	7
After 1 wash	30	15
After 3 washes	39	18
After 5 washes	48	Not determined

[a] Each wash for 20 minutes at 60°C.

TABLE 107

WATER-REPELLENCY EFFECTS AFTER DRY CLEANING AND SOLVENT TREATMENT

No. of treatments	Amount of water taken up	
	After dry cleaning (%)	After treatment with solvent (%)
One	80	23
Three	Not determined	26
Five	Not determined	28

are treated with a pure solvent, the water repellency is retained better (Table 107).

The finishing of acrylic fabrics with silicones to obtain an acid-repellent effect is particularly worth noting, as the resulting fabrics can be used for acid-protective clothing.

10.8.1.6.2 Handle, fiber smoothness, and slip resistance. One of the most important side effects of silicone impregnations is that the goods acquire a soft and supple handle due to the surface smoothness of the individual fibers. It is not possible to obtain the same effect with other water-repellency agents. This fact has contributed greatly to the increasing use of silicone impregnation. A finishing treatment with polydimethylsiloxanes is particularly striking in this respect. In certain cases, however, the resulting smoothness of the fibers is a disadvantage—particularly in the finishing of smooth, light-weight fabrics (e.g., lining materials), the slip resistance of which is greatly lowered. Consequently, the use of a mixture of polymethyl-*H*-siloxanes and polydimethylsiloxanes is recommended, this being a combination which leads to a better adherence of the fibers to one another (*70*).

The simultaneous use of softening agents and silicone water-repellency agents is possible but generally involves an impairment of the water-repellency effect.

10.8.1.6.3 Wet-soil-repelling effect. Fabrics made water repellent with silicones have a repelling effect on soil from aqueous media. However, it is impossible to speak of a soil-repellent effect with respect to dry dust or of an oil- and fat-repellent behavior. In fact, under certain conditions, fats and oils are retained by silicone-impregnated fabrics more tenaciously than by untreated fabrics. On the other hand, sticky substances show a surprisingly low adherence to silicone-impregnated fabrics (*78*).

10.8.2 Use of Silicones for Coating Fabrics

Textile fabrics can be coated with heat-vulcanizing silicone rubber (see Sections 8.1 and 9.6) applied from dispersions or solutions in organic solvents. The fabrics must be singed, cleaned, and smoothed before the application. To apply the silicone rubber mass and coating, machines of the usual kind with air, roller, or rubber blanket spreaders may be used. The viscosity of the coating mass is adjusted with an organic solvent.

For textiles, 20 to 80 gm/m^2 of the coating is necessary according to the purpose in mind, and this can be applied in one operation. Thicker coatings must be applied by repeated spreading operations. Under these conditions, care must be taken to evaporate the solvent between individual spreading processes at a temperature below 90°C since otherwise vulcanization of the lower layer will impair adhesion of the subsequent layer.

Silicone rubber films adhere best to cotton. Their adherence to synthetic fiber fabrics and even to smooth fabrics of regenerated cellulose fibers is inadequate, although it can be improved by the use of various adhesives. An immersion process is usually used for the glass fiber fabrics, the fabric being freed from organic finishes by heating to 300 to 350°C before immersion in a diluted dispersion of silicone rubber.

Silicone rubber coatings exhibit the properties specific for silicones, such as resistance to heat and cold, outstanding aging properties, and resistance to many chemicals and atmospheric agents; their *release effect* with respect to sticky articles is of particular importance. This behavior makes the silicone rubbers suitable coating materials for *liner fabrics* in the rubber industry, where they have already found acceptance (see Section 10.4.2).

10.9 Silicones in the Leather Industry*

Leather is a porous fibrous material; its fibers, irregularly felted with one another, are hydrophilic. In contact with water in liquid or vapor form, leather can take up considerable amounts of moisture with swelling. The degree of water uptake depends on the type of the leather itself, and on its method of preparation. In addition, leather can also absorb water by capillary action and transport it. The water uptake and the water permeability of leather is almost always disadvantageous in practical use. On the other hand, a permeability to water vapor and a certain amount of water-vapor absorption are very desirable for comfort since leather thereby prevents an excessive accumulation of body moisture under the clothing or in the shoes. It is therefore an important problem to find preparations decreasing the water permeability of leather and its uptake of liquid water without impairing its water-vapor permeability and its capacity for taking up water vapor.

The water permeability of leather can be reduced by contents of 15 to 25% of fat. However, leather stuffed in this way has many disadvantages: the water-vapor permeability falls sharply, since all the pores are more or less blocked with fat; the heat-insulating power is greatly reduced; under certain conditions the leather tends to form fatty spews; the leather becomes stiff when cold and tends to give off oil when hot; and the capacity for taking up water vapor is reduced.

The use of silicones on leather was proposed immediately following their success in other fields. Consequently, the industrially available organic polysiloxanes were tested on leather, at a relatively early stage, predominantly in

* Section 10.9 was written by Dr. B. Zorn, Leverkusen, West Germany.

the form of emulsions. These initial experiments were necessarily unsuccessful because polyorganosiloxanes alone, in technically and economically acceptable amounts, give a relatively small water-repellency effect on leather, and ordinary emulsifiers cannot be destroyed on leather sufficiently completely to eliminate their wetting effect; often, moreover, unsuitable leather was used for the impregnation. These difficulties have now been overcome, and silicone-treated leather goods such as shoes, clothing, bags, and technical leather articles are being used to a relatively large extent.

10.9.1 Leather-Impregnating Agents and Leather-Impregnating Processes Based on Silicones and Silanes

As in the impregnation of textiles (see Section 10.8), polymeric siloxanes with or without functional groups and silanes from which, partly by reactions with the fibrous materials, siloxane films are produced can be used for the impregnation of leather. Organoalkoxyhalosilanes (*14*), polyorganosilazanes (*39*, *41*, *42*), and organosilanols or polyorganosiloxanols (*40*, *109*) have been recommended as the starting materials. However, in general such products have not gained acceptance, partly because they give rise to harmful by-products in processing and partly because the effects that can be obtained with them are inadequate.

The silicone products recommended today for the water-repellency treatment of leather contain substituted or unsubstituted polymethylsiloxanes, probably almost exclusively, as silicone components.

They are used either as *solutions* in organic solvents or as *aqueous emulsions*.

The impregnation of leather with liquid polydimethylsiloxanes dissolved in *organic solvents* does indeed, under certain circumstances, give very good results when so-called static methods of testing (*180*, *237*, *297*) are used as the criterion of water resistance, but not with dynamic testing methods in which the leather is moved and worked in contact with water. Such dynamic methods have been described, for example, by Maeser (*196*) ("Maeser Machine"), Kauppi (*160*), and Baumann (*16*) ("Bally Penetrometer"). They simulate the stressing of the shoe upper leather in the walking folds during use; the results obtained with the Penetrometer, at least, are in harmony with those obtained by tests in wear.

Leather found to be satisfactory on dynamic testing is obtained by impregnation with a silicone oil and a mixture of polymethylsiloxanes made from $(CH_3)_3SiO_{1/2}$ and $SiO_{4/2}$ units (*160*). A polymethylsiloxane of this type can be obtained, for example, by the cohydrolysis of trimethylchlorosilane with silicon tetrachloride or ethyl silicate in suitable molar ratios, and is used in organic solution. Samples of leather impregnated in this way and subjected to Kauppi's dynamic test allow a drop of water to penetrate only after 3000 to

10,500 flexures, while this took place after only 500 to 650 flexures with an untreated leather used for comparison.

A further improvement in the water-tightness of leather is achieved by combining polymethylsiloxanes in solution with hydrolyzable titanium compounds, particularly tetrabutyl titanate and also tetramethyl and tetradodecyl titanates (*56*). Components suitable for this process are, for example, mixtures of a polymethylsiloxane copolymerizate derived from $(CH_3)_3SiO_{1/2}$ and $SiO_{4/2}$ units with a CH_3:Si ratio of 1.2:1 and a polymethylsiloxane derived from 97 mole-% of $(CH_3)_2SiO$ and 3 mole-% of $(CH_3)_3SiO_{1/2}$ units. At an optimum ratio of the components, which is about 1 : 1 : 1, samples of leather treated with mixtures of the three components in 75% solution in gasoline allow water to penetrate in a dynamic test only after about 50,000 flexures.

Work has frequently been carried out on the improvement of such systems. The light sensitivity of titanium-containing polysiloxane solutions observed in many cases can be reduced by the addition of aliphatic nitro compounds, e.g., nitroethane or nitrobenzene (*102*). The analogous zirconium compounds, which do not show such light sensitivity, have also been recommended (*139*, *140*).

Most of the silicone leather-impregnating agents on the market today are probably formulated on the basis of mixtures of polysiloxanes with esters of titanic or zirconic acid.

The water-repellency effects can be increased by special measures. Thus, it has been reported that complexes of zirconium with β-dicarbonyl compounds, e.g., acetylacetone, enhance the effect of polysiloxanes on leather (*103*). Organic solutions of polysiloxanes with the condensation products of alkoxides of polyvalent metals and phosphoric acid esters (*130*) have a similar effect, and so have mixtures of polysiloxanes with the reaction products of titanic acid esters and formaldehyde (*220*).

The sweating-out of the silicone oils when the impregnated leather is dried can be counteracted by additions of polyisobutylene, butyl rubber, or styrene-butadiene copolymers (*208*). The use of siloxanes in admixture with phenol-formaldehyde and glycerol-phthalic acid resins (*312*) can impart to the leather an impermeabilizing effect besides a water-repellency effect.

Since the use of solvent-containing impregnating agents requires certain precautionary measures (in view of their combustibility or toxicity) and may necessitate the use of relatively extensive solvent-recovery plants on the large scale, attempts have also been made to use the polyorganosiloxanes in the form of *emulsions*. In general, the emulsification of titanic esters in an aqueous phase fails because of the sensitivity of these products to hydrolysis, but emulsions of the siloxanes in water can be combined with triethanolamine titanate (*266*) or products of the reaction of titanium tetrachloride with polyols and alkanolamines (*317*, *318*). Very recently, it has also become possible, by the application of combinations of solvents and emulsifiers, to use polysiloxane emulsions together with hydrolyzable alkyl titanates in aqueous dispersion without a premature hydrolysis of the titanate taking place (*337*).

On the other hand, the water-repellency effect of emulsions of polyalkylsiloxanes or polyalkyl-*H*-siloxanes may also be improved by combination with Zr salts or mixtures of Zr and Zn salts (acetates) (*206*, *219*). A similar effect is shown by mixtures of polyalkyl-*H*-siloxanes with chromiun(III) complexes of the higher aliphatic carboxylic acids (*171*), or by water-soluble coordination complexes of chromium(III) salts with polyorganosiloxanes containing carboxyalkyl or hydroxyalkyl groups (*101*).

In addition to polymethylsiloxanes, certain silicic esters are capable of imparting water repellency to leather (*338*).

10.9.2 Application and Action of Silicone-Impregnating Agents on Leather

The application and mode of action of silicone-impregnating agents on leather may be illustrated below on the basis of an example.*

The air-dry pieces of leather are briefly immersed—an immersion time of 0.5 to 2 minutes is generally sufficient—in solutions of the impregnating agent in organic solvents, particularly hydrocarbons or chlorohydrocarbons. Immersion gives the most uniform impregnation and the most satisfactorily reproducible results, which is particularly important for laboratory work on comparative tests. With careful working and the application of a sufficient amount, almost equally uniform and effective impregnation can be achieved by application with brushes, rollers, or plush. However, the tendency toward the formation of fatty spots is greater with this process than with the immersion process for many leathers containing a large amount of unbound fat (crude oil, natural fat). The most convenient method of application, spraying, is recommended only with relatively large pieces of leather. The impregnation obtained by this process is generally very superficial, but is sufficient in many cases. Particularly with small pieces of leather, a considerable loss of the impregnating medium may occur by spray missing the target and by spray rebounding. In this case, it is preferable to use higher boiling diluents than with the preceding methods, e.g., those having a boiling range of 150 to 200°C.

After impregnation, the leather is predried in a current of air and is then heated (50 to 60°C).

In principle, all full chrome-tanned leathers and slightly or moderately strongly retanned leathers are suitable for impregnation. However, it must be borne in mind that the leather must possess a structure as dense and fine as possible and that all wetting agents, emulsifiers, loosely bound or nonbound materials (e.g., tanning agents, dyes), and large amounts of fat in the leather

* Xeroderm S-100, Farbenfabriken Bayer AG, Leverkusen, West Germany.

may impair the action of the impregnating agent or eliminate it entirely. Train-oil-containing fatting agents should not be used on any account. Besides chrome leathers, leathers prepared with vegetable (in combination with synthetic) tanning agents can also be impregnated satisfactorily, provided that they satisfy the above requirements. However, higher amounts of impregnating agent are necessary than in the case of chrome-tanned leathers.

There is an optimum amount of impregnating agent for each type of leather and each batch, which it is uneconomic to exceed. If the uptake of water after a given time is plotted as a function of the number of flexures necessary to lead to the penetration of water with various concentrations of the impregnating agent, a curve is obtained (Fig. 100) which shows that the uptake of water does

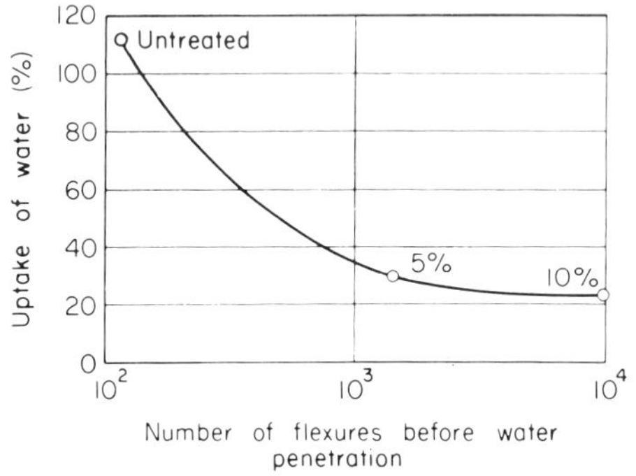

FIG. 100. Example of the uptake of water in per cent of the weight of the leather of a non-dressed boarded leather treated with silicone-impregnating agents of different concentrations (5 and 10%) as a function of the number of flexures.

not fall to zero in proportion to the increase in the concentration of the impregnating agent and the number of flexures but approaches a limiting value. With boarded leather, the optimum concentration of the impregnating agent is between 7 and 10% and with calf between 2 and 3%.

The effect of good silicone-impregnating agents does not fall off even during prolonged storage of the leather.

The air permeability of calf suede leather impregnated with a 10% solution of impregnating agent by immersion does not fall to a measurable extent; the water-vapor permeability falls only by about 10%. As is well known, the air permeability* of boarded leather is markedly depressed when this is dressed with filmlike drying dispersions of organic polymers. Comparative investigations have shown that the air permeability decreased under these conditions by more than 90% as an average of several tests, while an impregnation with a

* Determined according to DIN 53334.

10% solution of silicone-impregnating agent by immersion reduced the air permeability by only about 60%; the water-vapor permeability in each case fell by about 20%.

In combining silicone-impregnating agents and binder finishes, it must be borne in mind that the latter often bring emulsifiers onto the leather which can nullify the effect of the silicone impregnation.

Silicone-impregnated leathers are not only water repellent but they also exhibit a wet-soil-repelling action, can easily be cleaned by washing with water, have a relatively high abrasion resistance, are less sensitive to heat and cold, and retain a good appearance for a longer time.

Silatrioxanes of types (I) and (II) (*83*) and epoxide-group-containing

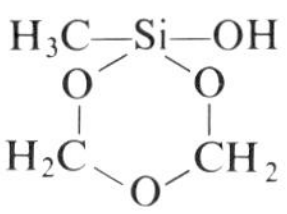

I

and

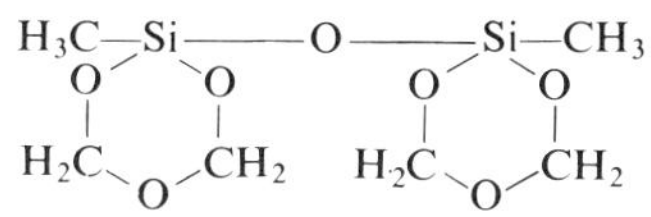

II

siloxanes of formula (III) (*3*) with $n=2$ or 3 act as *tanning agents* on leather.

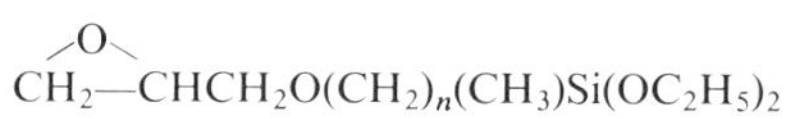

III

However, the leathers obtained have so far shown no decisive technical advantages over those made with conventional agents. It is interesting that heterocycles containing Sn, Si, O, and CH_2 bridges such as

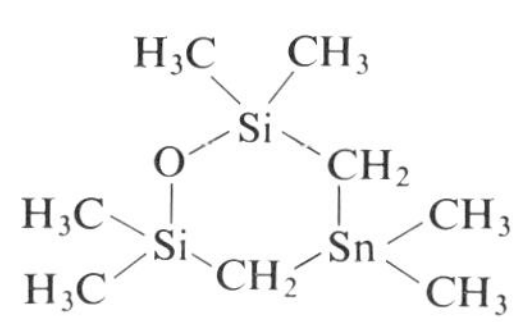

are antibacterial with respect to gram-positive organisms and fungicidal with respect to a causative agent of interdigital mycosis, *Aspergillus flavus*, and others, and are therefore suitable for the antibacterial and antimycotic finishing of leather.

10.10 Silicones in the Paper Industry

In the paper industry, as in the textile industry, silicones are used to achieve water-repellency and nonadhesion effects. The processes by means of which these problems are solved are also similar to those used in the treatment of textiles.

In the very first years of the development of silicones it was proposed to make paper water repellent by treatment with organochlorosilanes (*233*, *243*) or with alkali-metal methylsiliconates (*82*). Since the acid liberated from the chlorosilanes by hydrolysis or the alkali of the siliconates had a deleterious effect on the fibers, attempts were made to counter this danger by rapid neutralization; on the other hand, organoacyloxysilanes (*13*), which give the less aggressive carboxylic acids on hydrolysis, or polyorganosilazanes (*117*), proved to be more suitable for water-repellency treatment. However, these processes lost their interest after it had been learned that good effects could also be obtained with siloxanes, which are more readily accessible and easier to handle.

Paper can be made water repellent or abhesive especially with polymethyl-*H*-siloxanes, which can be fixed on it even at relatively low temperatures; siloxanes with other functional groups, e.g., OH or OR, are also suitable. The best effects are obtained from combinations of these types of siloxanes. A large number of such combinations, which may also contain resinlike components or fluoroalkylsiloxanes, have been described, e.g., refs. *161*, *162*, and *188*.

The silicones are applied either from solutions or from aqueous emulsions, but do not draw substantively onto the paper. If siloxane emulsions containing the oleates or stearates of morpholine or triethanolamine as emulsifiers are applied to paper previously treated with aluminum triformate, the emulsions are rapidly broken (*267*).

The usual curing catalysts, such as the zinc, tin, or lead salts of carboxylic acids, or aluminum chelate compounds, accelerate the drying, improve the adhesion and, particularly important with release papers, prevent the migration of silicone oil into the adhesive layer. The addition of boron compounds is particularly effective in the last-mentioned sense (*59*).

Suitable papers are kraft papers, parchment, artificial parchment, and glassine; because of their smooth surface, the latter are preferred for the preparation of release papers. Sizing does not modify the silicone effect. Impregnation is carried out either by immersing the paper web in dilute solutions or emulsions of the silicone preparations and, if desired, squeezing out the excess with rolls or by spreaders. As usual, the time and temperature of condensation are in inverse relationship to one another. In continuous operation, most preparations require a condensation time of 1–2 minutes at 120–150°C, though recent developments have made it possible to reduce the time to 15 seconds at 120°C.

An amount of 0.5 to 2 g of silicone per m^2 is necessary to obtain a sufficient water-repellent and release effect, depending on the porosity of the paper. In view of the cost relationship between silicone and paper, this markedly increases the cost of the latter, so that silicones have not yet replaced paraffin waxes completely as general water-repellency agents in this field in spite of their numerous advantages (complete absence of odor and taste, retention of the nature of the paper).

On the other hand, there has been considerable interest in the fact that such papers can be used as release materials. They are used particularly as covering papers and interlayers for adhesive foils and tapes; even tenaciously adhering films are readily detached from silicone-treated paper, without—when the silicone treatment has been carried out correctly—loss of adhesive power even after long contact with the silicone. Silicone-impregnated papers are used in many cases as release papers to prevent baked goods from sticking to the tins; however, in this case their use is restricted to materials with a low content of sugar. The application of silicones protects the paper in deep-freeze packages from the melting ice water and prevents the contents from freezing to it. Similarly, silicone-impregnated papers have repellent properties with respect to unvulcanized rubber, asphalt, pitch, waxes, electrical insulating tapes, adhesive plaster, meat, and many other substances, and therefore form an ideal packaging material which makes it possible to remove the contents easily and completely.

Special auxiliaries are today available for the gluing of silicone-impregnated papers (*311*).

If thin tissue paper is impregnated with liquid polymethylsiloxanes (CH_3:Si ratio = 1.75 to 2.25), in such a way that the siloxane content is between about 5 and 12%, a silicone film can be transferred to glass with this material (*47*).

Side by side with attempts to apply the typical effects of classical silicones, especially their abhesivity, to paper treatment, more recent attempts have also been made in the opposite direction: to make paper surfaces adhesive by means of organofunctional silicon compounds. The situation is quite similar to that relating to glass (Sections 10.7.2 and 10.7.3). Thus, it has been found that polyaminoalkylsilanes, particularly

$$H_2N—CH_2—CH_2—NH(CH_2)_3—Si—(OCH_3)_3$$

when applied or hydrolyzed onto paper, make it possible to line the paper with a polyethylene foil (*261*).

10.11 Silicones in the Cleaner and Polish Industry

Cleaners and polishes for the treatment of paints, metals, wood, or plastics have hitherto contained as the active principle a wax component which is distributed over the surface of the object in the form of a film by rubbing. Owing to their capacity for giving very thin films by rapid and uniform spreading, their water-repelling action and weather resistance, their resistance to resinification, and their general chemical and physiological inertness, silicone oils appear to be particularly suitable for the production of such protective films; in fact, they were very quickly accepted in the cleaner and polish industry. Today, they are sometimes used instead of waxes (but generally together with them) to make very effective preparations for the cleaning and preservation of car paint, furniture, metals, glass, and floor coverings.

The chief advantage of silicone-containing cleaners and polishes over preparations based on wax is that because of the spreading power of the silicone film a uniform coating of the surface and a high gloss are obtained quickly and with very little effort. Since the silicone film is effective even in extremely thin layers, the amount necessary is extremely small, and therefore the general nature of the surface remains unchanged when the polishing process is carried out correctly. Surfaces treated with silicone preparations are highly water repellent and can be washed and cleaned more easily than untreated surfaces, probably because the dust particles lie only loosely on the silicone film. Silicone films applied from polishes possess no repellent effect against dry dust.

The observation that polishes containing silicone and wax can be applied very much more satisfactorily than preparations containing wax only is rather remarkable. The phenomenon is probably due to the fact that the silicone acts as a slip agent for the wax crystals, and that these can be smeared and distributed over a film of silicone oil without difficulty. Immiscibility of the silicone oil and the wax are prerequisites for this effect to occur.

The commercial preparations are either solutions of the silicone oil and the wax in organic solvents or emulsions of such solutions of the oil-in-water type. The majority of the preparations are designed for the care and maintenance of car lacquers. Pastes are also marketed for special purposes. Many preparations simultaneously contain highly dispersed fatty materials that act as slip agents and facilitate the cleaning and freshening up of the surface. Suitable formulations have been given repeatedly and have been patented; here a reference to some of the first patent specifications may suffice (*11*, *50*, *55*, *97*, *306*). In detail, the recommendations differ partly in accordance with the nature of the silicone oil used (polydimethylsiloxane oils with viscosities between 100 and 1000 cSt are preferred), the type of wax component (carnauba wax is recommended predominantly) or other additives, and the choice of the emulsifiers for the emulsion types.*

Surfaces that have been treated with preparations containing polydimethylsiloxane oil can, in general, because of the release effect of the silicones, be coated with adhesive films only after the film of silicone oil has been effectively

* For systematic investigations of the influence of the silicone and wax components on the quality of the polish see refs. *211* and *313*.

removed. This must be borne in mind particularly in the respraying of a car body that has been treated with silicone preparations.

When using silicone cleaners and polishes, it should be remembered that it is not in general the purpose of a surface treatment to make the surface water repellent. For example, in contrast to earlier ideas (*271*), experience has shown that automobile windshields should be hydrophilic so that the water runs off them in a uniform sheet which can be easily removed by the wipers. Consequently, special preparations, generally containing wetting agents, which permit the unavoidable water-repelling film of oil and dirt to be removed, are suitable for cleaning windshields.

10.12 Silicones in the Protection of Masonry*

10.12.1 Influence of Moisture on Building Materials

Most silicate building materials are more or less porous, as a result of which they are relatively good thermal insulators and permit the exchange of gases necessary for a healthy living climate. The moisture of living rooms is transported to the exterior through the capillaries, generally in the vapor phase, but under very unfavorable conditions also by capillary migration. On the other hand, it is a natural disadvantage of porous building materials that when they are subjected to prolonged wetting from the outside, i.e., by rain, an inward transport of liquid may take place. If, however, the pores of the material are filled with water, the heat-insulating effect is impaired, and this effect increases with increasing porosity. In addition, once it is taken up, water evaporates only very slowly, and the heat of evaporation is taken from the wall material. In order to retain a sufficient heat-insulating effect, therefore, greater wall thicknesses are often used than would be necessary on static grounds (*214*).

Prolonged wetting leads to the moisture damage familiar with new buildings. The external walls are highly stressed by the alternation of wetting and drying, and the "working" of the outer wall zones during swelling and drying gives rise to stresses which can produce fissures and splits in the materials, which are generally very brittle. This increases the absorption capacity still further, until severe damage occurs (*183*). Frequent wetting also accelerates soiling of the buildings, since dust particles drawn in with the water float into the pores and fissures and become anchored there.

The above difficulties can be substantially avoided by preventing the ingress of moisture from the outside without suppressing the outward transport of moisture formed inside. It is frequently proposed to add to the mortars or plaster mixtures materials which have a water-repellent effect but do not block the pores and therefore do not obstruct the exchange of gases (*184*, *248*, *345*).

* Section 10.12 was written by Dr. W. Krauss, Leverkusen, West Germany.

Silicones satisfy these conditions ideally. Because of the very low surface tension of the silicones and the resulting spreading power, silicone films applied to porous substrates become distributed over the pore walls without sealing the passages. Air-permeability measurements, which will be reported in detail in Section 10.12.3.2, clearly show that a silicone treatment does not change the pore cross section of a material. A further advantage is that, because of the chemical relationship between silicates and silicones, a much better anchoring of a silicone film on a silicate carrier material is achieved than is the case, for example, with purely organic water-repellency agents such as paraffin wax and oils.

There are at present two methods, differing in their principle, available for making building materials water repellent by means of silicones:

1. Production of a silicone resin film by the application of (a) an aqueous solution of sodium or potassium methylsiliconate, or (b) a solution of a silicone resin dissolved in organic solvents.

2. Mixing of the solid plaster constituents with an insoluble methylsilsesquioxane powder.

10.12.2 Suitable Products

10.12.2.1 *Water-Soluble Organosilane Derivatives*

In principle, the mode of action of these products is based on the production of a water-insoluble silicone from a water-soluble monomeric compound after the evaporation of the water by reactions at the normal temperature. In practice, the only products used today are alkali-metal methylsiliconates (alkali-metal methylsilanolates), e.g., the sodium salt to which the formula

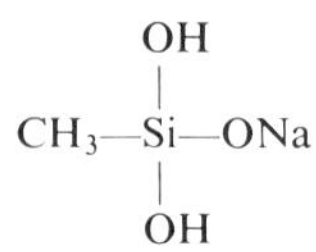

can be ascribed only as a simplification; in actual fact, because of partial intermolecular condensation, siloxanes such as

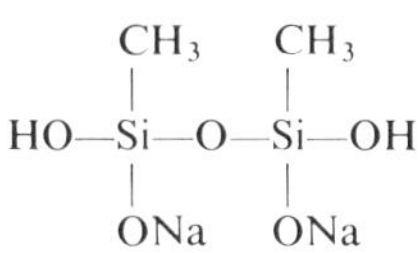 or 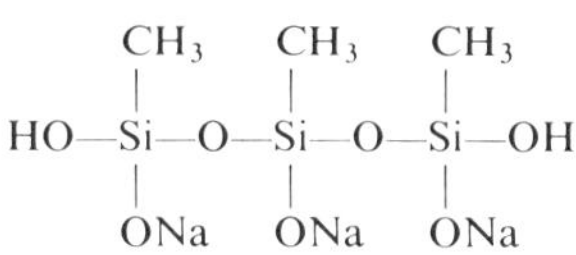

are also present. The addition of acids to aqueous solutions of these salts gives rise to methylsilanols which change spontaneously into polymethylsiloxanes by intermolecular condensation. Atmospheric carbon dioxide acts in this direction:

$$2CH_3Si(OH)_2ONa + CO_2 + H_2O \rightarrow 2[CH_3Si(OH)_3] + Na_2CO_3$$

$$n[CH_3Si(OH)_3] \rightarrow (CH_3SiO_{3/2})_n + 3/2nH_2O$$

The sodium carbonate is washed out almost completely by the first rain, so that there is no risk of efflorescence caused by the carbonate. Potassium salts have the advantage that the potassium carbonate arising in reaction with the carbon dioxide of the air is hygroscopic and can be washed out still more easily, but their silicon content is lower than that of sodium salts.

All the advantages and disadvantages of water-repellency treatment with water-soluble methylsiliconates can be deduced from the chemistry of the formation of a hydrophobic silicone layer.

An advantage of importance in practice is certainly the solubility of the siliconate in water; even a still-damp substrate can be treated with the aqueous solution. A disadvantage is the relatively slow reaction with carbon dioxide, since the applied material remains water soluble for some time and can therefore be washed out by rain. The reaction with CO_2 is suppressed almost completely if free lime is still present simultaneously, since calcium hydroxide traps the carbon dioxide. A pre-formed layer of polysiloxane can be decomposed by calcium hydroxide with the formation of calcium methylsiliconate, which can be washed out by rain even though its solubility is low (*250*). Consequently, on a lime-containing substrate, effective siliconization can extend only to the depth of the surface zone in which a carbonation of the free calcium hydroxide has taken place. This can be seen on fresh fracture surfaces of wetted lime cement mortar prisms in Fig. 101 (*231*).

In contrast to sodium methylsiliconate, a silicone resin passes this boundary and remains active even in the presence of free lime. The decomposition of a polysiloxane layer by calcium hydroxide therefore appears to be limited to the lower polymers, such as the first condensation products of methylsilanetriol. Since carbonation increases with increasing age of the lime-containing material, the action of alkali-metal methylsiliconates on such substrates rises with increasing time of aging before impregnation. In practice, setting times of 2 to 3 weeks appear to offer a sufficient guarantee of adequate carbonation of the surface zones (*34*, *121*, *178*).

The waiting time before impregnation can be substantially shortened by pretreating the substrate with lime-binding agents. Particularly recommended for this purpose are hexafluorosilicic acid and its salts (*350*), and also soluble carbonates, especially ammonium carbonates (*231*). With this artificial aging, however, it must not be forgotten that in addition to the hardening of the lime there is, in the former case, a separation of colloidal silica which may have a pore-blocking effect (*66*, *270*).

The solubility of the siliconates only in water is another disadvantage, since they cannot be used for the further water-repellency treatment of a material that has been previously treated and is therefore already water repellent.

In principle, the reaction products of alkylalkoxysilanes with ethylene glycol (*173*) or with hydroxyl-containing polyethers (*150*) are suitable as nonalkaline impregnating agents to be used in aqueous solutions. By heating with ethylene glycol, methylsilsesquioxane can be converted into a water-soluble product which is also suitable for use in impregnation (*174*).

FIG. 101. Relationship between the depth of penetration of silicone-impregnating agents and carbonation of the lime shown on fracture surfaces of mortar blocks. Center: Non-impregnated sample; the lime-containing core has been stained with phenolphthalein and appears dark, the border shows the CaO-free zone. Left: Impregnated with sodium methylsiliconate and wetted; the nonwetted edge part corresponds approximately to the thickness of the CaO-free zone. Right: Impregnated with silicone resin solution and wetted; the nonwetted edge zone is wider than the CaO-free zone.

10.12.2.2 *Impregnating Agents Soluble in Organic Solvents*

Hydrophobic films on building materials can also be produced with substantially precondensed polyorganosiloxanes; such products are soluble only in organic solvents and must be applied in them. The formation of the hydrophobic film no longer depends on the occurrence of chemical reactions but only on the rate of evaporation of the solvent used, so that the activity of these impregnating agents is much less affected by external conditions than is the case with the alkali-metal methylsiliconates.

After the solvent has evaporated, the silanol groups still present in these resins condense eventually to give new siloxane bonds, so that the resin film formed becomes continuously harder. Particularly rapid hardening occurs with polymethylsiloxanes containing Si—OCH_3 groups (*172*); the use of even a proportion of these more active resins with other components can be advantageous (*15*) and so can a combination with polypropylsiloxanes containing ethoxy groups (*207*).

The fact that an incompletely condensed silicone resin can give a silicone film with completely satisfactory properties immediately after evaporation of the solvent appears to contradict what has been said on the properties of the silicone resins with respect to paints and lacquers (Sections 8.2 and 9.4), but it must be borne in mind that the silicone films produced on building materials are extremely thin—perhaps only a few layers of molecules thick. Since, moreover, the silicone resins used here are highly cross-linked, they dry rapidly even at the normal temperature to nonsticky invisible films.

Silicone-impregnating agents dissolved in organic solvents not only have the advantage of the very rapid formation of a hydrophobic film but also the advantage that their action is less affected by the presence of free calcium hydroxide; in contrast to the aqueous siliconate solutions, they can be applied repeatedly to a substrate previously treated with silicones; the new solvent redissolves the film present in each case and transports it deeper into the material, depending on the amount of new material applied. The disadvantage of these products is the restriction to organic solvents. Attempts have been made to overcome this problem by using aqueous emulsions, but such emulsions are very rapidly broken at the surface of plasters. Thus, any action in depth is lost and the danger exists that the hydrophobic film will be destroyed if the surface is mechanically damaged to a slight extent. Consequently, emulsions are little used as protective agents for buildings, but they are used for the water-repellency treatment of glass fibers, slag wool fibers, and asbestos fibers (cf. Section 10.12.4).

10.12.2.3 *Solid Products*

Mortar and plaster mixtures can be made very effectively water repellent by the addition of solid highly disperse methylsilsesquioxanes (*227*, *228*). In contrast to the methods of application of silicone-impregnating agents mentioned above, the water-repelling effect exists even during the setting of the plaster. This may affect the transport of dissolved salts, as shown in certain circumstances by an increased tendency to effloresce which is of course particularly noticeable on dark plasters. As far as can be seen, this phenomenon occurs particularly at low temperatures and high humidities, so that when the solid silicone products are used the conditions of processing for plaster mixtures must be observed with particular care.

Extremely finely divided *methylsilsesquioxanes* are obtained by spraying methylchlorosilanes into a moist atmosphere (*62*). As can be seen from Fig. 102 the primary particles have a spherical shape and an average diameter of 0.2 to 0.5 μ. In itself, methylsilsesquioxane does not differ essentially in its chemical constitution from the polymethylsiloxanes described in the preceding section, but here the condensation has proceeded so far that the product is insoluble in organic solvents. Correspondingly, its resistance to chemicals (e.g., calcium hydroxide) is higher, so that direct processing together with noncarbonated lime as is required by its mixing in plaster becomes possible. In contrast to the film-forming liquid impregnating agents, the addition of highly disperse

FIG. 102. Electron micrograph of the primary particles of a highly disperse methylsilsesquioxane used as an additive for mortar and plaster.

methylsilsesquioxane gives particles of silicone distributed uniformly through the whole depth of the plaster layer without the formation of a closed film. However, the number of silicone particles per unit area is so high that the same effect as that of a closed film is obtained with respect to drops of water. This is responsible for the important requirement that silicones must be distributed very carefully in the plaster mixture and that a powder with the smallest possible primary particles should be used (*228*).

10.12.3 Physical Properties

10.12.3.1 *Water Repellency*

In capillaries immersed in liquids, the level of the liquid rises or falls according to the magnitude of the contact angle to an extent given by the formula:

$$H = \frac{2\sigma}{d \cdot r} \cos\theta$$

where H is the height of rise, σ the surface tension, d the density of the liquid, and r and θ denote, respectively, the radius of the capillary and the contact angle (cf. Section 9.1.6). For $\theta > 90°$, the rise is negative, and the capillary suction therefore turns into a capillary "pressure". In this case, before the liquid can penetrate into the capillary it must overcome a force which is directly proportional to the contact angle and inversely proportional to the capillary radius. The essence of the water-repellency treatment consists in coating the surface of the material with the thinnest possible layer which increases the contact angle with respect to water to such an extent that H in the above formula becomes negative. Since water forms a contact angle of up to about 105° at the silicone and air interface (Section 9.1.6), the treatment of porous building materials with a silicone product eliminates the capillary suction and in fact actually gives an "anticapillarity," preventing the entry of water into the pores.

The most interesting question now is whether driving rain beating against a water-repellent wall with sufficient force is capable of penetrating into the pores and therefore whether the "anticapillarity" achieved by siliconization of the capillaries present in the various building materials is sufficiently great. Heavy rain supplies large amounts of water in a very short time, so that even the most porous material cannot absorb it immediately and the excess runs off on the outside of the wall. This curtain of water is now acted upon simultaneously by the wind, so that the wall surface and therefore the pores are under a continuous pressure of water. From the wind pressure formulas of DIN 1055 (*344*), we obtain for some wind velocities the surface pressures given in Table 108, when the wind blows at right angles to the surface.

In order to include the most severe conditions, laboratory tests are generally carried out with a pressure of 5 to 10 cm of water, and the uptake of water of the material is measured as a function of time. Very good differentiation is also possible by determining the amounts of water absorbed by test specimens immersed to a depth of 3 mm (*217*).

TABLE 108

SURFACE PRESSURE OF THE WIND[a]

Wind velocity (km/hr)	Surface pressure at normal incidence (cm of H_2O) (approx.)
50	1.2
102	5.0
140	9.5

[a] According to DIN 1055.

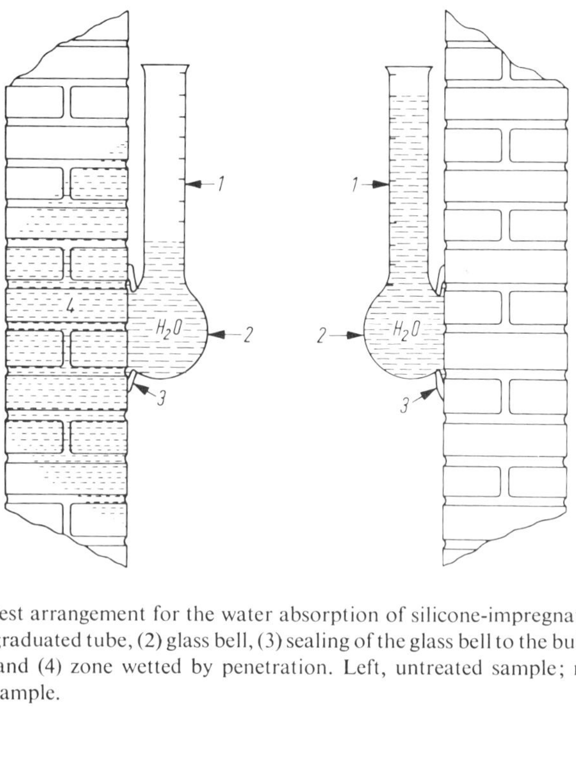

FIG. 103. Test arrangement for the water absorption of silicone-impregnated plaster (cf. Fig. 107): (1) graduated tube, (2) glass bell, (3) sealing of the glass bell to the building material to be tested, and (4) zone wetted by penetration. Left, untreated sample; right, silicone-impregnated sample.

The following method, which can also be used on the finished structure, is very suitable for approximate tests of the water repellency of a material in the laboratory and in practice (*120*, *158*) (Fig. 103). A glass bell provided with a graduated tube is cemented to the material to be investigated and is then filled to a height of 10 cm. The amount of water that must then be added to keep the

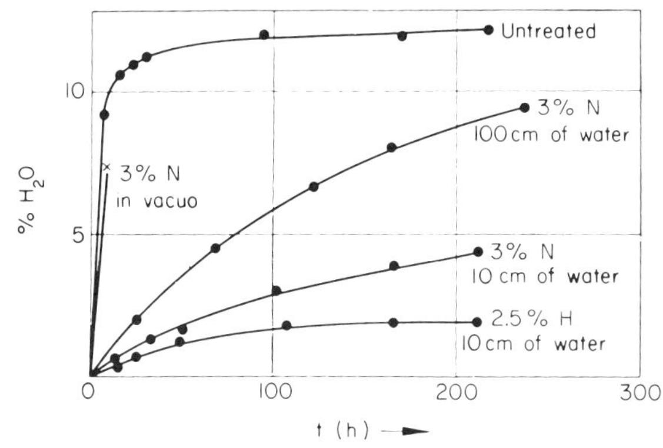

FIG. 104. Rate of water uptake by silicone-impregnated bricks at various pressures (cm of water). The percentage figures refer to the silicone content of the impregnating solution used; N=solution of sodium methylsiliconate, H=resin solution.

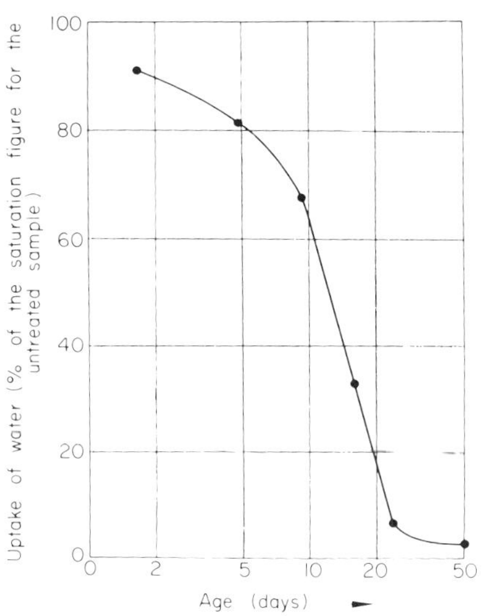

FIG. 105. Activity of sodium methylsiliconate on a lime-containing substrate as a function of its age after setting.

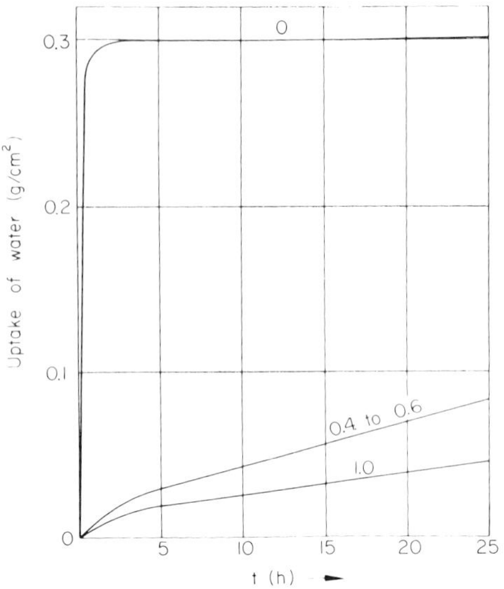

FIG. 106. Uptake of water of a plaster, made hydrophobic with methylsilsesquioxane powder, at a pressure of 5 cm of water. The figures on the curves give the amount of methylsilsesquioxane added in per cent on the dry plaster.

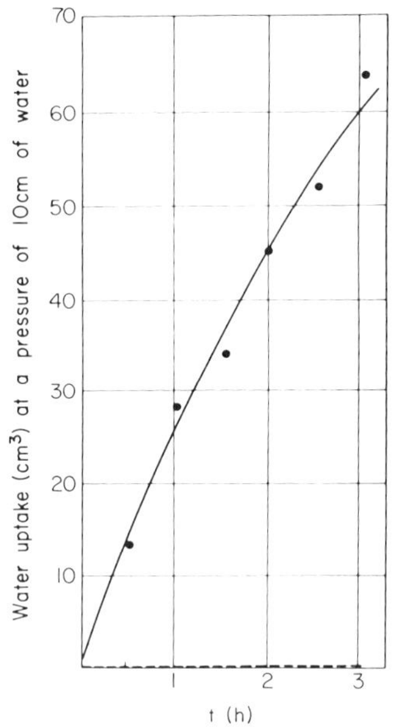

FIG. 107. Water uptake of a plaster five years old impregnated with resin solution, measured with the device according to Fig. 103: (——) untreated; (– – –) siliconized.

level of the water constant is a measure of the absorption power of the investigated material.

Figures 104 to 107 show graphically the results of some tests of the water absorption of various building materials and plasters. It can be seen from Fig. 104 that the hydrophobic effect of silicone impregnation falls off markedly when a certain pressure of water is exceeded and, as was to be expected, when very high pressures are used it can no longer be detected. Consequently, foundations or water containers cannot be sealed with silicone products. Figure 104 also illustrates a superiority of the silicone resin over sodium methylsiliconate with respect to the hydrophobic effect, which is often observed and which may perhaps be ascribed to the greater condensation of the resin, or to soluble reaction products (Na_2CO_3) of the salts which have not been washed out completely.

Figure 105 shows the rise in the hydrophobic effect with increasing age of the lime-containing substrate coated with sodium methylsiliconate. According to this, no reduction of sodium methylsiliconate activity by free lime is to be feared after about 20 days.

The water-repellent effect of a solid methylsilsesquioxane on a plaster mixture is shown in Fig. 106. When this product is used, no coherent film is formed; the separate particles of silicone form a "hydrophobic network" which, with sufficiently small meshes, has the properties of a coherent silicone film. It can be seen that the particle network is quite closely linked when as little as 0.4 to 0.6% is added to the dry plaster.

The measurements in Fig. 107 were obtained as in Fig. 103, on a facing plaster five years old that had been partially impregnated with a resin-based silicone-impregnating agent.

10.12.3.2 *Permeability to Air and Water Vapor*

In addition to a good water-repellency effect, it is required that an impregnating agent should not, or not substantially, affect the air and water-vapor permeabilities of building materials. It has been shown by various measurements that silicone impregnation does not reduce the permeability to air (*231, 250, 323*).

The water-vapor permeability is generally determined in accordance with DIN 53 122 (*347*). The sample to be investigated is cemented onto a test vessel filled with a drying agent. The unit is then brought into a constant-humidity room. The vapor transmission can be easily calculated from the increase in the weight of the unit and the surface area of the sample. A modification in which the test vessel is filled with water, stored in air of defined humidity, and the decrease in weight is measured, seems to be more in accord with practice (*217*).

Impregnation with silicone films obtained from an aqueous phase does not affect the water-vapor permeability. However, very variable results were obtained when solid polysiloxanes in powder form were added to the mortar (*181*, *217*).

Here, the drying of the fresh plaster apparently exerts an essential influence; the results are better for samples in which part of the making-up water was removed by suction—the fresh plaster being applied to an absorbent substrate, such as is always the case in practice—than for samples from which all the making-up water had to be removed by evaporation.

The rate of drying out for moist building materials, which is important for the removal of the moisture of the materials and moisture due to living conditions, is closely related with the water-vapor permeability. It has been shown in experiments on lime-sand bricks (*323*) that only in very few cases—particularly with materials having a very high moisture content in the saturated state—is the rate of drying out appreciably reduced by a silicone treatment. Such states of saturation are seldom reached in practice.

In the case of building materials which store a very large amount of moisture and therefore exhibit very low rates of drying as compared with their moisture capacity, the moisture content continuously increases with prolonged treatment with water. If such a material is impregnated with silicones, the entry of water is prevented and at the same time drying from the inside out is not affected. An example of this is given by results obtained with Ytong bricks weathered for seven months (*181*).

10.12.3.3 *Prevention of Efflorescence*

Since silicone impregnation suppresses the capillary transport of water to the surface of a building material, no salts can be transported from the interior of this material to the outside, so that efflorescence is suppressed. This phenomenon can be shown very strikingly (Fig. 108) if one half of a brick is treated with a silicone-impregnating agent and the brick is then placed in a salt solution. In the nonimpregnated half, the salt solution is sucked up copiously, migrates in the direction of a concentration gradient to the surface, and there deposits the salt in the form of a crystalline layer when the water evaporates, while the treated side of the brick remains unchanged. The prevention of efflorescence is particularly important in the case of decorative masonry.

10.12.3.4 *Improvement of the Frost Resistance of Building Materials*

In the final analysis, the damage caused to building materials by frost depends on their capacity for taking up water, which is sometimes very high. It must therefore be assumed that a silicone treatment which prevents, or at least greatly reduces, the uptake of water will also bring about an improvement in the frost resistance (*34*). Experiments carried out in accordance with DIN 105 (*343*) and measurements on bricks led to the results shown graphically

FIG. 108. Prevention of efflorescence on a brick by means of silicone treatment: left-hand side, untreated; right-hand side, siliconized.

in Fig. 109. In evaluating the damage, a distinction was made between small fissures (hair fissures), large (gaping) fissures, and complete disintegration.

Similar trials on roofing tiles (*20*) have shown that here again their sensitivity to the action of frost is reduced by a suitable silicone treatment. The extent of the effect depends not only on the nature of the impregnating process but also essentially on the structure of the tile. Under extreme conditions (e.g., under a very high pressure of snow) the results may in certain circumstances be reversed (*131*), or other disadvantages such as the formation of water of condensation under the tiles may occur (*5*).

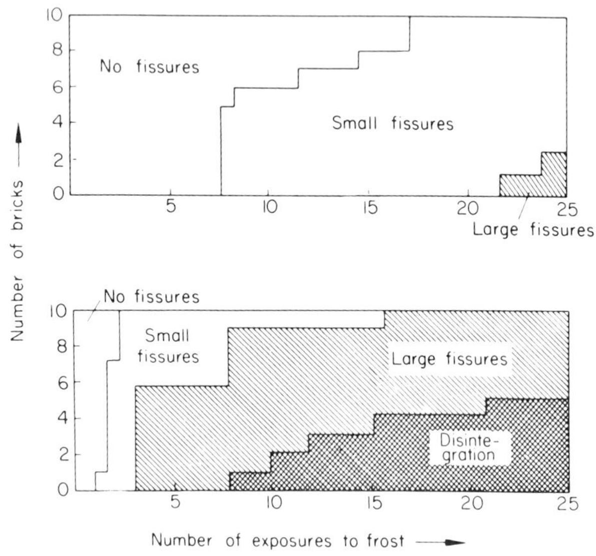

FIG. 109. Frost resistance of bricks: top, bricks impregnated with sodium methylsiliconate (time of immersion 60 seconds); bottom, nonimpregnated bricks.

10.12.3.5 *Dust and Soil Repellency*

Silicone films have a repelling action against moist dirt and soil, as can be seen from Fig. 110. Here an experimental plate of facing plaster was impregnated in checker fashion with a polymethylsiloxane and was then exposed to a very dusty industrial atmosphere for a year. The nonimpregnated areas are much more severely soiled than the impregnated areas. The effect is probably due to a less firm anchorage and smaller depth of penetration of the soil particles, and to the fact that they can be washed off more easily by rainwater.

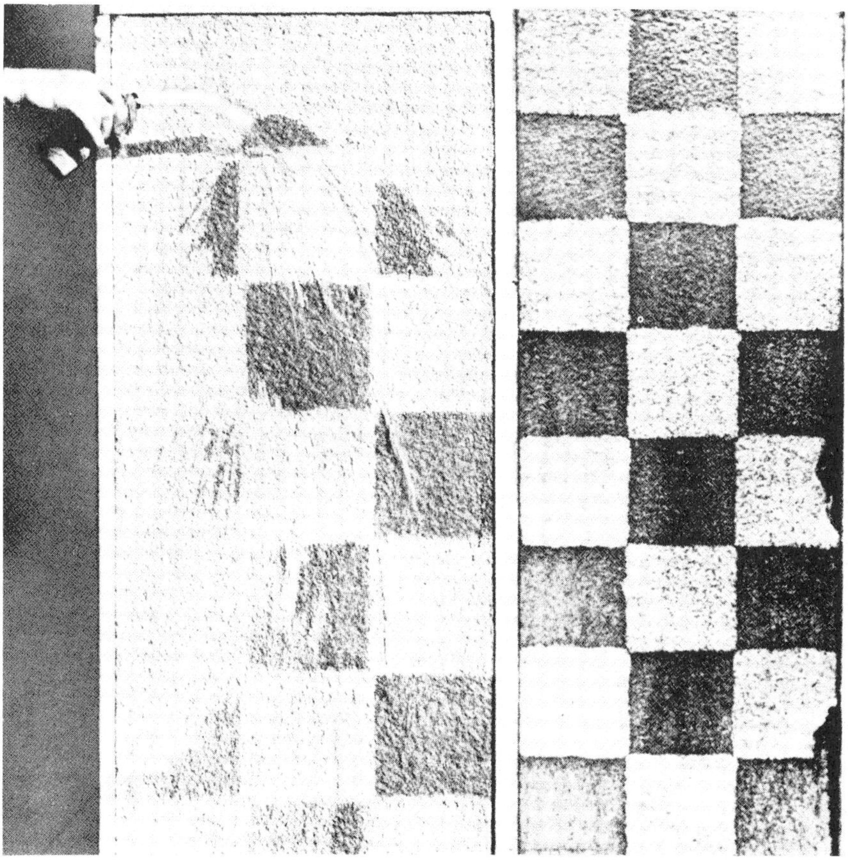

FIG. 110. Dust and soil repellency of silicone-impregnated materials: left, facing plaster sheet impregnated in checker fashion wetted to show siliconization; right, the same, dry, after exposure to an industrial atmosphere for 1 year.

10.12.3.6 *Durability of Silicone Impregnations*

The question of the durability of a silicone impregnation cannot be answered precisely. Data on this point (*6, 350*) vary between 5 and 10 years. The degree of durability depends primarily on the conditions during processing. A silicone layer penetrating into a material to only a small extent is more highly exposed to mechanical degradation, and such a layer therefore quickly loses its activity. Layers which are too thin arise when excessively diluted solutions of impregnating agents are used or when $Ca(OH)_2$-containing substrates are impregnated. Insufficient penetration must also be feared when excessively damp walls are impregnated. According to experience so far, the durability of an impregnation with silicone resins appears to be better than one with sodium methylsiliconate, probably because the sodium methylsiliconate or its low-molecular-weight hydrolysis products may be subjected to too many harmful influences before they have reached a state of condensation sufficiently resistant to the action of water and chemicals.

In the evaluation of the durability of a water-repellency treatment it is important to use as a criterion not only the formation of beads by water but also and primarily the uptake of water, which can be tested, for example, by the tube test (Section 10.12.3.1). The bead effect can disappear after only a few weeks in a highly contaminated atmosphere or can be destroyed by mechanical damage to the surface of the material.

10.12.4 Further Possibilities of Application

In the previous section, the various silicone-impregnating agents were discussed primarily as water-repellency agents for masonry, plaster, and roofing tiles; however, their range of application is much greater, and extends, strictly speaking, to all materials that have a certain degree of capillary structure and are therefore absorbent. Such materials include natural and artificial stone, slag wool, and asbestos cement products.

The water-repellency treatment of *asbestos cement sheets*, which are being increasingly used for roofings or as walling materials, is particularly important in practice. The main problem here is less the achievement of good water repellency than the suppression of efflorescence, which is particularly deleterious for colored sheets. This object is achieved by treating the surface with a mixture of polysiloxanes and polymethacrylic esters (*7*, *94*). Of the pure silicones, amylmethylsiloxane has been found to be particularly effective on asbestos cement (*118*) while methylsilicones alone gave unsatisfactory effects.

Good results have been obtained from the water-repellency treatment of concrete floors, bridge ramps, etc. (*260*). This improves the resistance, particularly the frost resistance, of the layer of concrete. The waterproofing of the silicone-impregnated concrete remains almost unaffected after heavy rain, while with untreated concrete it falls to half (*36*). Furthermore, concrete floors treated in this way are easier to free from adherent ice.

10.13 Silicones in Medicine, Pharmacy, and Cosmetics*

The properties of the silicones, particularly their physiological inertness, which has been outlined in Section 9.7, have also interested physicians in this class of materials. Methylsilicone oils and methylsilicone rubber have proved

* Section 10.13 was written by Prof. G. Hecht, Wuppertal-Elberfeld, West Germany, and supplemented by the author.

especially valuable as auxiliaries in medicine, and also in pharmacy and cosmetics.

10.13.1 Dermatological Applications including Cosmetics and Industrial Protection

Methylsilicone oils are used to a large extent as components of ointment bases and as water-repellent principles of ointments and pastes for the industrial, cosmetic, and clinical protection of the skin. The reasons for their favorable acceptance are their excellent compatibility with human skin, the ease with which they form extremely thin films, and the highly water-repellent properties which they impart to the substrate even when used in very small amounts.

The inertness of these substances with respect to skin shown by the animal experiments sketched in Section 9.7 has been confirmed in extended clinical investigations (*141*, *305*). It is repeatedly stated that in this respect silicone oils are superior to most ointment bases used previously, particularly paraffin oil and petroleum jelly.

10.13.1.1 *General*

The satisfactory results on the industrial and clinical protection of the skin were the starting points for the introduction of silicone oils into cosmetics. At the present time, silicone oils are used in agents for skin care and in protective skin creams. They have also been proposed as additives for nail varnishes, lipsticks, powder-creams, and anti-sunburn preparations, performing in each case the function of a water-repellent agent (*189*, *193*, *322*). This applies particularly to anti-sunburn preparations, since silicone oils have no specific uv absorption (*334*, *335*). Silicone oils may also be incorporated in toothpastes and other preparations for the care of the teeth and give the teeth a water- and stain-repellent coating (*77*) which also preserves astringent and antibacterial components on the gums (*189*). Because of their hydrophobic properties, they are also of great interest for preparations designed to keep the hair in place, regardless of humidity. However, the results obtained so far show that this problem cannot be solved with normal polydimethylsiloxane oils, but that the silicone must be fixed on the hair by means of functional groups. Some proposals have been made (e.g., refs. *38* and *300*), but the problem cannot be regarded as finally solved.

10.13.1.2 *Silicone-Containing Ointment Bases*

In ointment bases, silicone oils are combined with other components such as fatty alcohols, polyethylene glycols, and polyethyleneglycol addition products. Methylsilicone oils are miscible neither with these substances nor with paraffin waxes or petroleum jelly, and we must consider this in formulating the ointment. Therefore, even anhydrous ointment bases need the inclusion of constituents which impart stability to the dispersion. However, special emulsifiers are particularly important with water-containing ointment bases (*252*). Low-molecular-weight polymethylphenylsiloxane oils miscible with paraffin oils, petroleum jelly, etc., have recently been produced (cf. Section 9.2.11). Anhydrous pastes stable in storage are obtained by dispersing calcium stearate in liquid polydimethylsiloxanes (*223*).

Many dermatologically active materials can be incorporated in silicone-containing ointment bases to give stable compositions (*216*). Anhydrous ointment bases based on silicones are recommended for the incorporation of water-sensitive active materials, such as, in particular, penicillin (*165*, *284*). Surprisingly, resorption of the active materials through the skin takes place (*165*) in spite of the hydrophobic nature of the silicone component.

10.13.1.3 *Skin Protection with Silicones*

The silicone film produced on the skin from silicone-containing ointments ensures protection against numerous skin-damaging agents, particularly acids, mineral oils, salt-containing waters, and lubricating greases, and also against metal, wood, fertilizers, and building-material dust. Correspondingly, silicone-containing skin-protecting creams have found wide acceptance as industrial skin protectives for the prevention of industrial dermatoses. Their successful use is confirmed by a series of reports (*137*, *141*, *147*, *182*, *187*, *191*, *274*, *305*, *327*). Effective skin protection against modern detergents and other cleaning and washing agents has also been found, with the suppression of the possible domestic and industrial allergies. Better cleansing of the skin is a particular advantage of treating the skin with silicone-containing creams (*182*). For these reasons, such creams are widely used, even outside the field of industrial skin protection.

The skin-protecting action of the silicones can be shown by an impressive test. Optical brighteners are taken up copiously by the skin from optical solution and are firmly retained, as can easily be seen from the fluorescence which occurs. A pretreatment of the skin with a silicone-containing paste substantially reduces this uptake (*278*).

In the medical field, the use of silicone ointments or pastes on patients whose skin is endangered by continuous wetting accompanying their disease has proved successful (*86*, *195*, *307*). Even babies' skin is protected against diaper rash by silicone-containing creams (*166*, *186*). Skin irritations by artificial limbs and contact eczemas can be prevented by silicone ointments. Moreover, such preparations have been successfully used because of their good skin compatibility in the treatment of dyshidrosiform eczemas and in neurodermatitis, and also, because of their pleasant cooling effect on first-degree burns

(*187*). Excellent results have been obtained with silicone ointments as coating agents in surgery, e.g., with fistulas, draining channels, prolapses, ulcerations, sanious wounds, and those healing by second intention (*8*). They are also suitable for the prophylaxis of industrial eczemas in dental practice (*128*).

10.13.2 Silicones in Dental Medicine

In dentistry, room-temperature-vulcanizing silicone rubber is increasingly used as a replica medium for molding the jaw (with or without teeth) in the manufacture of dentures (*79*, *93*, *126*, *129*, *201*, *336*). Depending on the case, the material is used as pastes of various consistencies into which a curing agent is mixed before use. The transition from a pasty spreadable consistency to the rubber-elastic state takes place at body temperature within about 4 to 5 minutes. After this time, because of the elasticity of the material, the replica can be removed from the mouth without difficulty and can then be used for casting a working model, e.g., with plaster of paris.

The advantage of the silicone replica materials, as compared with the plaster used previously, lies in the elasticity of the cured material, its resistance to breakage, and its good release from the jaw. Silicone replica pastes also possess advantages over the alginates which are much used at the present time. Unlike the latter, they are not sensitive to loss of water, which may be the cause of shrinkage phenomena, and possess a better molding accuracy. The dimensional stability of silicone replicas is outstanding and is retained even on prolonged storage in the air.

Recently, room-temperature-vulcanizing silicone rubber products have also been proposed for the production of soft and elastic lower linings of dental plates (*333*).

10.13.3 Silicones as Auxiliaries in Surgery and General Medicine

In surgery and in various other fields of medicine, silicone rubber and silicone oils are now valued for certain special properties. Their use has been partly determined by their heat resistance, which permits heat sterilization, partly by their slip and release properties, and partly by their surface activity (foam suppression). However, it was the physiological inertness of the methylsilicones and, in the case of silicone rubber, the freedom of the vulcanizates from any added plasticizers, which gave the first impulse to the use of the products.

10.13.3.1 *Catheters, Implantations, Lubricants*

Artificial arteries made of nylon fabric are compatible with body tissue and are characterized by flexibility and low blood permeability after treatment with silicones. The walls cannot crack or adhere.

Silicone rubber articles play an important part in surgery, particularly as tubes and drains. They withstand hot sterilization, to which the rubber articles used previously could not be subjected. Even after repeated sterilization, silicone rubber articles undergo no change in their physical properties (*76*).

Silicone rubber tubes have been used as implantations or as permanent catheters and have given good results (*67*, *190*). A silicone tube implanted as a replacement for a resected portion of the urethra healed in without giving a reaction (*67*). As permanent catheters, silicone rubber tubes appear to have the advantage over ordinary rubber tubes of evoking no tissue reactions.

Because of their lubricating properties, silicone oils are used as lubricants for catheters in urology; however, silicone oil can cause burning of the mucous membranes similar to that described in Section 9.7 for the ocular conjunctiva (*246*).

10.13.3.2 *Retardation of Blood Clotting*

Blood clotting is retarded at surfaces of silicone products or on walls of vessels coated with silicone films (*88*, *146*). Silicone rubber tubes are therefore used for blood transfusions; siliconized glass vessels are used for storing blood (*258*). In other cases, silicone materials are also used successfully to retard blood clotting (*190*).

10.13.3.3 *Miscellaneous*

Silicone oils have been recommended for the *hot sterilization* of surgical instruments and syringes. They withstand the necessary temperatures and heating times without change or evaporation, and do not attack the instruments; on the contrary, the silicone coating adhering to them after sterilization has an advantageous effect, since it saves subsequent oiling and gives some protection against corrosion (*52*, *53*, *304*).

Silicone oils have also proved valuable aids in separating blood constituents by centrifuging, such as is carried out for the *detection of tumor cells*. On centrifuging, a silicone oil with a density of 1.075 separates the heavier erythrocytes and granulocytes from the lighter lymphocytes and tumor cells and thus makes it possible to isolate the tumor cell layer more cleanly; this can then be sent for cytological diagnosis (*111*).

Silicone antifoaming agents based on methylsilicone oils are occasionally used in medicine. Thus, in gastroscopy, optical study of the interior of the stomach is extraordinarily difficult when foam is present, and is thus facilitated by the previous administration of suitable silicone preparations (*95*, *133*). According to animal experiments, the treatment of severe pulmonary edemas due to irritant gases is possible only if the edematous foam is broken; however, according to more recent investigations, life is not prolonged (*127*).

10.13.4 Veterinary Medicine

Successful experiments have been made to combat colics occurring in ruminants and horses by the administration of silicone antifoaming agents. The silicone can be given prophylactically with the fodder (*205*) or be injected therapeutically into the rumen (*9, 65*). Siliconized glass vessels are used for the artificial insemination of cattle (*28*).

10.13.5 Pharmacy and the Foodstuffs Industry

The physiological inertness of the silicones has also interested the *pharmaceutical industry*. Silicone rubber stoppers are used in glass vessels containing valuable medications, especially antibiotics.

Aqueous liquids can be completely removed from siliconized bottles and ampuls. Siliconized ampuls (cf. Section 10.7) have been used widely for storing penicillin suspensions or penicillin powder to be dispersed as an aqueous suspension in the ampuls before injection. The suspensions run rapidly off the walls of such glass vessels so that the doctor can easily follow the emptying process. For the same reason, it is possible to remove the contained dose accurately, and the overfilling previously required becomes unnecessary. Finally, the silicone film on the wall of the vessel has a foam-suppressing action on the suspension, and this again facilitates the emptying process and its observation.

The storage properties and stability of sensitive drugs are said to be increased by coating tablets, pills, gelatine capsules, etc., with thin films of silicone resin (*332*). In the manufacture of tablets, added silicones act as release agents. A silicone oil emulsion is suitable for moist granulation and siliconized talc for dry granulation. The addition of silicones also has the remarkable effect of preventing the troublesome "post-hardening" of the tablets on storage (*203*). Because of their release effect, coatings of silicone resins on drying sheets, and in dishes and pans, are also of advantage in the pharmaceutical industry when sticky materials are being processed.

Foam-preventing and foam-suppressing agents based on silicones are valuable auxiliaries in the manufacture of pharmaceutical preparations because of the physiological inertness of the silicones and the very small amounts that generally have to be added (cf. Section 10.14.1).

It may also be mentioned that there are some problems in the *foodstuffs* industry corresponding to these uses of the silicones in the pharmaceutical industry. Here, however, the relevant legislation, which may vary from country to country, must be considered. German law permits certain silicone oils, silicone resins, and silicone elastomers for articles in use (*32*) but there are as yet no permissive regulations for silicones as food additives. In the United States the regulations of the Federal Register apply. For example, in contrast to the German legislation, they permit the addition to foodstuffs of liquid polydimethylsiloxanes with viscosities greater than 300 cSt in amounts not exceeding one part per million.

10.14 Miscellaneous Applications

10.14.1 Silicones as Antifoaming Agents

Silicone antifoaming (defoaming, foam-suppressing) agents are available either in the undiluted form or as aqueous emulsions. The pure type of antifoaming agent consists of silicone oils with additives which facilitate the dispersion of the silicone, and thereby enhance the foam-suppressing effect (*54*). This type is usually diluted before use with organic solvents, since, in view of the small amounts used, it is too difficult to measure out in concentrated form. The antifoaming emulsions differ mainly in the type of emulsifier used in their preparation, and therefore often have different activities in different fields of application. Such relationships between the nature of the emulsion and its effect become understandable when one considers that the formation of a surface film of silicone oil is necessary for the foam-suppressing effect (cf. Section 9.1.7). An oil film can arise only when the emulsion is broken, for example, by dilution in the system to be defoamed, by a change in the pH or in the concentration of foreign ions, or by a rise in temperature. It is therefore natural that there is a certain antagonism between the stability of the emulsion and its foam-suppressing action; the more labile the emulsion, the greater its tendency to react to factors such as those mentioned above, and therefore to suppress foam with the formation of the superficial film of silicone. For technical products in general, compromise solutions are made between the requirements set by the stability of the emulsion to transport and storage, on the one hand, and its defoaming power on the other. Very effective and stable defoamer emulsions are obtained by adding polyether siloxanes, which enhance the defoaming action of the silicone defoamer emulsion without reducing its stability. The effect is the more surprising, because polyether siloxanes themselves possess no defoaming action (*230*).

The advantages of the silicone antifoams consist in their activity in extraordinarily low concentrations; according to the system, the latter vary from something like 10^{-5} to 10^{-1}%. Owing to the unusually low amount of foam suppressor needed, no significant "contamination" of the system to be defoamed need be feared, particularly since silicone foam suppressors are physiologically inert (Section 9.7). Silicone antifoams do not, in general, lose their activity and are not decomposed even at relatively high temperatures.

The difficulty of making general statements about antifoaming agents prevents a detailed account of the range of applications. Consequently, we generally remark that silicone antifoams have proved satisfactory and have been introduced in, for example, the following processes: in resin boiling in the paint industry, in the preparation of latex in the rubber industry, for defoaming fine dispersions of polyvinyl acetate, in the manufacture of paper and glue, for the surface defoaming of mineral oils, in the oil industry, in the boiling of linseed oil, in the cooking of bitumen, to prevent the foaming and "spitting" of water in steam boilers, in the processing of printing inks, in the preparation and processing of dispersion dyestuffs, in pharmaceutical chemistry, and in chemical technology.

Silicone antifoaming agents are capable both of eliminating foam that has already formed and also of preventing the appearance of foam.

It is occasionally observed that silicones have *foam-producing* effects; these effects are limited to systems in which silicones are soluble (cf. Section 9.1.7). As surface-active substances in dissolved form, they can lower the surface tension and therefore become foaming agents.

10.14.2 Silicones as Ore Flotation Reagents

The fact that silicones can be used as ore flotation reagents is explained, like their foam-suppressing effect, by their surface activity (Section 9.1.7).

It might be expected that, because of their silicate affinity, in mixtures of minerals containing silicate and sulfide or oxide components, the silicones would wet the silicate minerals. Indeed, the first investigations pointed in this direction (*74*, *277*). By treating mixtures of minerals with chlorosilanes and subsequent flotation, the silicate gangue was floated while sulfide and carbonate minerals, together with noble metals, remained in the sludge.

In contrast to this, later experiments showed that galena could be separated from sand by silicone oils through flotation of the sulfide mineral (*325*). However, the working conditions were such that it was impossible to speak of a foam flotation process in the true sense of the words.

Proof that silicones can be used as true ore flotation reagents that are active through the formation of molecular films, that work in correspondingly low doses, and that are capable of a selective separation of minerals, was first given for systems consisting of galena, zinc blende, and silicate gangue (*106*, *107*). The selection of suitable silicones and emulsifiers for the polyorganosiloxane to be used in emulsion form is important. Short-chain polybutylmethylsiloxanes and nitrogen-containing emulsifiers proved to be the most effective combination. By means of this and some measures already known in flotation technique—suppression of the zinc blende with zinc sulfate solution, revival of the zinc blende with copper sulfate solution—galena can be separated selectively from zinc blende and the latter, in turn, from the gangue. The course of the recovery in the treatment of a relatively high intergrown lead-zinc ore is shown by Fig. 111 as a function of the flotation time. The degree of enrichment is good and is comparable with that from commercial flotation reagents. The low consumption of reagents is surprising, being about 4 gm of silicone per ton of crude ore, as compared with a xanthate requirement of

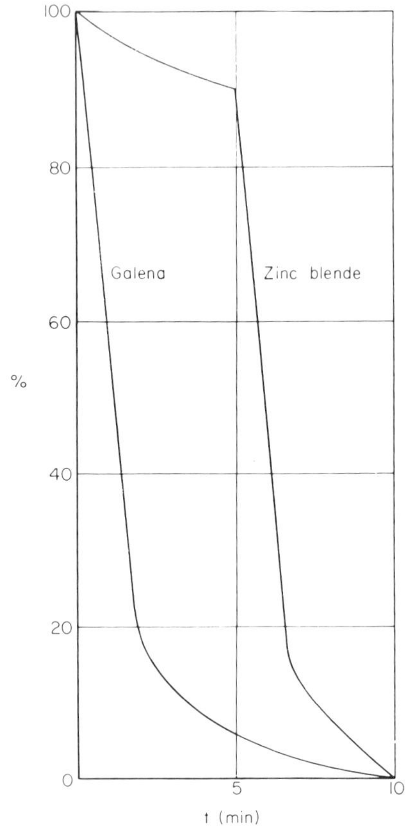

FIG. 111. Selective flotation of crude ore containing galena and zinc blende with a silicone preparation. Decrease in the metal content in the sludge, in per cent of the initial content, as a function of the time of flotation.

about 150 to 200 gm/ton. The difference can be understood in that silicones possess only a very small volume solubility and are transferred to the mineral particles through surface solutions (Section 9.1.7); in their case, therefore, a saturated surface solution is achieved with substantially less material than when water-soluble collectors are used.

10.14.3 Molding Materials Based on Room-Temperature-Vulcanizing Silicone Rubber

As already mentioned in Section 8.1.2.3, certain types of silicone rubber can be rapidly converted by catalysts at room temperature into rubberlike elastic vulcanizates. Apart from the purposes of electrical insulation (Section 10.2), such materials have been used mainly as *molding materials.* They satisfy all the requirements set for such materials. They can be made of any desired consistency—kneadable, castable, or stretchable—and can be vulcanized in a few minutes or hours depending on the catalyst added; because of the creep capacity of the silicone, they fill completely even the most complicated molds; they can easily be removed from the substrate without damage because of the release properties of the silicones and their rubberlike elasticity; they exhibit no appreciable dimensional shrinkage, even on prolonged storage; and, finally, they are physiologically inert. They have therefore been introduced, *inter alia*, as dental molding materials and have proved eminently suitable. We have already mentioned the possibility of using them even for molding casting resins (Section 10.6). Moreover, they are of interest wherever the replicas must have high precision and good permanence: in art work, in research on the history of art, and also in police work and in many other fields.

Recently, however, products based on room-temperature-vulcanizing silicone rubber have also acquired special importance as *sealing materials.* These are materials which, when applied in pasty consistency on the site, spontaneously change into a rubberlike elastic material. The development of one-component systems (Section 8.1.2.3.2), in particular (i.e., compositions which contain all components and "vulcanize" merely on access of atmospheric humidity), opened the door to new applications.

The advantages of the end product are its elasticity, its outstanding weather resistance, its light-fastness, its water repellency, its capacity for being colored in different shades, and its good adhesion to many process and constructional materials, particularly glass, wood, and metal, which, if necessary, can be increased still further by adhesives. Consequently, a wide variety of applications are indicated in the building industry and in the construction of machines, cars, aircraft, and ships. In addition, when supplied to the layman in the form of "do-it-yourself products," they promise to become practical and universal aids in the small technical needs of everyday life.

10.14.4 Silicones in Laboratory Technique

Finally, silicones are also useful in some aspects of laboratory work.

Silicone oils, particularly liquid polymethylphenylsiloxanes with viscosities of about 100 to 300 cSt/20°C are suitable for replacing sulfuric acid as *liquid baths* in melting point apparatus. They do not become discolored, and remain clear and transparent, thus facilitating observation. Suitable types of polymethylphenylsiloxanes can be heated at 250°C for long periods, and even above 300°C for short periods.

The same oils are becoming increasingly important as heat-transfer agents in thermostats working at relatively high temperatures. The figures given in Section 9.2.2 give an idea of the temperature to which they can be exposed. Their heat stability in air is always lower than under an inert gas, so that it is advisable (e.g., in thermostats working above 250°C) to arrange a nitrogen blanket above the oil phase. In open systems (i.e., those at atmospheric pressure) it is advisable to use a reflux condenser at the breathing point to prevent the diffusion loss of thermal degradation products formed. Special reference may be made here to the possibility of achieving peak temperatures at 450°C in closed systems with certain types of silicone oils at relatively low pressures of the order of magnitude at 10 atm (Section 9.2.2).

Low-molecular-weight representatives of the polymethylphenylsiloxanes (see Section 9.2.3) are outstanding *diffusion-pump oils*. High-quality working fluids based on these products give, with water cooling, final vacuums of the order of magnitude of 10^{-8} torr. They are very resistant to air leakage; tests have shown that even after 300 air leakages at intervals of 18 to 20 minutes, each lasting for 1 minute, there was not only no impairment, but even a slight improvement, of the final vacuum (*288*).

Silicone oils are also suitable for the preparation of *cooling mixtures*. Acetone or methanol mixed with dry ice is replaced by a not too viscous methylsilicone oil (*163*). Since silicone oil has no appreciable vapor pressure, no topping-up is required. The use of silicone oil here also appears advantageous because of its high ignition point and the absence of unpleasant odor.

Liquid silicones have also proved suitable as *solvents* for cryoscopic *molecular weight determinations* (*265*). Octamethylcyclotetrasiloxane has proved an almost ideal solvent for the molecular weight determination of silanes and siloxanes; its solvent power is good and its cryoscopic constant, 11.2, is about twice as high as that of benzene. Another advantage over benzene is its low solvent capacity for water. Apart from being used in siloxane chemistry, it also seems suitable for determinations of organic compounds (*265*).

Silicone oils are used in *gas chromatography* as stationary liquid phases for special separating problems (*17*, *287*). Their suitability is determined not only by their good heat stability and chemical inertness, but also by the fact that as nonpolar compounds they exhibit no selectivity with respect to polar materials and, in general, separate them only according to their boiling points (*17*, *287*).

Furthermore, the excellent storage stability of the silicone oils, especially the constancy of their viscosity even during many years of storage at room temperature, makes them valuable *calibrating standards* for *viscometers*.

Silicone greases have been generally introduced for the lubrication of glass joints; their viscosity changes little with rising or falling temperature and they withstand temperatures at which organic types of grease are decomposed. Their resistance to chemicals is generally sufficient, owing to the small area of attack which they offer to reagents in a joint. Silicone greases or methylsilicone oils also facilitate the fitting of rubber tubes onto glass tubes. A film of silicone prevents the rubber adhering firmly at high temperatures and has no effect on the rubber.

The *surface water-repellency treatment* of glass apparatus for analytical purposes has often been described and discussed (*99*, *100*). In siliconized burettes and pipettes, water and dilute aqueous solutions have convex menisci and run out without wetting the wall; consequently, they should—with recalibration—allow more rapid and precise work. However, they have not realized the hopes expected of them, since in the long run the silicone film deteriorates under chemical stresses, particularly the action of alkaline solutions, and then loses its activity. Only in special cases has a surface water-repellency treatment of glass proved satisfactory as an aid in measuring techniques. As an example of this, we may mention that the quality of a polarographic dropping electrode is improved by siliconization, which gives a greater constancy of the rate of flow, the drop weight, and the dropping time of the mercury (*293*).

REFERENCES

(1) AEG, unpublished. (2) AEG-Fabrik, Mülheim/Ruhr, unpublished. (3) R. Alsgaard and J. W. Gilkey, Dow Corning, U.S. Pat. 2970881 [15.12.1958]. (4) J. Alvergnat, *Rev. gén. Electric.* **69**,393 [1960]. (5) E. Amrein, *Schweizerische Tonwarenindustrie* **59**, No. 10, 8 [1956]. (6) F. O. Anderegg, *Progr. Architect.* **32**, 94 [1951]. (7) H. Arens, Th. Goldschmidt, German Appl. (West German) 1080010 [20.11.1957]. (8) R. Asche, *Therap. Ber.* **29**, 184 [1957]. (9) J. A. Austin and K. A. Ratcliff, Jensen-Salsbery Laboratories, U.S. Pat. 2635981 [23.3.1949].

(10) J. S. Axelson and F. M. Precopio, *ASTM-Bull.* **244**, 40 [1960]. (11) M. E. Baer and L. E. Concannon, U.S. Pat. 2584413 [30.6.1948]. (12) D. L. Bailey and A. S. Pater, Union Carbide, Belg. Pat. 618786 [U.S. Prior. 12.6.1961]. (13) A. J. Barry, Dow Chemical, U.S. Pat. 2405988 [10.12.1943]. (14) —, —, U.S. Pat. 2469625 [10.12.1943]. (15) S. L. Bass, Dow Corning, U.S. Pat. 2706723 and 2706724 [21.6.1951]. (16) E. Baumann, *Leder* **8**, 8, 185 [1957]. (17) E. Bayer, Gas chromatography, Elsevier, New York, 1961. (18) S. H. Bell, *J. of the Oil and Colour Chemists' Association* **35**, No. 386, 373 [1952]. (19) L. I. Berberich and K. N. Mathes, *Trans. Amer. Inst. Electr. Engr.* (Power App. and Syst.) **77** (35), 69 [1958].

(20) K. Bergmann, *Ziegelindustrie* **21**, 47 [1956]. (21) C. A. Berridge and R. Treat jr., *Mater. Design Engng.* **51**, No. 166, 128 [1960]. (22) H. H. Bertram, *Gummi, Asbest, Kunststoffe* **15**, 1182 [1962]. (23) —, *Rubber Chemistry and Technology* **36**, No. 4, 1148 [1963].

(24) E. BEYELER, *Schweizer Maschinenmarkt* **18**, 255 [1958]. (25) L. P. BIEFELD, Owens-Corning Fibreglas, U.S. Pat. 2723210 [2. 10. 1952]. (26) A. R. BLANK: An environmental high performance, rugged silicone cable design for missile application: 11th Annual Wire and Cable Symposium, Asbury Park, New Jersey, 28–30 November 1962. (27) S. A. BRADY, *Off. Digest* **30**, 958 [1958]. (28) H. BRINKMANN, Dissertation, Tierärztl. Hochschule Hannover 1955. (29) British Thomson-Houston, Brit. Pat. 593727 [U.S. Prior. 1.7.1944].

(30) F. K. BROCHHAGEN, *Kunststoffe* **44**, 555 [1954]. (31) D. T. BUMPSTEED, *J. Soc. Glass Techn.* **42**, Trans. 204 [1958]. (32) *Bundesgesundheitsblatt* 1961, No. 8, p. 120–122, Springer-Verlag, Berlin, Göttingen, Heidelberg. (33) A. BUTCHART, *J. Soc. Glass Techn.* **42**, Trans. 189 [1958]. (34) H. L. CAHN, *Concrete Construction Mag.* July 1958. (35) —, *Paint Industry Mag.* **76**, No.4, 8 [1961]. (36) —, R. V. MACHEY JR., *ASTM-Bull.* **235**, 37 [1959]. (37) P. C. CANOVAI JR., Union Carbide, U.S. Pat. 2934459 [2.8.1957]. (38) R. CHARLE, R. RITTER and G. KALOPISSIS, Soc. Monsavon-l'Oreal, U.S. Pat. 2944942 [French Prior. 4.8.1956]. (39) N. D. CHERONIS, U.S. Pat. 2564674 [7.8.1947]; U.S. Pat. 2579416 [25.1.1946]; U.S. Pat. 2579417 [7.8.1947].

(40) —, U.S. Pat. 2568384 [13.11.1947]. (41) —, U.S. Pat. 2635059 [26.10.1948]. (42) —, U.S. Pat. 2579418 [24.1.1949]. (43) H. A. CLARK, Dow Corning, U.S. Pat. 2762717 [18.4.1955]. (44) — and E. P. PLUEDDEMANN, *Mod. Plastics* **40**, No. 10, 133 and 195 [1963]. (45) S. R. COCKETT, J. G. EVANS and G. LANDELLS, Bradford Dyers' Association, U.S. Pat. 2774690 [Brit. Prior. 16.4.1951]. (46) T. J. COLLIER, Owens Corning Fiberglas, U.S. Pat. 2974062 [14.2.1956]. (47) W. R. COLLINGS, Dow Corning, German Pat. (West German) 849737 [French Prior. 29.3.1949]. (48) Companie des Prod. Chim. et Electrometallurgie, Alais Froges et Camarque, Brit. Pat. 666820 [French Prior. 16.9.1948]. (49) Commission Électrotechnique International, Genf 1957, Publication 85: *Recommendations rélatives à la classification des matières destinées à l'isolement des machines et appareils électriques en fonction de leur stabilité thermique en service.*

(50) H. H. COOKE and D. RUSSO, Standard Oil, U.S. Pat. 2626870 [5.12.1949]. (51) Corning Glass Works, Brit. Pat. 588762 [26.5.1944]. (52) F. W. CROWE, *J. Amer. Med. Assoc.* **149**, 1464 [1952]. (53) M. C. CROWLEY and F. D. OSTRANDER, *Science* [Washington] **108**, 542 [1948]. (54) C. C. CURRIE, Dow Chemical, U.S. Pat. 2632736 [22.8.1946]. (55) —, Dow Corning, U.S. Pat. 2523281 [26.9.1949]. (56) —, —, U.S. Pat. 2672455 [4.8.1952]. (57) M. DAHM and W. SIMMLER, Farbenfabriken Bayer, German Appl. (West German) 1122698 [10.6.1960]. (58) —, —, —, German Appl. (West German) 1150516 [Brit. Prior. 9.6.1961]. (59) K. DAMM and H. E. GÖTTGENS, —, Belg. Pat. 616399 [German Prior. (West German) 25.4.1961].

(60) —, H. HORNIG, G. SINN and H. STEINBACH, —. French Pat. 1361771 [German Prior. (West German) 6.7.1962]. (61) — and R. MÜLLER, *Gummi, Asbest, Kunststoffe* **15**, 302 [1962]. (62) —, K. SCHNURRBUSCH, W. OETKE, R. ERDMENGER and K. STADE, Farbenfabriken Bayer, Belg. Pat. 572412 [German Prior. (West German) 31.10. and 25.11.1957]. (63) J. H. DAVIS, *Electr. Rev.* **1957**, 709. (64) —, D. W. REES and J. H. RILEY, *Proc. Inst. Electr. Engrs.* (G.B.) Paper 3627 [International Conf. on Components and Materials used in Electronic Engineering] June 1961. (65) R. DE R. BARONDES, W. D. JUDGE and C. G. TOWNE, *Military Surgeon* **106**, 379 [1950]. (66) G. DE JONGE, *Lack- u. Farben-Chem.* [Zofingen] **6**, 76 [1952]. (67) R. R. DE NICOLA, *J. Urology* **63**, 168 [1950]. (68) F. L. DENNETT, Dow Corning, German Pat. (West German) 886591 [Brit. Prior. 11.5.1948]. (69) —, —, U.S. Pat. 2588365 [10.11.1949].

(70) —, —, U.S. Pat. 2588366 [9.3.1950]. (71) —, *Amer. Dyestuff Reporter* **39**, No. 1, 63 [1950]. (72) —, Dow Corning, U.S. Pat. 2588367 [4.5.1950]. (73) —, —, German Pat. (West German) 1046564 [U.S. Prior. 29.4. and 7.9.1954]. (74) A. DESNOES and R. TESTUT, *Rev. Ind. minérale* **35**, 211 [1954]. (75) J. F. DEXTER, *Electr. Engr.* **75**, 258 [1956]. (76) S. DIMANT, *Lancet* **267**, 533 [1954]. (77) Dow Corning, Brit. Pat. 686429 [9.1.1951]. (78) P. DUGGAN and F. FORTESS, *Mod. Textile Mag.* **34**, No. 3, 31 [1953]. (79) A. EFFINGER, *Dtsch. Zahnärztl. Z.* **13**, 256 [1958].

(80) G. EHLERS, *Elektrotechn. Z.* [A] **74**, 553 [1953]. (81) G. E. EILERMAN, Pittsburgh Plate Glass, U.S. Pat. 3019122 [15.12.1958]. (82) J. R. ELLIOTT and R. H. KRIEBLE, General Electric, U.S. Pat. 2507200 [10.2.1945]. (83) H. H. ENDER, Union Carbide, U.S. Pat. 3078293 [23.9.1960]. (84) H. ENDERS and H. DEINER, Chemische Fabrik Pfersee, German Appl. (West German) 1060347 [6.4.1957]. (85) J. G. EVANS, G. LANDELLS and W. S. MEALS, Bradford Dyers' Association, German Appl. (West German) 1058013 [Brit. Prior. 4.1.1955]. (86) L. FINKE, *Therap. Ber.* **29**, 225 [1957]. (87) L. G. FITZGERALD-LEE, *Electr. Rev.* (G.B.) **167**, No. 19, 771 [1960]. (88) R. F. FLETCHER, *Lancet* **270**, 509 [1956]. (89) D. FORKEL, *Konstruktion* **8**, 452 [1956].

(90) F. FORTESS, *Ind. Engng. Chem.* **46**, 2325 [1954]. (91) H. K. FRENSDORFF, *Rubber Age* **83**, 812 [1958]. (92) C. L. FREY, Dow Corning, German Appl. (West German) 1111182 [U.S. Prior. 10.9. and 28.10.1954]. (93) H. FULDE, *Zahnärztl. Praxis* **10**, 68, [1959]. (94) A. GANCBERG, Union Chimique Belge, German Appl. (West German) 1102629 [Belg. Prior. 14.11.1955]; German Appl. (West German) 1065551 [Belg. Prior. 23.12.1955]. (95) M. GASTER, J. O. WESTWATER and W. E. MÒLLE, *Gastroenterology* [Baltimore] **27**, 652 [1954]. (96) N. J. GAYNES, *Ind. Finishing* [Indianapolis] **38**, No. 6, 106 [1962]. (97) H. C. GEEN and J. D. QUIST, S. C. Johnson and Simoniz, U.S. Pat. 2812263 [10.3.1949]. (98) P. G. GEMEINHARDT and J. W. BRITAIN, Mobay Chemical, German Appl. (West German) 1121802 [U.S. Prior. 3.11.1958]. (99) P. T. GILBERT JR., *Science* [Washington] **114**, 637 [1951].

(100) —, *Chemist-Analyst* **41**, 52 [1952]. (101) J. W. GILKEY, Dow Corning, U.S. Pat. 2894967 [2.8.1956]. (102) —, —, Belg. Pat. 561185 [U.S. Prior. 28.9.1956]. (103) —, —, Belg. Pat. 564179 [U.S. Prior. 23.1.1957]. (104) M. A. GLASER, *Ind. Engng. Chem.* **46**, 2334 [1954]. (105) —, R. E. CAPLAN and R. W. CLOPE, *Official Digest* **33**, 1197 [1961]. (106) A. GÖTTE, W. NOLL, H. STEINBACH and A. ZAPPEL, Farbenfabriken Bayer, German Pat. (West German) 1156724 [1.10.1958]. (107) — and H. STEINBACH, *Aachener Blätter* **1959**, No. 1. (108) R. GOLDMAN, Premo Pharmaceutical Lab. U.S. Pat. 2504482 [17.6.1949]. (109) Th. Goldschmidt, French Pat. 1298127 [German Prior. (West German) 29.8.1960].

(110) H. J. GOLDSTEIN and P. V. STROTHER, Texas-U.S. Chemical, U.S. Pat. 3073789 [26.5.1959]. (111) F. GRAEBER, *Krebsarzt* **16**, 473 [1961]; *Therapeutische Ber.* **32**, 247 [1960]. (112) P. A. GRIFFIN, Paint Manufact. **30**, 345 [1960]. (113) —, ibid. **31**, 411 [1961]. (114) C. GÜNDEL, *Silikattechnik* **8**, 533 [1957]. (115) C. J. GUILLISSEN and A. GANCBERG, U.S. Pat. 2732320 [Belg. Prior. 31.1.1951]. (116) —, —, Union Chimique Belge, German Pat. (West German) 975736 [Belg. Prior. 26.2.1951]. (117) C. P. HABER, General Electric, U.S. Pat. 2553314 [1.7.1944]. (118) T. F. HART and R. S. WISHART JR., Union Carbide, U.S. Pat. 2833673 [25.3.1954]. (119) G. HAUSCHILD, Dissertation, TH Aachen 1954.

(120) L. HAVENITH, *Das Baugewerbe* **41**, No. 20, 1006 [1961]. (121) —, ibid. **41**, No. 22, 1097 [1961]. (122) S. D. HAYDEN, *Electronic Engng.* **1956**, No. 2, 58. (123) B. F. W. HAYWORD, ibid. **1956**, No. 3, 115. (124) R. C. HEDLUND, *Jb. Chem. Ind.* **1957/58**, 66. (125) — and C. R.

OLSON, *Paint Varnish Production* **53**, No. 5, 40 [1963]. (126) A. HENNICKE, *Dtsch. Zahnärzteblatt* **12**, 833 [1958]. (127) D. HENSCHLER, F. AMEND and A. JÜTTNER, *Arch. exp. Pathol. Pharmakol.* **239**, 288 [1960]. (128) H. HERRMANN, *Dtsch. Zahnärzteblatt* **10**, 325 [1956] (129) —, ibid. **13**, 348 [1959].

(130) R. HEYDEN and J. PLAPPER, Böhme Fettchemie, Swiss Pat. 362483 [German Prior. (West German) 14.5.1959]. (131) H. HIENDL, *Die Ziegelindustrie* **14**, 428 [1961]. (132) M. HILBERT, K. DAMM and K. SCHNURRBUSCH, Farbenfabriken Bayer, Belg. Pat. 622877 [German Prior. (West German) 28.9.1961]. (133) B. L. HIRSCHOWITZ, R. L. BOLS and H. M. POLLHARD, *Gastroenterology* [Baltimore] **27**, 5 [1954]. (134) F. HOSTETTLER, Union Carbide, German Appl. (West German) 1091324 [U.S. Prior. 25.9.1957]. (135) D. A. HUGHES, *J. Soc. Glass Techn.* **42**, Trans. 214 [1958]. (136) D. R. HULL, DuPont, U.S. Pat. 2729538 [30.11.1951]. (137) K. HUMPERDINCK, *Med. Klin.* **48**, 15 [1953]. (138) M. J. HUNTER, Dow Corning, U.S. Pat. 2645624 [3.3.1950]. (139) — and C. F. DUDLEY, —, German Appl. (West German) 1048667 [U.S. Prior. 21.6.1954].

(140) —, —, —, U.S. Pat. 2728736 [21.6.1954]. (141) J. HUSSONG, *Die Medizinische* **42**, 1477 [1955]. (142) J. F. HYDE, Corning Glass Works, U.S. Pat. 2439689 [11.6.1943]. (143) —, Dow Corning, French Pat. 991088 [U.S. Prior. 11.8.1948]. (144) —, —, German Pat. (West German) 865576 [U.S. Prior. 23.2.1950]. (145) J. IGLISCH, Dissertation, Stuttgart 1958. (1946) L. B. JACQUES, E. FIDLAR, E. T. FELSTED and A. G. MACDONALD, *Canad. Med. Assoc. J.* **55**, 26 [1946]. (147) H. JÄRNECKE, *Dtsch. med. Wschr.* **80**, 755 [1955]. (148) H. JEBSEN-MARWEDEL and G. MARWEDEL, *Farbe und Lack* **66**, 314 [1960]. (149) M. H. JELLINEK, Union Carbide, German Appl. (West German) U 2099 [U.S. Prior. 18.3.1952].

(150) —, —, Belg. Pat. 554172 [U.S. Prior. 18.1.1956]. (151) —, —, U.S. Pat. 2965515 [23.5.1955]. (152) R. JOLY, *Rev. gén. Electr.* **70**, 341 [1961]. (153) H. KALLAS, *Elektro-Anz.* **1955**, 494. (154) —, *Elektrotechn. und Maschinenbau* **74**, 335 [1957]. (155) —, *Elektrotechn. Z.* [B] **9**, 392 [1957]. (156) —, ibid. [B] **11**, No. 10, 398 [1959]. (157) —, 2. Internationales Drahtlack-Kolloquium vom 2. u. 3. Sept. 1964 in Hamburg. (158) R. KARSTEN: *Bauchemie, Straßenbau, Chemie und Technik*, Verlagsgesellschaft m.b.H., Heidelberg 1960, p. 381. (159) W. S. KATHER, General Electric, U.S. Pat. 2713545 [27.5.1952].

(160) T. A. KAUPPI, Dow Corning, U.S. Pat. 2678893 [21.2.1951]. (161) J. W. KEIL, —, German Appl. (West German) 1057442 [U.S. Prior. 22.12.1955]. (162) —, —, U.S. Pat. 3050411 [16.9.1959]. (163) H. KIRCHER, Farbenfabriken Bayer, unpublished. (164) D. X. KLEIN, DuPont, U.S. Pat. 2532559 [3.10.1946]. (165) P. KLEIN, Hygien. Inst. d. medizin. Akad. Düsseldorf 1955, unpublished. (166) P. KNIPE, *Brit. med. J.* **1956**, I, 1044. (167) T. KOCH, N. V. Onderzoekingsinstituut Research, Arnheim, German Appl. (West German) p 51555 [Dutch Prior. 28.10.1948]. (168) G. J. KOOKOOTSEDES and J. L. SPEIER, Belg. Pat. 625526 [U.S. Prior. 1.12.1961]. (169) M. KRAMER, *Melliand Textilber.* **36**, 806 [1955].

(170) —, *Textilpraxis* **13**, 1050 [1958]. (171) —, W. NOLL and O. PANTKE, Farbenfabriken Bayer, German Appl. (West German) 1080061 [26.7.1957]. (172) K. W. KRANTZ, General Electric, U.S. Pat. 2810704 [19.4.1954]. (173) —, —, U.S. Pat. 2916461 [30.4.1956]. (174) —, —, U.S. Pat. 2887467 [3.2.1958]. (175) W. KRAUSS, 6. Fatipec Kongress, Wiesbaden, Mai 1962, Verlag Chemie, Weinheim/Bergstr, p. 332. (176) — and G. MENNICKEN, Farbenfabriken Bayer, German Appl. (West German) 1092585 [15.1.1958]. (177) —, —, —, German Appl. (West German) 1111320 [7.11.1958]. (178) — and H. WEISSBACH, *Baugewerbe* **38**, 333 [1958]. (179) B. H. KRESS and H. A. HOPPENS, *Paintindia* **2**, No. 9, 35 [1952].

(180) V. KUBELKA and V. NEMEC, *Collegium* [Darmstadt] **1933**, 311. (181) H. KÜNZEL, *Dtsch. Bauztschr.* 7, 449 [1959]. (182) I. KWOCZEK and R. SCHAEFER, *Dermatol. Wschr.* **1958**, No. 7, 177. (183) K. LADE and A. WINKLER: *Ursachen der Putz- und Anstrichschäden*, Carl Maurer Verlag, Geislingen/Steige 1956. (184) —, —,: *Putz, Stuck, Rabitz*. Julius Hoffmann Verlag, Stuttgart 1955. (185) W. LANGMANN and W. LEHMANN, Farbenfabriken Bayer, German Appl. (West German) 1154066 [21.11.1959]. (186) F. LANGNER, *Therap. Ber.* **29**, 278 [1957]. (187) M. LAUMANNS, *Therap. d. Gegenwart* 3, 109 [1958]. (188) H. J. LEAVITT, General Electric, U.S. Pat. 2985545 and U.S. Pat. 2985546 [26.3.1958]. (189) O. LEBERL, *Kosmetik-, Parfüm-, Drogen-Rundschau* 7, No. 1/2, 3/4, 5/6 [1960].

(190) F. LEISS and K. H. PETER, *Arzneimittelforsch.* **4**, 614 [1954]. (191) P. LEVAN, *New-corner Calif. Med.* **81**, 210 [1954]. (192) W. LIEBING, *AEG-Mitt.* **48**, 336 [1958]. (193) G. LINDEMANN, *Riechstoffe und Aromen* **5**, 179 [1955]. (194) G. LOEW, Farbenfabriken Bayer and Mobay Chemical, German Appl. (West German) 1101752 [U.S. Prior. 4.9.1958]. (195) J. J. LUBOWE, *Soap, Perfum., Cosmet.* **28**, 1003 [1955]. (196) M. MAESER, *J. Amer. Leather Chemists' Assoc.* **42**, 8, 390 [1947]. (197) J. MARSDON and R. H. SAVAGE, *Amer. Inst. Electr. Engr., Proceedings*, T-Section 8211, 67, 1 [1948]. (198) J. T. MARSH, G. BELL and A. G. THOMPSON, Tootal Broadhurst Lee, Brit. Pat. 695703 [31.10.1949]. (199) K. N. MATHES, General Electric, U.S. Pat. 2523037 [3.8.1946].

(200) R. R. MCGREGOR: *Silicones and their Uses*, McGraw-Hill, New York 1954, p. 79. (201) J. W. MCLEAN, *Brit. Dental J.* **104**, 441 [1958]. (202) R. N. MEALS and F. M. LEWIS: *Silicones*, Reinhold Publishers, New York 1959. (203) W. MERZ, *Deutsche Apotheker-Zeitung* **98**, No. 10, 223 [1958]. (204) Midland Silicones, Brit. Pat. 780553 [U.S. Prior. 18.10.1954]. (205) —, Brit. Pat. 778356 [30.11.1955]. (206). —, Brit. Pat. 809766 [U.S. Prior. 28.8.1956]. (207) —, Brit. Pat. 860522 [U.S. Prior. 4.11.1957]. (208) R. G. MITTON and K. G. WYATT, British Leather Manufacturers' Research Assoc., Brit. Pat. 848838 [7.5.1957]. (209) V. M. MONTSINGER, *J. Amer. Inst. electr. Engr.* **49**, 293 [1930].

(210) B. E. MOODY, *J. Soc. Glass Techn.* **42**, Trans. 177 [1958]. (211) A. MORTON and T. W. WATSON, *Paint Manufacture* **29**, 238 and 258 [1959]. (212) F. NAUSCH, *Elin-Zeitschr.* **12**, No. 2, 110 [1960]. (213) H. T. NELSON and R. A. PANCHERI, United States Rubber, U.S. Pat. 2705690 [5.6.1953]. (214) H. NEUBERT, *Dtsch. Bauztschr.* 7, 441 [1959]; DIN 4108: Wärmeschutz im Hochbau. (215) A. NEUNTEUFEL, *Elektrotechn. Z.* [B] **9**, 491 [1957]. (216) F. NEUWALD and K. H. ADAMS, *Dtsch. Apotheker-Ztg.* **96**, 131 [1956]. (217) P. NEY, *Dtsch. Bauztschr.* 7, 450 [1959]. (218) P. O. NICODEMUS, General Electric, U.S. Pat. 2442613 [22.5.1944]. (219) S. NITZSCHE and E. PIRSON, Wacker-Chemie, German Appl. (West German) 1064192 [21.11.1956].

(220) —, —, —, U.S. Pat. 3002949 [German Prior. (West German) 13.11.1957]. (221) S. J. NIZINSKI, *Wire and Wire Products* **37**, 628, 633, 634 and 677 [1962]. (222) W. NOLL, Farbenfabriken Bayer, French Pat. 1072136 [German Prior. (West German) 4.3.1952]. (223) —, — German Appl. (West German) 1000570 [15.9.1953]. (224) —, —, French Pat. 1156600 [German Prior. (West German) 3.8.1955 and 11.4.1956]. (225) —, —, German Appl. (West German) 1106070 [13.11.1959]. (226) —, *Chimia* **16**, 245 [1962]. (227) —, K. DAMM and H. WEISSBACH, Farbenfabriken Bayer, Belg. Pat. 572974 [German Prior. (West German) 26.11.1957]. (228) —, —, —, —, Belg. Pat. 572727 [German Prior. (West German) 8.11.1957; 26.11.1957; 27.8.1958]. (229) —, O. GLENZ and W. LANGMANN, —, Brit. Pat. 915787 [German Prior. (West German) 13.11.1959].

(230) — and K. SCHNURRBUSCH, —, Belg. Pat. 626473 [German Prior. (West German) 30.12.1961]. (231) — and H. WEISSBACH, *Zement-Kalk-Gips* **9**, 476 [1956]. (232) F. J. NORTON, General Electric, U.S. Pat. 2386259 [30.7.1942]. (233) —, —, U.S. Pat. 2412470 [22.2.1943]. (234) P. NOWAK, *Elektrotechn. Z.* [B] **7**, 354 [1955]. (235) — and J. BOLLIG, Licentia-Patent-Verwaltungs G.m.b.H., German Appl. (West German) 1096521 [31.10.1958]. (236) E. H. OLSON, DuPont, U.S. Pat. 2719073 [30.11.1951]. (237) G. OTTO, *Collegium* [Darmstadt] **1941**, 158. (238) Owens-Corning Fiberglas, Brit. Pat. 750135 [U.S. Prior. 17.8.1951]. (239) H. A. PACE, Goodyear, French Pat. 1151710 and 1153994 [U.S. Prior. 21.12.1955].

(240) H. PÄRSCH, *Dtsch. Elektrotechn.* **11**, 59 [1957]. (241) A. S. PATER, Union Carbide, German Appl. (West German) 1141085 [U.S. Prior. 30.6.1958]. (242) W. J. PATNODE, General Electric, U.S. Pat. 2306222 [16.11.1940]. (243) — and F. J. NORTON, Thomson-Houston, French Pat. 948923 [U.S. Prior. 16.11.1940 and 30.7.1942]. (244) J. PETER and G. FROMANDI, Farbenfabriken Bayer, German Pat. (West German) 922490 [17.7.1952]. (245) CH. L. PETZE, *SPE-J*, Vol. **14**, No. 12, 35 [1958]. (246) L. PFAU, *Z. Urol.* **47**, 426 [1954]. (247) E. H. PHREANER, *Rubber Age* **80**, 255 [1956]. (248) W. PIEPENBURG: *So macht man guten Putz*, Bauverlag, Wiesbaden-Berlin, 2nd Ed. (249) R. M. PIKE and E. L. MOREHOUSE, Union Carbide, French Pat. 77928/1184198 [U.S. Prior. 12.10.1956].

(250) E. PIRSON, *Kunststoffe* **47**, 597 [1957]. (281) — and M. WICK, *Fette, Seifen, Anstrichmittel* **55**, 792 [1953]. (252) J. B. PLEIN and E. M. PLEIN, *J. Amer. pharmac. Assoc.*, sci. edit. **42**, 79 [1953]. (253) E. P. PLUEDDEMANN, *J. Chem. Engng. Data* **5**, 59 [1960]. (254) —, Dow Corning, U.S. Pat. 2946701 [12.11.1957]. (255) —, —, Belg. Pat. 613364 [U.S. Prior. 2.2. and 19.5.1961]. (256) — and H. A. CLARK, —, Belg. Pat. 613466 [U.S. Prior. 6.2. and 19.5.1961]. (257) J. P. POOLE, *The Glass Industry* **1956**, 195. (258) W. PRIBILLA, *Die Medizinische*, No. 27/28, 928 [1953]. (259) M. PUSCHNER, *Kjellberg-Esab-Schriften* **5**, 15 [1958].

(260) C. E. REED: *The Industrial Chemistry, Properties and Application of Silicones*, Edgar Marburg Lecture 1956, publ. by the American Soc. for Testing Materials, Philadelphia. (261) L. H. REINKE, Dow Corning, U.S. Pat. 3046155 [6.10.1960]. (262) A. REISCHL, W. SIMMLER and E. MÜLLER, Farbenfabriken Bayer, German Appl. (West German) 1114632 [27.5.1960]. (263) —, H. HOLTSCHMIDT, W. NEUMANN and W. SIMMLER, —, German Pat. (West German) 1190176 [25.4.1961]. (264) —, W. GÖBEL, H. RÖHR and W. SIMMLER, —, French Pat. 1427660 [German Prior. (West German) 11.7.1963]. (265) H. REUTHER and E. ROSENBAUM, *Chem. Techn.* **7**, 333 [1955]. (266) Rhône-Poulenc, French Pat. 1087484 [25.8.1953]. (267) —, Brit. Pat. 792390 [French Prior. 13.4.1955]. (268) —, Brit. Pat. 872411 [French Prior. 5.8.1958]. (269) L. RIEDEL, *Glas-Email-Keramo-Technik* **1957**, 176.

(270) L. RISTER, *Österreichische Bauztg.* **9**, 254 [1933]. (271). E. G. ROCHOW: *An Introduction to the Chemistry of the Silicones*, 2nd Ed., New York, John Wiley and Sons, Inc., 1951, p. 135. (272) H. ROTHERT, *Elektrotechn. u. Maschinenbau* **74**, 326 [1957]. (273) F. RUTTKOWSKI, *AEG-Mitt.* **48**, 363 [1958]. (274) SALOMON, *Ind. Med. Surgery* **26**, 474 [1957]. (275) T. R. SANTELLI, I. G. MOHR and C. A. CALENDINE, Libbey-Owens-Ford Glass, French Pat. 1097128 [U.S. Prior. 24.3.1953]. (276) J. H. SAUNDERS and P. G. GEMEINHARDT, Mobay Chemical, German Appl. (West German) 1104690 [U.S. Prior. 4.9.1958]. (277) K. F. SCHEU and C. H. SCHEU, U.S. Pat. 2564546 [23.7.1947]. (278) W. SCHNEIDER, Augsburg 1955, unpublished. (279) K. SCHNURRBUSCH and W. NOLL, Farbenfabriken Bayer, Belg. Pat. 621194 [German Prior. (West German) 9.8.1961].

(280) H. A. Schuyten, J. W. Weaver, I. D. Reid and J. F. Jurgens, *J. Amer. Chem. Soc.* **70**, 1919 [1948]. (281). H. E. Schwiete, T. Langauer and G. Auchter, *Glastechn. Ber.* **35**, 65 [1962]. (282) P. W. Sherwood, *Paint Technology* **27**, No. 8, 57 [1963]. (283) E. G. Shur et al., *Paint, Oil and Chem. Rev.* **114**, No. 23, 59 [1951]. (284) G. Siebert, *Die Medizinische* **1952**, 1631. (285) Siemens-Schuckert-Werke, Belg. Pat. 535431 [German Prior. (West German) 30.4. and 30.6.1954]. (286) S. I. Sil'vestrovich and J. A. Boguslavaskii, *Proc. Acad. Sci. USSR, Chem. Technology Sect.* (*Engl. transl.*) **129**, 179 [1959]. (287) W. Simmler, *G-I-T-Fachzeitschrift für das Laboratorium* **6**, 101, 189, 278 [1962]. (288) — and W. Bächler, *Vacuumtechnik* **8**, 155 [1959]. (289) —, H. Niederprüm, M. Dahm and H. W. Patzelt, Farbenfabriken Bayer, German Appl. (West German) 1197616 [30.7.1963].

(290) G. Sinn, H. Hornig and W. Simmler, —, French Pat. 1339069 [German Prior. (West German) 17.10.1961]. (291) R. Smith-Johannsen, General Electric, U.S. Pat. 2601336 [17.2.1949]. (292) L. A. Spitze and D. O. Richards, *J. Appl. Physics* **18**, 904 [1947]. (293) M. von Stackelberg and V. Toome, *Leybolds polarograph. Ber.* **1**, No. 4, 55 [1953]. (294) H. Stamm and H. Arens, Th. Goldschmidt, German Pat. (West German) 857188 [21.11.1950]. (295) —, —, —, German Pat. (West German) 903621[22.8.1952]. (296) J. A. Stapp, *Amer. Ceram. Soc. Bull.*, Febr. **1959**, 64. (297) F. Stather and H. Herfeld, *Collegium* [Darmstadt] **1935**, 13. (298) H. Steinbach, Farbenfabriken Bayer, German Pat. (West German) 1161419 [23.12.1961]. (299) —, —, French Pat. 1364448 [German Prior. (West German) 28.7.1962].

(300) —, K. Damm and W. Simmler, —, Belg. Pat. 630579 [German Prior. (West German) 6.4.1962]. (301) R. Steinmann, Libbey-Owens-Ford Glass, French Pat. 1072614 [U.S. Prior. 7.1.1952]. (302) S. Sterman and I. G. Marsden, *Mod. Plastics* **40**, No. 11, 125 [1963]. (303) E. Stolte, *VDE-Fachberichte* **21**, I, 20 [1960], VDE-Verlag, Berlin. (304) E. Stutz, *Schweiz. Mschr. Zahnheilkunde* **61**, 438 [1951]. (305) R. R. Suskind, *Archiv. Ind. Hyg. and Occup. Med.* **9**, 101 [1954]. (306) R. G. Swanson, DuPont, U.S. Pat. 2614049 [27.6.1949]. (307) J. R. Talbot, J. K. McGregor and F. W. Crowe, *J. Invest. Dermat.* **17**, 125 [1951]. (308) M. E. Thayer, Dow Corning, U.S. Pat. 2474704 [1.8.1948]. (309) R. Theumann, *Jb. Chem. Ind.* **1957/58**, 70.

(310) R. N. Thomas, *Paint Industry Mag.* **76**, No. 3, 10 [1961]. (311) J. F. Thompson, *Paper, Film and Foil Converter* **28**, 22 [1954]. (312) U. J. L. Thuau, French Pat. 1316679 [22.12.1961]. (313) P. E. Tucker and T. W. Watson, *Paint Manufact.* **28**, 12 [1958]. (314) L. J. Tyler, Dow Corning, U.S. Pat. 2611774 [9.1.1950]. (315) Ulmer Preßwerk F. Zwick, German Pat. (West German) 955095 [9.4.1954]. (316) Union Carbide, Belg. Pat. 544555 [U.S. Prior. 21.1.1955]. (317) Union Chimique Belge, Belg. Pat. 551234 [24.9.1956]. (318) A. Verheyden, R. Leclercq and A. Gancberg, Union Chimique Belge, Belg. Pat. 541400 [19.9.1955]. (319) Wacker-Chemie, Belg. Pat. 607563 [German Prior. (West German) 26.8.1960].

(320) K. Wark, *AEG-Mitt.* **50**, 10/11, 460 [1960]. (321) J. A. C. Watt, *J. Textile Inst.* **48**, Transact. 175 [1957]. (322) K. Weigel, *Parfümerie u. Kosmet.* **37**, 14 [1956]. (323) H. Weissbach, *Zement-Kalk-Gips* **11**, 357 [1958]. (324) G. Weissenberg, German Pat. (West German) 925226 [4.1.1951]. (325) M. Weist, VEB Silikonchemie, German Pat. (East German) 5812 [4.12.1951]. (326) H. Wessel, *Silikattechnik* **9**, 201 [1958]. (327) A. Wessel, *Therap. Ber.* **30**, 100 [1958]. (328) C. L. Whipple, Dow Corning, Belg. Pat. 603552 [U.S. Prior. 13.5.1960]. (329) L. C. Whitman and P. Doigan, *Trans. Amer. Inst. electr. Engr.* (I) (Commun. and Electronics) **73**, 194 [1954].

(330) M. WICK, *Elin-Zeitschr.* **12**, No. 2, 106 [1960]. (331) W. WUNDER, *Melliands Textilber.* **43**, 1089 [1962]. (332) E. C. YEN and F. E. STIRN, American Cyanamid, U.S. Pat. 2512192 [26.5.1948]. (333) W. ZECH, *Dentaldienst* **10**, 10 [1958]. (334) M. ZINGSHEIM and R. LANGE, *J. med. Kosm. u. Sexologie* **1952**, 5. (335) —, —, *Strahlentherapie* **90**, 638 [1953]. (336) R. ZÖLLICH, *Zahnärztl. Welt u. zahnärztl. Reform* **60**, No. 12, 369 [1959]. (337) B. ZORN and K. DAMM, Farbenfabriken Bayer, Brit. Pat. 923067 [German Prior. (West German) 28.2. 1959]. (338) — and W. ZEISS, —, German Appl. (West German) 1027360 [27.4.1956]. (339) AIEE-(Amer. Inst. Electr. Engr.) Standard No. 510: *Test procedure for evaluation of systems of insulating materials for random-wound electric machinery*, New York, Nov. 1956.

(340) ibid. No. 511: *Proposed test procedure for evaluation of systems of insulating materials for electric machinery employing form-wound pre-insulated coils*, published for trial use, New York, Nov. 1956. (341) Bayer-Kunststoffe, *Leverkusen* **1955**, p. 268. (342) *Bayer-Mitt. für die Gummi-Industrie* No. 24, Dez. 1956, p. 29. (343) DIN 105, March 1957. (344) DIN 1055, November 1953, sheet 4. (345) DIN 18550: *Putz, Baustoffe und Ausführung*, October 1955. (346) DIN 46453, December 1954. (347) DIN 53122, February 1958. (348) *Electr. Times* **1956**, 44. (349) General Electric: *Silicone Rubber for Electronic Wire Insulation* (Publication).

(350) *Österreichische Malerztg.* **9**, 53 [1958]. (351) Silicones for traction-motor insulation, *Electr. Energy* **1958**, No. 5, 188. (352) Silicone insulation for traction motors, *Engineer* **1958**, 528. (353) VDE 0170/September 1957 and VDE 0171/September 1957. (354) VDE 0283/October 1952. (355) VDE 0530, *Bestimmungen für elektrische Maschinen*, part 1/1.66, § 13, Table 3. (356) VDE 0530, *Bestimmungen für elektrische Maschinen*, Part 1/1.66, § 26, Table 6, No. 9 and note 6. (357) VDE 0532/July 1955, § 12, column 6.

CHAPTER 11

Esters of Silicic Acid (Organooxysilanes and Polyorganooxysiloxanes)*

Owing to the ease with which these compounds can be made by the reaction of silicon tetrachloride with alcohols or phenols, and with water in the case of siloxanes—the first silicic acid ester was probably synthesized in 1846 (*20*)—this field was opened up very early and was explored much more intensively than that of the more difficultly obtainable organosilicon compounds. Large numbers of silicic acid esters were already known at the beginning of the twentieth century, when the organosilicon compounds were scarcely more than curiosities. The interest of chemists in them has been retained up to the present time, but the field has not experienced such rapid industrial development as that of the organosilicon compounds.

A comprehensive review of the chemistry of silicic acid esters cannot be given here (for special literature see refs. *31* and *100*). We shall merely make an attempt to develop parallels and divergencies between the organosilicon and the organooxysilicon compounds, and to show what directions of development have become technically fruitful in this region closely related to that of the chemistry of the silicones.

11.1 Chemistry and Constitution

In contrast to the organosilicon compounds, which contain at least one Si—C bond in the molecule, in the organooxysilanes and polyorganooxysiloxanes the organic groups are attached to the silicon only via oxygen atoms. Compounds which contain *both* C—Si and O—Si linkages, i.e., organoorganooxysilanes and organoorganooxysiloxanes, have already been treated in the preceding chapters as belonging to the chemistry of the silicones.

* For the nomenclature see Section 1.2.

The constitution of the silicic acid esters is closely related to that of the silicones. While in the latter it is the number of Si—C bonds which characterizes the siloxane units, in the former it is the number of Si—OC bonds (Table 109).

TABLE 109

SILOXANE UNITS OF THE ORGANOOXYSILANES AND POLYORGANOOXYSILOXANES (SILICIC ACID ESTERS) AS COMPARED WITH THOSE OF THE POLYORGANOSILOXANES (SILICONES)

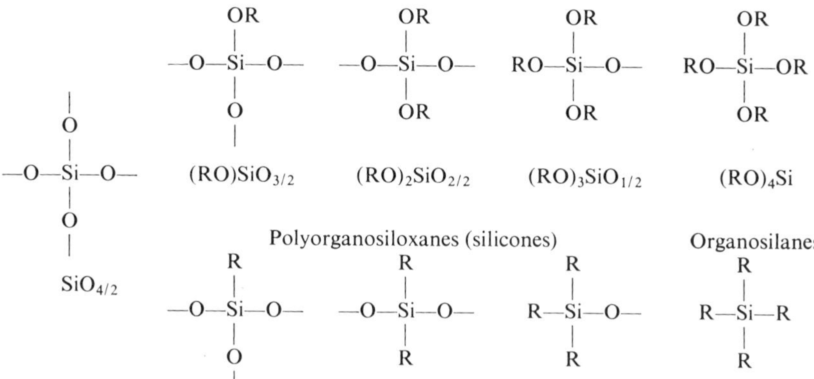

Q	T	D	M	
	Polyorganooxysiloxanes (Polysilicic acid esters)			Organooxysilanes (orthosilicic acid esters)
$SiO_{4/2}$	$(RO)SiO_{3/2}$	$(RO)_2SiO_{2/2}$	$(RO)_3SiO_{1/2}$	$(RO)_4Si$
	Polyorganosiloxanes (silicones)			Organosilanes
	$RSiO_{3/2}$	$R_2SiO_{2/2}$	$R_3SiO_{1/2}$	R_4Si

In parallel with the true silicones (polyorganosiloxanes) $R_nSiO_{(4-n)/2}$, therefore, there is a series of polyorganooxysiloxanes (polysilicic acid esters) $(RO)_nSiO_{(4-n)/2}$ with $n=1$, 2, and 3. As with the former, $SiO_{4/2}$ units (with $n=0$) can also be included as constituents.

The counterpart to the organosilanes R_4Si are organooxysilanes (orthosilicic acid esters). While, however, the organosilanes have acquired practically no importance in the chemistry of the silicones, the monomeric organooxysilanes are an important member of the silicic acid series—in fact, industrially perhaps the most important.

The analogies in their constitutions permit us to expect that polymers similar to those of silicone chemistry, and with similar forms, can be built up in the chemistry of the silicic acid esters. Here again, a multiplicity of polymers is conceivable which, starting from the limiting form of silica glass, pass via resinlike more or less intensively cross-linked polymers to the region of liquid siloxanes with chain or ring structures. The monomeric organo-oxysilanes, with their properties that can be varied in many directions, are also of interest.

TABLE 110

SOME REACTIONS OF THE $\equiv Si-OR$ BOND

$2\equiv Si-OR + H_2O$	$\rightarrow$	$\equiv Si-O-Si\equiv + 2ROH$
$\equiv Si-OR + HO-Si\equiv$	$\rightarrow$	$\equiv Si-O-Si\equiv + ROH$
$\equiv Si-OR + R'OH$	$\rightarrow$	$\equiv Si-O-R' + ROH$
$\equiv Si-OR + R'O-Si\equiv$	$\rightarrow$	$\equiv Si-O-Si\equiv + ROR'$
$2\equiv Si-OR + 2HCl$	$\rightarrow$	$\equiv Si-O-Si\equiv + 2RCl + H_2O$
$\equiv Si-OR + Cl-Si\equiv$	$\rightarrow$	$\equiv Si-O-Si\equiv + RCl$
$2\equiv Si-OR + 2R'COOH$	$\rightarrow$	$\equiv Si-O-Si\equiv + 2R'COOR + H_2O$
$\equiv Si-OR + R'COO-Si\equiv$	$\rightarrow$	$\equiv Si-O-Si\equiv + R'COOR$
$2\equiv Si-OR + R'CHO$	$\rightarrow$	$\equiv Si-O-Si\equiv + R-\underset{RO\ \ \ OR}{C}-H$
$\equiv Si-OR + NaOH$	$\rightarrow$	$\equiv Si-O-Na + ROH$
$\equiv Si-OR + HO-B<$	$\rightarrow$	$\equiv Si-O-B< + ROH$
$\equiv Si-OR + R'MgCl$	$\rightarrow$	$\equiv Si-R' + Mg(OR)Cl$

* Page number of chapter references.

In principle, the whole spectrum of silicone polymers could be reproduced again in the chemistry of the silicic acid esters, but there are considerable difficulties in applying principles of definite polymer construction to silicic esters. This is inherent in the properties of the Si—OR bonds, the discussion of which will also lead to an understanding of the technological possibilities in the chemistry of the silicic acid esters and their limits.

The important point to remember is that the Si—OC bond is in general substantially *more reactive* than the Si—C bond (cf. also Section 3.3.3). It takes part in various reactions which lead to siloxane compounds (Table 110) and is, *inter alia*, subject to hydrolytic, acidolytic, and thermal degradation. In comparison with the Si—C bond, therefore, the Si—OC grouping can be regarded in general as *functional*, and as such it plays a certain part in the chemistry of the silicones (Section 3.3).

This somewhat generalized statement obviously requires a limitation; on the one hand, the Si—C bond can be so weakened by electronegative substituents that it too becomes sensitive to hydrolysis (Section 4.5.4), and on the other hand, as will be shown below, the sensitivity of the Si—OC bond to attack can be greatly reduced. The differences between the reactivities of the two types of bond are most strongly pronounced in silanes and siloxanes with only unsubstituted aliphatic radicals bound to silicon through direct Si—C bonds and in those with only organooxy groups having a small number of C atoms, especially methoxy and ethoxy groups.

Because of the reactivity of the Si—OC bond, silicic acid esters might be regarded as fairly unpromising from the point of view of polymer chemistry, since they seem to offer little hope of forming substances resistant to atmospheric humidity. This point of view has now changed, and more and more examples have recently become known of special Si—OR bonds characterized by a sometimes astonishing stability. The common principle of all these types of bond appears to be a *bulkiness* of the structure of the R group and a *steric screening* of the Si—OC bond.

In the series of aliphatic silicic acid esters, the methyl and ethyl esters are most susceptible to hydrolysis. The sensitivity to hydrolysis falls with the increasing number of carbon atoms and with the increasing branching of the hydrocarbon residue (*28*, *68*, *70*).* Thus, for example, it falls to a very pronounced extent in the sequence:

$$\gt Si—O—n\text{-}C_4H_9 > \ \gt Si—O—s\text{-}C_4H_9 > \ \gt Si—O—t\text{-}C_4H_9$$

Silicic acid esters of phenols (I) are already less sensitive to hydrolysis than the esters of the lower aliphatic alcohols. The resistance to hydrolysis can be

* The statements are fairly qualitative. Numerical comparisons are difficult since the sensitivity of hydrolysis has not yet been investigated by standardized methods.

increased considerably by the introduction of ethyl groups into the benzene ring in the *o,o′* position (II) (*127*):

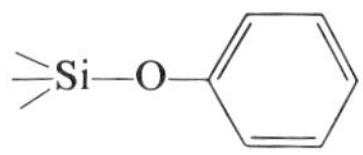

I

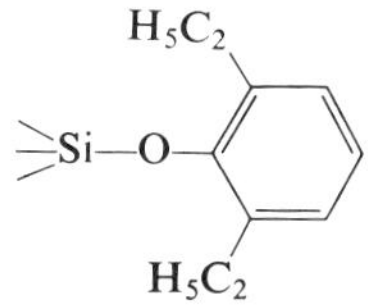

II

On passing from silicic acid esters of phenols to those of *o*-hydroxybiphenyl and of *o,o′*-dihydroxybiphenyl, the resistance to hydrolysis increases in the sequence (*122*):

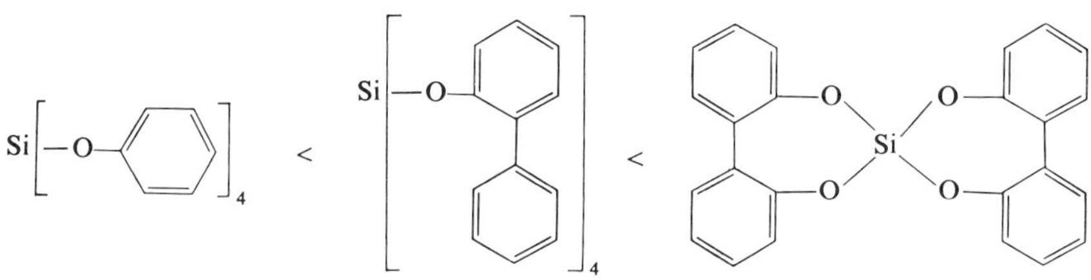

By making use of this knowledge, it has today become possible to prepare silicic acid esters that can be used industrially as synthetic oils or as resin components.

These efforts are in contrast to the attempts to make use of the high reactivity of the lower alkoxy groups bound to the silicon atom for the preparation of silicic acid, so that developments in this field are characterized by two completely divergent tendencies: on the one hand, toward the stabilization of the Si—OR bond and, on the other hand, toward making it as reactive as possible (cf. Section 11.3).

11.2 Preparation

11.2.1 Tetraorganooxysilanes and Spiroorganooxysilanes

11.2.1.1 *Preparation from Silicon Tetrachloride*

The commonest process for the preparation of organooxysilanes is the reaction of *silicon tetrachloride* with *alcohols* or *phenols*.

The performance of the reaction depends in detail on the nature of the alcohol or phenol. Account must be taken of possible side reactions, among which esterification of the alcohol by the hydrogen chloride formed in the reaction plays a particular role:

$$\mathrm{ROH+HCl \rightarrow RCl+H_2O}$$

This reaction becomes particularly noticeable in the preparation of organooxysilanes with *lower aliphatic* OR groups; it leads to a reduction in the yield of tetrasubstituted silanes and causes the formation of corresponding amounts of polymeric organooxysiloxanes.

The side reactions can be suppressed to a large extent by a suitable sequence of combining the reactants. The alcohol should preferably be added to the chlorosilane, so that there is always an excess of silicon tetrachloride and the hydrogen chloride formed finds no excess of alcohol. In this case, the reaction mixture is cooled by the removal of the heat of evaporation of the hydrogen chloride, which also supresses the side reactions. For the preparation of tetramethoxysilane, this method of working is not only recommended but necessary to obtain useful yields (*88*); in addition, it is advisable to use solvents and to cool the reaction mixture (*109*). From propanol onward, with an increasing number of carbon atoms the yield becomes less sensitive to the side reactions mentioned, so that figures of 90% can easily be achieved. The esterification always requires heating for several hours with a small excess of alcohol after the components have been mixed. The alkoxysilane is finally isolated from the reaction mixture by distillation.

The preparation of *tetraaryloxysilanes* presents no difficulties if the melting points of the phenols are below the boiling point of silicon tetrachloride. Otherwise, suitable organic solvents are used.

If *polyalcohols* are treated with silicon chloride, *polymers* with groups such as

$$\mathrm{-O-\overset{|}{\underset{|}{Si}}-O-R-O-\overset{|}{\underset{|}{Si}}-O-}$$

are generally produced. However, with certain difunctional alcohols monomeric *spirocyclic* silanes are also formed, e.g., with pinacol, butane-1,4-diol, diethyleneglycol, and 1,1′-dihydroxy-1,1′-bicyclohexyl (cf. Table 111). The process is carried out in inert solvents such as benzene or chloroform, and in the presence of a tertiary amine (pyridine) as the hydrogen chloride acceptor (*34*, *106*).

The preparation of silicic acid esters is also possible as a continuous process, either by passing the silicon tetrachloride and alcohol together in the same direction or by passing the former as a vapor countercurrent to the liquid alcohol (*21*, *24*).

Alkali-metal alkoxides react with tetrahalosilanes like the alcohols; they give tetraorganooxysilanes even when steric hindrance is present and the

TABLE 111
TETRAORGANOOXYSILANES[a]

Compound	ν (cSt) at	t (°C)	Mp (°C)	Bp (°C) at	p (mm Hg)	n_D^{20}	d_4^t (gm/cc) at	t (°C)
$Si(OCH_3)_4$	—	—	2 (*78*)	121 (*116*)	764	1.3681 (*116*)	1.0339 (*104*)	20
$Si(OC_2H_5)_4$	0.800 (*102*)	20	−77 (*12*)	168.5 (*108*)	760	1.3831 (*116*)	0.9346 (*116*)	20
$Si(n\text{-}OC_3H_7)_4$	1.66 (*102*)	20	< −80 (*102*)	224–225 (*35*)	750	1.40107 (*104*)	0.9105 (*104*)	20
$Si(i\text{-}OC_3H_7)_4$	1.28 (*73*)	25	−22 (*74*)	185–186 (*86*)	742	1.38520 (*104*)	0.8754 (*104*)	20
$Si(n\text{-}OC_4H_9)_4$	2.33 (*102*)	20	< −80 (*102*)	273–277 (*35*)	760	1.41333 (*104*)	0.8981 (*104*)	20
$Si(i\text{-}OC_4H_9)_4$	—	—	—	250–251 (*86*)	741	1.4090 (*30*)	0.884 (*86*)	20
$Si(s\text{-}OC_4H_9)_4$	~3.0 (*23*)	20	~ −105 (*23*)	87 (*30*)	2	1.4070 (*30*)	0.883 (*30*)	20/20
$Si(t\text{-}OC_4H_9)_4$	—	—	46–48 (*7*)	92–95 (*7*)	9	—	—	—
$Si(O{-}CH_2{-}CH(C_2H_5){-}C_4H_9)_4$	6.83 (*55*)	37.6	< −73 (*55*)	191–192 (*55*)	9	—	—	—
$Si(OC_6H_5)_4$	6.6 (*94*)	55	52–53 (*8*)	400 ± 5 (*15*)	760	1.5583[b] (*94*)	1.1653 (*94*)	55/55
$Si(o\text{-}OC_6H_4{-}CH_3)_4$	31.5 (*73*)	25	—	445 (*73*)	760	—	1.114 (*73*)	25
$Si(m\text{-}OC_6H_4{-}CH_3)_4$	—	—	—	284–286 (*90*)	15	—	1.120 (*90*)	16
				443–446 (*37*)	720			
$Si(p\text{-}OC_6H_4{-}CH_3)_4$	—	—	69–70 (*37*)	442–445 (*37*)	~760	—	—	—
				290–291 (*90*)	15			
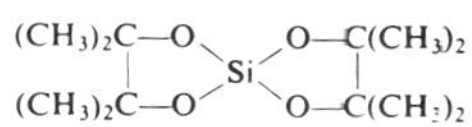	—	—	112 (*34, 106*)	238 (*34, 106*)	758	—	—	—
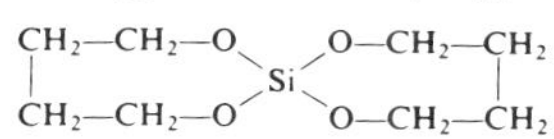	—	—	46 (*34, 106*)	—	—	—	—	—
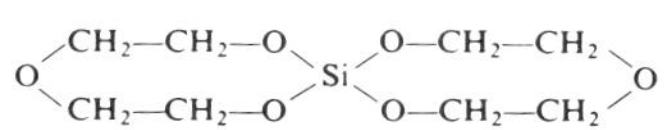	—	—	84–85 (*34, 106*)	108 (*34, 106*)	3	—	—	—
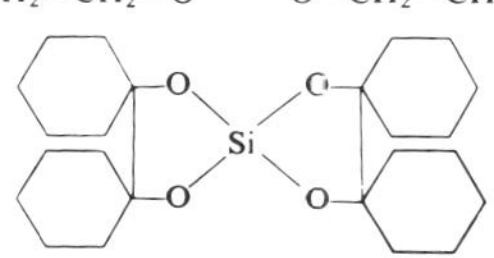	—	—	282 (*34*)	—	—	—	—	—

[a] Numbers in parentheses indicate reference numbers.
[b] 54.7°C.

reaction of the chlorosilane with alcohol stops at the stage of di- or trisubstitution by RO groups. Thus, with *t*-butanol in pyridine, silicon tetrachloride gives tri-*t*-butoxychlorosilane which can subsequently be converted by means of sodium isopropoxide into tri-*t*-butoxy-isopropoxysilane in 63% yield (*2*); on heating under pressure, it also reacts with sodium *t*-butoxide (*118*). At 220°C with sodium *t*-butoxide, tri-*t*-butoxybromosilane gives tetra-*t*-butoxysilane (*7*). The reaction of Si—Cl bonds with *alkali-metal phenoxides* in inert organic solvents also takes place with good yields (*51*).

Even the *Si—F bond*, which hardly reacts with alcohols, can be caused to react smoothly with *alkoxides*. On heating with aluminum alkoxides to 260 to 300°C, silicon tetrafluoride and sodium hexafluorosilicate give tetraalkoxysilanes in more than 80% yield (*40*, *81*, *82*).

The reaction of *ethers* with silicon tetrachloride does not give tetraorganooxysilanes but alkoxychlorosilanes, which in turn can be condensed further to siloxanes (cf. Section 3.2.3). Of the reactions with *ketones*, it is known that *benzophenone* gives a spirocyclic alkoxysilane with the reaction product of silicon tetrachloride and magnesium in ether (*45*):

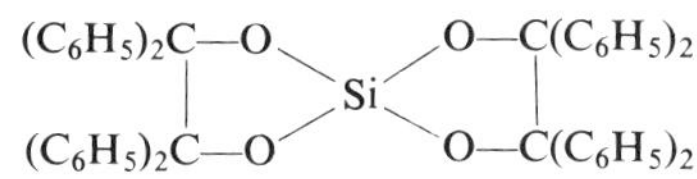

Epoxides react with $SiCl_4$ with ring rupture and the formation of ω-haloalkoxysilanes; with silicon tetrachloride, on boiling under reflux, ethylene oxide gives an almost 75% yield of tetra-(2-chloroethoxy)-silane (*67*):

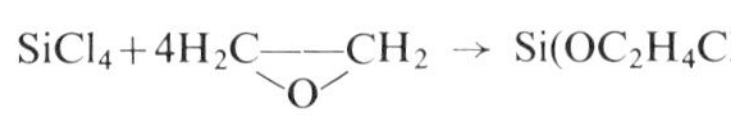

11.2.1.2 *Preparation by Transesterification*

Organooxysilanes with small aliphatic R groups, particularly ethoxy- and methoxysilanes, can often be *transesterified* with higher aliphatic alcohols in a smooth reaction. Such transesterification reactions have acquired particular importance for the preparation of silanes which are difficultly accessible because of steric hindrance or because of side reactions; they are also important for the preparation of mixed substituted alkoxysilanes. Under pressure, the reverse reaction can take place, with a higher alkoxy group being replaced by a lower one (*25*). The reaction is catalyzed by hydrogen chloride, preferably obtained by the addition of small amounts of silicon tetrachloride to the reaction mixture (*72*, *76*), and, particularly, by alkali-metal alkoxides (*6*, *29*, *36*, *50*). The reaction with phenols is also catalyzed by sodium or aluminum alkoxides or by sodium phenoxide (*50*, *117*).

Transesterification can also be carried out with aryloxysilanes if there is a sufficient difference in boiling points between the phenols to be exchanged. Thus, tetraphenoxysilane reacts with hydroquinone in accordance with the equation (*80*):

$$2Si(OC_6H_5)_4 + HOC_6H_4OH \rightarrow (C_6H_5O)_3Si{-}OC_6H_4O{-}Si(OC_6H_5)_3 + 2C_6H_5OH$$

11.2.1.3 *Preparation from Silicon or Silicon Alloys*

In order to avoid the roundabout route via the chlorination product of silicon, $SiCl_4$, attempts have been made to obtain silicic esters directly from silicon and alcohols or phenols.

Alcohols react with silicon in the presence of copper at elevated temperatures with moderate or poor yields to give tetraalkoxysilanes (*77*) (cf. Section 2.2.4):

$$Si + 4ROH \rightarrow Si(OR)_4 + 2H_2$$

In the case of tetramethoxysilanes, the yield amounts to 40% on the alcohol and ~5% on the silicon. Higher yields (63% on the alcohol, 40% on the silicon) are obtained by heating $CaSi_2$ with methanol in the presence of small amounts of copper and chlorine at 230°C (*105*). The reaction with magnesium silicide takes place even at room temperature and gives yields of about 40% on the silicon (*33*):

$$Mg_2Si + 8CH_3OH \rightarrow Si(OCH_3)_4 + 2Mg(OCH_3)_2 + 4H_2$$

The reactions of silicon with phenols take place more readily. With silicon-copper alloys at 300 to 400°C under (hydrogen) pressure, phenol gives tetraphenoxysilane with a yield of almost 80% (*4*, *128*). 2,2′-Dihydroxybiphenyl reacts even at 130°C (*4*, *128*).

11.2.1.4 *Preparation from SiS_2, SiO_2, and Other Compounds*

Besides the processes mentioned above, there are a number of other possibilities of preparing organooxysilanes; however, in general they have acquired no practical importance. Thus, alcohols and phenols can react readily and smoothly with silicon disulfide (*14*, *15*):

$$4ROH + SiS_2 \rightarrow Si(OR)_4 + 2H_2S$$

The reaction takes place via the formation of alkoxythiosilanols, which may also be present in the final product under certain circumstances (*124*).

In the form of silica sol or highly disperse SiO_2, or even as quartz, silica reacts with *o*-dihydroxyarenes, particularly catechol, in a weakly alkaline medium, even at room temperature, to give stable silicic acid esters (*123*). It is less reactive with respect to alcohols. The addition of alkali-metal hydroxide or of alkali-metal silanolates and heating at 100 to 200°C at up to about 30 atm are necessary for the reaction to take place. The water formed according to the equation

$$4ROH + SiO_2 \rightarrow Si(OR)_4 + 2H_2O$$

has to be removed continuously. Under such conditions tetraethoxysilane, for example, is obtained in almost quantitative yield (*3*). Under normal pressure and at room temperature, this reaction takes place only in traces.

Alcohols, particularly those with a small number of carbons, also react with the *Si—H bond* in trichlorosilane (*26*, *35*):

$$SiHCl_3 + 4ROH \rightarrow Si(OR)_4 + H_2 + 3HCl$$

In other cases, the reaction must be catalyzed, for which purpose alkali-metal alkoxides, for example, are suitable (*59*, *69*). Silane reacts with lower alcohols—apart from methanol—only in the presence of a lithium alkoxide:

$$SiH_4 + 4ROH \xrightarrow{LiOR} Si(OR)_4 + 4H_2$$

but the catalytic effect can be observed only when the alkoxide contains the residue of the alcohol used (*69*).

Finally, *amines* can react with silanols to give alkoxysilanes. Primary amino groups on tertiary carbon atoms react with the silanol bond at elevated temperatures and in the presence of basic catalysts (*125*, *126*):

$$\diagdown\!\!-Si-OH + H_2NR \rightarrow \diagdown\!\!-Si-OR + NH_3$$

The reaction is suitable for introducing tertiary alkoxy residues into already alkoxy-substituted silanes in which steric hindrance impedes further substitution.

11.2.2 Polyorganooxysiloxanes

11.2.2.1 *Preparation by the Hydrolysis of Monomeric Silanes and Condensation*

To construct a polymer chemistry of the polyorganooxysiloxanes with a systematic completeness similar to that of the silicones, we would need mono-, di-, and trifunctional organooxysiloxane units which, in turn, again in analogy with silicone chemistry, would be made from correspondingly functional organooxysilanes:

$$(RO)SiX_3 \quad (RO)_2SiX_2 \quad (RO)_3SiX$$

The fact that this problem is very much more difficult than in silicone chemistry, is due to the above-mentioned reactivity of the Si—OR bond, which is affected to a greater or smaller extent any measures employed for splitting the functional Si—X bond; in other words, during the treatment of the monomers, their functionality is at least partially increased, and the desired polymers cannot therefore be obtained with the desired certainty.

Nevertheless, some investigations are available which show that the route sketched out is possible in principle. Definite polymers have been obtained by the hydrolysis of *alkoxychlorosilanes* when the process was carried out in the presence of *acid acceptors* which protected the Si—OC bond from the attack of the aggressive hydrogen chloride, or if the hydrolysis was carried out with *salt hydrates* (*62*).

Triethoxychlorosilane can be converted into hexaethoxydisiloxane in 60% yield with the calculated amount of water in the presence of pyridine (*71*):

$$2(RO)_3SiCl + H_2O \xrightarrow{\text{pyridine}} (RO)_3Si{-}O{-}Si(OR)_3 + 2HCl$$

Under analogous conditions, tri-*n*-butoxychlorosilane gives hexa-*n*-butoxydisiloxane (*38*); and triphenoxychlorosilane reacts with half a mole of water, added in the form of sodium sulfate decahydrate, to give hexaphenoxydisiloxane (*79*).

Cyclic polymers containing 3 to 8 siloxane units in the ring have been prepared from di-*n*-butoxydichlorosilane by hydrolysis in the presence of a tertiary amine (*38*) (cf. Table 112):

$$(RO)_2SiCl_2 \xrightarrow[\text{amine}]{\text{water}} [(RO)_2SiO]_n \qquad (n = 3 \text{ to } 8)$$

It is an interesting fact that practically no high-molecular-weight linear polysiloxanes are found in this case, in contrast to the situation in the hydrolysis of dimethyldichlorosilane.

On analogous hydrolysis, *n*-butoxytrichlorosilane gave a viscous liquid which solidified to a friable gel on heating in vacuum (*38*). Mixtures of mono- and di-*n*-butoxychlorosilanes have given resinlike masses. The hydrolysis of mixtures of mono-*n*-butoxy and tri-*n*-butoxychlorosilanes has actually given small amounts of high-boiling oils which possibly resemble polycyclic or spherocyclic siloxanes in their structure (cf. Table 43) (*38*).

The hydrolysis and condensation of di-*t*-butoxydichlorosilane takes place particularly simply, since the Si—OC bond in it is protected sterically. The silane hydrolyzes and condenses through a series of definite isolable intermediates, of which the chlorosilanols are more reactive than the silanols. The cyclic trimeric siloxane predominates among the condensation products, obviously being preferentially formed under the steric constraint of the bulky *t*-butoxy groups (*83*).

The *Si—H grouping* has also been used as a functional bond for the formation of alkoxysiloxane units and their polymers. It is converted by dilute caustic soda into the siloxane bond, but it is impossible to prevent the Si—OC bond from being involved to a smaller or greater extent as well (*41*, *75*).

TABLE 112

LINEAR AND CYCLIC POLYORGANOOXYSILOXANES[a]

Compound	Mp (°C)	Bp (°C) at	p (mm Hg)	n_D^{20}	d_4^{20} (gm/cc)
(a) Linear polymers					
$(CH_3O)_3SiOSi(OCH_3)_3$	—	118.5 (*112*)	30	1.38058 (*104*)	1.1233 (*104*)
$(C_2H_5O)_3SiOSi(OC_2H_5)_3$	—	230–232 (*41*)	760	1.39160 (*104*)	0.9979 (*104*)
$(n\text{-}C_3H_7O)_3SiOSi(On\text{-}C_3H_7)_3$	—	205–208 (*1*)	25	1.40643 (*104*)	0.9583 (*104*)
$(i\text{-}C_3H_7O)_3SiOSi(Oi\text{-}C_3H_7)_3$	—	278–280 (*86*)	764	1.39241 (*104*)	0.9260 (*104*)
$(n\text{-}C_4H_9O)_3SiOSi(On\text{-}C_4H_9)_3$	—	245–250 (*1*)	20	1.41718 (*104*)	0.9362 (*104*)
$(i\text{-}C_4H_9O)_3SiOSi(Oi\text{-}C_4H_9)_3$	—	310–312 (*86*)	742	—	0.926 (*86*)
$(s\text{-}C_4H_9O)_3SiOSi(Os\text{-}C_4H_9)_3$	—	139–141 (*126*)	1	1.4132 (*126*)	—
$(t\text{-}C_4H_9O)_3SiOSi(Ot\text{-}C_4H_9)_3$	~260 (*2*) decomp.	160–170 (*2*) sublimes	2	—	—
$H(n\text{-}C_4H_9O)_2SiOSi(On\text{-}C_4H_9)_2H$	—	146–150 (*75*)	6	1.4104 (*75*)[b]	0.9151 (*75*)[c]
$HO(t\text{-}C_4H_9O)_2SiOSi(Ot\text{-}C_4H_9)_2OH$	77 (*83*)	140 (*83*)	8	—	—
$(C_6H_5O)_3SiOSi(OC_6H_5)_3$	—	308–310 (*79*)	3	—	—
$(CH_3O)_3SiO[(CH_3O)_2SiO]Si(OCH_3)_3$	—	95–96 (*112*)	2	1.38640 (*104*)	1.1658 (*104*)
$(C_2H_5O)_3SiO[(C_2H_5O)_2SiO]Si(OC_2H_5)_3$	—	268–273 (*84*)	760	—	—
$(i\text{-}C_3H_7O)_3SiO[(i\text{-}C_3H_7O)_2SiO]Si(Oi\text{-}C_3H_7)_3$	—	318–329 (*86*)	745	—	1.007 (*86*)
$(i\text{-}C_4H_9O)_3SiO[(i\text{-}C_4H_9O)_2SiO]Si(Oi\text{-}C_4H_9)_3$	—	345–348 (*86*)	741	—	0.949 (*86*)
$(s\text{-}C_4H_9O)_3SiO[(s\text{-}C_4H_9O)_2SiO]Si(Os\text{-}C_4H_9)_3$	—	187–188 (*126*)	1	1.4178 (*126*)	—
$(C_2H_5O)_3SiO[(C_2H_5O)_2SiO]_2Si(OC_2H_5)_3$	—	290–295 (*84*)	760	—	—
$(C_2H_5O)_3SiO[(C_2H_5O)_2SiO]_4Si(OC_2H_5)_3$	—	300–310 (*84*)	760	—	—
(b) Cyclic polymers					
$[(C_2H_5O)_2SiO]_4$	2.5 (*110*)	138–140 (*110*)	1.5	1.4001 (*110*)	1.0946 (*110*)
$[(C_3H_7O)_2SiO]_4$	—	237–240 (*86*)	15	—	1.053 (*86*)
$[(n\text{-}C_4H_9O)_2SiO]_3$	—	185–190 (*38*)	3	1.4196 (*38*)	—
$[(n\text{-}C_4H_9O)_2SiO]_4$	—	220–225 (*38*)	3	1.4220 (*38*)	—
$[(n\text{-}C_4H_9O)_2SiO]_5$	—	245–250 (*38*)	3	1.4228 (*38*)	—
$[(n\text{-}C_4H_9O)_2SiO]_6$	—	275–280 (*38*)	3	1.4230 (*38*)	—
$[(n\text{-}C_4H_9O)_2SiO]_8$	—	300–326 (*38*)	3	1.4240 (*38*)	—
$[(i\text{-}C_4H_9O)_2SiO]_4$	—	256–260 (*86*)	15	—	0.977 (*86*)
$[(t\text{-}C_4H_9O)_2SiO]_3$	158–159 (*83*)	201 (*83*)	5	—	—
$[(C_6H_{11}O)_2SiO]_3$	216 (*98*)	—	—	—	—

[a] Numbers in parentheses indicate reference numbers [b] n_D^{25}. [c] d_4^{25}.

It has been suggested that *silanol condensation* actually occurs in the production of hexamethoxydisiloxane by the hydrolysis of tetramethoxysilane, trimethoxysilanol being formed as an intermediate (*46*). Condensations between *alkoxysilanols* and *other functional alkoxysilanes* are also possible, in analogy with the corresponding reactions of organosilane derivatives (*11*, *32*, *126*) (see Section 5.2):

$$\equiv\mathrm{Si{-}OH} + \mathrm{X{-}Si}\equiv \;\rightarrow\; \equiv\mathrm{Si{-}O{-}Si}\equiv + \mathrm{HX} \qquad (\mathrm{X} = \mathrm{Cl}, \mathrm{NH_2})$$

The reaction of trialkoxysilanols containing higher and branched alkoxy groups with dialkoxydichlorosilanes has given octaalkoxytrisiloxanes (*56*):

$$\begin{array}{c} \mathrm{OR} \\ | \\ \mathrm{(RO)_3Si{-}O{-}Si{-}O{-}Si(OR)_3} \\ | \\ \mathrm{OR} \end{array}$$

It is substantially more difficult to obtain definite polymers from *tetraalkoxysilanes* than from those alkoxysilanes which contain a functional group other than the alkoxy group. Partial hydrolysis of these silanes gives, in general, mixtures of siloxanes of different functionalities (*86*, *111*). On partial hydrolysis, tetraisopropoxy- and tetraisobutoxysilanes are converted into linear hexaisopropoxy- and hexaisobutoxydisiloxanes and into cyclic octaisopropoxy- and octaisobutoxytetrasiloxanes (*86*). Tetraisopropoxysilane is also partially hydrolyzed with sodium carbonate containing water of crystallization (*63*). In general, such partial hydrolyses are used for the preparation of high-molecular-weight polyorganooxysiloxanes. They can be controlled better when the starting materials are *mixed* alkoxysilanes containing easily hydrolyzable alkoxy groups together with difficult hydrolyzable ones, and these are hydrolyzed in the presence of the difficultly hydrolyzable alcohols (*61*).

Another kind of "degradation" of tetraalkoxysilanes to polyalkoxysiloxanes consists in treating them with the calculated amount of a *carboxylic acid*, particularly acetic acid, so that the desired molar proportion of alkoxy groups is split off as carboxylic acid esters. This principle can be used for a *one-stage* synthesis of polyalkoxysiloxanes by allowing silicon tetrachloride, alcohol, and a carboxylic acid (particularly acetic acid) to react simultaneously (*101*).

The *total hydrolysis* of tetraalkoxysilanes gives *silicic acid gels*, particularly rapidly in the presence of acid or alkali. Stable *dispersions of silicic acid* are obtained in the presence of water-soluble amines, especially triethanolamine (*95*, *99*).

However, polyalkoxysiloxanes can be obtained more simply than by the roundabout route via definite monomeric alkoxysilanes and their hydrolytic degradation, by combining the *esterification* of silicon tetrachloride with a simultaneous *partial hydrolysis*; thus, silicon tetrachloride is treated with a mixture of alcohol and water, in which the water content is so adjusted that the desired number of Si—O—Si bonds is produced in the polymer (*22*). The resulting polymers are obviously less defined and uniform than in the hydrolytic degradation of tetraorganooxysilanes, but as a one-stage process it is of interest for industrial purposes.

11.2.2.2 *Alkoxylation of Polysiloxanes*

Instead of forming the Si—OR bond in a first stage and the Si—O—Si bond in a second stage, or both simultaneously, it is also possible to start from polymers already containing the siloxane bond and, in addition, reactive substituents which can be replaced by alkoxy groups.

Suitable polymers for this purpose are *polychlorosiloxanes*, which can be obtained, for example, by the partial hydrolysis of silicon tetrachloride. Their reaction with alcohol takes place in a manner fundamentally similar to that of chlorosilanes. Possibly, owing to disproportionation, tetraalkoxysilanes are obtained as well as polyalkoxysiloxanes (*1*). In this way, various hexa-alkoxydisiloxanes have been prepared from hexachlorodisiloxane, e.g., with propanol and butanol (*1*), 1,3-dimethylbutanol (*10*), 1-ethylpentanol (*10*), hexanol (*65*), cyclohexanol (*85*), 2-ethylbutanol (*126*), 2-ethylhexanol (*10, 65*), phenol (*57*), and perfluoroalkanols (*52*).

Likewise, with ethanol (*115*) and phenol (*57*), octachlorocyclotetrasiloxane can give the corresponding polyorganooxysiloxanes. With ethanol, octachlorotrisiloxane, decachlorotetrasiloxane, and tetradecachlorohexasiloxane have given octaethoxytrisiloxane (*84*), decaethoxytetrasiloxane (*50, 84*), and tetradecaethoxyhexasiloxane (*50, 84*); octaisopropoxytrisiloxane (*63*) has been prepared with isopropanol.

Furthermore, poly-*H*-siloxanes can be used as starting materials; their hydrogen atoms can be replaced by organooxy groups by reaction with alcohols and phenols in the presence of zinc chloride (*119*).

Lower alkoxy groups in the polysiloxanes can also act as functional groups; for example, if polymethoxy- or polyethoxysiloxanes are treated with higher alcohols, *transesterification reactions* take place, such as are known with the monomers (*44*).

The fact that *low-molecular-weight* silicic acids *rich in hydroxyl groups* can be esterified (*39*) is of particular interest:

$$\equiv Si{-}OH + HOR \rightarrow \equiv Si{-}OR + H_2O$$

Aqueous solutions of low-molecular weight silicic acids, obtained by the acidification of solutions of sodium silicate, are used for this purpose, and the silicic acid is extracted from them with a suitable alcohol. Esterification is then brought about by removing the water from the alcoholic solution in the form of the azeotrope. Even *solid silicic acid* can be esterified by heating with alcohols above 100°C to such an extent that alkoxylated silicic acids are produced (*5*, *9*). With amorphous preparation of silicic acid, tetraalkoxysilanes give solid polyalkoxysiloxanes (*58*).

Polyalkoxysiloxanols have been condensed with alkoxychlorosilane derivatives of methylisobutylmethanol and 2-ethylhexanol in the presence of pyridine to give polysiloxanes in accordance with the equation (*56*):

$$\geq Si{-}OH + X{-}Si\leq \;\rightarrow\; \geq Si{-}O{-}Si\leq + HX$$

Finally, in analogy with the synthesis of tetraorganooxysilanes from SiS_2 and alcohols, polyalkoxysiloxanes can be prepared from *silicon oxysulfide*—which can be obtained by heating a mixture of silicon, silica, and sulfur—by boiling with alcohols (*42*, *54*).

11.3 Properties and Applications

The properties of silicic esters can be modified in many ways because of the large number of possible substitutions of the organooxy groups. The literature is therefore very rich in publications dealing with the use of silicic acid esters for various purposes.

There is hardly a conceivable application for which silicic acid esters have not been proposed or stated to be suitable. Many of these proposals have led to no technical success. We shall therefore confine ourselves to pointing out some directions of development which have actually acquired technical importance or which could do so.

As already outlined in Section 11.1, the technical interest in silicic acid esters is based on two facts: first, on the *stabilizability* of the Si—OR bond and therefore the prospect of producing polymers of adequate resistance to moisture at low or high temperatures; and second, the *susceptibility of silicic acid esters to hydrolysis*, particularly the esters of the lower aliphatic alcohols, and therefore the possibility of using these as starting materials for the production of silicic acid. In analogy with the silicones, stabilized silicic acid esters are of interest as hydraulic fluids, heat-transfer media, and resin components, while readily hydrolyzable silicic esters are of interest as raw materials for the production of heat-resistant binders.

11.3.1 Silicic Acid Esters as Hydraulic Fluids

Since the polyorganooxysiloxanes contain the same Si—O—Si linkage as the silicones, it was logical to ascertain whether typical properties of the silicones are also to be found in the polysilicic acid esters.

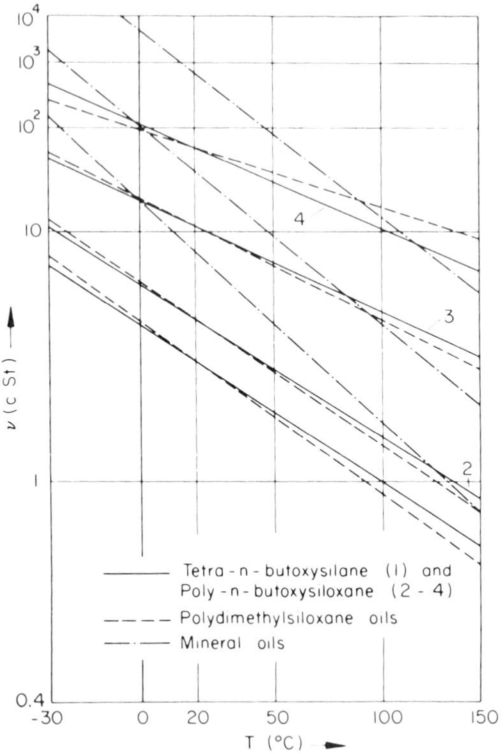

FIG. 112. Temperature-viscosity behavior of poly-*n*-butoxysiloxanes and tetra-*n*-butoxysilane compared with that of methylsilicone oils and mineral oils.

In fact, it has been found that certain liquid polyorganooxysiloxanes show a similar dependence of the viscosity on the temperature and—again in analogy with the silicones—have unusually low pour points. However, it is particularly interesting that tetraorganooxysilanes, i.e., orthosilicic acid esters, which contain no Si—O—Si but only Si—O—C bonds, exhibit the same characteristics (*102*).*

* From this it should be concluded that the helical structure of the siloxane chain (see Section 6.2.1) is not in itself a necessary condition for the low temperature-dependence of the viscosity. The principle common to the silicones and silicic acid esters appears rather to be the low strength of the intermolecular forces.

Of the aliphatic silicic acid esters, poly-*n*-butoxysiloxanes and tetra-*n*-butoxysilane have a very good temperature-viscosity behavior. The shape of the ν-t curves in the Ubbelohde diagram resembles that of the methylsilicones (Fig. 112) (*60*). Consequently, in this respect the butoxy group in the silicic

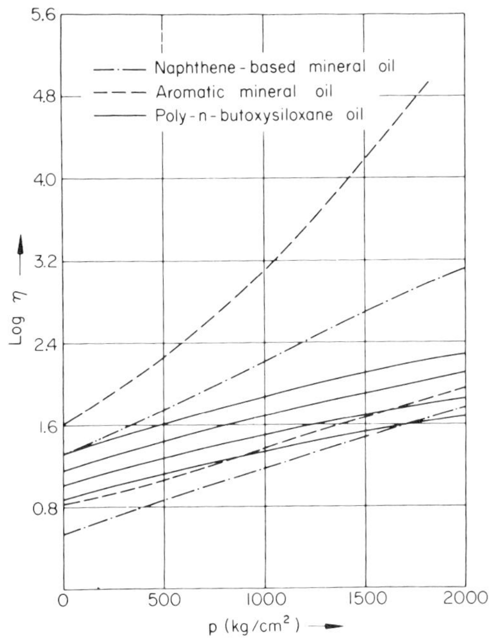

FIG. 113. Pressure dependence of the viscosity [ηP] of a poly-*n*-butoxysiloxane oil in the range from 0 to 2000 kg/cm^2 [after Kuss (*60*)].

acid esters proved to be equivalent to the methyl group in methylsilicones. In relation to the pressure-dependence of the viscosity, moreover, these silicic acid esters behave similarly to silicone oils (Fig. 113): The viscosity changes only relatively slightly with pressure in the range from 0 to 2000 kg/cm^2. With respect to the pour points, poly-*n*-butoxysiloxanes are far superior to the silicone oils: their pour points are round about $-100°C$. Their lubricating properties are better than those of methylsilicone oils (*60*).

Such polysilicic acid ester oils could also appear attractive as hydraulic fluids, particularly since they are considerably easier to obtain than silicone oils. However, in evaluating their properties it must be borne in mind that even the Si—O*n*-C_4H_9 bond is sensitive to hydrolysis to a certain extent, and is by no means so stable at high temperatures as a CH_3—Si bond. Considerable advances have been achieved in attempts to discover silicic acid esters having good stability to moisture as well as temperature-viscosity curves as flat as possible that are suitable for use at very low and elevated temperatures (*23*, *27*, *44*, *66*).

However, not even the most stable Si—OR bonds attain the hydrolytic and thermal stability of a CH_3—Si bond. Consequently, in technical use, these oils must be protected both against the entry of atmospheric humidity and against excessive overheating. It is not difficult to satisfy these two requirements in many technical applications, as in shock absorbers, electrohydraulic hoists, brake equipment, door-closing devices, etc. In general, the use of these products is of interest in all cases where their special properties can be suitably utilized: where the hydraulic devices have to operate uniformly even with large variations in temperature and remain capable of functioning at extremely low temperatures. They are consequently very useful in aircraft construction (*27*).

A very remarkable property of liquid poly-*s*-butoxysiloxanes is that they are miscible in any proportion with refrigerants based on difluorochloromethane and trifluorochloromethane, even at low temperatures (*23*, *49*, *103*, *107*). For this reason, and because of their outstanding temperature behavior, they are suitable as lubricants for refrigerators using these refrigerants.

11.3.2 Silicic Acid Esters as Heat-Transfer Media

The organooxysilanes include many compounds with astonishingly high boiling points (400 to 450°C/760 mm, cf. Table 111). Attempts have therefore been made to use them as heat-transfer media, and a large number of specifications deal with developments in this direction. In certain cases, suitable eutectic mixtures which lower the setting point of the mixture to below room temperature may be sought, while in others the aim is to improve the resistance to hydrolysis and the heat stability of the products by substitution with suitable organooxy groups (see, for example, refs. *16*, *18*, *43* and *53*). Esters containing aromatic and aliphatic organooxy groups simultaneously have low pour points but not such high boiling points (*17*, *114*). The investigations in this direction are not yet complete and the prospect of being able to transfer heat through liquid phases at temperatures up to 400°C without the application of pressure is still attractive. A complicating factor is the practical requirement that the fluids must be resistant to atmospheric humidity and atmospheric oxygen even at high temperatures and in the presence of iron, since the action of oxygen on the aromatic esters is catalyzed by the walls of iron vessels.

11.3.3 Silicic Acid Esters as Resin Components

Although silicic ester resins of relatively high stability have been prepared by suitable substitution (*97*, *121*), the question of whether technical resins and paints based on silicic acid esters *alone* will satisfy the stringent requirements that must be laid down for them is still open.

On the other hand, it is interesting that aromatic silicic acid esters, when incorporated into organic polyester resins, are capable of improving certain properties of the latter (*64*, *87*, *89*, *96*, *120*). In particular, the heat resistance is increased in such a way that it comes somewhere between those of good alkyd resins and silicones, although it is obviously nearer to that of the alkyd resins than that of the silicones; the thermal losses of material are reduced in comparison with those of alkyd resins. The resistance of such resins to hydrolysis is at least not worse than that of the alkyd resin components, and the Si—OC bond has proved to be resistant to the attack of water, perhaps because it is embedded in the network of the alkyd resin and is sterically screened and protected by this. The reaction products of polyester resins with 2,2′-biphenylenedioxydiisopropoxysilane are particularly outstanding in their behavior (*91*, *92*). Resins from this group have been tested as insulating materials for the electrical industry but have not yet succeeded in replacing or displacing silicone resins.

11.3.4 Silicic Acid Esters as Starting Materials for the Production of Silicic Acid as a Bonding Agent

Numerous applications of silicic acid esters are based exclusively on the tendency of the Si—OC bond to undergo hydrolytic cleavage and therefore on the possibility of preparing silicic acid from silicic acid esters by saponification or by thermal condensation. Such a silicic acid, when obtained correctly, exhibits good binding properties with respect to many inorganic oxides, silicates, carbides, etc., and can therefore be used as a heat-resistant binder for such materials. Here the silicic acid ester assumes the role of a silicic acid carrier relatively rich in SiO_2, free from alkali and easy to handle (in contrast to water-glass), from which the silicic acid can be formed *in situ* in a controllable manner.

The highest content of silicic acid would be obtainable in polymethoxysiloxanes; in addition, the silicic acid esters of methanol are also the easiest to decompose by hydrolysis. However, their use is limited by the fact that methoxysilanes are not unobjectionable physiologically (see Section 9.7.1), and that work with these substances can only be carried out with special precautions.

Consequently, the silicic esters of ethanol are in general preferred as binder carriers; they are used either as monomeric ethoxysilane or as polymeric polyethoxysiloxanes. According to their content of silica, technical products of this type have names such as "ethyl silicate 28" or "ethyl silicate 40." Liquid ethyl silicates with still higher contents of silica, e.g., more than 53% have been prepared by special measures (*19*). Polybutoxysiloxanes containing about 45% of silica are also suitable for obtaining silicic acid with binding properties.

Ethyl silicates are generally used by first preparing an aqueous alcoholic solution, in which the alcohol component acts as a solubilizing agent between ethyl silicate and water. Figure 114 shows the homogeneity region in the ethyl silicate-ethanol-water tertiary system. A mixture lying outside the region of homogeneity forms an emulsion which, if its composition is suitable, may also become homogeneous because of the liberation of alcohol by hydrolysis of the

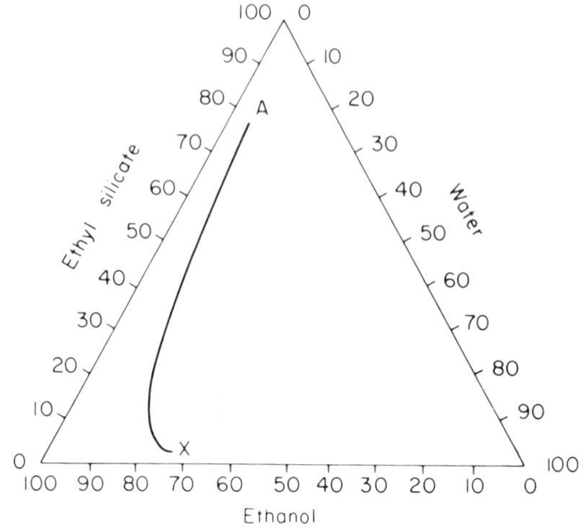

FIG. 114. Limits of miscibility in the ternary system of ethyl silicate, ethanol, and water. Miscibility exists to the left of the line shown.

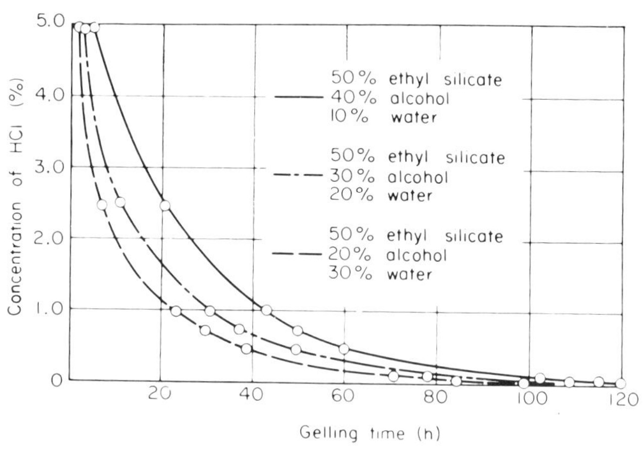

FIG. 115. Dependence of the gelling time of ethyl silicate solutions on the amount of HCl added (*47*).

ester. The rate of this hydrolysis is increased by acids or bases. Even small amounts of acids considerably reduce the gelling time (Fig. 115). The control of the hydrolysis of the silicic acid ester in the individual case is a matter of experience and special formulations (*13*, *47*, *93*, *113*).

After the completion of hydrolysis, the granular aggregate of the substrate to be bound is cemented with silica gel. In order to achieve the optimum

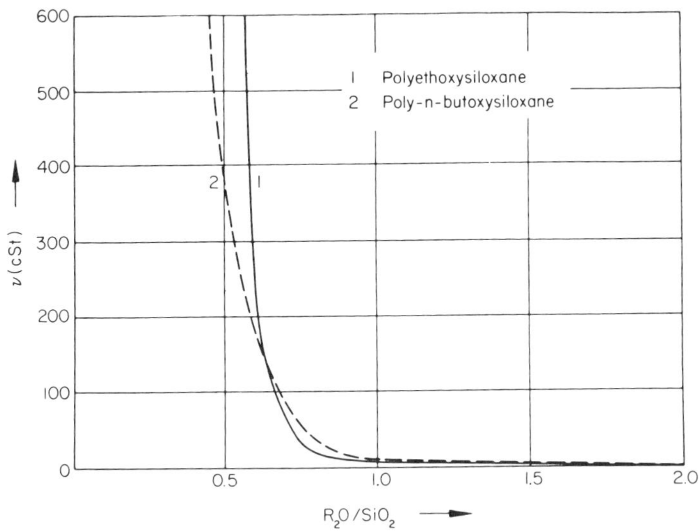

FIG. 116. Change in the viscosity of polyalkoxysiloxanes as a function of the R_2O/SiO_2 ratio.

mechanical strength and thermal stability, the system is heated, the temperatures being raised to values determined by the structure and composition of the mass but being as a rule far above 1000°C. Under these conditions, the silica gel is transformed into cristobalite, if new phases do not arise by reaction with the surface of the grains to be bonded. Thus, the formation of mullite in the binder has been demonstrated in the bonding of corundum with ethyl silicate (*47*).

To understand the mode of action of silicic acid esters as binders, it is interesting to consider their change in viscosity as a function of the R_2O/SiO_2 ratio; this is shown in Fig. 116 (*60*). If this ratio is in the range of about 0.6 to 0.7, the viscosity of the polyethoxy- and polybutoxysiloxanes investigated increases in sudden jumps, so that when the hydrolysis or condensation is carried out suitably a filler binder system solidifies very rapidly.

One of the most important fields of application of the hydrolyzable and condensable silicic acid esters is their use as binders for sand mixtures from which investment molds for precision casting are made (*93*, *113*). The pasty mixture of sand, magnesia, brick meal, etc., with ethyl silicate, water, and, where required, alcohol, can be hardened at the desired rate by suitable catalysis; the mold dimensions are reproduced well, and the shells can easily be detached from the model, are stable up to 1600°C, and are therefore fairly universally applicable in the casting of metals.

Silicic acid esters of the lower aliphatic alcohols are also suitable for the bonding of other refractory masses, e.g. sillimanite, silicon carbide, corundum (*47*), asbestos, and mica (*48*). In addition, silicic acid esters are used to fill the pores of coarse rocks such as sandstone with silica gel, thus making the material of the rock stronger and more resistant to weathering (*48*).

REFERENCES

(1) E. W. ABRAHAMSON, I. JOFFE and H. W. POST, *J. Org. Chem.* **13**, 275 [1948]. (2) H. J. BACKER and H. A. KLASENS, *Rec. Trav. chim. Pays-Bas* **61**, 500 [1942]. (3) D. L. BAILEY and F. M. O'CONNOR, Union Carbide, German Appl. (West German) 1023753 [U.S. Prior. 17.5.1955]. (4) E. BEANLAND, Hygrotherm Engineering, Brit. Pat. 913346 [1.6.1960 and 24.2.1961]. (5) K. L. BERRY and R. M. JOYCE, Du Pont, U.S. Pat. 2757098 [6.3.1952]. (6) D. C. BRADLEY, R. C. MEHROTRA and W. WARDLAW, *J. Chem. Soc.* [London] **1952**, 4204. (7) H. BREEDERVELD and H. I. WATERMAN, *Rec. Trav. chim. Pays-Bas* **73**, 871 [1954]. (8) British Thomson-Houston, Brit. Pat. 670476 [U.S. Prior. 15.12.1948]. (9) E. C. BROGE, Du Pont, U.S. Pat. 2736668 [9.5.1952].

(10) California Research, Brit. Pat. 703420 [U.S. Prior. 24.2.1950]. (11) —, Brit. Pat. 709345 [U.S. Prior. 28.8.1950]. (12) H. D. COGAN and C. A. SETTERSTROM, *Chem. Engng. News* **24**, 2499 [1946]. (13) —, —, *Ind. Engng. Chem.* **39**, 1364 [1947]. (14) J. B. CULBERTSON, H. DE W. ERASMUS and R. M. FOWLER, Union Carbide, U.S. Pat. 2569455 and 2569746 [8.6.1949]. (15) —, —, —, —, U.S. Pat. 2569746 [8.6.1949]. (16) E. DA FANO, J. B. Pierce Foundation, U.S. Pat. 2584334 [10.6.1950]. (17) —, —, U.S. Pat. 2641581 [Brit. Prior. 6.12.1948]; French Pat. 989579 [U.S. Prior. 17.8.1948]; U.S. Pat. 2684945 [17.5.1952]. (18) —, —, Brit. Pat. 706738 [30.1.1952]. (19) D. P. DODGSON, Monsanto Chemical, Brit. Pat. 722425 [23.10.1951].

(20) EBELMEN, *Liebigs Ann. Chem.* **57**, 319 [1846]; *Ann. Chimie Physique* **3**, 16 [1846]; *J. prakt. Chem.* **37**, (6), 347 [1846]. (21) H. G. EMBLEM, K. HARGREAVES and N. A. HURT, *Ind. Chemist* **39**, 576, 641 [1963]. (22) —, A. E. MEADOWCROFT and G. E. STOCKWELL, Monsanto Chemical, Brit. Pat. 674137 [17.9.1949]. (23) Farbenfabriken Bayer, Publication: Fluisile®. (24) H. FEICHTINGER, H. NOESKE, H. J. TOMUSCHAT and O. ROELEN, Ruhrchemie, German Appl. (West German) 1180359 [15.5.1962]. (25) C. FRIEDEL and J. M. CRAFTS, *Liebigs Ann. Chem.* **136**, 203 [1865]. (26) — and A. LADENBURG, ibid. **143**, 118 [1867]. (27) N. W. FURBY, R. L. PEELER and R. I. STIRTON, *Trans. Amer. Soc. Mechan. Engr.* **79**, 1029 [1957]. (28) L. GALIMBERTI, *Ann. Chim. applicata* **38**, 55 [1948]. (29) P. D. GEORGE and J. R. LADD, *J. Amer. Chem. Soc.* **75**, 987 [1953].

(30) W. GERRARD and J. V. JONES, *J. Chem. Soc.* [London] **1952**, 1690. (31) *Gmelins Handb. der anorgan. Chemie*, System-Nr. 15, Teil C: *Organische Silicium-Verbindungen*, Verlag Chemie, Weinheim/Bergstr. 1958, 8th Ed. (32) A. GOLDSCHMIDT and J. R. WRIGHT, California Research, U.S. Pat. 2758127 [30.12.1953]. (33) C. P. HABER, General Electric, U.S. Pat. 2445576 [19.6.1946]. (34) W. HAHN, *Makromolekulare Chem.* **11**, 51 [1953]. (35) M. E. HAVILL, I. JOFFE and H. W. POST, *J. Org. Chem.* **13**, 280 [1948]. (36) B. HELFERICH and W. REIMANN, *Chem. Ber.* **80**, 163 [1947]. (37) J. HERTKORN, *Ber. dtsch. chem. Ges.* **18**, 1679 [1885]. (38) R. K. ILER, *Ind. Engng. Chem.* **39**, 1384 [1947]. (39) — and P. S. PINKNEY, ibid. **39**, 1379 [1947].

(40) H. JENKNER, Kali-Chemie, German Pat. (West German) 952263 [23.2.1953]. (41) I. JOFFE and H. W. POST, *J. Org. Chem.* **14**, 421 [1949]. (42) I. B. JOHNS, Monsanto Chemical, U.S. Pat. 2673870 [27.4.1951]. (43) L. H. JOHNSTON, U.S. Pat. 2335012 [12.4.1941]. (44) W. S. KATHER and T. WIENER, Hodges Research, U.S. Pat. 2711418 [6.1.1950]. (45) F. S. KIPPING and J. T. ABRAMS, *J. Chem. Soc.* [London] **1944**, 81. (46) E. KONRAD, O. BÄCHLE and R. SIGNER, *Liebigs Ann. Chem.* **474**, 276 [1929]. (47) R. KRANNICH, *Silikattechnik* **12**, 78 [1961]. (48) W. KRINGS and W. DITTRICH, *Sprechsaal f. Keramik, Glas, Email* **93**, 126 [1960]. (49) H. J. LÖFFLER, *Kältetechnik* **9**, 135 [1957].

(50) L. MALATESTA, *Gazz. chim. ital.* **78**, 747 [1948]. (51) M. MARCHAND, German Pat. (East German) 12115 [19.4.1951]. (52) E. T. MCBEE and O. R. PIERCE, Purdue Research, U.S. Pat. 2777870 [7.2.1955]. (53) A. E. MEADOWCROFT, Monsanto Chemical, German Pat. (West German) 894839 [Brit. Prior. 11.4.1950 and 20.3.1951]. (54) Monsanto Chemical, Brit. Pat. 706288 [U.S. Prior. 27.4.1951]. (55) C. R. MORGAN, California Research, U.S. Pat. 2643263 [22.9.1950]. (56) —, W. F. OLDS and A. L. RAFFERTY, *J. Amer. Chem. Soc.* **73**, 5193 [1951]. (57) V. MORRILL JR., American Stove, U.S. Pat. 2416531 [26.10.1943]. (58) R. F. NICKERSON, Monsanto Chemical, French Pat. 1130254 [U.S. Prior. 6.5.1954]. (59) S. NITZSCHE and E. PIRSON, Wacker-Chemie, German Pat. (West German) 896414 [23.5.1950].

(60) W. NOLL, *Angew. Chem.* **66**, 41 [1954]. (61) —, H. JONAS and P. SIMONS, Farbenfabriken Bayer, German Pat. (West German) 865448 [13.1.1951]. (62) R. OKAWARA, *J. chem. Soc. Japan* **55**, 678 [1952]. (63) —, T. TANAKA and I. ISHIMARU, *Bull. chem. Soc. Japan* **27**, 45 [1954]. (64) M. M. OLSON, Pittsburgh Plate Glass, U.S. Pat. 2877202 [7.10.1953]. (65) B. A. ORKIN, Socony-Vacuum-Oil, U.S. Pat. 2626957 [1.9.1948]. (66) Oronite Chemical, Publication: Oronite High Temperature Hydraulic Fluids, 1957. (67) W. I. PATNODE and R. O. SAUER, General Electric, U.S. Pat. 2381137 [14.5.1942]. (68) W. PAWLOWSKI and J. MLODECKI, *Przemysl chem.* **12**, [36], 687 [1956]. (69) J. S. PEAKE, W. H. NEBERGALL and YU TI CHEN, *J. Amer. Chem. Soc.* **74**, 1526 [1952].

(70) G. W. PEDLOW JR. and C. S. MINER JR., Minnesota Mining, U.S. Pat. 2566364 and U.S. Pat. 2566365 [15.6.1946]. (71) D. F. PEPPARD, W. G. BROWN and W. C. JOHNSON, *J. Amer. Chem. Soc.* **68**, 70 [1946]. (72) —, —, —, ibid. **68**, 73 [1946]. (73) J. B. Pierce Foundation, Brit. Pat. 670630 [U.S. Prior. 17.8.1948]. (74) —, Austrian Pat. 171159 [U.S. Prior. 9.2.1950]. (75) A. F. REILLY and H. W. POST, *J. Org. Chem.* **16**, 383 [1951]. (76) D. RIDGE and M. TODD, *J. Chem. Soc.* [London] **1949**, 2637. (77) E. G. ROCHOW, General Electric, U.S. Pat. 2473260 [26.6.1946]. (78) —, *J. Amer. Chem. Soc.* **70**, 2170 [1948]. (79) W. RODZIEWICZ and J. PREJZNER, *Roczniki Chem.* **31**, 525 [1957].

(80) D. S. P. ROEBUCK, Monsanto Chemical, Brit. Pat. 879572 [15.3.1957]. (81) M. SCHMEISSER and H. JENKNER, Kali-Chemie, German Pat. (West German) 892438 [17.9.1951].

(82) —, —, *Z. Naturforsch.* **7b**, 191 [1952]. (83) G. SCHOTT and H. U. KIBBEL, *Z. anorg. allg. Chem.* **311**, 53 [1961]. (84) W. C. SCHUMB and D. F. HOLLOWAY, *J. Amer. Chem. Soc.* **63**, 2753 [1941]. (85) —, —, ibid. **63**, 2853 [1941]. (86) R. SCHWARZ and A. KESSLER, *Z. anorg. allg. Chem.* **263**, 15 [1950]. (87) —, — and K. KNOLL, Siemens-Schuckert-Werke, German Pat. (West German) 861612 [24.6.1949]. (88) — and K. G. KNAUFF, *Z. anorg. allg. Chem.* **275**, 176 [1954]. (89) — and W. KUCHEN, Siemens-Schuckert-Werke, German Appl. (West German) 1022725 [16.12.1952].

(90) —, —, *Chem. Ber.* **86**, 1144 [1953]. (91) —, —, Siemens-Schuckert-Werke, German Appl. (West German) 1191968 [5.4.1958]. (92) —, — and F. WEIGEL, —, German Appl. (West German) 1125023 [9.4.1958]. (93) H. F. SCOTT and H. G. EMBLEM, Refractories J. **27**, 286 [1951]. (94) O. F. SENN and A. P. BRADY, U.S. Atomic Energy Comm. Techn. Inform. Serv. No. NP 4592 **1953**, 1. (95) C. SHAW and W. E. LANGRISH-SMITH, Union Chimique Belge, French Pat. 968562 [Brit. Prior. 1.7.1947]. (96) Siemens-Schuckert-Werke, Brit. Pat. 717408 [German Prior. (West German) 6.7.1950]. (97) —, Belg. Pat. 553340 [German Prior. (West German) 17.3.1955]. (98) R. SIGNER and H. GROSS, *Liebigs Ann. Chem.* **488**, 56 [1931]. (99) Silicon (organic) Developments, Brit. Pat. 647528 [1.7.1947].

(100) W. SIMMLER: *Organische Derivate der Kieselsäure, in Houben-Weyl:* Methoden der organischen Chemie, Vol. VI (2), 71 [1963]. (101) P. SIMONS, H. JONAS and W. NOLL, Farbenfabriken Bayer, German Pat. (West German) 897627 [20.3.1951]. (102) —, —, — and M. ZIMMERMANN, —, German Pat. (West German) 916340 [26.8.1950]. (103) —, —, —, —, —, German Pat. (West German) 865172 [13.12.1950). (104) B. E. F. SMITH, *Svensk. kem. Tidskr.* **61**, 213 [1949]. (105) F. SPERR and E. WIBERG, Union Carbide, U.S. Pat. 2909550 [2.6.1955]. (106) H. STAUDINGER and W. HAHN, German Pat. (West German) 929189 [21.7.1952]. (107) H. STEINLE, *Kältetechnik* **8**, 12 [1956]. (108) D. R. STULL, *Ind. Engng. Chem.* **39**, 517 [1947]. (109) G. SUMRELL and G. E. HAM, *J. Amer. Chem. Soc.* **78**, 5573 [1956].

(110) T. TAKATANI, *J. chem. Soc. Japan, pure Chem. Sect.* **72**, 765 [1951] (C.A. **1952**, 11102). (111) —, ibid. **73**, 409 [1952]. (112) T. TANAKA and T. WATASE, *J. Chem. Physics* **22**, 1268 [1954]. (113) D. F. B. TEDDS, Metal Ind. **75**, 27 [1949]. (114) W. W. TRIGGS, J. B. Pierce Foundation, Brit. Pat. 676633 [26.5.1949]. (115) L. TROOST and P. HAUTEFEUILLE, *Bull. Soc. chim.* **19**, 255 [1873]. (116) M. G. VORONKOV and B. N. DOLGOV, *J. appl. Chem. USSR (Engl. transl.)* **24**, 103 [1951]. (117) — and G. B. KARPENKO, *J. Gen. Chem. USSR (Engl. transl.)* **27**, 361 [1957]. (118) —, A. N. LAZAREV and A. K. BAIGOZHIN, ibid. **26**, 3421 [1956]. (119) Wacker-Chemie, French Pat. 1040941 [German Prior. (West German) 27.7.1950].

(120) F. WEIGEL, Siemens-Schuckert-Werke, German Appl. (West German) 1019725 [8.1.1953]; German Pat. (West German) 945936 [24.9.1952]. (121) —, —, German Pat. (West German) 949311 [21.5.1954]. (122) —, —, NUREMBERG, personal communication, 1958. (123) A. WEISS, G. REIFF and A. WEISS, *Z. anorg. allg. Chem.* **311**, 151 [1961]. (124) W. WOJNOWSKI and R. PIEKOŠ, ibid. **314**, 189 [1962]. (125) J. R. WRIGHT, California Research, U.S. Pat. 2727054 [24.8.1954]. (126) —, R. O. BOLT, A. GOLDSCHMIDT and A. D. ABBOTT, *J. Amer. Chem. Soc.* **80**, 1733 [1958]. (127) W. ZEISS and W. NOLL, Farbenfabriken Bayer, German Pat. (West German) 955062 [23.7.1955]. (128) J. J. ZUCKERMAN, *J. Chem. Soc.* [London] **1962**, 873.

CHAPTER 12

Analytical Methods

The analytical chemistry of the silicones abounds in problems which have not yet been solved, or not in a satisfactory manner. Thus, the determination of the constitution of relatively complex polymeric siloxanes is a problem which can be solved only by the use of various methods, and often not unambiguously. The determination of the nature and the amount of small proportions of functional groups in polymeric siloxanes, which can substantially affect the properties of the material as a whole, also generally presents great difficulties. The desire often expressed in the industry of determining small concentrations of silicones in solid or liquid technical products cannot always be satisfied.

12.1 Chemical Methods

12.1.1 Elementary Analysis

In principle, the elementary analysis is carried out by classical procedures.

The most certain and universally applicable method for the *determination of silicon* consists in digesting the substance in a Parr bomb, dissolving the melt in water, and either separating the silicon from the solution as silica and weighing it, or determining it as the 8-hydroxyquinoline-silicomolybdic acid complex (*22*, *36*).

If low-boiling organosilicon compounds are to be digested in the Parr bomb, sealable tubes made of silica-free glass (*19*) or of Cellit (*23*) are used for weighing out the substance; in the latter case, the tube is heated and closed with crucible tongs. The digesting agent can be modified by the addition of potassium nitrate and cane sugar (*62*) or of ethylene glycol (*61*).

A simpler procedure is to decompose the organosilicon substance by digestion with fuming sulfuric acid (*36*), possibly with the addition of a little nitric acid, followed by elimination of the sulfuric acid, and weighing the SiO_2 residue. However, this method gives reliable results only when the starting material is free from low-molecular-weight siloxanes and when, in addition, such compounds cannot be formed and volatilized during the digestion.

Carbon and *hydrogen* are determined in the usual combustion apparatus by the micro or semimicro process. In this process, care must be taken that the organosilicon material is completely oxidized and that no silicon carbide is formed. The process is therefore preferably carried out with a large excess of oxygen, or, particularly in the case of resinlike substances, in such a way that the sample is heated to just below incandescence; this ensures that the substance oxidizes in layers, the SiO_2 formed flakes off, a new surface is continually exposed, and therefore the formation of SiC is avoided (*43*).

It must be borne in mind that small amounts of silica are entrained by the stream of oxygen as smoke, and can be transferred into the absorption tubes. Consequently, relatively long combustion tubes are used and the gases are filtered through a plug of glass wool (*43*) or a layer of pumice kept at about 200°C outside the combustion furnace by means of a special heating device (*32*). In the combustion tube itself, oxidation is carried out catalytically by means of fine heated platinum gauze (*43*) or nickel turnings and platinized asbestos (*32*). Low-boiling organosilicon compounds are passed into the combustion tube by means of a stream of nitrogen (*28*). It has in fact been recommended to determine the content of SiO_2 from the increase in weight of the combustion tube (*28*, *43*), but in any case digestion in the Parr bomb is more reliable for the determination of the silicon.

Halogens can be determined in the combustion process by absorption with powdered silver according to Dennstedt (*9*, *28*), or, better, by fusion in the Parr bomb. *Nitrogen* can be determined in the Dumas combustion apparatus or in suitable cases (e.g., when present as silazane nitrogen) by Kjeldahl's method (*5*).

If small contents of silicones in *mixtures of substances* and in *technical products* such as pastes, emulsions, polishes, and creams, or on paper, textiles, and the like, are to be detected and determined quantitatively, an extraction with organic solvents is carried out first, and an attempt is then made to obtain an idea of the nature of the silicones by physical methods, particularly by infrared spectroscopy. For quantitative determination, use is generally made of the Si content, which is determined by digestion in the Parr bomb, and the silicone content is deduced, at least approximately, if the type of silicone is known. Only with organic carrier materials which inherently contain no silicon can a direct determination of the SiO_2 be made by ashing the carrier. Special prescriptions for analyses of this type have been worked out [polishes (*25*), textiles (*42*), leather (*10*), and paper (*7*)].

12.1.2 Determination of Functional Groups

The *Si—Cl-bond* is easy to cleave hydrolytically. Consequently, the chlorine bound to silicon in silanes or siloxanes has generally been determined by hydrolyzing these with water or aqueous alkalis and titrating the resulting halogen ion argentometrically or acidimetrically in aqueous or alcoholic solution. In this process, it must be borne in mind that the rate of hydrolytic rupture of the Si—Cl bond may depend on the organic substituent; where necessary, therefore, the hydrolysis must be accelerated and completed by heating.

Recently, it has been shown in a very painstaking investigation that very reliable figures can be obtained rapidly if the chlorosilane is dissolved in isopropanol, which is a good solvent for the silane and for the hydrogen chloride formed by alcoholysis, and if the acid is then titrated with ethanolic potassium hydroxide using tetrabromophenolphthalein ethyl ester as the indicator until the color changes from yellow to blue. The very sharp color change agrees with the figures obtained on potentiometric determination in many organic solvents including isopropanol (*35*).

The *Si—H bond* is cleaved by aqueous sodium hydroxide; half the amount of the hydrogen evolved, which is produced by the reaction

$$\geq Si—H + NaOH \rightarrow H_2 + \geq Si—ONa$$

and is determined gas-volumetrically, shows the content of Si—H bonds (*54*). The decomposition proceeds very smoothly with piperidine-water mixtures (*59*). However, steric factors can greatly retard or completely suppress this reaction. Thus, tri-*o*-tolylsilane and tri-*o*-xenylsilane split off no hydrogen even after being heated for 24 hours under reflux with mixtures of piperidine and water (*59*). In certain polysilmethylenes the Si—H bond can be hindered sterically to such an extent that it cannot be determined by this method. In such cases, the analysis has been effected by converting the Si—H bond into a Si—Cl bond by chlorination, and determining the halogen by hydrolysis (*16*).

It must be borne in mind that the Si—Si bond also reacts with alkalis with liberation of hydrogen:

$$\geq Si—Si \leq + 2NaOH \rightarrow 2 \geq Si—ONa + H_2$$

and can itself be determined with 3*N* sodium methoxide solution by Zeriwitinoff's method (*55*). Where alkalis or alkali-metal alkoxides reacted too slowly, faster decomposition was achieved with piperidine, probably because of wetting factors (*29*). Hydrogen is liberated more rapidly by the action of piperidine/water on Si—H bonds than on Si—Si bonds (*59*).

* Page number of chapter references.

The Si—H group can also be determined quantitatively by cleavage with bromine in accordance with the equation

$$\rangle Si—H + Br_2 \rightarrow \rangle Si—Br + HBr$$

This reaction takes place completely in a few minutes even at room temperature. It is carried out with a solution of bromine in glacial acetic acid of known bromine content in such a way that no bromine escapes and the excess can be determined titrimetrically with $N/10$ $Na_2S_2O_3$ after the addition of KI (*13*, *15*).

Following the same principle the Si—C_6H_5 bond can be broken by bromine:

$$C_6H_5—Si\langle + Br_2 \rightarrow \rangle Si—Br + C_6H_5Br$$

and can therefore be estimated essentially as above. However, it reacts quantitatively only with *boiling* solutions of bromine in glacial acetic acid, so that by adjusting the reaction conditions it is possible to determine the Si—H and the Si—C_6H_5 bonds in a phenylsilane separately (*13*, *15*). If the phenylsilane contains several Si—H groups then, after the latter have been brominated, the phenyl groups are bound so firmly that their splitting off with bromine no longer takes place quantitatively. This difficulty in the determination of phenyl groups can be circumvented by adding water and boiling after the bromination of the Si—H groups with bromine in glacial acetic acid. This forms siloxane bonds, which strengthen the Si—phenyl bond less than bromine. In most cases, this method of determination, which is easy to perform, gives good results (*15*).

A further process suitable for the determination of the Si—H bond in volatile silanes consists in the reduction of aqueous solutions of $HgCl_2$ to Hg_2Cl_2:

$$\rangle Si—H + 2HgCl_2 + H_2O \rightarrow \rangle Si—OH + 2HCl + Hg_2Cl_2$$

The Hg_2Cl_2 formed is determined quantitatively by oxidation with I_2, and from this the content of Si—H bonds is deduced (*12*). The principle of the reduction of Hg^{2+} solutions is also exploited analytically by determining the reduced and nonreduced amounts of mercury ions polarographically in the form of the iodide complexes. It has been shown that this method enables Si—H bonds to be determined even in the presence of Si—Si bonds and of silanes containing unsaturated groups (*6*).

As a functional group, the *Si—OH bond* plays an important part in many technical silicone products. Its determination is complicated by the fact that its amount is low and that water is often present in amounts of about the same order of magnitude. The OH groups linked to silicons can be determined—together with any water that may be present—by the Zeriwitinoff method (*21*,

24, *34*, *39*, *52*, *53*) with lithium aluminum hydride (*27*, *53*), or with phenyl isocyanate (*8*):

$$2\,{\geqslant}Si{-}OH + 2C_6H_5NCO \rightarrow {\geqslant}Si{-}O{-}Si{\leqslant} + C_6H_5NH{-}CO{-}NHC_6H_5 + CO_2$$

While in the two cases mentioned first the gas liberated is measured volumetrically, in the last method the consumption of phenyl isocyanate is determined titrimetrically (*8*).

A distinction can be made between the silanol groups and water by means of a suitable modification of the Karl Fischer method (*8*, *40*), since water reacts very much more rapidly with the reagent than the OH in silanol groups. The differences become very pronounced and are capable of useful quantitative distinction when care is taken that only a very small excess of reagent is present during the titration. An automatic titration apparatus should preferably be used, operated by the *dead-stop method*. When an excess of reagent is used in the titration, the silanol groups can also be determined quantitatively (*20*, *40*).

The *Si—OR bond*, which, as a functional group in silicone chemistry, is found predominantly as $Si{-}OC_2H_5$, can be determined quantitatively by Zeisel's method by decomposition with hydriodic acid. The procedures which apply to organic compounds must be changed for alkoxysilanes, since for silicon compounds special solvents must be used which on the one hand form a homogeneous mixture with them and the hydriodic acid and, on the other hand, neither react with the OR group under the reaction conditions nor form an alkyl iodide with HI. The use of phenol and red phosphorus has proved suitable (the liberated iodine is converted into HI). The alkyl iodide formed can be determined by gas chromatography (*56*) or by conversion into iodate

$$RI + Br_2 \rightarrow RBr + IBr$$

$$IBr + 3H_2O + 2Br_2 \rightarrow HIO_3 + 5HBr$$

and subsequent titration of the latter (*14*).

Organofunctional groups are determined, on the whole, by classical organic methods. If the organic function is in the neighborhood of the silicon atom, this can limit the choice of processes under certain conditions. Thus, for example, experience has shown that the determination of the OH groups in the $-\overset{|}{\underset{|}{Si}}-\overset{|}{\underset{|}{C_n}}-OH$ grouping with $n = 1$–3 is more reliable with acetic anhydride and pyridine than with isocyanate (*47*).

However, the special features of the strength of the Si—C bond as a function of the substituents carried by the carbon atom also suggest specific methods of determination. Thus, the fact that the alkaline hydrolysis splits the Si—C bond in chloromethylchlorosilanes and gives different hydrolysis products:

$$\geq Si{-}CCl_3 \xrightarrow[OH^{\ominus}]{H_2O} CHCl_3$$

$$\geq Si{-}CHCl_2 \xrightarrow[OH^{\ominus}]{H_2O} CH_2Cl_2$$

$$\geq Si{-}CH_2Cl \xrightarrow[OH^{\ominus}]{H_2O} CH_3Cl(+Cl^{\ominus})$$

which can be determined by gas chromatography, can be used for quantitative group analysis (*17*).

The group detection of the organic function is however carried out most elegantly by means of nuclear resonance spectrometry (Section 12.2.4).

12.1.3 Functionality of the Siloxane Units

There is no simple and universally applicable method for the determination of the functionality of the siloxane units in a polymeric organosiloxane.

The most general method is probably the one in which the polymeric siloxane is decomposed with anhydrous hydrogen fluoride (*3*, *4*, *41*), the resulting monomeric organofluorosilanes corresponding to the siloxane unit (cf. Section 5.5.3). Their separation by quantitative fractionation permits certain deductions to be drawn concerning the structure of the polymer. However, the method is time-consuming and difficult.

The special problem of determining the proportions of T and M units in linear polysiloxanes can be partially solved by the Roedel test (*44*). If T units in D siloxanes are to be determined, increasing amounts of monofunctional units are gradually added to the siloxane polymers to be investigated, these are equilibrated, and the reduction in the capacity for gelling is investigated. The amount of added M units that just prevents the gelling of the oil is a function of the content of T groups. Conversely, the content of M siloxanes in D siloxanes is determined by the addition of T groups. According to another variant, resort is made to viscosity measurements and the proportion of T units is determined after the equilibration of M groups, or conversely by comparing the viscosities obtained with calibration curves of siloxanes with known M, D, and T contents.

12.2 Physical and Physicochemical Methods

12.2.1 Density and Refraction

When the group relationship of an organic silicon compound is known, certain conclusions concerning its composition can often be based on simple measurements of the density and the refractive index. These figures are always given as characteristics of the compounds listed in tables in the preceding chapters. Measurements of the density and the refractive index can be particularly valuable for monitoring the preparation of organosilicon monomers; e.g., the density of mixtures of dimethyldichlorosilane and methyltrichlorosilane changes linearly with the ratio in which these substances are present (*1*). Consequently, by a simple density measurement at a given temperature, it is possible to determine the purity of the substances.* Phenyl groups in polymethylphenylsiloxanes increase the density and refactive index as compared with the polymethylsiloxanes, so that when it is known that a siloxane belongs to this group, measurements of the density and refractive index can already give valuable information on its composition. Examples of the value of these simple determinations could be multiplied at will.

The familiar principle according to which the molecular refraction

$$MR = \frac{n^2-1}{n^2+2} \cdot \frac{M}{d}$$

(where n is the refractive index, M the molecular weight, and d the density) is composed additively of the bond and group refractions and can therefore be used to determine the composition and structure of a liquid phase from measurements of d and n, can also be applied to organosilanes and organosiloxanes (*46*, *58*). A systematic study of the situation in investigations independently carried out by two authors has led to the bond and group refractions in Table 113.

Additivity was shown with a large number of compounds, and, among other things, it was demonstrated that it is possible in this way to detect even small proportions (e.g. 5 mole-%) of trifunctional units in difunctional siloxanes.

TABLE 113

BOND AND GROUP REFRACTIONS IN ORGANOSILICON COMPOUNDS (cm^3/mole)

Bond	According to Sauer (*46*)	According to Warrick (*58*)
Si—O	1.751	1.75
Si—C (aliphatic)	—	2.50
Si—C (aromatic)	—	2.56
Si—H	3.22	3.20
Si—Cl	7.18	7.20
$Si—CH_3$	7.573	7.57
$Si—C_6H_5$	27.43	27.39

* For the use of Krutzsch's method for the analysis of mixtures of methylchlorosilanes see ref. *26*.

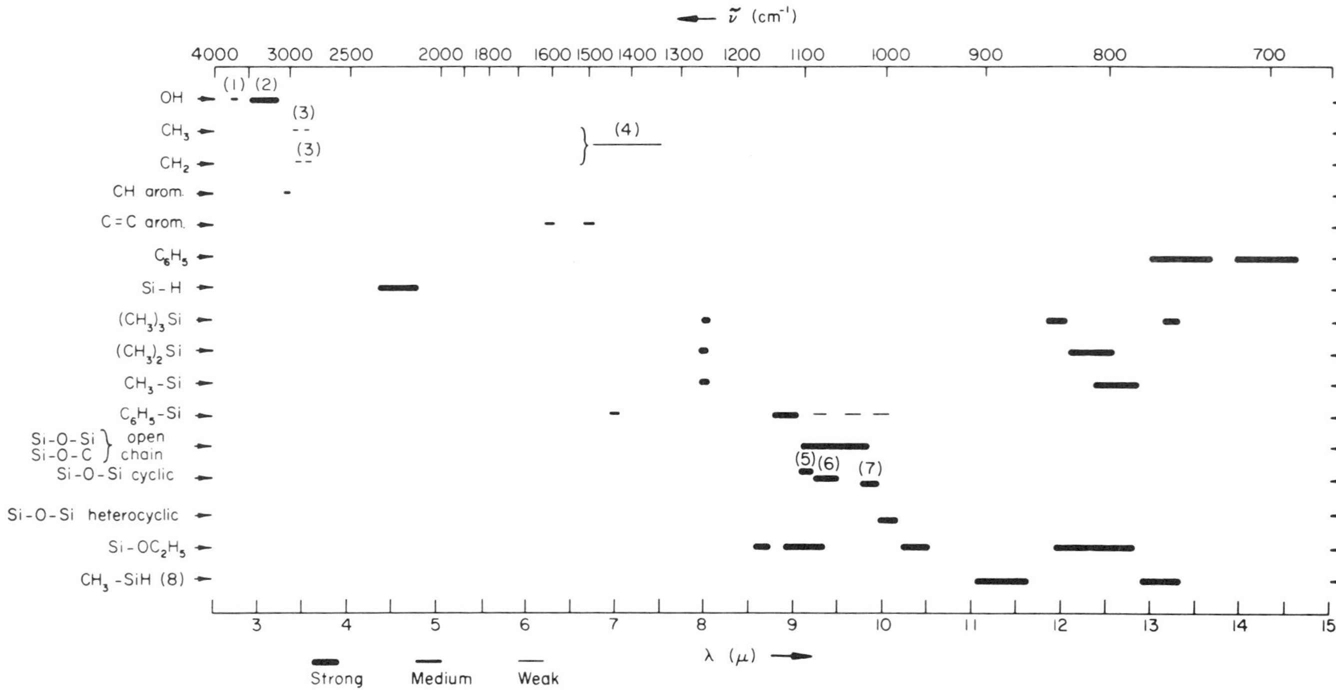

FIG. 117. Schematic representation of the positions and intensities of some infrared absorptions important for the chemistry of the silicones: (1) free OH groups; (2) associated OH groups (hydrogen bonding); (3) weak with Si—C bonds, strong with C—C bonds; (4) weak with Si—C bonds, medium-strong with C—C bonds; (5) in cyclotetrasiloxane; (6) in cyclic polysiloxanes with more than 4 siloxane units; (7) in cyclotrisiloxane. (8) Compounds with this configuration also show the absorption given under Si—H and Si—CH_3; with these, only the Si—CH_3 vibration at 12.5 μ is absent.

12.2.2 Infrared and Raman Spectroscopy

Infrared spectroscopy provides a preliminary idea of the composition and—to a certain extent—of the molecular structure of a siloxane; under certain conditions, it can also give semiquantitative information. The recording technique presents no particular difficulties. If the compounds must be investigated in the dilute state, solvents such as carbon tetrachloride and carbon disulfide are used; highly condensed cross-linked siloxanes such as hardened silicone resins are examined spectroscopically by the KBr method. Dissolved, condensable siloxanes such as still unhardened silicone resins can be spread as thin films on plates of rock salt and baked on.

Figure 117 gives a list of the infrared frequencies most important for silicone chemistry that can be determined spectroscopically with rock salt prisms (*2*, *37*).

In addition to the absorption that can be ascribed to the linkages of atoms or groups to silicon, the most frequent absorption frequencies of those vibrations in which silicon is not involved are also given. The positions of the vibrations in substituting groups are not affected by the linkage to silicon atoms; e.g., the C—H vibrations of the alkyl groups appear at the usual frequencies. However, the intensities of the bands of those C—H groups that are attached in the alpha position to silicon are considerably weakened.

The infrared absorptions of the bonds starting from silicon are relatively strong. Under certain circumstances, the position of the Si—R vibrations is affected by other substituents on the silicon atom. Thus, it has been observed (*38*) that in the $CH_3{-}\overset{|}{\underset{|}{Si}}{-}H$ group the characteristic absorption of Si—CH_3 at 12.5 μ is always absent. Instead, two strong absorption bands are observed, at about 11.2 and 13.2 μ. Similar displacements of the Si—CH_3 vibration in the long-wave region are found when Si—OC_2H_5 bonds are present as well.

In general, the spectra of silanes and siloxanes are relatively simple and it is easy to distinguish between methyl-, methylphenyl-, methyl-*H*-, and methylalkoxysilanes or siloxanes. This follows from Figs. 118 to 121, in which the spectra of some selected siloxanes are shown and the absorptions important for identification are marked.

More difficult than the identification of the substituents bound to the silicon atom is the determination of the *molecular size* and *siloxane functionalities*. The siloxane bond is characterized by an absorption in the region of 9 to 10 μ. Only in low-molecular-weight siloxanes with a small number of siloxane units is the absorption fairly sharp, as in hexamethyldisiloxane (Fig. 118*a*) and in hexamethylcyclotrisiloxane and octamethylcyclotetrasiloxane (Fig. 117). In cyclic and linear polysiloxanes, it is broadened when the number of Si atoms rises above 4 (*60*). Linear polysiloxanes with about 10 Si atoms already exhibit an absorption band whose nature no longer changes when the chain is extended.

In cyclotrisiloxanes, cyclotetrasiloxanes, and ring systems with a larger number of members, the absorption maximum is somewhat different (Fig. 117). With the increasing degree of branching in the high-molecular-weight siloxanes, the maximum of the usually broad bands shifts toward smaller wavelengths as compared to that of linear polymers (*38*). This effect has been analyzed in greater detail with definite low-molecular-weight siloxanes (*31*).

The absorption of the Si—O—Si grouping in heterocycles is quite sharp and occurs at somewhat higher wavelengths than in normal siloxanes, namely, at

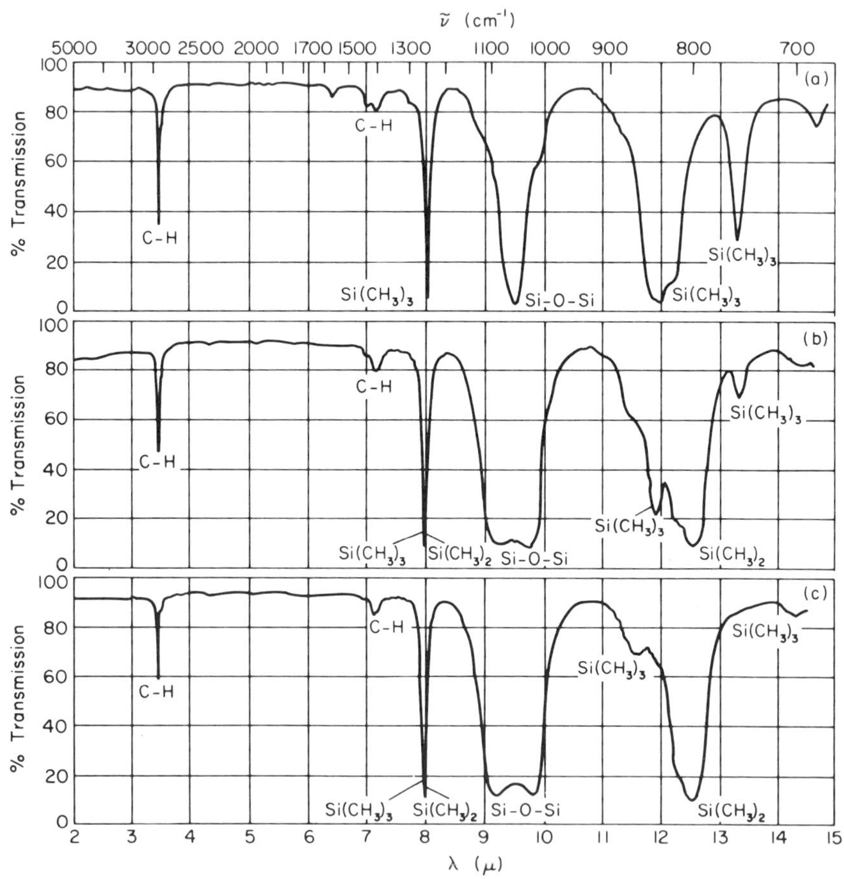

FIG. 118. Infrared spectra of three liquid linear polymethylsiloxanes MD_nM: (a) hexamethyldisiloxane M_2; (b) α,ω-bis-(trimethylsiloxy)-polydimethylsiloxane with $n \approx 10$; (c) α,ω-bis-(trimethylsiloxy)-polydimethylsiloxane with $n \approx 100$.

10.2 μ (*48*). This has been demonstrated in detail for six-membered rings of the type

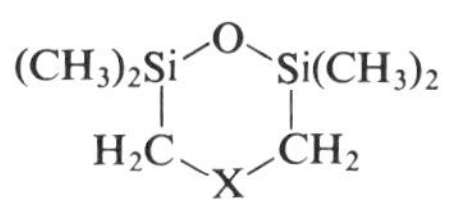

with X=O, NR, S, CHR, CR′R″

The functionality of the siloxane units can be deduced from the Si—R absorptions only to a certain extent. Mono- and difunctional groups can be more

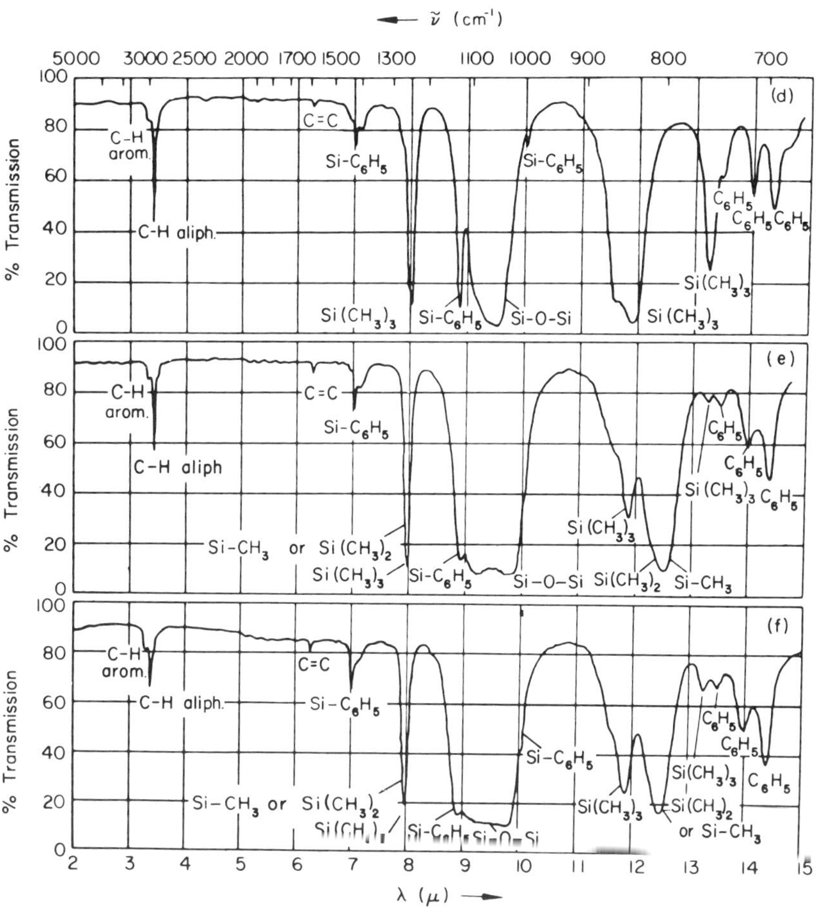

FIG. 119. Infrared spectra of three liquid methylphenylsiloxane polymers: (d) low-molecular-weight branched siloxane $C_6H_5Si[OSi(CH_3)_3]_3$; (e) linear α,ω-bis-(trimethylsiloxy)-polymethylphenylsiloxane with a low phenyl content; (f) linear, α,ω-bis-(trimethylsiloxy)-polymethylphenylsiloxane with a higher phenyl content.

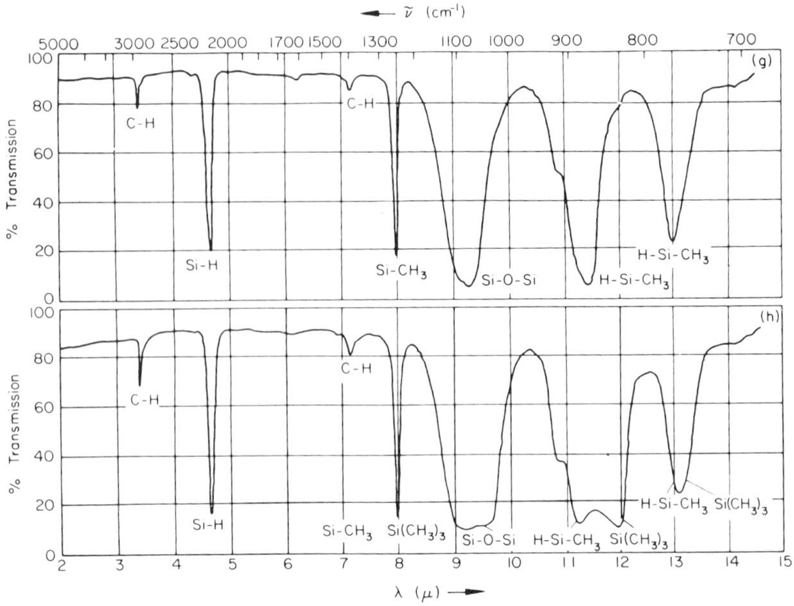

FIG. 120. Infrared spectra of two polymethyl-*H*-siloxanes: (g) 1,3,5,7-tetramethylcyclotetrasiloxane and (h) linear α,ω-bis-(trimethylsiloxy)-polymethyl-*H*-siloxane.

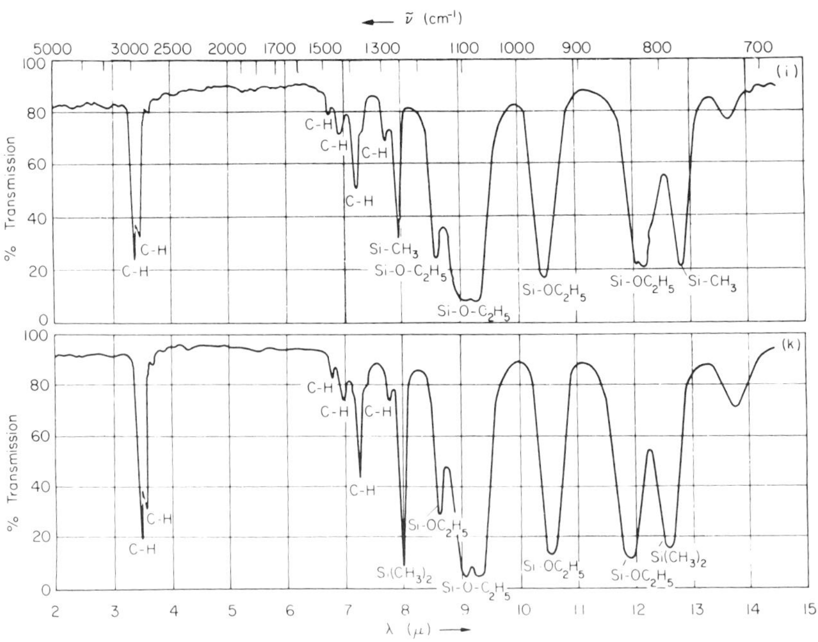

FIG. 121. Infrared spectra of methylethoxysilanes: (i) methyltriethoxysilane and (k) dimethyldiethoxysilane.

easily recognized in this way than trifunctional groups (Fig. 117). The position of the Si—H stretching vibrations clearly indicates whether the hydrogen is included in a mono-, di-, or trifunctional siloxane unit, as can be seen from the following data (*31*):

	$\nu_{Si—H}$
$(CH_3)_2Si(H)O_{1/2}$	2138 ± 8 strong
$CH_3Si(H)O$	2157 ± 9 strong
$Si(H)O_{3/2}$	2210 ± 4 medium

Information about *silanol groups* can generally be obtained from infrared spectra only when the content of these groups amounts to at least a few tenths of a per cent.

Vibrations available for this purpose are the stretching vibration at 2.7 μ, the deformation vibration at about 11 μ, and the combination vibration at 2.2 μ. This last vibration has so far been successfully used only in the analysis of silicate glasses, silica gels, and fumed silicas (*46a*, *59a*). It permits easy differentiation between these materials and any water that may also be present, the combination vibration of which is situated at 1.9 μ.

It is preferred in silicone chemistry to use the stretching vibration, which occurs in a readily accessible region that can be studied easily. However, since the OH absorption at 2.7 μ is subject to interference by water and other compounds containing OH groups, these must be eliminated first. The deformation vibration at 11 μ has the advantage that it does not coincide with the absorptions of other substances containing OH groups. However, the position of this absorption is not particularly constant, and it falls in a spectral region in which many other bands occur, so that it can be easily masked by these (*38*).

The stretching vibration has been used in the infrared spectroscopic study of, e.g., the processes involved in the condensation of silicone resins containing OH groups. It is clear (*8*) from the series of spectra in Fig. 122 how, on heating, the associated groups (which can be recognized by a broad absorption at about 3 μ) are first driven out, and then the silanol groups, whose contents can be inferred from the sharp absorption at 2.7 μ, are slowly condensed out.

As compared with the great practical importance of infrared spectroscopy in analytical problems of organosilicon chemistry, the role of *Raman spectroscopy* recedes into the background. This is due partly to the greater demands on apparatus and time, and partly to the fact that Raman spectroscopy requires extremely pure nonfluorescing preparations which, in general, can be obtained only from distillable low-molecular-weight compounds. Consequently, Raman spectroscopy has not hitherto succeeded in finding acceptance as an auxiliary for the analysis of high-molecular-weight siloxanes.

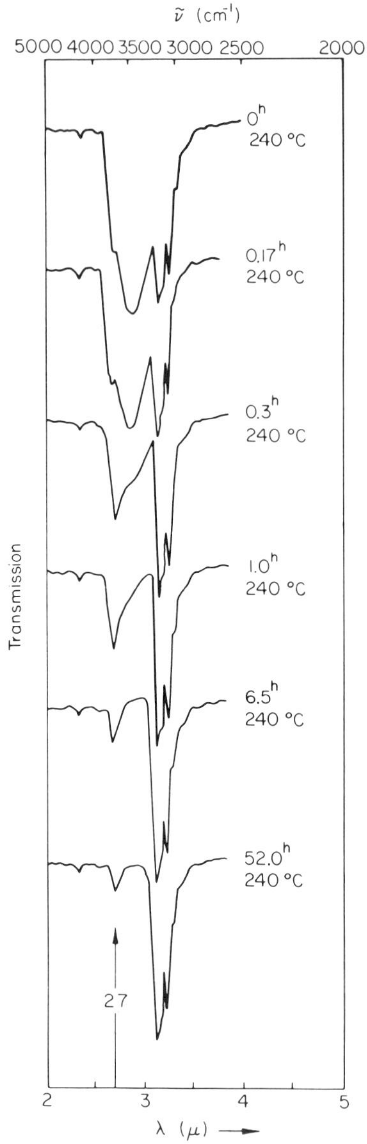

FIG. 122. Decrease in the OH content during the thermal condensation of a silicone resin shown in the infrared spectrum by the behavior of the 3 and 2.7 μ absorptions; see text.

12.2.3 Mass Spectrometry

Mass spectrometry offers an outstanding method for the rapid and reliable analysis of methylchlorosilanes (*63*). The peaks that can be used for the analysis of various mixtures of chlorosilanes and the fragmentary products which they show are collected in Table 114.

An investigation of the fragmentation processes taking place in the mass spectrometer showed that molecules of organochlorosilanes decompose with the splitting off of those substituents that are present in greatest number on the Si atom. Thus, in methyltrichlorosilane a chlorine atom is split off first, while in trimethylchlorosilane it is a methyl group. In dimethyldichlorosilane, the Si—CH_3 and not the Si—Cl bond is broken, obviously because of the greater thermal stability of the Si—Cl bond (85.7 kcal) as compared with the Si—C bond (69.3 kcal). The probability of rupture of the Si—R bond decreases with the increasing size of the R radical (*51*).

The analysis of mixtures of high-boiling silanes such as the phenylchlorosilanes is also possible with the mass spectrometer in principle if a provision is made for the inlet section to be specially heated. However, the expense of gas-chromatographic analysis is substantially lower (see Section 12.2.5).

12.2.4 Nuclear Resonance Spectrometry

In special cases, proton resonance spectrometry can be used to solve the analytical problems presented by organosilicon chemistry. Protons in CH_3—Si and C_6H_5—Si groups exhibit clearly distinct resonance frequencies. The ratio of the areas of the absorptions is equal to the ratio of the number of methyl and phenyl protons in the compound investigated, so that a differentiation and determination of the content of methyl and phenyl groups in silanes and siloxanes is possible (*30*).

Proton resonance spectrometry has acquired particular importance for the group diagnosis of certain functional groups in C-substituted silicon compounds. Thus, particularly in the case of C_α substitutions, a rapid distinction can be made between the following groups (*49*):

CH(CN); C(COOH)(CN); SR; NR; Br; Cl; $S^{\oplus}$R; $N^{\oplus}$R; OR; OOCR

from the chemical shift which methylene protons undergo on substitution (see Section 6.4).

12.2.5 Gas Chromatography

The rapid development of chromatographic methods of investigation in recent times has also been of benefit for the analytical problems of silicone chemistry. Some of the problems for which mass spectroscopy was previously reserved have today been taken over by gas chromatography. It is used predominantly for the analysis of silane mixtures, but low-molecular-weight

TABLE 114

PEAKS USED IN THE MASS-SPECTROMETRIC ANALYSIS OF METHYLCHLOROSILANES (*63*)

Mass	Fragment	Fragment of	System	For analysis of
168	$Si^{35}Cl_4^+$	$SiCl_4$	$(CH_3)_2SiCl_2$, CH_3SiCl_3, $SiCl_4$	$SiCl_4$
148	$CH_3Si^{35}Cl_3^+$	CH_3SiCl_3	$(CH_3)_2SiCl_2$, CH_3SiCl_3, $SiCl_4$; $(CH_3)_3SiCl$, $(CH_3)_2SiCl_2$, CH_3SiCl_3	CH_3SiCl_3
135	$Si^{37}Cl^{35}Cl_2^+$	CH_3SiCl_3	$(CH_3)_3SiCl$, $(CH_3)_2SiCl_2$, CH_3SiCl_3	CH_3SiCl_3
133	$Si^{35}Cl_3^+$	CH_3SiCl_3	$(CH_3)_3SiCl$, $(CH_3)_2SiCl_2$, CH_3SiCl_3	CH_3SiCl_3
128	$(CH_3)_2Si^{35}Cl_2^+$	$(CH_3)_2SiCl_2$	$(CH_3)_2SiCl_2$, CH_3SiCl_3, $SiCl_4$; $(CH_3)_3SiCl$, $(CH_3)_2SiCl_2$, CH_3SiCl_3	$(CH_3)_2SiCl_2$
113	$CH_3Si^{35}Cl_2^+$	$(CH_3)_2SiCl_2$	$(CH_3)_4Si$, $(CH_3)_3SiCl$, $(CH_3)_2SiCl_2$	$(CH_3)_2SiCl_2$
108	$(CH_3)_3Si^{35}Cl^+$	$(CH_3)_3SiCl$	$(CH_3)_3SiCl$, $(CH_3)_2SiCl_2$, CH_3SiCl_3	$(CH_3)_3SiCl$
93	$(CH_3)_2Si^{35}Cl^+$	$(CH_3)_2SiCl_2$, $(CH_3)_3SiCl$	$(CH_3)_4Si$, $(CH_3)_3SiCl$, $(CH_3)_2SiCl_2$	$(CH_3)_3SiCl$
73	$(CH_3)_3Si^+$	$(CH_3)_3SiCl$, $(CH_3)_4$ Si	$(CH_3)_3SiCl$, $(CH_3)_2SiCl_2$, CH_3SiCl_3; $(CH_3)_4Si$, $(CH_3)_3SiCl$, CH_3SiCl_3	$(CH_3)_3SiCl$, $(CH_3)_4Si$

TABLE 115

STATIONARY PHASES FOR THE GAS CHROMATOGRAPHY OF MIXTURES OF METHYLCHLOROSILANE[a]

Stationary phase	Stability	Separation of critical pairs $(CH_3)_2SiCl_2/CH_3SiCl_3$	$(CH_3)SiHCl_2/(CH_3)_3SiCl$	$(CH_3)_3SiCl/SiCl_4$
Dibenzyl ether (*57*)	Good	+	+	Unknown
Nitrobenzene (*11*, *33*)	Poor	+	+	+
Poly-γ-trifluoropropylmethylsiloxane (*50*, *57*)	Very good	+	+	−
Polydimethylsiloxane, high molecular weight (*57*)	Very good	−	+	+
Poly-γ-cyanopropylmethylsiloxane (*45*)	Good	+	+	+
Diethyl phthalate + polydimethylsiloxane (*18*)	?	+	Unknown	+

[a] Numbers in parentheses indicate reference numbers.

siloxanes, silalkylenes, and the like can also be analyzed by gas chromatography (*13*).

An important practical problem that can be solved satisfactorily by means of gas chromatography is the analysis of mixtures of methylchlorosilanes. The separation of such mixtures by distillation is very laborious, and becomes impossible when only small amounts are present, because of the very

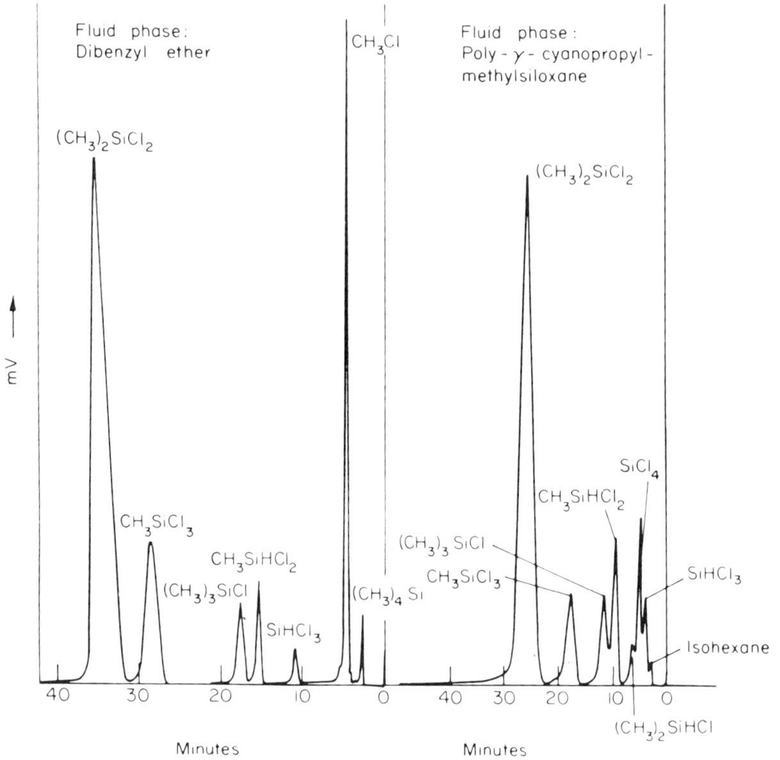

FIG. 123. Gas chromatograms of mixtures of silanes (reaction products of direct synthesis) (*57*): 3 meter column; temperature 50°C (left) and 30°C (right); carrier gas H_2 and He, pressure 1.0 and 1.5 atm gage, respectively.

closely adjacent boiling points. For this purpose, columns between one and four meters long are used. As carrier for the liquid phase they preferably contain, instead of the kieselguhr used previously, chemically treated and prepared fireclay which is available on the market under various names.

Numerous liquids have been tested as stationary phases, particularly for the separation of methylchlorosilanes. They differ in respect to their stability and separating power, in which connection the separation of the pairs of substances $(CH_3)_2SiCl_2/CH_3SiCl_3$, $CH_3SiHCl_2/(CH_3)_3SiCl$, and $(CH_3)_3SiCl/SiCl_4$ must be regarded as particularly critical (Table 115). Chromatograms of mixtures of methylchlorosilanes obtained using dibenzyl ether and poly-γ-cyanopropylmethylsiloxane as liquid phases are shown in Fig. 123.

Higher boiling silanes or silicon compounds can be satisfactorily analyzed by means of polydimethylsiloxane oils, particularly types with very high molecular weights, as the stationary phase. Phenylchlorosilanes (*57*) and

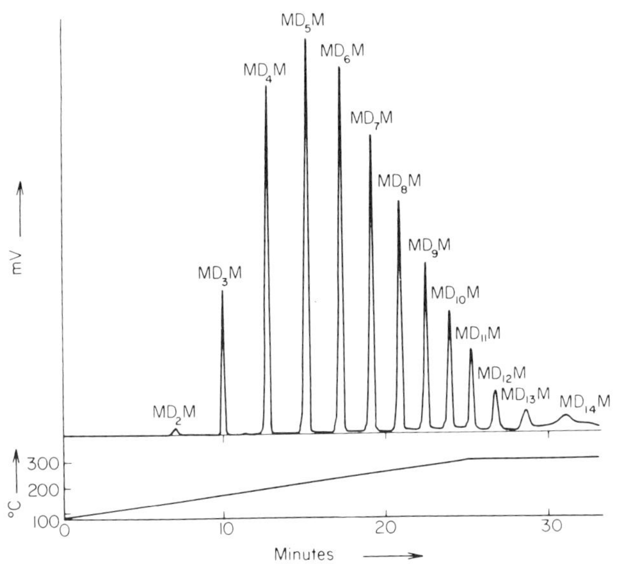

FIG. 124. Gas chromatogram of a low-molecular-weight polydimethylsiloxane oil ($\nu \sim 5$ cSt/20°C) (*57*): 1.8 meter column; carrier gas: He; atmospheric pressure; stationary phase: high-molecular-weight polydimethylsiloxane.

chloromethylchlorosilanes, as well as disilmethylenes (*18*), can be separated in this way.

Finally, Fig. 124 illustrates the analysis of a low-molecular-weight polydimethylsiloxane oil. This was obtained by equilibration of D_4 and M_2 by means of bleaching earth as catalyst and the low-boiling components were distilled off by heating to 145°C/15 torr. The chromatogram very impressively shows the distribution of the homologous short-chain siloxanes in the polymer starting with MD_2M up to $MD_{14}M$ and forming a bell-type curve.

REFERENCES

(1) E. W. BALIS, W. F. GILLIAM, E. M. HADSELL, H. A. LIEBHAFSKY and E. H. WINSLOW, *J. Amer. Chem. Soc.* **70**, 1654 [1948]. (2) L. J. BELLAMY: Infra-Red Spectra of Complex Molecules, John Wiley and Sons, Inc., New York, 1958, 2nd Ed. (3) H. S. BOOTH and M. L. FREEDMAN, *J. Amer. Chem. Soc.* **72**, 2847 [1950]. (4) C. A. BURKHARD, E. G. ROCHOW, H. S. BOOTH and J. HARTT, *Chem. Reviews* **41**, 97 [1947]. (5) A. BYGDÉN: *Silicium als Vertreter des Kohlenstoffs in organischen Verbindungen*, Dissertation University of Uppsala 1916, p. 74. (6) J. ČERMÁK and P. DOSTÁL, *Collect. Czechoslov. chem. Commun.* **28**, 1384 [1963]. (7) M. CHÈNE and O. MARTIN-BORRET, ATIP Bull. No. 1/1958, p. 7; Abstract: *Allg. Papier- Rundschau* **1959**, No. 6, 275. (8) K. DAMM and W. NOLL, *Kolloid-Z.* **158**, 97 [1958]. (9) M. DENNSTEDT and W. UTERMARK: *Anleitung zur vereinfachten Elementaranalyse nach der Makro- und Mikromethode.* Otto Meissner-Verlag, Hamburg 1947, 5th Ed.

(10) R. G. H. ELLIOTT, *J. Soc. Leather Trades' Chem.* **48**, 3, 98 [1964]. (11) K. FRIEDRICH, *Chem. and Ind.* **1957**, 47. (12) G. FRITZ, *Z. anorg. allg. Chem.* **280**, 134 [1955]. (13) —, *Fortschr. Chem. Forschg.* **4**, 459 [1963]. (14) — and H. BURDT, *Z. anorg. allg. Chem.* **307**, 12 [1960]. (15) —, —, ibid **317**, 35 [1962]. (16) — and J. GROBE, ibid. **299**, 302 [1959]. (17) —, — and D. KSINSIK, ibid. **302**, 175 [1959]. (18) — and D. KSINSIK, ibid. **304**, 241 [1960]. (19) W. F. GILLIAM, H. A. LIEBHAFSKY and A. F. WINSLOW, *J. Amer. Chem. Soc.* **63**, 801 [1941].

(20) H. GILMAN and L. S. MILLER, ibid. **73**, 2367 [1951]. (21) F. O. GUENTHER, *Analytic. Chem.* **30**, 1118 [1958]. (22) A. HORNIG, in Houben-Weyl: *Methoden der organischen Chemie, Analytische Methoden*, Verlag G. Thieme, Stuttgart, 1953, 4th Ed., p. 65. (23) — and W. NOLL, in Houben-Weyl: ibid. p. 63. (24) J. F. HYDE, *J. Amer. Chem. Soc.* **75**, 2166 [1953]. (25) L. IVANOVSKY, *Soap, Perfum. Cosmet.* **28**, 187 [1955]. (26) M. KAHLERT and H. CLAUSS, *Chem. Techn.* **6**, 14 [1954]. (27) S. W. KANTOR, *J. Amer. Chem. Soc.* **75**, 2712 [1953]. (28) H. KAUTSKY, G. FRITZ, H. P. SIEBEL and D. SIEBEL, *Z. analyt. Chem.* **147**, 327 [1955]. (29) F. S. KIPPING and J. E. SANDS, *J. Chem. Soc.* [London] **119**, 848 [1921].

(30) O. KOCH and E. KÖSTER, Farbenfabriken Bayer, 1958/59, unpublished. (31) H. KRIEGSMANN, *Z. Elektrochem.* **65**, 342 [1961]. (32) H. LÄMMER, *Chem. Techn.* **4**, 491 [1952]. (33) B. LENGYEL, G. GARZO and T. SZÉKELY, *Acta Chim. Hung.* **37**, 37]1963]. (34) G. R. LUCAS and R. W. MARTIN, *J. Amer. Chem. Soc.* **74**, 5225 [1952]. (35) J. A. MAGNUSON and P. W. BAILLARGEON, *Analytica Chim. Acta* (Amsterdam) **32**, 186 [1965]. (36) J. A. MCHARD, P. C. SERVAIS and H. A. CLARK, *Analytic. Chem.* **20**, 325 [1948]. (37) W. MEISE, Farbenfabriken Bayer; his own results and literature data were used. (38) —, —, 1959, unpublished. (39) R. MÜLLER and L. KLENK, *J. prakt. Chem.* **4**. Reihe, **1**, 129 [1955].

(40) W. NOLL, K. DAMM and R. FAUSS, *Kolloid-Zeitschrift*, **169**, 18 [1960]. (41) W. H. PEARLSON, T. J. BRICE and J. H. SIMONS, *J. Amer. Chem. Soc.* **67**, 1769 [1945]. (42) J. PETERTILL, *Textil* [Prague] **12**, 422 [1957]. (43) E. G. ROCHOW: An Introduction to the Chemistry of the Silicones, John Wiley and Sons Inc., New York, 1951, 2nd Ed. p. 162. (44) G. F. ROEDEL, *Analytic. Chem.* **20**, 705 [1948]. (45) H. ROTZSCHE, *Z. anorg. allg. Chem.* **328**, 79 [1964]. (46) R. O. SAUER, *J. Amer. Chem. Soc.* **68**, 954 [1946]. (46a) H. SCHOLZE, *Naturwiss.* **47**, 226 [1960]. (47) W. SIMMLER, Farbenfabriken Bayer, 1961, unpublished. (48) —, W. MEISE and H. WALZ, *Z. anorg. Allg. Chem.* **314**, 245 [1962]. (49) —, H. WALZ and H. NIEDERPRÜM, *Chem. Ber.* **96**, 1495 [1963].

(50) A. D. SNEGOVA, L. K. MARKOV and V. A. PONOMARENKO, *Zh. analit. Khim.* **19**, 610 [1964]; *Zbl.* **1965**, No. 13, Ref. 1993. (51) N. N. SOKOLOV, K. A. ANDRIANOV, and S. M. AKIMOVA, *J. Gen. Chem. USSR* (*Engl. transl.*) **25**, 647 [1955]. (52) L. H. SOMMER and L. J. TYLER, *J. Amer. Chem. Soc.* **76**, 1030 [1954]. (53) M. M. SPRUNG and F. O. GUENTHER, ibid. **77**, 3990 [1955]. (54) A. STOCK and C. SOMIESKI, *Ber. dtsch. chem. Ges.* **52**, 695 [1919]. (55) U. TAKAHASHI, *Bull. chem. Soc. Japan* **28**, 443 [1955]. (56) S. VERTALIER and F. MARTIN, *Chim. analytique* **40**, 80 [1958]. (57) E. WALASCHEWSKI, Farbenfabriken Bayer, 1957, unpublished. (58) E. L. WARRICK, *J. Amer. Chem. Soc.* **68**, 2455 [1946]. (59) H. J. S. WINKLER and H. GILMAN, *J. Org. Chem.* **26**, 1265 [1961]. (59a) G. WIRZING, *Naturwiss.* **50**, 466 [1963]; **51**, 211 [1964].

(60) N. WRIGHT and M. J. HUNTER, *J. Amer. Chem. Soc.* **69**, 803 [1947]. (61) B. WURZSCHMITT, *Mikrochimica Acta* **36**/**37**, 769 [1951]; *Chemiker-Ztg.* **74**, 356 [1950]. (62) —and W. ZIMMERMANN, *Fortschr. chem. Forsch.* **1**, 485 [1950]. (63) P. D. ZEMANY and F. P. PRICE, *J. Amer. Chem. Soc.* **70**, 4222 [1948].

Subject Index

B

C

D

N

O

R

T